# 16th International Workshop on Advanced Computing and Analysis Techniques in Physics Research

# (ACAT2014)



Journal of Physics: Conference Series Volume 608

## Prague, Czech Republic
## 1-5 September 2014

ISBN: 978-1-5108-0794-5
ISSN: 1742-6588

**Printed from e-media with permission by:**

Curran Associates, Inc.
57 Morehouse Lane
Red Hook, NY 12571

**Some format issues inherent in the e-media version may also appear in this print version.**



Printed by Curran Associates, Inc. (2015)

For permission requests, please contact the Institute of Physics
at the address below.

Institute of Physics
Dirac House, Temple Back
Bristol BS1 6BE UK

Phone: 44 1 17 929 7481
Fax: 44 1 17 920 0979

techtracking@iop.org

**Additional copies of this publication are available from:**

Curran Associates, Inc.
57 Morehouse Lane
Red Hook, NY 12571 USA
Phone: 845-758-0400
Fax: 845-758-2634
Email: curran@proceedings.com
Web: www.proceedings.com

## *Table of contents*

## Volume 608

**16th International workshop on Advanced Computing and Analysis Techniques in physics research (ACAT2014)**
**1–5 September 2014, Prague, Czech Republic**

**Accepted papers received: 31 March 2015**
**Published online: 22 May 2015**

### Preface



### Papers

### T1 - Computing Technology for Physics Research

## T3 - Computations in Theoretical Physics: Techniques and Methods

Journal of Physics: Conference Series **608** (2015) 011001 doi:10.1088/1742-6596/608/1/011001

# 16th International Workshop on Advanced Computing and Analysis Techniques in Physics Research (ACAT2014)

**"bridging disciplines"**

## Preface

This volume of IOP Conference Series is dedicated to scientific contributions presented at the 16th International Workshop on Advanced Computing and Analysis Techniques in Physics Research (ACAT 2014), this year with motto "bridging disciplines", which took place on September 1-5, 2014, at the Faculty of Civil Engineering, Czech Technical University in Prague, Czech Republic.

The 16th edition of ACAT explored the boundaries of computing system architectures, data analysis algorithmics, automatic calculations, and theoretical calculation technologies. It provided a forum for confronting and exchanging ideas among these fields, where new approaches in computing technologies for scientific research were explored and promoted.

This year's edition of the workshop brought together over 140 participants from all over the world. The workshop's 16 invited speakers presented key topics on advanced computing and analysis techniques in physics. During the workshop, 60 talks and 40 posters were presented in three tracks: Computing Technology for Physics Research, Data Analysis - Algorithms and Tools, and Computations in Theoretical Physics: Techniques and Methods. The round table enabled discussions on expanding software, knowledge sharing and scientific collaboration in the respective areas.

ACAT 2014 was generously sponsored by Western Digital, Brookhaven National Laboratory, Hewlett Packard, DataDirect Networks, M Computers, Bright Computing, Huawei and PDV-Systemhaus.

Special appreciations go to the track liaisons Lorenzo Moneta, Axel Naumann and Grigory Rubtsov for their work on the scientific program and the publication preparation. ACAT's IACC would also like to express its gratitude to all referees for their work on making sure the contributions are published in the proceedings. Our thanks extend to the conference liaisons Andrei Kataev and Jerome Lauret who worked with the local contacts and made this conference possible as well as to the program coordinator Federico Carminati and the conference chair Denis Perret-Gallix for their global supervision.

Further information on ACAT 2014 can be found at http://www.particle.cz/acat2014.

L. Fiala, M. Lokajicek, N. Tumova
Institute of Physics of
the Czech Academy of Sciences

Journal of Physics: Conference Series **608** (2015) 011001 doi:10.1088/1742-6596/608/1/011001

# Committees

## Local organizing committee

- Charles University in Prague, Faculty of Mathematics and Physics
  - Tomas Davidek
  - František Knapp
- Czech Technical University in Prague, Faculty of Nuclear Sciences and Physical Engineering
  - Vojtech Petracek
- Institute of Physics, Academy of Sciences of the Czech Republic
  - Lukas Fiala
  - Milos Lokajicek (chair)
  - Nina Tumova
- Nuclear Physics Institute, Academy of Sciences of the Czech Republic
  - Dagmar Adamova
  - Michal Sumbera

## Programme committee

- Track 1
  - Dagmar Adamova (NPI, Czech Republic)
  - Jiri Chudoba (FZU, Czech Republic)
  - Axel Naumann, liaison (CERN, Switzerland)
  - Niko Neufeld (CERN, Switzerland)
- Track 2
  - Alina Grigoras (CERN, Switzerland)
  - Lorenzo Moneta, liaison (CERN, Switzerland)
  - Martin Spousta (Charles University, Czech Republic)
- Track 3
  - Radja Boughezal (Argonne National Laboratory, USA)
  - Michal Malinsky (Charles University, Czech Republic)
  - Grigory Rubtsov, liaison (INR, Russia)

# International Advisory and Coordination Committee

- Andrej Arbusov (BLTP JINR Russia)
- Pushpalatha Bhat (Fermilab USA)
- David Britton, advisor for Track 1 (Glasgow University UK)
- Federico Carminati (CERN Switzerland)
- Denis Oliveira Damazio, advisor for Track 2 (BNL USA)
- Bruce Denby (U. Pierre et Marie Curie France)
- Junpei Fujimoto, advisor for Track 3 (KEK Japan)
- Fabrizio Gagliardi (Supercomputing Centre Barcelona Spain)
- Clara Gaspar, advisor for Track 1 (CERN Switzerland)
- Gudrun Heinrich, advisor for Track 3 (MPI-Munich Germany)
- Andrei Kataev, advisor for Track 3 (INR Russia)
- Christian Kiesling (MPI-Munich Germany)
- Alexander Kryukov (SINP Russia)
- Jerome Lauret, advisor for Track 1 (BNL USA)
- Milos Lokajicek (FZU Prague Czech Republic)
- Daniel Maitre, advisor for Track 3 (Durham University UK)
- Axel Naumann (CERN Switzerland)
- Denis Perret-Gallix, Chair (IN2P3/CNRS France)
- Fons Rademakers (CERN Switzerland)
- Sudhir Raniwala (Rajasthan University India)
- Tord Riemann, advisor for Track 3 (DESY Germany)
- Grigory Rubtsov (INR Russia)
- Jose Seixas (Rio de Janeiro Federal University Brazil)
- Thomas Speer, advisor for Track 2 (Brown University USA)
- Liliana Teodorescu (Brunel University UK)
- Peter Uwer, advisor for Track 3 (Humboldt-University Germany)
- Jianxiong Wang (IHEP China)
- Monique Werlen (LAPTH France)

# Sponsors

| | | | |
|---|---|---|---|
| *Platinum* | WD | BROOKHAVEN NATIONAL LABORATORY | hp |
| *Gold* | DataDirect NETWORKS | M computers | |
| *Silver* | Bright Computing | HUAWEI | |

Journal of Physics: Conference Series **608** (2015) 011002 doi:10.1088/1742-6596/608/1/011002

# Peer review statement

All papers published in this volume of *Journal of Physics: Conference Series* have been peer reviewed through processes administered by the proceedings Editors. Reviews were conducted by expert referees to the professional and scientific standards expected of a proceedings journal published by IOP Publishing.

# Massive affordable computing using ARM processors in high energy physics

**J W Smith, A Hamilton**
Department of Physics
University of Cape Town, Rondebosch, South Africa
E-mail: `joshua.wyatt.smith.cern.ch`

**Abstract.** High Performance Computing is relevant in many applications around the world, particularly high energy physics. Experiments such as ATLAS, CMS, ALICE and LHCb generate huge amounts of data which need to be stored and analyzed at server farms located on site at CERN and around the world. Apart from the initial cost of setting up an effective server farm the cost of power consumption and cooling are significant. The proposed solution to reduce costs without losing performance is to make use of ARM® processors found in nearly all smartphones and tablet computers. Their low power consumption, low cost and respectable processing speed makes them an interesting choice for future large scale parallel data processing centers. Benchmarks on the Cortex™-A series of ARM® processors including the HPL and PMBW suites will be presented as well as preliminary results from the PROOF benchmark in the context of high energy physics will be analyzed.

## 1. Introduction

High energy physics (HEP) at the Large Hadron Collider (LHC) [1] creates an enormous amount of data that need to be stored for later analysis. Dedicated server farms have been built at CERN and Hungary (Tier 0) as well as around the world (Tier 1's and Tier 2's) and are connected through The Worldwide LHC Computing Grid (WLCG). The initial setup costs of these server farms can be immense and the costs to maintain such a server farm can be even larger. Cooling and the power required to maintain such servers at their peak performance make server farms an expensive venture. In early 2013, CERN's Tier 0 had a power capacity of 3.5MW.

The proposed solution is to make use of ARM® processors from here on referred to as ARM. These are found in smartphones and tablet computers where the combination of low power consumption and high performance is the top priority. Significant savings might be achieved if ARM processors are able to cope with the huge amount of data processed. This letter serves to explore the capabilities of ARM processors through parallel processing benchmarks.

## 2. ARM Processors

The ARM processor was designed to perform only a few instructions at once. This reduces the need for hardware such as transistors and thus minimizes power consumption. This letter addresses three system on chip (SoC) setups in the Cortex™-A range, namely the A7 MPCore™, A9 MPCore™ and A15 MPCore™ [2, 3, 4], from here on referred to as the A7, A9 and A15 respectively. Table 1 summarizes the different processors used. An obvious advantage to the A9 is the number of cores, however this is irrelevant because the number of cores is an intrinsic property of the SoC. The hardware *Floating-Point Unit* (FPU) generates results for the speed at which multiplication and addition operations are carried out. v3/4 refers to the respective FPU version. A traditional Intel® computer is also used (Hep405). This serves to provide some reference to the reader. Power measurements for the A7, A9 and A15 were taken using a Fluke 289 Digital Multimeter. Measurements on Hep405 were taken using the Intel® Power Gadget. However, if the power usage characteristics and temperatures vary from those used in Intel's calibration then there will be errors between estimated and actual power usage. For this reason, when referring to Hep405's results the reader must take care in remembering that these values serve as more of an estimate.

**Table 1.** The different setups with key features.

| Setup | Processor | Cores | RAM | Cache | FPU | OS |
|---|---|---|---|---|---|---|
| Cubietruck | AllWinner A20, 1.2GHz | A7 dual core | 2GiB DDR3 | 512 KiB L2 | VFPv4 | Archlinux, hard float |
| Wandboard-Quad | Freescale i.MX6 Quad, 996MHz | A9 quad core | 2GiB DDR3 | 32KiB L1, 1 MiB L2 | VFPv3 | Archlinux, hard float |
| ArndaleBoard-K | Samsung Exynos 5250, 1.7GHz | A15 dual core | 2GiB DDR3 | 32 KiB L1, 1MiB L2 | VFPv4 | Fedora 19, hard float |
| Hep405 | Intel® Core i7-2600, 3.4GHz | quad core | 16GiB DDR3 | 256KiB L1, 1MiB L2, 8MiB L3 | - | Scientific Linux 6 |

## 3. Benchmarks

### *3.1. HPL*

The High-Performance LINPACK (HPL) benchmark is the parallel version of the LINPACK benchmark [5] which is synthetic and CPU intensive. *Flops* of a system are calculated by splitting the large matrix into blocks that are then solved on different cores. Figure 1 shows the relationship between matrix size, block size, *Gflops*, power consumption and *Gflops*/watt for the HPL benchmark. The grey area is an "envelope function" which splits the data into blocks along the time axis and then finds the average power value for that window. It gives a representation of the average power consumption which is indicated by the blue line. A trend we see for the A7 and A9 is that as the size of the matrix block that is passed to the separate cores increases, the average power consumption decreases. This is because slower computation means that less work in the form of communication has to happen between the processors. The average power consumption remains fairly constant with the A15 because of the faster clock speed of 1.7GHz. The *Gflops*/watt is calculated by taking the average power consumption for the time window where there is a HPL measurement. We can see that the A9 is the most efficient provided the block size is relatively large. The A7 does not have a fast enough processor and the A15 uses to much power for the output it delivers. The A9's *Gflops*/watt is comparable to the output of Hep405.

### *3.2. PMBW*

The Parallel Memory Bandwidth Benchmark (PMBW) [6] is a suite that measures bandwidth capabilities of a multi-core computer. This is an important test because more cores result in the floating point performance increasing in a linear fashion. However, if the memory bandwidth

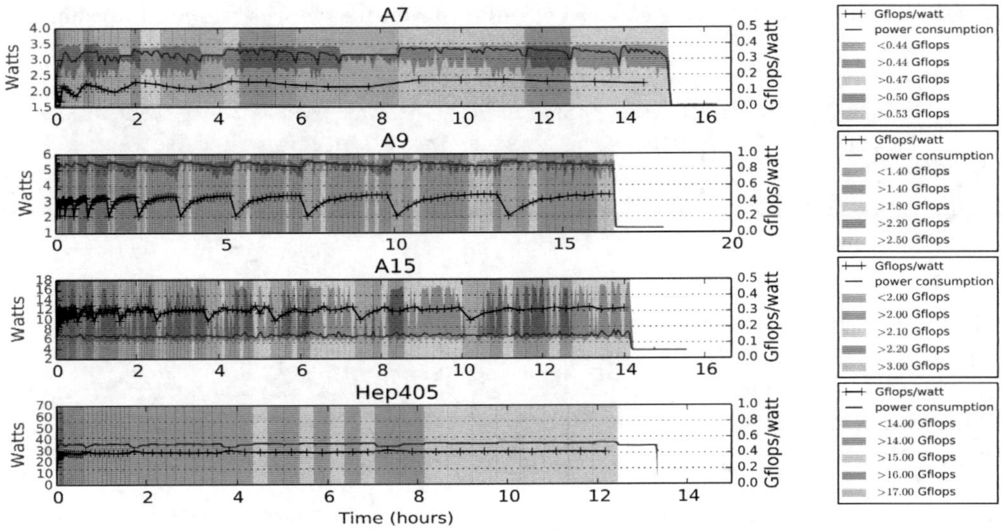


**Figure 1.** The HPL benchmark applied to the 3 different ARM processors and 1 Intel processor.

is not capable of processing the data fast enough those processors will stall. Unlike floating point units the memory bandwidth does not scale with the number of cores running in parallel. The code was developed in assembler language which means compile flags for the SoC become unimportant, thus there are no optimizations. The code uses two general synthetic access patterns, namely sequential scanning and pure random access. The benchmark outputs an enormous amount of data and so only the results for multiple threads, read/write scan operations with each clock-cycle transferring 32-bits (ARM) and 256 bits (HEP405) are shown in Table 2. It is accurate to say that none of these ARM processors perform very well when it comes to memory bandwidth.

**Table 2.** Summary of the important results for the HPL and PMBW benchmarks.

| | A7 | A9 | A15 | Intel® i7 |
|---|---|---|---|---|
| HPL: Average *Gflops*/W | 0.109 | 0.408 | 0.323 | 0.421 |
| PMBW: Max read (GiB/s) | 9.505 | 24.611 | 21.472 | 298.264 |
| PMBW: Max write (GiB/s) | 9.035 | 21.428 | 21.546 | 163.123 |

*3.3. Dedicated PROOF Cluster*

The Parallel ROOT Facility, PROOF [7], is an extension to the well known framework, ROOT [8]. PROOF strives to introduce massive parallelism by splitting up files among workers. The dedicated PROOF cluster at the University of the Witwatersrand (Wits) in Johannesburg, South Africa uses 7 A9 boards sharing a NFS directory hosted by a traditional server. The PROOF benchmark suite [9] was used to determine performance. It is a further add-on designed to test the topology and scalability of clusters by running cycle- (CPU) and data-driven (I/O)

Journal of Physics: Conference Series **608** (2015) 012001 doi:10.1088/1742-6596/608/1/012001

processes. Results for vfpv3-d16, neon and default are shown. These correspond to compile flags used in the compilation of ROOT that are specific to the A9.

*3.3.1. CPU benchmark* The default TSelHist selector was used with the output shown in Figure 2. The unit of measurement is the number of events per unit time.We can see that the cluster scales linearly for CPU-intensive tasks. This is to be expected as the only significant constraint is the overhead produced when the master has to pool the results and create the histograms.

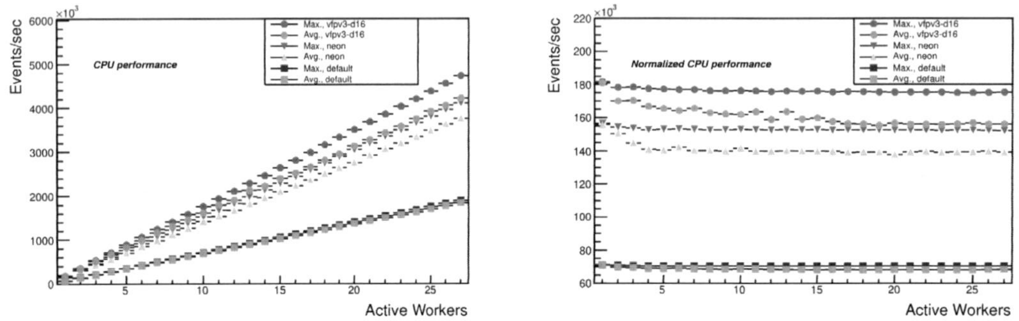


**Figure 2.** PROOF CPU benchmark for ARM cluster.

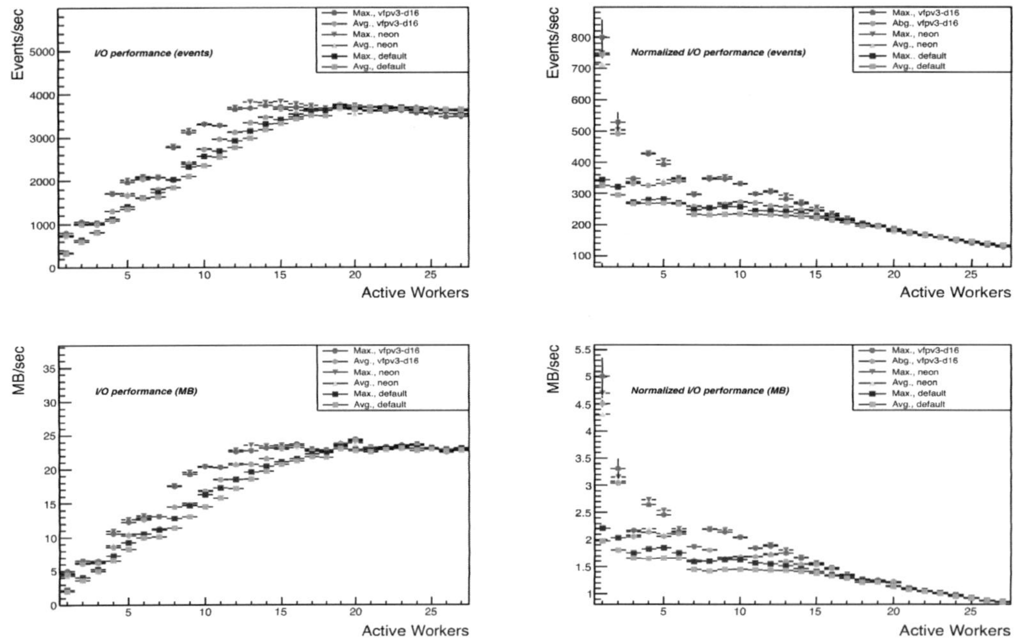


**Figure 3.** PROOF I/O benchmark for ARM cluster.

*3.3.2. I/O benchmark* The default TSelEvent selector was used with the output shown in Figure 3. This benchmark shows the bottleneck that the NFS server creates. Scalability saturates at around 12-13 workers which equates to reading events at a total of around 23 MB/sec. Its clear that a shared NFS directory reduces a lot of the advantage that one might gain from a PROOF cluster.

## 4. Conclusions

ARM processors use significantly less power than traditional processors. However, they are also significantly slower and so parallel computing for the right type of problem must be exploited. HEP data analysis provides tasks that are easily parallelizable and with the correct frameworks in place can be successful. PROOF provides one such framework, however communication bottlenecks have to be overcome. The performance per watt for the A7 and A15 shows poor results when compared to the A9 and Intel machines. The A9's advantage over the other ARM processors lies in the quad core SoC. This processor is still slow compared to traditional computers due in part to the age of the technology. The newer A15 quad core processors should show a dramatic improvement in overall performance and specifically performance per watt. Furthermore, the newly developed 64 bit architecture along with their respective SoCs will have a large impact on the CPU performance and memory bandwidth of these ARM processors.

## Acknowledgments

The authors would like to thank the National Research Foundation of South Africa and the University of Cape Town for funding as well as the Massive Affordable Computing (MAC) team at the University of the Witwatersrand for building and providing access to the A9 ARM cluster.

## References

[1] 2008 *Journal of Instrumentation* vol 3 (IOP and SISSA)
[2] ARM Ltd 2013 *Cortex$^{TM}$-A7 MPCore$^{TM}$ Technical Reference Manual* r0p5 ed
[3] ARM Ltd 2012 *Cortex$^{TM}$-A9 MPCore$^{TM}$ Technical Reference Manual* r4p1 ed
[4] ARM Ltd 2013 *Cortex$^{TM}$-A15 MPCore$^{TM}$ Processor Technical Reference Manual* r4p0 ed
[5] Dongarra J Linpack benchmark Computer Science Technical Report CS - 89 – 85 University of Tennessee
[6] Bingmann T 2013 pmbw - Parallel Memory Bandwidth Benchmark / Measurement website
[7] Ballintijn M, Brun R, Rademakers F and Roland G 2003 *CHEP03 Intl. Conf.*
[8] Brun R, Rademakers F An object orientated analysis framework URL `http://root.cern.ch`
[9] Ryu S, G Ganis 2012 *J. Phys.Conf. Ser. 368 012020 IOPscience*

Journal of Physics: Conference Series **608** (2015) 012002 doi:10.1088/1742-6596/608/1/012002

# The Massive Affordable Computing Project: Prototyping of a High Data Throughput Processing Unit

**Mitchell A. Cox and Bruce Mellado**

School of Physics, University of the Witwatersrand. 1 Jan Smuts Avenue, Braamfontein, Johannesburg, South Africa, 2000

E-mail: mitchell.cox@cern.ch

**Abstract.** Scientific experiments are becoming highly data intensive to the point where offline processing of stored data is infeasible. High data throughput computing or High Volume throughput Computing, for future projects is required to deal with terabytes of data per second. Conventional data-centres based on typical server-grade hardware are expensive and are biased towards processing power rather than I/O bandwidth. This system imbalance can be solved with massive parallelism to increase the I/O capabilities, at the expense of excessive processing power and high energy consumption. The Massive Affordable Computing Project aims to use low-cost, ARM System on Chips to address the issue of system balance, affordability and energy efficiency. An ARM-based Processing Unit prototype is currently being developed, with a design goal of 20 Gb/s I/O throughput and significant processing power. Novel use of PCI-Express is used to address the typically limited I/O capabilities of consumer ARM System on Chips.

## 1. Introduction

Projects such as the Large Hadron Collider (LHC) at CERN and the Square Kilometer Array (SKA) in South Africa generate enormous amounts of raw data which presents a serious computing challenge. After planned upgrades in 2022, the data output from the ATLAS Tile Calorimeter (TileCal) will increase by 200 times to over 40 Tb/s (Terabits/s) [1].

A simple plot, shown in Fig. 1, of the increase in CPU processing power in MIPS (Million Instructions Per Second) and hard drive read-write speed in MB/s (MegaByte/s) over many years clearly demonstrates the fact that hard drive I/O rates are insufficient, and will not become sufficient in the near future, to store the entirety of raw data from modern scientific experiments such as the SKA and LHC [2].

The increase in Ethernet throughput, however, is at a similar rate to the increase in CPU processing power. Based on Amdahl's Laws it has been recommended that approximately one compute instruction per bit of data is required for a balanced system and this relationship is clear when comparing CPU and Ethernet in Fig. 1 [3]. It appears upon inspection that CPU performance and Ethernet throughput are well balanced but in reality high-end Ethernet and other external I/O interconnects in not commonly available except on very high-end and therefore expensive systems. For example, 1 Gb/s Ethernet from 2002 is suitably balanced with a 2002 performance CPU. It is imbalanced when it is coupled with a modern CPU with an order of magnitude higher performance, however this is a very common situation since high-end CPUs are more prevalent than cutting-edge external I/O.

It is important to remember that the balance between I/O and CPU processing power is highly dependant on the workload. For some high-level triggering tasks and event reconstruction in the LHC, for example, the algorithms are very CPU intensive. In other situations, simpler computations such as Optimal Filtering takes place and the system becomes I/O bound.

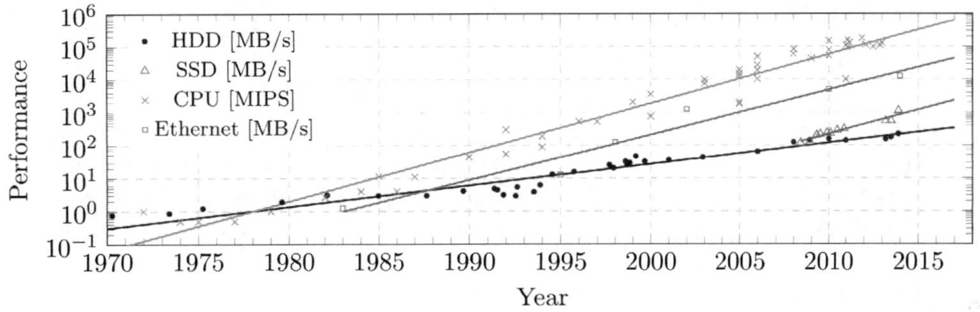


Figure 1: Hard drive (HDD) and Solid State Drive (SSD) and Ethernet 802.3 data throughput and CPU performance on a log scale with time [2, 4].

A specialised triggering and data acquisition system is currently employed by the LHC to reduce the amount of data produced to a manageable quantity for offline storage. This solution is not always suitable and so a paradigm shift is necessary to deal with future workloads and new projects.

The cost, energy efficiency, processing performance and I/O throughput of the computing system to achieve this task is vitally important to the success of future big science projects. Current x86-based microprocessors such as those commonly found in personal computers and servers are often biased towards processing performance and not I/O throughput and are therefore less-suitable for cost-effective high data throughput applications due to the necessity for massive parallelism.

High Volume throughput Computing (HVC) provides a suitable paradigm for high data throughput streaming applications [5]. HVC is a datacenter based computing paradigm where the focus is on loosely-coupled throughput-oriented workloads in terms of either requests (service type applications), processed data (big data applications) or the maximum number of simultaneous subscribers (interactive real-time applications). The definition does not include data-intensive MPI workloads since these are suitably covered by High Performance Computing (HPC).

One of the first steps to the development of an effective HVC system is a high data throughput Processing Unit (PU). This PU should be well balanced in terms of CPU performance and I/O throughput and latency to maximise energy efficiency and cost. The primary aim of the Massive Affordable Computing Project at the University of the Witwatersrand, Johannesburg in South Africa is to develop such a PU.

ARM System on Chips (SoCs) are found in almost all mobile devices due to their low energy consumption, high performance and low cost and are the basis for the PU under development [6]. Section 2 provides a brief overview of the performance of four different ARM-based platforms with various ARM SoCs as well as their specifications. In Section 3 an overview of an architecture for the PU is presented. Finally, a PCI-Express test setup is described and preliminary results are given in Section 4 and Section 5 concludes.

Table 1: CPU benchmark results of the ARM platforms at a fixed clock frequency of as close to 1 GHz as possible.

| Core Revision | **Cortex-A7** r0p4 | **Cortex-A9** r2p2 | **Cortex-A15** r3p2 | **Cortex-A15** r3p3 |
|---|---|---|---|---|
| System on Chip | AllWinner A20 | Freescale i.MX6Q | Samsung 5410 | NVIDIA TK1 |
| Clock (MHz) | 1008 | 996 | 1000 | 1090 |
| Cores | 2 | 4 | 4 | 4 |
| Feature Size (nm) | 40 | 40 | 28 | 28 |
| SP GFLOPS | 1.76 | 5.12 | 10.56 | 11.70 |
| DP GFLOPS | 0.70 | 2.40 | 6.04 | 6.26 |
| CoreMark | 4858 | 11327 | 14994 | 16689 |
| Load Power (W) | 2.85 | 5.03 | 7.48 | 6.11 |
| Idle Power (W) | 1.16 | 2.02 | 1.97 | 2.26 |
| Calc. Power (W) | 1.69 | 3.01 | 5.51 | 3.85 |
| DP GFLOPS/W | 0.42 | 0.80 | 1.10 | 1.63 |
| Ethernet (Mb/s) | 100 | 1000 | 100 | 1000 |
| PCIe (Gb/s) | - | 5 | - | 40 |

## 2. ARM System on Chips

ARM System on Chips (SoCs) are low cost, energy efficient and high performance which has led to their extensive use in mobile devices. Several ARM platforms have been tested by the group and a summary of the specifications and of the CPU benchmark results is presented in Tab. 1.

It is clear that the CPU performance per Watt is increasing with newer generation ARM SoCs. This is naturally driven by consumer demand for faster mobile devices. The exact cost of the SoCs is not known because it is either under Non-Disclosure Agreement (NDA) or is based on order quantity. It can be deduced from the cost of development platforms and the commodity nature of the SoCs presented that the cost is low.

The Freescale i.MX6Q and NVIDIA Tegra-K1 are particularly suitable for a high data throughput PU because of their PCI-Express I/O interface. The available Ethernet interfaces on commodity ARM SoCs are not feasible for multi-gigabit I/O. The Tegra-K1 also has excellent CPU performance and power efficiency due to the small silicon feature size.

## 3. Processing Unit

A diagram of the PU architecture that is currently being prototyped is shown in Fig. 2. Several ARM SoCs will be integrated onto a single circuit board to minimise duplicated components and form a compact solution. The SoCs will be connected via a PCI-Express switch for a high data throughput interconnect. Finally, a network processor SoC, such as one typically found in high-end network routers, will be used to bridge the internal PCIe network to an external, industry standard Ethernet-based network. The external network can also be another PCIe interface if the PU is integrated closely with an FPGA, for example.

Multiple 10 Gb/s XAUI or SFP+ Ethernet connections via a single server-grade (and therefore more expensive) SoC will enable high data throughput I/O. The use of affordable commodity SoCs will enable significant computational power. The combination of high data throughput I/O and affordable, energy efficient computing forms an excellent candidate for a HVC PU.

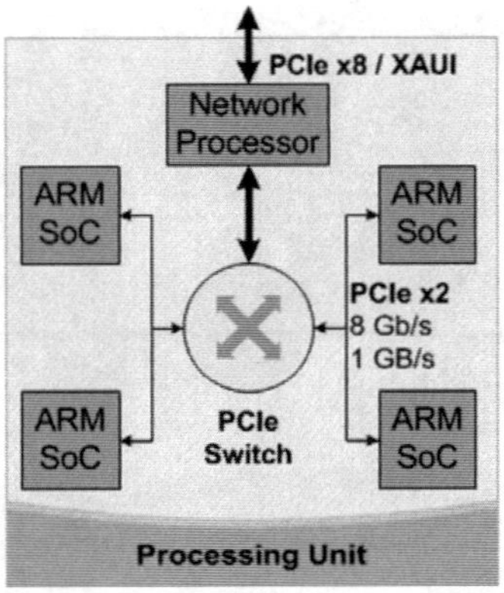


Figure 2: Block diagram of PU architecture showing four hypothetical ARM SoCs, each with PCIe x2 connectivity and an external XAUI (10 Gb/s Ethernet) or PCIe interface.

## 4. PCI-Express Testing

PCI-Express throughput tests have been performed on a pair of Freescale i.MX6 quad-core ARM Cortex-A9 SoCs clocked at 1 GHz, located on Wandboard development boards [7]. The results are presented in Tab. 2 and a photo of the custom test setup designed by the author is in Fig. 3. Three tests were run to ascertain the maximum data throughput that can be obtained from the i.MX6 SoC: a simple CPU based memcpy command and two Direct Memory Access (DMA) transfers, initiated by the Endpoint (EP) or slave and the Root Complex (RC) which is the host. Unfortunately the i.MX6Q SoC does not have a DMA unit on the PCIe controller and so the Image Processing Unit DMA unit was used instead. This is a workaround provided by the manufacturer.

Table 2: PCI-Express throughput results of a i.MX6 (Wandboard) pair.

| | **CPU memcpy** | **DMA (EP)** | **DMA (RC)** |
|---|---|---|---|
| Read (MB/s) | $94.8 \pm 1.1\%$ | $174.1 \pm 0.3\%$ | $236.4 \pm 0.2\%$ |
| Write (MB/s) | $283.3 \pm 0.3\%$ | $352.2 \pm 0.3\%$ | $357.9 \pm 0.4\%$ |

The theoretical maximum throughput for the PCI-Express Gen 2 x1 link that was used is 500 MB/s. The best result is using DMA initiated by the RC but it is only 72% of the theoretical maximum. The RC-mode drivers are more optimized than the EP-mode drivers due to limited manufacturer support for EP-mode. The read results are lower than write because of overheads to initiate the read. The PU architecture will take these differences into account and use a data push rather than a pull based approach.

A Wandboard to standard PCI-Express x1 slot adapter has also been designed and manufactured by the author. A photo of this is shown in Fig. 4. This will enable future testing of multiple SoCs connected via a PCI-Express switch to closely prototype the proposed PU architecture from Section 3.

## 5. Discussion, Conclusions and Future Work

High data throughput computing, or more formally High Volume throughput Computing (HVC), is required for projects such as the LHC and SKA which produce enormous amounts of raw

Journal of Physics: Conference Series **608** (2015) 012002 doi:10.1088/1742-6596/608/1/012002

Figure 3: PCI-Express test setup for a pair of i.MX6 SoCs (Wandboards).

Figure 4: Photo of two Wandboard (via the EDM Connector) to PCI-Express adapters without Wandboards attached.

data. A general purpose ARM System on Chip based processing unit is being developed at the University of the Witwatersrand, Johannesburg which hopes to enable affordable and energy efficient HVC.

In the current prototype, a PCI-Express x1 interface will be used for the raw data transfer between several ARM Cortex-A9 SoCs and a single high-end network processor. This network processor bridges the internal PCIe network to multiple standard external 10 Gb/s Ethernet connections. PCI-Express is superior to Ethernet in energy efficiency, I/O throughput and latency, especially in light of the fact that commodity ARM SoCs do not support Ethernet faster than 1 Gb/s, however PCI-Express is not suitable for longer distance communications.

Initial throughput measurements presented for a pair of Freescale i.MX6 quad-core Cortex-A9 SoCs are 72% of the theoretical maximum 500 MB/s for the available x1 link. Twelve of these SoCs would therefore be connected in parallel to provide 40 Gb/s throughput at a power consumption of less than 100 W. As a proof of concept the Cortex-A9 prototype aims to provide 20 Gb/s aggregated throughput.

The more recent NVIDIA Tegra-K1 SoC supports 40 Gb/s PCI-Express and significantly higher processing power, as well as integrated CUDA GPU processing support. Several of these SoCs will likely be used in the final prototype in future.

The next stage of research by the author will be to test a small PCIe cluster of Cortex-A9 SoCs. An adapter board to connect the Cortex-A9 SoC development board (Wandboard) to a standard PCI-Express x1 connector has been designed and manufactured. A PCI-Express switch development motherboard will be used to connect up to eight devices together. Linux driver development to emulate and Ethernet device over PCIe is currently in progress.

The use of multiple energy efficient commodity ARM SoCs interconnected via PCI-Express and a single higher-end SoC for external communications via multiple 10 Gb/s Ethernet connections is theoretically well suited as a HVC Processing Unit. Future big science experiments may be jeopardised by prohibitive data processing costs but the processing unit presented in this paper may be a possible solution to this problem with its high data throughput, energy efficient and affordable computing capabilities.

**Acknowledgements**
The financial assistance of the National Research Foundation (NRF) towards this research is hereby acknowledged. We would also like to acknowledge the School of Physics, the Faculty of Science and the Research Office at the University of the Witwatersrand, Johannesburg. The Massive Affordable Computing Project team and specifically Robert Reed is acknowledged for some of the results presented in this paper.

## References

[1] Carrió F *et al.* 2014 *Journal of Instrumentation* **9** C02019–C02019 ISSN 1748-0221 URL `http://stacks.iop.org/1748-0221/9/i=02/a=C02019`
[2] Dursi J 2012 Parallel I/O doesn't have to be so hard: The ADIOS Library Tech. rep. SciNet URL `http://wiki.scinethpc.ca/wiki/images/8/8c/Adios-techtalk-may2012.pdf`
[3] Szalay A S *et al.* 2010 *ACM SIGOPS Operating Systems Review* **44** 71 ISSN 01635980 URL `http://dl.acm.org/citation.cfm?id=1740390.1740407`
[4] Wikipedia 2014 IEEE 802.3 — Wikipedia, The Free Encyclopedia accessed: 16 September 2014 URL `http://en.wikipedia.org/wiki/IEEE_802.3`
[5] Zhan J *et al.* 2012 High Volume Throughput Computing: Identifying and Characterizing Throughput Oriented Workloads in Data Centers *2012 IEEE 26th International Parallel and Distributed Processing Symposium Workshops & PhD Forum* (IEEE) pp 1712–1721 ISBN 978-1-4673-0974-5 URL `http://ieeexplore.ieee.org/lpdocs/epic03/wrapper.htm?arnumber=6270846`
[6] Rajovic N *et al.* 2013 *Journal of Computational Science* **4** 439–443 ISSN 18777503 URL `http://www.sciencedirect.com/science/article/pii/S1877750313000148`
[7] Wandboardorg 2012 Wandboard - Freescale i.MX6 ARM Cortex-A9 Opensource Community Development Board accessed: 18 February 2014 URL `http://www.wandboard.org/`

# Adaptive track scheduling to optimize concurrency and vectorization in GeantV

**J Apostolakis[1], M Bandieramonte[2], G Bitzes[3], R Brun[1], P Canal[4], F Carminati[1], J C De Fine Licht[5], L Duhem[6], V D Elvira[4], A Gheata[1], S Y Jun[4], G Lima[4], M Novak[1], R Sehgal[7], O Shadura[8], S Wenzel[1]**

[1] European Organization for Nuclear Research (CERN) - Geneva, Switzerland
[2] University of Catania and INAF (IT)
[3] University of Athens (GR)
[4] Fermi National Accelerator Laboratory (US)
[5] University of Copenhagen (DK)
[6] Intel Corporation
[7] Bhabha Atomic Research Center (IN)
[8] National Technical University of Ukraine, "Kyiv" Politechnic Institute

Andrei.Gheata@cern.ch

**Abstract**. The *GeantV* project is focused on the R&D of new particle transport techniques to maximize parallelism on multiple levels, profiting from the use of both *SIMD* instructions and co-processors for the CPU-intensive calculations specific to this type of applications. In our approach, vectors of tracks belonging to multiple events and matching different locality criteria must be gathered and dispatched to algorithms having vector signatures. While the transport propagates tracks and changes their individual states, data locality becomes harder to maintain. The scheduling policy has to be changed to maintain efficient vectors while keeping an optimal level of concurrency. The model has complex dynamics requiring tuning the thresholds to switch between the normal regime and special modes, i.e. prioritizing events to allow flushing memory, adding new events in the transport pipeline to boost locality, dynamically adjusting the particle vector size or switching between vector to single track mode when vectorization causes only overhead. This work requires a comprehensive study for optimizing these parameters to make the behaviour of the scheduler self-adapting, presenting here its initial results.

## 1. GeantV project in the simulation context of LHC

One of the striking facts for the LHC computing was that more than half of the available CPU resources of the experiments were used for producing simulated samples. Assuming a linear dependency of simulation needs with the acquired data, the implications from the perspective of the LHC upgrade schedule become sizeable as far as computing resources are concerned. As the integrated luminosity will increase by factors in the near future and by orders of magnitude for the high-luminosity LHC era, we are looking at a tremendous CPU power to cover future simulations.

The "why" part of the observation above is very important for understanding what can be done in the future besides just scaling up the resources, but also using them more efficiently. This is what triggered the investigation and prototyping work that started in 2011 and which is the base for the current R&D behind the *GeantV* project [1]. Looking at the main reasons for the low number of instructions per cycle of about 0.8 in particle transport simulations and HEP code in general, we can identify a large range of sources, starting with abusing object-oriented design in data access patterns and ending with poor data and code locality caused by the sequential nature of the classical particle transport approach.

This simulation optimization initiative has grown during the last years into a project looking at many different performance dimensions, which can potentially speed up simulation by important factors on the architectures available today. The main idea behind the first version of the *GeantV* prototype was changing the basic work unit for transport simulation from a single particle to a vector of particles fulfilling geometry locality criteria. This was motivated by the observation that most of the simulation time is spent in only 5-10% of all logical volumes of a HEP experiment, and in addition geometry locality means material locality, which partially implies physics locality. Adding multithreaded concurrency on top of this model allowed building the initial version of the prototype which sets the basis for exploring new features like: SIMD vectorization, fine grained parallelism, transparent usage of resources (CPU/GPU) or re-design of the code in terms of specialization and re-usability.

The project is evolving in three main directions which are complementary enough to allow for independent development and which share core data structures and features into a single simulation engine, which currently is still in the prototyping phase.

- The geometry modeling development aims to re-design the main features of the standard geometry modelers such as the *Geant4* [2] and *ROOT* [3] ones using techniques like template specialization and generic programming, supporting vectorization at library level and generating optimized code for both CPU and GPU. This is evolving into the next generation geometry tool, triggered by the *GeantV* project but usable well beyond it. This geometry development, known as *VecGeom* [4], recently started to be merged with the *USolids* library [5], which was already crossbreeding the features of *ROOT* and *Geant4* shape primitives.
- The physics development aims to refactor the *Geant4* physics engine and algorithms into vectorized kernels usable on both CPU and GPU, using the same specialization techniques whenever possible and allowing for "fast" features like generalized tabulated cross sections or inclusion of fast simulation techniques.
- The kernel and scheduler development aims to globally optimize the flow of the simulation application, managing concurrency and deploying vectors of tracks matching appropriate locality criteria to both physics and geometry algorithms. The major challenge of the scheduler is to keep a positive balance between the gains from locality and vectorization and the overheads coming from the extra data management required by the vector approach. The following sections of the paper will focus more on these areas.

## 2. The data model: vectors of tracks

One of the important requirements allowing compilers to vectorize the code is the contiguity and alignment of the data to be processed. Any indirections using pointers or indexing represent major overheads or even blockers for vector architectures, besides the obvious negative impact on caching. Since the presence of a virtual function table for a type and the use of virtual function generally prevent vectorization, plain old data like types are to be favored in a vector approach, as well as the usage of structures of arrays (*SOA*) versus the classical arrays of structures (*AOS*). As rule of thumb, arranging the data in a way that avoids collecting at run-time from many places improves performance but is generally hard to achieve.

In *GeantV*, we implemented a custom *SOA* approach, which can automatically handle vectors of tracks expanding them into internally managed aligned arrays corresponding to each track data member. Memory allocation is performed at initialization time and expanded as needed. This structure (called *GeantTrack_v*) can pass pointers to arrays of positions, momenta or other fields to any vectorized client. Due to the size of the track object (currently 192 bytes), an *SOA* approach can have important caching overheads, so we are also investigating a hybrid *AOSOA* (array of structures of arrays) to alleviate this effect.

### *2.1. Reshuffling overheads*

Providing a vector container is far from being enough to percolate vectors down into non-trivially branched algorithms. Conditional code is harmful for both SIMD and GPU processing but in many cases can be partially avoided by using masking techniques [6] at the level of single algorithms. On the other hand, successive calls of algorithms that involve track selections or which are changing individual track states require container reshuffling or compacting operations.

A simple example is propagating a vector of particles in a single volume. Some of the particles will reach the volume boundary or the point where a physics process was sampled earlier than others, leaving "holes" in the container. Continuing the propagation in vector mode requires compacting the fields of the remaining tracks in contiguous memory and possibly moving the stopped tracks to a different vector container. Another common technique is to apply a vector operation to a subset of the tracks, which requires reshuffling those matching the criteria for the vector operation into a contiguous block at the beginning of the container.

All these operations involve costly memory copying, which is an overhead compared to a scalar approach. The challenge is to minimize these overheads as much as possible in order to keep them below the performance gains due to locality and vector processing.

## 3. Track scheduling model

The *GeantV* scheduler controls the simulation workflow. In the current implementation, a single thread controls the flow, monitoring the work queue and other critical parameters such as the vector size to trigger the appropriate corrective actions.

The prototype transports particles in a continuous flow, which is fed either from an external generator invoked on demand, or from the particles transported by a previous propagation step. The picture is very different from the standard sequential particle stack approach and mixes together tracks coming from different events. The *GeantV* scheduler design allows for an arbitrary number of filter modules to select and group together from the input flow all particles matching certain locality conditions, as shown in the top part of figure 1.

Locality means in our terminology the repeated execution of a sequence of instructions for all (or most of) the vector of particles, without any logical branching that could interrupt this sequence. The local code creates the premises for better caching and parallel execution, and is allowed to call inline functions or methods supporting a vector input signature on their turn.

The locality criteria are strongly dependent on the tasks to be executed within the track propagation (stepping) procedure, presented in the bottom of figure 1. For example, computing the distance to the next object boundary becomes local when called for a vector of particles contained by the same logical volume, allowing computing the distance in a vectorized fashion without any extra data gathering. The corresponding geometry filter will therefore group particles by this criterion. The physics sampler on the other hand requires particles of the same type and within the same energy range to become local. In the current version of the prototype, geometry locality is the only implemented filter, but we are extending the concept to other use cases as well.

An important future feature of *GeantV* will be the possibility to combine full and fast parametric simulation in the same session, in an arbitrary proportion. The use of fast simulation models will be triggered by user-defined filters, which would group particles by the criteria allowing these models to be applicable. Using the same filtering idea, additional resources such as general-purpose graphics processing units can be used transparently by grouping particles for which a given transport task would take profit from running on such devices.

Journal of Physics: Conference Series **608** (2015) 012003 doi:10.1088/1742-6596/608/1/012003

Every locality filter fills its current vector of tracks up to a threshold, and after generates a *basket*, which is put in a common work queue. The baskets produced by a given filter are only used by the computing task they are filtered for. For example, a geometry filter generates baskets of tracks located in the same volume, which are processed by the *GeantV* vector propagator.

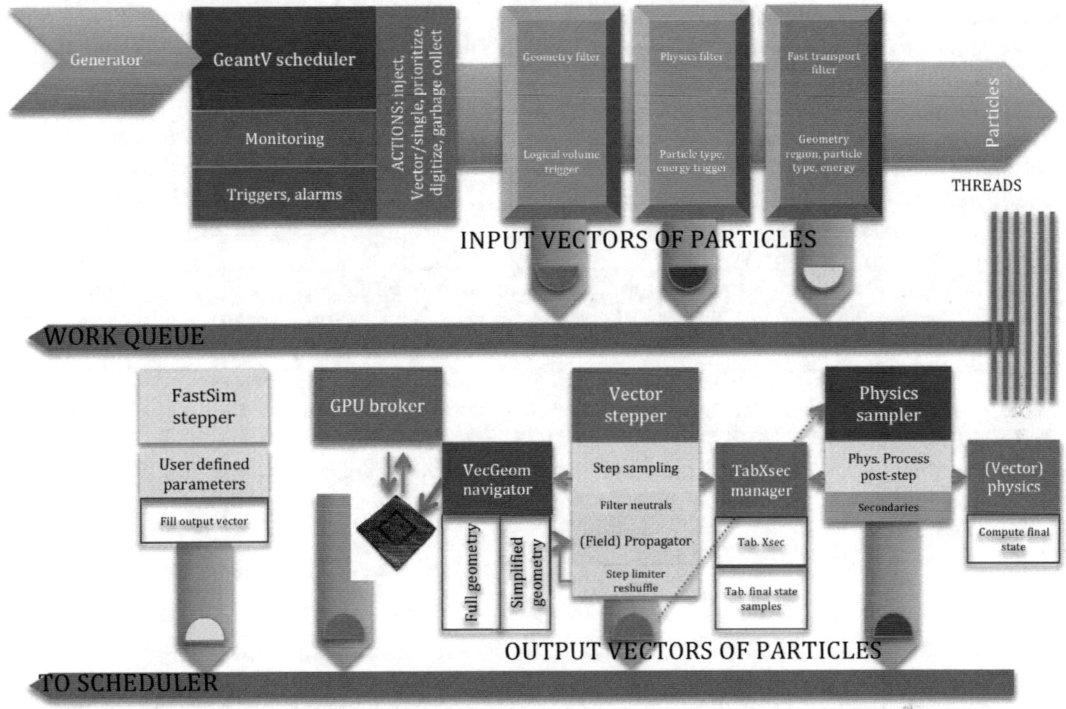


Figure 1. The *GeantV* scheduler aims to improve code locality and allow for SIMD vectorization in a concurrent environment by dispatching vectors of tracks to be transported. The design is based on a set of mutually exclusive filters grouping particles according certain locality criteria, that can be exploited in a vector manner by the different tasks executed in the propagation procedure, shown in the bottom part of the picture.

While the vector stepper operates on a basket, the input vector gets updated with the proposed steps sampled by a physics manager. This uses cross sections tabulated from a *Geant4* physics list for all the processes involved by the simulation. The vectorized geometry navigator queries the geometry for the geometrical step limits for each track. After invoking the field propagator, the track vector gets gradually smaller as particles reach their proposed steps, so the compacting mechanism has to be called several times for the vector container. The code can automatically switch to a scalar approach when the vectors become inefficient for vector processing. An important mechanism implemented in the scheduler is the possibility to postpone single particles (not yet fully propagated) to be regrouped later with others matching the same locality criteria. Postponing is triggered systematically when geometry locality can be easily fulfilled (e.g. after firing a new event), but much less in the "sparse" regime when particles are spread in the full detector, having a wide range of energies. At the end of a propagation step, all particles (primaries as well as generated secondaries) are copied in the output vector of the basket.

*3.1. Workload balancing and queue control*

An important concurrency feature is automatic workload balancing. In *GeantV* this is achieved by using a single queue where any worker thread can retrieve the next basket to be processed. In this approach, an empty queue implies wasted idle resources waiting for work, but an over-populated queue points to inefficient vector size and additional overheads due to inter-thread communication in the "basketizing" phase. The scheduler has therefore to constantly monitor the work queue load and take actions to correct the "bad" work regimes.

The parameters used to control the queue sanity are the low and high watermarks for the number of contained baskets, which are subject to optimization. These are now chosen in a more or less arbitrary manner, and reaching them triggers specific actions. The low watermark triggers a special garbage collection by event range, to transport with priority tracks belonging to "old" events. Once an event and its corresponding data structures get flushed and digitized, the generator produces a new event, increasing the queue size. The high watermark triggers a re-evaluation of the thresholds used by the locality filters to "basketize" tracks, generating larger vectors and therefore decreasing the queue size at the profit of vectorization.

The algorithm for choosing the "appropriate" vector size is an important scheduling parameter to optimize. This depends on many factors, like the filter type, physics, or geometry complexity. These dependencies and correlations with other scheduling parameters are not well understood yet, but we are currently doing scans of the performance for different parameter values. In figure 2 we can observe that the performance can vary by more than 30% when changing the value for the particle vector size, with an optimal value of 64 tracks per basket. This corresponds however to a very simple geometry setup and just proves the principle. One possible approach that we are considering for optimizing such critical model parameters is based on genetic algorithms [7], as described in section 5.

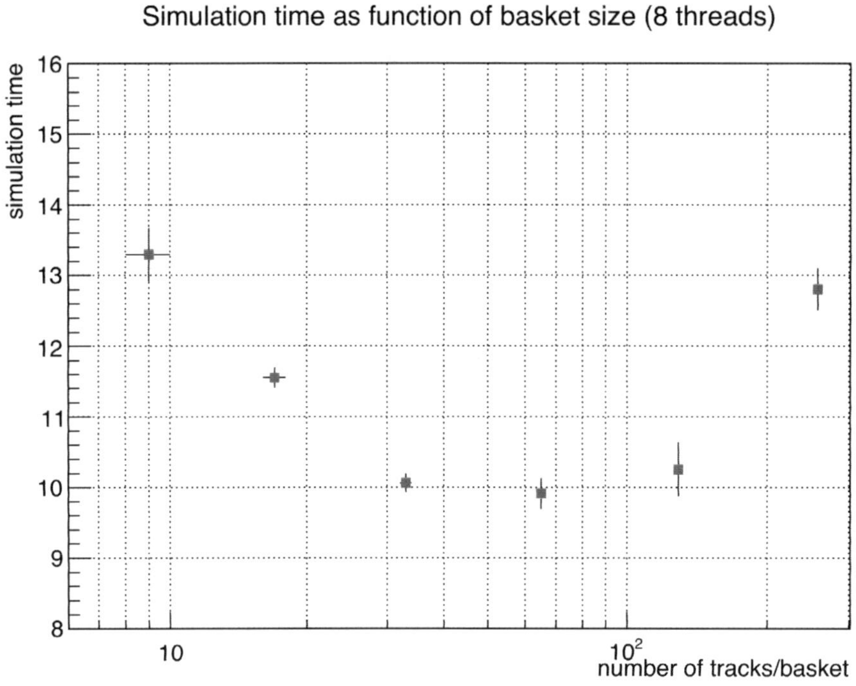


Figure 2. The simulation time has an optimum against the track vector size. Small vectors imply inefficient vectorization and dispatching overheads, while long vectors are penalized by frequent garbage collections in the "sparse" regime when locality is hard to enforce. The plot was produced on X86_64 apple-darwin13 Core i7 supporting AVX2.

## 4. Scalability and performance

Scalability is a very important requirement for the *GeantV* scheduler in order to minimize the memory footprint and maximize cache coherency. This is difficult to achieve due to the fine-grained parallelism, which requires concurrent data scatter and gather between propagation steps.

The first non-vectorized version of the prototype only needed to copy track pointers to new baskets, which was a lightweight operation that could be done for up to 12 workers by a single scheduling thread. The current vectorized version has to copy the full track data, which can easily bottleneck when executed by a single thread. An improved version of the scheduler performs now concurrent re-basketizing, but still suffers from an important serial bottleneck due to concurrent copy operations of track data onto a single receiving track container. To alleviate this, we are investigating an approach with multiple filtering threads managing alone the copy operations for subsets of the total basket types.

We are constantly monitoring the scalability of the vector prototype using the simple geometry setup presented in figure 3a, imported from one of the *Geant4* novice examples. This allows testing and comparing regularly different approaches for concurrency. Most of the tests are performed on an Intel® Xeon® CPU E5-2695 v2 @ 2.40 GHz machine with 24 native threads supporting AVX2 instructions. It is clear that more work needs to be done to improve this property, and the current goal is to achieve a good scalability up to at least 12 workers. One of the important milestones of the project is verifying that the new transport scheduling approach brings indeed performance improvements compared to the classical approach. This requires the ability to reproduce in a realistic way the physics (at least electromagnetic) for a complex setup and compare the performance with a standard *Geant4* run. Since we do not have yet all the ingredients to do that, we were aiming for a more simple check to hint for performance improvement or degradation.

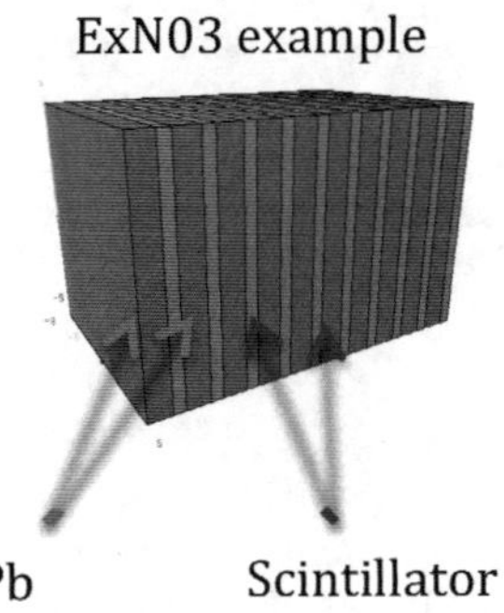


Figure 3a. Simple *ExN03* setup used for the nightly scalability tests run on.

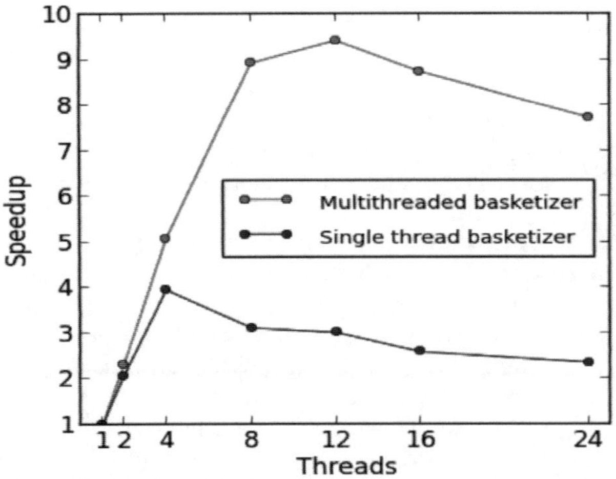


Figure 3b. Scalability plots of the speedup against number of worker threads. The blue line corresponds a single thread performing track re-basketizing between tracking steps. The red line corresponds to the current concurrent basket filling.

To reproduce the physics using a standard *Geant4* physics list we tabulated the cross sections for all the processes, particles and elements involved in the simple setup from Figure 3a. We have also tabulated final states coming from these processes, and implemented a special tabulated cross section process to have the exact same behavior in *Geant4*. Without entering into the full details of this procedure, we were able to compare the results for the following three configurations: *Geant4* simulation using the tabulated cross section process, *GeantV* using the scalar *ROOT* geometry for navigation, and *GeantV* using the vectorized *VecGeom* geometry to check the impact of vectorization.

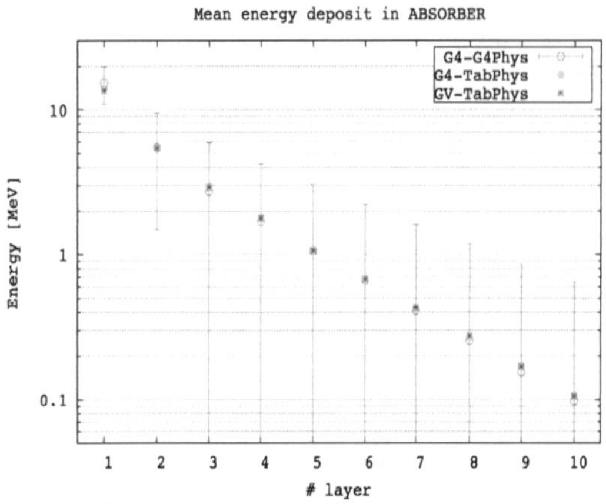


Figure 4a. The *GeantV* simulations based on tabulated physics and the corresponding native *Geant4* simulation produce compatible physics results (here energy deposit per layer in the *ExN03* setup)

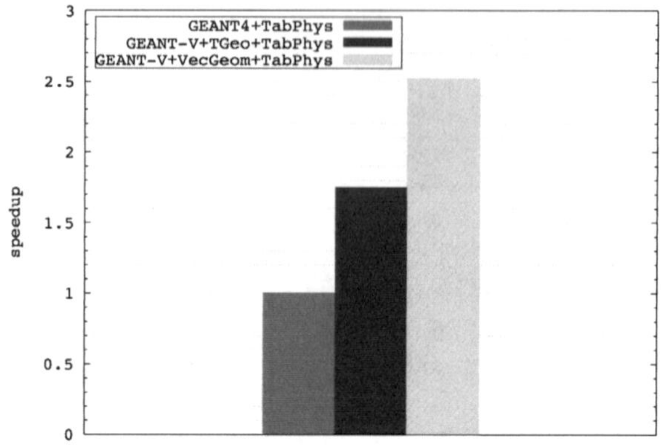


Figure 4b. Performance profile of the vector prototype for the *Geant4 ExN03* example. The performance improvements in case of tabulated physics are mainly due to geometry vectorization and better caching, despite the data gather overheads.

While getting statistically compatible results for the energy deposit and track length in the scintillator layers for incident electrons with energies ranging from *30 MeV* to *30 GeV*, we can observe a sizeable improvement of the total transport time even for this very simple setup compared to the *Geant4* + tabulated physics case. While an important contributor appears to be the vectorized geometry, the caching gains due to vector treatment have clearly a major impact. We will soon test a more complex setup, where we hope that the increased computational complexity of the vectorized code will enhance the benefit of the vectorized treatment of tracks.

## 5. Plans for optimising scheduling parameters

The *GeantV* scheduling strategy is based on a model that has a certain complexity and a considerable number of critical parameters that need to be tuned. The tuning procedure itself is not obvious and requires the understanding of single parameter behaviors, work that has already started.

We are currently investigating the possibility of using genetic algorithms as one possible tool for the optimization procedure. This will use as model "chromosomes" the states or ranges of the current model parameters, like the work queue thresholds, the vector sizes per locality filter or the number of worker threads. The evolution of the population of chromosomes will be done by mutation or crossover of these properties, and aim to improve the total simulation run time while keeping memory under controlled limits.

Since it is likely that the phase space for this procedure repeated for different geometry and physics setups can hardly be covered by a single set of parameters for all cases and even from the beginning to the end of the same simulation, we are foreseeing an adaptive behavior of the *GeantV* prototype, based on a very short training time of the model and a parameterization extracted from the genetic algorithm procedure.

## References

[1] Apostolakis J, Brun R, Carminati F and Gheata A 2012 *J.Phys: Conf. Ser.* **396** 022014 http://iopscience.iop.org/1742-6596/396/2/022014

[2] S Agostinelli et al 2003 *Nuclear Instruments and Methods* **A 506** (53pp)

[3] http://root.cern.ch
[4] Apostolakis J, Brun R, Carminati F, Gheata A and Wenzel S 2014 *J. Phys.: Conf. Ser.* **513** 052038
http://iopscience.iop.org/1742-6596/513/5/052038
[5] Marek Gayer et al 2012 *J. Phys.: Conf. Ser.* **396** 052035
http://iopscience.iop.org/1742-6596/396/5/052035
[6] Bilk A, Girkar M, Grey P and Tian X 2002 *Int. J of Parallel Programming* **30**, No. 2
[7] Goldberg G 1989 Genetic Algorithms in Search, Optimization, and Machine Learning **ISBN-13**: 078-5342157673

# The Error Reporting in the ATLAS TDAQ System

**Serguei Kolos**
University of California Irvine, USA
E-mail: `serguei.kolos@cern.ch`

**Andrei Kazarov**
CERN, Switzerland, on leave from Petersburg Nuclear Physics Institute, Kurchatov NPI, Gatchina, Russian Federation

**Lykourgos Papaevgeniou**
CERN, Switzerland

**Abstract.** The ATLAS Error Reporting provides a service that allows experts and shift crew to track and address errors relating to the data taking components and applications. This service, called the Error Reporting Service (ERS), gives to software applications the opportunity to collect and send comprehensive data about run-time errors, to a place where it can be intercepted in real-time by any other system component. Other ATLAS online control and monitoring tools use the ERS as one of their main inputs to address system problems in a timely manner and to improve the quality of acquired data.

The actual destination of the error messages depends solely on the run-time environment, in which the online applications are operating. When an application sends information to ERS, depending on the configuration, it may end up in a local file, a database, distributed middle-ware which can transport it to an expert system or display it to users. Thanks to the open framework design of ERS, new information destinations can be added at any moment without touching the reporting and receiving applications.

The ERS Application Program Interface (API) is provided in three programming languages used in the ATLAS online environment: C++, Java and Python. All APIs use exceptions for error reporting but each of them exploits advanced features of a given language to simplify the end-user program writing. For example, as C++ lacks language support for exceptions, a number of macros have been designed to generate hierarchies of C++ exception classes at compile time. Using this approach a software developer can write a single line of code to generate a boilerplate code for a fully qualified C++ exception class declaration with arbitrary number of parameters and multiple constructors, which encapsulates all relevant static information about the given type of issues. When a corresponding error occurs at run time, the program just need to create an instance of that class passing relevant values to one of the available class constructors and send this instance to ERS.

This paper presents the original design solutions exploited for the ERS implementation and describes how it was used during the first ATLAS run period. The cross-system error reporting standardization introduced by ERS was one of the key points for the successful implementation of automated mechanisms for online error recovery.

## 1. Introduction

ATLAS [1] is one of the four major experiments at the Large Hadron Collider accelerator at CERN. The ATLAS Trigger and Data Acquisition (TDAQ) [2] system selects and transports physics data from the 1600 detector read-out links to the mass storage. The system is composed of about 20K applications distributed over 3K computers. Controlling and monitoring such system requires flexible and reliable services for error reporting and handling both inside individual applications and between applications. The Error Reporting Service (ERS) gives software applications the opportunity of collecting and reporting comprehensive data about run-time issues as well as subscribing to the messages produced by the other applications. The ATLAS online control and monitoring tools use ERS as one of their inputs for getting information about the TDAQ system problems and reacting to them in a timely manner for improving the quality of acquired data. Figure 1 shows the main flow of the ERS messages in the TDAQ system.

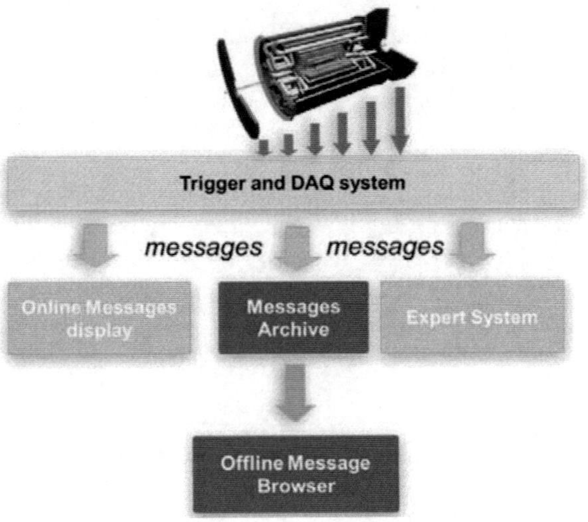


**Figure 1.** The flow of the ERS messages in the TDAQ system.

ERS is available in all programming languages used by the ATLAS software: C++, Java and Python. While the system design is the same for all of them, the implementations have their peculiarities which are defined by the specific features of the given programming language as discussed in the following sections.

## 2. The Error Reporting Interface

ERS interface is build around three main concepts: the Output Stream, the Issue and the Input Stream. An Issue contains description of a specific problem which can be reported to the Output Stream and can be retrieved, in the context of another application, via the Input Stream.

### *2.1. The Output Streams*

The Output Stream is a simple interface with a single function for sending arbitrary messages to that stream. There are six types of streams corresponding to different levels of severity of the reported issues: `DEBUG`, `LOG`, `INFO`, `WARNING`, `ERROR` and `FATAL`. The severity of any issue is established by the type of stream to which it has been sent. While ERS provides several default stream implementations, which can be used out of the box, it is an open framework which allows plugging in new implementations at any moment. Thus, new message destinations can be added without touching the frameworks code. Each Output Stream interface implementation is responsible for:

Journal of Physics: Conference Series **608** (2015) 012004 doi:10.1088/1742-6596/608/1/012004

- taking an action which is specific to that stream. This can be sending the given issue to an appropriate output device, e.g. standard output, database or a mobile phone,
- deciding whether the given issue has to be propagated further through the chain of stream implementations or immediately suppressed. This feature allows implementing the issue filtering.

### *2.2. The Output Streams Configuration*
The destination of the ERS messages depends solely on the run-time environment, in which the online applications are operating. Depending on the actual configuration, the messages which applications send to ERS may end up in a local file, a database, a message passing middle-ware, or in any other output which is supported by ERS. New output devices can be added to ERS as plugins without touching the ERS code. Stream configuration is very flexible and can be defined for each particular stream type at the level of an individual software process by setting the appropriate environment variables. A value of such variable is a comma-separated list of tokens where each token is a well defined ID of a stream implementation class optionally followed by the list of initialization parameters for that implementation. For example the Error stream can be configured via application environment in the following way:

```
TDAQ_ERS_ERROR="stderr, filter(Tile), abort"
```

In this case all errors will be sent to the standard error stream and any issue originated from the Tile Calorimeter subsystem will immediately abort the running application.

### *2.3. The Issue Class*
The main class for reporting problems is called Issue. This class inherits from the std::exception in order to provide compatibility with the standard C++ library. Any custom ERS exception can be caught as std::exception.

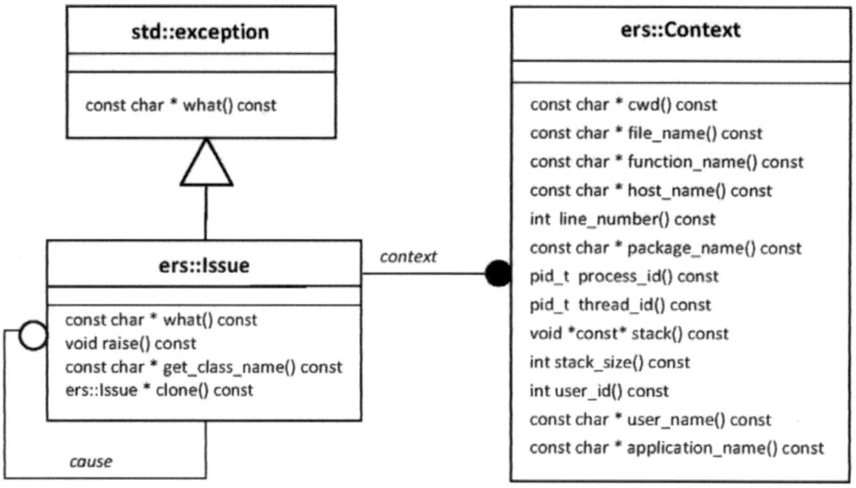


**Figure 2.** The main ERS public API classes.

The Issue class is abstract. Any software package which has to report an error must declare a specific issue class by inheriting it from the abstract ers::Issue or alternatively reuse an existing issue class from another package. In this case each kind of problem is described by the corresponding C++ class, which facilitates the application of the expert system techniques for errors analysis and decisions taking procedures. As shown in Figure 2, each issue has the Context attribute which describes the place in the code where the issue has occurred. In C++ this information is provided by the special macro `ERS_HERE`, which must be the first argument to each Issue constructor. In Java and Python such information is extracted in the Issue constructor itself, so no additional arguments is used. An issue may contain a pointer to another issue which provides more detailed information on the cause of the problem. Such chain of issues may have arbitrary depth thus giving any required level of details for the problem description.

### *2.4. The Input Stream*

Within the same process ERS issues are passed using the programming language exception mechanism, e.g. `try...catch` in C++ and Java and `try...except` in Python. For the inter-process issue handling ERS defines the asynchronous `Input` stream interface for receiving ERS Issues reported by other applications. The idea is that for each `Output` stream implementation one can implement a corresponding `Input` stream so that it will be capable of receiving the issues reported to its `Output` counterpart. For a pair of such streams a user can simply register a callback function with the `Input` stream implementation to get all issues reported by all processes to the corresponding `Output` stream.

## 3. C++ ERS API

Having well defined and strictly typed errors is highly desirable for simplifying the system maintainability and absolutely indispensable for replacing human operator with expert system. On the other hand writing issues classes declarations would have been tedious and error prone. To overcome this problem ERS uses the mind breaking BOOST Preprocessor [3] package. Despite conventional opinion the usage of macro constructs in C++ may be extremely useful and convenient, drastically reducing the amount of code which has to be written and improving the code quality and maintainability. The following two macros have been designed for the C++ ERS API and can be used for declaring component specific exception classes.

```
ERS_DECLARE_ISSUE(namespace, class_name, message, class_attributes)

ERS_DECLARE_ISSUE_BASE(namespace, class_name, base_class_name,
                       message, class_attributes, base_class_attributes)
```

The first macro can be used for declaring classes inheriting the ers::Issue class, while the other one can declare classes derived from the other macro-defined ERS exceptions. The most interesting part in these macros is the ability of passing arbitrary number of the class attributes via the `class_attributes` and `base_class_attributes` parameters. This is where the BOOST Preprocessor package, which implements advanced parsing and looping functions for standard C macro, has been exploited. A class attributes is the sequence of the type and name macro tuples which looks like:

```
((attribute1-type) attribute1-name)
((attribute2-type) attribute2-name)
...
((attributeN-type) attributeN-name)
```

The macro generate C++ classes with the constructors containing a given number of parameters. A C++ developer will need to provide appropriate values for all the attributes when constructing the corresponding exception. Those values will be stored in the exception and inserted into the main message of the issue, which is declared only once at the issue definition time. This C++ class also provides accessor functions, like `get_attributeX-name()` for all declared attributes. For example placing the following declaration into a C++ file:

```
ERS_DECLARE_ISSUE(
  io,
  FileIssue,
  "Basic issue with '" << file_name << "' file",
  ((const char *)file_name ))
```

will produce the C++ code like:

```
namespace io {
  class FileIssue {
    FileIssue(const ers::Context & context , const char * file_name );
    FileIssue(const ers::Context & context , const char * file_name,
              const std::exception & cause );

    const char * get_file_name() const;
  };
}
```

The `ERS_DECLARE_ISSUE_BASE` macro is similar to the `ERS_DECLARE_ISSUE` one but allows declaring hierarchies of exceptions classes. For example one can declare the `io::PermissionDenied` and `io::CantOpenFile` exceptions which inherit the `io::FileIssuein` the following way:

```
ERS_DECLARE_ISSUE_BASE(
  io,
  PermissionDenied,
  FileIssue,
  "Insufficient privileges for '" << operation << " '" << file_name
          << " file which has '" << permissions << "' permissions",
  ((const char *)file_name ),
  ((int)permissions) ((const char*)operation))

ERS_DECLARE_ISSUE_BASE(
  io,
  CantOpenFile,
  FileIssue,
  "Cant open file " '" << file_name << "' file",
  ((const char *)file_name ),
  ERS_EMPTY )
```

### *3.1. Catching and Reporting Issues*
There are six functions for reporting issues to different ERS streams, e.g. ers::debug(), ers::log(), ers::info(), ers::warning(), ers::error() and ers::fatal(). The last three of them accept only strictly typed messages, i.e. objects of a type inherited from the ers::Issue. In this way the important messages are imposed to be strictly typed, allowing the TDAQ controlling and monitoring systems to rely on their content and meaning. An example of the usage of the three functions for reporting issues which accept only strictly typed messages is shown below.

```
try {
  open_file( file_name );
}
catch ( io::PermissionDenied & ex ) {
  ers::warning(io::CantOpenFile( ERS_HERE, file_name, ex ));
}
catch ( io::FileIssue & ex ) {
  ers::error( ex );
}
catch ( std::exception & ex ) {
  ers::fatal(io::FileIssue( ERS_HERE, file_name, ex ));
}
```

The other three streams accept both the ers::Issue instances and arbitrary dynamically constructed information. For the latter ERS provides three macros: `ERS_DEBUG`, `ERS_LOG`, `ERS_INFO`, which can be used to pass arbitrary information to ERS without defining a new issue type. For example:

```
ERS_DEBUG( 1, "test debug macro " << 12345 );

ERS_LOG( "So far " << event_number << " events have been collected");

ERS_INFO( "The run " << run_number << " has been started" );
```

## 4. Java ERS API

Java ERS API supports exceptions which are derived from the `ers.Issue` as well as the standard Java exceptions, which can be reported by third party libraries, e.g. derived from the `java.lang.Exception`.Contrary to C++ Java does not have macro preprocessor so exception classes have to be declared manually, e.g.:

```
public class ConnectionIssue extends ers.Issue {
  public String reason ;
  public int port_number ;

  ConnectionIssue (String reason, int port_number) {
    super("Connection to %p port failed because of %s", port_number, reason);
  }

  ConnectionIssue (String reason, int port_number, ers.Issue cause) {
    super("Connection to %p port failed because of %s", port_number, reason, cause);
  }
};
```

Note that one has to declare two constructors: one takes another exception as its last argument to allow chains of the issues and the other one has only the actual issue specific parameters to be used for creating the first level exception.

### *4.1. Reporting Issues*

Reporting of ERS Issues in Java is done using static functions declared in the `ers.Logger` class. The *Debug*, *Info* and *Log* functions support plain strings as their arguments, other methods only accept `ers.Issue` or `java.lang.Exception` objects. For example:

```
...
catch { java.net.SocketException ex } {
  ers.Logger.error(new ConnectionIssue("Socket is not available", port, ex)));
}
catch { java.lang.StackOverflowError ex } {
  ers.Logger.fatal(ex) ;
}

ers.Logger.debug(2, "Verbose debug message");
ers.Logger.log("A message to be logged in log file");
```

### *4.2. Log4J Mapping to ERS*

Most of third-party java libraries widely used in TDAQ software use de-facto standard Java logging framework `log4j`(http://logging.apache.org/log4j/2.x/) for reporting messages. To allow log4j messages to be routed over ERS, Java ERS library includes an implementation of log4j *appender*, which converts log4j messages to ERS issues and redirects them to the appropriate ERS streams. For example the following `log4j` configurations routes all `DEBUG` messages to ERS stream.

```
<log4j:configuration xmlns:log4j="http://jakarta.apache.org/log4j/">
   <appender name="ERS" class="ers.log4j.ERSAppender">
    <layout class="org.apache.log4j.SimpleLayout"/>
   </appender>

   <root>
    <level value="DEBUG" />
    <appender-ref ref="ERS"/>
   </root>
</log4j:configuration>
```

## 5. Handling Issues in the Distributed Environment

In order to report and receive ERS issues in the distributed TDAQ system, two special `Input` and `Output` streams have been implemented. They are based on the Message Transfer System (MTS) [4], which is the CORBA [5] based distributed message passing middle-ware developed in the ATLAS TDAQ project.

## 6. Conclusion

During the Run 1 ERS has proved to be extremely useful for both error detection and problems analysis. In total about 30M ERS messages have been produced by the TDAQ system processes and reported via ERS. All those messages have been archived to the Oracle [6] database. A dedicated application called Log Browser has been provided for browsing the content of that database. At run time about 10 different types of ERS issues have been used by the ATLAS online expert system for error detection and handling during data taking.

For the upcoming data taking period many more ERS messages are going to be used by the online expert system to improve the automation of the data taking. In addition to that the archived system back-end will be changed to Splunk [7], which is an easy-to-use web interface and powerful enterprise platform for analyzing machine generated data, like log files, message archives, etc. This allows having access to the archived messages from all over the world without the need of installing any ATLAS specific software.

Journal of Physics: Conference Series **608** (2015) 012004 doi:10.1088/1742-6596/608/1/012004

## References

[1] ATLAS Collaboration, The ATLAS experiment at the CERN large hadron collider, Journal of Instrumentation, vol. 3, no. 08, p. S08003, Aug. 2008. Available: http://iopscience.iop.org/1748-0221/3/08/S08003

[2] ATLAS Collaboration, ATLAS high-level trigger, data-acquisition and controls, CERN, Technical Design Report ATLAS-TDR-016 CERN-LHCC-2003-022, 2003. Available: http://cdsweb.cern.ch/record/616089/

[3] Boost Preprocessor library home page http://www.boost.org/doc/libs/release/libs/preprocessor/

[4] A. Kazarov et al., A Scalable and Reliable Message Transport Service for the ATLAS Trigger and Data Acquisition System, *Proc. of the 19th Real-Time IEEE Conf., Nara, Japan, May 2014*. Available: http://cds.cern.ch/record/1703434

[5] CORBA home page http://www.corba.org

[6] Oracle home page http://www.oracle.com

[7] Splunk home page http://www.splunk.com

# Data-flow Performance Optimisation on Unreliable Networks: the ATLAS Data-Acquisition Case

**Tommaso Colombo on behalf of the ATLAS Collaboration**
CERN, CH-1211 Geneva 23, Switzerland and Institut für Technische Informatik (ZITI), Universität Heidelberg, B6 26, 68131 Mannheim, Germany

E-mail: Tommaso.Colombo@cern.ch

**Abstract.** The ATLAS detector at CERN records proton-proton collisions delivered by the Large Hadron Collider (LHC). The ATLAS Trigger and Data-Acquisition (TDAQ) system identifies, selects, and stores interesting collision data. These are received from the detector readout electronics at an average rate of 100 kHz. The typical event data size is 1 to 2 MB. Overall, the ATLAS TDAQ system can be seen as a distributed software system executed on a farm of roughly 2000 commodity PCs. The worker nodes are interconnected by an Ethernet network that at the restart of the LHC in 2015 is expected to experience a sustained throughput of several 10 GB/s.

A particular type of challenge posed by this system, and by DAQ systems in general, is the inherently bursty nature of the data traffic from the readout buffers to the worker nodes. This can cause instantaneous network congestion and therefore performance degradation. The effect is particularly pronounced for unreliable network interconnections, such as Ethernet.

In this paper we report on the design of the data-flow software for the 2015–2018 data-taking period of the ATLAS experiment. This software will be responsible for transporting the data across the distributed Data-Acquisition system. We will focus on the strategies employed to manage the network congestion and therefore minimise the data-collection latency and maximise the system performance. We will discuss the results of systematic measurements performed on different types of networking hardware. These results highlight the causes of network congestion and the effects on the overall system performance.

## 1. Introduction: the ATLAS Trigger and Data-Acquisition system in 2015–2018

ATLAS [1] is one of the experiments installed at the Large Hadron Collider (LHC), CERN, Geneva, Switzerland. It observes proton-proton collision events delivered by the LHC at a design frequency of 40 MHz. Each event corresponds to a data size of 1 to 2 MB. The ATLAS Trigger and Data-Acquisition (TDAQ) system [2] is responsible for the selection of interesting events, reducing the initial frequency of 40 MHz to $\sim 1$ kHz of stored events. This requires an overall trigger rejection factor of the order of $10^4$ against minimum bias events, while retaining the rare new physics processes, such as Higgs boson decays. The TDAQ system, outlined in Figure 1, is based on the combination of a hardware-based first-level trigger and a software-based High-Level Trigger.

The first-level trigger (Level-1) is a synchronous pipelined electronics system, with a guaranteed maximum latency of 2.5 μs. It selects events using coarse-grained data from the calorimeters and the muon spectrometers. It triggers the detector readout at a maximum rate of 100 kHz, imposing the necessary dead-times to protect the detector front-end buffers. As part

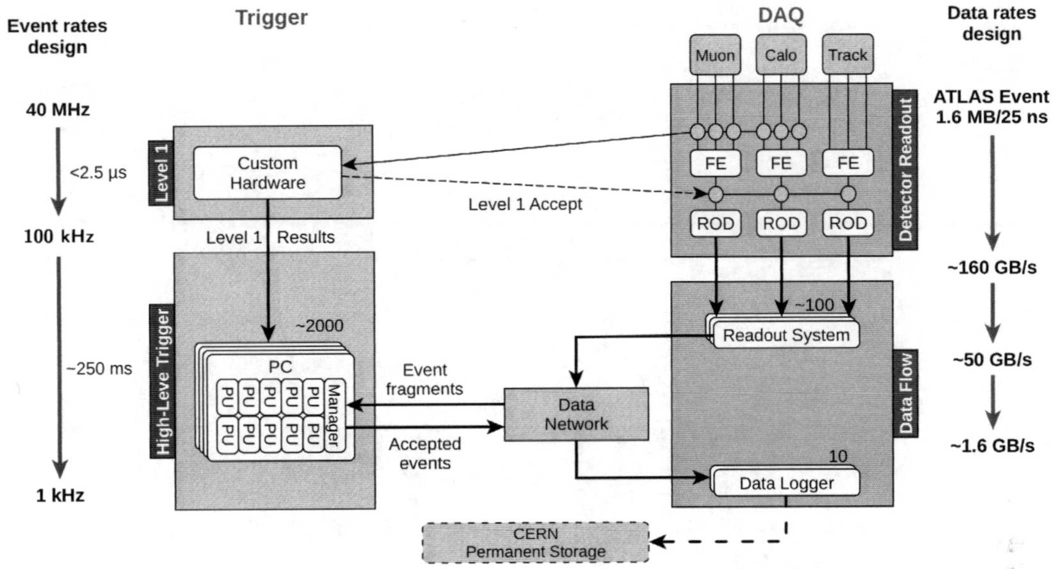


**Figure 1.** Outline of the ATLAS Trigger and Data-Acquisition system, stating the event and data rates foreseen for the 2015–2018 data-taking period.

of the selection process, it identifies a so-called region-of-interest (RoI): a part of the detector that recorded interesting signals that is used as a seed for the High-Level Trigger.

The High-Level Trigger (HLT) is a distributed software system running on around 2000 PCs interconnected by an Ethernet network. When an event is accepted by the Level-1, its data fragments (1860 fragments of variable sizes around 1 kB) are distributed, using custom optical links, to hardware buffers in the Readout System (ROS) nodes. At the same time, region-of-interest information is assigned to one of the HLT Processing Units. Each HLT worker node hosts one HLT Processing Unit per CPU core. Using the region-of-interest as a starting point, the Processing Unit incrementally retrieves and analyses event fragments, until a decision can be taken. The event can be rejected even without analysing all its fragments, thus limiting the fraction of data to be retrieved. Fragments of rejected events are deleted from the ROS buffers, while accepted events are transferred to one of the Data Logger nodes for storage. The rate of events that can be accepted is mostly determined by the availability of offline computing and storage resources, and is foreseen to be around 1 kHz for the 2015–2018 data taking period.

The physical layout of the Data-Acquisition and High-Level Trigger system is represented in Figure 2. The core of the system consists of two large network routers with a maximum capacity of several hundred 10GbE ports. ROS nodes are directly connected to both core routers with 4 10GbE links. HLT nodes are organised in racks of at most 40 nodes. Each node in a rack is connected to an aggregation switch with a GbE link. The rack switches have 10 GbE links to both core routers. For the rationale behind this design, please refer to [3].

## 2. Performance challenges: network congestion

The latency of the data transfer between the ROS and the HLT Processing Units (i.e. the data collection latency) is critical to the performance of the whole system: each of the HLT Processing Units operates exclusively on an event assigned to it, with the HLT selection proceeding iteratively starting from the region-of-interest identified by the Level-1, collecting data fragments incrementally as needed. A Processing Unit is blocked while it waits for the data fragments to

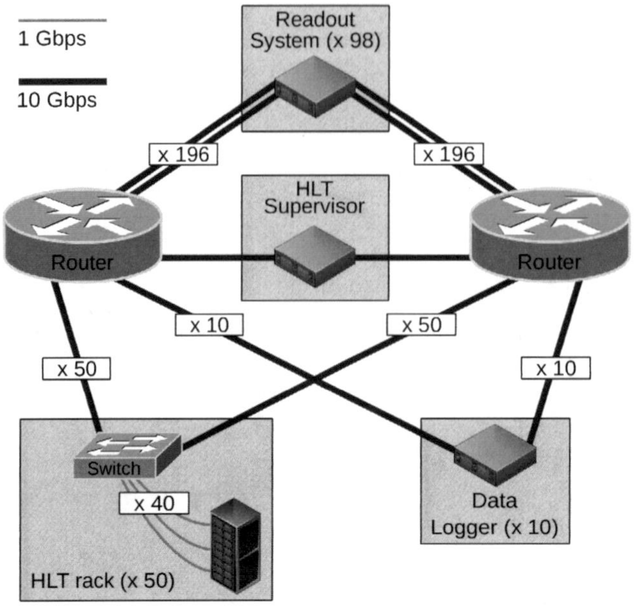


**Figure 2.** Network architecture of the ATLAS Data-Acquisition and High-Level Trigger system.

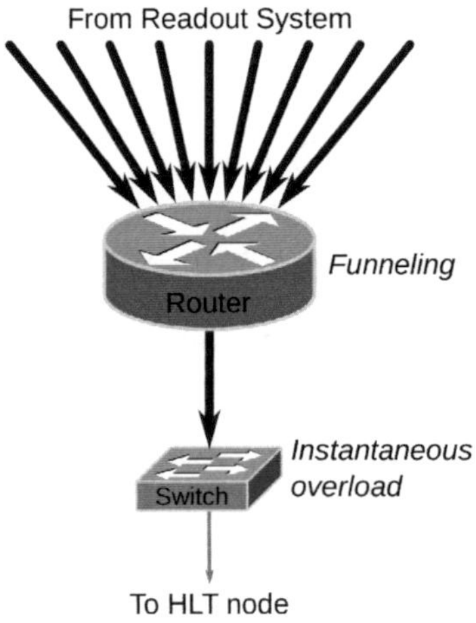


**Figure 3.** Visualisation of the network congestion.

be collected, which means that the data collection latency effectively translates into lost CPU time.

Data transfers from the ROS are naturally bursty. Event fragments are striped over all the ROS nodes, since each node buffers data from a specific region of the detector. A single HLT Processing Unit usually requests fragments from multiple ROS nodes at the same time. As the fragments are already available in the ROS buffers and are sent as soon as the request reaches the ROS, many nodes will start sending fragments at the same time to the same destination, thus creating instantaneous network congestion (see Figure 3).

As mentioned in Section 1, the network technology used throughout the HLT system is

Ethernet. The messaging among applications is based on the TCP protocol. Ethernet provides an "unreliable" transport for network packets. In particular, packets can be dropped in response to a variety of conditions, including network buffer overflows caused by congestion. TCP provides reliable connections by detecting dropped packets and retransmitting them. However, this comes at a price: TCP has both an active packet drop detection mechanism, in which the receiver detects the lost packet and causes the sender to react very quickly, and a passive mechanism, which is based on relatively long timeouts ($\geq$ 200 ms) on the sender side [4]. Unfortunately, in cases of network congestion, the passive mechanism plays a much more important part than the active mechanism. Hence, network congestion can radically impact the data collection latency and therefore the overall performance of the system.

## 3. Preventing packet drops: a client-side traffic shaping algorithm

Smoothing the rate of data requests generated by a HLT node can alleviate the network congestion by controlling the maximum size of the traffic burst from the ROS. Obviously such a smoothing mechanism imposes a trade-off: excessive smoothing can increase the data collection time by unnecessarily delaying the requests for data, whereas insufficient smoothing will not eliminate packet drops.

For the measurements presented here, a credit-based traffic shaping algorithm was used. Its basic rules are as follows.

- Each HLT node has a fixed number of credits available.
- Each data request to a ROS node uses as many credits as the number of fragments it asks for.
- Each response returns the credits used by the corresponding request.
- If all available credits are used, further requests are blocked until the necessary credits become available.

Since all event fragments are similar in size, the number of fragments in a request gives a rough estimation of the size of the corresponding response. Therefore, this algorithm effectively limits the maximum burst size of data transfers directed to the same HLT node.

## 4. Measuring the effects of congestion

Detecting buffer overflows in an Ethernet network is a relatively simple task: most switches publish cumulative counts of dropped packets per port and the Linux kernel provides the total number of packet retransmissions that have occurred in a TCP connection. However, the relationship between the dropped packets count and the overall application performance is rather complex, especially when using a protocol with congestion control features, such as TCP. The parameter space to investigate is very large and the monitoring information (e.g. instantaneous buffer occupancies) scarce or unavailable. Therefore, generating synthetic traffic patterns, which can be tightly controlled and are known in advance, proves instrumental in investigating the performance profile of the network. The tests described in the following paragraphs employ this approach.

The key application performance metric in the ATLAS HLT case is the total data collection latency per event: as mentioned in Section 2, data collection directly translates to an inefficiency source for the HLT farm.

### *4.1. Test set-up*

The tests were performed using some of the hardware available in the ATLAS TDAQ system. At the time of these tests, the TDAQ infrastructure was still under consolidation. Hence the test set-up is slightly different from the one described in Section 1. In particular:

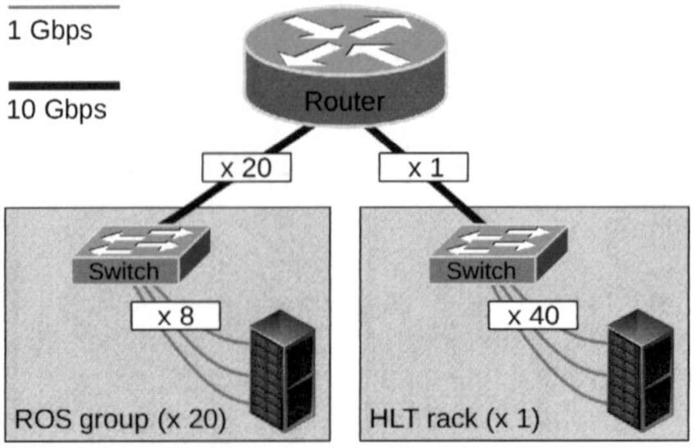


**Figure 4.** Test set-up.

- ROS nodes were not directly connected with 10GbE links to the core routers; they were instead connected with GbE links to an intermediate aggregation switch with a 10GbE uplink to the core (the intermediate switch was kept under-subscribed so that no congestion could appear there).
- HLT racks are only connected to one of the two core routers.

Given these constraints, the following test set-up, shown in Figure 4, was chosen:

- 20 ROS groups:
  - Each group consists of 8 nodes connected to one switch (160 total).
  - 12 event fragments of 1.1 kB are served by each node (1920 total fragments, 2112 kB full event size).
- 1 HLT rack:
  - It consists of 39 PCs connected to one switch.
  - Each PC hosts 24 HLT Processing Units (936 total).
- The uplinks from the switches connect to a single core router.

This configuration was chosen because it provides with a realistic model of the expected network congestion phenomena in the final system topology. In particular, the congestion at the rack-level switch is well represented.

The core router uses input-buffering with a peculiar implementation of switch-level virtual output queueing: input ports are grouped in modules of 8 ports at most and each module maintains multiple, distinct queues to every output port on the router. Buffer space is abundant: 1.5 GB per module, shared among the queues.

The top-of-rack switches use output-buffering. Two different models were tested:

- a switch with small (600 kB), per-port dedicated buffers
- a switch with a small (a few MB) buffer, shared among all ports
- a switch with a large (1.25 GB) buffer, shared among all ports

The results obtained with such a set-up should scale reliably with the higher number of HLT racks in the complete system: the network congestion phenomena individually affect the output buffers of the network ports that are actually in use.

### *4.2. Simple test scenario: full event building*

A very simple traffic pattern is the so-called full event building:

- Events are assigned to HLT Processing Units at a constant rate.
- The Processing Units immediately collect all fragments of assigned events.

It should be noted that this pattern is the harshest in terms of generated traffic bursts: since the data corresponding to an event is collected all at once, the maximum burst size corresponds to the event size.

The assignment rate parameter for this test was selected with the goal of utilising roughly 90% of the available bandwidth of the HLT rack uplink: at the chosen 500 Hz assignment rate, the total rack input bandwidth was $\sim$ 1.1 GB/s.

The results of the measurements are shown in Figure 5. The total data collection time per event (shown in the plots on the left) is influenced both by the network conditions and by the traffic shaping mechanism: with few traffic shaping credits available, the large data-collection time is due to collection inefficiency because the HLT nodes cannot fully utilise the network bandwidth; with many traffic shaping credits the network congestion determines the performance.

It is possible to estimate the minimum possible value for the data collection latency to use as a reference point. For the sake of simplicity, the time it takes for data requests to reach the ROS nodes and the time it takes for the ROS to prepare the data can be neglected[1]. The dominant component is given by the so-called packetisation delay of the data, i.e. the amount of time it takes to transmit the data on the slowest link in the path. In this set-up, this corresponds to the amount of time it takes to transmit a full event on the GbE link from the rack-level switch to the HLT node, which is $\sim$ 18 ms.

When the switch with small per-port buffers is used, each buffer can be though of as belonging to the HLT node connected to its switch port. As the traffic bursts directed to an HLT node get bigger than the size of the buffer, packets are dropped: the maximum and average round trip-times (shown in the plots on the right) increase steeply and the data collection time is accordingly influenced. Since fixed-size fragments were used for these tests, the maximum number of credits assigned to each HLT node translates directly to the size of the maximum allowed traffic burst (e.g. 200 credits, with a 1.1 kB fragment size, translate to a burst size of 220 kB plus protocol overhead). As expected, the minimum data collection latency results from a configuration with 500 credits (i.e. a maximum burst size of 550 kB plus protocol overhead) in which the buffers are used almost up to their maximum capacity, but are never overflowed.

The results using the switch with a small shared buffer is less easily interpreted: the burst size is not the only relevant parameter. The shared buffer is bigger than the $\sim$ 2 MB full event size, so, in principle, for a single burst no overflows can occur. However, since the buffer is shared, the dynamics of the system must be considered: the shared buffer is filled with event data by a 10GbE input and emptied by multiple GbE outputs, handling one event each. Therefore, multiple traffic bursts, corresponding to different events targeting different outputs, will overlap in the shared buffer, requiring a buffering capacity that is several times higher than the maximum burst size. As a consequence, also in this case the minimum data collection latency results from a configuration with a maximum burst size reduced using traffic shaping.

## 5. Summary

A particular challenge to the performance of the ATLAS TDAQ system is the inherently bursty nature of the data traffic from the ROS buffers to the HLT nodes. On the TDAQ Ethernet network, this can lead to packet drops which cause very high data collection latencies, an inefficiency source for the system. Traffic shaping techniques can mitigate these phenomena.

[1] Measurements show that both time intervals are smaller than 0.5 ms.

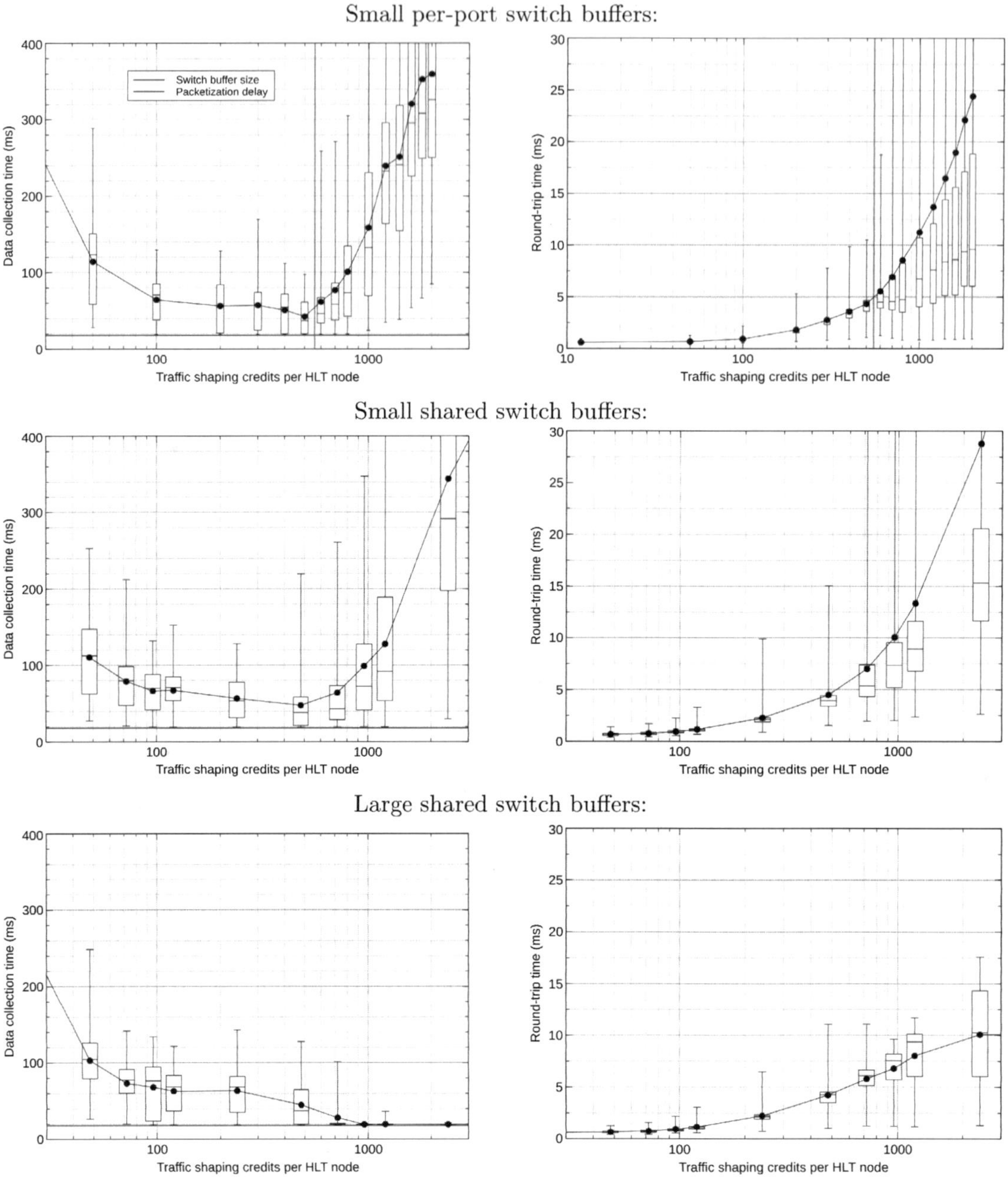


**Figure 5.** Data-Collection time and round-trip time as a function of the number of traffic shaping credits for the three different switch models under study. The test conditions are detailed in the text. The bullets represent the average values. The horizontal box lines represent the first quartile, the median, and the third quartile. The box whiskers represent the first and the 99th percentile.

As shown in this paper, the buffering space available in the network device plays a crucial role. When a switch with large buffers is used, packets are never dropped, even without traffic shaping: the data collection latency can get close to the minimum value allowed by the network topology. If a switch with buffers too small to handle the bursts is used, traffic shaping is effective in reducing the data collection latency by preventing packet drops. However, even in the best case, the average data collection latency does not reach its minimum possible value.

As a significant price premium is placed on network devices with large buffers, optimising the system performance becomes a balancing exercise among the cost of a better performing network, the cost of developing better traffic shaping techniques, and the cost of the inefficiencies introduced by network congestion.

## 6. Further investigations

The behaviour of an Ethernet network subject to congestion can be studied by injecting controlled traffic patterns and monitoring the relevant application performance metrics. However, gaining a deeper understanding of the congestion effects on a commercial system is a daunting task, as many useful metrics, such as queue occupancies, are either unavailable or too coarse. An alternative solution is the development of simulation models, which allow deeper inspection. Currently, models of the ATLAS TDAQ system are being developed, with the goal of enabling further investigations without having to deal with the limitations of testing on a production system. Once the models are proven capable of reproducing the known behaviour of the system, their results will be used to guide the optimisation of the network usage patterns.

## References

[1] ATLAS Collaboration 2008 *JINST* **3** S08003
[2] ATLAS Collaboration 2003 *ATLAS High-Level Trigger, Data-Acquisition and Controls Technical Design Report* ATLAS-TDR-016 CERN-LHCC-2003-022 CERN
[3] Pozo Astigarraga M E *in these proceedings*
[4] Allman M, Paxson V, and Stevens W *TCP Congestion Control* RFC 2581 IETF

# Evolution of the ATLAS Trigger and Data Acquisition System

**M E Pozo Astigarraga, on behalf of the ATLAS collaboration**

CERN, CH-1211 Geneva 23, Switzerland

E-mail: mpozoast@cern.ch

**Abstract**. ATLAS is a Physics experiment that explores high-energy particle collisions at the Large Hadron Collider at CERN. It uses tens of millions of electronics channels to capture the outcome of the particle bunches crossing each other every 25 ns. Since reading out and storing the complete information is not feasible (~100 TB/s), ATLAS makes use of a complex and highly distributed Trigger and Data Acquisition (TDAQ) system, in charge of selecting only interesting data and transporting those to permanent mass storage (~1 GB/s) for later analysis. The data reduction is carried out in two stages: first, custom electronics performs an initial level of data rejection for each bunch crossing based on partial and localized information. Only data corresponding to collisions passing this stage of selection will be actually read-out from the on-detector electronics. Then, a large computer farm (~17 k cores) analyses these data in real-time and decides which ones are worth being stored for Physics analysis. A large network allows moving the data from ~2000 front-end buffers to the location where they are processed and from there to mass storage. The overall TDAQ system is embedded in a common software framework that allows controlling, configuring and monitoring the data taking process. The experience gained during the first period of data taking of the ATLAS experiment (Run I, 2010-2012) has inspired a number of ideas for improvement of the TDAQ system that are being put in place during the so-called Long Shutdown 1 of the Large Hadron Collider (LHC), in 2013/14. This paper summarizes the main changes that have been applied to the ATLAS TDAQ system and highlights the expected performance and functional improvements that will be available for the LHC Run II. Particular emphasis will be put on the evolution of the software-based data selection and of the flow of data in the system. The reasons for the modified architectural and technical choices will be explained, and details will be provided on the simulation and testing approach used to validate this system.

## 1. Introduction

ATLAS (A Toroidal LHC ApparatuS) is a multipurpose particle detector located in the Large Hadron Collider (LHC) at CERN [1]. When the first proton beams began to circulate in the LHC, by the end of 2009, the period known as Run 1 commenced. From that moment until the beginning of 2013, the ATLAS detector operated successfully recording proton-proton collisions at a center of mass energy of 8 TeV. In 2013, the Long Shutdown 1 period, or LS1, was scheduled in order to do the necessary maintenance and upgrade operations on the different systems of the apparatus. In particular, the Trigger and Data Acquisition (TDAQ) system of the ATLAS detector experienced important changes some of which have been described in [2] and [3].

The ATLAS TDAQ system is responsible for the readout, selection and sending to permanent storage of the physics events and plays a fundamental role in the ATLAS operation. This document presents the current status of the Data Flow, Read-Out System and Data Collection network during the

LS1 evolution. In the first quarter of 2015, a new period named Run 2 will begin in the LHC, with collisions expected at a center of mass energy of 13 TeV, inaugurating a new era for the HEP experiments.

## 2. The Data Flow upgrade

Protons bunches traveling nearly at the speed of light collide in the center of ATLAS several millions of times per second. In order to reduce the data volume generated in the ATLAS subdetectors multiple stages of event filtering are performed. The Level 1 (L1) Trigger performs a first rejection based on simple calorimetry and muon tracking information reducing the rate by more than a factor hundred. The events passing the L1 Trigger are read-out from the subdetectors' Front-End electronics and the data is stored in large memory buffers in the Read-Out System for further real-time analysis in the so called High Level Trigger (HLT) farm. A set of applications, libraries and communication protocols transporting and delivering the information from the Read-Out System (ROS) to the processing units in the HLT farm is known as the Data Flow.

During Run 1, two separated computer farms were in charge of the filtering of the events, performing the Level 2 (L2) and Level 3 (L3) trigger respectively. In the first farm, known as Data Collection farm, the triggering algorithms filtered the events at 75 KHz based on a partial reconstruction of the collision data. Once an event passed the L2 filter, a dedicated set of nodes, known as Sub-Farm-Input (SFI), made the full reconstruction of the events (1.5 MB) by pulling the fragments from the Read-Out System. In the second farm, the Event Filtering farm, the processing nodes filtered the events at 3 KHz rate executing the HLT algorithms on the full event pulled from the SFI nodes. Finally, the events retained were pushed to the Sub-Farm-Output (SFO) for temporary storage until the CERN Storage System could accept them for permanent storage (200 Hz).

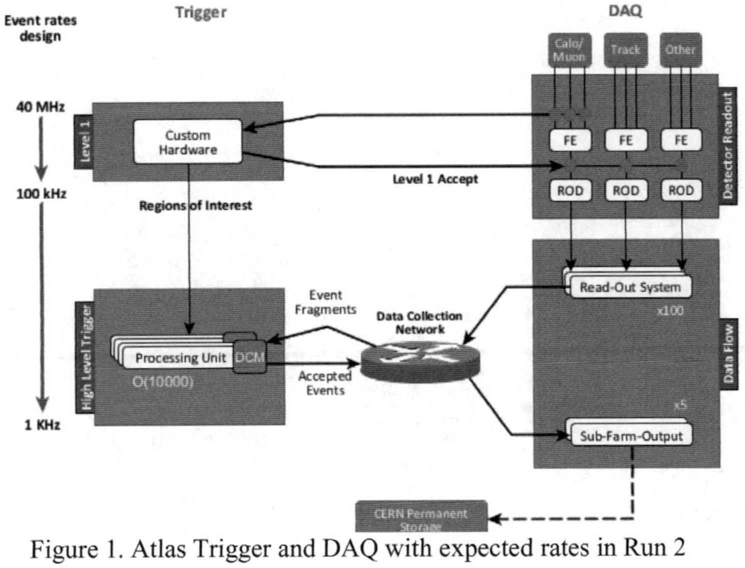


Figure 1. Atlas Trigger and DAQ with expected rates in Run 2

The Data Flow has been completely reviewed during LS1 in order to cope with the higher physics event rates and sizes expected in Run 2: detector read-out of ~2 MB events at 100 KHz. The HLT algorithms use often more data than originally foreseen for the Level 2 selection but very rarely need the full event to take the final decision. Driven by the HLT Supervisor nodes, a new Data Flow component running in each of the processing nodes, the Data Collection Manager (DCM), requests the event fragments progressively to the Read-Out-System as they are needed by the HLT algorithms. In the same way as for the original design, the events passing the HLT filter are sent to the Sub-Farm-Output where they are stored until they can be sent to the CERN permanent storage for Offline analysis.

This new architecture maximizes the flexibility of data access by the HLT algorithms, thus optimizing the selection efficiency; it profits from the advances in technology that allow the deployment of higher throughput and port density networks.

## 3. The Read-Out System upgrade

The Read-Out System receives and stores event fragments after the L1 accept. The fragments are requested by the processing nodes in the HLT farm for filtering and removed from the memory buffers only when a decision has been taken.

During Run 1, a hundred and fifty PCs ensured the correct behavior of the Read-Out-System. Each PC hosted from four to five PCI cards named ROBINs, and each ROBIN had three Read-Out-Links (ROLs) which connected the detector read-out to the ROS PCs. The event fragments stored in the ROBIN's buffers were requested by the processing nodes of the HLT farm thanks to the two 1 Gbps Ethernet links on a dual-port network interface card.

For Run 2, the design goal for the Read-Out System is 50% of read-out at 100 KHz. The increase in luminosity expected for Run 2 required a 20% increase in number of ROLs [4]. In addition, the limited memory capacity of the ROBINs and the obsolescence of the electronics made necessary a review and upgrade of the cards. For this purpose, a new PCIe card named RobinNP was developed with substantial improvements on the resources and performance characteristics. The connectivity of the ROS PCs to the HLT farm is provided by four 10 Gbps Ethernet ports on two dual-port network interface cards.

## 4. The Network architecture upgrade

The DAQ system makes use of two independent computer networks: the Data Collection Network and the Control Network.

### *4.1. The Data Collection Network*

The Data Collection (DC) network connects the HLT processing nodes with the Read-Out System and the Sub-Farm-Output. The DC network has evolved to provide network connectivity as required by the new Data Flow architecture: the old Data Collection and Back-End networks have been merged into a single Ethernet network, the ROS PCs have been directly connected to the network cores and the Sub-Farm-Input removed (see Figure **2**. Data Collection network).

On the ROS side, the removal of the aggregation layer had important consequences on the traffic patterns. This change was motivated by the current state of the technology that allowed only the aggregation from 10 to 40 Gbps Ethernet. Because of the high cost of this type of network devices compared to the low aggregation factor they provide (4:1), it was decided that the aggregation layer could be removed. Each ROS PC has currently quadruple active-active bonded connections to the network routers and because the over-dimensioning of the link capacity it can support almost transparently the failure of any of these links.

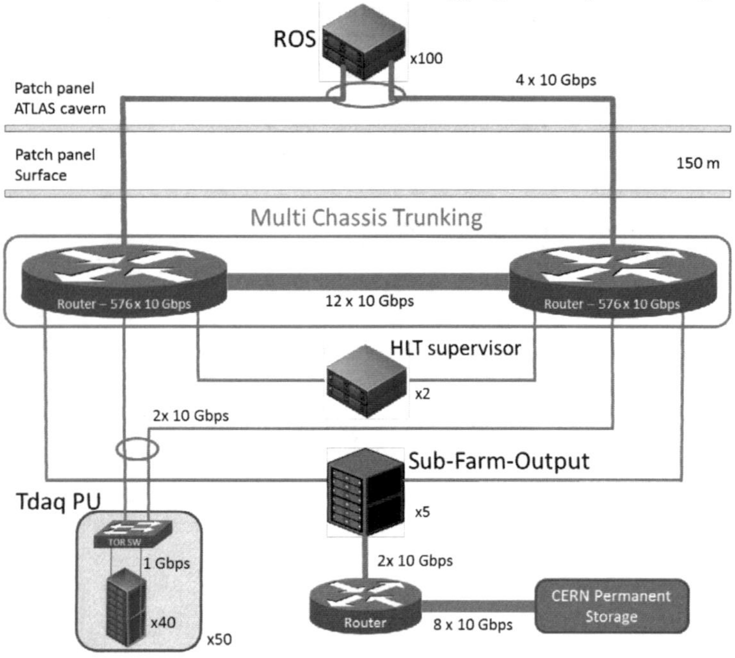

Figure 2. Data Collection network

The main change on the network side came from the replacement of the two network core routers that provided redundant connectivity to the different sub-systems. These devices were at the end of their lifetime and have been replaced by newer generation ones which are optimized to operate at 10 Gbps Ethernet. The internal architecture of the new routers incorporates a major improvement in the packet forwarding capacity by introducing the Virtual Output Queuing (VOQ) technique, which assigns different packet queues to each output port. Thanks to the VOQ technique, the packets addressed to a given output port can be forwarded independently of the traffic going to the other output ports, removing the head-of-line blocking phenomenon[5]. The consequence is that the device backplane has now a 100% packet forwarding efficiency.

The router manufacturer proposed several types of 10 Gbps Ethernet linecards for the routers with eight and twenty-four ports and different CAM memory sizes. The final purchase of the linecards, representing 60% of the total price, was delayed to the second half of the LS1 period in order to evaluate the various models. A simple software tool was developed to simulate the traffic throughput and patterns expected in the final system allowing the measurement of the switching capacity of the linecards and the backplane capacity of the routers. In the Figure 3. 8 port linecard (left) Vs 24 port linecard (right) maximum full-duplex traffic throughput we can find the comparison between the throughput for the eight and twenty-four port linecards when bidirectional TCP traffic is forwarded by the linecards' ASIC. The eight port linecard showed a performance of 93% of the theoretical values while the twenty-four port one performed at 87%. In order to measure the backplane capacity of the router, traffic was sent to the output ports in different linecards showing a 93% of efficiency for both models. Further tests with the ROS PCs were carried out using the Data Acquisition framework and modifying the traffic burstiness in the Data Flow software. This was possible by tuning the traffic shaping mechanism in the Data Collection Managers (DCM) that controls the number of event fragments that are consecutively requested to the Read-Out-System. All these tests proved that the linecard with twenty-four 10 Gbps Ethernet ports was the best cost-effective solution and that due to the unidirectional nature of the traffic, seven ports in a group of eight could be used safely in non-blocking mode, even if the specifications stated that only six should be used.

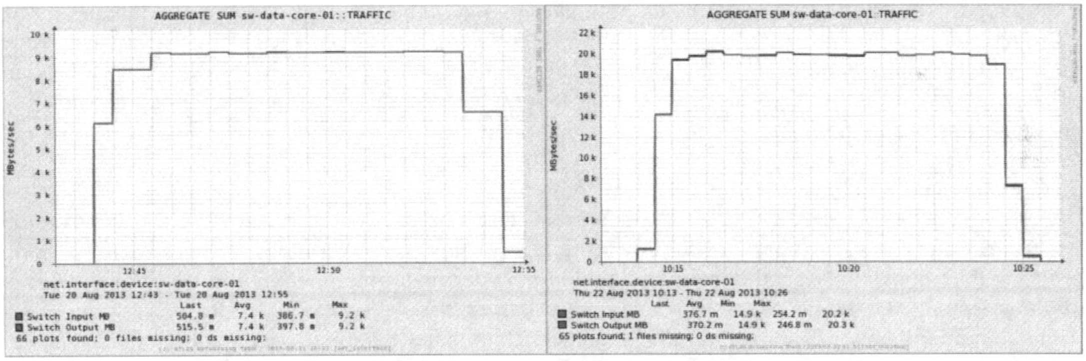


Figure 3. 8 port linecard (left) Vs 24 port linecard (right) maximum full-duplex traffic throughput

An additional enhancement of the new generation linecards is the deep packet buffers capable of absorbing few hundreds of milliseconds of packet bursts, covering most of the worst case scenarios. As a consequence, the network packets drops occur almost exclusively to the Top-Of-the-Rack switches in the processing nodes racks.

An interesting feature of the new network routers known as *Multi Chassis Trunking* provides both load balancing and link redundancy to the network (see Figure **2**. Data Collection network). This is achieved creating aggregated links on the devices connected to the routers which are perceived as a single virtual network device. A proprietary protocol running between the two routers ensures the Routing and Forwarding table synchronization between the two network cores needed for an optimal forwarding of the traffic.

Finally, several Top-Of-the-Rack switches for processing nodes racks were studied during LS1; this allowed us to clearly identify the required characteristics for future replacements. Many different candidate switches were evaluated taking into consideration the new traffic patterns and the traffic shaping policies that the DCMs can apply to the fragment requests. The results of the tests showed that without any traffic shaping, a switch with a buffer size bigger than a full event size per output port is needed to avoid packet drops (~2 MB per output port). However, applying an intelligent traffic shaping policy on the DCM, it was possible to obtain reasonably good results with much smaller shared buffers per device. Nowadays the data center switch market is moving into that direction and cheaper switches can be obtained. This subject is still under study and a decision about a suitable switch replacement will be taken in the following months.

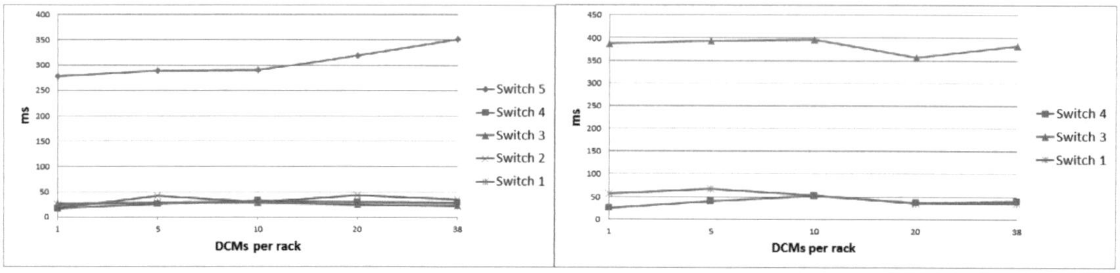


Figure 4. Event Building time (ms) with traffic shaping (left) and without (right)

### 4.2. The Control Network

The Control Network is a parallel network to the DC network in which the traffic for the control, configuration and monitoring of the TDAQ infrastructure flows. There are not strong performance requirements for the control network so the main upgrade activities concerned the redundancy and fail-over techniques. As part of the redundancy improvements, an Active-Backup setup has been installed for all important nodes in the system: Online and Monitoring PCs, ROS, HLT supervisor and SFOs. Several fail-over tests have been carried out to validate the technical choices, trying to anticipate any potential issue.

## 5. Conclusions

The right functioning of the TDAQ system has a direct impact on the operation of the ATLAS experiment and the achievement of its Physics goals. In this paper we have shown how the TDAQ architecture has been reshaped during LS1 period in order to profit from the technological progress and to maximize the flexibility and efficiency of the data selection process.

The preliminary tests carried out so far have demonstrated the correct behaviour of the deployed system and validated our technical choices. The overall TDAQ has nevertheless not yet been completely installed, thus more measurements and tests will be needed before the start of data taking with the LHC in 2015.

## References

[1] ATLAS Collaboration. 2008. *Journal of Instrumentation* 3 S08003

[2] A. Krasznahorkay et al. 2013. The evolution of the Trigger and Data Acquisition System in the ATLAS experiment. ATL-DAQ-PROC-2013-018, https://cds.cern.ch/record/1604503

[3] N. Garelli et al. 2013. The Evolution of the Trigger and Data Acquisition System in the ATLAS Experiment. ATL-DAQ-PROC-2013-029, https://cds.cern.ch/record/1622797

[4] W. Vandelli et al. 2014. Evolution of the ReadOut System of the ATLAS experiment ATL-DAQ-PROC-2014-012 https://cds.cern.ch/record/1710776

[5] N. McKeown, A. Mekkittikul; V. Anantharam, J.Walrand. 1999. "Achieving 100% Throughput in an Input-Queued Switch". *IEEE Transactions on Communications* 47 (8): 1260–1267

# Intelligent operations of the data acquisition system of the ATLAS experiment at LHC

**G Anders, G Avolio, G Lehmann Miotto and L Magnoni**

CERN, CH-1211 Geneva, Switzerland

**Abstract.** The ATLAS experiment at the Large Hadron Collider at CERN relies on a complex and highly distributed Trigger and Data Acquisition (TDAQ) system to gather and select particle collision data obtained at unprecedented energy and rates. The Run Control (RC) system is the component steering the data acquisition by starting and stopping processes and by carrying all data-taking elements through well-defined states in a coherent way. Taking into account all the lessons learnt during LHC's Run 1, the RC has been completely re-designed and re-implemented during the LHC Long Shutdown 1 (LS1) phase. As a result of the new design, the RC is assisted by the Central Hint and Information Processor (CHIP) service that can be truly considered its "brain". CHIP is an intelligent system able to supervise the ATLAS data taking, take operational decisions and handle abnormal conditions. In this paper, the design, implementation and performances of the RC/CHIP system will be described. A particular emphasis will be put on the way the RC and CHIP cooperate and on the huge benefits brought by the Complex Event Processing engine. Additionally, some error recovery scenarios will be analysed for which the intervention of human experts is now rendered unnecessary.

E-mail: Giuseppe.Avolio@cern.ch Gabriel.Anders@cern.ch

ATL-DAQ-PROC-2014-028
25 September 2014

## 1. Introduction

The Trigger and Data Acquisition (TDAQ) system [1] of the ATLAS detector [2] at the Large Hadron Collider (LHC) at CERN is composed of a large number of distributed hardware and software components (about 2000 machines and more than 15000 concurrent processes at the end of LHC's Run I) which provide, in a coordinated manner, the data-taking functionality of the overall system.

The Run Control (RC) and the Central Hint and Information Processor (CHIP) are key components of the Online Software framework that encompasses the software to configure, control and monitor the TDAQ system. The RC system steers the data acquisition by starting and stopping processes and by carrying all data-taking elements through well-defined states in a coherent way. Given the size and complexity of the TDAQ system, errors and failures are bound to happen and must be dealt with. The data acquisition system has to recover from these errors promptly and effectively, possibly without the need to stop data taking operations. During LHC Run 1, the detection and handling of problems was based on an embedded rule-based forward-chaining expert system (CLIPS [3]), which was deeply integrated with the RC system. Even though the system performed well, it had major disadvantages: new rules could not be tested without reproducing the error conditions in the production environment and monitoring of system resources used by specific rules was not possible. This made the development and debugging of new rules difficult. Additionally, the expert system was lacking the natural support for detections of temporal patterns.

Journal of Physics: Conference Series **608** (2015) 012007 doi:10.1088/1742-6596/608/1/012007

During the LHC Long Shutdown 1 (LS1) the RC has been completely re-designed and re-implemented in order to address the problems mentioned beforehand. The new RC is now assisted by the CHIP that can be truly considered as its "brain". CHIP is an intelligent application having a global view on the TDAQ system. It supervises the ATLAS data taking, takes operational decisions and handles abnormal conditions. Furthermore, CHIP automates complex procedures and performs advanced recoveries.

## 2. The Run Control system

### *2.1. Overview*

Applications in the ATLAS TDAQ systems are organized in a tree-like hierarchical structure (the run control tree, see figure 1) following the functional de-composition into systems and sub-systems of the ATLAS detector. Each application in the run control tree is managed by a parent Controller. The root node of the tree is the Root Controller. Controller applications are responsible to keep the system in a coherent state by starting and stopping their child applications and by sending them the proper commands needed to reach a state suitable for data-taking. The composition of the run control tree is completely specified in the TDAQ configuration database [4], where applications and parent/child relationships are described.

Operations across the run control tree are synchronized using Finite State Machine (FSM) principles. FSM transitions are usually initiated by the human operator via a graphical user interface: commands are sent directly to the Root Controller and then automatically propagated throughout the tree of leaf Controllers. Once an application completes the execution of a command (or changes its internal status by any reason) it notifies the parent Controller which in this way can evaluate when a coherent state is reached.

Controller applications are also the RC elements interacting with CHIP. Controllers inform CHIP about any change in the status of their controlled children. CHIP, in its turn, is able to detect any anomaly in the system analysing the status of all the applications and can notify the Controllers about proper actions to be taken. Examples of actions are setting a simple error flag or restarting/ignoring offending applications.

Controllers are able to supervise applications of any kind: "simple" applications not implementing RC functionalities are not expected by their Controllers to follow the FSM but are just started and stopped at the right time (*i.e.*, as described in the configuration database).

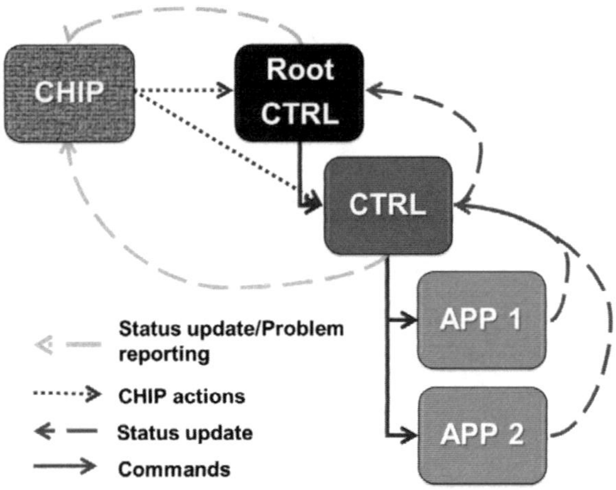


**Figure 1.** Schema of a simple Run Control tree: the Root Controller (Root CTRL), a child controller (CTRL), two leaf applications (APP 1 and APP 2) and CHIP are shown.

### *2.2. Communication schema*

The communication schema has been designed and implemented starting from the requirement that at any moment in time each Controller needs to know the state of its children, and that CHIP must have a complete and coherent view of the status of all the applications in the system. In such a way Controllers shall be able to determine when a child application has completed the execution of a command or has been properly started/stopped or has changed its state, and CHIP to detect abnormal situations and take recovery actions. As a result:

- any child application or Controller shall be able to notify its parent about any status change;
- any Controller shall be able to notify CHIP about any change in its children's status (the Root Controller being the only Controller sending CHIP information about itself);
- CHIP shall be able to notify any Controller about actions to be taken to resolve an abnormal situation.

Since CHIP, Controllers and child applications run on different hosts of the TDAQ computing farm, they communicate each other via a CORBA based remote Inter Process Communication (IPC) system.

### *2.3. Architecture of a Controller*

The Controller's design is strongly based on the definition of clear responsibilities for all of its components. This strategy allows easy possible future extensions. A Controller is made up of several logical sub-controllers that cooperate together in order to properly deal with incoming commands:

- *Main Controller* – It orchestrates all the operations performed by the Controller; it receives commands from other applications and takes care of their execution delegating, when needed, the responsibility to other components;
- *FSM Controller* – It is responsible for the FSM operations completely encapsulating all the FSM's implementation details;
- *Application Controller* – Its responsibility is to interact with the *ProcessManager* [5] system in order to properly start and stop child applications and monitor its life-time;
- *Command Controller* – It executes commands as requested by the *Main Controller*, keeps track of all the commands currently in execution and enforces that only compatible commands are executed concurrently;
- *DVS Controller* – It interacts with the *DVS* [6] sub-system (implementing the Test Management functionalities) in order to verify the functional status of the controlled child applications.

*2.3.1. Commands as resources.* The *Command Controller* uses a simple mechanism in order to both keep track of all the commands being executed at any moment in time and to manage the concurrent execution of multiple commands. Commands are treated as if they were resources:

- every command declares a list of other commands it should not be executed concurrently with;
- before every command is executed, the *Command Controller* checks whether some "incompatible" command is already being executed and, if not, a resource for the current command is allocated (the resource being the command itself);
- when a command execution is completed, the corresponding resource is freed.

### *2.4. Performances*

From a performance point of view it is important to keep low the overhead introduced by the RC system in dispatching commands and receiving their acknowledgments. In order to evaluate such an overhead, the time needed by a controller application to fully perform a FSM state transition is measured as a function of the number of child applications. With about one thousand child applications the time needed to perform a state transition is less than 180 ms[1] (see figure 2). Taking into account that transition actions performed by real-life applications during physics runs take tens of seconds and that during the LHC Run 2 a single controller will supervise O(100) children, the controller's performance is fully satisfactory.

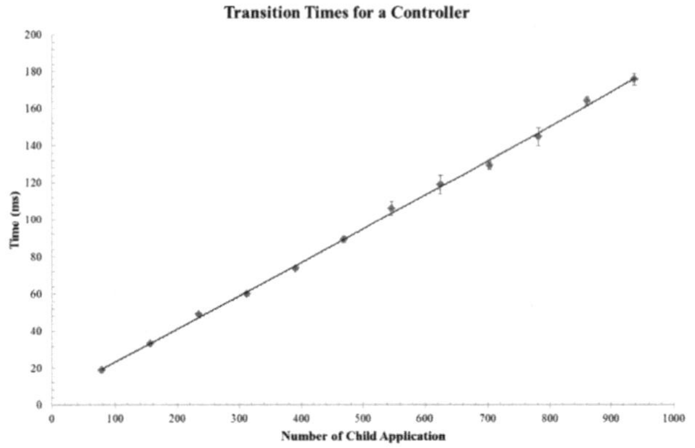


**Figure 2.** Plot showing the time needed by a Controller to perform a FSM state transition as a function of the number of child applications (evenly distributed on a rack of 39 computing nodes). Child applications are configured to not execute any action during state transitions (*i.e.,* they just receive commands from the parent controller and notify it when the command execution starts or completes).

## 3. The CHIP

### *3.1. Overview*

The application CHIP gathers various information and employs a Complex Event Processing (CEP) engine in order to aggregate, correlate and analyze this information. Its knowledge base consists of a set of rules loaded from text files. After evaluation of the various CEP engines available on the market, CHIP was based on ESPER [7]. This engine was chosen because of its comprehensive documentation, its open-source license and because it is based on Java which is a well-known and widely used programming language. ESPER allows:

- *Efficient handling of very high information update rates* – The peak rate is typically several tens of thousands of events per second. These peaks coincide with global state transitions of the RC system. Since errors are more likely to occur during state transitions than at other times, it is important that the engine can timely react to failures also in these conditions.
- *Rule testing* – The correct logic of new rules can be verified by artificial injection of events in a unit test. This makes the development of new rules easier, since the behaviour of rules can be tested without creating special conditions in the production environment.
- *Metrics analysis* – It is possible to monitor the CPU usage of individual rules. This is a powerful feature because it allows to identify CPU intensive rules which then can be revised and optimized. In this way potential CPU bottlenecks can be circumvented.
- *Configuration of threading model* – It is possible to change the size of the various engine inherent thread pools in order to tune the engine for best performance.
- *Natural support for temporal correlations* – It is often of high interest to correlate events happening at different points in time. An example for this may be the detection of applications, which, even though automatically restarted after failure, repeatedly crash in a given time window. The frequent crashes may indicate a severe underlying problem and should be investigated.

---

[1] Tests have been executed on nodes equipped with two Intel Xeon E5645 CPUs, 24 GB of RAM and GbE link connection.

- *Sophisticated anomaly detection* – The ability to perform complex correlations of data from various information providers is one of the main characteristics of CEP engines.

Besides processing the information coming from the RC Controllers, CHIP makes use of the information available in the logging service, the operational data service and the configuration service. For each type of information a corresponding adapter was implemented which makes the information available to the CEP engine (see figure 3). Furthermore CHIP interacts with the Test Management service. Amongst others this allows to detect DAQ problems originating from hardware or network failures.

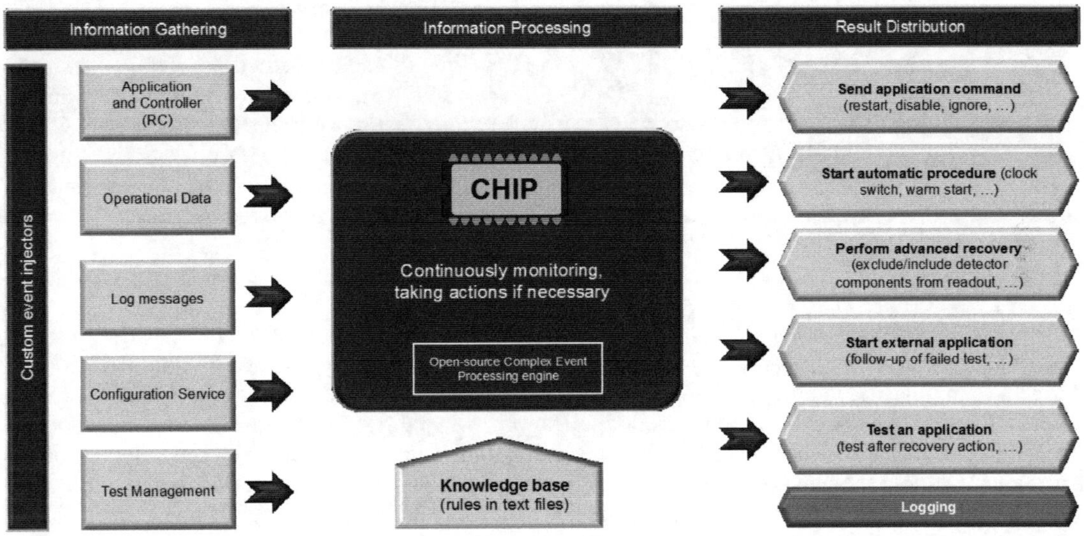


**Figure 3.** Architectural overview of CHIP. The flow of information is from left to right and can be divided into three phases: gathering of information using various event injectors, processing of the information within the CEP engine and distribution of results.

*3.2. Knowledge base*

At present CHIP handles about twenty different kind of recoveries, some of which are very generic, e.g. for crashed applications, others are very sophisticated, e.g. for recovering failing high level trigger applications. Additionally, CHIP automates about ten procedures, e.g. the raise of high voltage of detector components when collisions are established and beam conditions are safe, or the selection of the ATLAS reference clock depending on current detector and beam conditions. The knowledge base is thought of as a non-constant set of rules which is gradually extended over time when new type of errors occur which can be resolved by well-defined recovery procedures.

*3.3. Required resources and performance*

CHIP runs on a dedicated 16-core machine and uses about 1GB of memory[2]. The CPU utilization follows the incoming event rate and typically varies between <1% and 300% (*i.e.*, the equivalent of 3 CPU cores out of 16). The CPU utilization of the different rules is regularly measured in dedicated test sessions based on the CEP engine's built-in metrics (see figure 4). The responsiveness of CHIP was verified in tests during which the information update peak rate was at least twice as high as expected for LHC Run 2.

[2] Tests have been executed on a node equipped with two Intel Xeon E5540 CPUs, 24 GB of RAM and GbE.

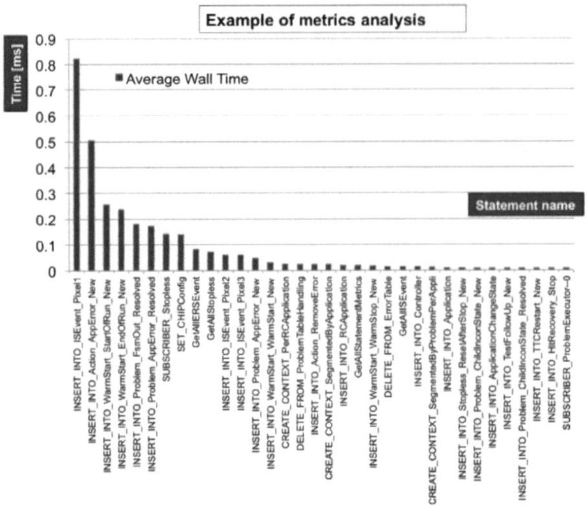


**Figure 4.** This diagram shows the CPU wall time needed for the evaluation of different ESPER statements (not all existing statements shown). The evaluation time of each statement is averaged over the duration of an approximately 15 minutes long test session, during which the RC was used to cycle the DAQ system through various states and during which various failures were provoked.

### *3.4. Error recovery example*

In case the data taking is blocked due to a faulty sub-detector component, CHIP can dynamically disable the corresponding read-out links without the need for stopping the data taking session. The procedure involves obtaining the read-out link endpoints from the configuration database and communicating with the concerned read-out applications. Once the sub-detector issue has been looked into and resolved, CHIP can re-enable the corresponding read-out components and integrate them back, again without the need of stopping the data taking session. The advantages of this procedure are two-fold: firstly, the recorded data may still be suited for physics analyses depending on the disabled sub-detector channels and secondly, the DAQ system down-time caused by this procedure is much smaller than the down-time caused by a full stop and restart of the complete data taking session. Thus less integrated luminosity is lost for physics analyses.

## 4. Conclusions

The re-design of the RC system, which is assisted by the application CHIP, achieves the clear separation between the steering and supervising functionality. Whereas the former is taken care of by the RC system, the latter is implemented in CHIP. The decision to base CHIP on a CEP engine allows advanced anomaly detection and error recognition. With unit tests the correct behavior of rules can be verified by injection of artificial events without the need for recreating the error conditions in the production environment. Dedicated test sessions and metrics analyses have shown that CHIP is responsive even during peaks of the incoming information updates. The peak rates during those test sessions were larger than anticipated for LHC Run 2.

## References

[1] ATLAS Collaboration ATLAS high-level trigger, data-acquisition and controls : Technical Design Report (Geneva : CERN) https://cds.cern.ch/record/616089?ln=en

[2] ATLAS Collaboration The ATLAS experiment at the CERN Large Hadron Collider JINST **3** (2008) S08003

[3] CLIPS http://clipsrules.sourceforge.net/

[4] Almeida J, Dobson M, Kazarov A, Lehmann Miotto G, Sloper JE, Soloviev I and Torres R The ATLAS DAQ system online configurations database service challenge *Proc. CHEP 2007,* J.Phys.: Conf.Ser. 119 (2008) 022004

[5] Avolio G, Dobson M, Lehmann Miotto G and Wiesmann M The ProcessManager in the

ATLAS DAQ system *Proc. Real-Time Conference, 2007, 15th IEEE-NPSS,* IEEE Trans.Nucl.Sci. 55 (2008) 399-404

[6] Avolio G, Corso-Radu A, Kazarov A, Lehmann Miotto G, Papaevgeniou L, Soloviev I, and Unel G A dynamic test management framework for the ATLAS experiment *ATLAS conference slide ATL-DAQ-SLIDE-2014-226* (https://cds.cern.ch/record/1703297?ln=en)

[7] EsperTECH http://esper.codehaus.org

# The performance and development for the Inner Detector Trigger algorithms at ATLAS

**Ondrej Penc on behalf of the ATLAS Collaboration**
Academy of Sciences of the Czech Republic, Institute of Physics, Prague, CZ
E-mail: ondrej.penc@cern.ch

**Abstract.** A redesign of the tracking algorithms for the ATLAS trigger for LHC's Run 2 starting in 2015 is in progress. The ATLAS HLT software has been restructured to run as a more flexible single stage HLT, instead of two separate stages (Level 2 and Event Filter) as in Run 1. The new tracking strategy employed for Run 2 will use a Fast Track Finder (FTF) algorithm to seed subsequent Precision Tracking, and will result in improved track parameter resolution and faster execution times than achieved during Run 1. The performance of the new algorithms has been evaluated to identify those aspects where code optimisation would be most beneficial. The performance and timing of the algorithms for electron and muon reconstruction in the trigger are presented. The profiling infrastructure, constructed to provide prompt feedback from the optimisation, is described, including the methods used to monitor the relative performance improvements as the code evolves.

## 1. Introduction

The ATLAS detector [1] is one of the two general purpose detectors thanks to which the Higgs boson was discovered [2].

To build on this success from Run 1 of the LHC operation which ended in spring 2013, the hardware and software for the ATLAS detector is being upgraded in readiness for LHC Run 2, due to commence in spring 2015.

Towards the end of Run 1, the LHC was routinely delivering instantaneous luminosities of $7.7 \times 10^{33}\,\mathrm{cm}^{-2}\mathrm{s}^{-1}$ with proton-proton collisions at 8 TeV centre-of-mass energy. Following the upgrade, the centre-of-mass energy will be increased to 13 TeV at a luminosity to $2\times10^{34}\,\mathrm{cm}^{-2}\mathrm{s}^{-1}$. The expected parameters of LHC after the upgrade are summarized in Table 1. The upgraded LHC will allow the measurement of the properties and branching ratios of the Higgs boson, and also improvements in the measurements of many Standard Model physics processes and searches for physics beyond the Standard Model.

The time between bunch crossings will be reduced from 50 ns to the 25 ns from the original design, which has the potential, with the increased instantaneous luminosity, to more than double the input rate of physics triggers. This increased collision rate and increased average number of separate proton-proton pileup interactions per bunch crossing will place a significant burden on the hardware and software subsystems of the ATLAS Trigger.

For the upgrade, the ATLAS High Level Trigger (HLT) architecture has been simplified such that both the Run 1 software stages of the HLT – the Level 2 (L2) and Event Filter (EF) – have been merged in to a single HLT stage to run on single CPU nodes.

**Table 1.** Comparison of the LHC parameters at the end of Run 1 and updated parameters for Run 2.

| Machine parameter | Run1 | Run2 | Unit |
|---|---|---|---|
| Centre-of-mass energy | 8 | 13 | TeV |
| Luminosity | $7 \times 10^{33}$ | $2 \times 10^{34}$ | $cm^{-2}s^{-1}$ |
| Bunch crossing spacing | 50 | 25 | ns |
| L1 input frequency | 20 | 40 | MHz |
| Average pileup | 21 | $\sim 55$ | events per bunch crossing |

This simplifies the data-flow, removing the need for network communication between the L2 and EF trigger levels, and removing the need for detector data for L2-accepted events to be also requested from the DAQ system again also for the separate EF processing. The trigger processing still consists of similar algorithms as in the old L2 and EF processing, still within an RoI, but instead of then following the L2 processing with the Event Building on a separate CPU for events that have been accepted by L2, and then starting the EF processing from scratch, the Event building just becomes an additional step in the processing of algorithms on the single HLT CPU node. This means the positioning of the full Event Building in the full processing chain is more flexible, and can be deferred until much later in the processing chain. In addition, the L2 and EF algorithms can be combined more efficiently, so that their equivalent algorithms can avoid repetition of reconstruction steps so reducing the processing time [3].

Together with the unification of L2 and EF on the CPU nodes, and the related network changes, the L1 hardware is also being upgraded by the addition of a new Topological Processor (TP) and enhanced Trigger Central Processor using FPGAs instead of custom electronics [4]. The TP uses the topological information on jet or muon directions to achieve additional background rate reduction. In addition, a new hardware stage, the Fast Tracker (FTK) [5], is being added that will reconstruct tracks in the silicon tracker following a Level 1 accept, but before the HLT processing, using massively parallel processing.

## 2. The HLT tracking

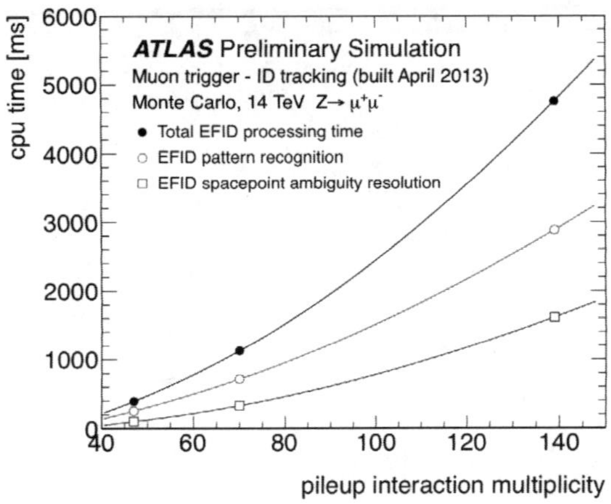


**Figure 1.** The Event Filter Inner Detector (EFID) trigger track reconstruction time for the ATLAS muon trigger as a function of the pileup interaction multiplicity. Shown are the times for the main pattern recognition algorithm (red) and Ambiguity Solver (blue) together with the total EFID tracking time (black) [6].

In both Run 1 and Run 2 the HLT tracking system runs as two stages. During Run 1, these were a fast tracking stage at L2, and an independent, precision tracking stage running a modified version of the offline tracking for the EF. In both of these stages, the independent pattern recognition was performed. During Run 1, this EF tracking was one of the most time consuming stages of the HLT processing. For Run 2, the same overall algorithm structure is retained - a fast tracking algorithm, followed by a precision tracking stage - but information from the pattern recognition of the fast tracking stage can be reused by the precision tracking, to avoid running the pattern recognition again for the precision tracking.

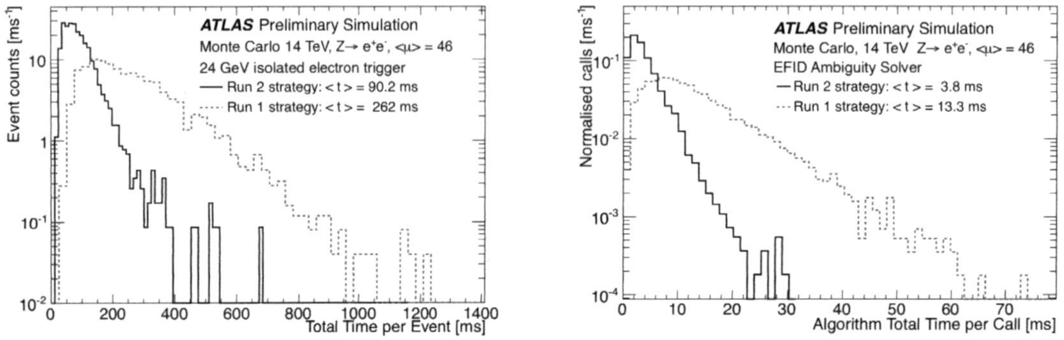


(a) (b)

**Figure 2.** The distribution of total processing time for the complete processing for the 24 GeV isolated electron trigger (a) and processing time for the Ambiguity Solver (b). Both plots compares the tracking algorithms that were executed during Run 1 (dashed red line) and algorithms that are supposed to be executed during Run 2 (solid black line) [6].

The high pileup multiplicity and high rate expected for Run 2 highlight the need for a new tracking strategy and software optimisation. The CPU time consumption in simulated Run 2 events as a function of the pileup interaction multiplicity is illustrated in Figure 1. Shown are the times for the two most time consuming stages of the EF Inner Detector (ID) trigger strategy as used during Run 1. The time consumption can clearly seen to grow non-linearly with the pileup multiplicity, clearly indicating that this would quickly become too costly to run in the HLT at high multiplicities.

These long processing times have led to an optimisation of the tracking code for the Precision Tracking, and a redesign of the trigger tracking strategy to be used for Run 2. The new strategy also had to include the data from the new Insertable B-layer (IBL) [7] – the new layer of high precision pixel detectors in the centre of the ATLAS Inner Detector.

For Run 2 the fast tracking stage is based on the pattern recognition algorithms used in L2 during Run 1 [8], but has been completely rewritten to provide a new Fast Track Finder stage. This identifies hits in the layers of the ID and makes up a pattern that later on becomes a track using hits in Regions of Interest (RoI), identified by the L1 trigger system. The RoIs determine the parts of the detector that are worth further investigation. The clusters within an RoI from the silicon tracking detectors are combined with the geometrical information to provide three-dimensional space-points. The subsequent pattern recognition is then based on space-point triplet finding within the subset of three different ID layers followed by the merging of triplets that are consistent with the same track trajectory [9]. This pattern recognition is followed by a preliminary track fit, rejection of outlying hits, and extension of the tracks into the Transition Radiation Tracker (TRT) - a straw tube tracker, which surrounds the silicon detectors.

In the Run 2 strategy the task of the precision tracking second stage is to perform the final track fit, and to resolve ambiguities between close or crossing tracks, but without first running an independent pattern recognition stage. During Run 1, the EF tracking ran the pattern recognition anew, even after the L2 tracking. With the new single node HLT strategy for Run 2, the pattern recognition stage of the Precision Tracking can be omitted, instead using the pattern recognition from the Fast Track Finder. The execution times for the Run 1 and Run 2 strategies on simulated Run 2 data can be seen in Figure 2(a). The *Run 1 strategy* runs those algorithms that were executed during Run 1 data taking, which consisted of the fast tracking stage, followed by the modified version of the full offline tracking, running the pattern recognition and Ambiguity Solver stages. In contrast, the *Run 2 strategy* also runs a fast tracking stage, but then performs the Precision Tracking by directly seeding the Ambiguity Solver stage from the output of this fast tracking stage rather than running the offline pattern recognition. Comparison of the Ambiguity Solver part of the reconstruction running in both strategies is seen in Figure 2(b). The Run 2 HLT code will also be able to seed the Precision Tracking with alternative track seeds such as tracks from the FTK. The IBL data was fully integrated into the complete reconstruction process during the redesign of the trigger tracking code.

## 3. ATLAS code optimisation

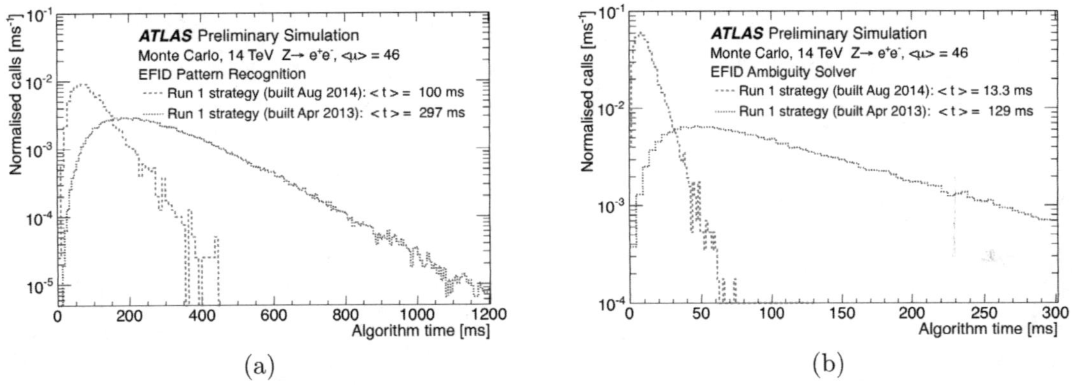


**Figure 3.** The distribution of processing times per call for the pattern recognition stage (a) and for the Ambiguity Solver stage (b) of the Event Filter Inner Detector trigger tracking. Shown are the times for the tracking strategy used during the ATLAS Run 1 data taking in 2012 but implemented in different versions of the ATLAS code; one built in April 2013 (blue dashed line) and a more recent build from August 2014 (red dashed line) [6].

The performance of the upgraded LHC machine will place significant demands on the ATLAS trigger system and will need to deal with very busy events containing many individual pileup interaction in each bunch crossing. Profiling and optimisation of the new tracking code has been performed as part of the overall redesign to identify the code where further optimisation will be most beneficial. For the Run 2 Fast Track Finder, the profiling tool Callgrind [10] is being used to determine the most often called, or computationally expensive parts of the code. Even small and simple changes in the code can lead to significant reduction of the execution time of an algorithm, such as precalculation of a constant variable before a *for*-cycle, or avoiding precalculation of a variable, if there is the possibility that the variable will not be used. Other optimisations, such as including modifications to avoid branch misprediction or cache misses, have also been studied previously [8].

An example of successful optimisation is depicted in Figure 3. This shows the processing times for the versions of the pattern recognition and Ambiguity Solver stages of the Run 1 strategy from two different releases of the ATLAS software, one from 2013 and the other from August 2014. The code from 2014 is significantly faster than that from 2013 as the offline tracking algorithms used in the Event Filter tracking underwent significant optimisation prior to the later build. The second example, Figure 3(b), shows that optimised code executes an order of magnitude faster. This is significant, since the Ambiguity Solver is still executed in the Run 2 trigger tracking, whereas the pattern recognition is not.

## 4. Muon efficiency in the Run 2

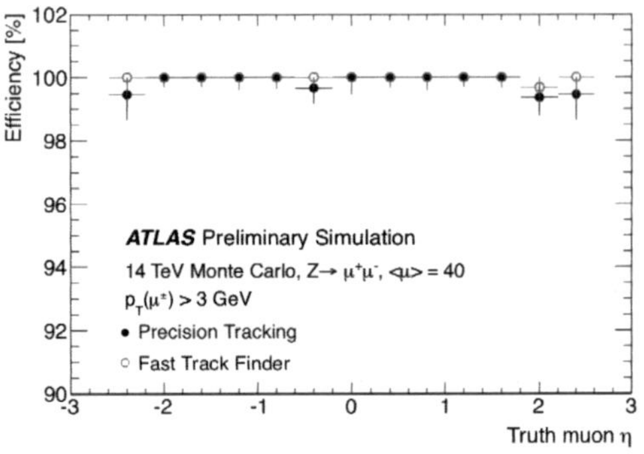


(a)

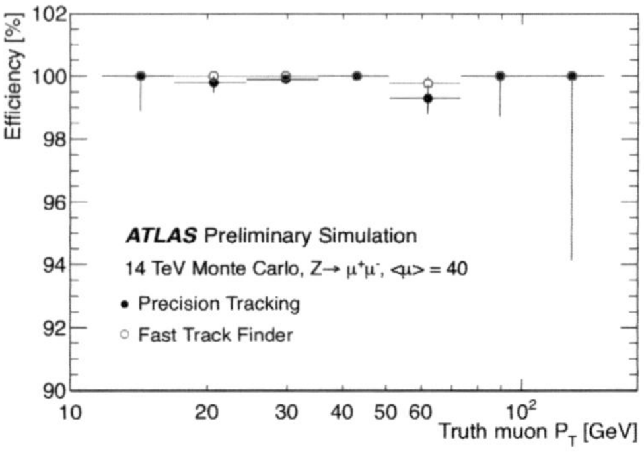


(b)

**Figure 4.** The muon finding efficiency for the ATLAS Inner Detector trigger tracking with respect to the true muon pseudorapidity (a) and transverse momentum (b). The efficiency is shown for the Run 2 strategy, for both the Fast Track Finder (hollow red points), and the Precision Tracking (solid black points) [6].

The outstanding performance of the ATLAS trigger during Run 1 is also required for Run 2. Figure 4 shows the efficiencies of the Fast Track Finder, and the Precision Tracking in the Run 2 strategy with respect to true muons estimated using simulated Run 2 Monte Carlo $Z \to \mu^+\mu^-$ events with a mean number of 40 additional pileup interactions per bunch crossing. The efficiency is above 99 %, illustrating that the ATLAS trigger performance has been preserved. The Fast Track Finder serves to seed the Precision Tracking which performs additional rejection of fake tracks using the Ambiguity Solver algorithm, and refits the track. The apparent small loss of the efficiency for the Precision Tracking at high pseudorapidity is due to the rejection of poorly reconstructed, mismeasured, or split tracks which result from the looser requirements of the fast tracking which is optimised for efficiency rather than purity. The simulated data also contain hits from the IBL.

## 5. Summary

The ATLAS trigger performed exceptionally well during Run 1. To deal with the significantly larger rates and occupancies expected during Run 2, the ATLAS tracking trigger code has been redesigned and significantly improved. The high efficiencies of the ID trigger tracking for the muons in $Z \to \mu^+\mu^-$ Monte Carlo show that the trigger performance has been preserved. This will allow ATLAS to continue the previous successes from Run 1 also in to Run 2.

The significant reduction of the time required by the tracking, as illustrated by the reduction by a factor of three in the processing time for the complete trigger for the 24 GeV isolated electron trigger for Run 2, when compared with Run 1, promises that the more challenging requirements for the trigger during Run 2 will be met. The successful installation of the IBL and the inclusion of the data from it in the tracking trigger code, together with provisions to include the FTK in the HLT provide a very bright outlook for the ATLAS tracking trigger and will allow ATLAS to rise to the challenge of data taking for the coming years.

## 6. References

[1] ATLAS Collaboration 2008 The atlas experiment at the cern large hadron collider *Journal of Instrumentation* **3** S08003 URL http://stacks.iop.org/1748-0221/3/i=08/a=S08003

[2] ATLAS Collaboration 2012 Observation of a new particle in the search for the standard model higgs boson with the atlas detector at the lhc *Physics Letters B* **716** 1 – 29 ISSN 0370-2693 URL http://www.sciencedirect.com/science/article/pii/S037026931200857X

[3] Bartoldus R S *et al.* 2013 Technical Design Report for the Phase-I Upgrade of the ATLAS TDAQ System Tech. Rep. CERN-LHCC-2013-018. ATLAS-TDR-023 CERN Geneva final version presented to December 2013 LHCC.

[4] Simioni E *et al.* 2012 Topological and Central Trigger Processor for 2014 LHC luminosities Tech. Rep. ATL-DAQ-PROC-2012-041 CERN Geneva

[5] Shochet M, Tompkins L, Cavaliere V, Giannetti P, Annovi A and Volpi G 2013 Fast TracKer (FTK) Technical Design Report Tech. Rep. CERN-LHCC-2013-007. ATLAS-TDR-021 CERN Geneva aTLAS Fast Tracker Technical Design Report

[6] ATLAS Collaboration 2014 ID Trigger timing and efficiency results Tech. Rep. ATL-COM-DAQ-2014-088 CERN Geneva

[7] Capeans M, Darbo G, Einsweiller K, Elsing M, Flick T, Garcia-Sciveres M, Gemme C, Pernegger H, Rohne O and Vuillermet R 2010 ATLAS Insertable B-Layer Technical Design Report Tech. Rep. CERN-LHCC-2010-013. ATLAS-TDR-19 CERN Geneva

[8] ATLAS Collaboration 2013 Studies for the development of the Inner Detector trigger algorithms at ATLAS Tech. Rep. ATL-DAQ-PUB-2013-002 CERN Geneva

[9] ATLAS Collaboration 2012 Performance of the atlas trigger system in 2010 *The European Physical Journal C* **72** 1849 ISSN 1434-6044 URL http://dx.doi.org/10.1140/epjc/s10052-011-1849-1

[10] Nethercote N and Seward J 2007 *Proceedings of the 2007 ACM SIGPLAN Conference on Programming Language Design and Implementation* PLDI '07 (New York, NY, USA: ACM) pp 89–100 ISBN 978-1-59593-633-2 URL http://doi.acm.org/10.1145/1250734.1250746

# Latest evolution of EOS filesystem

**Geoffray Adde and Belinda Chan, Dirk Duellmann, Xavier Espinal, Alessandro Fiorot, Jan Iven, Lukasz Janyst, Massimo Lamanna, Luca Mascetti, Joaquim M. Pereira Rocha, Andreas J. Peters, Elvin A. Sindrilaru**

CERN, IT Deparment, Data Storage and Services group, Geneva, Switzerland

E-mail: geoffray.adde@cern.ch

**Abstract.** EOS is an open source distributed filesystem developed and used mainly at CERN. It provides low latency, high availability, strong authentication, multiple replication schemas as well as multiple access protocols and features. Deployment and operations remain simple and EOS is currently being used by multiple experiments at CERN providing a total raw storage space of 86PB. A brief overview of EOS's architecture is given then its main and latest features are reviewed and some operations facts are reported. Finally, emphasis is laid on the new infrastructure aware file placement and access engine.

## 1 Introduction

EOS is an open source distributed filesystem (*DFS*) developed at CERN. It follows some simplified POSIX semantics. It runs on commodity hardware and it is mainly focussed on low latency, high availability, ease of operation and low total cost of ownership. It is built on top of the XRootD framework and thus offers a native XRoot access to the data.

Some of the closest major DFS's to EOS are DPM, GPFS, Lustre and CephFS. First, only the three latter are strictly POSIX compliant. Compared to GPFS and CephFS, EOS proposes a simpler architecture which makes it easier to set up and to operate. It also reduces the latency as no blocking inter-component communication is required to serve requests. As a consequence of this architecture, the hardware of the head node is the scalability bottleneck of the system. GPFS is commercial software, which in many cases, is a major distinction per se. GPFS relies mainly on RAID storage whereas EOS is based exclusively on RAIN storage. CephFS is a filesystem implemented on top of the Ceph object storage system. This additional layer can also provide block device access. EOS does not propose such a unifying approach and remains simpler. Lustre does not provide data replication of any type. This is delegated to dedicated hardware/subsystem. DPM is more focussed on small instances and less oriented toward low latency access.

EOS uses an agile development cycle which leads to numerous releases. In the first part of this paper, we expose the latest evolution of EOS, including new features that are already in operation and those which will be used soon.

Recently, the environment in which EOS is operated at CERN went through a major change. A second computer centre was opened in Hungary, and now, EOS runs over a WAN. In the second part of this document, we make a detailed focus on an evolution which was driven by this change : the new infrastructure-aware file placement and access engine[1].

## 2 Architecture

EOS is an open source project developed at CERN in the Data Storage Services group of the IT department. It is written in C++ and released under GPLv3 license. It is exclusively based on open source technologies and does not have any dependency on commercial software including database.
EOS's architecture has been designed and kept simple in order to allow easy administration of the instances and to keep the latency low. An instance is merely composed of three types of components. These software components are just plug-ins into the same XRootD server daemon thus being given different roles.

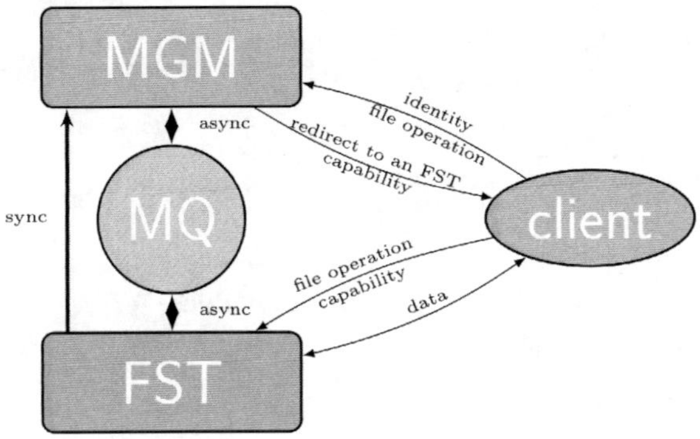


**Figure 1.** Architecture of EOS

*MGM*: This is the management component. This component usually runs on a dedicated machine (optionally two with one in slave mode for fast fail-over). This machine is the head node of the instance. This component provides:

- authentication and identity management
- permissions and quota accounting
- namespace and file placement/access
- instance management
- node balancing and draining

This component is not involved in any data transfer.

*MQ*: This is the message queuing component. It is actually a message broker that allows the coordination of the asynchronous messaging in the instance. Commonly, it runs on the same machine as the MGM.

*FST*: This is the file storage component. It physically stores the files (replicas or stripes). It transfers data to and from the clients. It is also in charge of checksumming and disk verification. This component runs on many servers - currently, up to a few hundred.

Figure 1 gives a schematic view of how these components interact in a typical file operation from a client.

[1] In this, paper, *file placement and access* is also referred as *file scheduling*

## 3 Operations
EOS has been in operations at CERN since 2011. Today there are 6 instances running. They sum-up to the following facts:

- 195M files
- 86PB of raw space
- 35000 disk drives
- 1 lost files per million and per year over the two last years

The biggest instance has 70 million files and 26PB of raw storage in 10800 drives scattered in 421 nodes. Peak reading rate in operations was reached at 15GB/s. This is far from the limits of the instance - most of nodes having a 10Gb/s link - but this is constrained by the physics analysis use case. Some scalability tests have been run and it has been reported that within the limits of the current hardware, this instance can double in size without hitting any scalability issue.

## 4 Features and Evolutions
EOS targets mainly the physics analysis use case at CERN. In average, it is defined by

- write once, read many
- possibly highly concurrent access of a bunch of hot files
- simplified POSIX access including scattered short reads, mainly through *ROOT*.

It drives most of its evolutions. In this section, we give an overview of the main features of EOS and of their latest developments.

### *4.1 Authentication, user accounting and data access*
EOS provides strong authentication via Kerberos 5, X509 and Simple Shared Secret. User/group mapping comes together with UNIX permissions and ACLs. An advanced quota system runs along it and provides user and group quotas. Lately, a new project quota has been added. It is attached to a subtree of the namespace. A configurable recycle bin feature has also been added and now provides an additional data safety measure in operation especially against erroneous client-side individual or bulk deletions.

### *4.2 Core features*
EOS has two strong traits in its architecture.
The first one is low latency. It is helped by a fast in-memory namespace whose memory footprint has been reduced as well as its boot-up loading time. Dramatic improvements in case of scattered IO have been brought by the implementation of vector reads; they should hit operations soon.
The second trait is tunable physical layout. In EOS, the storage unit is a file and data placement/replication is managed at this level. While the standard replica layout is still the most widely used, newer RAIN layouts using erasure encoding allow to set more accurately the desired trade-off between file read availability, write speed and redundancy. As the access and replication requirements about a file are likely to evolve, a converter engine has been implemented and is now used in operations. It allows layout changes to be carried out upon individual requests or under the supervision of the new LRU engine that watches the last access time stamps of files. Physical placement and access of the stripes/replicas has recently received infrastructure awareness. Taking into account networking, racks and locations increases speed and availability of the whole instance while it decreases internal data transfers.
EOS's potential of scalability has recently been levered by allowing delegation of the authentication to dedicated servers. It can offload this burden from the MGM node as it has

been identified as one of the most CPU consuming tasks. It is ready to use but not yet in production.

*4.3 Protocols and client features*
EOS is being operated in multiple ecosystems. To that end, it offers native access to the data through XRootD and HTTP(S)/WebDav.
Interfaces are provided for SRM and more recently for S3 through HTTP. A Dropbox-like client has also been added. A GridFTP DSI plug-in is also provided to run gateways between GridFTP and EOS. Lately, it has been rewritten to allow interaction with latest versions of XRootD and Globus software.
EOS features a multi-user FUSE client which allows to mount EOS instances onto a client file tree. It has been rewritten. It is more stable and now provides better caching, usage of vector reads as well as parallel streams for RAIN layouts.
Finally, a ROOT integration is provided. It has been vastly improved thanks to caching optimization and vector reads.

*4.4 Availability, Sustainability and Operating Costs*
In this field, EOS keeps evolving towards higher availability and simpler operations.
EOS has the ability to run in dual master/slave MGM node configuration. The slave MGM then follows all the operations and takes over in case of failure of the master MGM. The fail-over mechanism has been reimplemented and is now instantaneous. Being based on XRootD 4, latest versions of EOS benefit from the updates of the framework, including IPv6 compliance.
Everyday operations are eased by automatic disk/node draining/balancing for which policies have been improved. These work in close collaboration with a background disks scrubbing process and a real-time filesystem check feature. In addition, a new periodic direct-IO file checksum scan has been implemented to check the integrity of the data actually stored on disk. It features an optional auto-repair mechanism able to recreate replica/stripes if enough data is still available.

## 5 New feature closeup : infrastructure aware file scheduling

*5.1 Previous Architecture*
So far, EOS has been aware of the infrastructure it runs on. This information has mainly been incorporated in the way scheduling groups are designed. A scheduling group is a bunch of machines and a scheduling subgroup is a list of filesystems, each one being taken from each machine in the group (fig. 2). Each replica/stripe being allocated in the same subgroup, it avoids collocation of any stripe/replica of the same file and then avoids a common point of failure (CPOF) related to files having two replicas/stripes hosted by the same machine. In this context, file placement is carried out by a Round-Robin selection of the subgroup followed by a random selection of the target remote filesystems, the sampling being weighted with the network and disk activity of each remote FS. The file access is also done using a weighted random selection but among the remote FS hosting a replica/stripe of the file being accessed.

*5.2 New Architecture*
Bringing infrastructure awareness to EOS (fig. 3) brings three major types of improvement to this initial design.

- skip many more types of CPOF like racks, routers, rooms and sites.
- optimize file access by directing the client to the closest available source.
- optimize internal data moves triggered by balancing and draining.

These assets are especially important when operating in a WAN.

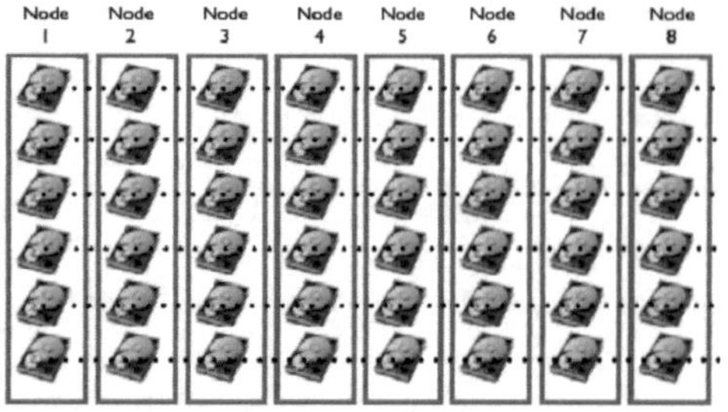


**Figure 2.** an EOS Scheduling group and its subgroups (horizontal slices)

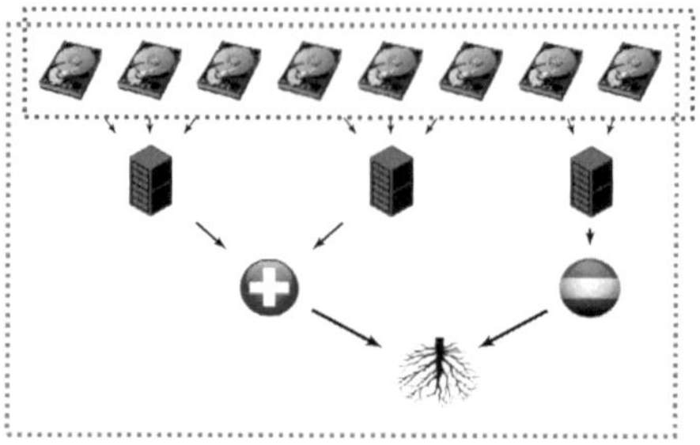

**Figure 3.** Scheduling tree for a scheduling subgroup

In this new design, the concepts of scheduling groups and subgroups (fig. 2) are still valid. The subgroup gets enhanced from a flat vector of remote FS to a tree structure of the underlying infrastructure (fig. 3).
File placement starts the same way with round-robin selection of a subgroup. Selecting the target FS can then be carried out following one of multiple policies.

- spread out replicas/stripes as much as possible in the tree to strengthen data safety/availability by minimizing the number of CPOF
- concentrate replicas/stripes as much as possible around a given node in the tree to maximize the file access capacity for a given group of client machines.
- any kind of intermediate policy between the two above, possibly specific to the physical layout of the file.
- any policy involving any kind of distinction in the branches of the tree (*e.g.* remote site, SSD rather than HDD).

File access algorithm is also modified. First, a copy of the scheduling tree is populated with the available stripes/replicas for the file being accessed. Then, in the scheduling tree, the closest node to the client is computed. The closest replica/stripe to that node is then selected.
Other types of information can be integrated to the selection on the remote FS in the tree. It can lead to convenient behaviours like online balancing of the storage nodes, network traffic management, etc. For instance, the current implementation skips saturated nodes as much as possible and provides an optional inline storage balancing feature.

*5.3 Implementation*

Compared to the previous one, the new architecture is "stateful" because trees and internal structures need to be kept up-to-date. It was implemented with two major constraints in mind:

- No significant latency should be added to the file placement/access operations.
  This was carried out by using two distinct types of structures : trees that are flexible and easy to update and snapshots that are compact and very fast (typically Mops/s). The latter retain the structure of a tree and some data attached to it and can be efficiently copied and used as a support for further computations.
- The new subsystem should scale linearly with the number of threads.
  To ensure this, the implementation makes sure that only a small amount of data needs to be concurrently updated by the serving threads - which can be many. When these updates are needed, they are processed using atomic instructions without any mutexing.

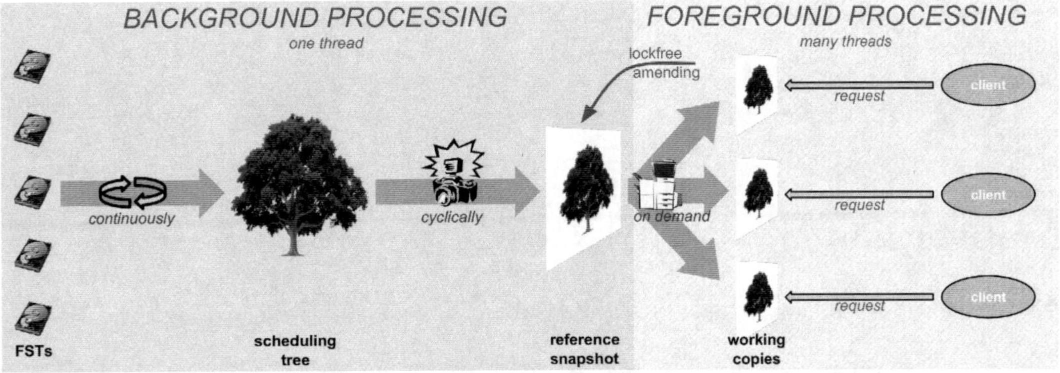


**Figure 4.** Infrastructure-aware implementation overview

Figure 4 shows an overview of the new architecture. The continuous update of the scheduling tree is fed with asynchronous messages from the FST nodes periodically reporting their state. A reference snapshot of the tree is also made periodically and then used to carry out all the placement/access requests. As a result, there is a latency in the state of the instance as it is seen by the system. This could lead to overloading some FSTs because they are seen as idle while they actually have been selected to serve many requests. To bypass this latency, every time a request is served, the reference tree is amended with an approximation of the overhead caused by the operation. To avoid mutexing, this is made using atomic instructions. As a side-effect of this design, a tiny part of these updates is also lost on the way. It is not an issue as we already deal with an approximation of the correction of the instance's state. This approximation is not extremely precise and does not need to be as errors don't get accumulated.

Tests were run on a small instance with 8 FSTs with 8 FSs each. The MGM node had 4 core i7 2.2Ghz cores and 8GB of RAM. Concurrent read and write of small files (a few bytes) were issued to/from the instances. Half of them with the former algorithm, half of them with the new algorithm. Execution time of the placement/access procedure were recorded and averaged (fig. 5).

| | 2 replicas | 3 replicas | raid6 |
|---|---|---|---|
| placement | 250/150 | 270/155 | 320/160 |
| access | 100/135 | 140/135 | 270/140 |

**Figure 5.** Execution time in $\mu s$. Flat alg./Tree alg.

In most cases, latency is reduced. The order of magnitude remains the same : $0.1ms$ per operation. Given the architecture and these measurements, the derived processing capacity is $\mathcal{O}(10\text{k schedulings/thread}/s)$ scaling linearly with the number of cores.

## 6 Conclusion

Latest enhancements of EOS have been reviewed. The new tree-based file placement and access algorithm has been presented. It offers several assets to optimize EOS operations in a WAN. This system will likely be introduced into the balancing and draining algorithms. The whole system will become operational in a near future.

# An overview of the DII-HEP OpenStack based CMS data analysis

**L Osmani[1], S Tarkoma[1], P Eerola[2], M Komu[3], M J Kortelainen[4], O Kraemer[4], T Lindén[4], S Toor[4] and J White[4]**

[1] Department of Computer Science, P.O.B. 68, FI-00014 University of Helsinki, Finland, [2] Department of Physics, P.O.B. 64, FI-00014 University of Helsinki, Finland, [3] Ericsson Research, 02420 Jorvas, Finland [4] Helsinki Institute of Physics, P.O.B. 64, FI-00014 University of Helsinki, Finland

E-mail: `tlinden@cc.helsinki.fi`

**Abstract.** An OpenStack based private cloud with the Gluster File System has been built and used with both CMS analysis and Monte Carlo simulation jobs in the Datacenter Indirection Infrastructure for Secure High Energy Physics (DII-HEP) project. On the cloud we run the ARC middleware that allows running CMS applications without changes on the job submission side. Our test results indicate that the adopted approach provides a scalable and resilient solution for managing resources without compromising on performance and high availability.

To manage the virtual machines (VM) dynamically in an elastic fasion, we are testing the EMI authorization service (Argus) and the Execution Environment Service (Argus-EES). An OpenStack plugin has been developed for Argus-EES.

The Host Identity Protocol (HIP) has been designed for mobile networks and it provides a secure method for IP multihoming. HIP separates the end-point identifier and locator role for IP address which increases the network availability for the applications. Our solution leverages HIP for traffic management.

This presentation gives an update on the status of the work and our lessons learned in creating an OpenStack based cloud for HEP.

## 1. Introduction

The vast amounts of data that needs to be recorded, processed and analyzed by the experiments at the Large Hadron Collider (LHC) has required the creation of a worldwide distributed computing system. The Worldwide Large Hadron Collider Computing Grid (WLCG) has been very successful as it was an necessary and important part enabling the discovery of the Higgs boson in 2012. The future LHC runs will pose even larger computing challenges in terms of data complexity and the amounts of data needed to be handled. Cloud computing can provide a more flexible framework for building distributed computing systems using virtual machines, virtual networks and virtualized storage. Adapting public or private clouds for High Energy Physics (HEP) simulation and data analysis requires some work to provide a seamless and smooth transition from grid technology to cloud technology. The creation of secure distributed computing system poses an important challenge in the cloud environment.

The Datacenter Indirection Infrastructure for Secure HEP Data Analysis (DII-HEP) project is a collaboration between the Helsinki Institute of Physics and the Department of Computer Science at the University of Helsinki funded by the Academy of Finland for the years 2012 –

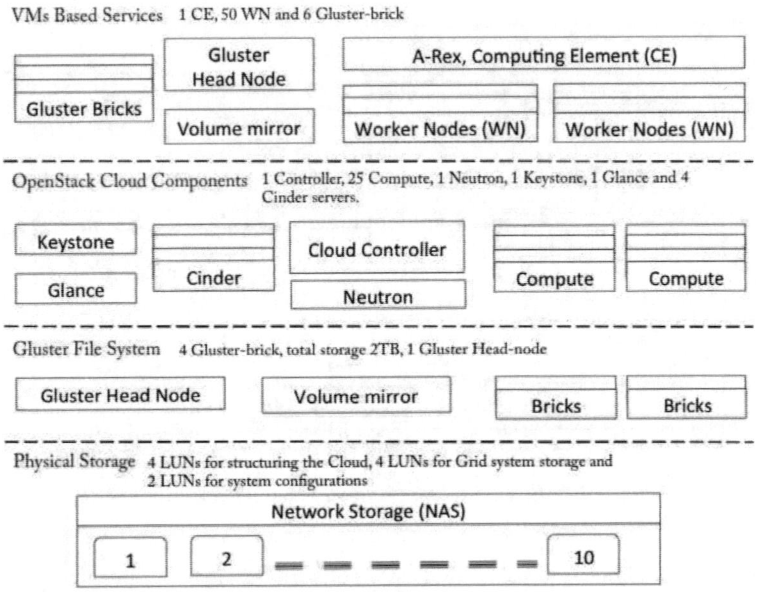


**Figure 1.** Overall system architecture based on OpenStack and GlusterFS.

2014. The goals of the project are to explore the latest software stacks for distributed computing infrastructures and construct a secure and scalable setup for scientific applications using the CMS analysis and production framework as a test case.

## 2. Initial setup

Helsinki Institute of Physics runs a *distributed* CMS Tier-2 site T2_FI_HIP [1] based on the Advanced Resource Connector (ARC) middleware [2]. The main CPU resource is the 768 core Linux cluster and ARC Compute Element (CE) Jade shared between CMS and ALICE. The Madhatter Storage Element is using dCache and is located next to Jade. On the Kumpula campus some 10 km from Jade there are two Linux clusters Korundi (400 cores) and Alcyone (892 cores) both with two ARC CEs. Korundi is shared between the Departments of Physics, Chemistry and the Helsinki Institute of Physics. Alcyone is shared between the Departments of Physics, Chemistry, Computer Science and the Helsinki Institute of Physics.

In the DII-HEP project the requirement was to *add and integrate cloud resources to the existing grid production environment* so that all *jobs could be submitted unchanged.* The use of the grid (ARC) interface also on the cloud resource allows all applications to run unchanged. The setup is **cloud based** and **grid enabled**. The cloud resource is OpenStack [3] based and uses the Gluster File System as shown in Fig. 1. It runs on a part of the Ukko cluster (1920 cores) also on the Kumpula campus. ARC is used to control the execution of jobs and it runs on a virtual ARC CE nodeslab-0002. HTCondor [4] is used as the local batch system and CERN VM File System (CVMFS) [5] is used for application software distribution. CMSSW jobs read data from the local SE using the xrootd protocol [6]. Monitoring is done with Graphite, Site Availability Monitoring- (SAM) and Nagios jobs.

OpenStack is an Open Source infrastructure for public and private clouds. It consists of several different components, *Nova* (Compute), *Quantum* / Neutron (Network), *Swift* (Object based storage), *Cinder* (Block storage), *Keystone* (Identity management) and *Glance* (Image management). The DII-HEP project started with the **Folsom** release and the current setup is

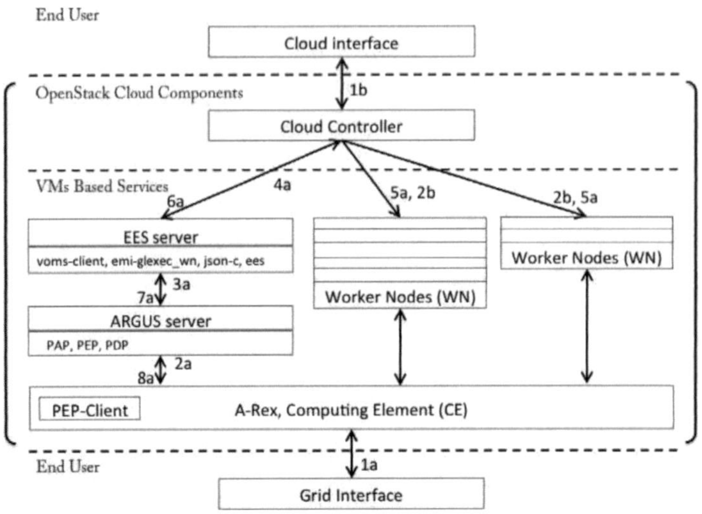


**Figure 2.** How EES is used with Argus for elastic VM-provisioning.

running on **Havana**, testing has been done also with the **IceHouse** release. GlusterFS provides a shared filesystem for VM-instance booting and provides the shared area for the data of the running jobs, see Fig. 1. OpenStack has been deployed on Ubuntu 12.04 LTS and the VMs are based on Scientific Linux CERN 6.4.

So far the system has run a maximum of 200 concurrent jobs. The average CPU efficiency of all CMS jobs over the previous year is 83 % from the Swedish Grid Accounting System (SGAS). There has been 147290 jobs run for 9163 walltime days for the period August 2013 to August 2014. The CPU efficiency loss due to virtualization was measured to be 4 % by running the HEPSPEC 2006 benchmark on bare metal and on VMs. This setup provides efficient resource management, but semi-statically as the *VMs are manually added or removed.* This initial *semi-static* setup has been presented at the CHEP 2013 conference [7].

## 3. Elastic setup status

To achieve *elastic VM-provisioning* and improve on the initial semi-static setup we have used the Argus Execution Environment Service (EES) [8] which is part of the European Middleware Iniative (EMI) release. EES is a service that may co-exist with Argus [9] to start VMs or send jobs to a cloud, see Fig 2. EES has been tested with Argus Policy Execution Point demon (PEPd) and it has a plugin written for OpenNebula. A prototype EES OpenStack plugin has been developed in the DII-HEP project to implement elastic VM-provisioning.

The EES OpenStack plugin controls the start/stop of VMs in the OpenStack cloud and communicates with the OpenStack controller using the JSON API. It is written in C and uses the standard JSON-C library to parse the OpenStack JSON responses. The C Curl library is used to send and receive JSON objects to/from OpenStack. XACML (eXtensible Access Control Markup Language) obligations are handled by the EMI saml2-xacml2 library. For more details on the EES OpenStack plugin see this article [10].

## 4. Host Identity Protocol

The Host Identity Protocol (HIP) [11] has been designed for mobile networks and standardized in the following Request for Comments (RFC) documents 4423, 5201–5207, 6092. It provides

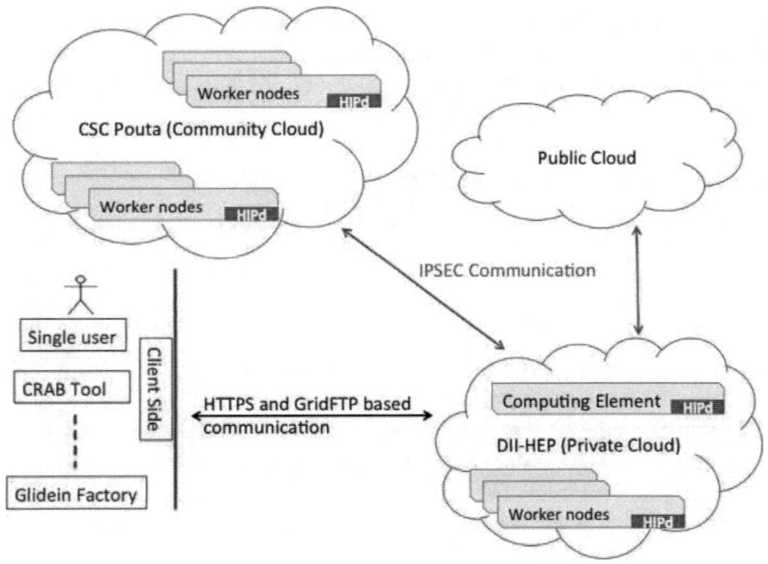


**Figure 3.** Hybrid cloud environment based on DII-HEP and cPouta clouds.

persistent cryptographic identifiers and supports both IPv4 and IPv6 addressing. The namespace is secure in the sense that HIP identifiers cannot be spoofed. The Host Identifiers (HI) are not routable, so they are translated into routable addresses (locators) between network and transport layer. The mapping from HI to network layer locators is dynamic, which enables end-host mobility and multihoming. *This can be useful for creating a hybrid cloud where the IP-address spaces can be different in different clouds.* The dynamic HI mapping can also be useful for VM live migration and site renumbering. The HIP connections are typically protected with IPSec. *Application software can be used unchanged with the HIP protocol.* HIP can in a way be viewed as a Virtual Private Network (VPN), but without a gateway.

A study of the impact of using the HIP protocol was made by comparing the performance for CMS CPU intensive CRAB (CMS Remote Analysis Builder) jobs in three different cases: no specific security mechanism, HIP protocol and SSL/TLS based security. The jobs run for about 170 minutes and each test was running for a about a week on 200 concurrent job slots. The summary of these tests is that the HIP protocol creates negligble CPU and network overhead on the CE. For more details on the HIP protocol use in DII-HEP see this paper [12].

## 5. Secure Hybrid Cloud setup

The HIP protocol has been used in DII-HEP to simplify the network configuration in a hybrid cloud setup. An additional benefit of the HIP protocol is that the communication is encrypted which is useful inside a multitenant cloud as well as on WAN connections.

CSC, The Finnish IT Center for Science has a new data center in Kajaani, more than 500 km north of Helsinki. The supercluster Taito (initially 9 216 and currently 19 048 cores) is located in Kajaani. The part of Taito dedicated for cloud computing is called cPouta, which is an Infrastructure as a Service resource. OpenStack Grizzly is in use on cPouta before the upgrade to the IceHouse release. The DII-HEP setup in Kumpula, Helsinki currently runs 40 cores on OpenStack. HTCondor can join together several HTCondor pools by *flocking* them together. On cPouta there are several VMs as worker nodes and one special HTCondor node that is used for the pool flocking connecting the cPouta worker nodes to the Kumpula cloud resources. Fig. 3 shows how the the Host Identity Protocol is used for *secure cloud bursting* from the Kumpula DII-HEP cloud to the Kajaani Pouta cloud. Tests with 44 cores running on cPouta and connected to the OpenStack in Helsinki are currently in progress.

## 6. Future work

Work on making the EES OpenStack plugin production quality is continuing as well as work on scaling up the intra-cloud setup based on HIP. The Jade cluster is planned to be replaced by a cloud setup as presented here on cPouta. The presented cloud setup could also be used by other scientific applications, like in the Finnish Grid and Cloud Infrastructure consortium.

## 7. Summary

A scalable infrastructure for CMS jobs has been created which is **cloud based** and **grid enabled**. The ARC grid interface has been used to provide the interface for CMS applications, so that all CMS jobs can run unchanged. The cloud setup itself does not rely on grid tools. The same setup could be used for other applications as well. The cloud provides flexibility in managing the infrastructure. Up to 200 simultaneous jobs have run so far. An EES OpenStack plugin prototype has been written. Initial results show that the HIP protocol can be very helpful in resource mobility and security. Cloud bursting has been demonstrated using the HIP protocol and HTCondor flocking for 44 cores so far over a distance of more than 500 km.

## Acknowledgments

The authors thank the CSC technical staff and the CMS collaboration for support. This work is funded by the Academy of Finland, grant numbers 255932 (CS) and 255941 (Physics).

## References

[1] Edelmann E, Happonen K, Koivumäki J, Lindén T and Välimaa J 2011 *J. Phys.: Conf. Ser.* **331** 062037

[2] Ellert M, Grønager M, Konstantinov A, Kónya B, Lindemann J, Livenson I, Nielsen J L, Niinimäki M, Smirnova O and Wäänänen A 2007 *Future Generation Comput. Syst.* **23** 219–240

[3] OpenStack Cloud Software `http://www.openstack.org`

[4] Thain D, Tannenbaum T and Livny M 2005 *Concurrency - Practice and Experience* **17** 323–356

[5] Skipsey S, Doherty T and Turner A G D 2012 *J. Phys.: Conf. Ser.* **396** 042054

[6] Dorigo A, Elmer P, Furano F and Hanushevsky A 2005 *WSEAS Transactions on Computers* **4** 348–353

[7] Toor S, Osmani L, Eerola P, Kraemer O, Lindén T, Tarkoma S and White J 2014 *Journal of Physics: Conference Series* **513** 062047

[8] EMI Execution Environment Service `https://twiki.cern.ch/twiki/bin/view/EMI/ARGUS_EESv0_0_10Details`

[9] EMI Argus `http://www.eu-emi.eu/kebnekaise-products/-/asset_publisher/4BKc/content/argus`

[10] White J, Toor S, Eerola P, Lindén T, Kraemer O, Osmani L and Tarkoma S 2014 *Proceedings of Science* **ISGC2014** 019

[11] Moskowitz R and Nikander P 2006 Host identity protocol (hip) architecture, rfc 4423 (informational) `http://www.ietf.org/rfc/rfc4423.txt`

[12] Osmani L, Toor S, Komu M, Kortelainen M J, Lindén T, White J, Khan R, Eerola P and Tarkoma S 2015 Secure cloud connectivity for scientific applications submitted

# A development of an accelerator board dedicated for multi-precision arithmetic operations and its application to Feynman loop integrals

**S Motoki[1], H Daisaka[2], N Nakasato[3], T Ishikawa[1], F Yuasa[1], T Fukushige[4], A Kawai[4], J Makino[5]**

[1] High Energy Accelerator Research rganization KEK 1–1 ho Tsukuba Ibaraki 3 5– 1 Japan
[2] Hitotsubashi niversity 2–1 aka Kunitachi Tokyo 1 6– 1 Japan
[3] niversity of Aizu Aizu akamatsu Fukushima 65– 5 Japan
[4] K&F Computing Research Co. 1–21–6–4 Ko imacho Chofu Tokyo 1 2– 26 Japan
[5] RIKE Advanced Institute for Computational Science –1–26 inato ima minami machi Chuo ku Kobe Hyogo 65 – 4 Japan

E mail smotoki@post.kek.jp

**Abstract.** Higher order corrections in perturbative uantum eld theory are re uired for precise theoretical analysis to investigate ne physics beyond the Standard odel. This indicates that e need to evaluate Feynman loop diagrams ith multi loop integrals hich may re uire multi precision calculation. e developed a dedicated accelerator system for multi precision calculations RAPE PX . e present performance results of our system for the case of Feynman t o loop bo and three loop selfenergy diagrams ith multi precision.

## 1. Introduction

With the discovery of Higgs particle at the CERN Large Hadron Collider, precision measurements are expected at a future International Linear Collider to explore new physics beyond the Standard Model. In tandem with experiments, an accurate theoretical prediction is required. To meet such a demand, higher order correction in perturbative quantum eld theory becomes more and more important, and methods to evaluate multi-loop integrals precisely should be provided.

We have been developing DCM (Direct Computation Method) for loop integrals [1]. This is a fully numerical method of the combination of multi-dimensional integration and the extrapolation technique. It is known that the accurate numerical evaluation of the integral is a hard problem due to its divergent nature. Yuasa et al. [2] reported that the numerical evaluation is numerically unstable with double-precision operations. A solution to this difficulty is to compute the integral in multi-precision arithmetic, that is, it needs more bits for mantissa than double-precision. In addition, we have a case that needs more bits for exponent. With Double Exponential Formulas for numerical integration (DE) [3], we can estimate that 15-bit wise exponent is required to reduce a relative error smaller than a reference value, although the exponent of double-precision is only 11-bit wise.

There are several ways to accomplish multi-precision arithmetic. One way is to use the speci c softwares, for examples, `GMP` [4], `MPFR` [5], `ARPREC` [6], `MPFUN90` [6], `DD/QD` [6], and `quadmath` on a new version of `GNU` compiler. These softwares are easy to use. However, high performance can not be expected, since one multi-precision arithmetic operation is realized with, at least, more than 10 double-precision arithmetic operations in these softwares. Another way is to design a new, dedicated hardware which has circuits for calculation with precision higher than double-precision. We have been developing such a hardware based on the fundamental idea of GRAPE originally developed for gravitational $N$-body simulations [7]. We designed GRAPE-Multi-Precision (MP) processor which consisted of a number of processing elements (PE) and memory components with dedicated circuits for quadruple, hexuple, octuple-precision arithmetic. We implemented this processor on a structured ASIC (Application Speci c Integrated Circuit) which was called GRAPE-MP [8] , and on an FPGA (Field Programmable Gate Array) board with a control processor, which was called GRAPE-MPX [9]. On these systems, we measured the performance of quadruple, hexuple, octuple-precision calculation and obtained the higher performance compared to a software implementation.

In this work, we developed a new system called GRAPE9-MPX. It is similar to GRAPE-MPX but is a rather larger system. This means that larger number of PEs can be implemented in the new system, which will be suited for large scale simulations. In addition to designing a hardware, we have been developing a programming interface named Goose and LSUMP which enable us easily to use GRAPE9-MPX.

This paper is organized as follows. Next section covers a brief explanation of the design and the implementation of the new system from hardware and software viewpoints. In section 3, we brie y explain Feynman loop integrals used for the performance measurement. The results are shown in section 4. The last section is devoted to the summary and our future prospects.

## 2. GRAPE9-MPX system

### *2.1. Hardware*

**Figure 1.** Schematic diagram (left) and picture (right) of GRAPE9-MPX. Note that 4 FPGA boards are included in the system.

Figure 1 shows a schematic diagram and a picture of GRAPE9-MPX system. It is an accelerator which is connected to a host computer. It consists of FPGA boards (up to 8 boards available) which are housed on a PCIe extender board (G9MG). For the FPGA board, we used Altera Arria V board from Altera Co. [10]. On this FPGA, we implemented the MP processor and its Control Processors (CP). The MP processor consists of PEs and memory components for quadruple/hexuple/octuple-precision arithmetic (named as MP4/MP6/MP8), which forms a Single-Instruction-Multiple-Data (SIMD) processor. Each PE has a  oating-point multiply unit and an add unit which perform in every clock cycle. Note that for MP4, we used the numerical representation compatible with IEEE-754-2008:binary128 format. For details of the PE, CP, and the numerical representations, see Daisaka et al. [8] and Nakasato et al. [9].

One of applications suited for our system is an interaction type calculation as $f_i = \sum_{j=1}^{n_j} f(X_i, Y_j)$, where $X_i$ is $i$-th element of $X$, $Y_j$ is $j$-th element of $Y$, and $n_j$ is the number of elements of $Y$. The function $f(X_i, Y_j)$ describes an interaction form of $X_i$ and $Y_j$. Note that data $X_i$ is set on $i$-th PE, whereas data $Y_j$ is set on all PE, then each PE calculates $f(X_i, Y_j)$ and sums up from $j = 1$ to $j = n_j$ in accordance with instructions. An example of this type calculation is gravitational interactions among particles. We should note that a multi-dimensional integration can be expressed in a similar form by *loop fusion* technique that merges multiple loops into a single loop. This indicates that we can accelerate the calculation of Feynman loop integrals e ectively by using GRAPE9-MPX.

Table 1 summarizes our current implementation of MP4/MP6/MP8 processors on the FPGA board. The peak performance of MP4, for example, can be estimated as 2 (operations) $\times$ 36 (PEs) $\times$ 92 (MHz) $=$ 6.6 G ops per board. It adds up to 26.4 G ops for the system with 4 boards.

| | MP4 | MP6 | MP8 |
|---|---|---|---|
| Number of PEs | 36 | 20 | 12 |
| Clock Speed(MHz) | 92 | 81 | 70 |
| Theoretical Peak(G ops) | 6.6 | 3.2 | 1.6 |

**Table 1.** Number of PEs, clock speed, and theoretical peak performance for MP4/MP6/MP8 implemented on the FPGA.

### *2.2. Programming interface*

We have also developed Goose compiler which provides a programming interface for GRAPE9-MPX system. Figure 2 shows the ow of the compile process for our system, and the part of a sample program for one-loop box integral. Goose is a directive base compiler like OpenMP, in which a directive is inserted in an original code. By the directive (#PRAGMA GOOSE PARALLEL seen in the sample code), Goose extracts the loop next to the directive and generates an intermediate representation which is like an assembler code. Goose also generates API calls in which functions of I/O interfaces for GRAPE9-MPX are embedded. In order to convert the intermediate representation to machine instructions (kernel code) for the MP processor, Goose uses LSUMP backend [11] which is a domain speci c language (DSL) compiler specially developed for the MP processor. The kernel code is read by an executable le which is generated from the API calls by C/C++ compiler. For more detail, see Nakasato [11].

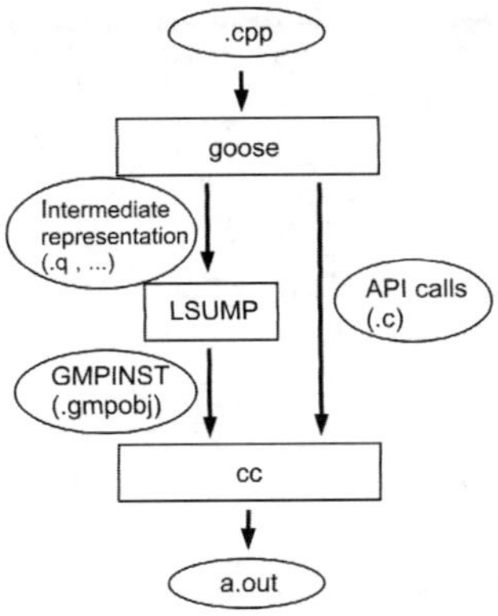


```
#pragma goose parallel for loopcounter(ixy, iz)
for(ixy = 0; ixy < ni; ixy++) {
  sumzG[ixy] = 0.0;
  for(iz = 0; iz < nj; iz++) {
    xx = dev_xx[ixy];
    yy = dev_yy[ixy];
    zz = x30_1[iz] * dev_cnt4[ixy];
    d = - xx * yy * s -
    tt * zz * (one - xx - yy - zz) +
    (xx + yy) * lambda2 +
    (one - xx - yy - zz) * (one - xx - yy) * fme2+
    zz * (one - xx - yy) * fmf2 ;
    sumzG[ixy] += gw30[iz] / (d * d);
  }
}
```

**Figure 2.** Flow of the compile process (left) and a sample code with the directive (right) for our system.

Journal of Physics: Conference Series **608** (2015) 012011 doi:10.1088/1742-6596/608/1/012011

### 3. Feynman loop integrals

Figure 3 shows the two-loop crossed box diagram and the three-loop selfenergy diagram used for the performance measurement. The number of dimensions of the loop integral is six for the

**Figure 3.** Two-loop crossed box diagram and three-loop selfenergy diagram used for the performance measurement. In these diagrams, $p_i$ and $x_i$ denote a momentum of an external line and a Feynman parameter, respectively.

two-loop crossed box, and seven for the three-loop selfenergy, respectively. We consider the case where all masses of the internal lines are 1 and set the Mandelstam variables $s = t = 1$ and the external momenta $p_i^2 = 1 (1 \le i \le 4)$ for the two-loop crossed box diagram and $s = 1$ and $p^2 = 1$ for the three-loop selfenergy diagram, respectively. We choose these values to compare numerical results to ones in the previous studies [12, 13].

With these parameters, the integrals of the two-loop crossed box with variable transformations from $\{x_i\}$ to $\{\xi_i\}$ to change the integration domain to the unit cube from the unit simplex is expressed as

$$I_{2loop} = 2\int_0^1 d\xi_1\, d\xi_2\, d\xi_3\, d\xi_4\, d\xi_5\, d\xi_6\, (\xi_1^2 {\xi'_1}^3 \xi'_2 \xi_4 \xi'_4)\frac{C}{D^3}, \qquad (1)$$

where

$$\begin{aligned} C &= \xi'_1(\xi_1 + \xi'_1\xi_4\xi'_4), \qquad (2) \\ D &= \xi'_1((\xi_1 + \xi'_1\xi_4\xi'_4) \\ &- (\xi_1(\xi_1\xi'_2\xi'_2\xi_3\xi'_3 + \xi'_1\xi_4\xi'_4(\xi'_2(\xi_3\xi'_5\xi_6 + \xi'_3\xi_5\xi'_6) - \xi_2\xi_5\xi_6)) \\ &+ \xi_1\xi_2(\xi'_1\xi_4\xi'_4(-\xi_5\xi_6 + \xi'_5\xi'_6) + (\xi_1\xi'_2\xi_3 + \xi'_1\xi_4\xi'_4\xi_5) + (\xi_1\xi'_2\xi'_3 + \xi'_1\xi_4\xi'_4\xi_6)) \\ &+ \xi'_1\xi_4(\xi_1\xi_5(\xi_2 + \xi'_2\xi'_3)(\xi_4\xi'_5 + \xi'_4\xi_6) + \xi_1\xi'_2\xi_3\xi'_5(\xi_4\xi_5 + \xi'_4\xi'_6) + \xi'_1\xi_4\xi'_4\xi_5\xi'_5) \\ &+ \xi'_1\xi'_4(\xi_1\xi_6(\xi_2 + \xi'_2\xi_3)(\xi_4\xi_5 + \xi'_4\xi'_6) + \xi_1\xi'_2\xi'_3\xi'_6(\xi_4\xi'_5 + \xi'_4\xi_6) + \xi'_1\xi_4\xi'_4\xi_6\xi'_6))), \end{aligned}$$

with $\xi'_i = 1 - \xi_i$. Note that the integration domain for three-loop selfenergy diagram, $I_{3loop}$, can be changed by the similar way.

In order to evaluate $I_{2loop}$ and $I_{3loop}$, we used DE [3] iteratively for the multi-dimensional integration. Computation time for the numerical integration depends on the total number of evaluation points, $N_{total}$. In order to take advantage of the hardware characteristics of GRAPE9-MPX, we divide $N_{total}$ into two parts, $i$-loop (outer loop) part and $j$-loop (inner loop) part, so as to satisfy $N_{total} = N_i \times N_j$. This is realized by *loop fusion* technique, in which the outer 3-dimensional integration in Eq. (1) can be performed in $i$-loop and the most inner 3-dimensions in $j$-loop. In the performance measurements, we set $N_i$ to be $2^{18}$ for two-loop crossed box and $2^{23}$ three-loop selfenergy, respectively. On the other hand, we vary $N_j = 2^k$ where $k = 4, 5, 6, ..., 13$. Hereafter, we call the maximum problem size for $N_j = 2^{13}$. In the maximum problem size, $N_{total}$ is $2^{31}$ for the two-loop crossed box, and $2^{36}$ for the three-loop selfenergy, respectively.

The numerical results in quadruple precision obtained by GRAPE9-MPX in the maximum problem size are $I_{2loop} = 0.008536$ and $I_{3loop} = 0.279609$, which are within the relative errors of $10^{-3}$ for the two-loop crossed box and $10^{-5}$ for the three-loop selfenergy compared to the previous works [12, 13]. It seems that the error is slightly larger because of smaller number of $N_{total}$, but the calculation with this error level is achieved in a very short time, as seen later.

## 4. Performance results

We measured the performance of GRAPE9-MPX with up to 4 FPGA boards. It is connected a Linux PC (Scienti c Linux 6.5) with Intel Xeon 2687W (3.4GHz) CPU on ASRock Xtreme11 motherboard (X79 chipset). The performance results are plotted in Fig.4. From this gure, we can see that for the two-loop crossed box integral, the e ective performance in the maximum problem size was 2.4 G ops with a single board, 4.7 G ops with 2 boards, and 9.1 G ops with 4 boards, respectively. On the other hand, the e ective performance for the three-loop selfenergy integral in the maximum problem size was 8.7 G ops with 4 boards.

This gure also shows that the speed up by multiple boards is well scalable in the maximum problem size. It becomes 3.8 times faster using 4 boards, and 1.9 times faster using 2 boards than using a single board. On the other hand, the scalability is not good in smaller $N_j$. This is because the overhead by communication between the GRAPE9-MPX boards and the host computer via PCI express is not negligible for smaller $N_j$.

We compared the efficiency of the e ective performance for the two-loop crossed box integral and for the three-loop selfenergy integral. We show that it is about 34% of the theoretical peak performance for the two-loop crossed box, whereas about 33% for the three-loop selfenergy. The reason why the latter is slightly lower than the former comes from the di erence of the number of instructions. From LSUMP, the operation count of the instruction is 91 for two-loop crossed box, whereas it is 89 for three-loop selfenergy.

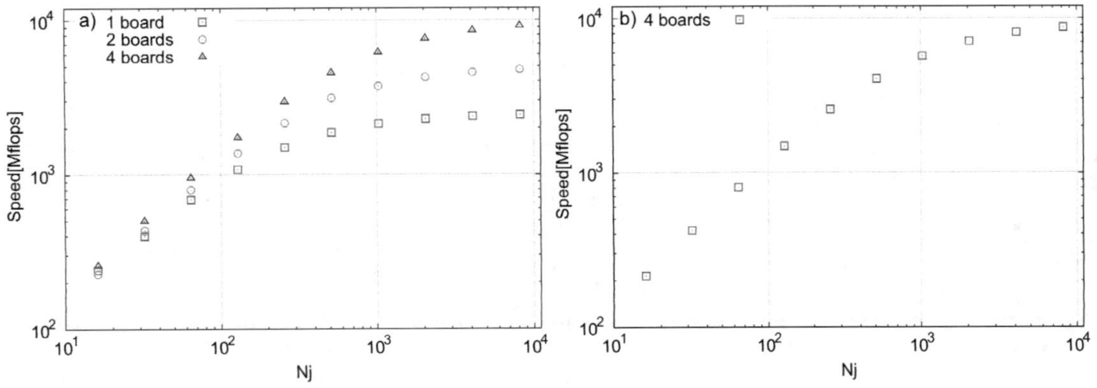


**Figure 4.** Performance results for two-loop crossed box diagram (left) and three-loop selfenergy diagram (right).

We also compared the computation time for the two-loop integral in the maximum problem size measured in GRAPE9-MPX of MP4 with 4boards, and the time measured in software implementations with the same numerical format (15-bit exponent and 112-bit mantissa). The measured calculation time is 21.3 sec for GRAPE9-MPX, 1074.7 sec for `quadmath` on gcc4.6.3, and 6086.7 sec for `GMP-MPFR`. Although we used `OpenMP` for the multi-threaded computing and the number of threads used was 16, GRAPE9-MPX system was 52.9 times faster than `quadmath` and, 319.5 times faster than `GMP-MPFR`. Thus, we could show the advantage of our system in the performance.

## 5. Summary

We have developed GRAPE9-MPX system, an accelerator system for multi-precision arithmetic operations which can accelerate the calculation of Feynman loop integrals. Our system consists of multiple FPGA boards in which a lot of PEs are implemented and forms SIMD way. Each PE has a dedicated logic for multi-precision operations. We measured the performance of the system in quadruple precision (15-bit exponent and 112-bit mantissa) for the two-loop crossed box and the three-loop selfenergy integrals, and got the e ective performance of 9.1 G ops and of 8.7 G ops with 4 boards in the maximum program size, respectively.

One of the advantages of our system is that we have developed not only the hardware with high performance in multi-precision calculation, but also the programming interface which enables us to port applications easily for our system. The combination of Goose and LSUMP provides a simple way to use GRAPE9-MPX. All we have to do is to insert the directive in the original code. This simple programming model plays an essential role in treating various kinds of loop integrals in the automatic computation of Feynman diagrams.

Although we mainly presented the results of MP4 in this paper, we have measured the performance of MP6 and MP8 for the same Feynman loop integral. For the case of the two-loop crossed box, we achieved 4.4 G ops for MP6 and 2.3 G ops for MP8 with 4 boards, respectively. The detail of the results will appear in near future.

## Acknowledgment

We acknowledge the support from Grants-in-Aid for Scienti c Research 23540328 and 24540292.

## References

1 de Doncker E Shimizu Y Fu imoto J and Yuasa F 2 4 *Computer Physics Communications* **159** 145–156
2 Yuasa F de Doncker E Fu imoto J Hamaguchi Ishika a T and Shimizu Y 2 *PoS(ACAT)* vol RL `http://pos.sissa.it/cgi-bin/reader/conf.cgi?confid=50`
3 ori 2 5 *Publications of the Research Institute for Mathematical Sciences* **41(4)** – 35
4 P RL `https://gmplib.org/`
5 PFR RL `http://www.mpfr.org/`
6 High Precision Soft are Directory RL `http://crd-legacy.lbl.gov/~dhbailey/mpdist/`
Koike K 2 *Development of Control Processor for Multipurpose Computer GRAPE-DR and its Performance Evaluation* Thesis raduate niversity for Advanced Studies RL `http://id.nii.ac.jp/0201/00001514`
Daisaka H akasato akino J Yuasa F and Ishika a T 2 11 *Procedia Computer Science* **4** –
akasato Daisaka H Fukushige T Ka ai A akino J Ishika a T and Yuasa F 2 12 *Embedded Multicore Socs (MCSoC), 2012 IEEE 6th International Symposium on* 5– 3
1 Altera Co *Arria V Device Handbook* RL `www.altera.co.jp/literature/hb/arria-v/arriav_handbook.pdf`
11 akasato and akino J 2 *IEEE International Conference on Cluster Computing and Workshops* pp 1–
12 Laporta S 2 *International Journal of Modern Physics A* **15** 5
13 Yuasa F de Doncker E Hamaguchi Ishika a T Kato K Kurihara Y and Fu imoto J 2 12 *Computer Physics Communications* **183** 2136–2144

# The Long Term Data Preservation (LTDP) project at INFN CNAF: CDF use case

**S. Amerio[1], L. Chiarelli[2],L. Dell'Agnello[3], D. Gregori[3], M. Pezzi[3], P. Ricci[3], F. Rosso[3], S. Zani[3]**

[1] University of Padova, Italy
[2] GARR
[3] INFN-CNAF, Bologna, Italy

E-mail: michele.pezzi@cnaf.infn.it

**Abstract**. In the last years the problem of preservation of scientific data has become one of the most important topics inside international scientific communities. In particular the long term preservation of experimental data, raw and all related derived formats including calibration information, is one of the emerging requirements within the High Energy Physics (HEP) community for experiments that have already concluded the data taking phase. The DPHEP group (Data Preservation in HEP) coordinates the local teams within the whole collaboration and the different Tiers (computing centers). The INFN-CNAF Tier-1 is one of the reference sites for data storage and computing in the LHC community but it also offers resources to many other HEP and non-HEP collaborations. In particular the CDF experiment has used the INFN-CNAF Tier-1 resources for many years and after the end of data taking in 2011, it is now facing the challenge to both preserve the large amount of data produced during several years and to retain the ability to access and reuse the whole amount of it in the future. According to this task the CDF Italian collaboration, together with the INFN-CNAF computing center, has developed and is now implementing a long term future data preservation project in collaboration with Fermilab (FNAL) computing sector. The project comprises the copy of all CDF raw data and user level ntuples (about 4 PB) at the INFN-CNAF site and the setup of a framework which will allow to access and analyze the data in the long term future. A portion of the 4 PB of data (raw data and analysis-level ntuples) are currently being copied from FNAL to the INFN-CNAF tape library backend and a system to allow data access is being setup. In addition to this data access system, a data analysis framework is being developed in order to run the complete CDF analysis chain in the long term future, from raw data reprocessing to analysis-level ntuples production and analysis. In this contribution we first illustrate the difficulties and the technical solutions adopted to copy, store and maintain CDF data at the INFN-CNAF Tier-1 computing center. In addition we describe how we are exploiting virtualization techniques for the purpose of building the long term future analysis framework.

## Introduction

Long-term data preservation can be defined as the ability to provide continued access to digital materials and the capability to analyze them. Interest in the long term preservation of scientific data and their availability to general public is growing.

Data collected in High Energy Physics (HEP) experiments are the result of a significant human and financial effort. The preservation of HEP data beyond the lifetime of the experiment is of crucial importance to ensure the long term completion and extension of scientific programs, to allow cross collaboration analysis, analysing data from several experiment at once, to perform new analysis with new theoretical models and techniques and for education, training and outreach.

HEP data preservation poses many technical and organizational challenges: data preservation implies migration to new storage media when available, adjusting data access methods if needed; moreover data analysis capabilities must be preserved ensuring the experiment legacy software runs on new platforms, or on old ones with no security issues; validation systems have to be set up to regularly check data access and analysis framework; all the information needed to access and analyze data has to be properly organized and archived. For these reasons a data preservation project can be divided into two main areas: the first consists in "bit preservation", how we preserve the data, while the second part consists in the preservation of the analysis capabilities.

In this paper we describe a project to preserve at INFN-CNAF computing center in Italy a complete copy of the data of CDF experiment which run at the Fermilab Tevatron until September 2011. The project aims at having at CNAF a copy of all raw data and user level ntuples (4 PB) and providing users with data access and data analysis capabilities in the long term future.

## 2. Bit preservation and storage configuration

For the bit preservation, there are two important aspects to consider: first how the data is transferred and second where the data is stored.

As shown in Figure 1, we are using dedicated network resources for the CDF long term data preservation project.

At CNAF two servers with 10GE network interfaces are directly connected to the core switch/router of the Computing Center in order to guarantee the maximum network performance and the minimum round trip time (RTT).

In Local Area Network, servers involved in this project are isolated in a spacific VLAN and are using a dedicated IP network.

Geographical connectivity is provided by GARR and ESNET and consists on a dedicated 10 Gbps link from CNAF to FNAL.

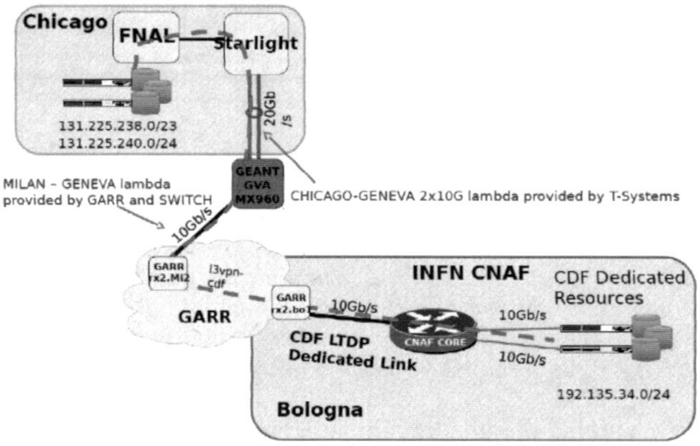


**Figure 1** : Layout of the FNAL-CNAF copy network.

For the data transfer Sequential Metadata Access (SAM)[1], a data handling tool developed at Fermilab, has been installed on a dedicated machine at CNAF and it is used to pilot the transfer. Figure 2 shows how the data transfer works.

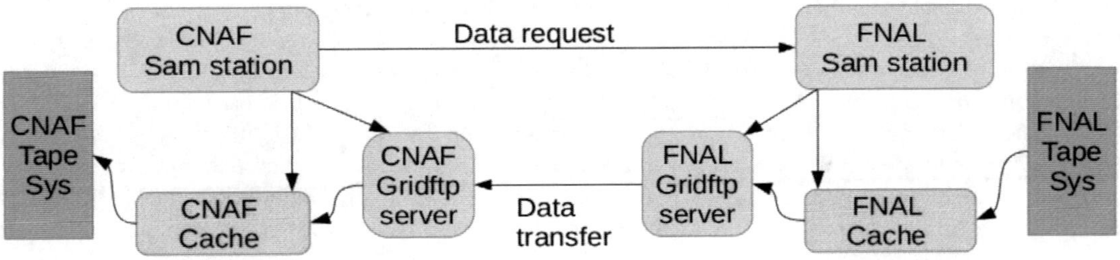


**Figure 2** : Data transfer layout.

Integrating the SAM and GridFTP transfer commands within the same bash script, the data transfer takes place in a semi-automatic way:
1) CNAF requests from FNAL data that are staged at FNAL cache.
2) Data are copied using GridFTP protocol, using a third party transfer.
3) To check data integrity the file checksum is computed and compared with the value stored in SAM database.
4) Once the data files are in the CNAF cache and if the data integrity check is successfull, they are automatically migrated to tape and the SAM bookeeping database is updated.

The adopted solution at CNAF for the "bit preservation" of CDF data is GEMSS[2], a well established system, in production since several years at CNAF. GEMSS consists of a pool of disks managed by GPFS (IBM General Parallel File System), a tape library infrastructure for the archive back-end managed by Tivoli Storage Manager[3] (TSM) and an integration system to transfer data from disk to tape and vice versa.

The storage layout system is composed by several elements as shown in Figure 3.

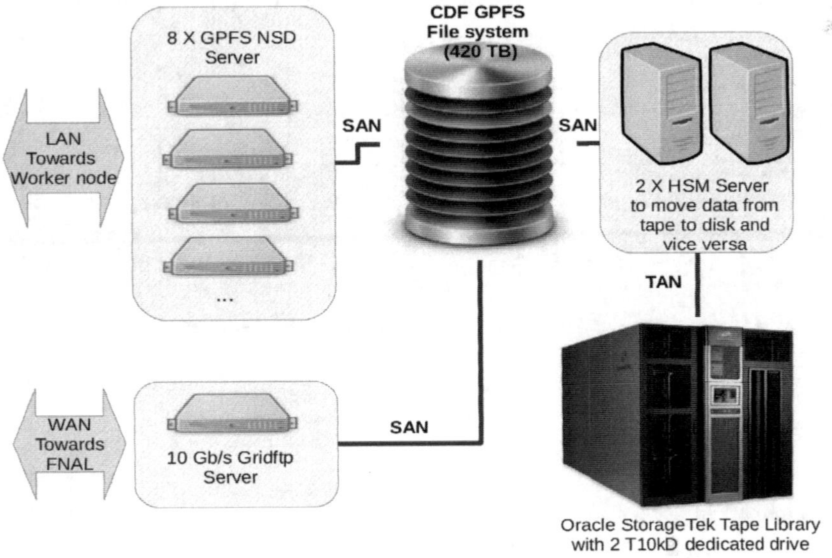


**Figure 3** : Storage system layout.

The transferred data are stored on a dedicated GPFS cluster file system. On this file system two different directories were created. The first directory (cache) is the actual SAM disk cache that is used during network transfer. After a data file is copied to the disk cache and its data integrity is checked, it is moved to a second directory (durable) from which files are regularly and automatically migrated to tape. Two dedicated HSM servers are used for moving data from tape to disk and vice versa using 2 tape drives through the Tape Area Network (TAN). During the copy phase, once a file is migrated to tape it is automatically removed from disk, in order to maintain a significant ($\sim$ 300 TB) disk buffer for the copy. In the short term future, once the copy is completed, we will keep a small ($\sim$ 100 TB) CDF-dedicated disk area to cache datasets requested by users for the analysis. In the long term future instead, when intermittent usage is foreseen, the disk cache will be allocated on demand. In any case a small disk cache (~ few tens of TB) is needed for periodic data read for integrity checks. In case of damage or loss, the file can be recovered by Fermilab, where two additional copies of the data are archived.

## 3. Long term future analysis framework

In order to access and process the data in the long term future it is essential to have an infrastructure that allows to use the experiment software. The goal is to make the software and data available and functional even when the experiment is no longer active. In the following we will briefly describe the main ingredients of the framework under development at CNAF.

In Figure 4 the main elements of CDF analysis frameworks are shown. The user submits a job from his/her user area. The job submission system contacts the SAM station to access the data. Data is sent to worker nodes together with a tarball containing the analysis code. If needed, e.g. for MC production, detector and run conditions are retrieved from a dedicated database. All these services are already installed at CNAF and for the long-term future we intend to replicate this system as much as possible.

We expect in the long-term future data will be accessed and processed very rarely. For this it is necessary to study a solution which is robust but at the same time allows to minimize the needed resources (number of physical machines within the framework). As shown in the Figure 4 the only real machine, except the storage system, is the database, which is currently located at FNAL. All other services are instantiated on virtual machines. The virtual machines will be handled by WNoDeS[4], an INFN-developed framework that makes possible to dynamically allocate virtual resources out of a common resource pool, in order to instantiate the virtual machines when a job is started.

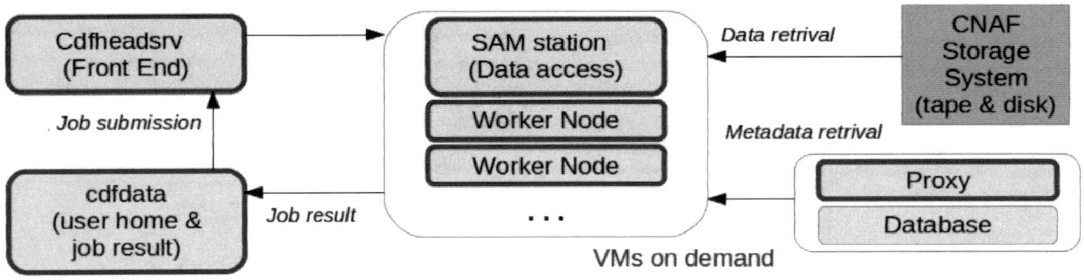


**Figure 4** : Long term analisys framework.

### *3.1 Data access*

We will keep using SAM to access data in the long term future. CNAF tape system is transparent for SAM: upon a request to access a file, the file is recalled from tape on a disk cache before being sent to the final destination (worker node or user area). As already stated in the previous section, in the long term future we foresee intermittent access to CDF data, so the necessary disk cache will be allocated on-demand using CNAF resources in opportunistic mode. We estimate that a user area of 50 TB will be sufficient for each analysis, as the majority of CDF datasets amounts to few TB and the biggest datasets do not exceed 20 TB. As described in section 2, CNAF used GPFS filesystem, and the tape library which is developed and maintained by IBM. SAM code is based on SL6 and will not be ported to new operating systems thus compatibility problems may arise if IBM will not support GPFS SL6 version in the long term future. A possible solution is accessing the data via NFS protocol, which is expected to be compatible with older operating system. The first tests of this access method were successful: CDF GPFS filesystem was exported by a server via NFS protocol and data accessed via standard SAM commands. Scalability tests will follow on extended datasets (hundreds of files).

*3.2 Data analysis*
Future analysis on CDF data will use a software legacy release based on Scientific Linux 6 (SL6) distributed through CVMFS. At CNAF as a first step squid proxy servers have been setup to access the CVMFS server located at FNAL. In a second phase the FNAL CVMFS server will be replicated at CNAF.

The legacy release has not yet been officially released by FNAL; in the meantime the development version is being tested on CNAF SL6 nodes. As far as job submission is concerned, the system currently installed for CDF at CNAF is based on glideinWMS [5] to exploit computing resources at CNAF and additional LCG resources at different Tier-2 sites in Italy and other European countries. In the short term future the current system will be maintained, upgrading it to use the legacy release. For the long term future instead a simpler job submission system is under study which will exploit only CNAF CPU resources and LSF workload management system, already used at CNAF by other experiments.

*3.3 Long term database issues*

Oracle databases for data bookkeeping and run conditions are currently located at FNAL. In order for the CDF analysis framework at CNAF to be independent from FNAL, they would need to be locally replicated. There are basically two options currently under review. As first straightforward approach, we could plan to setup the needed Oracle database instances at CNAF, and keep them synchronized with one of the high availability or disaster recovery features available with Oracle. Choosing this solution we could keep compatible with the CDF analysis framework, but it would be quite expensive in terms of Oracle licenses, especially in the long term. Therefore, as an alternative approach, we are investigating the idea to rewrite the database interface inside the CDF software, opening the path to use open-source RDBMSs such as PostgreSQL or MySQL. This second approach is under discussion, because it needs some assessment of the needed developing effort. However it could be worthwhile to reduce the fixed costs, especially considering that this is a long term data preservation project.

## 4. Conclusion
A project to preserve CDF data and analysis capabilities is being implemented at CNAF. The project is be divided into two areas: the "bit preservation"and the preservation of the analysis framework. For the bit preservation a system to copy the data from FNAL has been setup; the data is being copied and about 2 PB of data out of 4 PB are already on tape. For the analysis framework, a system is under development: the goal is to exploit as much as possible the CDF analysis framework already existing at CNAF, investigating solutions to keep it functional for many years to come, e.g. using NFS to access data and virtualization techniques to run CDF legacy code.

## References

[1] Stonjek S, Baranovski A, Kreymer A, Lueking L, Ratnikov F et al., *Deployment of SAM for the CDF Experiment*, 2005 1052–1054 2, Proceedings of 2004 CHEP conference
[2] D. Bonacorsi et al., *The Grid Enabled Mass Storage System (GEMSS): the Storage and Data management system used at the INFN Tier1 at CNAF*, 2012 J. Phys. Conf. Ser. 396 042051 Proceedings of 2012 CHEP conference.
[3] IBM website references for Tivoli Storage Manager info and documentation https://www.ibm.com/developerworks/wikis/display/tivolidoccentral/Tivoli+Storage+Manager

and http://www-03.ibm.com/software/tivoli/products/storage-mgr/
[4] D.Salomoni et all, WNoDeS, a tool for integrated Grid and Cloud access and computing farm virtualization 2010 J. Phys. Conf. Ser. 331 052017 Proceedings of 2010 CHEP conference
[5] S.Amerio et al, *Eurogrid: a new glideinWMS based portal for CDF data analysis*, Journal of Physics: Conference Series 396 (2012) 032001
[6] S. Amerio et al., *Long Term Data Preservation for CDF at INFN-CNAF* 2013 J. Phys. Conf. Ser. 513 (2014) 042011 Proceedings of 2013 CHEP conference

# The INFN-CNAF Tier-1 GEMSS Mass Storage System and database facility activity

**Pier Paolo Ricci[1,4], Alessandro Cavalli[1], Luca Dell'Agnello[1], Matteo Favaro[1], Daniele Gregori[1], Andrea Prosperini[1], Michele Pezzi[1], Vladimir Sapunenko[1], Giovanni Zizzi[1] and Vincenzo Vagnoni[2]**

[1] INFN CNAF, viale Berti Pichat 6/2 40127 Bologna, Italy.
[2] INFN Sezione di Bologna, via Irnerio 46, 40126 Bologna, Italy.

E-mail: pierpaolo.ricci@cnaf.infn.it

**Abstract.** The consolidation of Mass Storage services at the INFN-CNAF Tier1 Storage department that has occurred during the last 5 years, resulted in a reliable, high performance and moderately easy-to-manage facility that provides data access, archive, backup and database services to several different use cases. At present, the GEMSS Mass Storage System, developed and installed at CNAF and based upon an integration between the IBM GPFS parallel filesystem and the Tivoli Storage Manager (TSM) tape management software, is one of the largest hierarchical storage sites in Europe. It provides storage resources for about 12% of LHC data, as well as for data of other non-LHC experiments. Files are accessed using standard SRM Grid services provided by the Storage Resource Manager (StoRM), also developed at CNAF. Data access is also provided by XRootD and HTTP/WebDaV endpoints. Besides these services, an Oracle database facility is in production characterized by an effective level of parallelism, redundancy and availability. This facility is running databases for storing and accessing relational data objects and for providing database services to the currently active use cases. It takes advantage of several Oracle technologies, like Real Application Cluster (RAC), Automatic Storage Manager (ASM) and Enterprise Manager centralized management tools, together with other technologies for performance optimization, ease of management and downtime reduction. The aim of the present paper is to illustrate the state-of-the-art of the INFN-CNAF Tier1 Storage department infrastructures and software services, and to give a brief outlook to forthcoming projects. A description of the administrative, monitoring and problem-tracking tools that play a primary role in managing the whole storage framework is also given.

## 1. The storage system of the INFN-CNAF Tier-1

In this paper we describe the main activities and the consequent results obtained during the last years concerning mass storage resources at the INFN-CNAF Tier-1. Starting from the end of 2005 the Tier-1 [1] has become the Italian national reference point for the INFN computing activities. In particular, all of the LHC experiments (ALICE, ATLAS, CMS and LHCb) use our site as a primary Tier-1 computing centre. In addition, about 20 non-LHC collaborations (e.g. CDF, Kloe, AMS2, Argo, Auger, Glast, MAGIC, Pamela, Borexino, Darkside, Virgo, etc.) currently use the computing and storage resources of the centre with a guarantee level of service of 24/7 support (24h every day non-stop service availability). The storage resources comprise 15 Petabyte (PB) of net disk space accessible in a Storage Area Network (SAN), mainly composed of RAID-6 disk arrays. Furthermore,

[4] Corresponding Author

tape storage is provided by an Oracle STK SL8500 tape library robot, characterized by a capacity of 17 PB of uncompressed tape space served by several Oracle STK T10K tape drives. The computing farm is composed of ~1000 nodes, which have direct access to 12 PB of disk space shared over 11 IBM General Parallel File System (GPFS) [2] filesystems.

In order to access these storage resources, a very reliable and high performance disk-based data access system is required, as well as an efficient Mass Storage Systems (MSS). The MSS allows the archival of several PB of data per year on tape media, that should remain available over a time scale of several years. For these reason in 2008 we started to work on the integration of our IBM GPFS disk storage infrastructure with the IBM Tivoli Storage Manager (TSM) [3], aiming at realizing a full Hierarchical Storage Management (HSM) system called GEMSS (Grid Enabled Mass Storage System). GEMSS uses StoRM as Grid Storage Resource Manager (SRM) [4], providing the standard interface used in the WLCG framework. It is distributed with standard RPM packages implementing a software layer for the interaction and optimization between GPFS, TSM and StoRM, thus providing a complete solution for tape and disk storage. The entire storage resources at the INFN-CNAF Tier-1 have been managed by GEMSS since the start of its production phase at the end of 2010 [5].

In Figure 1, a schematic representation of the main GEMSS components is reported.

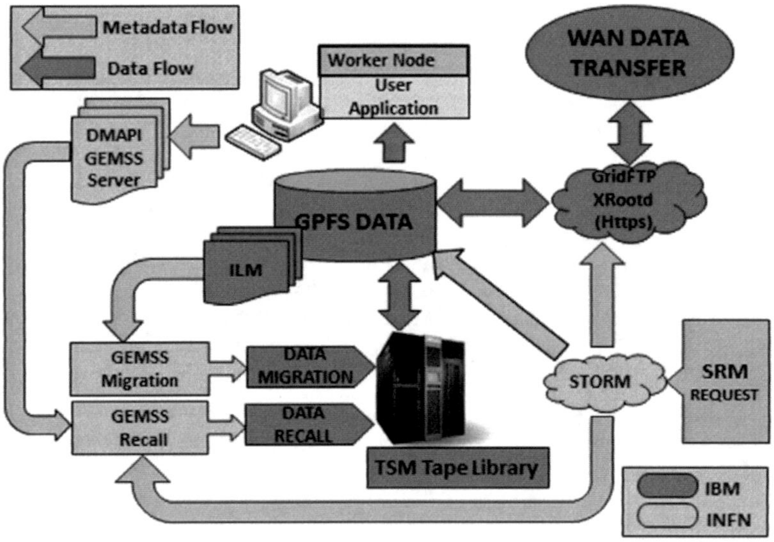


**Figure 1**. Logical schema of the main building blocks of the GEMSS system.

In the Figure, the five main components of the GEMSS system (some are developed by IBM and some by INFN) and the two main data and metadata flows (the pale blue and red arrows) are shown.

In particular it is possible to recognize the following main elements:

- GPFS DATA cluster and Information Lifecycle Management (ILM) engine which is the disk storage software main GPFS filesystem infrastructure with the utilization of ILM. By using ILM, the filesystem can be partitioned into a number of so-called GPFS "storage pools" implementing file placement policies and data migration rules from one pool to another according to some user-defined criteria.
- TSM (Tivoli Storage Manager) Tape Library. The tape library management system performs the data movement between disk and tape media using the "Data Migration" and "Data Recall" operations.
- StoRM, the SRM service.

- GridFTP/XRootD/(HTTPS) services. The LAN/WAN data transfer services and protocols which are used for transferring the data to and from the storage system. These protocols will be described in the next section.
- GEMSS migration and recall engine for the interaction between TSM and GPFS and for the general metadata management.

In addition, the DMAPI GEMSS server plays a central role when tape-resident data is directly accessed by User Application running on a worker node, without requesting in advance the tape recall via the SRM service. This is to allow the User Application to trigger tape recalls in an optimized and controlled way simply by accessing the files. The POSIX read requests are intercepted by the DMAPI GEMSS server and are submitted to the GEMSS system which groups and sorts the requests in an optimal way [5].

The GEMSS system is therefore the core of the INFN-CNAF Tier-1 storage infrastructure and it has been demonstrating stability, reliability and good performance over the last years of running activity. In the next section we will focus on the access to the MSS and on the specific performance monitoring tools implemented in order to improve the general service supervision.

## 2. MSS access and monitoring

The main entry point to data storage from the WAN at the INFN-CNAF Tier-1 is StoRM, a Grid middleware service that implements the SRM interface version 2.2 and is designed to support direct file access using native POSIX I/O calls to the storage. Actually StoRM is used by ATLAS, CMS and LHCb, and by other smaller Virtual Organizations (VOs). Data transfers are usually performed using the GridFTP service, that is also used without contacting first the SRM service by some non-LHC experiments using X509 certificates (DarkSide, Kloe, Theoretical physics groups, etc.). Besides the SRM protocol, the LHC experiments have also started considering other methods for accessing data. In particular, two interesting protocols are gaining popularity within the LHC world: XRootD [6] and HTTPS/WebDaV.

XRootD is now adopted by all LHC experiments for disk and tape data access. It is a protocol for data access where a user joined to a specific Federation can move, read and write data inside federated sites. User authentication is done using personal X509 certificates and the actual data transfer is done with checksum (adler32 algorithm) in order to grant the maximum reliability to the transfer operation. However, since tape-resident files are visible as ordinary files in a GPFS filesystem, a user could freely access such files causing massive recalls that would trigger uncontrolled operation and overload of the tape backend. For this reason, the tape recall over XRootD uses a specific ad-hoc "plugin" developed at CNAF to submit recall requests to GEMSS, when tape-resident files are directly accessed via XRootD. The specific "plug-in" is now in the XRootD development repository and it will be available to everyone in the standard XRootD release.

The HTTPS/WebDaV protocol implemented in StoRM has been used by the ATLAS and Xenon experiments. As well known in the literature, the WebDaV (Web Distributed Authoring and Versioning) protocol is an extension of the Hypertext Transfer Protocol (HTTP), and it makes the Web a readable and writable medium. It is a good framework for users to create, change and move files on a server (e.g. a web server or web share) and its main advantage lies in an intuitive and easy-to-use interface. At present, the utilization of this protocol at CNAF is somewhat limited to test activities or very specific use cases. Nevertheless, its diffusion is expected to increase quickly in the near future.

In order to monitor the performance and the correct behaviour of the whole storage system, two main monitoring tools are used in conjunction with other "minor" tool for specific monitoring of hardware storage boxes: Lemon and Nagios [7].

Lemon (LHC Era Monitoring) [8] is a CERN-developed server/client monitoring system available for Linux, where an agent running on the monitored machine gathers information using a set of standard and customized sensors (e.g. network traffic, memory and CPU usage, etc.). Information records are gathered by one central server that provides archiving over an Oracle database and a visual interface for organizing the information in a set of graphs over a web interface. Specific sensors have been developed at CNAF, and are currently used in production in addition to the default ones. In particular, information on the GPFS performance utilization, the Fibre Channel traffic over SAN interlinks, the tape drive throughput, and the TSM operations (e.g., number of successful or failed migrations or recalls operations) have proven to be very useful for identifying problems or data traffic bottlenecks. In particular the specific Lemon Fibre Channel sensor contacts the SAN switches using SNMP commands and obtains the counters values for the total traffic throughput at a specific timestamp. After a conversion these values are displayed as standard megabyte per second values. Regarding to the TSM operations, some specific Lemon sensors have been developed in order to analyse the latest GEMSS operation logs and summarize the useful information in a more readable form. Since the GEMSS logs include all the operation of the system and are rather verbose, some specific "keywords" are printed by the GEMSS daemons in the log files for operations such as recall file or mount a tape. The Lemon sensors search for these keywords in the last part of the log files and collect the information in standard plots (e.g. the number of successful and failed recall operations with a 5 minutes granularity) and in statistical average values on daily basis.

Nagios [9] is a widespread open source monitoring tool that can be easily adapted to specific local use. At our site, Nagios is used for alerting the site administrators via email or SMS notifications, when something actually goes wrong. Basically, the Nagios monitoring daemon runs intermittent checks on hosts and services using plugins which return status information to the central server. In case of failures, depending on the logics of the configured check, Nagios can also execute corrective actions like trying to restart the service or removing a machine from a cluster, or it can simply notify the failure. Custom Nagios checks have been developed for monitoring the GPFS cluster, TSM clients and server, GridFTP, and the StoRM services. Effectively some of these services (e.g. GridFTP and StoRM front-ends) use a multi-hosts definition list in the DNS alias name for redundancy and load balancing. In case of unrecoverable error on one specific host, the Nagios central server can interact with the local DNS server and remove the failed machine from the alias name of the service, leaving only the working machines in the list.

In addition to the Lemon and Nagios tools, specific monitoring tools have been installed and customized in order to obtain performance information with the finest granularity over the storage system boxes. In particular since the majority of the storage hardware at our site is from DataDirect Network (DDN), specific plugins have been provided by the vendor for accessing the whole set of information provided by their storage systems. A dedicated Cacti [10] customized tool service has been installed and configured at our site for monitoring the DDN hardware. It collects and provides detailed information on bandwidth performance (throughput reported in megabyte per second and rate of input/output operations per second) for each single logical and physical hard disk in the DDN storage boxes. This information provides a valuable help when debugging performance problems and throughput bottlenecks.

The INFN-CNAF Tier-1 storage infrastructure has shown good stability and excellent performance figures during the last years of service, also owing to its "parallel" and "clustered" layout. Throughout several stressful computing activities the Tier-1 storage system responded well, in some cases even exceeding expectations.

In Figure 2 a Lemon plot of the LHCb GPFS filesystem performance during three weeks in fall 2013 is reported. During the period of October-November 2013 the LHCb collaboration launched a massive stripping campaign, recalling from tape to disk at the INFN-CNAF Tier-1 a total of ~700 TB (divided in ~150k files from a total of roughly 300 tapes). As shown in the Figure, the sustained throughput from the tape backend has been between 300 and 450 MB/s (twice as fast as requested by the experiment) and this has reduced the total time for the whole operation to 23 days instead of the initial estimation of two months.

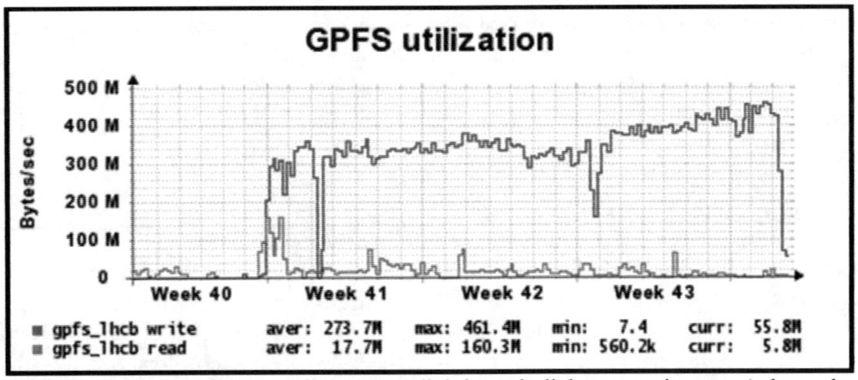


**Figure 2**. Lemon plot of the recall (tape to disk in red, disk to tape in green) throughput during the LHCb stripping campaign of oct-nov 2013.

## 3. The INFN-CNAF Tier-1 Oracle database services

The Oracle database service is one of the core services of the INFN-CNAF Tier-1. The Oracle database level of service has evolved during the last five years and now relies on different Oracle Real Application Clusters (RAC) for the maximum reliability and performance.

In Figure 3 a sketch of the infrastructure of the database service is shown. There are different Oracle instances in production and the connection with the mass storage is realized through the Tier-1 SAN. Some of the storage is currently phasing out due to hardware obsolescence and it is used only for testing purposes. A new storage box in production is now starting to replace the old disk volumes. Mixed Oracle 10g and 11g RAC clusters run over physical machines with Red Hat Enterprise Linux rel. 5 and 6 (the Oracle Linux operating system is under test). Oracle 12c is currently installed on a separate stand-alone testbed. The production storage hosts both the data files and the data of the Oracle Application and E-Business Suite related services. The services are monitored by a 10g clustered Oracle Grid Console, based on a RAC database for the repository and an active-passive console for total redundancy.

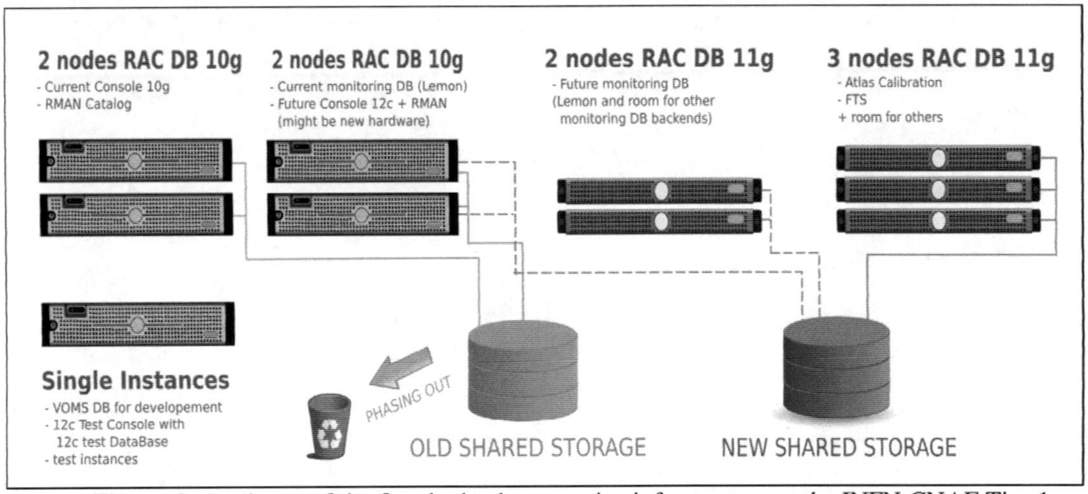


**Figure 3**. A schema of the Oracle database service infrastructure at the INFN-CNAF Tier-1.

At present the production services are operated on 2 RACs and a single instance:

- An Oracle 11g RAC composed of 3 nodes which runs:
    - the ATLAS Calibration database, in production within 2014;
    - the Grid File Transfer Service (FTS) database, available for WLCG and other users;
    - Oracle RMAN catalog for E-Business Suite related databases.
- An Oracle 10g RAC composed by 2 nodes which runs:
    - database for Lemon monitoring with history available for query and plot.
- A single Oracle 10g database dedicated to CNAF software developers.

Future plans and developments for the Oracle database service at CNAF include the installation of 2 nodes in a new Oracle 11g RAC cluster in order to migrate the current Lemon monitoring database and for other monitoring tool backends. Furthermore, the new Oracle 12c Cloud Control with 12c database is under test with the database service and the E-Business Suite for replacing the current 10g version. This will bring great advantages to the local monitoring and management of the RAC database, as it imples the "jump" of two major releases. In addition there are plans for a replica of the CDF FNAL database for the Long Term Data Preservation (LTDP) project [11].

All databases in production are backed-up using Oracle RMAN on separate disks for the "nearline" disk-backup, and the IBM Tivoli Data Protection (TDP) for archiving the backup over tape storage, in order to achieve the maximum data protection in case of disasters.

## 4. Conclusions

This paper is a site report of the INFN-CNAF Tier-1 concerning storage activities and services. Since the foundation of the Tier-1 centre, storage services have evolved in flexibility and stability with the introduction of the GEMSS storage system, that fulfilled the performance requests to the site, in addition to other clustered services like the Oracle Real Application Cluster (RAC), that helped increasing the database level of service. Furthermore, the use of monitoring tools, like Nagios, allowed tracking and automatically correcting failures of services or server hardware. The introduction of additional monitoring software packages (e.g., hardware specific ones) provided additional information that can be collected over the mass storage and that turned out to be very useful when trying to diagnose specific problems. The introduction and the wide distribution of additional methods for accessing data from the WAN (e.g. XRootD and HTTPS/WebDaV) enables the storage system to respond to the performance and flexibility requirements during the years to come. The disk and tape storage capacity in production at our Tier-1 are expected to increase quickly in the near future, and we are confident that the whole system will continue to exhibit great performance figures and reliability.

## References

[1] G. Bortolotti et al. *The INFN Tier-1* 2012 J. Phys. Conf. Ser. 396 042016 Proceedings of 2012 CHEP conference.

[2] *IBM General Parallel File System Administration and Programming Reference* Version 3 Release 2 SA23-2221-01.

[3] IBM website references for Tivoli Storage Manager info and documentation: https://www.ibm.com/developerworks/wikis/display/tivolidoccentral/Tivoli+Storage+Manager and http://www-03.ibm.com/software/tivoli/products/storage-mgr/.

[4] Info about StoRM available online http://www.italiangrid.it/middleware/storm/.

[5] D. Bonacorsi et al., *The Grid Enabled Mass Storage System (GEMSS): the Storage and Data management system used at the INFN Tier1 at CNAF*, 2012 J. Phys. Conf. Ser. 396 042051 Proceedings of 2012 CHEP conference.

[6] D. Gregori et al., *Xrootd data access for LHC experiments at the INFN-CNAF Tier-1* 2013 J. Phys. Conf. Ser. 513 (2014) 042023 Proceedings of 2013 CHEP conference.

[7] S. Antonelli et al., *INFN-CNAF Monitor and Control System* 2010 J. Phys. Conf. Ser. 331 (2011) 042032 Proceedings of 2010 CHEP conference.
[8] Info about Lemon available online http://lemon.web.cern.ch/lemon/index.shtml
[9] Info about Nagios available online http://www.nagios.org/.
[10] Info about Cacti available online http://www.cacti.net/.
[11] S. Amerio et al., *Long Term Data Preservation for CDF at INFN-CNAF* 2013 J. Phys. Conf. Ser. 513 (2014) 042011 Proceedings of 2013 CHEP conference.

# A self-configuring control system for storage and computing departments at INFN-CNAF Tier1

**Daniele Gregori[1], Stefano Dal Pra[1], Pier Paolo Ricci[1], Michele Pezzi[1], Andrea Prosperini[1], Vladimir Sapunenko[1]**

[1] INFN-CNAF, Viale Berti-Pichat 6/2, 40127 Bologna, Italy

E-mail: daniele.gregori@bo.infn.it

**Abstract.** The storage and farming departments at the INFN-CNAF Tier1[1] manage approximately thousands of computing nodes and several hundreds of servers that provides access to the disk and tape storage. In particular, the storage server machines should provide the following services: an efficient access to about 15 petabytes of disk space with different cluster of GPFS file system, the data transfers between LHC Tiers sites (Tier0, Tier1 and Tier2) via GridFTP cluster and Xrootd protocol and finally the writing and reading data operations on magnetic tape backend. One of the most important and essential point in order to get a reliable service is a control system that can warn if problems arise and which is able to perform automatic recovery operations in case of service interruptions or major failures. Moreover, during daily operations the configurations can change, i.e. if the GPFS cluster nodes roles can be modified and therefore the obsolete nodes must be removed from the control system production, and the new servers should be added to the ones that are already present. The manual management of all these changes is an operation that can be somewhat difficult in case of several changes, it can also take a long time and is easily subject to human error or misconfiguration. For these reasons we have developed a control system with the feature of self-configure itself if any change occurs. Currently, this system has been in production for about a year at the INFN-CNAF Tier1 with good results and hardly any major drawback. There are three major key points in this system. The first is a software configurator service (e.g. Quattor or Puppet) for the servers machines that we want to monitor with the control system; this service must ensure the presence of appropriate sensors and custom scripts on the nodes to check and should be able to install and update software packages on them. The second key element is a database containing information, according to a suitable format, on all the machines in production and able to provide for each of them the principal information such as the type of hardware, the network switch to which the machine is connected, if the machine is real (physical) or virtual, the possible hypervisor to which it belongs and so on. The last key point is a control system software (in our implementation we choose the Nagios software), capable of assessing the status of the servers and services, and that can attempt to restore the working state, restart or inhibit software services and send suitable alarm messages to the site administrators. The integration of these three elements was made by appropriate scripts and custom implementation that allow the self-configuration of the system according to a decisional logic and the whole combination of all the above-mentioned components will be deeply discussed in this paper.

## 1. Introduction

The departments Storage and Farming at the INFN-CNAF Tier1 administer around a thousand computing nodes and some hundreds of servers. A control system checking the status of services and hosts, able to send email alerts and run automatic recovery actions in case of an error is an essential tool. Given the large number of machines to be controlled it becomes very difficult to keep aligned the control system with the machines actually in production. For this reason, we decided to realize a control mechanism able of self configuration according to whether the machines are really in production and to their role in terms of the type of server (e.g. the IBM General Parallel File System GPFS, GridFTP, Worker Node, User Interface). For this purpose we have used a set of standard open source tools: Quattor[2], Nagios[3] and a Postgres database containing information on servers in the computing center named Docet[4]. All these tools have been integrated together by means of a series of Bash scripts that have the purpose of reading the information on the machines in production and creating the updated Nagios configuration. In the following sections, the tools used and the operating logic of integration will be described.

## 2. Adopted Tools

### *2.1. Quattor*

Quattor is a generic open-source tool-kit used to install, configure, and manage computers; it ensures that all machines that need to be monitored by Nagios are configured to perform the required checks.

### *2.2. Nagios*

Nagios is an open source computer system monitoring, network monitoring and infrastructure monitoring software application. Nagios offers monitoring and alerting services for servers, switches, applications, and services. It alerts the users when things go wrong and alerts them a second time when the problem has been resolved.

### *2.3. Docet*

Docet is a PostgreSQL database (DB) that contains all hardware information on all the machines in production at the Tier1. The information can be extracted from the DB and used to define which machines are actually in production and therefore which need to be controlled via Nagios.

## 3. Project Constraints

At the time of writing the system design has taken into account what are the specific requirements of each department. In particular for farming the large number of computing nodes $\sim 1000$ must be taken into account. For the storage unit, the configuration of the GPFS cluster is an element that must be taken into consideration in order to control the specific GPFS server nodes as quorum.

- The system must be able to control thousands of Worker Nodes and in the case of problems must be able to perform automatic recovery actions.
- The system must be able to automatically learn which machines (worker node or server) should be monitored and how the GPFS clusters are configured.
- The system must be quickly restored in case of failure of the control server (disaster recovery).

The Worker Nodes of the farming department must check a list of running daemons and mounted filesystems. In case of problems the control system must provide a restart of the services or mount any filesystem that is not mounted. Given the high number of nodes and controls to be done Nagios is not able to perform active controls (synchronous) via plugin NRPE[5]. For this purpose we employ passive controls (asynchronous) via plugin NSCA[6]. On the Worker Node a series of sensors to control the daemons and the file system are performed within an hour without setting the exact time using a random delay. The sensors send their results to the Nagios server on a named pipe, the Nagios server in this way reads the status value of the service

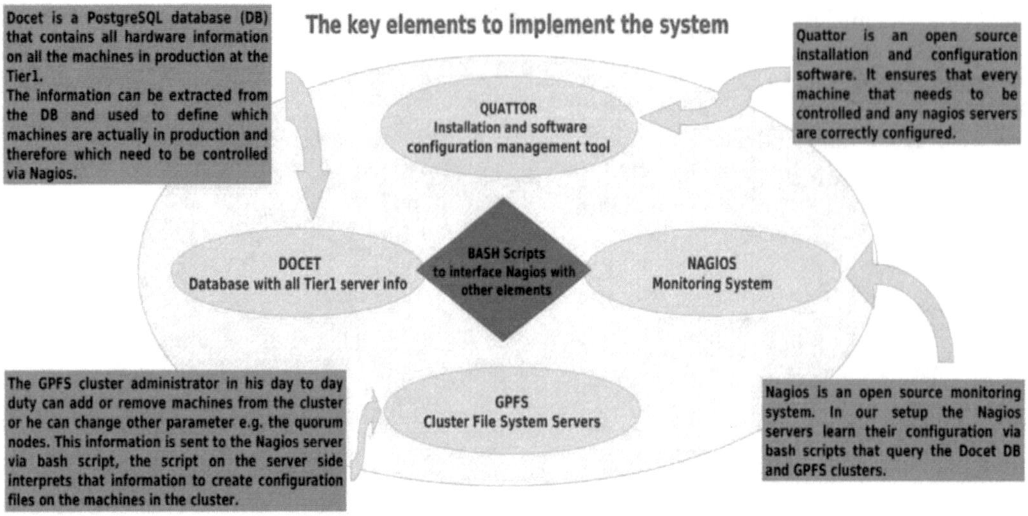


**Figure 1.** Overview of the key elements of the control system.

checked. If there is no status report sent for service for more than an hour Nagios status changes to "Unknown" and a mail is sent to the administrators of that service. If from a Worker Node the status "Error" is sent then Nagios will react immediately, proactively via NRPE plugin, requesting an attempt to restart the service and then send the new status. For this type of machine Nagios learns which of them should be checked through a query to the Docet DB. The query returns besides the list of Worker Nodes in production also the hostname, the types of hardware, whether they are physical or virtual machines, and the network switch to which they are connected. Based on this information, the Bash scripts construct the Nagios configuration and make an initial syntax checking and duplication of IP addresses. At this point the reload of the configuration is left to the Nagios administrator.

The requirements for a storage department are different. There are numerous GPFS clusters that must be controlled, a server within a cluster can be added or removed, or may change its role as quorum manager. All this information is not contained in the Docet DB, so in order to learn the cluster configuration one of the cluster servers must communicate with Nagios. The script `/opt/gpfs-nagios-info-sender.sh` is added to the cluster nodes via Quattor. This script sends the `mmlscluster`[7] command output directly to the Nagios server. In this way the Nagios family scripts can build configuration for all GPFS cluster, taking into consideration the quorum nodes and creating a cluster service for these particular machines.

To restore the server in case of disaster, Quattor is used. Quattor is able to reinstall the machine from scratch in about 30 minutes. Once the Nagios server is restored it can recreate his configuration: at first the machines of the GPFS cluster must send their configuration with the script `/opt/gpfs-nagios-info-sender.sh` then by querying the Docet DB to know the hardware characteristics. Overall, the reconstruction of the configuration takes a few minutes. In Figure 1 we present the schema of the self-configuring control system with characteristics of its elements.

## 4. The configuration system operation

Once the Nagios server has received the information from all the GPFS clusters, you can run the script that creates the configuration. There are two scripts that are executed in succession, the first is `/opt/nagios-gpfs/read_cluster_info.sh` and the second is `/opt/nagios-gpfs/create_all.sh`. The first reads the file sent by the GPFS cluster and creates an ordered list of servers per cluster, making sure that there are no machines belonging to more than one cluster, which is forbidden by GPFS. The second creates the configuration of Nagios itself, making sure that the servers are not in a blacklist, in which case they are ignored. It makes query to the database to check the hardware type of the individual server in order to add specific hardware controls (IPMI, RAID, POWER SUPPLY). Here are the steps that are followed:

(i) Control of information from the GPFS cluster and creation of a file containing the servers.

(ii) Save the current configuration for the case of recovery.

(iii) Through requests to the Docet DB other types of machines (GridFTP, GPFS, etc.) are added, for these machines the “type” information is contained in docet.

(iv) Verify that each server found in the preceding paragraphs is not present in the blacklist, in which case it is ignored.

(v) From the information gathered in the previous steps the script creates the configuration files of Nagios groups, hosts, commands and services.

(vi) A check is made of the correctness of the configuration with the command `nagios -v /etc/nagios/nagios.cfg`.

(vii) Nagios is ready to restart or reload the configuration.

For the farming departments the steps are similar to the above with the exclusion of the first point since there is no GPFS cluster to control and that the machines which require information from Docet are Worker Node, User Interface and other servers.

## 5. Some NagiosGraph tests

In order to increase the number of information available from the Nagios server we have created some sensors to measure the parameters of the network and memory of the clients. This data are sent to the Nagios server which stores it in a Round Robin DB and displays it via the plugin NagiosGraph[8]. With a granularity of 5 minutes for each quantity kept for one year hundreds MB are collected. In Figure 2 an example of a NagiosGraph plot is shown.

## 6. Conclusions

The automated system for Storage and Farming has been in production for over a year on two separate servers. During this period it has proven to be a reliable tool, which has significantly reduced the time spent managing Nagios and the possibility of manual errors. Possible developments are cross-checking of the two servers and in case of a failure an automatic activation of a third Nagios that takes the place of the server disabled by a hardrware failure.

## References

[1] D. Gregori et al, *INFN-CNAF Activity in the Tier-1 and Grid for LHC Experiments*, 2009, E-ISBN : 978-1-4244-3750-4

[2] `http://www.quattor.org/`

[3] `http://www.nagios.org/`

[4] S. Dal Pra, A. Crescente, *The data operation centre tool. Architecture and population strategies*, 2012, Journal of Physics: Conference Series 396(2012) 042014

[5] `http://exchange.nagios.org/directory/Addons/Monitoring-Agents/ /NRPE--2D-Nagios-Remote-Plugin-Executor/details`

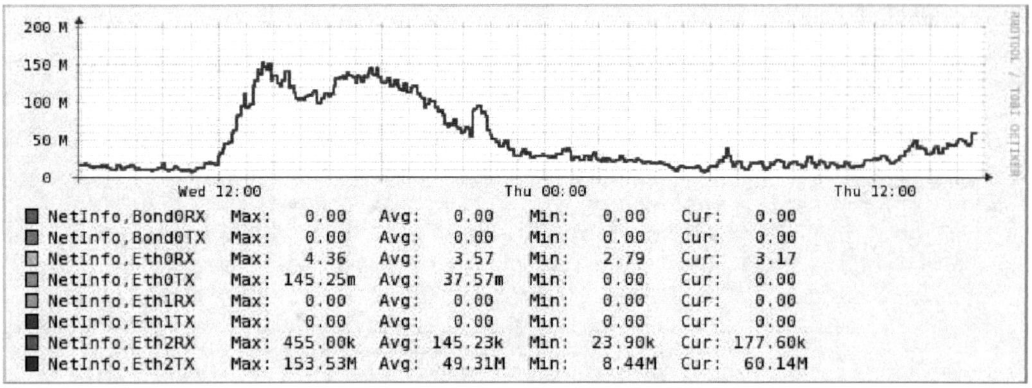


**Figure 2.** Network NagiosGraph plot.

[6] `http://exchange.nagios.org/directory/Addons/Passive-Checks/ /NSCA--2D-Nagios-Service-Check-Acceptor/details`

[7] *General Parallel File System (GPFS) Administration and Programming Reference Version 4 Release 1.0.4*, E-ISBN : SA23-1452-01

[8] `http://nagiosgraph.sourceforge.net/`

Journal of Physics: Conference Series **608** (2015) 012015 doi:10.1088/1742-6596/608/1/012015

# Multilevel Workflow System in the ATLAS Experiment

**M Borodin[1], K De[2], J Garcia Navarro[3], D Golubkov[4,5], A Klimentov[6], T Maeno[6] and A Vaniachine[7] on behalf of the ATLAS Collaboration**

[1] Department of Elementary Particle Physics, National Research Nuclear University "MEPhI," Moscow, 117513, Russia

[2] Physics Department, University of Texas Arlington, Arlington, TX 76019, United States of America

[3] Instituto de Fisica Corpuscular, Universidad de Valencia, E-46980 Paterna, Spain

[4] Experimental Physics Department, Institute for High Energy Physics, Protvino, 142281, Russia

[5] Big Data Laboratory, National Research Centre "Kurchatov Institute" Moscow, 123182, Russia

[6] Physics Department, Brookhaven National Laboratory, Bldg. 510A, Upton, NY 11973, United States of America

[7] High Energy Physics Division, Argonne National Laboratory, 9700 South Cass Avenue, Argonne, IL 60439, United States of America

E-mail: vaniachine@anl.gov

**Abstract.** The ATLAS experiment is scaling up Big Data processing for the next LHC run using a multilevel workflow system comprised of many layers. In Big Data processing ATLAS deals with datasets, not individual files. Similarly a task (comprised of many jobs) has become a unit of the ATLAS workflow in distributed computing, with about 0.8M tasks processed per year. In order to manage the diversity of LHC physics (exceeding 35K physics samples per year), the individual data processing tasks are organized into workflows. For example, the Monte Carlo workflow is composed of many steps: generate or configure hard-processes, hadronize signal and minimum-bias (pileup) events, simulate energy deposition in the ATLAS detector, digitize electronics response, simulate triggers, reconstruct data, convert the reconstructed data into ROOT ntuples for physics analysis, etc. Outputs are merged and/or filtered as necessary to optimize the chain. The bi-level workflow manager – ProdSys2 – generates actual workflow tasks and their jobs are executed across more than a hundred distributed computing sites by PanDA – the ATLAS job-level workload management system. On the outer level, the Database Engine for Tasks (DEfT) empowers production managers with templated workflow definitions. On the next level, the Job Execution and Definition Interface (JEDI) is integrated with PanDA to provide dynamic job definition tailored to the sites capabilities. We report on scaling up the production system to accommodate a growing number of requirements from main ATLAS areas: Trigger, Physics and Data Preparation.

## 1. Introduction

The multi-purpose nature of the ATLAS experiment [1] at the LHC resulted in continuous growth in use cases for Big Data processing, as more data and new requirements emerge. To process Big Data the ATLAS experiment adopted the dataset transformation approach, where software applications

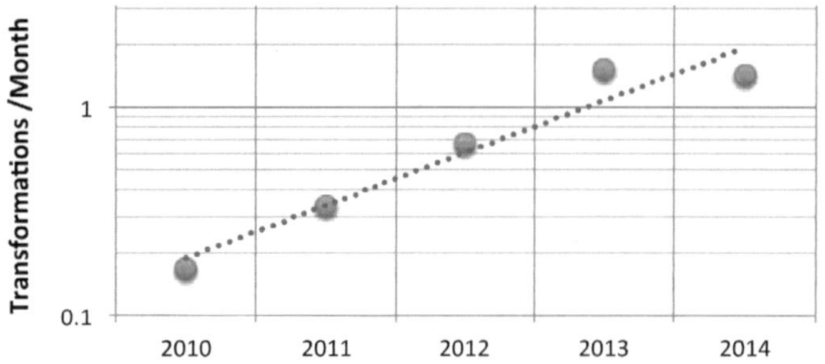


**Figure 1.** Continuous growth in the rate of data transformations added for Big Data processing in the ATLAS experiment.

transform the input datasets into the output datasets [2]. A success of this approach is evident from the exponential growth rate in the number of new data transformations used for Big Data processing in the ATLAS experiment (figure 1). Figure 2 shows the number of datasets produced during the major Run 1 simulation campaign dominated by datasets of SUSY physics group. The names of other physics and combined performance groups shown on the figure are: phys-exotics, phys-higgs, phys-gener (event generators), physics-sm (Standard Model), physics-top, soft-simul (software validation for the Monte Carlo simulations), perf-flavtag (flavour tagging), perf-jets, phys-beauty, perf-egamma (electron and gamma), and perf-tau.

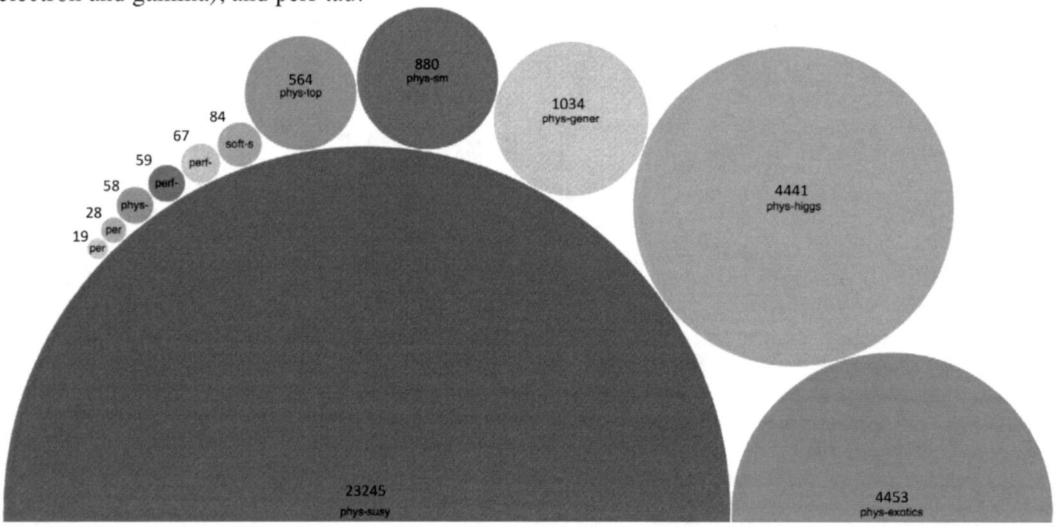


**Figure 2.** The number of datasets produced during the major Run-1 simulation campaign represents the scale and variety of Big Data processing requirements provided by ATLAS physics groups.

In ATLAS Big Data processing, each dataset transformation is represented by a task [3], having the individual data processing tasks organized into workflows with outputs merged and/or filtered as necessary. Figure 3 shows the Monte Carlo workflow steps: generate or configure hard-processes, hadronize signal and minimum-bias (pileup) events, simulate energy deposition in the ATLAS detector, digitize electronics response, simulate triggers, reconstruct data, convert the reconstructed data into ntuples for physics analysis, etc.

**Figure 3.** Monte Carlo workflow is composed of many steps.

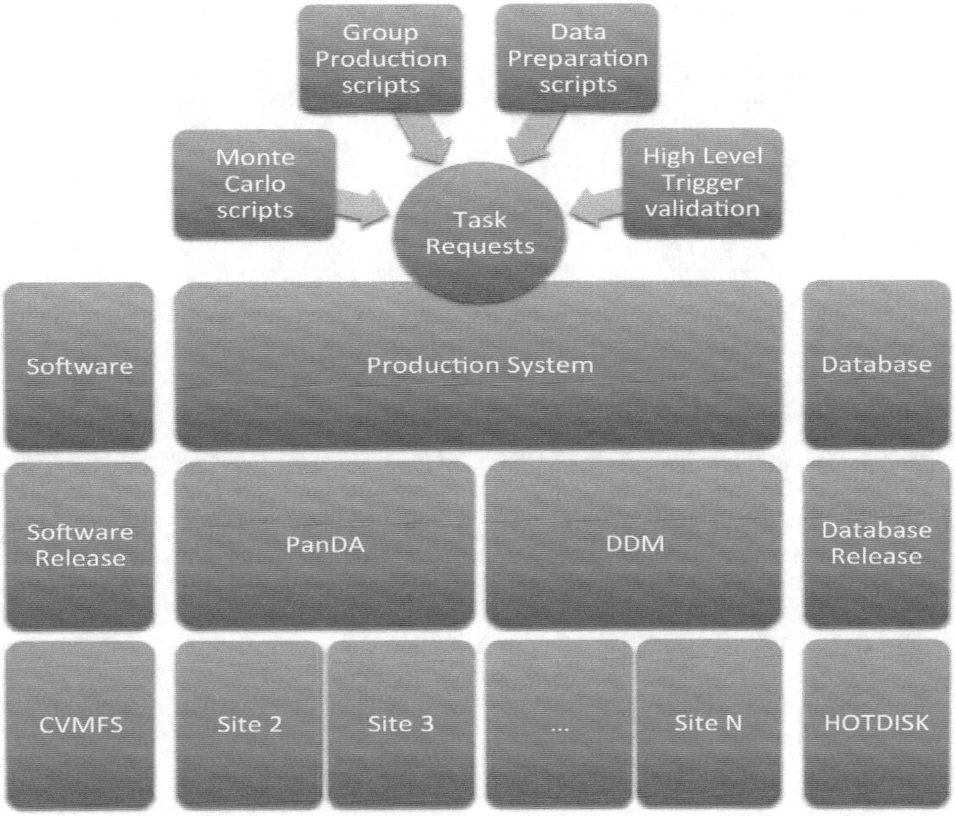


**Figure 4.** The ATLAS Big Data processing infrastructure during the LHC Run 1.

## 2. Big Data Processing

Figure 4 shows the multilevel Big Data processing infrastructure used by the ATLAS experiment during and after the LHC Run 1. The top management layer includes the Production System with the front-end Task Request interface [4]. In the middle are the workload management system PanDA [5] and the Distributed Data Management (DDM) system DQ2 [6]. The bottom layer shows Grid sites together with the software and database access technologies deployed at the sites. During Run 1, the infrastructure fully satisfied the requirements of ATLAS data reprocessing, simulations, and production by physics groups. As an example, we report below our experience with a representative use case of data reprocessing.

### *2.1. Run 1 Experience*

A starting point for ATLAS data processing is data reconstruction. During reconstruction, the raw detector data are processed with software algorithms to identify and reconstruct physics objects such as charged particle tracks. Following the prompt reconstruction at the computing centre at CERN (the Tier-0 site), the ATLAS data are reprocessed on the Grid, which allows reconstruction of the data with updated software and calibrations improving the quality of the reconstructed data for physics analysis. The collaboration completed four major reprocessing campaigns, with up to 2 PB of data being reprocessed every year. Automatic job resubmission avoids data losses at the expense of CPU time used by the failed jobs. Table 1 shows that failures have not presented a problem, as the fraction of CPU-time used for data recovery is not significant.

**Table 1.** Cost of recovery from transient failures for the reconstruction jobs.

| Reprocessing campaign | Input Data Volume (PB) | CPU Time Used for Reconstruction ($10^6 h$) | Fraction of CPU Time Used for Recovery (%) |
|---|---|---|---|
| 2010 | 1 | 2.6 | 6.0 |
| 2011 | 1 | 3.1 | 4.2 |
| 2012 | 2 | 14.6 | 5.6 |
| 2013 | 2 | 4.4[1] | 3.1 |

Another example of the success of our task-based data transformation approach is represented by the exponential growth of the rate of production tasks submission over the years [4]. As a result, the ATLAS production tasks count exceeded 1.6 million, with each task containing hundreds or thousands of jobs submitted by PanDA for execution on the Grid.

*2.2. Run 2 Preparations*

The LHC shutdown presented an opportunity to reengineer the ATLAS Big Data processing infrastructure, adding extra layers to further improve the system scalability and flexibility. To prepare the production for Run 2 challenges, PanDA has been upgraded with the Job Execution and Definition Interface (JEDI) [5], the production system enhanced with the Database Engine for Tasks (DEfT) [7], and the DDM system being upgraded to Rucio [8] (figure 5). On the production system upper level, the Database Engine for Tasks (DEfT) empowers production managers with templated workflow definitions. On the lower level, the Job Execution and Definition Interface (JEDI) is integrated with PanDA to provide dynamic job definition tailored to the sites capabilities. The data transformations will be configured using the ATLAS Metadata Interface (AMI) [9] harmonizing definitions of AMI tags between Tier-0 and the production system.

**Figure 5.** Multi-level architecture of the ATLAS Big Data processing infrastructure for Run 2.

In the bi-level production system, the JEDI layer is coupled with PanDA, while the DEfT layer implemented as the flexible database engine for bookkeeping. These two independent layers communicate via customized JSON protocol. During task execution, the JEDI layer defines the jobs

[1] In 2013 reprocessing, 2.2 PB of input data were used for selecting about 15% of all events for reconstruction, thus reducing CPU resources vs. the 2012 reprocessing.

tailored to the actual resources: disk space, CPU-time, memory, networks, etc. In contrast, the Run 1 production system employed the static job definition.

## 3. Conclusions

During the LHC Run 1, the ATLAS Big Data processing infrastructure supported a diverse range of workflows handling centrally ATLAS Big Data reprocessing (reconstruction of LHC data) and Monte Carlo production (full and fast simulations, digitization and reconstruction of simulated data). The production system scalability and flexibility has been demonstrated by managing double exponential growth in the number of task requests and data transformations. The total number of tasks exceeded 1.6 million; the data transformations count exceeded 120.

The LHC shutdown provided an opportunity for upgrading the production system with new capabilities, such as automatic recovery of lost data files and tailoring jobs to the capabilities of particular sites. As the ATLAS experiment continues optimising the use of Grid computing resources in preparation for the LHC Run 2 in 2015, the next generation production system is being integrated with other layers. The commissioning is in progress, scaling up the production system for a growing number of tasks and transformations that will process data for physics analysis and other ATLAS LHC Run 2 activities.

## Acknowledgments

We thank all our colleagues who contributed to the ATLAS Big Data processing infrastructure development and operations. This work was funded in part by the U. S. Department of Energy, Office of Science, High Energy Physics under Contract No. DE-AC02-06CH11357.

## References

[1] The ATLAS Collaboration 2008 The ATLAS experiment at the CERN Large Hadron Collider *J. Inst.* **3** S08003
URL http://iopscience.iop.org/1748-0221/3/08/S08003

[2] Stewart G A *et al.* 2014 ATLAS Job Transforms: A Data Driven Workflow Engine *J. Phys.: Conf. Ser.* **513** 032094
URL http://iopscience.iop.org/1742-6596/513/3/032094

[3] Vaniachine A V for the ATLAS Collaboration 2011 ATLAS detector data processing on the Grid *IEEE Nuclear Science Symposium and Medical Imaging Conference* pp 104-107
URL http://ieeexplore.ieee.org/xpl/articleDetails.jsp?arnumber=6154460

[4] Golubkov D *et al.* 2012 ATLAS Grid Data Processing: system evolution and scalability *J. Phys.: Conf. Ser.* **396** 032049
URL http://iopscience.iop.org/1742-6596/396/3/032049

[5] Maeno T *et al.* 2012 Evolution of the ATLAS PanDA production and distributed analysis system *J. Phys.: Conf. Ser.* **396** 032071
URL http://iopscience.iop.org/1742-6596/396/3/032071

[6] Garonne V *et al.* 2012 The ATLAS Distributed Data Management project: Past and Future *J. Phys.: Conf. Ser.* **396** 032045
URL http://iopscience.iop.org/1742-6596/396/3/032045

[7] De K *et al.* 2014 Task management in the new ATLAS production system *J. Phys.: Conf. Ser.* **513** 032078
URL http://iopscience.iop.org/1742-6596/513/3/032078

[8] Garonne V *et al.* 2014 Rucio – the next generation of large scale distributed system for ATLAS data management *J. Phys.: Conf. Ser.* **513** 042021
URL http://iopscience.iop.org/1742-6596/513/4/042021

[9] Fulachier J *et al.* 2014 Looking back on 10 years of the ATLAS Metadata Interface. Reflections on architecture, code design and development methods *J. Phys.: Conf. Ser.* **513** 042019
URL http://iopscience.iop.org/1742-6596/513/4/042019

# Monitoring of IaaS and scientific applications on the Cloud using the Elasticsearch ecosystem

**S Bagnasco[1], D Berzano[2], A Guarise[1], S Lusso[1], M Masera[1,3], S Vallero[1,3]**

[1]Istituto Nazionale di Fisica Nucleare, Via Pietro Giuria 1, 10125 Torino, IT
[2] CERN - European Organization for Nuclear Research, CH-1211 Geneva 23, CH
[3] Dipartimento di Fisica, Università degli Studi di Torino, Via Pietro Giuria 1, 10125 Torino, IT

E-mail: `svallero@to.infn.it`

**Abstract.** The private Cloud at the Torino INFN computing centre offers IaaS services to different scientific computing applications. The infrastructure is managed with the OpenNebula cloud controller. The main stakeholders of the facility are a grid Tier-2 site for the ALICE collaboration at LHC, an interactive analysis facility for the same experiment and a grid Tier-2 site for the BES-III collaboration, plus an increasing number of other small tenants. Besides keeping track of the usage, the automation of dynamic allocation of resources to tenants requires detailed monitoring and accounting of the resource usage. As a first investigation towards this, we set up a monitoring system to inspect the site activities both in terms of IaaS and applications running on the hosted virtual instances. For this purpose we used the Elasticsearch, Logstash and Kibana stack. In the current implementation, the heterogeneous accounting information is fed to different MySQL databases and sent to Elasticsearch via a custom Logstash plugin. For the IaaS metering, we developed sensors for the OpenNebula API. The IaaS level information gathered through the API is sent to the MySQL database through an ad-hoc developed RESTful web service, which is also used for other accounting purposes. Concerning the application level, we used the Root plugin TProofMonSenderSQL to collect accounting data from the interactive analysis facility. The BES-III virtual instances used to be monitored with Zabbix, as a proof of concept we also retrieve the information contained in the Zabbix database. Each of these three cases is indexed separately in Elasticsearch. We are now starting to consider dismissing the intermediate level provided by the SQL database and evaluating a NoSQL option as a unique central database for all the monitoring information. We setup a set of Kibana dashboards with pre-defined queries in order to monitor the relevant information in each case. In this way we have achieved a uniform monitoring interface for both the IaaS and the scientific applications, mostly leveraging off-the-shelf tools.

## 1. Introduction

The INFN-Torino computing centre hosts a private Cloud infrastructure, which provides IaaS services to different scientific computing applications [1]. The infrastructure is managed with the OpenNebula cloud controller [2]. The main stakeholders of the facility are a grid Tier-2 site for the ALICE collaboration at the CERN LHC, an interactive analysis facility for the same experiment and a separate grid site for the BES-III collaboration. Moreover, the centre offers virtualised batch farms on-demand to a number of smaller local tenants, such as a theory group

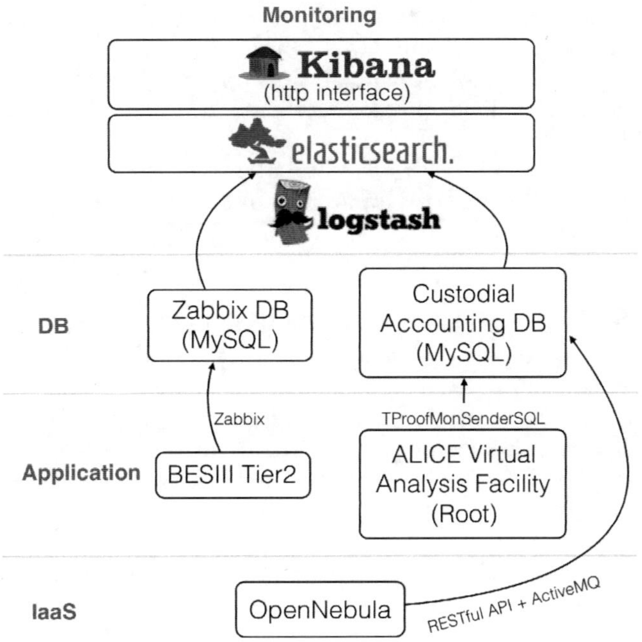


**Figure 1.** Set-up of the monitoring framework for IaaS and applications at the INFN-Torino private Cloud.

and the Compass collaboration, and single virtual machines tailored to the needs of specific use-cases (e.g. nuclear plant simulations). We are working to achieve the complete automation of resources allocation, which is already fulfilled in the case of the ALICE Virtual Analysis Facility (VAF) [3].

Within this scenario, a detailed monitoring and accounting of resource usage is both mandatory, not least for billing purposes, and challenging because of the heterogeneous data sources. For instance, we need to gather information at the IaaS level (from the cloud controller) and at the application level (e.g. Root [4]) and possibly exploit the data coming from other tools already in use to monitor some application on our Cloud (i.e. Zabbix [5]). Our goal is to achieve a uniform and user-friendly monitoring interface as a single entry point to these miscellaneous data. This paper illustrates how we set up a monitoring framework based on the Elasticsearch [6], Logstash [7] and Kibana [8] stack.

## 2. Implementation

In the current implementation, the heterogeneous accounting data is fed to different MySQL databases and sent to Elasticsearch via Logstash. We configured a set of Kibana dashboards with pre-defined queries in order to display the relevant information in each case. The set-up is illustrated in Figure 1 and explained in more detail in the following sub-sections.

### *2.1. The Elasticsearch ecosystem*

The Elasticsearch ecosystem is composed by Elasticsearch (ES), Logstash and Kibana and it is generally referred to as the 'ELK stack'.

ES is a search and analytics engine built on top of the Apache Lucene information retrieval library (Apache open source license). It is document-driven: entries are stored as JSON documents and all fields can be indexed and used in a single query. In our set-up, each application is indexed separately. ES also allows for full-text search on unstructured data, though in our specific case this feature is not fully exploited. Moreover, ES is API driven and can be interfaced with any RESTful API using JSON over http.

Logstash is an open source tool used to collect and parse events and logs to a central service. It can be easily customised via plugins for input, output and data-filters. In order to fit Logstash into our set-up we have developed only a simple plugin to retrieve input data from a MySQL database and a set of configuration files to customise data indexing (one for each application we wished to monitor).

Kibana is the official GUI to display and search ES data. The software could be exposed with any web server, we chose Apache, and no installation is required. Since Kibana does not provide any authentication/authorisation mechanism, these should be implemented at the level of the web server. Kibana has an easy search syntax to query ES. The creation of custom interactive dashboards can be achieved in few mouse clicks, without any prior GUI programming knowledge. This feature makes it a good candidate for provisioning monitoring-as-a-service to the Cloud tenants. Moreover, Kibana provides a set of useful pre-defined plot types like pies, histograms or trends.

The ELK stack was our first choice for implementing the monitoring infrastructure because it is open source, it provides most of the needed features out of the box and it is relatively easy to configure and use. So far, this solution has proven to be suited for our purposes.

### *2.2. Data preservation: the SQL backend*

Monitoring data are stored on a high-availability MySQL server prior to be inserted in the ES engine. This might be seen as a redundant step since data are partially duplicated between the database and ES. Indeed, ES itself can be configured as a fully-fledged (NoSQL) database cluster using its built-in high-availability features. However, in order to ensure flexibility and modularity to our monitoring system, we chose to use ES as a pure search engine with a simple single-node configuration and to delegate data preservation to the MySQL database. With the use of a standard and widely used back-end solution we will be able to move seamlessly away from the ELK stack without losing historical data, should it prove itself not to be the optimal tool. Moreover, the main application hosted at our Cloud besides the Tier-2 sites (i.e. the VAF) relies on Proof [4], which comes with a built-in plugin to send monitoring data to a SQL database at the end of each query. We preferred to avoid writing a custom plugin fitting explicitly our specific monitoring solution.

### *2.3. Sending data to the database*

For the IaaS metering, we developed sensors for the OpenNebula xml-rpc API. The IaaS level information gathered through the API is sent to the MySQL database through an ad-hoc developed RESTful web service, which is also used for accounting purposes. The service relies on the ActiveMQ messaging server to asynchronously and efficiently feed metering records to the database. A similar application is also being developed for billing. The relevant quantities that we feed to the database are the resource usage per tenant/group in terms of number of virtual machines, number of cores, memory and ephemeral disk space.

Concerning the application level, we use the Proof plugin *TProofMonSenderSQL* to collect accounting data from the interactive analysis facility. At the end of each user query, data are gathered from the worker-nodes to the Proof master and sent to the database with a standard MySQL client/server protocol. In this case, we also monitor some additional observables such as e.g. the number of workers, the datasets analysed or the number of events processed. Moreover, since the BES-III Tier-2 site used to be monitored with Zabbix, as a proof of concept we also

Journal of Physics: Conference Series **608** (2015) 012016 doi:10.1088/1742-6596/608/1/012016

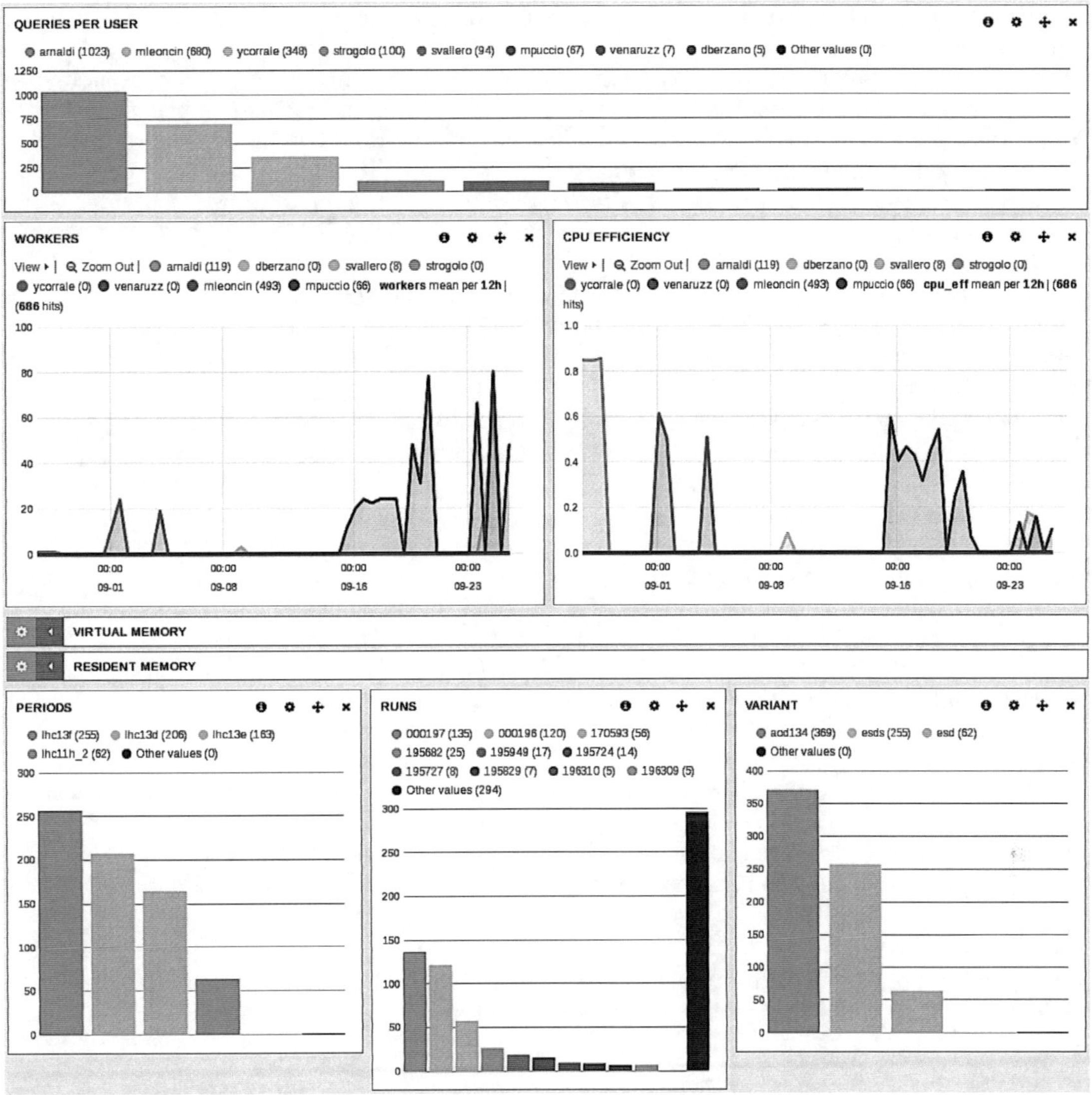


**Figure 2.** Partial view of the Kibana dashboard for the ALICE Analysis Facility at the INFN-Torino computing centre.

use the data from the Zabbix database. In this case we can get even more refined information such as resource consumption for specific processes, e.g. the BES-III Offline Software System (BOSS).

## 3. Custom dashboards

At present, the monitoring framework described in this paper is used to display data coming from the cloud controller (both from the accounting service and the test version of the billing service), the VAF and the BES-III Tier-2. Each use-case has its own Kibana dashboard.

As an example, in Figure 2 we show a snapshot of part of the VAF dashboard. The top-most

plot is a *terms* panel, which displays the results of an ES facet as a bar chart. By mouse-clicking on a user's bar, only entries for that user are displayed in all other panels.
The second top-most plots are *histogram* panels, showing the number of workers requested by each user and the average CPU efficiency as a function of the query time (similar plots for the memory usage on the master/workers are hidden in the figure). The plots' time-range can be easily selected by dragging with the mouse on the corresponding axis. This operation also sets the desired time-range on all other plots in the dashboard. In order to draw a separate line for each user in the plots, we have saved a set of pre-defined ES queries, one for each user, and configured Kibana in order to show them all in the same panel.
In the bottom-most row one can see some information concerning the datasets analysed: the LHC data-taking period flag, the run number and the type of file analysed. By selecting e.g. one data-taking period in the bottom-left panel, one automatically gets a list of users who analysed it in the top-most panel.

Similar dynamic dashboards can be set-up in few minutes without any programming, only a good knowledge of the back-end database schema is required.

## 4. Summary and outlook

At the INFN-Torino computing centre we have set-up a uniform monitoring interface for both the IaaS and the scientific applications, mostly leveraging off-the-shelf tools. The infrastructure relies on a SQL database back-end for data preservation and to ensure flexibility to choose a different monitoring solution if needed. We are now starting to consider dismissing the intermediate level provided by the SQL database and evaluating a NoSQL option as a unique central database for all the monitoring information.
For the IaaS accounting we have developed a custom RESTful web service to manage and query the database, which relies on a message queue to transport usage records.

We are currently working on a billing system for our private Cloud, which also relies on the ELK stack in its graphical interface.
The next step will be to define a model for monitoring-as-a-service, based on the tools described in this paper, which the Cloud tenants could easily configure to suit their needs.

## Acknowledgements

The present work is partially funded under contract 20108T4XTM of Programmi di Ricerca Scientifica di Rilevante Interesse Nazionale (PRIN), Italy.

## References

[1] Bagnsco S, Berzano D, Brunetti R, Lusso S and Vallero S 2014 *Journal of Physics: Conference Series* **513** 032100
[2] Moreno-Vozmediano R, Montero R S and Llorente I M 2012 *IEEE Computer* **45** 65-72
[3] Berzano D, Blomer J, Buncic P, Charalampidis I, Ganis G, Lestaris G and Meusel R 2014 *Journal of Physics: Conference Series* **513** 032007
[4] Brun R and Rademakers F 1997 *Nuclear Instruments and Methods in Physics Research* A **389** 81-86
[5] Zabbix [`http://www.zabbix.com/`]
[6] Elasticsearch [`http://www.elasticsearch.org`]
[7] Logstash [`http://logstash.net`]
[8] Kibana [`http://www.elasticsearch.org/overview/kibana`]

# Analyzing data flows of WLCG jobs at batch job level

**Eileen Kuehn, Max Fischer, Manuel Giffels, Christopher Jung and Andreas Petzold**

Karlsruhe Institute of Technology (KIT), Karlsruhe, Germany

E-mail: `eileen.kuehn@kit.edu`, `max.fischer@kit.edu`, `manuel.giffels@kit.edu`, `christopher.jung@kit.edu`, `andreas.petzold@kit.edu`

**Abstract.** With the introduction of federated data access to the workflows of WLCG, it is becoming increasingly important for data centers to understand specific data flows regarding storage element accesses, firewall configurations, as well as the scheduling of batch jobs themselves. As existing batch system monitoring and related system monitoring tools do not support measurements at batch job level, a new tool has been developed and put into operation at the GridKa Tier 1 center for monitoring continuous data streams and characteristics of WLCG jobs and pilots. Long term measurements and data collection are in progress. These measurements already have been proven to be useful analyzing misbehaviors and various issues. Therefore we aim for an automated, realtime approach for anomaly detection. As a requirement, prototypes for standard workflows have to be examined. Based on measurements of several months, different features of HEP jobs are evaluated regarding their effectiveness for data mining approaches to identify these common workflows. The paper will introduce the actual measurement approach and statistics as well as the general concept and first results classifying different HEP job workflows derived from the measurements at GridKa.

## 1. Introduction

The increase in opportunistic resource usage [1] and federated storage [2, 3, 4, 5] in different computing models of WLCG communities requires network-centric monitoring techniques. Different approaches dealing with this issue implemented user- and job-specific, or VO-specific monitoring [6, 7, 8]. Others deploy agents to data servers of federations itself [9]. Additionally there are mediating [10, 11] and aggregating [12] approaches for the different techniques. But none of the aforementioned approaches tracks the logical connections between a single job and its data transfers. This might be derived by accumulating data flow centric approaches measuring traffic for data federations with job-specific information. However, this is prone to errors but more importantly incomplete. The measurement of data flows requires knowledge about data servers, so opportunistic resources and data outside of data federations is not included in such measurements. Especially in data centers operating a batch system with commodity hardware, it is essential to know about the logical connections. As the network is shared by thousands of batch jobs and users, a differentiation and metric for job-specific network usage is desirable.

For this goal to be achieved a monitoring tool has been implemented at the GridKa data and computing center profiling data flows of WLCG batch jobs (section 2). In addition, first toy analysis show the relevance of measured data (section 3) to gain experience in identifying access patterns and estimating data transfers to finally understand data flows.

## 2. Monitoring the traffic of batch jobs

Existing monitoring tools in use at GridKa Tier 1 – including Cacti® [13], Icinga [14], Ganglia [15], and Univa Grid Engine [16] – deliver fine grained information about network traffic regarding different ports, the accumulated traffic by rack, or node and different services and alarms based on the data. But they are not capable of measuring network traffic on batch job level. As existing batch system monitoring and related system monitoring tools do not support measurements at batch job level, a new tool has been developed and put into operation at the GridKa Tier 1 center for monitoring continuous data streams and characteristics of WLCG jobs and pilots.

The design and implementation of the tool is based on C/C++. It has been modeled based on the OpenSource tool NetHogs [17]. NetHogs groups bandwidth by process instead of breaking traffic down per protocol or subnet. It does not rely on a special kernel module to be loaded but the Linux operating system and libpcap [18]. Libpcap is a system-independent interface for packet capturing in the user space. Network packets are copied into a buffer for further analysis. By utilizing libpcap, UDP and TCP network packets can be analyzed and due to procfs connections can be tracked. As NetHogs itself has not been implemented with monitoring in mind it can not be used directly for our purpose.

Important data to be analyzed is the destination IP and port as well as the source IP and port. Those are needed for matchmaking each connection with the sockets located in procfs. After the socket has been found the managing process needs to be identified. Therefore again the procfs is being utilized. Each process itself is tracked while monitoring the network packets for a fast identification and association to the batch job itself. Processes executed by root can also be tracked but do not necessarily belong to the traffic of jobs. Therefore different configurations for running the tool can be utilized.

### *2.1. Hardware and setup*

The tool is currently running on two racks consisting of 32 worker nodes. Each worker node has 24 job slots that need to be monitored in parallel. Long term measurements and data collection are in progress. These measurements already have been proven to be useful analyzing misbehaviors and various issues. Therefore we aim for an automated, realtime approach for anomaly detection.

### *2.2. Data format*

All data are time series regarding the actual network traffic (traffic rates, count of inbound and outbound packets, destination and source IP as well as ports), relevant unix process information (`pid`, `ppid`, `uid`) as well as information about the batch job itself (see figure 1). Every 20 seconds a new record for accumulated traffic information is created. Relevant process information are stored directly after the associated system event. For an improved data handling and analysis there is additional metadata associated to the measurements. These are stored in an additional database enabling a fast access to specific data.

## 3. Analyzing data flows of batch jobs

The actual measurements contain very detailed data as well as implicit tree structures of the executed unix processes per job. Each job is defined by a single multidimensional time series being correlated to the jobs on the same worker node and also to all other worker nodes inside the GridKa. Thus, the creation and collection of monitoring data is one important component of the monitoring chain. The other one is the data mining process to provide an insight into usage patterns and create statistics.

Before actually performing a clustering to identify patterns inside the dataset, there is the need to examine the general value and significance of the monitored data. For this purpose,

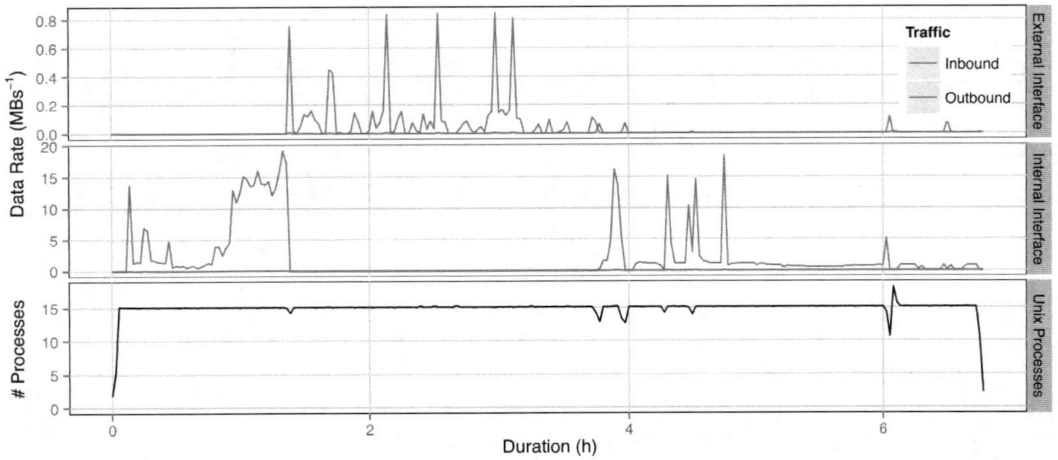


**Figure 1.** For each job detailed information about inbound and outbound traffic on internal and external interfaces, as well as running processes and metadata are available. The first plot visualizes incoming and outgoing traffic for the external interface. The second plot visualizes the same information but for the internal interface. The third plot displays the current number of running processes with regard to their proportion of the current timeframe.

a toy analysis to classify the executing VO of a batch job was designed. By performing the analysis it should be proven that the corresponding VO can be predicted by analyzing only the measured traffic profiles with an efficiency and purity greater than 90%. This use case can clearly be validated by verifying the monitored `uid`.

### *3.1. Preprocessing*

To ensure the best quality of the model a filtering of batch jobs is performed in preprocessing. Imbalanced data sets are a special case for classification problems. The class distribution is not uniform and therefore complicates the data mining process, since standard classification algorithms usually consider a balanced training set. To avoid biased results towards the majority class only the four main LHC experiments ALICE, ATLAS, CMS, and LHCb were chosen to be included in the final dataset. Additionally the matching jobs were evaluated for their completeness: Unfinished jobs, misbehaving jobs as well as jobs with missing data should be excluded. Missing data is recognized by reconstructing the unix process tree from the logs and assigning the appropriate traffic. As soon as a parent process can not be found and the current process is not the batch job shepherd itself, some data must be missing from the job. There are also jobs whose processes run longer inside the system than the batch jobs itself. As this behavior is not evaluated yet, those jobs also need to be excluded from the data set. Furthermore short jobs like Service Availability Monitoring (SAM) [19] tests or pilots running into a timeout are excluded by explicitly selecting jobs with a duration longer than 2.100 seconds.

From the whole variable space the following set of variables is included into the modeling: connection category (internal or external interface), incoming and outgoing volume (kB), incoming and outgoing rate (kB/s), duration and the job id itself.

The different variables are accumulated for every job so that each observation of a job consists of a six-dimensional feature set. In figure 2 pair-wise scatterplots and densities are visualized for selected features of the internal network interface.

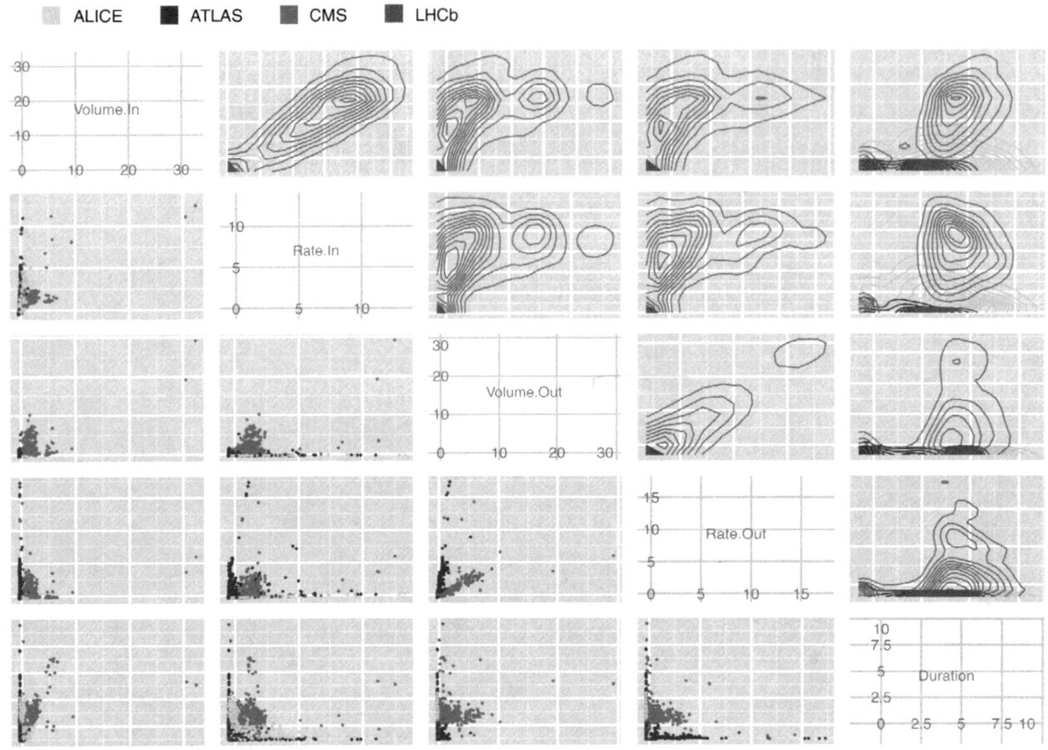


**Figure 2.** Scatter matrix visualizing selected scaled features for the internal network interfaces. Below the diagonal pair-wise scatter plots of variables are presented. Above the diagonal the density is shown. The colors represent the single experiments.

During data preprocessing care is taken that the data is normalized since the range of values of raw data varies widely. The majority of classifiers calculate distance between points by distance. If one of the features has a broad range of values, the distance will be governed by this particular feature. By applying normalization each feature contributes approximately proportional to the final distance. To this end, the standard value is calculated by applying (1) to each variable j. Each value of variable j is centred around 0 by subtracting the mean $\mu_j$ and is additionally weighted by the standard deviation $\sigma_j$.

$$z_j = \frac{x_j - \mu_j}{\sigma_j} \tag{1}$$

### *3.2. Training a decision tree classifier*

For the actual test analysis a decision tree classifier is used. The classifier is capable of performing multi-class classification by learning simple decision rules inferred from the data features. It was chosen because decision trees are simple to understand, interprete, and by using statistical tests the reliability of the resulting model can be validated. The analysis with the decision tree classifier is done with the software RapidMiner [20]. RapidMiner is an integrated environment for machine learning, data mining, predictive analytics, and business analytics. For measuring the quality and especially the stability of the selected data mining model the dataset is separated into three different and independent subsets: training data (60%), test data (20%), and validation

**Table 1.** Classification Performance Confusion Matrix visualizing the classification performance for the validation set by using the learned decision tree classifier based on traffic patterns of ALICE, ATLAS, CMS, and LHCb.

| | **ALICE** | **ATLAS** | **CMS** | **LHCb** | **Efficiency (%)** |
|---|---|---|---|---|---|
| **Pred. ALICE** | 206 | 2 | 6 | 8 | 92.8 |
| **Pred. ATLAS** | 2 | 330 | 3 | 0 | 98.5 |
| **Pred. CMS** | 0 | 0 | 121 | 0 | 100.0 |
| **Pred. LHCb** | 1 | 1 | 2 | 154 | 97.5 |
| **Purity (%)** | 98.6 | 99.1 | 91.7 | 95.1 | |

data (20%).

The model is built on the training set, refined with the test set and its quality is finally estimated on the validation set. The data has been taken over a timespan of 15 days on six different worker nodes. Three of them are located in an own rack respectively. The data set is split by utilizing stratified sampling to obtain the original distribution of VOs in the data. Stratified sampling is a sampling method that generates samples whilst maintaining the original proportion of each class. This way the resulting data sets still represent the original distribution of VOs. It is ensured that the validation set is entirely separate and distinct from the training and test set to get an independent validation of the resulting model.

### *3.3. Evaluation*
Table 1 displays the confusion matrix of classification performance on the validation set for the resulting decision tree. Although the results outperform our requirements of 90% efficiency and purity the effect of imbalanced data sets is still visible. The best results are gained for ATLAS as most data was available. For ALICE important jobs are not considered in the analysis as parts of analysis trains do not pass the preselection stop regarding the job duration. The specific influence of pilots also has to be measured. It might influence the results as well. E. g. some of the jobs of CMS are no regular jobs but pilots.

Summarizing, this first test analysis shows that traffic data of batch jobs seem to be a good indicator for further analysis. But to improve the performance and ensure comparability of different data sets substructures need to be recognized.

## 4. Discussion and final remarks
The actual measurements and analysis undertaken at the GridKa Tier 1 center show that network traffic profiles of WLCG batch jobs are a good indicator to predict the VO ownership and therefore there is some good indication that network patterns themselves might be a good indicator for clustering. The data being taken is dependent on the actual community needs and evolution of computing models in use. Changes might therefore invalidate previously discovered patterns. As a result, there is a need for incremental methods that may update existing models and still evolve and update with temporal changes in data.

Nonetheless, pilots are currently not taken into account correctly. They are treated as single jobs but instead they are local batch systems themselves [21]. By introducing some further analysis to detect actual jobs inside the pilots might even further improve the classifier.

## Acknowledgments
The authors wish to thank all people and institutions involved in the project Large Scale Data Management and Analysis (LSDMA), as well as the German Helmholtz Association and the Karlsruhe School of Elementary Particle and Astroparticle Physics (KSETA) for supporting and funding the work.

## References
[1] Kreuzer P *et al.* 2014 *Journal of Physics: Conference Series* **513** 062028
[2] Bauerdick L *et al.* 2012 *Journal of Physics: Conference Series* **396** 042009
[3] Gardner R *et al.* 2014 *Journal of Physics: Conference Series* **513** 042049
[4] Bonacorsi D 2012 CMS storage federations *Nuclear Science Symposium and Medical Imaging Conference (NSS/MIC), 2012 IEEE* (IEEE) pp 2012–2015
[5] Bloom K and The CMS Collaboration 2014 *Journal of Physics: Conference Series* **513** 042005
[6] Lorenz D *et al.* 2009 *Future Generation Computer Systems* **25** 308–314
[7] Arkhipkin D, Lauret J and Zulkarneeva Y 2014 *Journal of Physics: Conference Series* **513** 032002
[8] Legrand I *et al.* 2009 *Computer Physics Communications* **180** 2472–2498
[9] Andreeva J *et al.* 2014 *Journal of Physics: Conference Series* **513** 032004
[10] Cooke A W *et al.* 2004 *Journal of Grid Computing* **2** 323–339
[11] Aiftimiei C *et al.* 2008 *Journal of Physics: Conference Series* **119** 062003
[12] Andreeva J J *et al.* 2007 Experiment Dashboard: the monitoring system for the LHC experiments *GMW '07: Proceedings of the 2007 workshop on Grid monitoring* (New York, New York, USA: ACM) pp 45–49
[13] Cacti Group 2013 Cacti®- the complete rrdtool-based graphing solution. URL `http://www.cacti.net`
[14] Icinga Project 2014 Icinga — open source monitoring. URL `https://www.icinga.org`
[15] Massie M L, Chun B N and Culler D E 2004 *Parallel Computing* **30** 817–840
[16] Univa Corporation 2014 Univa products: Grid engine software for workload scheduling and management. URL `http://www.univa.com/products/grid-engine.php`
[17] Engelen A 2013 Nethogs: What program is using that bandwidth? URL `http://nethogs.sourceforge.net`
[18] 2014 Tcpdump/libpcap public repository URL `http://www.tcpdump.org`
[19] Duarte A *et al.* 2008 *Journal of Physics: Conference Series* **119** 052014
[20] RapidMiner 2014 Predictive analytics, data mining, self-service, open source - rapid miner. URL `http://rapidminer.com`
[21] Thain D, Tannenbaum T and Livny M 2005 *Concurrency and Computation: Practice & Experience* **17** 323–356

# Tier 3 batch system data locality via managed caches

**Max Fischer, Manuel Giffels, Christopher Jung, Eileen Kühn and Günter Quast**

Karlsruhe Institute for Technology, Kaiserstraße 12, 76131 Karlsruhe

E-mail: {`max.fischer, manuel.giffels, christopher.jung, eileen.kuehn, guenter.quast`}`@kit.edu`

**Abstract.** Modern data processing increasingly relies on data locality for performance and scalability, whereas the common HEP approaches aim for uniform resource pools with minimal locality, recently even across site boundaries. To combine advantages of both, the High-Performance Data Analysis (HPDA) Tier 3 concept opportunistically establishes data locality via coordinated caches.

In accordance with HEP Tier 3 activities, the design incorporates two major assumptions: First, only a fraction of data is accessed regularly and thus the deciding factor for overall throughput. Second, data access may fallback to non-local, making permanent local data availability an inefficient resource usage strategy. Based on this, the HPDA design generically extends available storage hierarchies into the batch system. Using the batch system itself for scheduling file locality, an array of independent caches on the worker nodes is dynamically populated with high-profile data. Cache state information is exposed to the batch system both for managing caches and scheduling jobs. As a result, users directly work with a regular, adequately sized storage system. However, their automated batch processes are presented with local replications of data whenever possible.

## 1. Introduction

The computing models of the LHC collaborations have been built around data storage [1, 2, 3, 4]. This is reflected in the hierarchical responsibilities in the LHC computing Grid, which defines the access and distribution of data. Processing of data in High Energy Physics (HEP) usually involves access from a dedicated computing cluster to a likewise dedicated storage cluster. Data locality, i.e. communication distance from data to processing, is typically provided at the granularity of an entire computing centre, housing both computing and storage resources.

In contrast, recent cluster computing architectures provide data locality at the granularity of individual computing nodes. Systems such as *GoogleFS* [5] or *Hadoop* [6] use mixed compute and storage resources: data is stored directly on compute nodes and processing is steered towards nodes hosting the appropriate data. By promoting local data access, data can be read at full capacity of the storage media while network communication is drastically reduced.

Due to the reduced network reliance, the mixed architecture generally scales better with increasing processing throughput: processing power can be added with little impact on existing resources. However, dedicated storage architectures are advantageous when the total data volume by far exceeds the momentarily relevant data volume: storage capacity can be increased without requiring an increase in computing capacity.

The 'High-Performance Data Analysis' (HPDA) design thus aims at combining the two paradigms. A top layer provides for fast local access, while a background layer serves as high-volume storage (see figure 1).

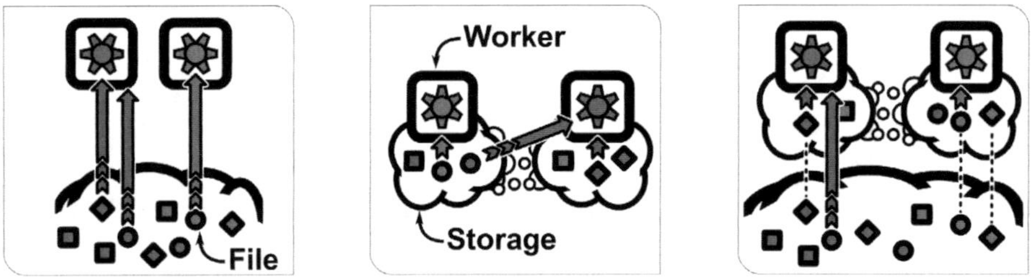


a) Dedicated Storage b) Integrated Storage c) HPDA Storage

Figure 1: Processing architecture illustration: Batch clusters in HEP usually use dedicated storage accessed via network from worker nodes (1a). Alternate setups such as *Hadoop* combine storage and worker nodes to allow local access (1b). HPDA provides local access to prominent data via caches overlaying classic high-volume dedicated storage (1c).

## 2. Conceptual Overview/Design Considerations

The workflows and data handling in HEP put several constraints on data usage. The HPDA design adopts two of these constraints as operation assumptions:

- Only a fraction of data is accessed on a regular basis with a high frequency.
- Non-local data access is efficient, provided it occurs with limited concurrency.

Therefore, data locality in the HPDA concept encompasses staging only the most relevant fraction of the total data locally on compute nodes. Any data can be fetched from remote storage, while prominent data is also kept available as a local copy; in essence, this is the functionality of a cache.

The deciding feature of HPDA caching is its scope: regular caching mechanisms for data from devices or hosts, such as Linux bcache [7] or CacheFS, operate individually on their single host. In the context of a pool of worker nodes however, all caches must likewise form a cache pool; state must be shared, e.g. to ensure even spread of data on all caches, to factor in imbalances of worker node capacities, or to decouple cache miss considerations from node availability.

In a cluster computing context such as HEP, data usage considerations extend beyond the current state of the system. For example, access to a portion of a data set implies an increased likelihood of subsequent access to the entire data set. Similarly, a history of repeated access to data suggests future usage beyond a classic cache expiration timespan dictated by data rate versus cache size (which would be low in the HEP context).

Thus, HPDA builds on the advantages of distributed caches [8], and extends it to actively coordinated caches. Caches are supervised globally, so data can be staged and released from caches asynchronously to current data accesses. The history of data accesses as well as features of external data are taken into consideration to formulate improved caching constraints.

## 3. Architecture and Implementation

While many individual features of the HPDA design have existing implementations, these are mostly part of specialized software or not compatible with HEP workflows. Hence, the HPDA core components are largely assembled as a custom implementation to ease flexibility and future extensions. The batch system itself, storage access and cache mapping rely on software and frameworks common to the HEP community.

All custom components are written purely in Python due to the language's extensive standard library and the proven accessibility. *HTCondor* [9] was chosen as the batch system since it offers extensive means for communication between nodes; in addition, it is well established in CMS computing and under investigation by German CMS groups for opportunistic resources usage. For transparent access to caches and remote filesystems, the logical filesystems of *xRootD* and *unionFS* are used.

### 3.1. Exposed Caches

The HPDA cache on the worker nodes serves two functions: maintaining cached files and exposing their state to the batch system. Several elements keep the cache up-to-date and give it the autonomy to interact with the pool (see figure 2).

Meta-data of caches and files is stored in a *catalogue*. Without the need to explicitly query local or remote storage, the cache state can be easily inspected for allocation and exposed to the batch system. For failure robustness, regular checkpoints are used in combination with incremental updates. This allows to quickly reconstruct the current state of the catalogue from a checkpoint by replaying stored updates.

A *worker* thread constantly maintains the cached files by validating the meta-data against remote storage. If new files are added to the catalogue, the worker asynchronously fetches them for caching. Catalogue and worker communicate only with generic, simple file state and move instructions. Plugins for storage and cache APIs allow supporting a range of storage systems without adjustments to the core elements.

An additional *allocator* thread provides limited autonomy for the cache. The allocator scores files based on their features, e.g. access frequency. If there are several local caches, this is used for deducing in which to place the file. When a new file is suggested for caching, the score is used to determine if it is accepted and which older files to evict.

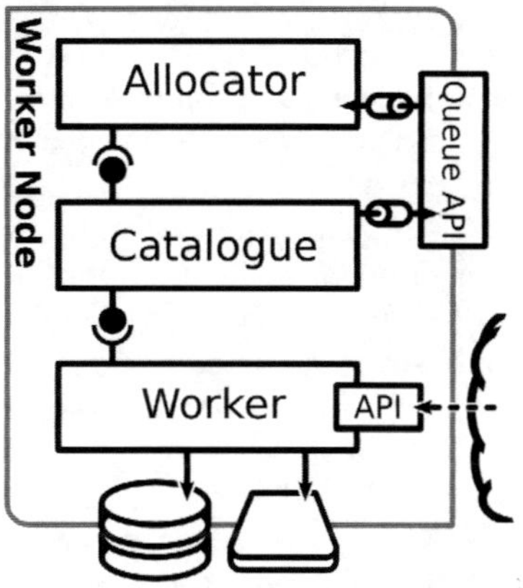


Figure 2: The HPDA cache consists of three main elements: The *catalogue* tracks the state of cached files. The *allocator* assigns files to cache media, creating fetching, movement and eviction tasks. The *worker* ensures validity of catalogued files and performs the allocator's tasks.

A generic API is used to exchange information with the batch system. The current content of the cache is regularly published to allow for data aware scheduling. Caching instructions are received via jobs and passed on to the allocator for validation; the job exit code is used to signal acceptance or rejection.

### 3.2. Coordinator

While the HPDA caches are able to share basic information, it is the responsibility of the coordinator to observe the big picture. It extracts resource requests from the job queue and history, and formulates staging orders to the cache pool if needed.

Information is extracted directly from the batch system, converting job ClassAds into attribute maps. Any feature of a job, such as expected I/O rate, can thus be considered for staging orders. The CMS grid overlay framework *glideinWMS* [10] has shown ClassAd extraction to be scalable to thousands of jobs even with a basic implementation.

Staging orders are dispatched in an anonymous fashion: An order encompasses the subject (i.e. file to stage), a score of importance, and resource constraints mimicking associated jobs. The assignment to distinct caches is the responsibility of the batch scheduler, while rebalancing and expiration is the responsibility of the caches. Thus, the coordinator does not need to be concerned with the topology of the worker and cache pool.

The optimal coordination strategy is of course a matter of optimisation. The current prototype implementation relies on manual staging orders; this is a logical side benefit of the need for manual management tools and sufficient for initial performance tests. Devising an optimal coordination strategy is the scope of future research.

### *3.3. Communication*

In order to avoid duplicated infrastructure, HPDA components rely on the existing batch system for information exchange across hosts. With the *HTCondor* batch system, two forms of communication are used: ClassAds and communication jobs.

The ClassAd [11] is *HTCondor*'s mechanism of describing entities, such as worker nodes or jobs, with key-value attribute maps. In addition to standard entities, generic ClassAds with arbitrary attributes can be published globally in the batch system. Using this method, HPDA components can globally expose information about themselves, e.g. for reporting free cache space.

Since direct communication in the cache-coordinator pool is for caching requests only, it always encompasses scheduling problem: a fitting host must be found. To promote consistency with job scheduling as well as modularity and scalability, the HPDA components do not address each other directly. Instead, the information is encapsulated in a batch job with constraints on fitting recipients. The batch system then directs the information to a suitable host, if any. Once the job starts, it addresses the local HPDA component via Inter-Process Communication and reports success or failure of the request via its exit code.

For both methods, a thin wrapper API allows using them similar to message queues. This makes HPDA independent of the choice of the batch-system, as long as an equivalent API, possibly using external resources, can be provided.

## 4. Experience

While the HPDA design has not been fully implemented yet, experiences with individual components have been promising so far.

Access to local caches on mass storage easily matches network performance. Hard disk drives (HDD) are comparable to 1 Gb/s network, while a single solid-state drive (SSD) reaches more than half the throughput of a 10 Gb/s network. The largest issues are concurrent reads for HDDs, with degrading performance above four concurrent accesses, and the limited capacity of SSDs.

Interfacing with the *HTCondor* batch system has proven to be easy and reliable. Automated job submission and information extraction was extensively tested during the development of the *grid-control* [12] job management tool. The global *glideinWMS* infrastructure demonstrates the scalability of ClassAd information extraction far beyond the needs of HPDA.

## 5. Summary and Outlook

The High-Performance Data Analysis (HPDA) design aims at merging dedicated storage with data local batch processing. An array of caches on worker nodes forms a cache pool that provides local access to high profile data. The pool is coordinated globally to include external data and adjust to predicted access patterns. For scalability, a modular, decoupled architecture is chosen that adds only low overhead of coordination on top of regular batch system activity.

On the infrastructure level, future research will focus on the optimal data locality. With modern network, node-local and rack-local access is comparable in performance for some applications [13]. Motivated by this, parallel filesystems distributed on machines of the same racks are candidates for future improvements.

For high-level development, primary field of interest is the optimization of cache coordination. Especially recognition of access patterns to anticipate data usage for pre-staging is a promising approach.

Data locality versus high performance remote access is not a clear cut decision. Future scaling tests will thus investigate the optimal environment for either approach. Since the underlying *HTCondor* at KIT spans a non-flat topology of desktop computers, opportunistic resources, as well as closely and loosely storage attached workers, we expect to be able to compare a wide range of different dependencies.

## Acknowledgments

The authors wish to thank all people and institutions involved in the project Large Scale Data Management and Analysis (LSDMA), as well as the German Helmholtz Association, and the Karlsruhe School of Elementary Particle and Astroparticle Physics (KSETA) for supporting and funding the work.

## References

[1] Grandi C, Stickland D, Taylor L et al. 2004 The CMS computing model preprint CERN-LHCC-2004-035/G-083
[2] Jones R and Barberis D 2008 The ATLAS computing model *J. Phys.: Conf. Series* **119** 072020 doi:10.1088/1742-6596/119/7/072020
[3] Brook N 2004 LHCb computing model CERN-LHCb-2004-119
[4] Carminati F et al. 2004 ALICE computing model CERN-LHCC-2004-038/G-086
[5] Dean J and Ghemawat S 2004 MapReduce: Simplified Data Processing on Large Clusters *OSDI'04: Sixth Symposium on Operating System Design and Implementation*
[6] The Hadoop project homepage **URL** http://hadoop.apache.org/
[7] The bcache project homepage **URL** http://bcache.evilpiepirate.org/
[8] Paul S and Fei Z 2001 Distributed caching with centralized control *Computer Communications* **24-2** 256-68 (doi: 10.1016/S0140-3664(00)00322-4)
[9] Thain D, Tannenbaum T and Livny M 2005 Distributed computing in practice: the Condor experience *Concurrency Computat.: Pract. Exper.* **17** 323–56 (doi: 10.1002/cpe.938)
[10] Sfiligoi I, Bradley D C, Holzman B, Mhashilkar P, Padhi S and Wurthwein F 2009 The Pilot Way to Grid Resources Using glideinWMS *WRI World Congress on Computer Science and Information Engineering* **2** 428–32 (doi:10.1109/CSIE.2009.950)
[11] Raman R, Livny M C and Solomon M 1998 Matchmaking: Distributed Resource Management for High Throughput Computing *Proceedings of the Seventh IEEE International Symposium on High Performance Distributed Computing* 140–6 (doi:10.1109/HPDC.1998.709966)
[12] The Grid-Control project homepage URL https://ekptrac.physik.uni-karlsruhe.de/trac/grid-control/
[13] Ananthanarayanan G, Ghodsi A, Shenker S and Stoica I 2011 Disk-locality in datacenter computing considered irrelevant *Proceedings of the 13th USENIX conference on Hot topics in operating systems* 12 (doi:10.1109/HPDC.1998.709966)

Journal of Physics: Conference Series **608** (2015) 012019 doi:10.1088/1742-6596/608/1/012019

# The ALICE analysis train system

**Markus Zimmermann for the ALICE collaboration**
CERN, 1211 Geneva 23, Switzerland
E-mail: m.zimmermann@cern.ch

**Abstract.** In the ALICE experiment hundreds of users are analyzing big datasets on a Grid system. High throughput and short turn-around times are achieved by a centralized system called the LEGO trains. This system combines analysis from different users in so-called analysis trains which are then executed within the same Grid jobs thereby reducing the number of times the data needs to be read from the storage systems. The centralized trains improve the performance, the usability for users and the bookkeeping in comparison to single user analysis. The train system builds upon the already existing ALICE tools, i.e. the analysis framework as well as the Grid submission and monitoring infrastructure. The entry point to the train system is a web interface which is used to configure the analysis and the desired datasets as well as to test and submit the train. Several measures have been implemented to reduce the time a train needs to finish and to increase the CPU efficiency.

## 1. Introduction

The ALICE collaboration is recording around 10 PBs every year. This amount of data is stored on different storage elements all over the world. To analyze a single dataset, on the order of $10^5$ files have to be read where those files containing data on the order of 50 TB. The creation and submission of the analysis jobs to the Grid can be done with the ALICE analysis framework [1]. These jobs can be monitored and resubmitted using a monitoring service (MonALISA [2]).
On top of these existing frameworks the LEGO (Lightweight Environment for Grid Operators) train system was created. It was developed to increase the CPU efficiency of the analysis jobs and thus to get more user analysis done with the same amount of computing resources. This is done by collecting multiple analysis tasks which should analyze the same dataset and by running them within the same analysis job. Following this procedure the data is read once, and then it is used for multiple analysis tasks. In comparison to that if the different user analysis would be done individually the data would be read for each analysis separately. This way more time would be used for loading the data and the CPU efficiency would be lower.
Another advantage of the LEGO trains is to hide the Grid complexity from the users. They just have to define their code on a web page which also provides the analysis results as soon as it is available. In between, the system computes the results fully transparent for the user. This process is fully automatized so that no one needs to supervise the jobs which are running on the Grid. Further improvements are implemented in the computing process to use the available resources as efficient as possible.
The train system contains several trains divided by physics working groups (jet analysis, particle correlations, ...) and different kind of datasets (pp-, pPb- or PbPb-collisions, data or Monte Carlo). The analysis of a dataset by a train is called a train run.

## 2. Running a LEGO train

Before running in a train a user has to submit the analysis code to a repository. Then the configuration of the code has to be defined on a web page in a so-called train wagon.
The trains are managed by train operators who define the datasets and configure the jobs of the train runs. On a regular time basis, which is usually once per day, the trains are started. To do so an operator composes a train run out of a set of wagons and a dataset. The train run is tested with an automatized procedure and only if the test is successful the train run can start on the Grid.
An example of a train test is shown in figure 1. The Base line tests the configuration of the train and does not contain any user code. The Full train contains all the train wagons. In between all wagons are tested individually. The test reports several measurements. The first one is the status if the test is finishing normal (OK) or if there is an exception in the code (Failed). The memory consumption is divided into the total consumption of the test and the growth per event. The later number indicates a memory leak if it is getting too big. To show this to the operator the number is highlighted in red if a memory leak is suspected. The timing information shows the used time per event. This allows a rough estimate of the train duration. In the last column the result of the merging test is reported.

| Wagon | Status | Memory consumption | | Timing | Merging |
|---|---|---|---|---|---|
| | | Total | Growth per event | | |
| **Base line**<br>stdout \| stderr | **OK** | 250 MB | 0.182 KB/evt | 34.10ms/evt | **No output** |
| **wagon 1**<br>stdout \| stderr | **OK** | 591 MB | **0.158 MB/evt** | 1.02ms/evt | **OK**<br>merge dir |
| **wagon 2**<br>stdout \| stderr | **Failed** | | /evt | ms/evt | **Not tested** |
| **Full train**<br>stdout \| stderr | **Failed** | | /evt | ms/evt | **Not tested** |

**Figure 1.** Example test of a train run. The Base line does not contain any user code, the wagons contain just their own code and the Full train contains all the wagon codes.

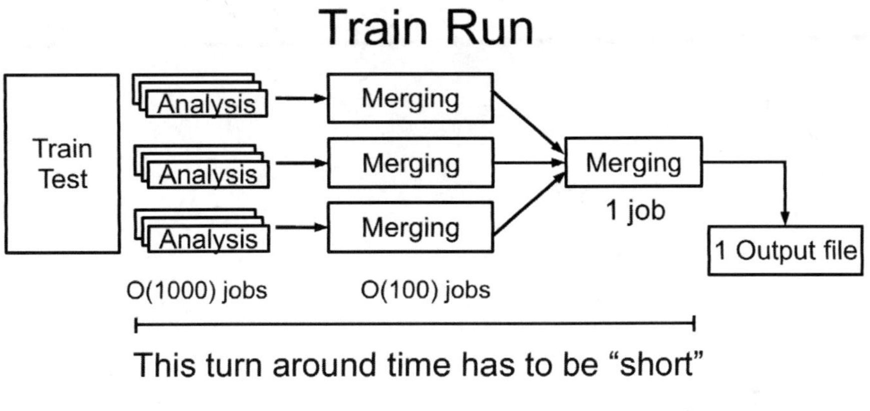


**Figure 2.** Procedure of a train run. After the train test the analysis jobs run. Then the results are merged into a single output file. The time to complete this procedure is called the turn around time.

If there is an issue in the test, the operator has to fix this or exclude the failing wagon(s). Afterwards the test has to be repeated. If the test is successful the train run can start. The chain of jobs run on the Grid is shown in figure 2. After the test, the analysis jobs are started. These jobs are grouped into sets which are merged as soon as all the jobs finished. The results of the merging jobs are merged again in a last merging job. The output of this last job is the result of the train run and is linked on the LEGO train web page. The time from starting the first analysis jobs until the merging is finished is called the turn around time. This is the time the user has to wait for the output and it should be as short as possible.
The configuration files of the train run and the final result can be downloaded from the web page by everyone inside ALICE. This way all the train runs are documented in the same way and it is easy to hand results to someone else within the experiment.

## 3. Statistics

In table 1 the usage statistics of the LEGO train system are shown. The system was set up in the beginning of 2012 and since then has continuously run. For 2014, the numbers are extrapolated from the measured values within the first 6 months of the year.
The line 'part of the user analysis done with the analysis trains' represents the total amount of computing time on the grid used by the trains divided by the total amount of computing time on the grid used for ALICE analysis. The value is increasing from 27% in 2012 to 70% in 2014. In the same time the number of users increased from 60 to 188. This shows that the LEGO trains are well accepted among the users and that they are now used for most of the analysis in ALICE.
For the users, the average turn around time is an important value. This value shows how long they have to wait on average to get the train results after the train has been submitted to the Grid. In figure 3, the development of the average turn around time per month is shown. It improves from 3 to 4 days at the beginning of 2012 to 14 hours in 2014. The reduction in July 2013 is achieved by putting the improvements discussed in section 4 into production.

**Table 1.** Usage statistics of the LEGO trains.

| | 2012 | 2013 | 2014 (extrapolated) |
|---|---|---|---|
| Users | 60 | 127 | 188 |
| Trains | 42 | 69 | 79 |
| Train runs | 1537 | 4794 | 7447 |
| Number of jobs | 12M | 26M | 34M |
| Train wagons/run | 14.9 | 10.1 | 8.9 |
| Part of the user analysis done with the trains | 27% | 57% | 70% |
| Processed data | — | 75PB | 130PB |
| Turn around time | 49h | 22h | 14h |

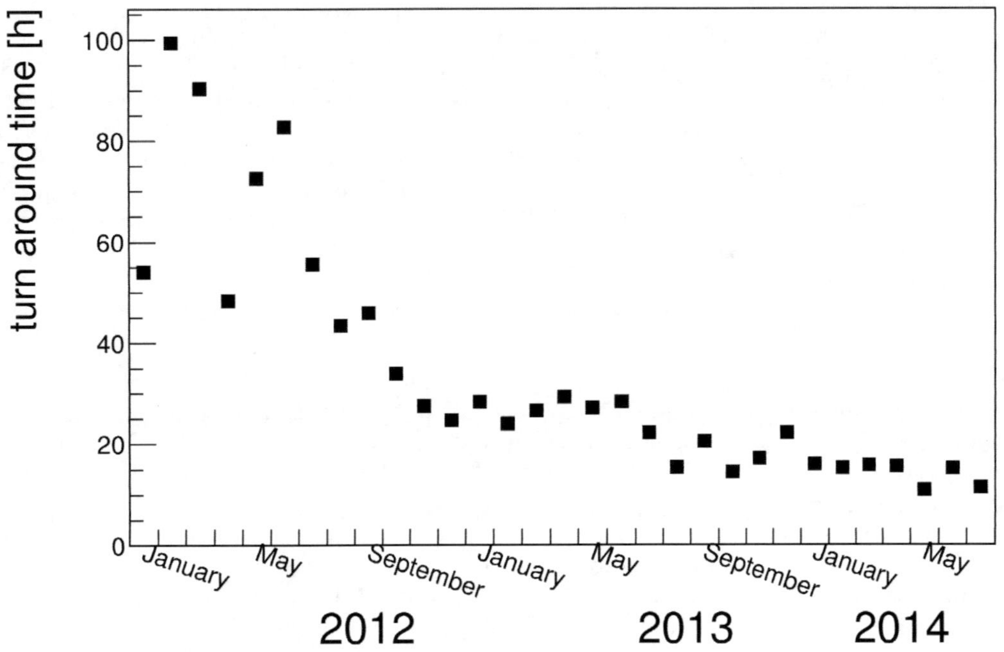


**Figure 3.** Average turn around time of the trains per month.

## 4. How to convince users to use the LEGO trains

Because the train system is collecting multiple analysis to run them together, the system works better when more people are using it. This section explains how the users were convinced to use the system.

The biggest advantage for the users is that they save time. The setup of a wagon is much easier and faster than submitting all the jobs for an analysis. Additionally, the resubmission of failed jobs, the submission of the merging jobs and the bookkeeping are done automatically by the system. As soon as a train finishes the user is notified by an email. Also new users can learn faster how to run an analysis via the train system than running it directly on the Grid.

The definition of the datasets is done by operators. This means that some datasets can already be divided into good and semi-good runs by a previous quality analysis. So the user just needs to choose on which dataset to run instead of defining it.

Some job parameters are predefined by the system or they are defined by the operators. This leads to a stable job management which has less failed jobs compared to user managed analysis. Additionally, trains make it easier for the operators and the Grid support personnel to find bugs in the jobs so that the users get faster support.

## 5. Improvements

One of the crucial parameters for the users is the turn around time of the trains. The objective of the improvements was to achieve a turn around time of one night. This way the user can define their analysis task in the evening and then the train runs over night. On the next morning, the user can continue with their work.

In this section some of the most effective improvements are discussed.

### *5.1. Job submission*

The ALICE computing model schedules the analysis jobs always to the computing sites which have the data on a local storage element (SE) to minimize the loading time. This can lead to the problem that some jobs have to wait for a long time because all job slots at the computing sites with a local copy of a certain file might be occupied. To solve this, the local data requirement is dropped for the train jobs as soon as 90% of the jobs are successfully finished. Then the remaining jobs can be analyzed on any computing site and the data is read remotely. Although this increases the time needed to read the data, it decreases the time until the last jobs are started. This speeds up the last jobs of the train and thus also the full train run itself. This is done at the cost of the CPU efficiency.

### *5.2. Dataset consolidation*

One job can only analyze multiple input files if they are saved at the same local SE. Most of the datasets (especially from MC productions) are spread over a lot of different SEs. So they end up having a lot of jobs which are processing just 1 input file. In addition to the physics analysis, each job produces some overhead for producing and starting the job.
The number of input files in a job can be increased by consolidating the dataset. This is done by copying some of the files to another SE so that more files can be executed within the same job. This is only done with the most used datasets. For some datasets, the consolidation reduces the number of jobs up to about 90%. Running fewer jobs means less overhead for the train run.

### *5.3. Run time of the jobs*

The introduction of a monitoring system to see the run time of the analysis jobs showed that most of the jobs are finishing within a quite short time. For most of the train runs, more than 95% of the jobs finish in half the time it takes the longest running job to finish. This offers a new way to speed up the train turn around time. The last 2% of the active jobs are killed and then the merging is started directly. This way some statistics is traded for a faster turn around time. This can only be done if the distribution of killed jobs is not introducing a bias on the data analysis.

## 6. Conclusions

The LEGO train system is a workflow for the organized analysis in ALICE. The system is commonly used and consumes 70% of the CPU time dedicated to user analysis. It hides the Grid complexity from the users and does the job submission, resubmission and bookkeeping automatically. The system saves computing resources by combining several user tasks in one analysis job. This results in a better CPU efficiency by loading the data less often.
By using the system, the Grid support personnel can find bugs easier and thus give faster support. This saves a lot of time for the users and the Grid support personnel compared to a situation in which all users do their own individual user analysis.

## References

**[1]** Gheata A 2008 ALICE analysis framework *Proc. XII Int. Workshop on Adv. Comp. and Analysis Techniques in Phys. Research (Erice) PoS(ACAT08)* p 028

**[2]** I. Legrand et al., MonALISA: An agent based, dynamic service system to monitor, control and optimize grid based applications, *Computer Physics Communication*, vol. 180, no. 12, pp. 2472 2498, 2004

# FAMoS - an information service on the usage of data files in AliEn

**A Abramyan[1,3], L Betev[2], P Buncic[2], C Grigoras[2], A Grigoryan[1], N Manukyan[1,3], M M Pedreira[2] and P Saiz[2]**

[1]A.I. Alikhanyan National Science Laboratory Foundation, 2 Brothers Alikhanyan str., 0036 Yerevan Armenia
[2] CERN, CH-1211, Geneve 23 Switzerland

E-mail: aabramya@yerphi.am nmanukya@yerphi.am

**Abstract**. The File Access Monitoring Service (FAMoS) leverages the information stored in the central AliEn file catalogue, which describes every file in a Unix-like directory structure, as well as metadata on file location and its replicas. In addition, it uses the access information provided by a set of API servers, used by all Grid clients to access the catalogue. The main functions of FAMoS are to sort the file accesses by logical groups, access time, user and storage element. The collected data identifies rarely used groups of files, as well as those with high popularity over different time periods. This information can be further used to optimize file distribution and replication factors, thus increasing the data processing efficiency. The paper describes the FAMoS structure and user interface and presents the results obtained in one year of service operation.

## 1. Introduction

The aim of ALICE [1], one of 4 Large Hadron Collider (LHC) experiments at CERN, is to explore the primordial state of matter that existed in the first instants of our Universe, immediately after the initial hot Big Bang. The ALICE detector has been built by a collaboration of over 1500 physicists and engineers from 151 Institutes in 37 countries. During data taking, the experiment produces tens of Petabytes of data annually. This data is accessed and analyzed by the members of ALICE Virtual Organization (VO) around the world, using the WLCG Grid infrastructure [2]. The ALICE collaboration deploys ALICE Environment (AliEn) [3] middleware across all Grid sites to facilitate the computing resources access.

The files created in the process of the data processing are stored in Storage Elements (SEs) distributed across the Grid sites. At present there are 82 SEs. Every recorded file and its replicas are annotated in the AliEn File Catalogue (section 2). Since the storage capacities of the sites are limited, the recorded data must be fully monitored for access and relevance, and some of it removed to free space for newly produced files. The FAMoS toolkit (section 3) is build to evaluate the data access and facilitate the removal of sparsely accessed data and replication of popular data sets.

[3] Our work was supported by the Grant of the World Federation of Scientists, within the Armenian H.E.P. National Scholarship Programme.

The service functionality can be summarized as follows:

1. Accumulation of file access attributes;
2. Definition of popularity by implementing optimization algorithms, determination of files to be removed, replicated and/or migrated from one site to another;
3. Definition of thresholds for various manipulations of files based on the analysis of information gathered in steps 1 and 2.

This paper describes our work on the 1st step, as well as presents the results obtained in one year of operation of the service (section 4).

## 2. File System and Data file and file set types of ALICE VO

### *2.1. File System of ALICE VO*

ALICE distributed data storage uses multiple types of storage systems and data transport protocols into one virtual file system called AliEn File Catalogue [4]. Unlike real file systems, the catalogue does not own the files; it only keeps an association between the Logical File Name (LFN) and (possibly more than one) Physical File Names (PFN) on disk or mass storage (tape) systems.

Below we give example of the LFN and PFN of a file named "testfile":

***LFN*** *=* */alice/cern.ch/user/n/nmanukya/testfile*

***PFN*** *=* *root://pcaliense04.cern.ch:1095//14/22310/e4b050e2...*

*protocol* *hostname* *port* *path of the file*

Figure1. Structure of LFN and PFN

In this example, the LFN is the address of the file in the catalogue. The PFN string contains the following fields:

- *protocol* to access the file remotely (e.g. gridFTP, root, rfio);
- *hostname* of the SE master node where the file is physically stored, the SE may consist of multiple file servers;
- *port* which is used by the protocol for the communication with SE;
- *path* shows the physical location of the file on the SE file system.

### *2.2. Data file and file set types in ALICE VO*

The following data containers are created and replicated in the processes of the data processing and Monte Carlo simulations:

- ESD (Event Summary Data) contain detailed information about characteristics of the reconstructed collisions. The ESDs are produced during event reconstruction from the raw data.
- For data analysis, the ESDs are filtered to smaller, AOD (Analysis Object Data) data containers.

In ALICE VO each ESD and AOD is stored two or more times on different SEs. The data are grouped in logical periods (categories), each corresponding to a specific set of data taking conditions or MC generation. The data replication is performed by logical periods.

## 3. FAMoS architecture

FAMoS accumulates file access information from log files of AliEn API and Authen services [5, 6]. The information is sorted by the attributes, shown in table 1.

Table 1. List of file access attributes

| Attribute | Description |
|---|---|
| File name | File LFN |
| SE name | SE from where the file was accessed |
| User name | User by whom the file was accessed |
| Access Time | Time and date of file access |
| Operation result | Successful or failure |

The monitoring is performed not only on the level of the files themselves, but also on the level of their combinations (categories), described in section 2.2.

FAMoS consists of the following components:

- ***Collector*** Bash script runs on daily basis on the API and Authen servers and uploads the log file of the last day accesses to the FAMoS server.
- ***Parser*** Perl module running on the FAMoS server and processing the log files of last day; aggregates the file attributes by categories, access time, user, etc. The Parser inserts the processed data into a database, called *Accesses*.
- ***Accesses*** MySQL database (section 3.1) contains access information with the time granularity of one day.
- ***Web interface*** (section 3.2) for statistical representation of categories access information, volumes, access time, user and SE.

### *3.1. The FAMoS database*

The structure of the database is presented in figure 2.

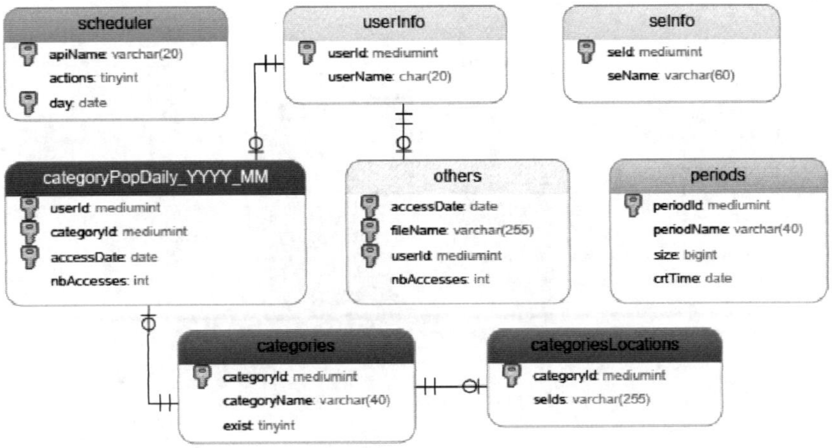


Figure 2. Accesses database structure

The database contains the following tables:

- ***scheduler*** - scheduling information (start time and state) for the operation of the FAMoS *Parser* module.
- ***userInfo*** and ***seInfo*** - values of user name and SE name attributes.
- ***categories*** and ***categoriesLocations*** - names of categories (extracted from the LFN), and the list of SE names from where a file of that category was accessed.
- ***periods*** - names of LHC periods, their volume and creation date. This information is gathered from the AliEn File Catalogue.

- ***categoryPopDaily_YYYY_MM*** - accesses to the categories. In order to minimize the reading time these tables are created for each month of accesses.
- ***others*** - access information for those files which cannot be grouped into a category.

*3.2. FAMoS Web interface*

FAMoS Web interface is an environment for graphical representation of the categories usage statistics as interactive charts and tables. To provide fast load of the web pages the interface design is simplified and page access from any device (from mobile phones to PCs) is done on the base of pure CSS. Currently it consists of the following components:

- *PHP pages* for the dynamic creation of database queries and collecting of their results, mostly in JSON format [7];
- *Highcharts API* [8] for visualization of the FAMoS data in the form of interactive charts;
- Google Chart API [9] for data representation in the form of interactive tables.

## 4. Statistics obtained in one year of operation of the FAMoS

FAMoS has been deployed and is in operation since August 2013. In this section we present several types of statistics provided by this service.

*4.1. Volumes of accessed or not accessed LHC periods*

Information on the volumes of used and/or unused periods/categories is critical for optimization of used space on the SEs. For example the plot in figure 3 presents the volumes of the periods which were not accessed during last year and plot in figure 4 presents their volumes grouped by creation year. The unused categories are deleted or archived, freeing space for new data.

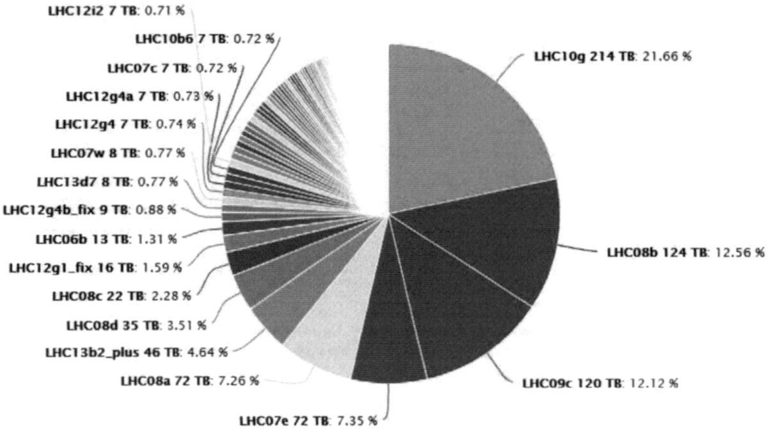


Figure 3. Volumes of the not accessed periods

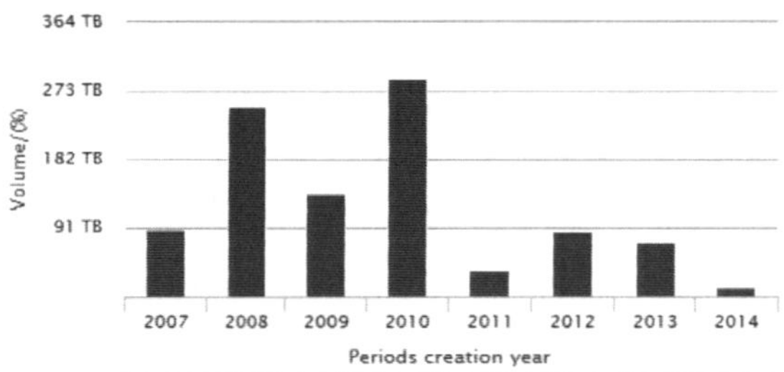


Figure 4. Volumes of not accessed periods grouped by their creation year

Journal of Physics: Conference Series **608** (2015) 012020 doi:10.1088/1742-6596/608/1/012020

### *4.2. Most and least accessed categories*

FAMoS identifies the frequently and rarely used categories over different periods of time. For instance figure 5 illustrates the most and least accessed categories during one year.

Currently we work on the analyses of this data for the precise definition of Popularity Index, by taking into account category volume, access time, number of users, which will be used for the optimization of distribution of the files/categories over SEs.

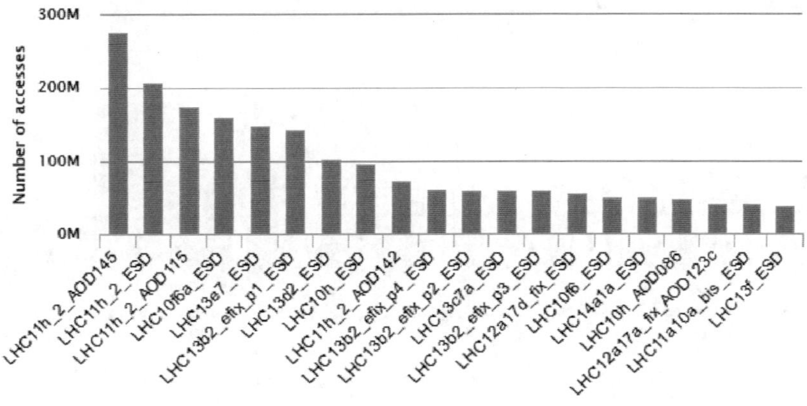


(a)

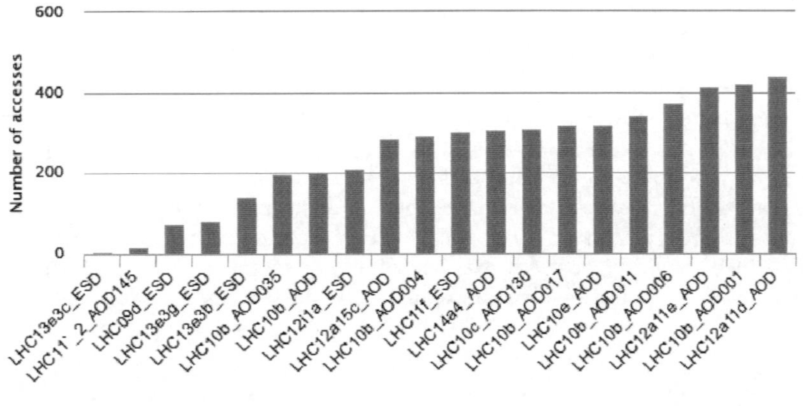


(b)

Figure 5. 20 most (a) and least (b) accessed categories during one year

### *4.3. Number of ESD and AOD accesses*

FAMoS provides statistics on the usage of the ESD and AOD files over time. Figure 6 shows the number of accesses to the ESD and AOD files, as well as the number of users who have accessed these files during the last year. This information will be used in the definition of Popularity Index for further location and number of replicas optimization.

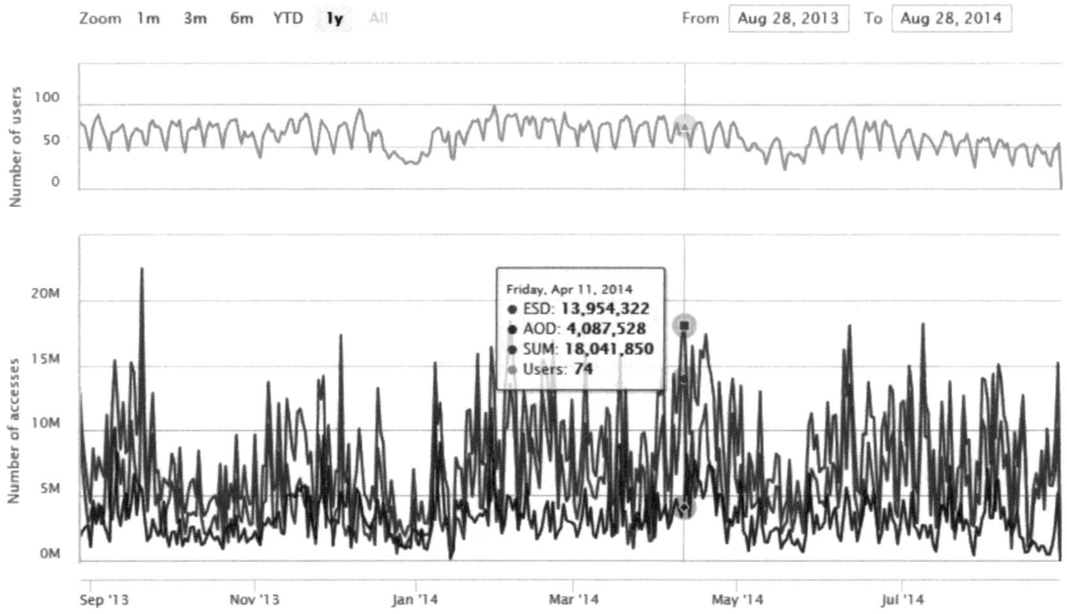


Figure 6. Evaluation of the number of ESD and AOD accesses during last year

## 5. Conclusions and future work

The File Access Monitoring Service has been deployed and is in operation since August 2013. It provides the information, necessary to optimize the file distribution over the Grid storage elements. The file Popularity Index calculated from the collected access parameters (described above) will determine which files should be removed, replicated and/or migrated from one site to another. The work on the distribution optimization algorithms and file Popularity Index is in progress.

## References

[1] ALICE Collaboration 'The ALICE Experiment at the CERN LHC' J. Instrum. 3, S08002 (2008); Web site URL http://aliweb.cern.ch
[2] Web site of WLCG infrastructure - http://wlcg-public.web.cern.ch/
[3] Web site of ALICE Grid Environment (AliEn) - http://alien.cern.ch
[4] P. Buncic, A. Peters, P.Saiz, The AliEn system, status and perspectives. Computing in High Energy and Nuclear Physics, 24-28 March 2003, La Jolla, California.
[5] Peters, Andreas-J. et al. – 'AliEnFS: A Linux file system for the AliEn grid services'. eConf C0303241 (2003) THAT005 cs/0306071 [cs-dc] CHEP-2003-THAT005
[6] J. Zhu et al, Enhancing the AliEn Web Service Authentication. 2011 J. Phys.: Conf. Ser. 331 062048;
[7] JavaScript Object Notation (JSON) - http://www.json.org/
[8] Highcharts charting API web site - http://www.highcharts.com/
[9] Google Chart API - https://developers.google.com/chart/interactive/docs/index

# Gaudi components for concurrency: Concurrency for existing and future experiments

**M Clemencic, D Funke, B Hegner, P Mato, D Piparo and I Shapoval**
CERN, CH-1211 Geneva 23, Switzerland

E-mail: {marco.clemencic, daniel.funke, benedikt.hegner, pere.mato, danilo.piparo, illya.shapoval}@cern.ch

**Abstract.**
HEP experiments produce enormous data sets at an ever-growing rate. To cope with the challenge posed by these data sets, experiments' software needs to embrace all capabilities modern CPUs offer. With decreasing $^{\text{memory}}/_{\text{core}}$ ratio, the one-process-per-core approach of recent years becomes less feasible. Instead, multi-threading with fine-grained parallelism needs to be exploited to benefit from memory sharing among threads.

Gaudi is an experiment-independent data processing framework, used for instance by the ATLAS and LHCb experiments at CERN's Large Hadron Collider. It has originally been designed with only sequential processing in mind. In a recent effort, the framework has been extended to allow for multi-threaded processing. This includes components for concurrent scheduling of several algorithms – either processing the same or multiple events, thread-safe data store access and resource management.

In the sequential case, the relationships between algorithms are encoded implicitly in their pre-determined execution order. For parallel processing, these relationships need to be expressed explicitly, in order for the scheduler to be able to exploit maximum parallelism while respecting dependencies between algorithms. Therefore, means to express and automatically track these dependencies need to be provided by the framework.

In this paper, we present components introduced to express and track dependencies of algorithms to deduce a precedence-constrained directed acyclic graph, which serves as basis for our algorithmically sophisticated scheduling approach for tasks with dynamic priorities. We introduce an incremental migration path for existing experiments towards parallel processing and highlight the benefits of explicit dependencies even in the sequential case, such as sanity checks and sequence optimization by graph analysis.

## 1. Introduction

Data analysis in HEP is a extremely compute intense task. Experiments have build elaborate software packages to reconstruct physical objects from the raw detector readout. These packages feature a wealth of algorithms for hit and track reconstruction, particle flow, analysis of physical objects, etc. A supporting framework is responsible for facilitating the execution of these algorithms, together with providing access to required data – such as detector geometry or magnetic field description – and services, e.g. histogramming and messaging. Conceived in the mid-1990s, these frameworks were designed to sequentially process one event, leveraging the embarrassingly parallel nature of HEP data through multiple jobs.

With higher collision energies, the complexity of the recorded events – and thus the time required for reconstruction – increases super-linearly. At the same time, CPU clock speeds have

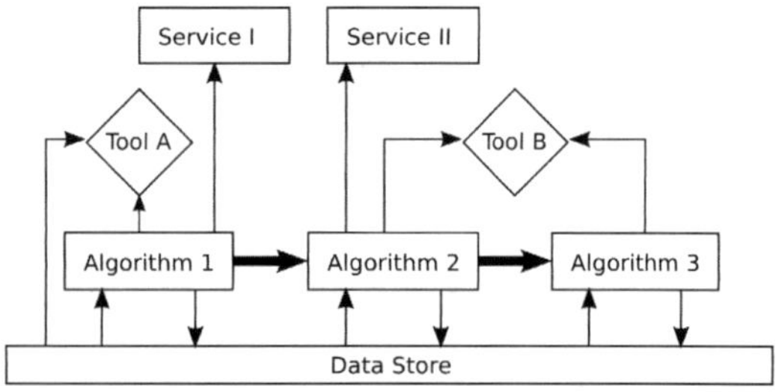


(a) Conceptual interaction of algorithms and tools with the data store.

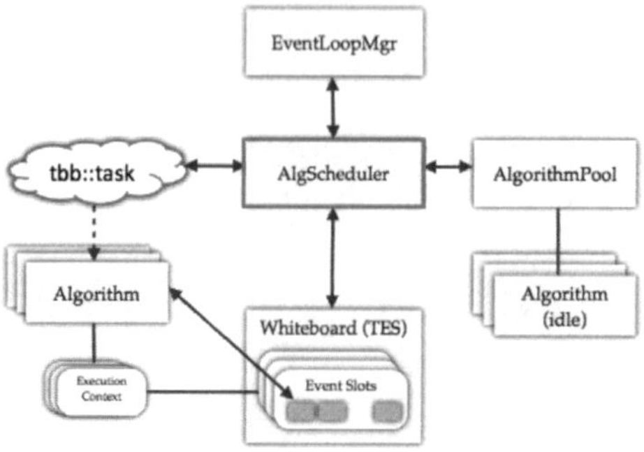


(b) Overview of Gaudi's components for concurrency [4].

Figure 1: Architectural overview of the Gaudi framework.

reached the power wall, putting an end to the "free-lunch" era [1]. Instead, multi-core processors need to be *actively* exploited, requiring a redesign of the executed software for parallel processing. Moreover, the $^{\text{memory}}/_{\text{core}}$ ratio is decreasing – particularly for many-core architectures, thus rendering the one-process-per-core approach unfeasible [2]. This gives rise to the need for more fine grained parallelization. This entails handling several events concurrently within one process – inter-event parallelism – as well as simultaneously executing independent algorithms for one event – intra-event parallelism. The treatment of intra-algorithm parallelism, i.e. concurrent processing of several physical objects, is beyond the scope of this work.

In this paper, we present concepts and components developed to allow inter- and intra-event level parallelism in the Gaudi framework [3].[1] Gaudi is an experiment independent framework offered by CERN, used for instance by the LHCb and ATLAS collaboration. Section 2 introduces the basic concepts of the framework. An overview of the developed components for concurrency is given in Section 3. Sections 4 and 5 give details on the implementation of data handles for automatic in- and output tracking and the scheduling of concurrent algorithms, respectively. The paper is concluded with a summary and an outlook on future work in Section 6.

## 2. The Gaudi Framework

Gaudi was originally developed as framework for the LHCb collaboration, but flexibility and adaptability to other experiments were among the primary drivers of all the design decisions involved [3]. The framework defines interfaces for *algorithms*, *tools* and *services*, that are loosely coupled encapsulations of work, communicating via data store(s), refer to Figure 1a. *Algorithms* constitute the main processing entity, transforming input data objects read from the data store to output objects placed in the store. Furthermore, they produce a binary decision, whether the data processed data is *accepted* or *rejected*. *Tools* encapsulate functionality reusable by many algorithms. A tool instance can either exclusively belong to one algorithm – private tool – or be owned by the tool service and shared among several algorithms – a public tool. Tools may retrieve or store objects in the data store. *Services* are managed by the framework and provide functionality to all algorithms and tools.

As mentioned above, algorithms are loosely coupled via the data store. They are arranged in sequences to ensure that (i) every algorithm finds its required input in the data store and (ii) only required algorithms are executed for an event. The latter is determined by the binary

[1] `http://proj-gaudi.web.cern.ch`

decision produced by the algorithms, which can lead to the premature termination of a sequence. Complex execution flows can be modeled by composing sequences with *and* or *or* behavior, with or without early termination. The configuration of an execution workflow is prescribed by a Python file, which defines the sequences and sets properties of the executed algorithms.

Gaudi was designed to leverage parallelism through multi-process execution. With more CPU cores available but decreasing memory (bandwidth) per core, the need arises for fine-grained inter- and intra-event level parallelism within one process. Thus, the *Concurrent Gaudi* project was started [5]. The following section gives an overview of the concepts and components introduced for parallel processing.

## 3. Gaudi Components for Concurrency

Goal of the Concurrent Gaudi project is the support of inter- and intra-event level parallelism. Different concepts need to be introduced to support these two levels of concurrency, refer to Figure 1b for an overview of the components described in the following:

*Inter-event level parallelism* requires a multi-slot data store to hold data products for simultaneously processed events – called "Whiteboard" in Concurrent Gaudi. To provide a minimally intrusive solution, algorithms interact with this data store using the same interface as for sequential processing, with the appropriate slot transparently selected by the framework prior to algorithm execution. Events are read by the event loop manager from the input file and handed over to the scheduler. For inter-event level parallelism, the scheduler can follow the sequential logic with the addition that executing an algorithm of a sequence requires the acquisition of an algorithm instance from the algorithm pool. The pool delivers and collects algorithm instances prior and after execution, respectively. For non-*clonable* algorithms, i.e. algorithms that use a unique resource, only one instance is present in the pool, thus creating a serialization point. Clonable algorithms have a configurable number of instances and therefore can be executed simultaneously for many events. Intel Threading Building Blocks (TBB) was chosen as multi-threading library for Concurrent Gaudi [6]. Therefore, the scheduler wraps the obtained algorithm instance in a TBB task and submits it to the TBB runtime. [4, 7] give a more detailed account of the presented components.

*Intra-event level parallelism* requires a more sophisticated scheduling strategy than inter-event level parallelism but promises increased speedups, as independent sub-workflows can be executed concurrently. The scheduler therefore needs to be aware of the relationships between algorithms. These manifest themselves in data dependencies and control flow restrictions, due to the sequence composition. In the sequential Gaudi framework, data dependencies were implicitly modeled by sequences. However, for parallel execution they need to be expressed explicitly. Therefore, data handles are introduced to the framework, described in detail in the following section. Section 5 discusses the scheduling strategies used for intra-event level parallelism.

## 4. Data and Tool Handles

The required inputs of an algorithm must be known to the scheduler after initialization. To facilitate an automatic tracking of read and written data objects, data handles are introduced to the Gaudi framework. These smart pointers contain the name and location of the required/supplied data product, read/write designator and optional flag. Furthermore, for read handles, alternate locations can be supplied. During the processing, one data object can be copied from one location to another, e.g. the raw detector readout might be copied from its original location in the data acquisition sub-tree of the data store to the reconstruction sub-tree, in order to have it available in files only containing the latter tree. With alternate locations,

Listing 1: Algorithm definition using data and tool handles.

```
class MyAlgorithm : public GaudiAlgorithm{

private:
  DataObjectHandle<LHCb::Tracks> m_tracks;
  DataObjectHandle<LHCb::Tracks> m_filteredTracks;

  ToolHandle<ITrackExtrapolator> m_extrapolator;
  ToolHandle<IMaterialLocator>   m_materialLocator;

public:
    MyAlgorithm( ... ) : GaudiAlgorithm( ... ) {
      declareInput("Tracks", m_tracks,
                   LHCb::TrackLocation::Default);
      declareOutput("FilteredTracks", m_filteredTracks,
                    "Analysis/FilteredTracks");

      declarePrivateTool(m_extrapolator, "TrackLinearExtrapolator");
      declareProperty("TrackExtrapolator", m_extrapolator); // optionally make it a property

      declarePublicTool(m_materialLocator, "DetailedMaterialLocator");
    }

    void execute() {
      LHCb::Tracks * tracks = m_tracks.get();
    }
};
```

Listing 2: Corresponding Python configuration for algorithm defined in Listing 1.

```
myAlg = MyAlgorithm('AnalysisFilter')
myAlg.Inputs.Tracks.Path = 'Skim/Tracks' # use pre-filtered tracks
myAlg = MyAlgorithm('AnalysisFilter')
myAlg.TrackExtrapolator.Iterations = 1 # rough estimate
```

a data object is first tried be found within the reconstruction sub-tree and – if unsuccessful – then in the data acquisition one, possibly requiring the opening of another file. All properties of a data handle can be customized in the Python configuration file of the workflow, therefore the final data flow can only be determined after the configuration has been processed.

When an algorithm is initialized, i.e. its configuration is processed, the data handle registers itself with the framework. The scheduler – being initialized after all algorithms – queries all algorithms for their in- and output handles and computes the data flow graph. Section 5 elaborates on the use of this graph for scheduling.

Besides algorithms, tools may also retrieve objects from the data store. Thus, an algorithm using a tool inherits its data dependencies. Therefore, we introduced tool handles to the Gaudi framework, to track the usage of tools by algorithms and to properly propagate the in- and outputs of tools to their callers. Accordingly, tools must also use data handles to interact with the data store to benefit from the automatic input/output tracking. Tool handles are smart pointers, containing name and type of the tool as well as a public/private designator. For private tools, properties of the tool can be configured through the parent algorithm in the Python configuration of the workflow. During the initialization of an algorithm, its tools – private and, if required, public – are initialized as well, therefore all input requirements are available at the time the scheduler computes the data flow graph.

Listing 1 exhibits a simple class, making use of the novel data and tool handles. The accompanying Listing 2 illustrates, how properties of data handles and private tools can be

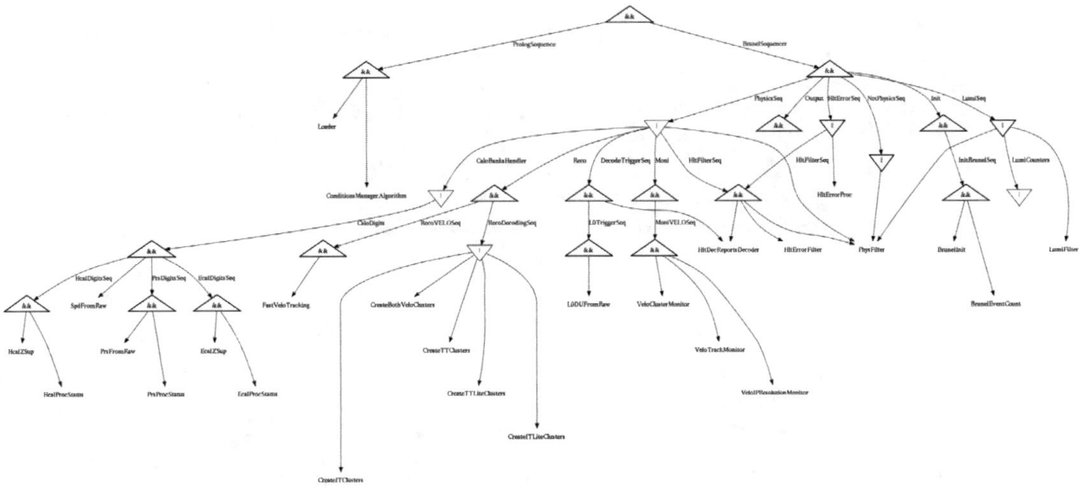

Figure 2: Sequence composition of the MiniBrunel workflow. Sequences are denoted by triangular shapes, the labels correspond to the appropriate C language operator.

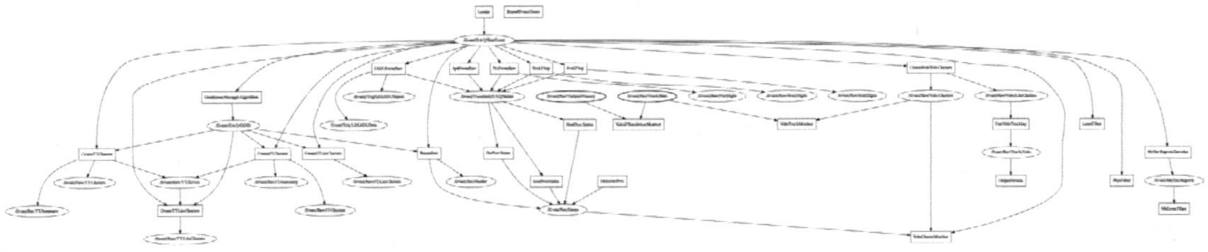

Figure 3: Data flow of the MiniBrunel workflow [7]. Algorithms are denoted by rectangular nodes, data objects by ellipses.

manipulated in the Python configuration of the worklfow. The syntax for declaring data and tool handles resembles the syntax for declaring properties of algorithms and tools and is therefore familiar to Gaudi users. To further ease migration, tools retrieved via the `tool<T>()` convenience method of `GaudiCommon`, which is widely used to access tools in the users' code base, also benefit from the automatic data dependencies propagation to the caller. For in/output tracking however, manual labor is required to introduce data handles to the user code.

The following section details the role of the data flow graph for intra-event parallel scheduling.

## 5. Scheduling

Scheduling of work is a crucial ingredient for multi-threaded applications. Gaudi's scheduling is twofold: (i) selection and order of algorithms to submit as task to the work queue and (ii) scheduling of work queue items on available processor cores. In this section, we detail the the former, while the latter is addressed by the TBB runtime [6].

As previously stated, a scheduler for parallel environments requires explicit knowledge of the workflow to be scheduled. The *control flow* stems from the composition of sequences of algorithms, as explained in Section 2. Figure 2 depicts the control flow for a subset of the LHCb Velo reconstruction, referred to as "MiniBrunel".[2] As seen in the figure, many sequences

[2] The full LHCb reconstruction is called "Brunel".

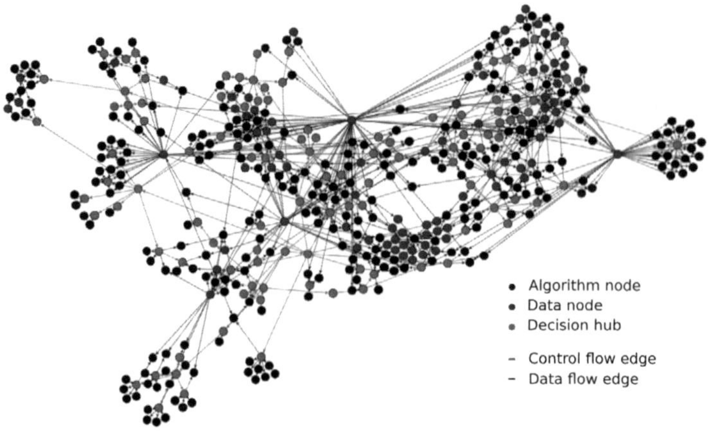


Figure 4: Unified execution flow graph of the entire Brunel workflow.[4]

are configured with early-return, thus the algorithms of a sequence are marked as executable one after the other. This allows for the first algorithm to act as a guard – or *filter* – for the subsequent ones; however, it severely limits the potential for parallel execution.

In addition to being marked as executable by the control-flow, an algorithm can only be scheduled if all its required input data objects are present. The data flow graph for the MiniBrunel sequence is shown in Figure 3. Both flows determine when a particular algorithm is executable, therefore we combine them into one unified execution flow graph, a precedence-constrained directed acyclic graph. Figure 4 shows the unified graph for the Brunel reconstruction sequence, with nodes for algorithms, data objects and decision hubs (stemming from sequences) as well as edges for control and data flow.

The scheduler maintains a list of all executable algorithms. After the execution of an algorithm, its binary decision is propagated along the control flow edges of the graph, updating decision and algorithm nodes accordingly. Additionally, its output data objects are marked as available, with the corresponding data nodes notifying their consumers along the outgoing data edges. If all inputs of an algorithm are available and its control flow requirements are fulfilled it is added to he list of executable algorithms.

As mentioned in the previous section, data handles can be marked as optional. For optional output data handles, the scheduler checks the presence of the object in the data store after the algorithm has finished.[3] For optional input objects, the algorithm execution is deferred as long as the optional object is still *producible*, i.e. its producing algorithms have not been executed yet, but are still reachable from an undecided decision hub.

The unified graph provides comprehensive insights into the workflow at hand, information that can be exploited for several purposes. The out-degrees of an algorithm's output data nodes determine its scheduling priority, as a high degree enables many further algorithms to run. If an algorithm is no longer required to be executed due to the control flow, its priority becomes zero. The critical path – i.e. the longest dependency chain – can be identified in the graph and extra priority given to the algorithms belonging to it, thus reducing the makespan. The graph also reveals the maximum concurrency in the workflow, i.e. the maximum number of algorithms executable at any given time. This bounds the number of occupiable threads, thus aiding in the resource planning in computing centers. However, pure graph analysis can *not* reveal the

[3] The check is omitted for non-optional output handles for performance reasons.

[4] brunel2012magdown workflow

obtainable speedup for a workflow, as the speedup depends on the relative time spent in each algorithm. Therefore, the number of threads that should *sensibly* be allocated to a process, can only be determined using runtime information. Further *static analysis* can be applied to identify unfulfillable data dependencies and superfluous control flow constructs.

## 6. Summary

The concurrent Gaudi project provides components and concepts to the experiments to introduce inter- and intra-event level parallelism in their software workflows. As non-intrusiveness is one of the primary goals of the project, all introduced components present an option to the user, while leaving their sequential production code intact. This allows for an incremental revision of the existing implementation and adoption of the components offered. In order to process data concurrently the user must (i) explicitly declare data in- and outputs for each algorithm, (ii) declare which tools are used by an algorithm, and (iii) revise thread-unsafe code, such as use of caches, back-channel communication, race conditions in the update of shared data, ...

The first two points can be achieved by using the newly introduced data and tool handles, respectively. To ease their use, the interfaces of the handles were designed according to components familiar to Gaudi developers. The third point requires a case by case analysis of the operations performed by the algorithms.

Having explicitly stated the algorithms' in- and outputs enables the automated deduction of the data flow among them. This allows for the computation of a unified execution flow graph, combining information from the algorithms' data dependencies and the control flow between them, stemming from their sequence composition. The graph is used to identify algorithms with both their data and control flow prerequisites fulfilled. These are submitted for execution by the scheduler, ordered by their priorities. The priority of an algorithm is determined by the connectedness of its output data objects in the execution flow graph, thus enabling maximal parallelism of the execution.

The graph structure also serves a secondary purpose, allowing the analysis of a given configuration statically for unfulfillable data dependencies and superfluous control flow constructs. Experiments can benefit from this analysis capabilities even without revising all their code for parallel execution by just introducing data and tool handles.

The gradual migration strategy and the benefits of static configuration analysis present a strong argument for the adoption of the components and concepts developed in the concurrent Gaudi project. Consequently, the LHCb collaboration has chosen to adapt these features into their production framework.

## References

[1] Sutter H 2005 *Dr. Dobb's Journal* **30** 202–210

[2] Borkar S 2007 *Proceedings of the 44th Annual Design Automation Conference* (New York, NY, USA: ACM) pp 746–749

[3] Barrand G *et al.* 2001 *Computer Physics Communications* **140** 45 – 55

[4] Clemencic M, Hegner B, Mato P and Piparo D 2014 *Journal of Physics: Conference Series* **513** 022013

[5] Hegner B, Mato P and Piparo D 2012 *2012 IEEE Nuclear Science Symposium and Medical Imaging Conference (NSS/MIC)* pp 2003–2007

[6] Willhalm T and Popovici N 2008 *Proceedings of the 1st International Workshop on Multicore Software Engineering* (New York, NY, USA: ACM) pp 3–4

[7] Clemencic M, Hegner B, Mato P and Piparo D 2014 *Journal of Physics: Conference Series* **513** 052028

Journal of Physics: Conference Series **608** (2015) 012022 doi:10.1088/1742-6596/608/1/012022

# Scalable cloud without dedicated storage

**D.V. Batkovich**[1], **M.V. Kompaniets, A.K. Zarochentsev**

Department of Theoretical Physics, St. Petersburg State University, Uljanovskaja 1, St. Petersburg, Petrodvorez, 198504 Russia

E-mail: batya239@gmail.com, mkompan@gmail.com, andrey.zar@gmail.com

**Abstract.** We present a prototype of a scalable computing cloud. It is intended to be deployed on the basis of a cluster without the separate dedicated storage. The dedicated storage is replaced by the distributed software storage. In addition, all cluster nodes are used both as computing nodes and as storage nodes. This solution increases utilization of the cluster resources as well as improves fault tolerance and performance of the distributed storage. Another advantage of this solution is high scalability with a relatively low initial and maintenance cost. The solution is built on the basis of the open source components like OpenStack, CEPH, etc.

## 1. Introduction

The Large Hadron Collider (**LHC**) is preparing for its second three-year run. Cool down of the vast machine has already begun in preparation for research to resume early in 2015 following a first long shutdown to prepare the machine for running at almost double energy in comparison to Run 1 . **ALICE** as the one of **LHC** experiments will need to process a greater amount of data. Even greater energy and hence data flow is planned for Run 3.

**ALICE** collaboration has plans to add a cloud infrastructure to the **GRID** facility before the start of Run 3. Cloud computing brings many positive aspects because it allows more efficient resource utilization and easy deployment methods. On one hand this allows one to balance the load between the projects, for example, you can easily balance the load between different LHC projects, or during **LHC** shutdowns you can use your cloud for different projects without having to re-install the software. On the other hand it is possible to easily maintain a number of different software versions for a long term data preservation.

At the present time **ALICE** infrastructure contains only **Tier1** and **Tier2** sites, this situation is connected with the fact that installation and configuration of **GRID** infrastructure is non trivial process that requires qualified personnel. At the same time there are a lot of small research groups which would like to contribute into **ALICE** IT infrastructure but do not satisfy even **Tier2** requirements. The main problem for such groups is that they have

- limited computational resources,
- limited finances,
- and lack of qualified IT personnel.

In this paper we offer a solution for **Tier3** sites [1] based on open-source software **OpenStack** [2] and **CEPH** [3]. This solution on one hand is intended to be easily installable,

[1] Speaker

maintainable and requires no highly qualified IT personnel, and on the other hand this solution is assumed to be scalable enough to be able to maintain growing amount of data.

## 2. Requirements for Tier3 clouds

The main requirement for all **TierX** clouds is to have a uniform interface. In our case **AMAZON EC2 API** [4] is chosen, due to the fact that main deployment method for **ALICE** IT infrastructure is assumed to be using **CernVM 3/CVMFS** [5, 6].

Because of the fact that **Tier3** sites have restrictions mentioned above (limited finances/computational resources), it is obvious that such a solution must be deployed on commodity hardware.

Another requirement is that all installation and maintenance procedures must be as easy as possible to allow non professionals to deploy the cloud.

The last requirement is that system must be at least minimally fault tolerant. **Tier3** sites are not supposed to store any significant data (only cache for data analysis), this means that cloud downtime and data loss must be minimized, but are not critical.

## 3. Storage configuration

One of the the cloud bottlenecks is a storage. The cheapest cloud solutions assume that there is some node which is used as a storage back-end (for example **OpenStack Cinder**[2] on **LVM**[3]). Such a solution cannot be considered as reliable one, because this is a single point of failure for storage and this configuration is not scalable enough due to concurrent IO from computation nodes.

As alternative one may consider some dedicated storage (hardware or software defined one). Hardware defined storage is very expensive and therefore not suitable for **Tier3** sites. On the other hand software defined storage shows reliable performance and fault tolerance. Another big advantage of the software defined storage is the number of available open-source solutions.

As a storage back-end for our solution we have chosen **CEPH** [3] based on following two considerations: first of all, **CEPH** is easily deployed software defined storage that provides distributed block devices which are extremely suitable for cloud infrastructure, second is that **CEPH** is already used in **CERN** as a storage back-end for **OpenStack**.

Additional positive aspects come from the fact that we want to maximally utilize resources and make our system as cheap as possible. As it was mentioned above, it is assumed that we run our cloud on commodity hardware (which implies that on each node we have the similar standard configuration including some CPU, RAM and HDD). The idea of the additional improvement comes from the fact that computation nodes are almost completely not using their HDD, while almost all storage nodes do not fully utilize their CPU. This leads us to the idea to combine computation and storage nodes. On one hand this allows us to utilize more disk space and make distributed storage more fast and fault tolerant, on the other hand this allows us to utilize CPU's from the storage nodes. All this allows one to utilize cluster resources more efficiently, but of course some careful tuning is still required to make computation and storage nodes work together on the same host without significant impact on performance. This leads us to the next layout of our configuration:

- Management servers: in most cases all of management responsibilities can be placed on one server. But due to the complex cloud network configuration and security requirements it was decided to setup the network management on a separate server. This server is also configured as a gateway between internal cloudnetwork and internet. As a result the management services are located on two separate hardware nodes.

[2] Cinder is a Block Storage service for OpenStack
[3] Logical Volume Manager. Cinder on LVM is most simple configuration of Block Storage service for OpenStack

- Worker(Computation/Storage) nodes: There must be at least 3 storage nodes (minimal **CEPH** configuration, see [3]). It is assumed, that when we setting up the worker nodes, we install both computation engine and storage engine, so that the number of computation nodes in the cloud infrastructure is the same as the number of storage nodes in the distributed storage. Of course, there is no reasonable limitation that requires to keep exactly the same number of computation and storage nodes. This configuration was considered in order to simplify the deployment procedure. In general case one can setup some nodes which run computation and storage engines simultaneously, and some dedicated nodes which run either computation or storage engine, but such a configuration is much more complicated than a homogeneous one.
- Network facilities: To setup cloud with **CEPH** back-end we need two separate networks. First is standard internal cloud network which is used as management network, network for VM's interconnect and as transport for accessing block devices. The second one is separate **CEPH** network for data replication. Additionally **CernVM3** requires access to **CVMFS** via caching proxy.

Among the existing cloud platforms we have chosen **CloudStack** and **OpenStack**. It should be noted that these solutions not just satisfied our requirements, but **OpenStack** was chosen because it is already used at **CERN** and has big community and support from corporate vendors, and **CloudStack** was chosen as a relatively simple and easily installable cloud solution. Both of them are free and do support **AMAZON EC2 API** and HTTPS (required to secure external connections to cloud infrastructure). Both cloud solutions have built-in **CEPH** support as well.

## 4. Choice of cloud platform

Historically **CloudStack** [7] was the first cloud platform we started to use with **CEPH**, but we found a lot of problems with setting it up in the required configuration. For example, it is not possible to configure **CloudStack AMAZON EC2 API** bridge in SSL mode. Additionally, automated deployment procedure (from XML-description) is non-functional since over a year now. Apart from that, there should be noted that CloudStack cannot be configured completely from the command line interface (CLI) and requires some manipulations in the web console. The main problem with **CloudStack** is that, in order to create a deployment script, one needs either to fix a number of **CloudStack** internal bugs or to write some nontrivial workarounds which allow one to setup CloudStack in required configuration without user intervention. Due to the facts mentioned above we have decided to continue to create our solution only with OpenStack as the cloud platform.

In contrary to **CloudStack**, **OpenStack** can be easily deployed from CLI, and installation procedure may be fully automated using **Packstack** [8] package based on puppet [9]. Additionally, there is no problem with configuring all external interfaces to use SSL.

**OpenStack** configuration we use consists from standard components **Glance** (image service), **Nova** (compute service), **Cinder** (block storage service), **Neutron** (networking service)[4] (see [2] for details). Neutron service can operate using different plugins, in our case we use default the Modular Layer 2 (**ML2**) plugin.

To estimate efficiency of proposed approach three different **OpenStack** configurations were created. All configurations has 3 compute nodes the difference is only in **Cinder** configuration:

(Conf. 1) Dedicated storage on **LVM**: **Cinder** is configured with **LVM** back-end on dedicated server

[4] It should be noted that OpenStack developers recommend to switch to Neutron networking service from legacy Nova networking.

(Conf. 2) Dedicated distributed storage: **Cinder** is configured with **CEPH** back-end running on additional 3 nodes

(Conf. 3) Distributed storage on compute nodes: **Cinder** is configured with **CEPH** back-end running on the compute nodes

**Packstack** includes modules for the most of **OpenStack** components. It means that at least 1st configuration (**Cinder** on **LVM**) could be deployed automatically. This can be easily done with one remark: current version of **ML2 Packstack** module has some bugs which do not allow to setup **ML2** networking correctly. To solve this problem we used custom shell scripts. It is also possible to install **Cinder** on **CEPH** using **Packstack**, but due to the fact that we want to setup computation and storage engine on the same worker node, we need a number of additional scripts which later on will be organized into a **Packstack** module.

## 5. Tests and results

As a performance test we have chosen IO benchmark **Bonnie++** [10]. It allows one to test disk and file system performance (block IO, latency, ...). We ran the tests for three **Cinder** configurations mentioned in the previous section[5]. Results are presented on Fig.1

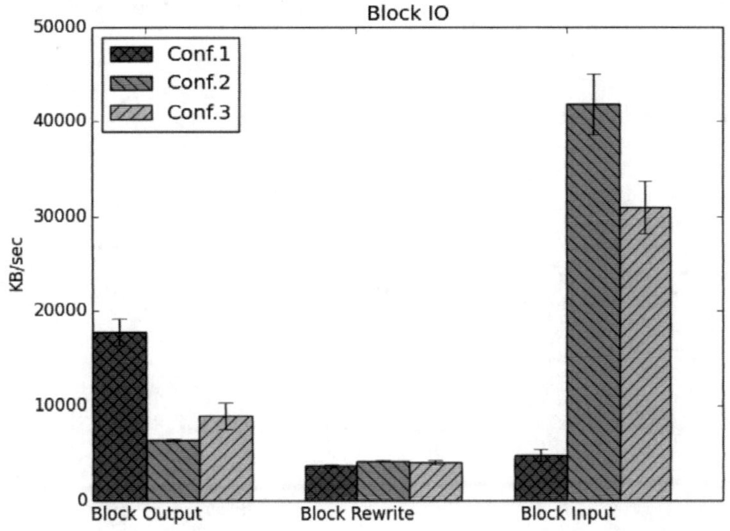


**Figure 1.** IO performance comparison for different configurations. Conf.1: Dedicated storage on **LVM**, Conf.2: Dedicated distributed storage, Conf.3: Distributed storage on compute nodes

At the y-axis on Fig. 1 one can see the IO speed in KB/sec. On the x-axis we have three groups: block output, block rewrite and block input. Bonnie++ block output is actually system's performance on write operations, block input - on read operations, and block rewrite is symmetric read and write test. Of course this is a synthetic test and it loads only disk IO (more realistic tests must load not only disk IO, but CPU and network also), but this benchmark shows some interesting tendencies.

[5] For tests we used Cisco Catalyst 4948 with 1Gb network and 4 Supermicro Twins. Each twin contains two servers with the following configuration: CPU 2 x Intel Xeon(R) E5420; RAM 16 GB; HDD Seagate Barracuda ES.2 250 Gb SATA-II.

At the "block output" section **Cinder** on **LVM**(conf.1) has best performance, due to the fact that in **CEPH** write operation is completed when data is replicated over whole **CEPH** cluster. While in "block input" we have directly opposite picture: **Cinder** on **CEPH** (conf.2 and 3) shows much better performance. And finally, for symmetric reads and writes (block rewrite) advantages of fast reads for **CEPH** seems to be compensated by slow writes.

As for different configurations with **CEPH**: dedicated distributed storage (conf.2) and distributed storage on compute nodes (conf.3) one can see that difference between performance is close to errors of measurements so we can say that these configurations have approximately the same IO performance.

As the summary we can say that with proper resource allocation between **CEPH** and **OpenStack** that runs on the same nodes we can achieve performance similar to the performance of standalone **CEPH** but with lower cost. This optimal ratio can be different for specific problems. Hence for the next step we have plans to deploy cluster following proposed scheme and to run real problem instead of synthetic **Bonnie++** tests. Moreover we must perform high load runs and test all of 3 configurations on different hardware.

## 6. Acknowledgments

The authors are grateful to Predrag Buncic for helpfull discussions and advises, to ACAT'14 Organizing Committee for support and hospitality, to Saint-Petersburg State University for a research grant 11.38.197.2014.

## References

[1] The Grid: A system of tiers `http://home.web.cern.ch/about/computing/grid-system-tiers`
[2] OpenStack `http://www.openstack.org/`
[3] CEPH `http://ceph.com/`
[4] Amazon EC2 `http://aws.amazon.com/documentation/ec2/`
[5] CernVM File System `http://cernvm.cern.ch/portal/filesystem`
[6] CernVM 3 and mCernVM Beta Release `http://cernvm.cern.ch/portal/ucernvm`
[7] CloudStack `http://cloudstack.apache.org/`
[8] Packstack `https://wiki.openstack.org/wiki/Packstack`
[9] Puppet `http://puppetlabs.com/`
[10] Bonnie++ `http://www.coker.com.au/bonnie++/`

# Towards a high performance geometry library for particle-detector simulations

**J Apostolakis[1], M Bandieramonte[2], G Bitzes[1], R Brun[1], P Canal[3], F Carminati[1], G Cosmo[1], J C De Fine Licht[1], L Duhem[4], V D Elvira[3], A Gheata[1], S Y Jun[3], G Lima[3], T Nikitina[1], M Novak[1], R Sehgal[5], O Shadura[6] and S Wenzel[1†]**

[1]European Organization for Nuclear Research (CERN) - Geneva, Switzerland
[2]University of Catania and INAF (IT)
[3]Fermi National Accelerator Laboratory (US)
[4]Intel Corporation
[5]Bhabha Atomic Research Center (IN)
[6]National Technical University of Ukraine, Kyiv Politechnic Institute
[†]corresponding author

E-mail: sandro.wenzel@cern.ch

**Abstract.** Thread-parallelisation and single-instruction multiple data (SIMD) "vectorisation" of software components in HEP computing has become a necessity to fully benefit from current and future computing hardware. In this context, the Geant-Vector/GPU simulation project aims to re-engineer current software for the simulation of the passage of particles through detectors in order to increase the overall event throughput. As one of the core modules in this area, the geometry library plays a central role and vectorising its algorithms will be one of the cornerstones towards achieving good CPU performance. Here, we report on the progress made in vectorising the shape primitives, as well as in applying new C++ template based optimisations of existing code available in the Geant4, ROOT or USolids geometry libraries. We will focus on a presentation of our software development approach that aims to provide optimised code for all use cases of the library (e.g., single particle and many-particle APIs) and to support different architectures (CPU and GPU) while keeping the code base small, manageable and maintainable. We report on a generic and templated C++ geometry library as a continuation of the AIDA USolids project. The experience gained with these developments will be beneficial to other parts of the simulation software, such as for the optimisation of the physics library, and possibly to other parts of the experiment software stack, such as reconstruction and analysis.

## 1. Introduction to VecGeom

Targeting the efficient use of vector- and in general SIMD microparallelism is becoming a necessity to fully exploit the capabilities of current and future computing hardware. This is mainly due to SIMD (single instructions multiple data) vector units becoming wider while the CPU clock rate is stalling at a more or less constant value. Unfortunately, only few existing software packages can directly benefit from this trend in hardware development (e.g., via a simple recompile statement) because in most cases the data is not organised in memory to be able to benefit from SIMD vector units, or worse there is no "vector" data to operate on in the first place. Fortunately, in particle-detector simulation we are in the first situation as there is an inherent “event-level” and ”track-level” parallelism thanks to which we can group data in “data vectors”. However, this requires a very substantial software development effort to recast existing simulation packages into a vector-friendly form. The Geant-Vector project (Geant-V) took up this challenge two years ago [1, 2]. The key idea of this R&D project is to group data from different events into vector data containers (named baskets) according to some locality criteria. One of the natural criteria is geometric locality – particles in the same (logical) volume are grouped into baskets, but other criteria might be applied as described in [3].

The concept of baskets can be seen as one of the necessary steps in simulation to benefit from vectorisation. A second, equally important, condition is to extend the interfaces and functionality of the major components of the simulation framework to provide an efficient basket processing. First of all, this implies the provision of new vector interfaces or API to the core functionality. Furthermore, it means the use of “horizontal” SIMD vectorisation (i.e., vectorisation over different particles) when processing the baskets.

One of the principal components of particle-detector simulation is the geometry modeler library. It provides the basic shape primitives used as building blocks to describe a detector and it offers basic geometry “ray-tracing” functionality required for the propagation of particles throughout the detector geometry model. The most important geometry tasks are a) collision detection between particles (rays) and shape primitives, b) determination if particles are inside or outside shapes, c) calculation of minimal (“safe”) distances to shape primitives and d) the determination of the distance to the exit point when leaving a shape primitive. Currently, a majority of software stacks in HEP uses one of the two wide spread libraries in form of the Geant4 [4] and ROOT(TGeo) [5] geometry modelers. Recently, the AIDA USolids project was started to converge to a single code base for the geometrical primitives in order to achieve better long-term maintainability of geometry algorithms and to foster further development of new code under a common umbrella to the benefit of both Geant4 and ROOT [6].

The present work, however, was motivated from the fact that none of these packages was designed with the requirement to process baskets, nor do any of those packages support compiling and executing geometry algorithms on a GPU. Hence, we are in a situation where those new features have to be developed almost from scratch. While this implies a major software development effort, it also brings with it the chance to modernise existing code, for example, by taking advantage of template features in C++ to increase performance and to write more generic and reusable code.

After our initial SIMD prototyping studies presented at CHEP2013 [7], our development called VecGeom (for vectorised geometry) is now merged with the AIDA USolids project [6] and seen as its evolution, as far as the geometric primitive code is concerned. Furthermore, VecGeom also offers geometry modeler components to actually describe hierarchical detectors and to perform scalar and vector navigation in these detectors. These parts go beyond the initial scope of the USolids project.

The following sections shortly report on the current progress of VecGeom starting with an overview of our software development pattern and a report of the implementation- and performance status thereafter.

## 2. Software engineering challenges

There are two major challenges to overcome in this project which are related to achieving good performance as well as good code maintainability. The first is the problem of how to achieve a reliable (multi-particle) SIMD vectorisation of our algorithms. In the past this often required platform specific code using assembly language or compiler intrinsics. However, in recent years this challenge is actually becoming a rather manageable task thanks to the availability of new high-level C++ (templated) wrapper libraries which hide away the complexity of SIMD hardware-instructions by providing user-friendly C++ types and operators [8–10].

Using such libraries of wrapper types, writing vectorised code is *often* no different to writing scalar code apart from using slightly different variable types (apart from some different treatment of conditional execution etc.). This is in particular true for multi-particle vectorisation for which we currently employ the Vc library [9], chosen for it's ease of use and the very good performance obtained in our initial vectorisation study [7].

The second challenge is related to software maintenance due to potential code duplication. Our two primary goals of providing new multi-particle vectorised geometry algorithms, as well as porting the geometry code to CUDA (or other GPU languages), in addition to the existing scalar code, could potentially lead to a code multiplication problem. This brings the danger of a significantly increased maintenance effort in the long term. To illustrate this, note that a brute force approach of just extending the code base in addition to already existing scalar code would imply having roughly 200 new functions to maintain for the approximately 20 supported primitive solids (box, sphere, tube,...).

Clearly, we do not want to follow this brute-force path. Rather, we make particularly extensive use of the generic programming pattern, possible in C++ with template classes and template functions. These template functions are then used to compile into specialised binary code for the scalar interface, the multi-particle interface, or the GPU kernel, as illustrated in Fig. 1. We refer the reader to our git repository for real code examples [11].

This approach is motivated by the following arguments:

- Scalar-particle and multi-particle code is very similar in many cases. A generic description of the algorithms in form of C++ template functions, using scalar or vector types as template parameters, is therefore able to describe both the scalar and multi-particle algorithms of the geometry library.
- Whenever the algorithms are not similar, either the explicit template specialisation approach, or the use of C++ traits and policy techniques [12, 13] which permit the specialisation of certain code regions, can be used.
- GPGPU computing follows the SIMD paradigm, so whenever we have developed a multi-particle algorithm, this should be the basis for a well performing GPU kernel.
- The CUDA nvcc compiler is now fully C++11 compliant having no problems to understand templates and is able to compile standard C++ code into GPU device kernels.[1]

## 3. Implementation status, new features and performance examples

### *3.1. Primitive solid types*

Following our initial vectorisation demonstrator that started mid 2013, our program of work consisted in refactoring the core geometry algorithms for the solid primitives using generic templated algorithms and we based our development mostly on the algorithms available in USolids [6].

This process involves a major development effort but the benefits are already clearly visible. Currently, we have achieved generic implementations for the following shape primitives used in detector construction: box, sphere, torus, tube(segments), cone(segments), trapezoids, symmetric trapezoids (trds), parallelepiped, paraboloid, hyperboloid. In all these cases, we are able to generate well performing SIMD kernels for the treatment of baskets and in all the above mentioned cases the same generic code is used for the scalar API with improved or equivalent performance compared to the original USolids code. Moreover, the same shape-primitives are now trivially available for detector construction on the GPU.

[1] The situation with OpenCL is not as good but more advanced standard and compilers (for example SYCL) are coming up which hopefully will make OpenCL accessible to us without the need to develop separate OpenCL code.

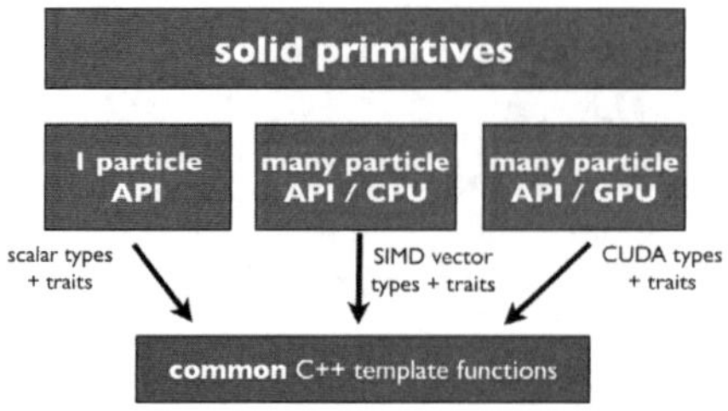


**Figure 1.** Diagram illustrating our software development approach. Templated C++ functions are written using abstract types and abstracted properties (traits). The different API then instantiate the template functions with concrete types and properties and the compiler produces different binary code in each case. In case of the multi-particle API, we currently inject SIMD vector types coming from the Vc library.

We should emphasise that next to these benefits, the process of refactoring the algorithms also has given us the opportunity to introduce some other optimisation or new constructs to the geometry library. Notably, these are:

- Improved algorithms, in particular via the elimination of many trigonometric functions often found in Geant4/ROOT algorithms.
- The use of template class specialisation to create specialised geometry objects from generic templates. Template shape specialisation is particularly useful for solid classes that can assume many different realisations. Traditionally, for example, the tube and cone shapes are each modeled by one C++ class but they can come in many different variations: hollow tubes and cones, full tubes and cones or hollow tubes/cones with a cut. This implies the presence of many conditional branches in the code. Template shapes specialisation helps avoiding these branches at runtime while keeping just one C++ implementation for all shape variations.

Taking all points together, our development effort has allowed us to maintain roughly the current code size in terms of the number of lines of code, to provide SIMD vectorised (or "GPUised") algorithms for multi-particle treatment *and* to increase the performance of the traditional scalar algorithms at the same time. This is for instance demonstrated in Fig. 2 which shows benchmark data for the case of the tube-segment shape which is one of the most important shapes in detector construction. From the point of view of the Geant-V project, the most important number is the speedup (generally between 2 and 3) due to the SIMD vectorised treatment of baskets which is promising. We see similar SIMD improvements for all our currently implemented shapes. Note that detailed benchmarks on the GPU are not yet included in this plot as this work is still in progress.

### *3.2. Other geometry components*

Beyond generic implementations at the shape level, VecGeom also includes a first portable implementation of a geometry navigator which is able to process single particles as well as baskets of particles on CPU and GPU. As a foundation layer to this, we provide an in memory representation of the detector which is synchronisable (copyable) between CPU and GPU. As a result, VecGeom can today serve as a minimalistic[2] replacement of the complete Geant4 or ROOT/TGeo geometry modellers. Indeed, we are able to do full particle-simulations in the

[2] In the sense that we do yet provide sophisticated navigation algorithms making use of spatial optimisations often used in computer graphics and game engines.

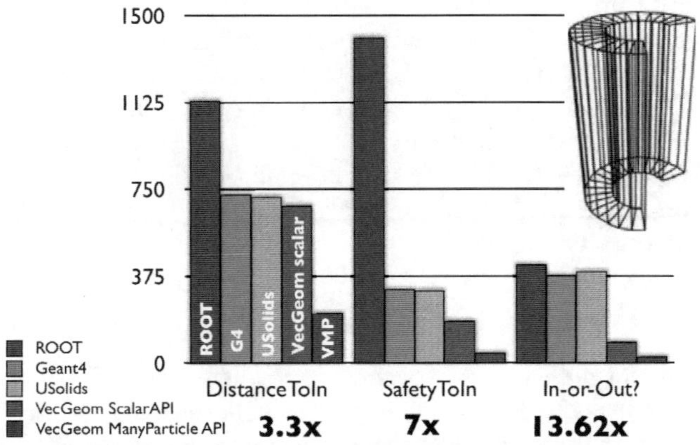


**Figure 2.** Performance numbers (CPU time in arbitrary units) for a tube-segment shape for three of the most essential algorithms from the Geant4, ROOT, USolids and the new generic VecGeom implementations. The two important features here are improved scalar performances of VecGeom over USolids (from better algorithms and template specialisation techniques) as well as excellent SIMD vectorisation results for the multi-particle treatment. The bold numbers indicate the total speedup as compared to the current implementation of the USolids library. The benchmark was performed using 1000 particles; We used gcc-4.7, ROOT5.34.18 and Geant4-10p01 in their release builds. The benchmark machine is a Intel(R) Core(TM) i7-3770 CPU@3.40GHz using the AVX instruction set.

Geant-V prototype either using TGeo or VecGeom. In a very simple detector simulation (based on the original Geant4 example N03, using only boxes), the total runtime improvement when using VecGeom instead of TGeo was measured to be roughly 40% (see [3] for more information). This is a first very promising number which is mainly due to a SIMD treatment of baskets in the geometry part of the simulation. This implies that together with a future vectorised treatment of the physics routines we hope to achieve considerable better overall performance gains. Of course, it is not possible to extrapolate this first result to realistic detector setups but we shall be able to provide those comparisons, together with the ability to represent more complex geometries, in the near future.

### *3.3. Ongoing work and outlook*

We are moving towards generic implementations of the more complicated shape primitives often used in HEP experiments. These are notably the polyhedra and polycone shapes. Due to their complex nature it might not be evident nor possible to achieve good SIMD vectorisation for the many-particle API. In this case, a good strategy seems to be to apply internal vectorisation to speed up the original algorithms due to the presence of internal loops over regular structures in those shapes. We have already achieved promising results for the polyhedra and generic trapezoid shapes.

Another area addresses complex shapes in form of boolean solids which follow the constructive solid geometry (CSG) approach to build more complicated solids out of primitive constituents. Here we made first prototypes replacing the current virtual function based CSG approach with a templated (policy like) approach allowing the compiler to perform better code optimisation for particular boolean solid instances. Concrete results will be presented elsewhere.

Journal of Physics: Conference Series **608** (2015) 012023 doi:10.1088/1742-6596/608/1/012023

## 4. Discussion and Summary

The present paper describes our effort to improve the performance of geometry libraries in detector simulations and to enable them for the new use cases in the context of multi-particle processing on CPU and GPU. We have made considerable progress towards this goal and we see that all our new implementations are currently faster (or equally fast) compared to all previous implementations, that they vectorise to the expected level and that they all use the same generic code implementation. We believe that a complete fully vectorised and otherwise optimised geometry library can be achieved soon.

One of the remaining challenges is how the vectorised solid algorithms can be used in the context of very complex detectors where optimised spatial tree-structures (or voxelisation) are used in the navigator.

Beyond the development in detector geometries, the present project has allowed us to gain invaluable experience in vectorising algorithms as well as in addressing software and portability issues, that will – without a doubt – help to tackle other developments in the Geant-V project and elsewhere.

## References

[1] Apostolakis J, Brun R, Carminati F and Gheata A 2012 *Journal of Physics: Conference Series* **396** 022014

[2] Apostolakis J, Brun R, Carminati F, Gheata A and Wenzel S 2014 *Journal of Physics Conference Series* **523** 012004

[3] Apostolakis J, *et al.* Adaptative track scheduling to optimize concurrency and vectorization in geantv (ACAT2014; this proceedings)

[4] Allison J, *et al.* 2006 *Nuclear Science, IEEE Transactions on* **53** 270–278 ISSN 0018-9499

[5] Gheata A URL `ftp://root.cern.ch/root/doc/18Geometry.pdf`

[6] Gayer M, Apostolakis J, Cosmo G, Gheata A, Guyader J M and Nikitina T 2012 *Journal of Physics: Conference Series* **396** 052035

[7] Apostolakis J, Brun R, Carminati F, Gheata A and Wenzel S 2014 *Journal of Physics Conference Series* **513** 052038

[8] Esterie P, Gaunard M, Falcou J, Lapresté J T and Rozoy B 2012 Boost.simd: generic programming for portable simdization *PACT* ed Yew P C, Cho S, DeRose L and Lilja D J (ACM) p 431 ISBN 978-1-4503-1182-3

[9] Kretz M and Lindenstruth V 2012 *Software: Practice and Experience* **42** 1409

[10] Fog A Vcl, c++ vector class library URL `http://www.agner.org/optimize/vectorclass.pdf`

[11] URL `http://git.cern.ch/pub/VecGeom`

[12] Myers N June 1995 "a new and useful template technique: Traits"

[13] Alexandrescu A 2001 *Modern C++ Design* (Addison-Wesley)

# Using Functional Languages and Declarative Programming to analyze ROOT data: LINQtoROOT

**Gordon Watts**

Professor of Physics
Box 351560
Department of Physics, University of Washington
Seattle, WA, 98195-1650
USA

E-mail: gwatts@uw.edu

**Abstract**. Modern high energy physics analysis is complex. It typically requires multiple passes over different datasets, and is often held together with a series of scripts and programs. For example, one has to first reweight the jet energy spectrum in Monte Carlo to match data before plots of any other jet related variable can be made. This requires a pass over the Monte Carlo and the Data to derive the reweighting, and then another pass over the Monte Carlo to plot the variables the analyser is really interested in. With most modern ROOT based tools this requires separate analysis loops for each pass, and script files to glue to the results of the two analysis loops together. A framework has been developed that uses the functional and declarative features of the C# language and its Language Integrated Query (LINQ) extensions to declare the analysis. The framework uses language tools to convert the analysis into C++ and runs ROOT or PROOF as a backend to get the results. This gives the analyser the full power of an object-oriented programming language to put together the analysis and at the same time the speed of C++ for the analysis loop. The tool allows one to incorporate C++ algorithms written for ROOT by others. A by-product of the design is the ability to cache results between runs, dramatically reducing the cost of adding one-more-plot and also to keep a complete record associated with each plot for data preservation reasons. The code is mature enough to have been used in ATLAS analyses. The package is open source and available on the open source site CodePlex.

## 1. Introduction

The LINQtoROOT project is a small piece of a new way of performing physics analysis, though it is technically the most difficult and central to realizing the vision. LINQtoROOT, an open source project, has been around for a few years undergoing sporadic development. This paper gives a brief overview of the LINQtoROOT project, recent developments motivated by its use in an ATLAS analysis, and finally a look towards the future of the project.

Physics analysis is still performed much the same way it was 15 years ago, though with larger datasets, better networking, and faster computers. The recent trend in software engineering has emphasised continuous integration, testability, keeping performance records over time, and repeatability. Physics analysis could benefit from this. If everything needed for a physics analysis were stored in a source code repository it could be automatically re-run each time a check-in occurred. The results could be automatically cached. There are a number of things that make this vision impractical – in particular the many hour run-times often required, the TB sized datasets that must be accessed, the

Journal of Physics: Conference Series **608** (2015) 012024 doi:10.1088/1742-6596/608/1/012024

complexity of running all parts of the analysis. Running in this mode was not an initial goal of the LINQtoROOT project, but it has turned out to be a powerful side-effect of it.

## 2. The LINQtoROOT Project

The project grew out of the author's general frustration with the steps required in modern collider physics analysis. Multi-pass runs on data files and Monte Carlo (MC) to derive scale factors that are later applied are not uncommon. Any change in selections cuts in an analysis will often require carefully re-running all the steps that use those selection cuts; especially tricky in a multi-dataset analysis.

The modern toolset to accomplish this requires familiarity with several different languages: bash and python to control the analysis, C++ to loop over the ROOT data files and generate plots, and CINT macros to produce the final files. LINQtoROOT attempts all of this with a single language.

Along with the goal of making analysis simpler, there were several requirements – mostly for practical reasons. First, all data files had to be in the ROOT format – no translation. Second, the processing had to be fast – the code must be C++. Third, it must be possible to run locally or remotely (e.g. PROOF).

This tool, written in Microsoft's C# language, was first presented at the CHEP 2012 conference [1]. The tool has developed quite a bit since then, however the conference paper detailing the inner workings of the tool is still accurate. An introduction to the tool is perhaps best served by a quick look at a small two pass analysis file:

```
var output = new FutureTFile(@"..\..\..\..\MCJ1-MCQCDStudyResults.root"));
var eventsMC = MCQCD.QueryableCollectionTree.CreateQueriable(dsMC);
var ptMC = eventsMC.SelectMany(e => e.Jets).FuturePlot("RawPtMC", "Raw Pt MC; pT",
     100, 0.0, 200.0, j => j.pt()/1000.0).Save(output);
eventsMC.SelectMany(e => e.Jets).FuturePlot("RawPtMC", "Raw eta MC; eta",
     100, 0.0, 200.0, j => j.eta()).Save(output);

var eventsDATA = MCQCD.QueryableCollectionTree.CreateQueriable(dsDATA);
var ptDATA = eventsDATA.SelectMany(e => e.Jets).FuturePlot("RawPtData", "Raw Pt Data”,
     100, 0.0, 200.0, j => j.pt()/1000.0).Save(output);
eventsDATA.SelectMany(e => e.Jets).FuturePlot("RawPtData", "Raw eta Data; eta",
     100, 0.0, 200.0, j => j.eta()).Save(output);

Helper.Divide(ptMC.Value, ptDATA.Value).Save(output);
```

Line 1 creates a ROOT output file where results will be written. Lines 2 and 5 open the MC and DATA datasets. These datasets can be multiple ROOT files chained together, or a dataset name on a PROOF server. Lines 3, 4, and 6 and 7 request that plots be made, and when they are done, that they be saved to the file. Note the `SelectMany(e => e.Jets)`: for each event this unrolls the jets, so it is the jet that is handed to the FuturePlot method. In FuturePlot's arguments, note the `j => j.pt()`. This is a C# lambda function that converts the jet into the jet's $p_T$ which is then plotted. The final line, 8, divides the histograms to create a ratio. The ratio could be used as input to another plot below if desired.

There is a subtly in this code that will be necessary to understand for discussion in later sections. The concept of futures, or promises, must be introduced to keep the analysis efficient [3]. Futures allow the system to accumulate many requests for plots without actually running a single one. When the actual value is requested, the LINQtoROOT library will process all known plots. This is key because almost all jobs are I/O bound, not CPU bound. When one considers a single pass through the file takes only 10 minutes, but one would need 500 passes to make 500 plots, it is easy to see why this some sort of facility like this must exist. Liens 3, 4, and 6, 7 do not make actual plots – just a promise to make a plot at some later time. Line 8's reference to the future's Value tells the backend it must now run over both datasets and produce an actual plot that can be used to calculate the ratio.

## 3. Using LINQtoROOT in a real analysis

This tool received extensive testing in an ATLAS physics analysis [4]. This tool was used by the author to accomplish a number of QCD background studies and cross checks in this analysis. None of the public plots were directly produced by this tool, but a good number appear it into the 200 page support note. Development since CHEP and the experiences described below are all from this experience.

## 4. General usage and performance

The performance of LINQtoROOT during the analysis is compared to the initial goals of the project. A good deal of further development was motivated by this analysis and that is described where appropriate.

### 4.1. Simplifying the boilerplate code in an analysis

There are several sources of boilerplate when building a physics analysis. At the lowest level there is the boilerplate that is required by the language: all the extra characters required to define functions, methods, classes, etc. Functional languages are perhaps the tersest, but this project required two language based features only available in C# when it was started (compile-time expression tree's and LINQ). The original paper contains extensive discussion over the language choice.

The second source of boilerplate is often required by the language and libraries themselves. For example, to declare, fill, and save a histogram using ROOT one ends up with source code spread throughout the analysis code: initialization to declare the histogram, code to fill it, and code to save it to the file. The design of LINQtoROOT means that all three occur in one line:

```
var allPlot = allJets
    .Select(j => j.Jet)
    .FuturePlot("ptAll", "AllMatchedJets", 100, 0.0, 200.0, j => j.pT/1000.0)
    .Save(output);
```

The *FuturePlot* method declares, books, and fills the histogram with the $p_T$ of each jet. Note that there is support to make a template histogram so every jet $p_T$ plot will have the same binning and limits. This was very successful and the code is much simpler than the standard C++ counterpart.

Efforts were also made to streamline the handling of input files. A separate text file lists the files and groups them under a single name – *test*, for example, or *mc* and *data*. That single name can be used to open the files in the code. By specifying a prefix like *proof* on the filename the LINQtoROOT backend would invoke the PROOF runner.

There are also numerous helper functions to help manipulate and post-process plots. This is necessary due to the Future mechanism. If one wants to manipulate a plot without forcing an actual run, some infrastructure is required. Languages that support monads natively will have the language syntax built in. As an example of what has to be done in C# to manipulate a plot without triggering evaluation:

```
FutureHelpers.Apply(gPTPlot, allPlot,
    (num, denom) =>
    {
        var h = num.Clone(num.Name + "_fraction") as ROOTNET.NTH1;
        h.Divide(denom);
        return h;
    })
    .Save(output);
```

Two future references to plots, gPTPlot and allPlot, will be passed to the lambda function that divides the plots, and saves the result to the output file. The lambda function will be executed when both gPTPlot and allPlot are already run, and not before. Though these functions and routines do remove a bunch of boilerplate, the difficulty, coined “monad hell”, of manipulating these futures in a non-monadic way remains awkward and can clutter the code.

### 4.2. Translating C# to C++

This requirement was one of the main reasons C# was chosen: all manipulation of the TTree data from the file to the filling of the plot is given to the LINQtoROOT library in the form of an Abstract Syntax Tree (AST) [5]. The AST is data, and can be manipulated, or optimized. An AST is a typical intermediate representation in a compiler like gcc, for example. The AST can be translated into C++ code, or even shipped to a different machine with a different version of ROOT to run on a PROOF cluster.

A great deal of work has been done to optimize multi-plot runs. For example, after a complex selection of jets, one often wants to plot several quantities of those same jets (e.g. $p_T$ and $\eta$). Hand coding one would make sure to select the jets once, and then plot all the variables. The LINQtoROOT library gets separate AST's for each plot and must intelligently combine them. Often it isn't obvious how to combine and optimize them. For example, expensive intermediate expressions (e.g. calculating $\eta$ from $p_x$, $p_y$, and $p_z$) can be lifted out of inner loops; complex sub-expressions are often calculated the same way in different parts of a plot's AST that can't obviously be combined – unless they are pulled out and the results are cached. This is by far the most complex part of the library. The author has gained new respect for compiler-optimizer writers. This is an ongoing area of work.

The AST used by LINQtoROOT is custom written by the author. However, the .NET framework contains a rather sophisticated and extensible AST. Utilities are starting to appear to help with optimizing these .NET AST's: a long term project is to replace the custom AST with the .NET one to take advantage of other people's optimization efforts.

### 4.3. Iterative analysis

Iterative analysis means being able to quickly add a plot to the analysis without being forced to wait hours for the new result. The final analysis, with 500 or so plots, takes close to 6 hours to run from scratch. A single additional plot often took less than 10 minutes to run on a local machine accessing the files across a fast network. Iterative means the ability to add a new plot to the existing 500 and have the results in about 10 minutes.

The AST, as described above, enables this feature. The AST plus any external input data (e.g. the input files), is turned into a cache key, with the file system as the cache. This has been remarkably successful. In no case has the cache-lookup time ever approached a typical run-time on the large files that are typical in an analysis. This enables numerous other scenarios. Changing a cut early in the analysis cut-flow will automatically re-run any dependent plots. If one then changes the cut back, the old results are still available in the cache. If a dataset doesn't need to be touched, then nothing will be run.

The caching hasn't been completely straight forward. For example, there is special code to prevent a re-run if one changes the title of a plot, though binning and limits remain the same. And while the look up is very fast compared to running over data, it isn't fast compared to some other operations, as described below. There is undoubtedly room for improvement: no effort has been made to optimize this yet.

### 4.4. A single programming language

This has been remarkably successful in eliminating the polyglot world physics analysis is performed in. One of the biggest aggravation points, having to run on multiple datasets with different script files, and then further scripts to glue the results together, is greatly mitigated, if not eliminated, by LINQtoROOT. Running over two large datasets is as simple as making plots from a single dataset. This plus the caching mechanism has eliminated a great deal of the "mess" involved in a traditional analysis.

The caching mechanism, however, is not as fast as it needs to be for the final stage of the analysis. The way the code is written and the design of C# leads to a very modular and composable layout of the code. Normally considered a good thing, but it makes running over matrices of cuts very easy, which can quickly lead to generating 500 or 1000 plots. The caching mechanism needs a finite amount of time to determine if each plot has been previously run. Somewhere between 500 and 1000 plots this is long enough that an analyser's attention will start to drift. This is problematic in the sense that near the end of an analysis one re-runs the analysis many times changing only plot formatting. With a normal dedicated script accessing ROOT files containing histograms from a big run, one can do this in about a second or so.

The second problem turns out to be the code itself. As mentioned in the previous section, monad hell makes writing the final plot mechanisms awkward (especially in combination with the non-functional design of the ROOT histogram library).

As a result the PlotLingo language was designed, as briefly described in Section 6.1.

## 5. Enabling a continuously updating analysis

Software developers over the last decade have discovered continuous integration. Each time there is a check-in to the source code repository a build of the software is done and tests are run. Why not with an analysis? One of the main reasons not to, up to now, has been the complexity and length of time it takes to run an analysis – many hours. In the software industry tests are designed to run in seconds or minutes. So the analyser keeps very careful control over when and how an analysis is re-run.

The LINQtoROOT library can potentially change this equation by virtue of its caching mechanism. Re-running the analysis with one new plot can be quite fast. The author used the TeamCity build robot to monitor check-ins to the ATLAS svn analysis repository, and re-run the analysis each time one occurred. This gave a complete record of the source code changes, captured the resulting plots, etc. As a result if a plot changed it was easy to go back in time and figure out exactly where the plot had changed. It also forces the analysis to be self-contained: there can be no hand-art involved if the build robot is to run the analysis. Given how successful this experience was, the author questions his ability to return to the old style of running the analysis.

## 6. Can a plot be self-describing?

A plot that sits in an electronic logbook or a computer desktop for a month often has questionable heritage. Exactly what selection cuts were made? What files is that actually from? Make a plot self-describing was not an original goal of this project, but was stumbled on as work continued, especially as the PlotLingo language developed.

### 6.1. The PlotLingo language

The PlotLingo package was designed to be a high speed plotting language to make the last step – the tedious manipulation of plots – fast. Second, it was designed to be intuitive and straight-forward. This was a result of two failings of the LINQtoROOT library: the time to determine 500 plots have been cached was long enough to slow down the process and the monad hell the future mechanism induced.

The reason this is a custom language and not just ROOT C++ code directly is because the ROOT plotting API's are, for the most part, very low level. For a plot to be useful even in an internal meeting in a collaboration like ATLAS it needs a lot of things – a legend, properly named axis, etc.

The core of the language itself is very simple. It is functional in nature, and meant to be very terse and single-minded – perhaps one plot per file. The language contains statements and functions and methods. The *for* loop and a *map* operation are implemented as a function or method call, and arguments can be code (e.g. lambda functions). It also contains a number of extensibility points – similar to Aspect Oriented Programming [6]. For example, the first time a plot is created, code is attached to the plot to automatically generate the legend, or perhaps normalize the histograms. There is a utility that will automatically assign plot colour depending on key words in the tittle to help maintain uniform colour assignment across multiple plots. Operators like plus, multiply, etc., can be overridden. A simple example PlotLingo script is shown here:

```
f = teamcity ("http://tc-higgs.XXX/repository/download/XXX/2250:id/dataJZ2W-
FirstJetStudies.root");
h1 = f.Get("EMFL05/JetPtFirstJetAll");
h2 = f.Get("EMFL05/JetPtSecondJetAll");
[h1,h2].plot().title("Jet p_T");
```

A few things to note about the script. The first line loads a file from a build robot that is running the analysis, the second and third lines will fetch two histograms. The fourth line plots them together with a plot title of jet $p_T$. Adding the legend is as simple as adding a call to the Legend method.

### 6.2. Plot history

All of the pieces are now in place to maintain a complete plot history of how the plot was made. The LINQtoROOT library keeps as part of its cache the AST that generates the plot, the input files, and the input data. This can be attached as meta-data to the histogram. Plot lingo is an interpreted script – there is no reason not to store the script along with the meta-data with the resulting plot. Finally, a small program can be written that will extract the information from the plot and dump it to a nicely formatted file.

The cache key, which is the AST, is generated as a by-product of the LINQtoROOT plot generation. It is currently text and quite ugly. Further, it does not have enough information to be compiled. The text, even for a fairly complex plot, is relatively short: 2 or 3 KB. There is an issue of how to attach the code to the histogram as it is written out to the ROOT file. The simplest thing has been to write an extra TObjString object alongside the histogram, but the ideal case would be to extend the histogram object itself. This isn't possible without losing compatibility with other programs that may want to use the result.

PlotLingo produces PDF or jpg output files. Each format is extensible: arbitrary blocks of data can be added to the image. PlotLingo thus needs only to add the plot meta-data and the PlotLingo script into a block of data inserted in the final image. A proof of concept has been added to the PlotLingo language for jpg files.

If one can achieve code quality of the data stored in the plot image, the plot image becomes more than just an image. Many cloud storage systems maintain versions of old files – it would now be possible to look back at the plot from a few days ago and see what selections changed. One could upload the plot to a webserver, the webserver could extract the plot code and run it nightly on new data. One could paste it into a log book, and come back 6 months later and realize it wasn't quite run on the proper dataset.

The code for these experiments has not been placed in the public repositories yet – now that the proof of principle code has been developed real implementation can start.

## 7. Conclusions and future work

The LINQtoROOT project has been more successful than the author predicted, though it has been a great deal of work. It, or a derivative of it, will almost certainly be used by the author in the future if ROOT continues to maintain compatibility with Windows.

The ability to do a continuous analysis and the possibility of "easily" recording plot history were pleasant surprises to this method. While there are a number of limitations that will likely prevent this tool from being adopted by the community (e.g. it is currently Windows only), the fact that these sorts of features are possible should drive all of us to find solutions on the platforms that we use regularly.

## References

[1] G. Watts, The LINQtoROOT Project, hosted on CodePlex (https://linqtoroot.codeplex.com/)
[2] G. Watts, Using Functional Languages and Declarative Programming to Analyze Large Datasets: LINQtoROOT, Computers in High Energy Physics 2012, Conference Proceeding
[3] Wikipedia, *Futures and Promises in Programming Languages*, http://en.wikipedia.org/wiki/Futures_and_promises.
[4] The ATLAS Collaboration, *Search for pair produced long-lived neutral particles decaying in the ATLAS hadronic calorimeter in pp collisions at sqrt(s) = 8 TeV*, ATLAS-CONF-2014-041.
[5] Wikipedia, *Abstract Syntax Tree*, http://en.wikipedia.org/wiki/Abstract_syntax_tree
[6] G. Watts, The PlotLingo Project, hosted on CodePlex (https://plotlingo.codeplex.com/)
[7] Wikipedia, *Aspect-oriented Programming*, http://en.wikipedia.org/wiki/Aspect-oriented_programming

# Distributed job scheduling in MetaCentrum

**Šimon Tóth**[1] **and Miroslav Ruda**[2]

CESNET a.l.e., Zikova 4, Prague 6, 160 00, Czech Republic

E-mail: [1] simon@cesnet.cz

E-mail: [2] miroslav.ruda@cesnet.cz

**Abstract.**
MetaCentrum – The Czech National Grid provides access to various resources across the Czech Republic. The utilized resource management and scheduling system is based on a heavily modified version of the Torque Batch System. This open source resource manager is maintained in a local fork and was extended to facilitate the requirements of such a large installation. This paper provides an overview of unique features deployed in MetaCentrum. Notably, we describe our distributed setup that encompasses several standalone independent servers while still maintaining full cooperative scheduling across the grid. We also present the benefits of our virtualized infrastructure that enables our schedulers to dynamically request ondemand virtual machines, that are then used to facilitate the varied requirements of users in our system, as well as enabling support for user requested virtual clusters that can be further interconnected using a private VLAN.

## 1. Introduction

The Czech National Grid – MetaCentrum is a highly geographically distributed grid providing both computational and storage resources.

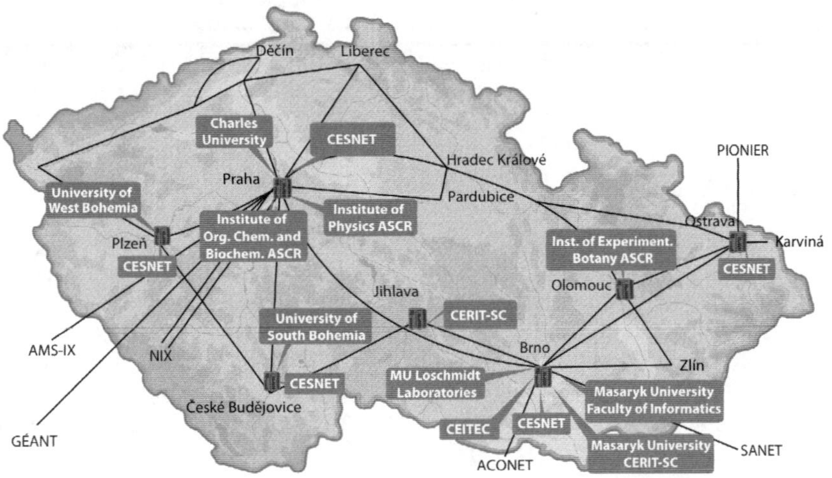

The grid contains a heterogeneous set of resources, currently encompassing approximately 10000 CPU cores distributed across 550 nodes, with varied memory capacity (12GB..6TB) and other capabilities (GPU cards, infiniband, high-speed NFS access, SSD scratch,...). Storage resources mainly consist of 1 PB permanent storage and 27 PB of long term (tape) storage. Heterogeneity of the resources is mirrored in the user base and workloads. The system commonly facilitates large job submissions (up to 10000 jobs), or very long jobs (several months of expected runtime).

This paper provides an overview of major features deployed in the Czech National Grid to deal with the varied provided resources.

## 2. Resource management

Original Torque [1] did not contain significant resource management features apart from managing CPU cores. We have extended Torque feature set significantly with additional resources, such as memory, GPU cards, scratch disk space and software licenses. These resources can all be requested by users in a very detailed manner. Users even have the option to request heterogeneous jobs with multiple machines, each with different resource allocation. For example a user can easily make a request for a job running on one master node with 1 CPU core and 48GB of memory, 10 slave nodes, each with 16 CPU cores and 32 GB memory, all nodes connected using infiniband. Additional features include negative requests, selecting machines based on their performance, exclusive full node allocation, etc.

All resource requests are processed by the scheduler and jobs are only executed on nodes strictly matching the jobs requirements, with all requested resources exclusively allocated to this job. Users can therefore rely on these resource being available throughout the entire jobs execution. For certain resources, in particular software licenses, we cannot provide a strict guarantee, as these resources can be exhausted without the intervention of the scheduler (from outside of the grid). For this style of resource we employ a heuristic to determine when a particular job requesting such resource can be safely executed.

To guarantee fair access to the provided resource, we are employing a complex fairsharing scheme [8] that takes into account utilization of CPU cores and memory [6]. The fairshare scheme is based on a dynamic ordering policy that orders jobs (across queues) based on their users historical resource consumption (with more recent consumption having a higher weight).

## 3. Distributed scheduling

MetaCentrum has experienced significant growth in the past five years (from 1500 CPU cores in 2009 to over 10000 CPU cores in 2014). Naturally this growth demanded changes in the scheduling infrastructure as the original centralized solution reached its scalability limits.

To address the scalability issue we went through multiple iterations of a distributed architecture [12, 13, 10], with the final version [11] targeting not only scalability, but also providing better resilience against local outages and giving resource providers better control over the local policies they want to enforce over their hardware.

The model is based on the notion of global queues, which can span across multiple servers (see Fig. 1) and are used to distribute information about jobs across these servers. Each scheduler is responsible only for the resources of its local server and follows all locally imposed policies. When a global queue is encountered it is treated the same way as any other queue, with the scheduler also pulling in information about jobs in the remote (on other servers) parts of the global queue. Remote and local jobs are all treated in the same manner.

Cross-server execution is handled in the background (between the servers), with the scheduler only issuing a standard run requests. For the convenience of the users, cross-running jobs have their execution information synchronized between the execution and the originating server.

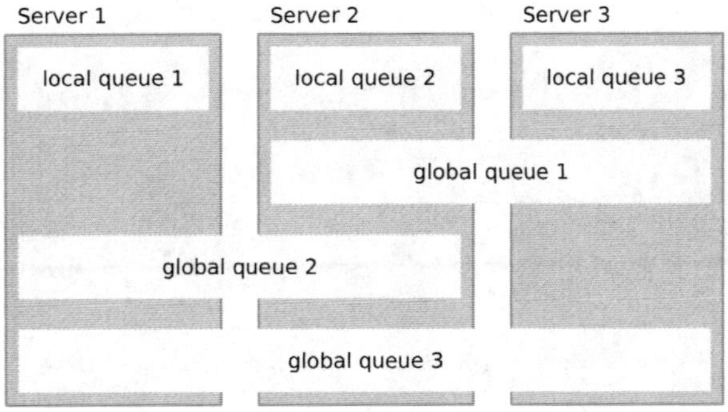


**Figure 1.** Distributed configuration utilizing global queues spanning across different servers.

This model puts very little constraints on the sites connected to the distributed system as they only need to run a compatible version of the Torque Batch System and can even use a different scheduler than the rest of the system as long as this scheduler is patched with the global queue support.

## 4. Virtualized infrastructure

Based on the Xen virtualization platform [2] and utilizing our Magrathea abstraction layer [4], the virtualized infrastructure was originally designed to provide full machine preemption. In the simplest case a physical machine is configured to host two virtual machines that both encompass the complete resource pool provided by the machine and are capable of switching resources between themselves (see Fig. 2). One of these machines is dedicated to low-priority preemptible jobs, one is dedicated to high-priority jobs and can preempt (take over the resources) from the low-priority virtual machine. This allows high-priority jobs immediate access to resources while heavily simplifying the scheduling process.

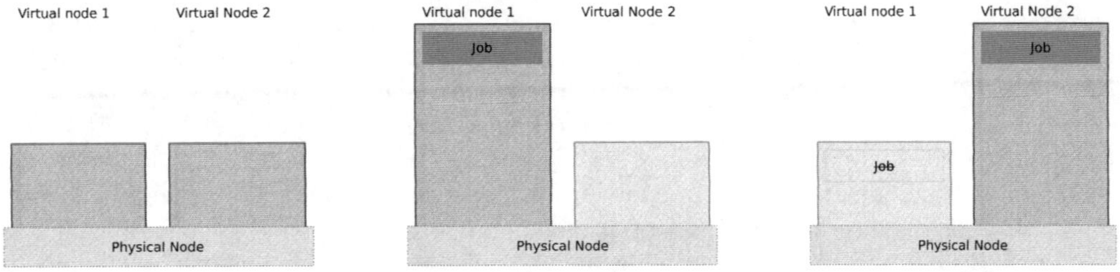


**Figure 2.** Example of a full machine preemption. Low-priority job claims all resources of the physical machine, but is then preempted by a high-priority job.

An extension of this basic model is the ability to start ondemand virtual clusters on top of our infrastructure. In this case, one of the virtual machines is left in an offline-bootable state and the scheduler is left with the option to start a new virtual machine with the desired (depending on the current overall users' demands) virtual image (see Fig. 3). This allows us to provide a more varied range of software configurations on the grid, without the need to statically dedicate a portion of the machine pool to a particular configuration.

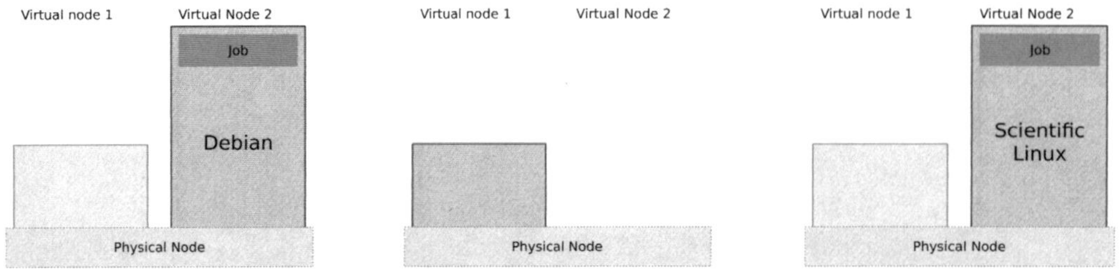


**Figure 3.** Example of an ondemand cluster. Once an instance with Debian is freed up, the scheduler boots up another instance with Scientific Linux.

Virtual clusters can also be requested directly by the users (including the possibility to connect the machines in the cluster using a VLAN). In this case the user can either request a virtual cluster that connects back to the resource management system, effectively creating a resource reservation, and specifying what users can access this prebuilt virtual cluster. For more complex use cases we are capable of providing a prebuilt image that connects to a preconfigured VLAN, extending the users computational infrastructure with more resources. This use case can cover even users that require access to Windows instances.

## 5. Management features

Extensive set of features, particularly important for the grid operator, was also implemented into Torque. This includes architectural features such as Kerberos [3] authentication with Kerberos based ACL for both job submission and system management and user oriented features such as the detection of invalid jobs that will never be executed under the current system configuration (these can be both the result of user error, or a maintenance event).

To maintain control over the systems behaviour, we have significantly extended the per-queue, per-group and per-user limits. Queues can be now limited to a particular subset of resources, can be limited in how many CPU cores they are allowed to occupy at any time, which extends to per-group and per-user limits.

To allow easy maintenance of machines, we are providing a dedicated maintenance queue, which disallows any other queues from accessing machines assigned to this queue. Furthermore, future maintenance can be specified for each node in the form of a deadline. No jobs that with runtime exceeding this specified deadline will be allowed to run on this particular node.

For testing and monitoring purposes it is sometimes required to submit small jobs to machines that are already fully utilized. Since we are heavily relying on exclusive resource allocation model, this would not be generally possible. To maintain this convenience, we have extended Torque with the support for admin slots. These represent an additional virtual CPU core that can be used to process small jobs (access to admin slots is limited to grid management and monitoring).

## 6. Conclusion

In this paper, we have presented an overview of important features that were implemented into our local fork of the Torque Batch System to facilitate the requirements of the Czech National Grid. Our resource management extensions allow users great control over what particular resources will be assigned to their jobs. Support for distributed scheduling allows us to divide the system into more manageable sites while still maintain global scheduling support and quality. Virtualized infrastructure gives our system the flexibility to deal with more exotic user requirements, be it infrequently utilized system configurations (through ondemand clusters) or dedicated virtual clusters. Our fork of the Torque Batch System is fully opensource and freely available at `https://cesnet.github.io/torque/`.

*Acknowledgements.* The work presented in this paper was conducted under the programme "Projects of Large Infrastructure for Research, Development, and Innovations" (LM2010005) funded by the Ministry of Education, Youth and Sports of the Czech Republic.

## References

[1] Adaptive Computing Inc. 2014 Torque batch system `http://www.adaptivecomputing.com/products/open-source/torque/`
[2] Barham P, Dragovic B and Fraser K, Hand S, Harris T, Ho A, Neugebar R, Pratt I and Warfield A 2003 Xen and the art of virtualization *SOSP*
[3] Neuman C, Yu T, Hartman S and Raeburn K 2005 The kerberos network authentication service *RFC*
[4] Ruda M, Šustr Z, Sitera J, Antoš D, Hejtmánek L and Holub P 2010 Virtual Clusters as a new service for MetaCentrum, the Czech NGI *CGW'09* pp 64–71
[5] Klusáček D, Rudová H and Jaroš M 2014 Multi resource fairness: problems and challenges *JSSPP'13* To appear.
[6] Klusáček D and Rudová H 2014 Multi-resource aware fairsharing for heterogenous systems *JSSPP'14* To appear.
[7] Tóth Š and Klusáček D 2014 User-aware metrics for measuring quality of parallel job schedules *JSSPP'14* To appear
[8] Klusáček D and Tóth Š 2014 On interactions among scheduling policies: finding efficient queue setup using high-resolution simulations *Euro-Par'14* pp 138–149
[9] Tóth Š and Klusáček D 2013 Tools and methods for detailed analysis of complex job schedules in the Czech National Grid *CGW'13* pp 83–84
[10] Tóth Š and Ruda M 2012 Practical experiences with Torque meta-scheduling in the Czech National Grid *Computer Science* **13** 2 pp 33-45
[11] Tóth Š 2013 Peer-to-peer scheduling in the Czech National Grid and beyond *CGW'13* pp 87–88
[12] Matyska L, Ruda M and Tóth Š 2011 Peer-to-peer cooperative scheduling architecture for national grid infrastructure *Data Driven e-Science* pp 105–118
[13] Tóth Š, Ruda M and Tóth Š 2011 Towards peer-to-peer scheduling architecture for the Czedch National Grid *CGW'11* pp 92–101

# Belle II distributing computing

**P. Krokovny**

Novosibirsk State University and Budker Institute of nuclear Physics, 630090, Novosibirsk, Russia

E-mail: `krokovny@inp.nsk.su`

**Abstract.** The next generation B factory experiment Belle II will collect huge data samples which are a challenge for the computing system. To cope with the high data volume and rate, Belle II is setting up a distributed computing system based on existing technologies and infrastructure, plus Belle II specific extensions for workflow abstraction. This paper describes the highlights of the Belle II computing and the current status. We will also present the experience of the latest MC production campaign in 2014.

## 1. Introduction

The existence of large matter-antimatter asymmetry (CP violation) in the $b\bar{b}$-quark system as predicted in the Kobayashi-Maskawa [1] theory was established by the B-Factory experiments [2, 3]. However, this cannot explain the magnitude of the matter-antimatter asymmetry of the universe we live in today. This indicates undiscovered new physics exists. The Belle II experiment [4], the next generation of the B-Factory, is expected to reveal the new physics by accumulating 50 times more data ( 50 ab$^{-1}$) than Belle by 2022. The Belle II computing system has to handle an amount of beam data eventually corresponding to several tens of PetaByte per year under an operation of the SuperKEKB accelerator with a designed instantaneous luminosity $810^{35}$ cm$^{-2}$s$^{-1}$. Under this situation, it cannot be expected that one site, KEK, will be able to provide all computing resources for the whole Belle II collaboration including the resources not only for the raw data processing but also for the MC production and physics analysis done by users. In order to solve this problem, Belle II employed a distributed computing system based on DIRAC, which provides the interoperability of heterogeneous computing systems such as grids with different middleware, clouds and the local computing clusters. Since last year, we performed the MC mass production campaign to confirm the feasibility and find out the possible bottleneck of our computing system. In parallel, we also started the data transfer challenge through the transpacific and transatlantic networks.

## 2. Computing model and distributed computing system

A consequence of the drastically increased data rate and volume compared to Belle is an increase in required computing resources by about two orders of magnitude. Because the requirements on storage, processing power, and network bandwidth will be at a similar scale as for the LHC experiments and because the collaboration has become more and more international, Belle II has decided to adopt a distributed computing model.

It has a similar hierarchical structure to the Worldwide LHC Computing Grid (WLCG), but is a bit simpler. The raw data is recorded and processed at KEK and copied only to one other site,

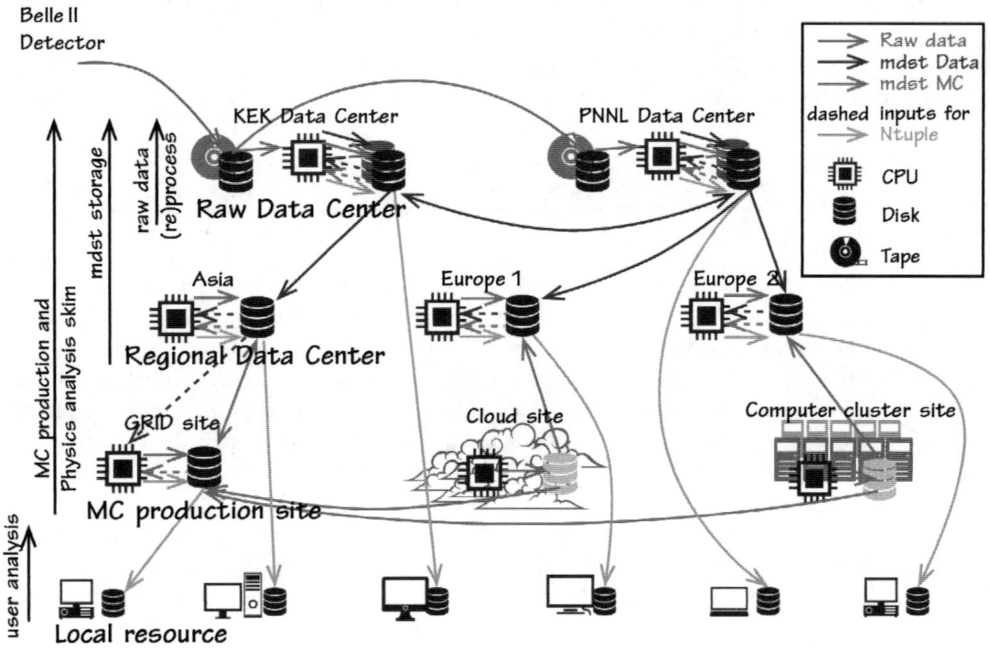


**Figure 1.** The Belle II computing model.

PNNL (Pacific Northwest National Laboratory). This provides a backup copy and the possibility to distribute reprocessing tasks over both sites. The output of the raw data processing are mDST files, a format containing all relevant information for physics analyses, which are copied to grid sites around the world. These sites also produce simulated data samples (Monte-Carlo, MC) in mDST format. Physicists process the mDST files on grid sites and transfer the output to local resources where the final analysis is performed. The Belle II computing model is illustrated in figure 1.

In the design of our distributed computing system we rely on existing and well-proven infrastructure and software. By choosing DIRAC [5], which was originally developed by LHCb, as the basis for our distributed computing system we can use resources provided via different middleware interfaces, including WLCG, Open Science Grid (OSG), and cloud resources. The location of output files is stored in the Logical File Catalog (LFC). For each file we record metadata in an AMGA database [6]. The installation of offline software releases is done conveniently via CVMFS (CERN Virtual Machine File System) [7], a system for mounting and caching a file system from a remote server, on most sites. The CVMFS stratum-0 server is kindly provided by CERN. A stratum-1 server is running at GridKa. At those sites without CVMFS support, mainly OSG sites, the software is distributed via installation jobs.

On top of these components Belle II has developed a user interface tailored to the needs of the Belle II collaborators. The interface is called gbasf2 and is designed to provide an easy transition from an analysis done with the offline software framework basf2 [8] to an analysis performed on the grid. A guiding design principle was to provide an abstraction of the workflow from single files and jobs to datasets and collections of jobs, called projects. The idea is that the user selects a set of input files based on metadata criteria and then processes them with the same analysis code. Instead of submitting multiple jobs for multiple files, the user starts just one project for the input dataset and then can refer to the collection of individual jobs via a chosen project name. This project-level management of jobs comprises the submission of jobs, the monitoring of jobs, the cancellation of jobs, and the retrieval of their output sandboxes. The workflow abstraction is also applied to the output files of a project which can be collectively referred to as an output dataset.

## 3. Production campaigns and data challenges

To evaluate the computing model and the distributed computing system, Belle II has performed a few MC production campaigns. The first campaign was performed in February-March 2013. In a first stage $B\bar{B}$ events were generated with EvtGen [9] and the detector response simulated based on a geometry implemented with Geant4 [10]. The output are raw data files. They are uploaded to storage elements and their metadata information is registered in AMGA. In a second stage the raw data files were the input to reconstruction jobs.

With the experience of the first MC production campaign at hand, the distributed computing and offline software were developed further to address the observed issues. Another assessment of the system was then done in a second MC production campaign from July 23rd to September 8th, 2013. In this campaign the event generation, the detector simulation, and the reconstruction were executed within one job. As the output was stored in mDST format, the ratio of output size to CPU usage is much lower than in the first campaign. A further difference with respect to the first campaign is that background data is mixed with the simulated signal events. This requires the import of background data files at the beginning of each job. The type of jobs executed in the second MC production campaign is similar to the type of jobs that will produce MC data samples for Belle II physics analysis later during the data taking phase. In total 630k jobs consumed 700 kHEPSpec06 days and produced 560M events, corresponding to an output size of 8.5 TB. Compared to the first campaign the used CPU resources could thus be increased by more than an order of magnitude as illustrated in figure 3. At the same time the job failure rate could be reduced considerably. While in the beginning it was at a level of 10% it reached about 1% at the end. Inefficiencies were caused by issues or downtimes of individual sites, the expiration of proxys, overloaded servers, and human errors like the submission of jobs with wrong parameters. It should be noted that no crash of the offline software was observed.

MC 3 and 3.5 were done to provide the first official MC samples to the physics group. It was started by April 3 2014 and finished in May 30 2014. In total 2.5M jobs consumed 2.2M HEPSpec06 days and produced 6000M events, corresponding to an output size of 45 TB. This sample is equivalent to 400 $fb^{-1}$ data. The MC sample equivalent 100 $fb^{-1}$ was open to user analysis.

Several sites around the world participated in the third MC production campaign. These are CoEPP (Australia), HEPHY (Austria), McGill HPC (Canada), CESNET in Czech Republich, DESY, GridKa, LRZ/RZG (Germany), INFN/CNAF (Italy), KEK-CRC, KMI (Japan), KISTI GSDC (Korea), Cyfronet, CC1 (Poland), NUSC, SSCC (Russia), SiGNET (Slovenia), ULAKBIM (Turkey), UA-ISMA (Ukraine), and OSG, PNNL (USA). A summary of the contributed CPU resources is shown in figures 2 and 3.

Like the support of the sites the support of shifters was another essential ingredient for the success of the production campaign. More than twenty volunteers kept production running by submitting jobs, monitoring them, and resubmitting failed jobs. Shifts were assigned in 8 hour blocks so that continuous coverage could be achieved by day shifts in Asia, Europe, and America. Log book information was recorded in a twiki. For communication, in particular at the hand over between shifts, it turned out that a video/audio conference with chat using SeeVogh/EVO provided a very efficient and productive solution.

In the Belle II computing model we rely on high-bandwidth network connections between sites, in particular between KEK and PNNL for the transfer of raw data. But also for the replication of MC datasets network connections between all sites are needed. To assess the

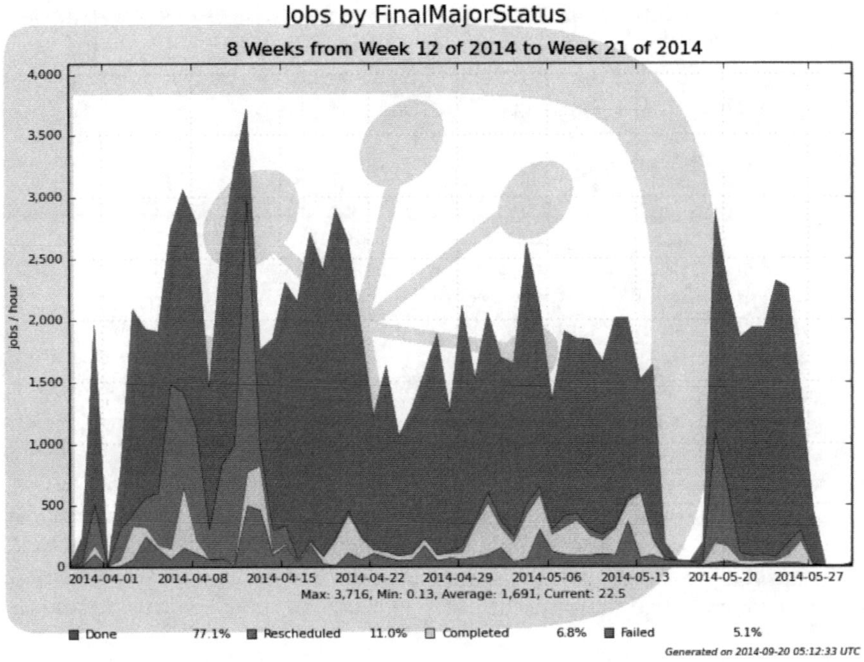


**Figure 2.** The number of running jobs for the third MC production campaign in April to May. Successfully completed jobs are shown in green and failed jobs in red. Light green indicates that the failover mechanism for the upload of the output was triggered.

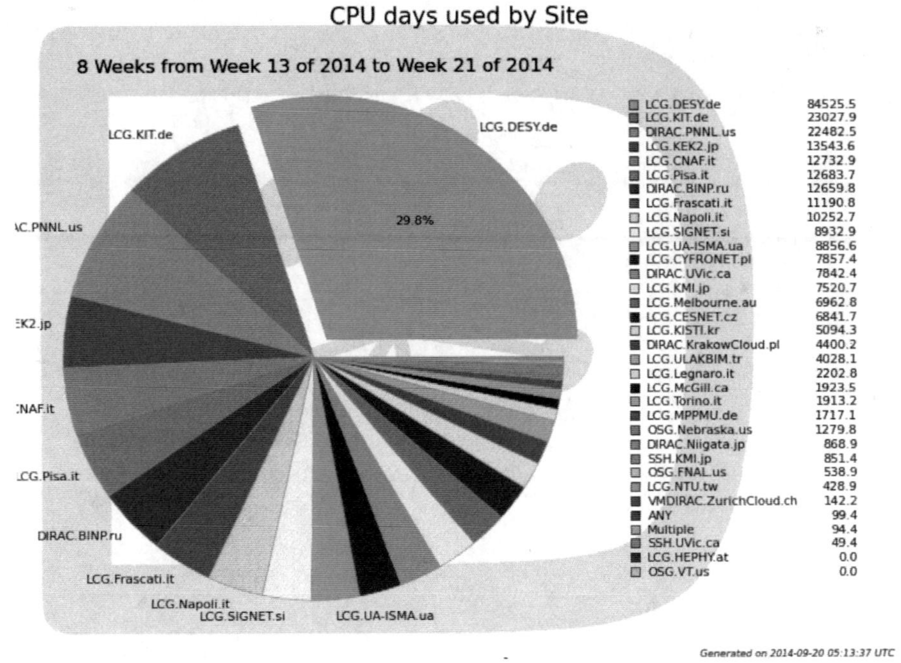


**Figure 3.** The contributions of sites to the third Belle II MC production campaign.

current status of the network resources available to Belle II, we exercised transfers in a data challenge in May 2013. The transfers were controlled by a FTS2 service running at GridKa. One of the conclusions from the data challenge was that the bandwidth for transfers from Japan to Europe is significantly lower than between Japan and the US.

## 4. Summary

The computing resources required to process and analyze the data of the next-generation B factory experiment Belle II will be of similar scale as the resources requirements of the LHC experiments. Therefore Belle II has adopted a distributed computing model which enables all collaborators to contribute. The distributed computing system is based on existing technologies and extended by Belle II specific features. An essential concept is the workflow abstraction from individual files and jobs to datasets and projects.

This concept and underlying infrastructure were successfully tested in three MC production campaigns. This success was only possible by the CPU, storage, and network resources kindly provided to us by sites and NRENs. The campaigns showed the issues and bottlenecks of our system and we managed to address several of them already during the production campaign so that a final job failure rate of about 1% was reached.

The fourth MC production campaign is started on September 2014. We include more sites and cloud resources. Another important use case of the system is the user analysis on the grid which may show completely different issues than a centrally managed MC production.

## Acknowledgments

We are grateful for the support and the provision of computing resources by CoEPP in Australia, HEPHY in Austria, McGill HPC in Canada, CESNET in Czech Republich, DESY, GridKa, LRZ/RZG in Germany, INFN/CNAF in Italy, KEK-CRC, KMI in Japan, KISTI GSDC in Korea, Cyfronet, CC1 in Poland, NUSC, SSCC in Russia, SiGNET in Slovenia, ULAKBIM in Turkey, UA-ISMA in Ukraine, and OSG, PNNL in USA. We acknowledge the service provided by CANARIE, Dante, ESnet, GARR, GEANT, and NII. We thank the DIRAC and AMGA teams for their assistance and CERN for the operation of a CVMFS server for Belle II. We acknowledge support from MinES of RF (grant 14.610.21.0002, identification number RFMEFI61014X0002).

## References

[1] Kobayashi M and Maskawa T 1973 Prog. Theor. Phys. **49** 652
[2] Abashian A *et al.* 2002 Nucl. Instrum. Meth. A **479** 117
[3] Aubert B *et al.* 2002 Nucl. Instrum. Meth. A **479** 1
[4] Abe T *et al.* arXiv:1011.0352 [physics.ins-det]
[5] Casajus A *et al.* [LHCb DIRAC Collaboration] 2010 J. Phys. Conf. Ser. **219** 062049; Tsaregorodtsev A *et al.* 2010 J. Phys. Conf. Ser. **219** 062029
[6] Ahn S *et. al.* 2010 Journal of the Korean Physical Society **57** issue 4 715
[7] Blomer J, Buncic P, Charalampidis I, Harutyunyan A, Larsen D and Meusel R 2012 J. Phys. Conf. Ser. **396** 052013
[8] Moll A 2011 J. Phys. Conf. Ser. **331** 032024
[9] Lange D J 2001 Nucl. Instrum. Meth. A **462** 152
[10] Agostinelli S et al. [GEANT4 Collaboration] 2003 Nucl. Instrum. Meth. A **506** 250

Journal of Physics: Conference Series **608** (2015) 012027 doi:10.1088/1742-6596/608/1/012027

# VISPA: Direct Access and Execution of Data Analyses for Collaborations

**Daniel von Asseldonk, Martin Erdmann, Robert Fischer, Christian Glaser, Gero Müller, Thorben Quast, Marcel Rieger, Martin Urban**

III. Physics Institute A, RWTH Aachen University, Aachen, Germany

E-mail: glaser@physik.rwth-aachen.de

**Abstract.** The VISPA project provides a graphical frontend to computing infrastructures. Currently, the focus of the project is to give an online environment for the development of data analyses. Access is provided through a web GUI, which has all functionality needed for working conditions comparable to a personal computer. This includes a new preference system as well as user configurable shortkeys. As all relevant software, data and computing resources are supplied on a common remote infrastructure the VISPA web framework offers a new way of collaborative work where analyses of colleagues can be reviewed and executed with just one click. Furthermore, VISPA can be extended to the specific needs of an experiment or other scientific use cases. This is presented in the form of a new GUI to the analysis framework Offline of the Pierre Auger collaboration.

## 1. Introduction

We are facing the situation that on the one hand our mobile computing devices become smaller and more portable whereas - on the other hand - more computing resources are needed and larger and larger data volumes need to be handled. Furthermore, it is getting more difficult to install software needed for data analyses on an ultrabook or tablet, even disregarding computing power and storage.

One solution is to use a software that all devices have in common: The web browser. Instead of installing software and transferring data onto a device, one opens a browser, connects to a web server and accesses a working environment by means of a web application.

The VISPA software provides this functionality. VISPA is a graphical front-end to infrastructures which makes software, data and computing resources available through the web.

The article is structured as follows: We first describe the software and its technical implementation in detail. After describing how VISPA can be extended by custom applications, current and future use cases are discussed.

## 2. The software

The VISPA software [1, 2] consists of two parts: The first part is the basic functionality which provides a GUI framework, user management and the communication between client (web browser), server and worker nodes. The second part are applications that use the VISPA framework. Several applications are shipped directly with the main VISPA package. These applications are a file browser, a terminal emulator and a code editor. However, the concept of

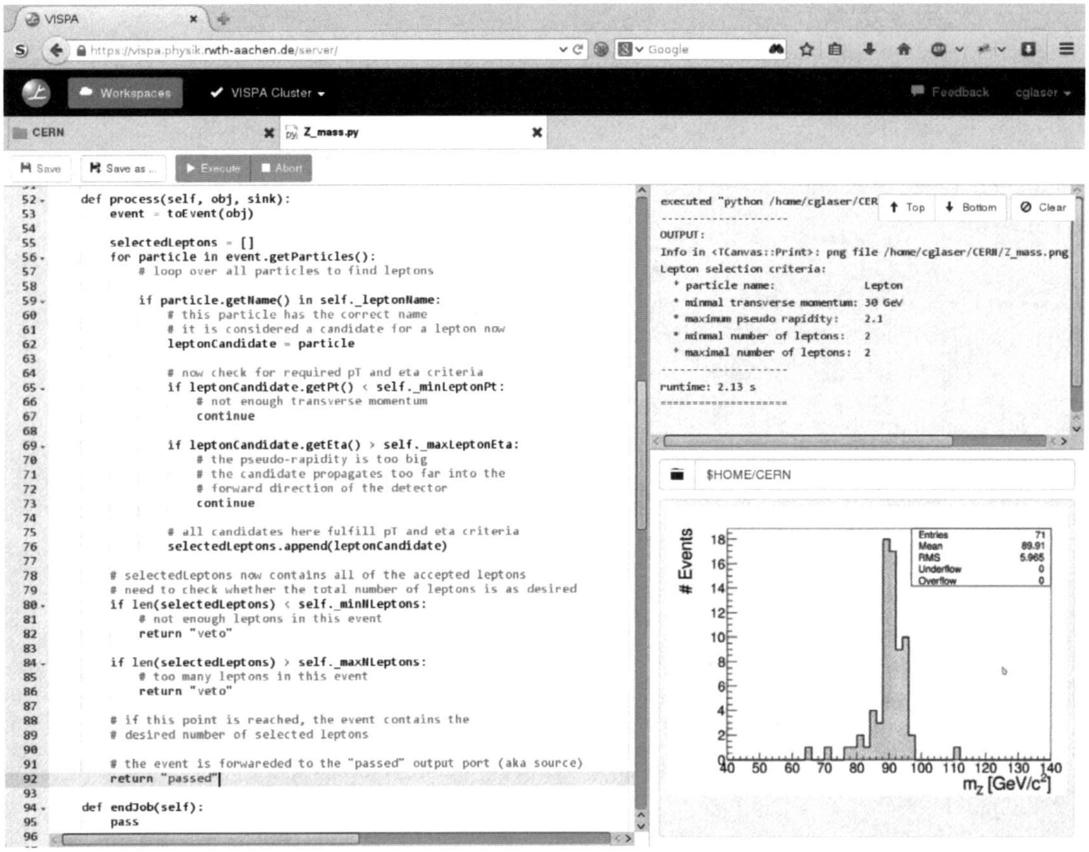


**Figure 1.** Screenshot of the code editor application.

VISPA is to provide a web framework that can then be extended with custom made applications by other groups themselves. This is demonstrated in section 4 by means of a graphical steering of the analysis framework of the Pierre Auger collaboration.

In the current version 1.0 of the VISPA platform, the GUI has been redesigned to optimize the workflow providing working conditions comparable to a local desktop environment. A preference system is used to satisfy users' individual tastes of the GUI appearance and working style. Frequently used operations are accessible with just one mouse click, or by using a shortkey respectively. Shortcuts are user configurable on the whole platform.

Figures 1, 2a and 2b show screenshots of the three basic VISPA applications. The main GUI elements are visible in the upper menu bar. A VISPA application is opened as a tab and multiple applications can be opened at the same time. The code editor application was extended to instantly execute python scripts. With a click on the green execute button the script is executed. The terminal output of the python job is visible in the upper right part of the application window. Output plots that are generated by the job are visible in the lower right part. A click on a plot will open a larger preview.

The VISPA platform consists of three tiers: Clients, the VISPA server and worker nodes (figure 3a). This separation enables increased scalability [3]. On the client side the web browser provides the graphical user interface. The VISPA server is a web server providing the content to the client, which it gathers from the backend servers. Worker nodes can be any computer

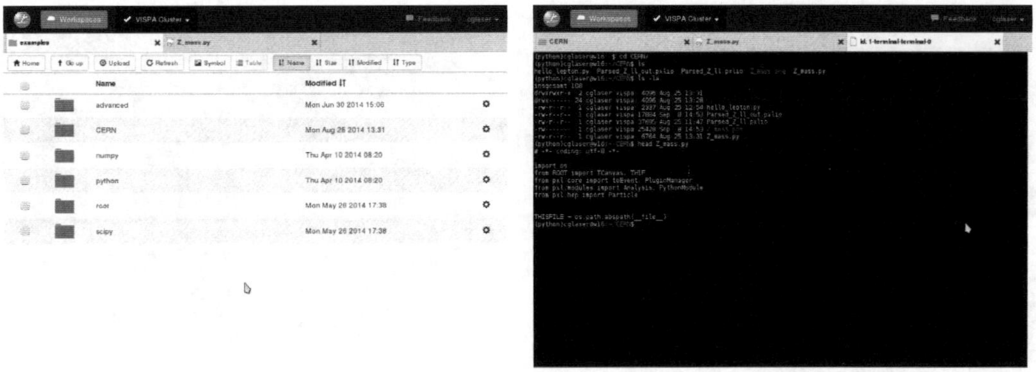


(a) (b)

**Figure 2.** Screenshot of the file browser (a) and the terminal (b) application.

ranging from a laptop to a computing cluster. The only requirements for the workers are SSH access and a Python [4] interpreter. Additional software may be required for custom VISPA applications.

The GUI (client code) is based on the HTML5 markup language, CSS3 and the JavaScript scripting language, which are supported by common web browsers [5]. Advanced functionality is implemented using the JavsSript library jQuery [6] and the Bootstrap3 framework [7]. Furthermore, most web content is generated by rendering HTML templates on server or client side using mako [8] or transparency [9].

Server and workspace code is written in Python [4]. The server is build on the web framework CherryPy [10]. This allows for a convenient integration of various scientific software packages written in Python, such as PXL [2], ROOT [11] and SciPy [12]. Persistent data, e.g. user management and preferences, are stored in databases using SQLAlachemy [13].

*Server - workspace communication* The communication between server and worker node (workspace) is established using a bootstrap approach: At the beginning of the communication, all relevant software is copied via ssh into the memory of the worker node. This makes a SSH access and a Python [4] interpreter the only requirement.

The first step is to create a ssh connection and execute a minimal Python code which makes the stdin and stdout pipe to exchange commands. Then all required Python packages are compressed into one zip archive on the server and copied and extracted into the workers memory. All further communication uses remote procedure calls. In our implementation we make use of the Python packages RPyC [14] and paramiko [15].

## 3. Extending VISPA with custom applications

VISPA is designed to be extended to custom needs of a user or a group. The VISPA framework provides all necessary functionality to utilize the client - server - worker system. Hence, the user needs to implement only the specific tasks of his application.

The VISPA source code [16] contains two well documented example applications, the so-called "dummy" and "demo" application. They demonstrate the implementation of a VISPA app and show most features of the VISPA framework: Utility functions are provided which simplify the server - client communication using Ajax, as well as the server - workspace communication. Furthermore, intercommunication between different apps is supported. For instance a file

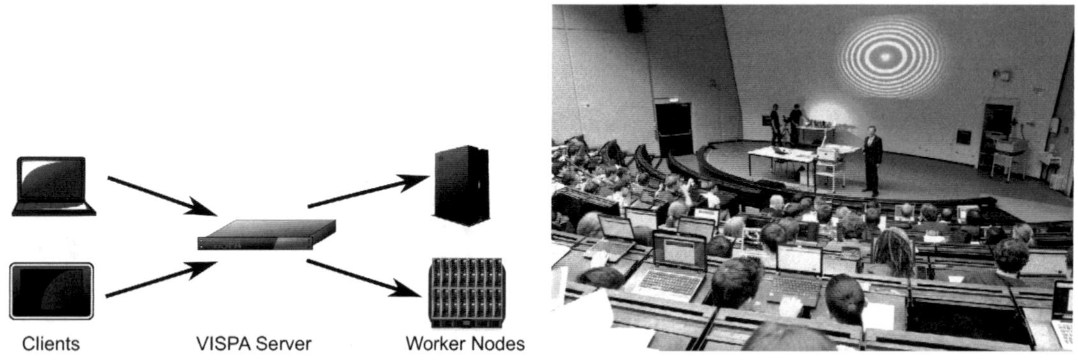


**Figure 3.** (a) Server setup. (b) Usage of VISPA in an experimental physics lecture. Picture: Peter Winandy (Aachen).

selection dialog can be opened as known from desktop applications.

## 4. Graphical steering of Pierre Auger analysis framework

An application which demonstrates the extensibility of the VISPA platform is the Auger Offline application which provides a graphical steering of the analysis framework Offline [17, 18] of the Pierre Auger collaboration. The reconstruction and analysis of raw data is steered by defining a sequence of modules and by optionally overriding the default module options. This is normally achieved by creating XML steering files which require knowledge of all necessary modules and module options including their correct spelling.

To simplify the usage of Offline, the VISPA framework was used to implement a GUI which can be seen in figure 4. In the left panel all available modules are displayed grouped by categories. In the middle panel, the module sequence can be defined. Via drag and drop modules can be moved from the left to the middle panel or sorted within the module sequence.

A click on one module opens the right panel where the module's options can be edited. To further assist the user, a tool-tip explaining the meaning of the option is displayed when hovering over an option. Furthermore, the user can take advantage of VISPA's job-management apps. With a few clicks one or multiple jobs can be created and submitted to different batch systems.

This VISPA application was written without modification of the Auger analysis framework Offline. All information is obtained dynamically by parsing XML configuration files that come with the Offline software. Hence, the VISPA app does not interfere with the development of Offline and will always be up to date to the installed version of the workspace.

The main advantages of using VISPA instead of a desktop GUI framework are: The application is usable from any device and one does not need to run the application on the computer where Offline is installed. The GUI is always up to date to the installed version on the computer or computing cluster that is accessed through VISPA and one can work directly on the remote files or submit a job. The device which steers Offline is decoupled from the device providing the computing resources.

## 5. Collaborative analyses

VISPA opens a new way of collaborative work. As the analysis as well as all relevant software and data are supplied on a common remote infrastructure, an analysis can be shared with colleagues by giving them access to the relevant files. The analysis can be reviewed or executed through

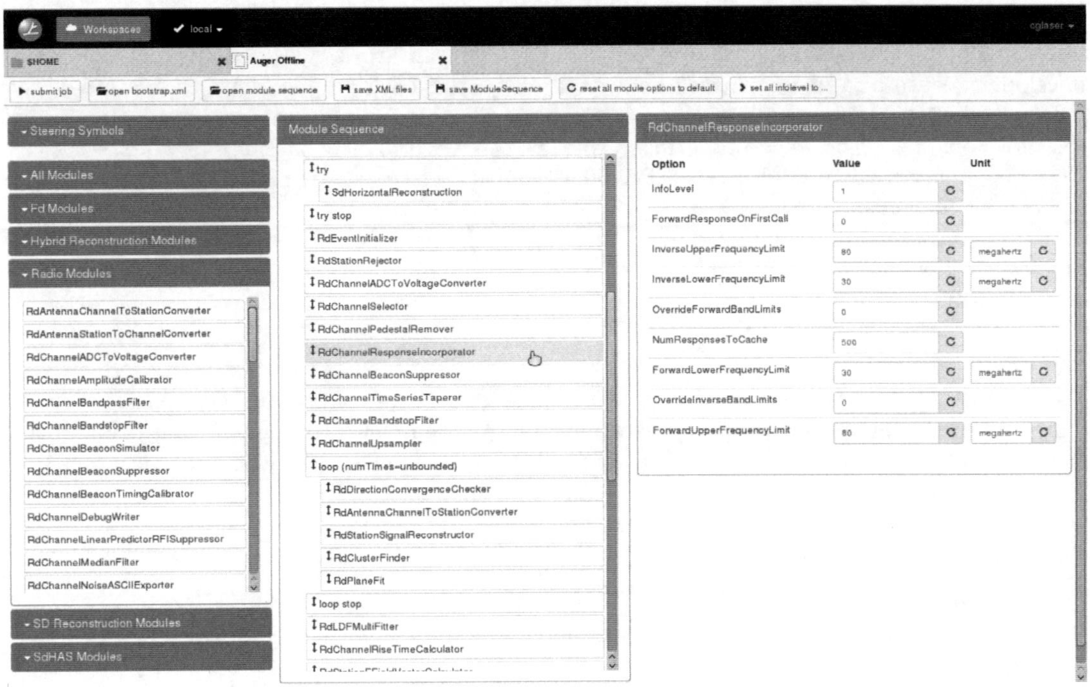


**Figure 4.** Screenshots of the VISPA application to steer the analysis framework Offline of the Pierre Auger collaboration.

the VISPA GUI giving a significant improve in comfort compared to remote ssh access. It is planned to add GUI interfaces to create working groups and to share analyses with other VISPA users as well as a GUI interface to common repository systems to facilitate shared work.

## 6. Application in education and outreach

Having a web browser as the only requirement for the user, VISPA is ideal to be used in education and outreach activities.

VISPA has successfully been used to supplement a 2nd and 3rd years experimental physics course at RWTH Aachen University in a blended learning initiative [19, 20]: Students analyzed lecture hall experiments themselves directly during the lecture (see figure 3b). After the experiment has been performed the data are uploaded on the VISPA server. A python analysis script is provided which should be extended by the students. Thereby the amount of what the students need to implement by themselves is adapted to the skills of the course. By using their own laptops, the students can log in to the VISPA server and analyze the measured data. An often task e.g. is to fit an analytic function to the data points and thereby determining fundamental physical constants such as the g-factor of an electron. More advanced analysis tasks are given as homework which are discussed in a weekly tutorial.

Furthermore, VISPA is used for CERN outreach activities for its 60th anniversary to perform a data analysis on public CMS data [21]. For example, the mass of the Z-boson can be reconstructed.

## 7. Conclusions

The VISPA web framework enables data analyses in a web browser. It provides a platform to develop, execute and share physics analyses. With the basic applications that are shipped with the VISPA package, all tools needed to develop data analyses are provided. As all relevant software, data and computing resources can be provided on a remote infrastructure that can be accessed through the VISPA GUI, analyses can be shared with colleagues keeping the comfortable graphical interface.

Furthermore, VISPA is designed to be extended with custom applications to satisfy the needs of specific user groups and unforeseen use cases. All necessary general functionality to write a web application is provided so that the user can focus on the specific task of his application.

## Achnowledgments

We wish to thank the organizers of the ACAT2014 conference for their kind support. This work is supported by the Ministerium für Wissenschaft und Forschung, Nordrhein-Westfalen, the Bundesministerium für Bildung und Forschung (BMBF) and the Deutsche Forschungsgemeinschaft (DFG).

## References

[1] VISPA web page `http://vispa.physik.rwth-aachen.de`
[2] Bretz H-P et al 2012 *J. Instrum.* **7** T08005 doi: 10.1088/1748-0221/7/08/T08005
[3] Erdmann M et al 2014 *J. Phys. Conf. Ser.* **513** 62034 doi:10.1088/1742-6596/513/6/062034
[4] van Rossum G Python language `http://www.python.org/`.
[5] Erdmann M et al 2014 *J. Phys. Conf. Ser.* **523** 012021 doi: 10.1088/1742-6596/523/1/012021
[6] jQuery `http://jquery.com/`
[7] Bootstrap `http://getbootstrap.com/`
[8] Mako `http://www.makotemplates.org/`
[9] Transparency `http://leonidas.github.io/transparency/`
[10] A Minimalist Python Web Framework `http://www.cherrypy.org`
[11] Antcheva I et al 2009 *Comput. Phys. Commun.* **180** 2499 doi: 10.1016/j.cpc.2009.08.005
[12] SciPy `http://www.scipy.org`
[13] SQL alchemy `http://www.sqlalchemy.org/`
[14] RPyC `http://rpyc.readthedocs.org/`
[15] Paramiko `http://www.paramiko.org/`
[16] VISPA source code `https://forge.physik.rwth-aachen.de/projects/vispa-web`
[17] Argirò S et al 2007 *Nucl. Instruments Methods Phys. Res. Sect. A Accel. Spectrometers, Detect. Assoc. Equip.* **580** 1485 doi: 10.1016/j.nima.2007.07.010
[18] Abreu P et al 2011 *Nucl. Instruments Methods Phys. Res. Sect. A Accel. Spectrometers, Detect. Assoc. Equip.* **635** 92 doi: 10.1016/j.nima.2011.01.049
[19] Bender H, Sauter A and Sauter W 2002 *Blended Learning: Effiziente Integration von E-Learning und Präsenztraining* Neuwied: Luchterhand.
[20] Erdmann M et al 2014 *Eur.J.Phys.* **35** 35018 doi: 10.1088/0143-0807/35/3/035018
[21] CERN outreach `http://opendata.cern.ch/`

# Planning for distributed workflows: constraint-based coscheduling of computational jobs and data placement in distributed environments

**Dzmitry Makatun[1,3], Jérôme Lauret[2], Hana Rudová[4] and Michal Šumbera[3]**

[1]Faculty of Nuclear Physics and Physical Engineering, Czech Technical University in Prague
[2]STAR, Brookhaven National Laboratory, USA
[3]Nuclear Physics Institute, Academy of Sciences, Czech Republic
[4]Masaryk University, Czech Republic

E-mail: dzmitry.makatun@fjfi.cvut.cz

**Abstract.** When running data intensive applications on distributed computational resources long I/O overheads may be observed as access to remotely stored data is performed. Latencies and bandwidth can become the major limiting factor for the overall computation performance and can reduce the CPU/WallTime ratio to excessive IO wait. Reusing the knowledge of our previous research, we propose a constraint programming based planner that schedules computational jobs and data placements (transfers) in a distributed environment in order to optimize resource utilization and reduce the overall processing completion time. The optimization is achieved by ensuring that none of the resources (network links, data storages and CPUs) are oversaturated at any moment of time and either (a) that the data is pre-placed at the site where the job runs or (b) that the jobs are scheduled where the data is already present. Such an approach eliminates the idle CPU cycles occurring when the job is waiting for the I/O from a remote site and would have wide application in the community. Our planner was evaluated and simulated based on data extracted from log files of batch and data management systems of the STAR experiment. The results of evaluation and estimation of performance improvements are discussed in this paper.

## 1. Introduction

Previous collaborative work between BNL (Brookhaven National Laboratory) and NPI/ASCR (Nuclear Physics Institute, Academy of Sciences of the Czech Republic) showed that the global planning of data transfers within the Grid can outperform widely used heuristics such as Peer-to-Peer and Fastest link (used in Xrootd)[1, 2]. Those results became the ground for continuation of research and extension of global planning to the entire data processing workflow, i.e., scheduling of CPU allocation, data transferring and placement at storage.

Long I/O overheads when accessing data from remote site can significantly reduce the application's CPUtime/WallTime ratio [3, 4]. For this reason, when setting up a data production at remote sites one has to consider the network throughput, available storage and CPU slots. When there are few remote sites involved in the data processing, the load can be tuned manually and simple heuristic may work, but, as the number of sites grows and the environment is constantly changing (site outage, fluctuations of network throughput and CPU availability), an automated planning of workflows becomes needed.

As an intuitive example of optimization let us consider a situation when a given dataset can be either processed locally, or can be sent to a remote site. Depending on transfer overhead it may appear to be optimal to wait for free CPU slots at the local site and process all the data there, or send a smaller fraction of the dataset for remote processing. Commonly used heuristics such as "*Pull a job when a CPU slot is free*" will not provide an optimization with respect to an overall processing makespan.

Another example arises from a workflow optimization which was done for inclusion of the ANL (Argonne National Laboratory) computational facility into the Cloud based data production of the STAR experiment [5]. In this case, and due to the lack of local storage at the site for buffering, the throughput of a needed direct on-demand network connection between BNL (New York) and ANL (Illinois) was not sufficient to saturate all the available CPUs at the remote site. An optimization was achieved by feeding CPUs at ANL from two sources: directly from BNL and through an intermediate site NERSC (National Energy Research Scientific Computing Center, California) having large local caching and with better connectivity to ANL. This example illustrates an efficient use of indirect data transfers which cannot be guessed by simple heuristics. A general illustration of distributed resources used for data production and their interconnection is given at Figure 1.

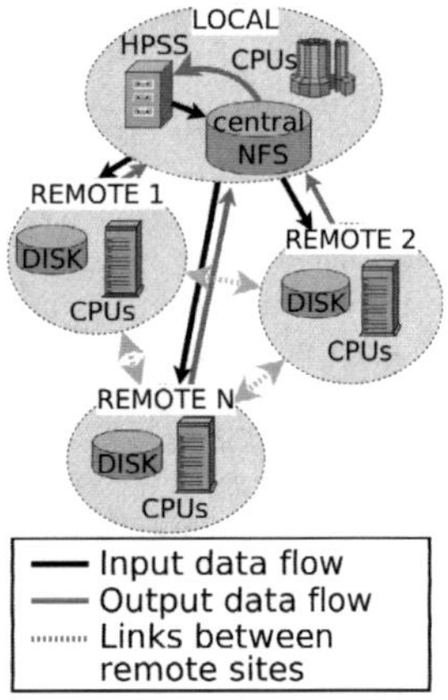


**Figure 1.** Schema of data production in the Cloud.

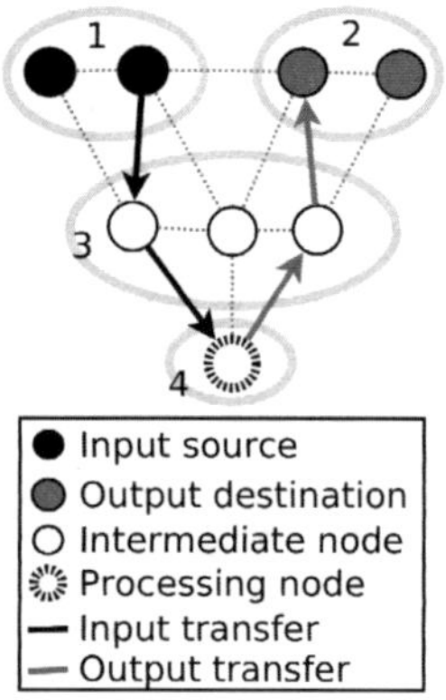


**Figure 2.** An example of a transfer path. Illustration for constraints 1-4 in section 2.

Scheduling of computational jobs submitted by users (user analysis) has even more degrees of possible optimization: selection between multiple data sources, grouping of jobs that use the same input files. This case becomes even more complex due to a poor predictability of the user analysis jobs. However, the main question for optimization remains the same as for the examples above: How to distribute a given set of tasks over the available set of resources in order to complete all the tasks within minimal time?

Problems of scheduling, planning and optimization are being commonly solved with the help of Constraint Programming (CP) [6]. It is a form of declarative programming which is widely used in scheduling, logistics, network planning, vehicle routing, production optimization etc... In the next sections we will introduce our Constraint Satisfaction Problem (CSP) formulation for a data production at multiple sites and provide a simulation-based evaluation of the proposed model.

## 2. Model formulation, assumptions and search approach

We will introduce only the core concepts of our CSP formulation and search algorithms, omitting detailed mathematical expressions. The following input parameters are necessary to define our CSP.

**Computational Grid** (see Figure 1) is described by directed weighted graph where nodes are computational sites $c$ with a given number of CPUs $cpu_c$ and storage space $disk_c$; edges are network links $l$ with weight $slowdown_l$ which is the time required to transfer a unit of data ($slowdown_l = \frac{1}{throughput_l}$). A dedicated storage facility, such as HPSS, can also be modeled as a node with $cpu_c = 0$.

**Set of jobs.** Each job $j$ has a $duration_j$, it needs one input file of $inputSize_j$, produces one output file of $outputSize_j$, input file is placed at $inputSourceNodes_j$ and output file must be transferred to one of $outputDestinationNodes_j$.

Our goal is to create a schedule of jobs at computational sites, transfers over links and a placement of files at storages for a given computational Grid and a set of jobs. In order to solve this problem the variables of our model define the *resource selection* and *timing* of each task:

**Resource selection variables** define a node $ProcessingNode_j$ where the job $j$ will be executed and a transfer path for each file $f$ (either input or output of a job). The transfer path is described by a set of boolean variables $X_{fl}$ where *true* means that a file $f$ will be transferred over a link $l$ and $false$ means the opposite.

**Time variables** are: $Js_j$ is a start time of a job $j$, $Ts_{fl}$ is a start time of a transfer of a file $f$ over a link $l$, $Fs_{fc}$ is a start time of a placement of a file $f$ at a node $c$, $Fdur_{fc}$ is a duration of a placement of a file $f$ at a node $c$.

In our model we assume that a network link can be modeled as an unary resource with no loss of generality. The measurements in [1] have shown, that a sequential transfer of a set of files does not require more time than a parallel transfer of the same set of files over the same link.

We use an incomplete search which can provide a suboptimal solution of required quality within a given time limit because the final goal is to create a planner that can process requests online. For a better search performance the overall problem is divided into two subproblems and the search is performed in two stages:

(i) Planning Stage: instantiate a part of variables in order to assign resources for each task.
  (a) Assign jobs to computational nodes.
  (b) Select transfer paths for input and output files.
  (c) Estimate a makespan for a given resource assignment $estMakespan$.
  (d) Find a solution for the subproblem with a minimal estimated makespan.

(ii) Scheduling stage: define a start time for each operation.
  (a) Define the order of operations.
  (b) Put cumulative constraints on resources in order to avoid their oversaturation at any moment of time.
  (c) Find a solution with a minimal *makespan* which is the end time of the last task.

At the planning stage we have to assign a transfer path for an input and an output file of each job which can be defined by the following constraints (see Figure 2):

**1.** An input file has to be transferred from one of its sources over exactly one link.

**2.** An output file has to be transferred to one of its destinations over exactly one link.

**3.** An intermediate node (neither source, destination nor selected for the job execution) either has exactly one incoming and outgoing transfer or is not on a transfer path:
$\exists$ incoming transfer $\Leftrightarrow \exists$ outgoing transfer.

**Table 1.** Variables and parameters used in cumulative constrains on resources.

| Task | Start | Duration | Usage | Limit |
|---|---|---|---|---|
| Job | $Js_{jc}$ | $duration_j$ | 1 | $cpu_c$ |
| Transfer | $Ts_{fl}$ | $size_f \cdot slowdown_l$ | 1 | 1 |
| File placement | $Fs_{fc}$ | $Fdur_{fc}$ | $size_f$ | $disk_c$ |

**4.** There must exist exactly one incoming transfer of an input file and exactly one outgoing transfer of an output file at the node which was selected for the job execution.

**5.** A file can be transferred from/to each node at most once.

In addition, we use constraints for loop elimination similarly as it is described in [7].

At the scheduling stage the problem is to assign a start time for each task. The following constraints on order of tasks are implemented:

- An outgoing transfer of a file from a node can start only after an incoming transfer to that node is finished. The first transfer of an input file from its source and the first transfer of an output file from the processing node are exceptions from this constraint.
- A job can start only after the input file is transferred to the selected processing node.
- An output file can be transferred only after the job is finished.
- A reservation of space for a file at a node is made when a transfer to that node starts.
- A file can be deleted from the start node of a link after the transfer is finished.
- A reservation of space for an output file is made at the processing node when the job starts.
- An input file can be deleted from a processing node after the job is finished.

Cumulative constraints are widely used in Constraint Programming for description of resource usage by tasks. Each cumulative constraint requires that a set of tasks given by *start times*, *durations* and *resource usage*, never require more than a *resource limit* at any time. In our case we use three sets of cumulative constraints: for CPUs, storages and links (see Table 1).

## 3. Simulations

The constraint satisfaction problem was implemented using MiniZinc [8] and Gecode [9] was used as a solver. The timelimit was set to 3 minutes for both planning and scheduling stages. The simulations were running under Windows 8 64-bit on a computer with Intel i5 (4 cores) 2.50 GHz processor and 6 GB of memory installed. The Gecode solver was running in a parallel mode using 4 threads.

The simulated environment consisted of 3 nodes: a central storage HPSS ($cpu_{HPSS} = 0$) which was the single source for input files and the single destination for output files, a local processing site and a remote processing site. The slowdown of links between the central HPSS and the local site was set to 0, which means that transfer overheads to/from the local site are negligible comparing to a job duration. The slowdown of the links to/from the remote site was increasing in each simulation proportionally to a slowdown factor. The parameters of jobs were taken from logging system of the STAR experiment's data production at computational site KISTI (Korea Institute of Science and Technology Information) [10]. The average job duration was 3,000 minutes and average time of transfer was 5 and 10 minutes to/from the remote site respectively (in the simulations where the slowdown factor = 1). Then, in further simulations the transfer times increase proportionally to the slowdown factor. In the simulated environment

80% of CPUs were available at the local site and 20% at the remote site. 2,000 of jobs were scheduled stepwise by subsets (chunks) of 200. Storage constraints were not considered in these simulations. Four different scheduling strategies were compared:

**Local:** All the jobs are submitted to the local site only. This strategy was used as a base line for comparison against other strategies.

**Equal CPU load:** Jobs are distributed between nodes with the goal to maintain an equal ratio of job duration per CPU. Each input file is transferred prior to the start of a job. At each node jobs are executed in input order.

**Data transferred by job:** Each CPU pulls a job from the queue when it is idle, then it has to wait for an input transfer before the job execution starts.

**Optimized:** This strategy is based on the model proposed in this paper.

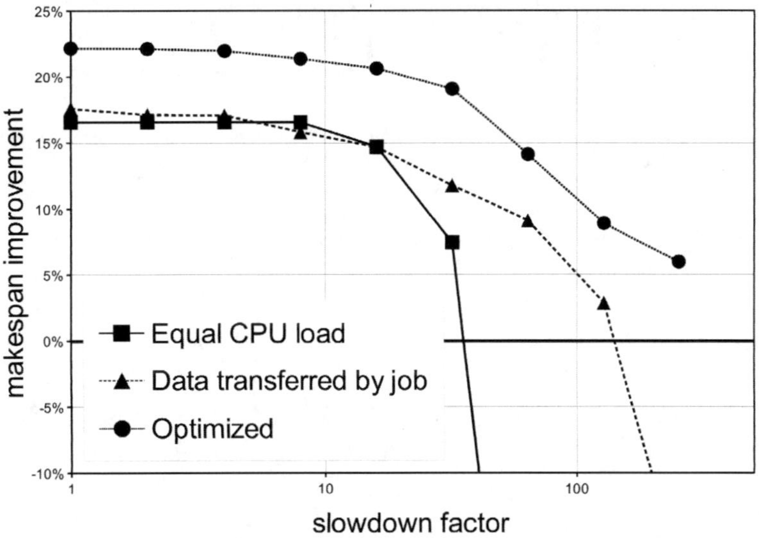


**Figure 3.** Results of simulations for real data production. Three strategies were evaluated and compared to a ideal local production. The optimized solution (our model) clearly provides the highest gain.

The plot at Figure 3 shows the gain in a makespan delivered by different scheduling policies compared to the job execution at the local site only. The curves shows the performance of the scheduling policies when an overhead of transfer to the remote site increases proportionally to the slowdown factor. When the transfer overhead becomes significant both heuristics ("Equal CPU load" and "Data transferred by job") fail to provide an efficient usage of the remote resources (the makespan improvement goes below zero). Negative makespan improvement means that, in this case, it would be faster to process all the data locally than to distribute it between several sites relying on the heuristic. The proposed global planning approach (Optimized) systematically provides a smaller makespan and adapts to the increase of transfer overheads better then the other simulated heuristics. It was able to provide a positive gain in makespan by using remote resources even when the transfer overhead is comparable to a job duration.

## 4. Conclusion

A model for scheduling of data production over Grid was formulated in form of constraint satisfaction problem and solved using constraint programming. The simulations based on data extracted from log files of batch and data management systems of the STAR experiment has shown that the proposed global planning approach systematically provides a smaller makespan and adapts to the increase of transfer overheads better then the other simulated heuristics. The proposed approach can provide an *optimization* and an *automatic adaptation* to fluctuating resources with no need for manual adjustment of a workflow at each site or tuning of heuristics. The future development of global planning for data processing in Grid is ongoing. In future we plan to test this approach on problems of larger size (more nodes, CPU's and links) and improve the search performance in order to enable online scheduling in real environment.

## Acknowledgments

This work has been supported by the Czech Science Foundation (GA-CR) and the Office of Nuclear Physics within the U.S. Department of Energys. We would like to acknowledge and thank Brookhaven National Laboratory for resources that they made available.

## References

[1] Zerola M, Lauret J, Barták R and Šumbera M 2012 One click dataset transfer: toward efficient coupling of distributed storage resources and CPUs *J. Phys.: Conf. Series* **368** 012022
[2] Makatun D, Lauret J and Šumbera M 2014 Study of cache performance in distributed environment for data processing, *J. Phys.: Conf. Series* **523** 012016
[3] Horký J, Lokajíček M and Peisar J 2013 *Influence of Distributing a Tier-2 Data Storage on Physics Analysis* (Beijing: 15th Int. Workshop on Advanced Computing and Analysis Techniques in Phys. Res.)
[4] Betev L, Gheata A, Gheata M, Grigoras C and Hristov P 2014 Performance optimisations for distributed analysis in ALICE *J. Phys.: Conf. Series* **523** 012014
[5] Balewski J, Lauret J, Olson D, Sakrejda I, Arkhipkin D, Bresnahan J, Keahey K, Porter J *et al.* 2012 Offloading peak processing to virtual farm by STAR experiment at RHIC *J. Phys.: Conf. Series* **368** 012011
[6] Rossi F, Beek P and Walsh T 2006 *Handbook of Constraint Programming* (Amsterdam: Elsevier)
[7] Troubil P and Rudová H 2011 Integer Linear Programming Models for Media Streams Planning. (Int. Conf. on Applied Operational Res.) *Lecture Notes in Management Sc.* **3** 509-22
[8] Nethercote N, Stuckey P, Becket R, Brand S, Duck G and Tack G 2007 MiniZinc: Towards a Standard CP Modelling Language *Lecture Notes in Comp. Sc.* **4741** 529-543
[9] Tack G 2008 *Gecode: an Open Constraint Solving Library* (Paris: OSSICP08 Workshop at CP-AI-OR08)
[10] Korea Institute of Science and Technology Information (KISTI) http://en.kisti.re.kr/

# High-speed zero-copy data transfer for DAQ applications

**Flavio Pisani [1,2], Daniel Hugo Cámpora Pérez[1] and Niko Neufeld[1]**

[1] CERN, CH-1211 Geneva 23, CH

[2] Dipartimento di fisica, Università degli Studi di Roma "La Sapienza", Piazzale Aldo Moro 5, 00185 Roma, IT

E-mail: `flavio.pisani@cern.ch`

**Abstract.** The LHCb Data Acquisition (DAQ) will be upgraded in 2020 to a trigger-free readout. In order to achieve this goal we will need to connect around 500 nodes with a total network capacity of 32 Tb/s. To get such an high network capacity we are testing zero-copy technology in order to maximize the theoretical link throughput without adding excessive CPU and memory bandwidth overhead, leaving free resources for data processing resulting in less power, space and money used for the same result. We develop a modular test application which can be used with different transport layers. For the zero-copy implementation we choose the OFED IBVerbs API because it can provide low level access and high throughput. We present throughput and CPU usage measurements of 40 GbE solutions using Remote Direct Memory Access (RDMA), for several network configurations to test the scalability of the system.

## 1. Introduction

The LHCb [1] experiment is one of the four big experiments placed in the Large Hadron Collider (LHC), situated in Geneva Switzerland. The detector collects data from proton-proton collisions at design center of mass energy of 14 TeV and a design rate of 40 MHz. To handle the big amounts of data produced by the detector the Data Acquisition System (DAQ) selects interesting events according to the underling physics in a multi-stage process usually called trigger.

In 2019-2020 the LHCb DAQ system will be upgraded in order to sustain the increasing amounts of data produced by the detector and to filter the data without a low level hardware trigger. Figure 1 shows the network design of the system, which follows a two-stage design. Readout Units (RUs) gather subdetector-specific incoming data fragments, and distribute them across Builder Units (BUs), following a many-to-one network flow pattern. The receiving BU is selected by a load-balanced distribution, managed by a centralized dispatcher named Event Manager (EM), and data is sent over a 100 Gb/s based event building network. In the next stage, the Filtering Units (FUs) receive the data from the BUs in a one to one pattern via a 10 Gb/s network. A more detailed description of the LHCb DAQ network upgrade can be found in [2].

We present a zero-copy implementation of a transport layer which is handling the event building traffic between RUs, BUs and EM. This transport layer is integrated in the Data Acquisition Protocol Independent Evaluator [2] (DAQPIPE) emulation software, which emulates an DAQ network traffic and makes possible to measure the performances of different network technologies.

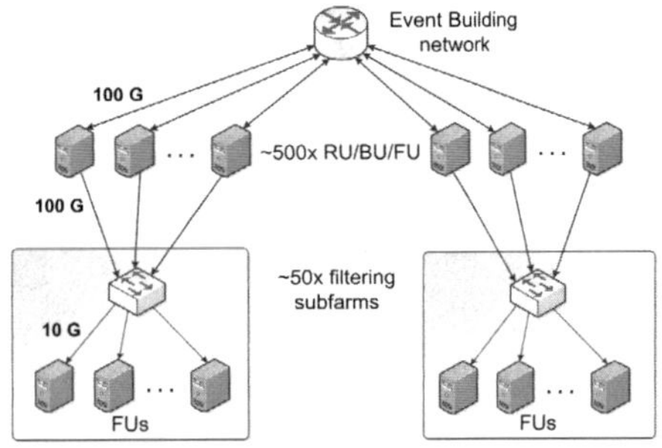


**Figure 1.** The LHCb upgrade network design.

## 2. Remote Direct Memory Access (RDMA)

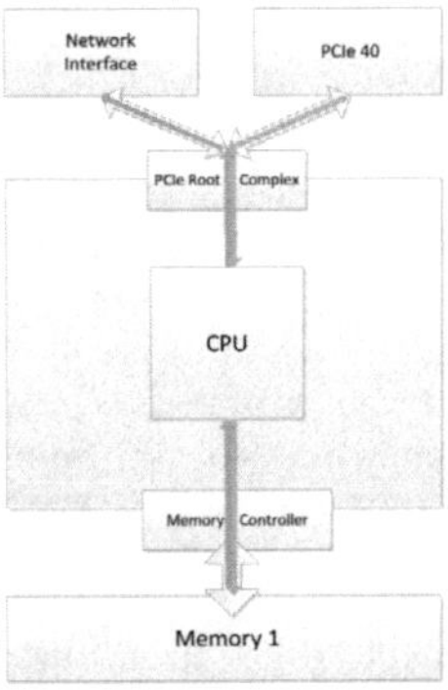


**Figure 2.** Example of non-RDMA data transfer from a PCIe read-out card to a network interface.

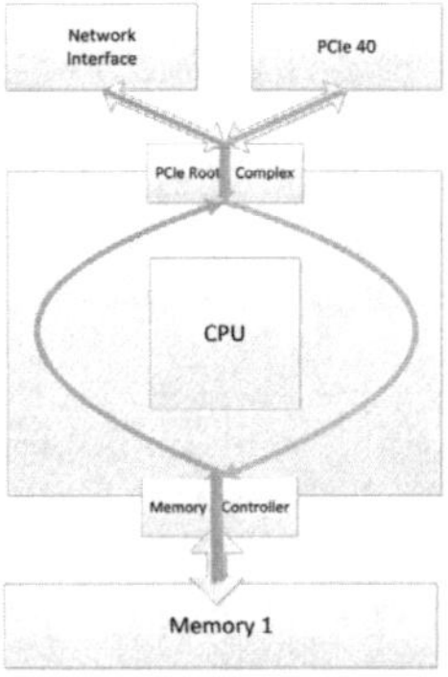


**Figure 3.** Example of RDMA data transfer from a PCIe read-out card to a network interface.

Traditional data transfer over the network requires, as shown in Figure 2, several operations to be performed by the CPU and the OS: moving the data between application space and kernel space; handling all the steps required by the protocol stack currently in use; moving the data between the memory and the network card. These operations produce CPU time and memory bandwidth overhead which is increasing with the increasing bandwidth that can be achieved with modern network technologies, for example 40/100 GbE or Infiniband [3].

Using RDMA data transfers, the data are written to or read from addresses in application space directly by the network card, as shown in Figure 3. For transferring data in this schema the user has to provide an available physical memory address space on both the local and the remote machine. This process, referred to as memory registration, is CPU intensive and locks the physical addresses on the RAM, on the other hand using RDMA removes all CPU and memory overhead introduced during the transmission itself. The address of the remote node needs to be exchanged over the network adding a small transmission overhead every time we want to transfer a message to a new destination memory address. Due to the nature of the overheads introduced by an RDMA transfer it is a suitable solution for transferring big chunks of data and reusing memory buffers where possible.

## 3. Protocol and API

In this section we briefly expose the essential parts of the IBVerbs API [4] and the protocol used in our DAQ implementation.

### *3.1. The IBVerbs API*

The IBVerbs library offers a user space API to access the RDMA capabilities of different devices across a variety of network technologies. The library allows the final user to abstract from the complexity of the hardware providing an hardware independent structure.

Below we discuss OFED-specific terminology [5] and aspects of the IBVerbs library which are necessary to understand the problems that may occur in a DAQ application and how to solve them. Further informations can be found in [6] and [7].

- **Work queue (WQ):** FIFO (First In First Out) queue used by the user to issue a work request to the network card. The work requests are then executed in a non-blocking way.
- **Completion queue (CQ):** FIFO queue used by the network card to notify the completion of a previously posted work request. Several WQs may share the same CQ.
- **Memory region (MR):** Chunk of memory ready to be used as source or destination of a data transfer.
- **Send op-code:** Sends data from a registered MR. There is no need to know the remote memory region but a *receive* request must be posted, on the receiving WQ by the receiver of the message, before the message is sent.
- **Receive op-code:** Writes data to a registered MR. Must be posted to the WQ before the *send* request.
- **RDMA write op-code:** Writes data from a locally registered MR to a remotely registered MR. The remote node does not receive any notification in the CQ when the transfer is over, the sender has to communicate this to the receiver.

Registering a new MR is a very time-consuming process which requires several operations to be performed by the OS and the driver [8]. In order to avoid registration and deregistration of the memory for every transfer, it is preferable to use designated memory buffers, which are registered only once.

The *Send/Receive* operations can be used to send small control messages to preregistered temporary MRs without introducing the heavy overhead of an *RDMA write* operation. To receive a *send* operation properly, a corresponding *receive* must have been posted to the WQ. To achieve this, one possible strategy, is to post enough *receive* request to receive the maximum number of messages that can be sent at once.

On the other hand the use of temporary buffers and the large number of work requests issued at the same time make *Send/Receive* a bad choice for transferring large data chunks to a specific memory location without any additional copy. The *RDMA write* is, at the contrary, a very suitable solution for sending large amount of data. To minimize the overhead introduced by memory registration a good practice, for a DAQ application, it is to define and register a large buffer in memory and to write the data to a subset of it.

### *3.2. Protocol*

Figure 4 depicts all the messages exchanged during the event building process for two different protocols, namely push and pull. The protocol determines the communication starter between RUs and BUs, in the push implementation the transfer is initiated by the RUs, in the pull implementation by the BUs. In both protocols the event is assigned to a specific BU by the EM every time there is a new trigger event. The trigger frequency can be configured at run time.

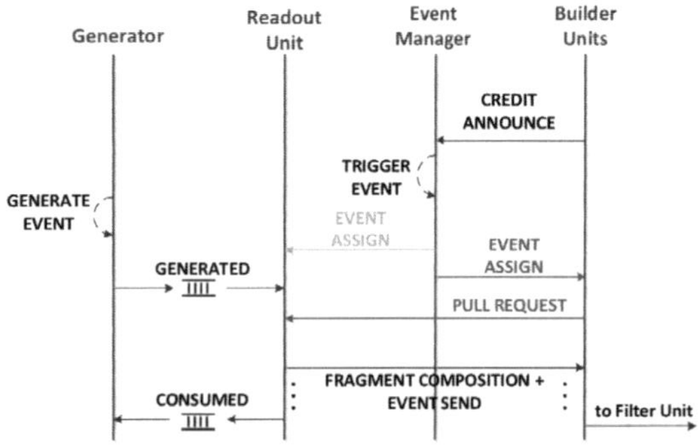


**Figure 4.** Protocol flow design. Push-specific messages are depicted in light green, and pull-specific ones are depicted in blue.

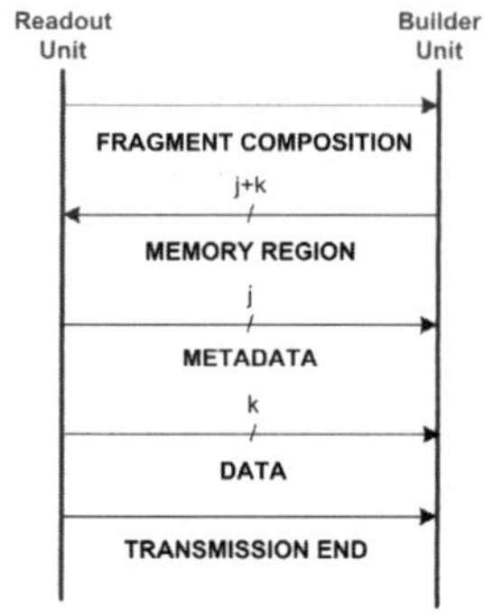


**Figure 5.** Fragment composition and event send flow design using RDMA transport layer.

The workload is balanced between the BUs using a credit system. The available resources are divided in slots, every time there is a free slot on a BU a *credit announce* message is generated. The EM assigns the events to the BUs according to the available slots. A more detailed description of the protocols and the messages exchanged is described in [2].

Figure 5 shows more in detail the messages exchanged between RUs and BUs in order to send the event fragments using an IBVerbs transport implementation. A full description of the messages follows:

(i) **FRAGMENT COMPOSITION:** The RU sends a fragment composition structure, which contains the number and the sizes of metadata and data fragments that are going to be sent, using a *send* operation.

(ii) **MEMORY REGION:** The BU sends, using a *send* operation, a valid MR for each fragment.

(iii) **METADATA and DATA:** The RU sends the metadata and the data fragments using and *RDMA write* operation.

(iv) **TRANSMISSION END:** The RU sends, using a *send* operation, a transmission complete message including the event ID of the completed event.

The need to send a MR before transferring the data and the metadata makes the protocol inherently pull, a possible way to avoid this is to use preregistered MR sent during the initialization of the system. Because each RU cannot calculate the exact amount of data collected by all the others RUs without gathering all the *fragment compositions*, to avoid dangerous overlaps in the data buffer, the RU must take into account the maximum possible size for every event resulting in higher memory usage.

## 4. Bandwidth and CPU usage Measurements

In order to test the performances of the transport layer implementation, we have built a test system consistent of 5 nodes with the hardware and software described in table 1. The nodes are connected through a Brocade ICX 7750 40 GbE switch and the network cards are configured for operating in Ethernet mode. In order to reduce the network overhead we enabled jumbo frames[1] on both the network cards and the switch.

[1] Jumbo frames are Ethernet frames with a payload greater that 1500 bytes.

**Table 1.** Server characteristics

| | |
|---|---|
| CPU | 2 x Intel Xeon CPU E5-2670 @ 2.60 GHz |
| Memory | 32 GB |
| Network card | Mellanox MT27500 Connectx3 |
| OFED version | Mellanox 2.2-1.0.1 |
| Linux version | slc 6.5 kernel 2.6.32-431.23.3 |

Table 2 shows the effective event building throughput achieved using both protocols for different number of nodes. In the different configurations tested the effective event building throughput is very close the theoretical maximum and scales up in a consistent way changing the number of nodes in the system for both push and pull.

**Table 2.** Effective event building throughput for different configurations

| Number of nodes | pull throughput [Gb/s] | push throughput [Gb/s] |
|---|---|---|
| 2 | 38.3 (95.7%) | 38.3 (95.7%) |
| 3 | 39.0 (97.6%) | 39.0 (97.4%) |
| 4 | 39.0 (97.5%) | 39.1 (97.6%) |
| 5 | 38.8 (96.9%) | 38.8 (97.0%) |

The total CPU usage, calculated using sar[2], for the DAQPIPE application using the IBVerbs transport is 0.94%. A full duplex transmission done by iperf reports a system CPU usage of 6.8%, we can provide an estimation of 86% lower CPU usage using RDMA transport.

## 5. Conclusions and future work

The Verbs implementation of the event building transport layer is performing according to the specifications in all the tested scenarios: push and pull protocol connecting up to 5 nodes through the ICX 7750 switch. The CPU usage is 86% lower using RDMA transport, leaving free resources for other tasks, like for opportunistic High Level Trigger (HLT) execution.

The scalability of the system up to 5 nodes is very promising, but larger scale tests will be performed in the short future. The many to one nature of the traffic makes the scalability to hundreds of nodes a challenging task, not only from an application point of view, but also from a networking perspective.

We will continue testing 40 Gb/s technologies looking forward to 100 Gb/s in the future, which should provide enough bandwidth for a complete trigger-free readout.

[2] The sar command writes to standard output the activity counters of the operating system. For more information see the sar manual.

## References

[1] The LHCb Collaboration et al 2008 *Journal of Instrumentation* **3** S08005
[2] Cámpora Pérez D H, Schwemmer R and Neufeld N *Protocol-Independent Event Building Evaluator for the LHCb DAQ System* (Proceedings of RT2014. Submitted to Transactions of Nuclear Science, pending review.)
[3] Infiniband Trade Association (IBTA) URL `http://www.infinibandta.org/`
[4] Mellanox Technologies 2001 *Mellanox IB-Verbs API (VAPI)*
[5] Open Fabric Alliance (OFED) URL `https://www.openfabrics.org/index.php`
[6] Grun P 2010 *Introduction to infiniband for end users*
[7] Mellanox Technologies 2014 *RDMA Aware Programming user manual*
[8] Mietke F, Baumgartl R, Rex R, Mehlan T, Hoefler T and Rehm W 2006 *Proceedings of Euro-Par 2006 Parallel Processing* (Springer-Verlag Berlin) pp 124–133 ISBN 3-540-37783-2

# Native Language Integrated Queries with CppLINQ in C++

**V Vassilev**
CERN, PH-SFT, Geneva, Switzerland
E-mail: `vvasilev@cern.ch`

**Abstract.** Programming language evolution brought to us the domain-specific languages (DSL). They proved to be very useful for expressing specific concepts, turning into a vital ingredient even for general-purpose frameworks. Supporting declarative DSLs (such as SQL) in imperative languages (such as C++) can happen in the manner of language integrated query (LINQ).

We investigate approaches to integrate LINQ programming language, native to C++. We review its usability in the context of high energy physics. We present examples using CppLINQ for a few types data analysis workflows done by the end-users doing data analysis. We discuss evidences how this DSL technology can simplify massively parallel grid system such as PROOF.

## 1. Introduction

Informally, imperative programming languages (such as C++) describe the necessary steps to get desired results, i.e the 'how' to perform the computations. On the other hand declarative programming languages (such as SQL) describe the desired result, i.e the 'what' the result should look like. Over the years, interoperability between both worlds has become vitally important to almost every general-purpose programming language. The interoperability in imperative languages is usually implemented through disabled type safety at compile time. Inter-process communication was done via compiler and library support mainly employing the type variants [1] [2] or type erasure. Another well-known approach to integrate domain-specific declarative programming language support is using strings, passed to bridging layers. For example, SQL is embeddable in a C++ application through string literals. They are passed to a driver, querying the database behind, producing results and finally wrapping them into C++ objects.

Comprehension syntax is very close to the syntax of a number of practical database query languages. A language of comprehensions can naturally and uniformly express operations on various collection types [3]. C# [4] introduced an incarnation of the comprehensions with its Language Integrated Queries (LINQ). LINQ's key concept is to provide a type-safe comprehension (SQL-like) syntax, operating on collections. The C# language provided rich language constructs, allowing LINQ implementation on top of the existing features. Moreover, LINQ introduced the new programming paradigm integrated into the type system of the underlying imperative language.

The newer C++ language specification standards introduced the necessary concepts to build type-safe, platform-independent comprehension syntax in C++. Since then a few open-source implementations of LINQ for C++ emerged. All of them are in early development stages, but definitely worth reviewing their power and usability in the C++ context. In particular we shall:

- describe the advantages of the computational abstraction;
- show the expressiveness of the comprehension syntax via concrete code snippets.

This paper is divided into sections as follows: Section 2, CppLINQ Overview, discusses in brief some of the advantages of the computation abstraction and how it can come naturally by introducing the comprehension syntax in C++. Section 3, Usability & Applications, shows the capabilities of CppLINQ and its possible usage scenarios in the field of the high-energy physics. Section 4, contains Conclusion & Future work.

## 2. CppLINQ Overview

The declarative programming paradigm helps to abstract out details such as computation control flow and tends to reduce side effects [5]. The paradigm puts less emphasis on the 'control' part of Kowalski's equation *algorithm* = *logic* + *control* [6]. Thus it leaves a lot of room for interpretation on how the logic semantics could be realized. The same logic can be expressed in many different ways, depending on the standpoints or the implicit requirements. Thus the system can delegate pinning down the control to a specialized layer, which performs further planning, optimizations and scheduling.

Going declarative can be too harsh for various reasons: some concepts are difficult to express; different algorithms exist only in an imperative language or are more efficient in an imperative language; the difficulty of educating developers; etc. These drawbacks lead to a hybrid approach – mixing a declarative language in an imperative one. The hybrid approach gives the flexibility to the developers of choosing the computational abstraction only where it makes most sense.

C# LINQ provides a declarative way to easily extract and process data from arrays, iterable classes, XML documents and relational databases. After its success in the .NET world it was ported to many unmanaged environments such as JavaScript and C++. CppLINQ is a common name of LINQ technology, implemented in C++. The implementation techniques vary, as a result there are a few CppLINQ dialects. Depending on the dialect, the same program logic could be expressed in different ways. Listing 1 introduces a typical CppLINQ program. The snippet illustrates invariant statements for every CppLINQ program. Writing code in CppLINQ is straightforward: include the CppLINQ-specific headers (line 1); add a using CppLINQ clause (line 7); use the 'from' statement (line 8) to enter the CppLINQ world.

```
1 #include "linq.h" // Include the necessary header files.
2 // Returns true if the parameter is a prime number.
3 bool is_prime(int x);
4
5 long sum_primes() {
6   auto xs = int_range(0, 100); // same as boost::counting_iterator.
7   using namespace cpplinq; // Add the using CppLINQ clauses.
8   return from(xs) // Switches to the CppLinq world.
9          .where(is_prime)
10         .sum();
11 }
```

Listing 1: A typical LINQ-style program in C++

### *2.1. Implementation Approaches*

Listing 1 demonstrate how to sum all prime numbers in the range of (0, 100) using CppLINQ. Some implementations rely on the 'operator.' and others on other operators, which are allowed to

Journal of Physics: Conference Series **608** (2015) 012030 doi:10.1088/1742-6596/608/1/012030

be overloaded. Usual criticism of the 'operator.' approach is that 'operator.' is not overloadable in C++. This makes user extensions very hard to implement. Alternatively, LINQ provides other hooks to handle third-party extensions. Another common implementation approach is using an 'overloadable' operator such as 'operator<<' or 'operator>>', at the cost of code readability.

One of the most mature and tested implementation of CppLINQ is by Microsoft Reactive Components research group [7]. It uses only standardized C++, making the library platform-independent. All references to CppLINQ in the paper relate to this particular implementation. Without limiting the generality of CppLINQ we will show a few examples in the context of high-energy physics.

## 3. Usability & Applications

The ROOT Framework [8] has become a standard for the field of high-energy physics. The tool is widely adopted for data analysis. A huge number of physicists are using ROOT on daily basis to perform data analysis on large scale data sets. A usual data analysis program starts with selection of the interesting pieces of data (interesting physics events). Physicists (data analysts) generally have a theory and expectations. They usually think in terms 'what' and not 'how', i.e they think about the logic and not the control of the Kowalski equation. For example, in order to check a theory or an expectation, one needs to 'select' all events, 'where' their properties match some criteria and 'order' them or 'group' them by certain properties. Translation of the process is straightforward to be done in terms of a comprehension syntax. Currently, the ROOT framework can be used only by an imperative manner (recommended is C++, but Python, Ruby and .NET bindings are available).

Lets consider a naive (and common) data layout, presented on Listing 2. A physics event data consists of a number of physics events, each one consists of number of particles.

```
//...
enum ParticleType { Photon = 0, Electron, Kaon, Pion };

struct Particle {
  ROOT::Math::XYZVector fPosition; // vertex position
  ROOT::Math::PtEtaPhiMVector fVector; // particle vector
  int fCharge; // particle charge
  ParticleType fType; // particle type
};

struct EventData {
  std::vector<Particle> fParticles; // particles of the event
};
std::vector<EventData> Events;
```

Listing 2: Physics event data sample

Our example data model is serialized on disk in the form of a ROOT tree in a ROOT file. Reading the data from disk in ROOT5 is tedious and cumbersome for various reasons. A major obstacle is ROOT5's non-conforming C++ interpreter and its lack of template support. This results in an awkward, non-type-safe implementation of data access. Moreover, iteration over the data entries with iterators is not possible.

Since the new C++ interpreter cling [9] came into service in ROOT6, ROOT can interact with very complex C++ constructs (for example, none of STL/Boost has to be hidden anymore, meaning ROOT can be exposed to heavily templated code). This allows implementing a strongly-typed ROOT file reader (aka TreeReader). The type safety comes through the

TTreeReaderValue (Listing3 line 4), where the user specifies the expected types and ROOT6 can report if there was a type mismatch.

```
1 void ReadEventData(std::vector<std::vector<Particle>>& v) {
2   std::unique_ptr<TFile> myFile = std::make_unique<TFile>(TFile::Open("eventdata_s99.root"));
3   TTreeReader tree("tree", myFile);
4   TTreeReaderValue<std::vector<Particle>> particles_value(tree, "fParticles"); //
5   while (tree.Next()) {
6      v.push_back(*particles_value); // particles_value is dereferenced as a 'normal' iterator.
7   }
8 }
```

Listing 3: Reading event data using TTreeReader class in ROOT6

The nested data structures can be flattened by using the TTreeReaderArray (line 4 on Listing 4) concept, which iterates type-safely over iterable containers.

```
1 void ReadEventDataFlat(std::vector<Particle>& v) {
2   std::unique_ptr<TFile> myFile = std::make_unique<TFile>(TFile::Open("eventdata_s99.root"));
3   TTreeReader tree("tree", myFile);
4   TTreeReaderArray<Particle> particles(tree, "fParticles"); //
5   while (tree.Next()) {
6     for (auto& p: particles)
7       v.push_back(p);
8   }
9 }
```

Listing 4: Reading event data elements using TTreeReaderValue class in ROOT6

ROOT6's TTreeReader helps the user express iterations over ROOT trees by implementing the std::iterator pattern. The inclusion of the TTreeReader in ROOT6 allows CppLINQ to operate on data coming directly from ROOT files in the form of tuples or trees. We will show three examples of event selection using the comprehension syntax. The new data reader is flexible and it is effortless to integrate it in new environments such as CppLINQ. All examples can run as ROOT6 macros or at the ROOT6 prompt [10].

Listing 5 shows how to open a ROOT file and prepares the TreeReader. Then using CppLINQ the number of events is calculated (line 7). Line 9 computes the count of all particles in all events in our data. Line 11 displays the particle count in the first 10 events.

```
1 root [0] #include "EventData.h"
2 root [1] TFile* f = TFile::Open("eventdata_s99.root");
3 root [2] TTreeReader t("tree", f);
4 root [3] TTreeReaderArray<Particle> particles(t, "fParticles");
5 root [4] #include "linq.hpp"
6 root [5] using namespace cpplinq;
7 root [6] from(t).count() //
8 (typename std::iterator_traits<iterator>::difference_type) 200
9 root [8] from(t).select([&](Long64_t){ return from(particles).count();}).sum() //
10 (size_t) 9729
11 root [9] from(t).take(10).select([&](Long64_t){ return from(particles).count();}).sum() //
12 (size_t) 492
```

Listing 5: CppLINQ at ROOT6's prompt

All code snippets use 'standard' LINQ operators [11]. Most of the operators take a closure as a parameter in the form of C++11 lambda function. The lambda functions are used to express the concrete rules to be held for every event selection.

Listing 6 shows how to select all events from a given ROOT tree, which have at least four particles with more than 15 GeV energy.

```
void Example1(const TTreeReader& tree, float Emin = 15.0) {
  TTreeReaderArray<Particle> particles(tree, "fParticles");
  typedef decltype(from(particles)) ParticleItr_t;
  auto goodParticles
    = from(tree)
      .select([&](Long64_t) { return from(particles); })
      .where([&](const ParticleItr_t &plist) {
         return plist.where([&](const Particle &p) { return p.fVector.E() > Emin; }).count() >= 4;});
  //...
```

Listing 6: Selection of events having at least 4 particles with energy more than N GeV

Listing 7 demonstrates mixing imperative concepts into CppLINQ. Alternatively, in the particular case one could use the default_if_empty operator. If there are cases where one needs to rely on the host programming language, it is possible to switch back to imperative mode.

```
void Example2(const TTreeReader& tree) {
  TTreeReaderArray<Particle> particles(tree, "fParticles");
  auto result = from(from(tree)
                    .where([&particles](Long64_t) {
                       bool hasAtLeast2 = from(particles)
                                         .where([](const Particle& p) { return p.fType == Pion; })
                                         .count() > 2;
                       auto electrons = from(from(particles)
                                             .where([](const Particle& p) { return p.fType == Electron; }))
                                         .select([](const Particle& p){ return p.fVector.Pt(); });
                       if (hasAtLeast2 && !electrons.empty())
                         return electrons.max();
                       return 0.;
                    }))
               .where([](double pt){ return pt != 0; }).count();
  //...
```

Listing 7: Select all events having at least 2 pions and show the momentum of the leading electron

```
void Example3(const TTreeReaderArray<Particle>& particles) {
  auto PosPtSum = from(tree).select([&](Long64_t) {
                         return from(particles)
                                .where([](const Particle& p) { return p.fCharge > 0; })
                                .select([](const Particle& p) { return p.fVector.Pt(); });
                                .sum();
                      });
}
 auto h = new TH1F("ptSum", "Sum p_T of events; p_T [GeV]", 200, 0, 500);
 from(PosPtSum).all([&h](double pt){h->Fill(pt); return true;});
 //...
```

Listing 8: Calculate the event momentum distribution for all positively charged particles

Listing 8 shows how CppLINQ works with the TTreeReaderArray by flattening the event data structure and iterating over all particles from in events in order to sum the momenta of the individual particles. More examples and demos can be found online [10].

## 4. Conclusion & Future work

CppLINQ is a platform-independent, open-source library, which uses well-standardized features of C++. It provides concepts to abstract out details of the algorithms. Combined with ROOT6. they can be helpful in the analysis of physics events. The library relies on iterator design pattern [12]. It works smoothly with STL and Boost.

Another complementary and fundamental concept of the technology is the delayed, on request, computation. They introduce parallelism through *expression trees* and will be a subject of a future paper. The technology is already adopted in cloud systems. They prove that the delayed computation and computational abstraction can be beneficial in many cases. Introducing the delayed computation concept from CppLINQ could enhance the current communication model between users and batch systems or between ROOT and PROOF, for example.

Abstracting out the control of the program opens a lot of opportunities. For example, it would be much easier to translate a data flow visual programming language for physics data analysis into C++. The data flow visual language constructs could be compiled into their CppLINQ representations, and vice-verse – CppLINQ programs could be visualized by a data flow visual representation, making them more understandable.

Certainly, the domain is rather new and opens a broad research and development horizon. While CppLINQ needs some more work to become more robust, mixing comprehensions with C++ is beneficial in many cases in the field of high-energy physics. It provides the necessary computational abstraction layer, which can be a mediator between programmers and tools. Such tools could perform powerful analyses, transformations, optimizations and parallelizations dynamically and interactively.

## References

[1] MSDN VARIANT structure URL `http://msdn.microsoft.com/en-us/library/windows/desktop/ms221627%28v=vs.85%29.aspx`

[2] Cantu M 2008 *Essential Pascal* (CreateSpace Independent Publishing Platform) chap 10

[3] Buneman P, Libkin L, Suciu D, Tannen V and Wong L 1994 Comprehension syntax *SIGMOD Rec.* **23** 87–96 ISSN 0163-5808 URL `http://doi.acm.org/10.1145/181550.181564`

[4] Hejlsberg A, Wiltamuth S and Golde P 2003 *C# Language Specification* (Boston, MA, USA: Addison-Wesley Longman Publishing Co., Inc.) ISBN 0321154916

[5] Lloyd J W 1994 Practical advantages of declarative programming *Joint Conference on Declarative Programming, GULP-PRODE* vol 94 p 94

[6] Kowalski R 1979 Algorithm = logic + control *Communications of the ACM* **22** 424–436 ISSN 0001-0782 URL `http://doi.acm.org/10.1145/359131.359136`

[7] Microsoft reactive components team reactive extensions URL `http://msdn.microsoft.com/en-us/data/gg577609.aspx`

[8] Brun R and Rademakers F 1997 ROOT – an object oriented data analysis framework *Nuclear Instruments and Methods in Physics Research Section A: Accelerators, Spectrometers, Detectors and Associated Equipment* **389** 81–86 ISSN 0168-9002 new Computing Techniques in Physics Research V URL `http://www.sciencedirect.com/science/article/pii/S016890029700048X`

[9] Vasilev V, Canal P, Naumann A and Russo P 2012 Cling – the new interactive interpreter for ROOT 6 *Journal of Physics: Conference Series* **396** 052071 URL `http://stacks.iop.org/1742-6596/396/i=5/a=052071`

[10] Vassilev V ROOT 6 examples with cpplinq URL `https://github.com/vgvassilev/RxCpp/tree/master/Ix/CPP/samples`

[11] MSDN The .NET standard query operators URL `http://msdn.microsoft.com/en-us/library/bb394939.aspx`

[12] Gamma E, Helm R, Johnson R and Vlissides J 1995 *Design Patterns: Elements of Reusable Object-oriented Software* (Boston, MA, USA: Addison-Wesley Longman Publishing Co. Inc.) ISBN 0-201-63361-2

# Recent Developments in the CernVM-File System Server Backend

**R Meusel, J Blomer, P Buncic, G Ganis, S Heikkila**
CERN PH-SFT
CERN, CH-1211 Geneva 23, Switzerland
E-mail: rene.meusel@cern.ch

**Abstract.** The CernVM File System (CernVM-FS) is a snapshotting read-only file system designed to deliver software to grid worker nodes over HTTP in a fast, scalable and reliable way. In recent years it became the de-facto standard method to distribute HEP experiment software in the WLCG and starts to be adopted by other grid computing communities outside HEP. This paper focusses on the recent developments of the CernVM-FS Server, the central publishing point of new file system snapshots. Using a union file system, the CernVM-FS Server allows for direct manipulation of a (normally read-only) CernVM-FS volume with copy-on-write semantics. Eventually the collected changeset is transformed into a new CernVM-FS snapshot, constituting a transactional feedback loop. The generated repository data is pushed into a content addressable storage requiring only a RESTful interface and gets distributed through a hierarchy of caches to individual grid worker nodes. Additonally we describe recent features, such as file chunking, repository garbage collection and file system history that enable CernVM-FS for a wider range of use cases.

## 1. Introduction

The CernVM File System (CernVM-FS) is a read-only file system [1] designed for accessing large centrally installed software repositories based on HTTP. The repository content is distributed through multiple layers of replication servers and caches to individual grid worker nodes where CernVM-FS repositories are usually mounted as a FUSE [2] file system. Both meta-data and file contents are downloaded on first access and cached locally. Nowadays CernVM-FS is a mission-critical tool for the distribution of HEP software [3, 4] in the World-wide LHC Computing Grid [5] and recently gains adoption in other fields [6]. The four LHC experiments sum up to some 110 million file system objects and 4 TB of file contents, which doubled in the last two years (Figure 1 shows the growth of the ATLAS software repository as an example).

Prior to distributing, the centrally installed software repository content is prepared by the CernVM-FS server tools. Each file is compressed and stored in a content-addressable storage based on a cryptographic hash of the file content. This allows for trivial consistency checks and yields de-duplication on file content level. Relevant file system meta information (i.e. path hierarchy, access flags, file names, ...) is stored in hierarchical file catalogs which are eventually put into the content-addressable storage along with the actual data.

Journal of Physics: Conference Series **608** (2015) 012031 doi:10.1088/1742-6596/608/1/012031

## 2. Usage Statistics and New Use Cases

In the last two years the amount of LHC experiment software accessible through CernVM-FS has more than doubled with a steady growth rate. In August 2014 CernVM-FS repositories under the domain `cern.ch`[1] were providing on-demand access to more than 7 TB of software and conditions data stored in about 125 million file system objects (i.e. files, directories, symlinks). All repositories hosted at CERN have been successfully migrated to the CernVM-FS 2.1.x repository scheme as of the 15th of September 2014 which is a crucial precondition to benefit from any of the discussed features in this work.

Beyond that, there are many more CernVM-FS repositories hosted by other institutions including but not exclusively DESY, NIKHEF and Fermilab. CernVM-FS's popularity inspired some new use cases and hence new challenges in the future development of the system. In this section we present statistics figures and a summary for future challenges due to these new use case scenarios.

### *2.1. Statistics for LHC Experiment Software Repositories*

The four LHC experiment software repositories are accounting for more than 4 TB of files and nearly 110 million file system objects. We have seen a steady growth for those repositories (see Figure 1) while the average file size increased only slightly. However, in Section 2.2 we take a look at other CernVM-FS repositories that exceed the average file size of the LHC experiment's repositories by several orders of magnitude (cf. Tables 1 and 2).

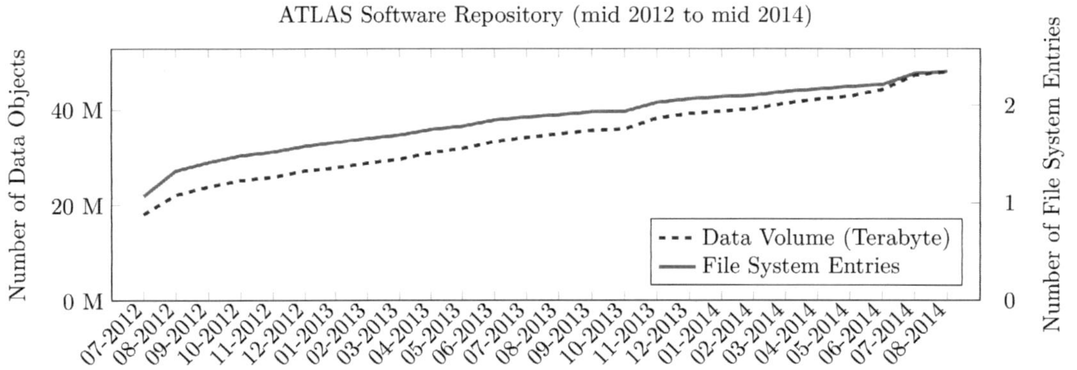


**Figure 1.** Growth of the ATLAS software repository (atlas.cern.ch) in the last 24 months. Showing a doubling in both file system entries (red) and available data volume (blue).

[1] alice-ocdb.cern.ch, alice.cern.ch, ams.cern.ch, atlas-condb.cern.ch, atlas.cern.ch, belle.cern.ch, boss.cern.ch, cernvm-prod.cern.ch, cms.cern.ch, geant4.cern.ch, grid.cern.ch, lhcb.cern.ch, na61.cern.ch, sft.cern.ch

| Repository | # File Objects | # Data Objects | Volume | Avg File Size |
|---|---|---|---|---|
| atlas.cern.ch | 48'000'000 | 3'700'000 | 2.2 TB | 68 kB |
| cms.cern.ch | 37'000'000 | 4'800'000 | 0.9 TB | 33 kB |
| lhcb.cern.ch | 15'800'000 | 4'600'000 | 0.5 TB | 43 kB |
| alice.cern.ch | 7'000'000 | 240'000 | 0.5 TB | 94 kB |
| **sum/average:** | 107'800'000 | 13'340'000 | 4.1 TB | 60 kB |

**Table 1.** File system statistics of conventional LHC experiment software repositories.
Effective: August 2014

### 2.2. *New Use Cases of CernVM File System Impose New Challenges*

CernVM-FS's scalability depends on the efficiency of its aggressive cache hierarchy. Therefore we assume that repository updates are infrequent (i.e. hours to weeks), individual distributed files are small ($< 500$ kiB) and the accessed working set of geographically collocated worker nodes is similar. These assumptions hold true for pure software repositories, hence the widespread and successful utilisation of CernVM-FS. Nevertheless, emerging use cases are relaxing some of these assumptions:

(i) Some LHC experiments began to distribute conditions databases[2] through CernVM-FS [3]. These files are accessed in very similar patterns as the experiment software itself, but the average *file size is orders of magnitude larger* (cf. Tables 1 and 2). CernVM-FS has to treat large files differently to avoid cache cluttering and performance degradation.

| Repository | # File Objects | # Data Objects | Volume | ∅ File Size |
|---|---|---|---|---|
| ams.cern.ch | 3'700'000 | 1'900'000 | 2.0 TB | 0.7 MB |
| alice-ocdb.cern.ch | 700'000 | 700'000 | 0.1 TB | 0.2 MB |
| atlas-condb.cern.ch | 8'400 | 7'800 | 0.5 TB | 60.7 MB |
| **sum/average:** | 4'400'000 | 2'600'000 | 2.6 TB | 20.5 MB |

**Table 2.** File system statistics of repositories containing experiment conditions data.
*Note:* ams.cern.ch does contain software also
Effective: August 2014

(ii) CernVM-FS persistently stores all historic snapshots of a repository, making it a good fit for long term analysis software preservation [7]. However, it lacks a means to easily manage and access these preserved snapshots.

(iii) It is planned to use CernVM-FS to distribute nightly integration build results to simplify development and testing of experiment software. While such repositories naturally experience large and frequent updates (i.e. LHCb would publish about 1'000'000 files summing up to 50 GB per day), the particular build results need to stay available only for a short period of time. Hence, the long term persistency mentioned before *quickly fills up the repository's backend storage* with outdated revisions and requires garbage collection.

(iv) With an increasing number of institutions hosting and replicating CernVM-FS repositories the ecosystem is evolving from a centralised and controlled software distribution service into a more mesh-like structure with many stakeholders. In this new environment configuration and distribution of public keys becomes increasingly challenging for individual clients.

[2] Repositories: atlas-condb.cern.ch and alice-ocdb.cern.ch

## 3. New and Upcoming Features in the CernVM-FS Server

The features presented here do not represent an exhaustive list but are a selection of changes relevant to the challenges we present in Section 2.2. Note that some of the features are already available in the current stable release (i.e. CernVM-FS 2.1.19) while others are scheduled for a release in the next 3 to 6 months.

### *3.1. File Chunking for Large Files*

CernVM-FS and its caching infrastructure were designed to serve many small files as efficiently as possible. Assuming that small files (f.e. binary executables or scripts) will be read entirely, CernVM-FS always downloads complete files on `open()` and stores them in a local cache. Larger files ($> \sim$ 50MiB) are much more likely to contain data resources (f.e. SQLite databases) that are usually read sparsely. Such access patterns in large files would unnecessarily clutter the cache and waste bandwidth.

Newer CernVM-FS Servers[3] therefore divide large files into smaller chunks, allowing for better cache exploitation and partial downloads. We use the XOR32 rolling checksum [8] for content aware cut mark detection to exploit de-duplication between similar large files in our content-addressable storage format. Hence, the chunk size is not static but varies in a configurable range (default: 4 MiB $<$ size $<$ 16 MiB).

This new feature is particularly interesting for repositories hosting large files like conditions databases. It is completely transparent for both the repository maintainer and the users of CernVM-FS and enabled by default.

### *3.2. Named CernVM File System Snapshots and Rollbacks*

Keeping track of historic snapshots in a CernVM-FS repository was cumbersome as they were identified only by a SHA-1 hash comparable to a commit hash in Git. With named file system snapshots[4] we now provide a means to catalogue these snapshots along with an arbitrary name and a description. The benefit of this history index is two-fold: facilitated mounting of ancient repository revisions on the client side and on the other hand the possibility to rollback revisions on the server side.

Nevertheless, we do not intend to transform CernVM-FS into a fully featured version control system. Named snapshots are meant to simplify the handling of old file system snapshots only. A server side rollback therefore invalidates all named snapshots that succeed the snapshot targeted by the rollback. The server side rollback is meant as an *undo*-feature. Thus we prevent the occurrence of diverging branches in a CernVM-FS repository.

### *3.3. Garbage Collection on Revision Level*

CernVM-FS is following an insert-only policy regarding its backend storage. Data is only added and references to it are updated, while preceding snapshots are never removed from the repository's backend storage. This yields a system where historic state remains reconstructible at the expense of an ever growing backend storage. In Section 3.2 we described how this can be utilised for long term preservation of software environments.

However, in use cases like the publishing of nightly integration build results this exhaustive history is not required or would even become prohibitively large in a short period of time.

In an experimental nightly build repository set up by LHCb, about one million new files per nightly build release summing up to some 50 GB were published. Due to de-duplication the backend storage (600 GB volume) was able to store the repository for about one month before

[3] Experimental file chunking as of version 2.1.7 (stable since 2.1.19)

[4] Named Snapshots are available since CernVM-FS 2.1.15

running full. Figure 2 plots the number of contained files (red) compared to referenced data objects (blue) and the stored data objects (brown) over the repository revisions.

Note that removing files from the CernVM-FS repository (see revisions 28 to 34 in Figure 2) leads to a decrease in the number of referenced data objects while the number of stored objects remains the same. The additionally stored data objects are referenced by historic snapshots but not in the latest revision of the repository. Since this use case does not require preservation of historic revisions, the yellow area describes the amount of garbage accumulated in the backend storage.

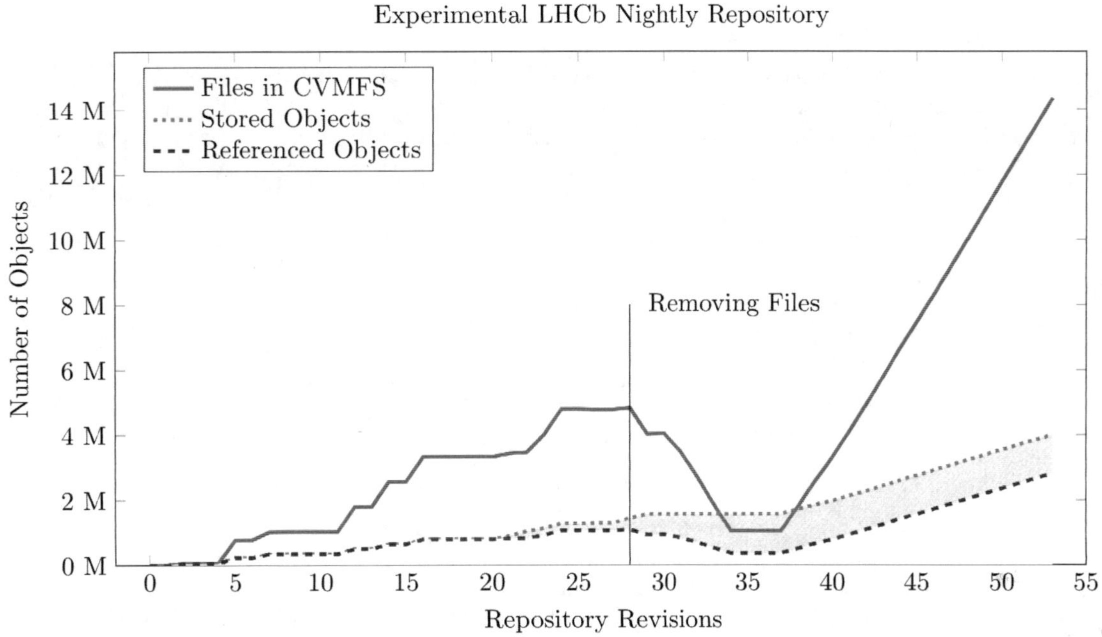


**Figure 2.** Statistics of an experimental LHCb nightly build repository comparing referenced files (red), referenced data objects (blue) and stored data objects (brown). Area in yellow depicts potential garbage in the repository's backend storage.

We are currently working on a garbage collector[5] for CernVM-FS on revision level that is striving to solve this problem. It will remove historic snapshots from a CernVM-FS repository including all data objects that are not referenced by any conserved snapshots. Due to the architecture of CernVM-FS's internal data structures (i.e. the file system catalogs) as a Merkle Tree [9] and a hash based content-addressable storage for all data objects; a simple mark-and-sweep garbage collector can efficiently detect such garbage objects. We already demonstrated the feasibility by a proof-of-concept implementation of such a garbage collector on the LHCb nightly repository.

### *3.4. Configuration Repository for Bootstrapping CernVM-FS Clients*

With a quickly growing number of CernVM-FS repositories hosted and replicated by various institutions the configuration of clients on both worker nodes and interactive machines becomes

[5] Planned for a release with CernVM-FS 2.1.20

increasingly complex. Therefore we are adding support for a special configuration repository[6] that is envisioned to contain both default settings and public keys for various scenarios and production repositories. This is meant to simplify the distribution of (default) configuration for many repositories through a central entry point.

Site administrators would point their CernVM-FS clients to this configuration repository instead of setting up repository and replication URLs and public keys manually. Hence, mounting a repository becomes a two-stage process: First the configuration repository is mounted to retrieve the default settings for the actual production repository which is mounted as a second step. As a result, the configuration of CernVM-FS will be drastically simplified.

## Summary
In this work we outlined use cases for CernVM-FS that have been emerging while the system became widely used along with the challenges they implied for the further development of CernVM-FS. In Section 3 we introduced new features that address these new requirements.

The new server components of CernVM-FS 2.1.x cope with a wider range of use cases while still keeping the focus on scalable software and conditions data distribution. In the next releases we will simplify the configuration of CernVM-FS clients as well as introduce a garbage collection feature allowing repository maintainers to prune unneeded historic repository snapshots.

## References
[1] Jakob Blomer, Predrag Buncic, and Thomas Fuhrmann. Cernvm-fs: Delivering scientific software to globally distributed computing resources. In *Proceedings of the First International Workshop on Network-aware Data Management*, NDM '11, pages 49–56, New York, NY, USA, 2011. ACM.
[2] FUSE: Filesystem in Userspace.
[3] Jakob Blomer, Carlos Aguado-Snchez, Predrag Buncic, and Artem Harutyunyan. Distributing lhc application software and conditions databases using the cernvm file system. *Journal of Physics: Conference Series*, 331(4):042003, 2011.
[4] J Blomer, P Buncic, I Charalampidis, A Harutyunyan, D Larsen, , and R Meusel. Status and future perspectives of cernvm-fs. *Journal of Physics: Conference Series*, 396(5):052013, 2012.
[5] LHC Computing Grid Technical Design Report.
[6] C Condurache and I Collier. Cernvm-fs beyond lhc computing. *Journal of Physics: Conference Series*, 513(3):032020, 2014.
[7] Dag Toppe Larsen, Jakob Blomer, Predrag Buncic, Ioannis Charalampidis, and Artem Haratyunyan. Long-term preservation of analysis software environment. *Journal of Physics: Conference Series*, 396(3):032064, 2012.
[8] Kendy Kutzner. *The decentralized file system Igor-FS as an application for overlay-networks.* PhD thesis, Karlsruhe Institute of Technology, 2008.
[9] Ralph C. Merkle. A digital signature based on a conventional encryption function. In *A Conference on the Theory and Applications of Cryptographic Techniques on Advances in Cryptology*, CRYPTO '87, pages 369–378, London, UK, UK, 1988. Springer-Verlag.

[6] `$CVMFS_CONFIG_REPOSITORY` is planned for CernVM-FS 2.1.20

# Techniques and tools for measuring energy efficiency of scientific software applications

**David Abdurachmanov[1], Peter Elmer[2], Giulio Eulisse[3], Robert Knight[4], Tapio Niemi[5], Jukka K. Nurminen[6], Filip Nyback[6], Gonçalo Pestana[5 6], Zhonghong Ou[6], Kashif Khan [5 6]**

[1] Digital Science and Computing Center, Faculty of Mathematics and Informatics, Vilnius University, Vilnius, Lithuania
[2] Department of Physics, Princeton University, Princeton, NJ 08540, USA
[3] Fermilab, Batavia, IL 60510, USA
[4] Research Computing, Office of Information Technology, Princeton University, Princeton, New Jersey 08540, USA
[5] Helsinki Institute of Physics, PO Box 64, FI-00014, Helsinki, Finland
[6] Aalto University, PO Box 11100, 00076 Aalto, Finland

E-mail: goncalo.pestana@aalto.fi

**Abstract.** The scale of scientific High Performance Computing (HPC) and High Throughput Computing (HTC) has increased significantly in recent years, and is becoming sensitive to total energy use and cost. Energy-efficiency has thus become an important concern in scientific fields such as High Energy Physics (HEP). There has been a growing interest in utilizing alternate architectures, such as low power ARM processors, to replace traditional Intel x86 architectures. Nevertheless, even though such solutions have been successfully used in mobile applications with low I/O and memory demands, it is unclear if they are suitable and more energy-efficient in the scientific computing environment. Furthermore, there is a lack of tools and experience to derive and compare power consumption between the architectures for various workloads, and eventually to support software optimizations for energy efficiency. To that end, we have performed several physical and software-based measurements of workloads from HEP applications running on ARM and Intel architectures, and compare their power consumption and performance. We leverage several profiling tools (both in hardware and software) to extract different characteristics of the power use. We report the results of these measurements and the experience gained in developing a set of measurement techniques and profiling tools to accurately assess the power consumption for scientific workloads.

## 1. Introduction

The Large Hadron Collider (LHC) [1] at the European Laboratory for Particle Physics (CERN) in Geneva, Switzerland, is an example of a scientific project whose computing resource requirements are larger than those likely provided in a single computer center. Data processing and storage are distributed across the Worldwide LHC Computing Grid (WLCG) [2], which uses resources from 160 computer centers in 35 countries. Such computational resources have enabled the CMS [3] and ATLAS [4] experiments to discover the Higgs Boson [5, 6], for example. The WLCG requires a massive amount of computational resources (250,000 x86 cores in 2012) and, proportionally, energy. In the future, with planned increases to the LHC luminosity [7], the dataset size will increase by 2-3 orders of magnitude, presenting even more challenges in terms of energy consumption.

In order to find and develop better solutions for improving energy efficiency in High Energy Physics (HEP) computing, it is important to understand how energy is used by the HEP systems themselves. We describe several tools and techniques that facilitate researchers to reach that goal.

As energy efficiency becomes a concern, new solutions have been considered to develop energy efficient systems. One potential solution is to replace the traditional Intel x86 architectures by low power architectures such as ARM. A comparison of the energy efficiency between ARMv7 and x86 Intel architecture is conducted in this article. The experiments use CMS workloads and rely on the techniques and tools described earlier to perform the measurements.

This article is structured as following. Firstly, we describe where is energy consumed in a HTC system and outline some of the tools and techniques available to measure and monitor energy consumption on HTC systems (Section 2). Secondly, we present the results of a comparison between ARMv7 and Intel Xeon architecture using CMS workloads (Section 3). Finally, we present IgProf, a general purpose, open source application performance profile. In addition, we describe its recent added energy profiling features and 64-bit ARM support.

## 2. Tools and techniques for energy measurement

When optimizing power usage, there are two granularities at which one can look at a computing system. The coarser granularity takes into account the behavior of the whole node (or some of its passive parts, e.g. the transformer) as part of a rack in a datacenter. This is usually investigated when engineering and optimizing computing centers. Alternatively, a more detailed approach is to look into the components which make up the active parts of a node, in particular the CPU and its memory subsystem since these are responsible for a sizeable fraction of the consumed power. They are also the place where the largest gains in terms of efficiency can be obtained through optimizations in the software.

If one is simply interested in the coarse power consumption by node, external probing devices can be used: monitoring interfaces of the rack power distribution units, plugin meters and non-invasive clamp meters (allowing measurement of the current pulled by the system by induction without making physical contact with it). They differ mostly in terms of flexibility. Their accuracy is typically a few percent for power, whereas their time resolution is in the order of seconds. This is more than enough to optimize electrical layout of the datacenters or to provide a baseline for more detailed studies.

A alternative approach takes into account the internal structure of a computing element of an HTC system, as shown in figure 1. Nowadays, every board manufacturer provides on-board chips which monitor energy consumption of different components of the system. These allow energy measurements of fine grained detail, as it is possible to individually monitor energy consumption of components such as the CPU, its memory subsystem, and others. An example of this chip monitors is the Texas Instruments TI INA231 [8] current-churn and power monitor which is found on the ARMv7 developer board which we used for our studies. It is quite common in the industry. Compared to external methods, these on-board components provide high accuracy and reasonably high precision measurements (millisecond level).

A special and slightly different case of these on-board monitors is a new technology called Running Average Power Limit (RAPL), provided by Intel beginning from the Sandy Bridge family of processors.

Contrary to other solutions, which are implemented as discrete chips, RAPL is embedded as part of the CPU package itself and provides information on the CPUs own subsystems. In particular RAPL provides data for three different domains: **package** (pck), which measures energy consumed by the system's sockets, **power plane 0** (pp0), which measures energy

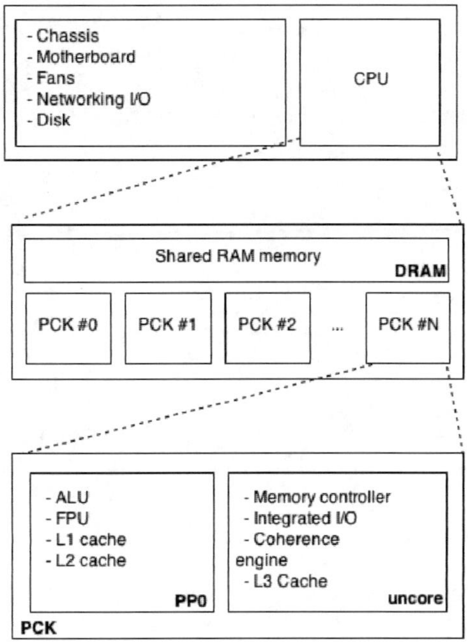


**Figure 1.** Components that contribute for power consumption in HPC

consumed by the CPU core(s), and **dram**, which accounts for the sum of energy consumed by memory in a given socket, therefore excluding the on-core caches [9]. As for the discrete components case, the timing resolution of measurements is in the millisecond range [10]. This is fine enough to permit exploiting such data to build an energy consumption sampling profiler for applications, similar to how performance sampling profilers work (see section 4). Finally, in addition to power monitoring of the sockets, RAPL can limit the power consumed by the different domains. This feature, usually referred as power capping, allows the user to define the average power consumption limit of a domain in a defined time window and allows more accurate independent measurements of the non limited components.

## 3. Power efficiency measurements with x86-64 and ARMv7

In this section, we demonstrate the potential of some of the tools we previously described. To that end, we perform several measurements of workloads from CERN, running on different architectures. The workloads used in the experiment run on top of Intel x86-64 architecture, traditionally used in HTC and data centers and 32 bit ARMv7 architectures (for similar studies for 64bit ARMv8 and Xeon Phi, please refer to [11]). The ARM architecture, initially developed for mobile devices, has been considered [12, 13] as a potential alternative to Intel in HTC, given its energy efficient computing. We also present a brief comparison between ARM and Intel architectures from the energy consumption perspective, based on the results obtained.

### *3.1. Tools and techniques*

For the Intel architecture, we used the RAPL technology to perform measurements of the energy consumed by the package, DRAM and cores (figure 1). The external measurements for the baseline were performed using a rack PDU, which provides an online API to gather the energy consumed by the system on the rack at a sampling rate of 1 second. For the ARM board, we used the Texas Instrument power monitor chip TI INA231 which allows reading of the energy consumed by the cores and dram at a sampling rate of microseconds. The chip was embedded in the board from the vendor. For the external measurements, we used an external plug-in power monitor with a computer interface for gathering and storing the results. In both cases we read the data as it was exposed to the system via the sysfs / devfs knobs. The machine specifications can be seen in figure 2.

| Name | Architecture | CPU | Cores | RAM Memory | Notes |
|---|---|---|---|---|---|
| Intel™ Xeon™ | Intel Sandy Bridge ™ | CPU E5-2650 | 32 | 252 GB | System on a rack |
| Exynos5 Octa Cortex™ | ARMv7 | A15 and/or A7 cores (big.LITTLE technology) | 4 | 2 GB | Development board |

**Figure 2.** Machine's specifications

*3.2. Experiment setup*

The workload used for the experiments was ParFullCMS, a multi-threaded Geant4 [14] benchmark application which uses a complex CMS geometry for its simulation. Using ParFullCMS, we ran simulation tasks on both the Intel and ARM machines (figure 3). The workflow was run several times with different number of threads in each machine. The number of threads run in each experiment is chosen according to the number of the cores of the machines.

*3.3. Analysis*

As expected, the ARMv7 architecture shows encouraging results from the energy efficiency perspective than Intel in all the experiments performed. Also as expected, both architectures do not perform better when overcommitted (more threads than the physical number of cores). Notice ARM results when overcommitted (8 threads) are much worse then Intel ones. This is due to the relatively modest amount of available DRAM (figure 2), causing the machine started swapping, greatly affecting performance. While this was expected, since the ARM system used is just a development board for mobile applications, this is a clear indication that when doing a final assessment of power efficiency for an architecture, one needs to have a full server-grade system in order to make a proper comparison.

## 4. Profiling for power efficiency

The hardware components described above provide measurements that are related to the full set of processes running on the machine. For the simple case where only a single benchmark application is running, these can be used to make comparisons between architectures. A further step is to try to see if there is a way to map the energy consumption measurements to functions and methods within an executing process. Such a mapping would allow for optimizations of the software itself. This kind of mapping can be done in two different ways, which we call *instrumentation* and *sampling profiling*.

In the *instrumentation* case, effective readings of profiled quantities (e.g. energy consumption), or quantities correlated with them (e.g. CPU power state transitions) are done at the beginning and the end of a profiled task and the difference between the two is used to estimate average power consumption over that period of time. By bookkeeping starting and stop values for monitored tasks one can get a fairly complete picture of what is happening to the system, provided the measured interval is large compared to the temporal precision of the measure being done. This is both to avoid a large error on the average estimation and to reduce performance overhead due to the measure itself.

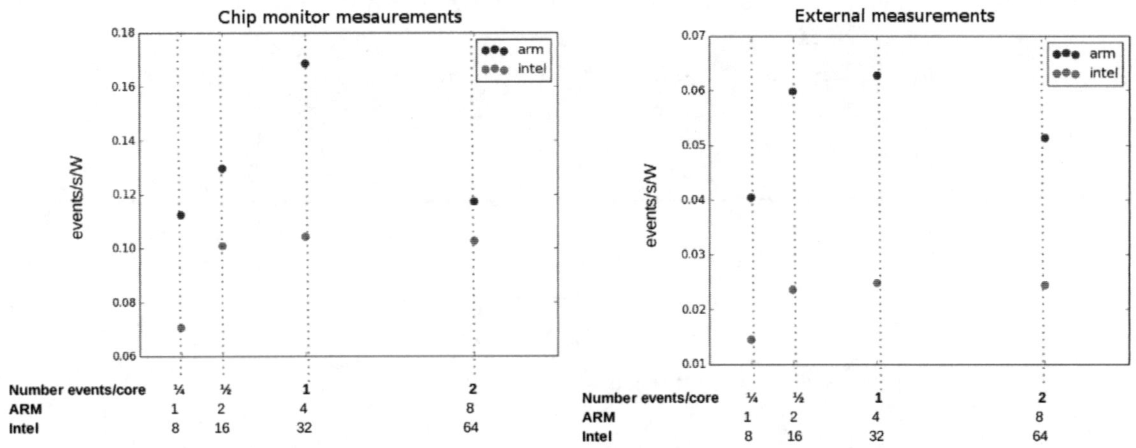


**Figure 3.** Chip monitor and external measurements results. The results are shown according to the relation events per number of cores of each machine and their absolute number of cores.

*Sampling profiling*, on the other hand, has a different approach where a given quantity is sampled regularly and at each sample the measured quantity is accumulated until it overflows a user provided limit. When this happens the profiler increments a counter for the process / function being executed in that precise moment. Assuming that the distribution of where time is spent in a system is constant over time (which is typically true for large data processing tasks), such a sampling algorithm converges to the actual distribution of the measured quantity. The advantage of this approach is that the fidelity of the measurement to first approximation depends only on the number of samples made, regardless of the error on the profiled quantity. This also allows minimizing the performance overhead by tuning the sampling period to be much larger than the measurement itself.

`IgProf` is a general purpose, open source application performance profiler. It was developed in HEP, but it is capable of profiling all types of software applications. The profiler has been available on the x86 and x86-64 platforms since many years [15, 16], and recently we have also ported it to ARMv7 and ARMv8. Moreover we have now added a statistical sampling energy profiling module which provides function level energy cost distribution [17]. Such a module uses the PAPI library to read energy measurements from the RAPL interface previously described.

To illustrate the new module, we use it to profile the memory benchmark `STREAM` [18]. Figure 4 compares the results from performance and energy profiling of the benchmarking tool. The X-axis describes the four main functions contributing to the execution time and energy consumption of the stream tool: Add, Copy, Triad and Scale. The left scale of the Y-axis and the perf_ticks series describe the execution time spent in each function, whereas the right scale of the Y-axis and the nrg_pkg, nrg_pp0 and nrg_pp1 series describe the amount of energy spent in each function. The energy consumption of the processor package domain and the power plane 0 (describing the CPU cores) seem to follow the time spent in the functions, whereas the energy consumption of power plane 1 seems to be fairly constant to zero (describing the unused GPU) .

As we would expect from a simple benchmark, the profiling results of a simple single-threaded application shows a correlation between the execution time and the energy spent in a function. While the energy profiling module is now fully functional within `IgProf`, further work needs to be done to tune the measurements and to gain experience with how to use the profiles obtained

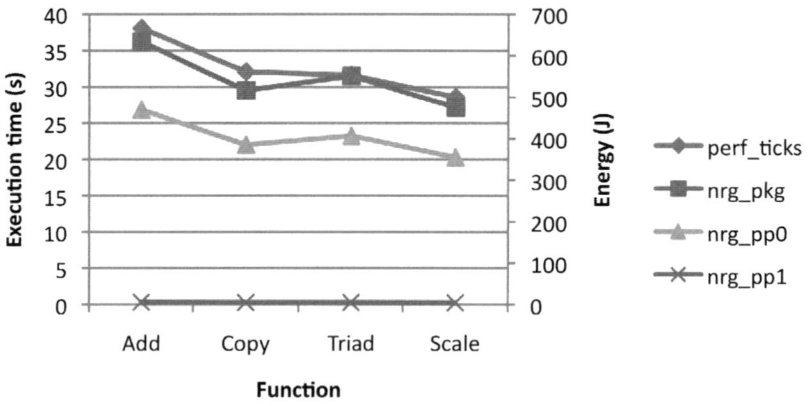


**Figure 4.** The results of performance and energy profiling of the `STREAM` tool.

for more complex applications.

## 5. Conclusions

Energy efficiency has become a major concern for HTC, given the large amount of computing resources - and thus energy - that recent experiments require. LHC computing is a prime example of the need for energy efficient facilities, given its present requirements and costs constraints. The need for energy efficiency drives an interest in accurately evaluating the different components of a HTC system to understand how and where energy is consumed and improve the overall efficiency. However, HTC systems are complex and composed of different components. In this paper we have presented a number of techniques and tools that provide insight into how and where energy is consumed from different perspectives and granularities. In addition, `IgProf`, an open source profiling tool, has been extended to run on 64-bit ARM and to provide function-level energy profiling capabilities. Using these tools and techniques we have also reported studies done to compare the energy performance of x86-64 and ARMv7 processors, confirming the potential of ARMv7 for efficient HTC systems should server grade systems be built around such chips.

## Acknowledgements

This work was partially supported by the National Science Foundation, under Cooperative Agreement PHY-1120138, and by the U.S. Department of Energy. ARMv8 and energy profiling support in IgProf was also supported by Google Summer of Code (GSoC 2014).

## References

[1] Lyndon Evans and Philip Bryant. LHC Machine. *JINST*, 3:S08001, 2008.
[2] Bird I. Computing for the Large Hadron Collider. *Annual Review of Nuclear and Particle Science*, 61, 2011.
[3] S. Chatrchyan et al. The CMS experiment at the CERN LHC. *JINST*, 3:S08004, 2008.
[4] Aad G et al (Atlas Collaboration). The Atlas Experiment at the CERN Large Hadron Collider. *JINST*, 3, S08003.
[5] S. Chatrchyan et al. Observation of a new boson at a mass of 125 GeV with the CMS experiment at the LHC. *Phys.Lett.*, B716:30–61, 2012.
[6] G. Aad et al. Observation of a new particle in the search for the Standard Model Higgs boson with the ATLAS detector at the LHC. *Phys.Lett.*, B716:1–29, 2012.
[7] L Rossi and O Bruning. High Luminosity Large Hadron Collider A description for the European Strategy Preparatory Group. Technical Report CERN-ATS-2012-236, CERN, Geneva, Aug 2012.
[8] Texas Instruments (TI) INA 231. `http://www.ti.com/product/ina231`. Last time accessed: 17-09-2014.

[9] *Intel 64 and IA-32 Software Developer Manuals - Volume 3.* www.intel.com/content/www/us/en/processors/architectures-software-developer-manuals.html.

[10] Balaji Subramaniam and Wu chun Feng. Towards energy-proportional computing for enterprise-class server workloads. *CPE '13 Proceedings of the 4th ACM/SPEC International Conference on Performance Engineering*, 2012.

[11] David Abdurachmanov, Brian Bockelman, Peter Elmer, Giulio Eulisse, and Robert Knight. Heterogeneous High Throughput Scientific Computing with APM X-Gene and Intel Xeon Phi. *to be published in ACAT 2014 proceedings*, 2014.

[12] David Abdurachmanov, Peter Elmer, Giulio Eulisse, and Shahzad Muzaffar. Initial explorations of arm processors for scientific computing. *Journal of Physics: Conference Series*, 523(1):012009, 2014.

[13] David Abdurachmanov, Kapil Arya, Josh Bendavid, Tommaso Boccali, Gene Cooperman, Andrea Dotti, Peter Elmer, Giulio Eulisse, Francesco Giacomini, Christopher D Jones, Matteo Manzali, and Shahzad Muzaffar. Explorations of the viability of arm and xeon phi for physics processing. *Journal of Physics: Conference Series*, 513(5):052008, 2014.

[14] S. Agostinelli et al. Geant4 a simulation toolkit. *Nuclear Instruments and Methods in Physics Research Section A: Accelerators, Spectrometers, Detectors and Associated Equipment*, pages 250–303. DOI: 10.1016/S0168-9002(03)01368-8.

[15] Giulio Eulisse and Lassi Tuura. IgProf Profiling Tool. *Proc. Computing In High Energy Physics (CHEP04) (Interlaken)*, 2004.

[16] IgProf, the Ignominous Profiler. `http://igprof.org/`. Accessed 2014-09-30.

[17] Vincent M. Weaver, Matt Johnson, Kiran Kasichayanula, James Ralph, Piotr Luszczek, Dan Terpstra, and Shirley Moore. Measuring energy and power with PAPI. *Proceedings of the International Conference on Parallel Processing Workshops*, pages 262–268, 2012. DOI: 10.1109/ICPPW.2012.39.

[18] John D. McCalpin. STREAM: Sustainable memory bandwidth in high performance computers. Web page. `http://www.cs.virginia.edu/stream/`. Accessed 2014-09-30.

# Heterogeneous High Throughput Scientific Computing with APM X-Gene and Intel Xeon Phi

**David Abdurachmanov[1], Brian Bockelman[2], Peter Elmer[3], Giulio Eulisse[4], Robert Knight[5], Shahzad Muzaffar[4]**

[1] Digital Science and Computing Center, Faculty of Mathematics and Informatics, Vilnius University, Vilnius, Lithuania
[2] University of Nebraska-Lincoln
[3] Department of Physics, Princeton University, Princeton, NJ 08540, USA
[4] Fermilab, Batavia, IL 60510, USA
[5] Research Computing, Office of Information Technology, Princeton University, Princeton, New Jersey 08540, USA

E-mail: David.Abdurachmanov@cern.ch

**Abstract.** Electrical power requirements will be a constraint on the future growth of Distributed High Throughput Computing (DHTC) as used by High Energy Physics. Performance-per-watt is a critical metric for the evaluation of computer architectures for cost-efficient computing. Additionally, future performance growth will come from heterogeneous, many-core, and high computing density platforms with specialized processors. In this paper, we examine the Intel Xeon Phi Many Integrated Cores (MIC) co-processor and Applied Micro X-Gene ARMv8 64-bit low-power server system-on-a-chip (SoC) solutions for scientific computing applications. We report our experience on software porting, performance and energy efficiency and evaluate the potential for use of such technologies in the context of distributed computing systems such as the Worldwide LHC Computing Grid (WLCG).

## 1. Introduction and Motivation

Processing the data produced by High Energy Physics (HEP) experiments like those at the Large Hadron Collider (LHC) [1] at the European Laboratory for Particle Physics (CERN) requires significant computing resources. The scale is beyond those available in a typical single computer center. The Worldwide LHC Computing Grid (WLCG) was established to provide the computing resources needed for the LHC experiments, and used, for example, by the CMS and ATLAS experiments for the discovery of the Higgs boson [2, 3]. It is a distributed computing resource across 170 computing centers in 40 countries. The CMS experiment, for example, used between 80,000 and 100,000 x86_64 cores from the WLCG for its processing needs in 2012. Planned luminosity upgrades of the LHC [4] will result in a 2-3 order of magnitude increase in dataset sizes over the next 15 years, requiring commensurate increases in processing capacity.

Intel is the leading company in general purpose server processors market. Alternative solutions, like ARMv8 64-bit, aim to provide high-density and energy efficient platforms for computing centers. These platforms delivered by multiple vendors are optimized for specific market segments while ensuring compatibility by running a common instruction set, i.e. a single ecosystem is shared. General purpose solutions provide benefits to a full software stack unlike graphics processing units (GPU), which require software redesign and only applicable for portions of the application.

In this paper we focus on two general purpose central processors: Applied Micro (APM) X-Gene 1 ARMv8 64-bit Server-on-Chip and Intel's Xeon Phi coprocessor. We investigate these two platforms as alternatives to the x86_64 family of processors used in the WLCG today. We base our results on performance (events per second) scalability over power (watts) usage. The power measurements we provide in this paper are for the silicon chip and not for a full computing node.

### *1.1. Market*

According to Gartner, Intel sold 92 percent of all server processors in 2013 [5]. Distributed High Throughput Computing (DHTC) is also dominated by x86_64. In contrast, ARM Holdings is a leader in low power and high energy efficiency processors market. Their business model differs from Intel as they provide license agreements for their intellectual property (IP) to partners. Instruction set (ISA) licenses allow partners to create custom silicon chips for wide variety of applications: mobile phones, notebook, servers and many others. In addition to providing ISA licenses ARM Holdings offers already designed and verified building blocks (CPU and GPU cores, on-chip interconnect and similar) for manufacturing of silicon chips solutions. Each solution is designed to meet specific market requirements, while keeping the compatibility by running a common ISA, like ARMv8 64-bit. ARM Holdings and partners introduced a new high energy efficient general purpose server products in 2014.

Intel announced Xeon Phi platform, also known as Many Integrated Cores (MIC), for HPC market in 2012. It is a many-core and long-vector machine combining benefits of CPU and GPU into a single product. Xeon Phi is available in a PCIe add-on card form factor and requires a host system with Intel Xeon processors. The paltform provides high computational power and high energy-efficiency solution for HPC market. We have previously reported results for the Xeon Phi [6] CMS Software (CMSSW) port (still incomplete due to issues with the Intel C++ Compiler).

We begin with a description of our efforts to port software and validate ARMv8 64-bit, and then describe a power and computational performance comparison between multiple architectures.

## 2. Software Port and Validation of ARMv8 64-bit

### *2.1. APM X-Gene*

APM is the first company to introduce server grade and custom designed ARMv8 64-bit Server-on-Chip solution, X-Gene 1 built on 40nm process by TSMC semiconductor foundry. A next generation X-Gene 2 built on 28nm processes is already being sampled. Princeton and CERN were provided with XC-1 development boards with a single X-Gene 1 APM883208 silicon chip for software porting efforts. It is primarily intended for speeding up development efforts. The specification of silicon chip is provided in table 1. We have been working with APM for more than a year to deliver CMS Software (CMSSW) and Open Science Grid (OSG) [7] software stack for ARMv8 64-bit.

### *2.2. Software Porting Issues*

We have previously reported work to port the CMS Software (CMSSW) to ARMv7 32-bit System-on-Chip silicon chips [8, 6]. It allowed us to understand many issues arising when porting to non-x86_64 architectures and to prepare for ARMv8 64-bit platforms. Indeed issues resolved for ARMv7 32-bit were often also relevant for ARMv8 64-bit, however some issues specific to ARMv8 64-bit were encountered:

- Update of `autoconf` package was required for successful detection of ARMV8 64-bit platform, `aarch64`.
- In cases where ARMv8 64-bit support was already present, software packages were updated to the newer version.
- We added ARMv8 64-bit support to the ROOT [9] software package. The ROOT (version 5) was extended with the `linuxarm64` target. One important feature, Reflex dictionary generation, is not supported for ARMv8 64-bit. ROOT uses GCCXML as the underlying implementation for data reflection, and GCCXML (based on GCC 4.2.1) is far too old for ARMv8 64-bit support. However, because the data model – LP64 – is the same between x86_64 and ARMv8 64-bit, we were able (as a workaround) to pre-generate dictionaries on x86_64 and then compile them on ARMv8 64-bit. We found that such dictionaries work successfully. More recent releases of ROOT (version 6) use LLVM/Clang in place of GCCXML for dictionary generation, but this still does not support ARMv8 64-bit due to the use of the old Just-in-Time (JIT) interface. This JIT interface will be removed from LLVM/Clang after the version 3.5 release and ROOT is scheduled to migrated to MCJIT interface in 2015 [10]. Until this migration happens `rootcling` will not work on ARMv8 64-bit systems.
- Oracle Instant Client is only provided as proprietary binary blobs and not as source code, and is not available for the ARMv7 32-bit and ARMv8 64-bit architectures. However, none of the standard CMS workflows run on the WLCG depend on Oracle. Only a few specialized applications used at CERN alone need Oracle, for example, loading detector calibrations.
- Toolchain (`GCC` and `binutils`) issues were found and reported upstream to the projects. The final `GCC` 4.9.1 and `binutils` (`bfd` linker) 2.24 is now capable of compiling CMSSW. However, one issue remains open with CERN Virtual File System (CVMFS) package used for CMSSW distribution across Grid sites. We had to modify the way CVMFS components (`libcvmfs_only.a`, `libz.a`, `libsqlite3.a`, `libcurl.a` and `libcares.a`) are merged into the final `libcvmfs.a` library.
- `IgProf` memory and performance profiler was ported to ARMv8 64-bit and `libunwind` library improved with fast backtracing capability for ARMv7 32-bit and ARMv8 64-bit [11].

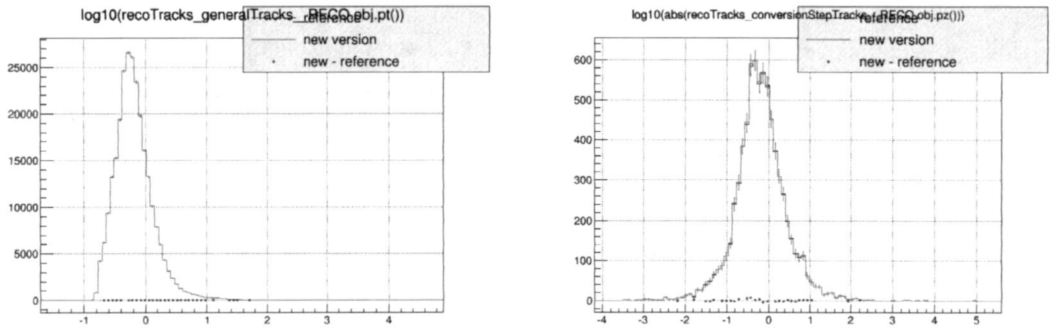


**Figure 1.** Example reconstruction validation distributions

*2.3. ARMv8 64-bit Validation*

To confirm that CMSSW provides comparable results on ARMv8 64-bit and x86_64, an initial validation was done using the reconstruction software workflow. The validation necessarily compared not only the two different architectures (ARMv8 64-bit to x86_64), but also two different Linux distributions (Fedora 19 to Scientific Linux CERN 6.5). Any discrepancies due to the latter, such as different math libraries, should be resolved in future validations. In addition, given that only a single board was available for ARMv8 64-bit, the validation was statistics limited to see any very subtle effects. With these minor caveats the observed differences are minimal (figure 1 shows two examples) and within physics validation limits.

### *2.4. Grid Software*

A new architecture such as ARMv8 64-bit would likely be initially deployed as a batch worker node. Thus we have also built the minimal set of required OSG software packages to support that: voms-clients, HTCondor and CVMFS. CVMFS is used by CMS to distribute software and small data packages around Grid sites. Thus no local installation of these packages is required. CMSSW for ARMv8 64-bit is available alongside with CMSSW for x86_64 in CVMFS. We have successfully used CMSSW release validation test suite on a local ARMv8 64-bit machine. It runs a number of predefined CMSSW (`cmsRun`) workflows with small number of events (10 or 100). The input for workflows was generated locally as the first step or fetched from a remote location. The CMSSW workflows successfully fetched their input data from a remote `xrootd` server [12, 13], which demonstrates a functioning voms-clients package on ARMv8 64-bit.

We installed and configured HTCondor on XC-1 development board to be a 8 slots (1-core and 2 GB memory) worker node at Princeton University. An additional x86_64 machine was configured with HTCondor as a master node for our ARMv8 64-bit worker node. The master node was able to communicate with the worker node and successfully execute the tasks.

## 3. Test Environments for Power and Performance Measurements

We now describe the test environments we have used to do power and performance measurements for an Intel Xeon processor, an APM ARMv8 64-bit X-Gene 1 Server-on-Chip and an Intel Xeon Phi coprocessor. As described above, we have demonstrated standard CMSSW applications running on the APM ARMv8 64-bit XC-1 board, however no three-way comparison of the platforms with CMSSW was possible due to the lack of a full CMSSW port on the Xeon Phi. We have thus used the Geant4 [14] benchmark, ParFullCMS, as a simple cross-platform test capable of running in multi-threaded mode. This benchmark uses a complex geometry (from CMS), but it is a standalone application distributed with Geant4. Table 1 shows the selected general purpose platforms used for the performance and energy efficiency benchmark.

**Table 1.** Silicon chips specifications

| | X-Gene 1 | E5-2650 | Xeon Phi SE10/7120 |
|---|---|---|---|
| Physical cores | 8 | 8 | 61 |
| Threads per core | 1 | 2 | 4 |
| Total threads | 8 | 16 | 244 |
| Frequency | 2.4 GHz | 2.0 GHz | 1.24 GHz |
| Memory | 16 GB (DDR3) | 256 GB (DDR3) | 16 GB (GDDR5) |

### *3.1. Intel Xeon*

Our reference platform is a dual-socket Intel Xeon CPU E5-2650 @ 2.00 GHz launched in Q1 2012. It is an 8 physical core Sandy Bridge CPU with hyper-threading (HT) and Turbo Boost

Journal of Physics: Conference Series **608** (2015) 012033 doi:10.1088/1742-6596/608/1/012033

(2.8 GHz) enabled. The particular system was equipped with 256 GB DDR3 memory. For silicon-to-silicon comparison we only measured power consumption of a single socket. For the test, the ParFullCMS benchmark was compiled with `GCC` 4.9.1.

Intel's Running Average Power Limit (RAPL) [15, 11] technology was used for measuring power usage of different parts of the silicon chip. It is available in modern Intel micro-architectures starting from Sandy Bridge. It is intended for controlling and limiting power usage in silicon chip. In addition, RAPL provides capabilities to measure energy and power usage. We used this capability to acquire readings from the silicon chip. The following RAPL domains are supported by server-grade silicon chips:

- PP0 (Power Plane 0) – processors cores subsystem.
- PKG (Package) – processor die.
- DRAM (Memory) – directly-attached DRAM.

The processors core subsystem consists of execution units, ALU, FPU, L1 and L2 caches. The uncore subsystem (`energy(PKG) - energy(PP0)`) consists of Intel QuickPath Interconnect (QPI), Last Level Cache (LLc), on-chip memory and I/O. RAPL provides ~1 ms [15] resolution measurements. For the test we only measured PKG power domain for a single socket. Reading RAPL sensor data was done via model-specific registers (MSR) exposed to Linux user-land via devfs.

### *3.2. Intel Xeon Phi*

The Xeon Phi card is a highly parallel machine with 61 physical cores and 4-way multi-threading running at 1.24 GHz. It includes GPU-class 16 GB GDDR5 memory, which provide ultra-high bandwidth compared to standard DRAM solutions. The card we used had only passive cooling. The card supports two execution modes: native and offloading. For simplicity our tests were done using native execution on the card itself. The Intel C++ Compiler (Intel Composer XE 2013 SP1 Update 2, 14.0.2 20140120) was used to compile the test as `GCC` does not provide Xeon Phi support.

The Xeon Phi card includes a number of sensors to acquire power usage from different domains on the card. We used `libmicmgmt` from Manycore Platform Software Stack (MPSS) 3.2 to read the sensor data for different domains (PCIe, 2x3, 2x4, VCCP, VDDG, VDDQ, instant power and more). There is drawback of non-direct sensor reading. `libmicmgmt` sends an interrupt over PCIe bus to uOS (Linux) forcing Xeon Phi to switch to higher power state. This method cannot be used for measuring idle power consumption. We measured idle power consumption of the card by physically removing it from the server enclosure. We found that idle power consumption was 17W. The resolution of sensors is 50 ms. Two distinct datasets were used for Xeon Phi: one for the whole card and another for VCCP + VDDG power domains. According to Performance API (PAPI) [16] source code, VCCP power rail is processors core subsystem and VDDG – uncore subsystem.

### *3.3. APM X-Gene*

APM X-Gene 1 is 8 physical core processor running at 2.4 GHz. The particular system, XC-1 development board, contains 16 GB DDR3 memory. The board provides 2 memory channels, but APM883408 is capable of addressing 512 GB of memory using 4 memory channels. Investigation done by APM showed that ParFullCMS test does not require high memory bandwidth. We would like to emphasize that the firmware for managing processor ACPI power states was not available in time for the test. We expect X-Gene 1 efficiency to increase once the firmware is available. XC-1 is running a custom build of Fedora 19 provided by Red Hat. The test was compiled with `GCC` 4.9.1. APM is working on X-Gene 1 optimizations for `GCC`, but changes were not integrated upstream in time for the benchmark.

The XC-1 development board includes sensors for reading the power usage of different domains of the silicon chip. We estimated power usage for the silicon chip by combining readings for PMD (core subsystem) and SOC (uncore subsystem) power domains. The XC-1 is an IPMI-enabled device, thus readings from the sensors can be acquired by `ipmitool` tool. We used IPMI before in previous research, but we found that executing `ipmitool` is expensive and restricts the measurements resolution. For XC-1 we read sensor data via I2C bus and avoided unnecessary overheads caused by execution of `ipmitool` tool.

## 4. Results of Power and Performance Measurements

### *4.1. Platform Efficiency*

Our energy efficiency benchmark provided three-way silicon level comparison between APM X-Gene 1 – the first on the market ARMv8 64-bit Server-on-Chip solution, Intel Xeon – the dominating processor family in server market and Intel Xeon Phi – a highly parallel long-vector processor family product intended for HPC workloads. We sampled power usage readings from sensors with one second resolution on all platforms.

First we compared absolute performance provided by different solutions by running on all available hardware threads (figure 2). We found that Intel Xeon Phi provided the best performance, 1.07 times higher performance to Intel Xeon. In the absence of the anticipated compiler optimizations, APM X-Gene 1 provided 2.48 times lower performance than Intel Xeon, but it was also operating at significantly lower power usage.

Then we measured how performance scales over power (figure 3, figure 4). We found that APM X-Gene 1 running at full capacity (8 threads) was drawing less power than Intel Xeon running a single thread and delivered 2.73 times higher performance. Intel Xeon was Turbo Boost enabled with up to 2.8 GHz for a single core. The benchmark is CPU bound and as expected Hyper-Threading (HT) on Intel Xeon did not deliver higher energy efficiency. Additional few percent of performance were gained at similar cost in power consumption (figure 3). We also overcommitted APM X-Gene 1 with 2 threads per physical core, but we did not observe significant change in energy efficiency. Intel Xeon Phi as expected required at least 2 threads to achieve high utilization of card.

## 5. Conclusions

We have built the software used by the CMS experiment at CERN, as well as portions of the OSG software stack, for ARMv8 64-bit. It has been made available in the official CMS software package repository and via the CVMFS distributed file system used by Grid sites. Our initial validation has demonstrated that APM X-Gene 1 Server-on-Chip ARMv8 64-bit solution is a relevant and potentially interesting platform for heterogeneous high-density computing. In the absence of platform specific optimizations in the ARMv8 64-bit GCC compiler used, APM X-Gene 1 shows excellent promise that the APM X-Gene hardware will be a valid competitor to Intel Xeon in term of power efficiency as the software evolves. However, Intel Xeon Phi is a completely different category of product.

As APM X-Gene 2 is being sampled right now, built on the TMSC 28 nm process, we look forward to extending our work to include it into our comparison.

## Acknowledgements

This work was partially supported by the National Science Foundation, under Cooperative Agreement PHY-1120138, and by the U.S. Department of Energy. We would like to express our gratitude to APM for providing hardware and effort benchmarking Geant4 ParFullCMS, and to TechLab at CERN for providing and managing the Intel Xeon and Intel Xeon Phi servers for the benchmarks.

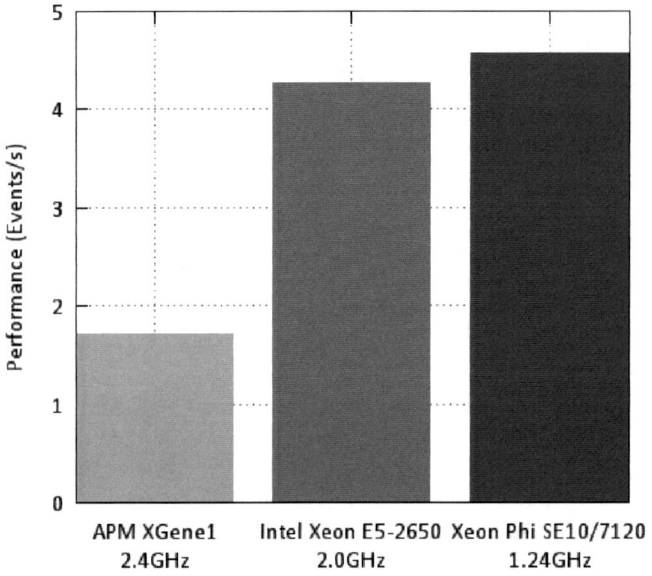


**Figure 2.** Absolute silicon chip performance

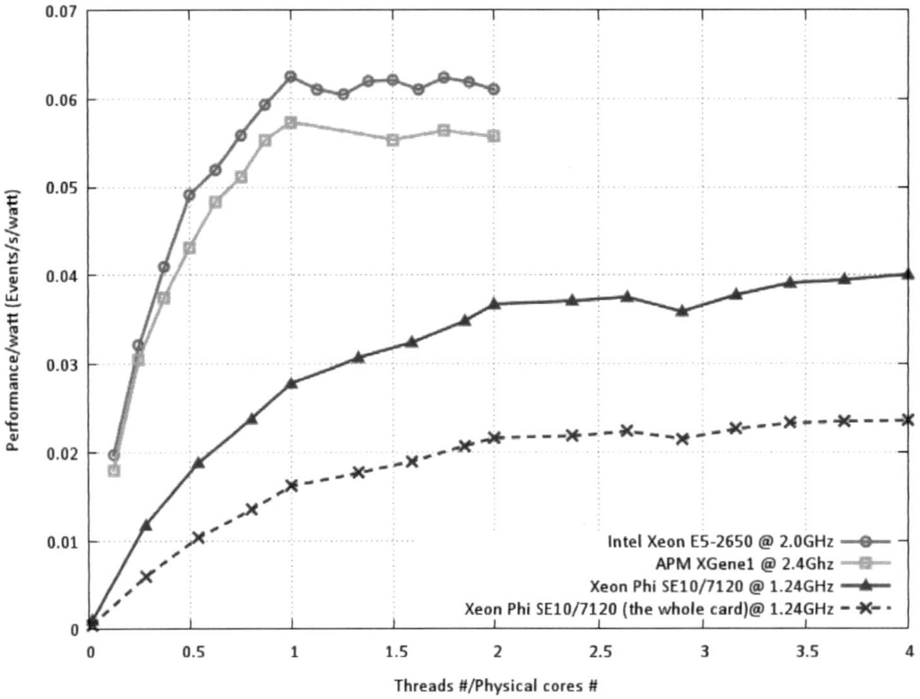


**Figure 3.** Energy efficiency scalability

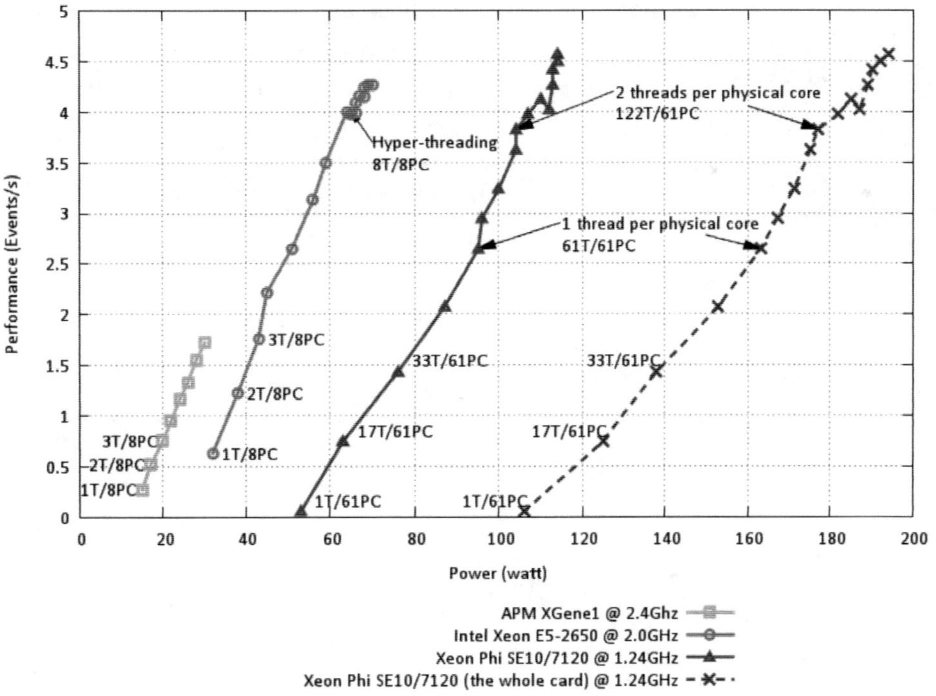


**Figure 4.** Performance scalability over power usage

## References

[1] Lyndon Evans and Philip Bryant. LHC Machine. *JINST*, 3:S08001, 2008.

[2] S. Chatrchyan et al. Observation of a new boson at a mass of 125 GeV with the CMS experiment at the LHC. *Phys.Lett.*, B716:30–61, 2012.

[3] G. Aad et al. Observation of a new particle in the search for the Standard Model Higgs boson with the ATLAS detector at the LHC. *Phys.Lett.*, B716:1–29, 2012.

[4] L Rossi and O Bruning. High Luminosity Large Hadron Collider A description for the European Strategy Preparatory Group. Technical Report CERN-ATS-2012-236, CERN, Geneva, Aug 2012.

[5] Michael J. Miller. Can AMD, ARM, or IBM dent Intel's server dominance? `http://forwardthinking.pcmag.com/none/323603-can-amd-arm-or-ibm-dent-intel-s-server-dominance`, 2014. [Online; accessed 21-September-2014].

[6] David Abdurachmanov, Kapil Arya, Josh Bendavid, Tommaso Boccali, Gene Cooperman, Andrea Dotti, Peter Elmer, Giulio Eulisse, Francesco Giacomini, Christopher D Jones, Matteo Manzali, and Shahzad Muzaffar. Explorations of the viability of ARM and Xeon Phi for physics processing. *Journal of Physics: Conference Series*, 513(5):052008, 2014.

[7] R Pordes, D Petravick, B Kramer, D Olson, M Livny, A Roy, P Avery, K Blackburn, T Wenaus, F Wuerthwein, I Foster, R Gardner, M Wilde, A Blatecky, J McGee, and R Quick. The open science grid. *Journal of Physics: Conference Series*, 78(1):012057, 2007.

[8] David Abdurachmanov, Peter Elmer, Giulio Eulisse, and Shahzad Muzaffar. Initial explorations of ARM processors for scientific computing. *Journal of Physics: Conference Series*, 523(1):012009, 2014.

[9] Rene Brun and Fons Rademakers. Root an object oriented data analysis framework. *Nucl.Instrum.Meth.*, A389(12):81 – 86, 1997. New Computing Techniques in Physics Research V.

[10] ARM Limited. ARM discloses technical details and partner support for ARMv7 architecture. `http://www.arm.com/about/newsroom/8513.php`, 2005. [Online; accessed 21-September-2014].

[11] David Abdurachmanov, Peter Elmer, Giulio Eulisse, Robert Knight, Tapio Petteri Niemi, Jukka Nurminen, Filip Nyback, Goncalo Marques Pestana, and Zhonghong Ou. Techniques and tools for measuring energy efficiency of scientific software applications. *To be published in ACAT 2014 proceedings*, 2014.

[12] A Dorigo, P Elmer, F Furano, and A Hanushevsky. XROOTD - A highly scalable architecture for data access. *WSEAS Transactions on Computers*, 4.3, 2005.

[13] L Bauerdick, D Benjamin, K Bloom, B Bockelman, D Bradley, S Dasu, M Ernst, R Gardner, A Hanushevsky, H Ito, D Lesny, P McGuigan, S McKee, O Rind, H Severini, I Sfiligoi, M Tadel, I Vukotic, S Williams, F Wuerthwein, A Yagil, and W Yang. Using Xrootd to Federate Regional Storage. *Journal of Physics: Conference Series*, 396(4):042009, 2012.

[14] S. Agostinelli et al. GEANT4: A Simulation toolkit. *Nucl.Instrum.Meth.*, A506:250–303, 2003.

[15] Intel. Intel 64 and IA-32 architectures software developers manual. combined volumes: 1, 2A, 2B, 2C, 3A, 3B and 3C, September 2014.

[16] `http://icl.cs.utk.edu/trac/papi/browser/src/components/micpower/README?rev=a7d7012`.

# Implementation of a Multi-threaded Framework for Large-scale Scientific Applications

**E Sexton-Kennedy[1], Patrick Gartung[1], CD Jones[1], David Lange[2]**
[1]Fermilab, P.O.Box 500, Batavia, IL 60510-5011, USA
[2] Lawrence Livermore National Laboratory, 7000 East Ave, Livermore, CA 94550

E-mail: sexton@fnal.gov , cdj@fnal.gov, gartung@fnal.gov, David.Lange@cern.ch

**Abstract.** The CMS experiment has recently completed the development of a multi-threaded capable application framework. In this paper, we will discuss the design, implementation and application of this framework to production applications in CMS. For the 2015 LHC run, this functionality is particularly critical for both our online and offline production applications, which depend on faster turn-around times and a reduced memory footprint relative to before. These applications are complex codes, each including a large number of physics-driven algorithms. While the framework is capable of running a mix of thread-safe and "legacy" modules, algorithms running in our production applications need to be thread-safe for optimal use of this multi-threaded framework at a large scale. Towards this end, we discuss the types of changes, which were necessary for our algorithms to achieve good performance of our multi-threaded applications in a full-scale application. Finally performance numbers for what has been achieved for the 2015 run are presented.

## 1.Introduction and Goals

It has become clear that to make use of the new hardware architectures coming to market in the next 10 years, our software must exploit concurrency. This has been described in HEP computing conferences in the past, for instance the CHEP conference in 2011 [1].

Traditionally, HEP computing has utilized the increasing CPU core count by increasing the number of single threaded jobs we run. Unfortunately, this does not scale for grid sites. As core counts double, grid sites are forced to double the number of batch slots, double the amount of memory on a machine, double the open file handles and double the number of network connections. By making the HEP applications concurrent, we can reduce these burdens on the sites and potentially target new sites whose single-threaded job resources were too limited for our single-threaded applications. Other goals include: 1. reduced processing latency for the larger event count files we expect to collect in Run 2, 2. more resource sharing between cores, including shared file handles, shared network connections, and sharing of infrequently updated memory holding calibrations, conditions, and I/O buffers 3. minimize changes to the existing framework. CMS' [2] challenge in switching to a concurrent framework is to avoid rewriting millions of lines of C++ legacy code to work in a new framework. In this paper we will briefly review the design of the new threaded framework; for a more complete description of the design see [3]. We will then discuss the thread safety tools and migration strategies we have found useful during the conversion. Finally performance results for the Run 2 reconstruction application will be presented.

## 2.Design

Journal of Physics: Conference Series **608** (2015) 012034 doi:10.1088/1742-6596/608/1/012034

### 2.1. *Legacy Design*

There are two main concepts that were retained from the legacy framework: system state transitions and data processing. Also retained from the legacy framework design are the user supplied algorithms which are encapsulated into modules. It is these modules that are informed when the framework (either the legacy or threaded) changes state. To aid migration to the new framework, we wanted to minimize any required changes to module code, therefor the legacy design heavily influences the new design.

### 2.2. *Threaded design*

The threaded framework design makes use of the naturally concurrent elements of the legacy framework. These elements occur at three different conceptual levels. At the highest level, the threaded framework runs multiple data streams concurrently. This is supported by introducing two new concepts to the framework: Stream and Global. The middle level supports running multiple modules concurrently while processing one event, run or lumi. To do that correctly, one must take into account the inter-module dependencies. At the lowest level, the threaded framework supports running multiple concurrent tasks within one module.

Eventually all three levels of concurrency will be implemented using Intel's Threaded Building Blocks [2] (TBB) library. In TBB, one breaks down concurrent work into *tasks* which TBB can then run in parallel. The framework therefore decomposes processing of transitions and modules into appropriate tasks, and then hands these tasks at the appropriate time to TBB. TBB can then appropriately schedule all three levels of tasks. Module code is allowed to directly call TBB however it must follow the rule that all tasks spawned by a module for a given transition call-back must complete before the module returns from that call-back.

#### 2.2.1. *Current Status*

As of the writing of this paper, the framework supports parallel execution at the highest and lowest levels. Non-event transitions are serialised as in Figure 1. At the highest level multiple events will execute in parallel in different streams according to the number specified in job configuration. At the lowest level developers are allowed to use TBB directly inside of their modules to add concurrent calculations.

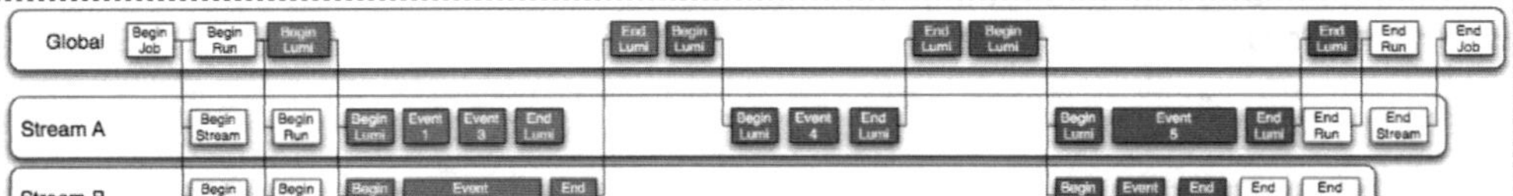


**Figure 1.** Note that only events are executed concurrently others are synchronised

## 3. Thread Safety

The legacy framework is composed of individual modules which communicate indirectly with each other by publishing data products to a shared container, the `edm::Event`, and then reading those published data products from the container. The threaded framework maintains the module and data product design but requires that both types obey some thread-safety rules.

In both the legacy and threaded frameworks, data products are passed to modules with the `const` modifier. This is done to enforce the rule that once published, data products cannot be changed. Therefore only const member functions of a data products are required to be thread-safe. Fortunately, this exactly matches the C++ 11 library guarantee that states that calling only const functions of standard library objects from different threads is thread-safe.

The majority of physicist defined code utilizes the module component of the framework. As such, the thread safety requirements of modules will be the requirements that most affect physicists. The threaded framework uses four different varieties of module, where each variety defines a different level of thread safety. The four module types are: Stream, Global, One and Legacy. The details of the thread safety requirements of each type of module is described in reference [3]. For the purposes of this paper, the work required of the collaboration was in converting modules from Legacy modules to Stream modules. In January when we started there were 20 conforming modules. In June we had 200 converted in time for the first release of the multi-threaded framework. By the time of the ACAT conference we had 230, which is significant because this is the snapshot of the CMS software in which the performance measurements reported later were collected with.

### *3.1. Required changes for Stream modules*

All modules in the multi-threaded framework are presumed to be thread unsafe and therefor are declared to the new frame work as Legacy modules. In order to convert them to a Stream module the author has to remove the use of non-const static member data, change the inheritance to `stream::Producer<>,` and remove any use of begin/end _job callbacks. If those are still needed they can be converted to begin/end stream callbacks, however this was rare.

### *3.2. Thread safety patches for 3rd party software*

In order to successfully run the full CMS reconstruction application in the multi-threaded framework, we had to make sure that all of the 3rd party software that we use was thread-safe. We used git repositories forked off of the official repositories of products like ROOT, CLHEP and fastjet to make minimal changes to them that allowed us to use them in our thread-safe framework.

## 4.Tools

In order to meet the challenge of migrating a large body of code like CMSSW into thread-safe implementations, we have developed tools to automate the discovery of non-conforming code. In addition, we found the use of run time tools for finding errors very important.

### *4.1. Static Code Analysis: Clang*

The Clang tool suite [3] has defined a framework that gives users the ability to extend it's compile time checking by adding user define rules to the default set provided with the tool. We have extended the clang static analyzer with customizations that find coding patterns that violate the rules of the threaded framework. The checkers fall into two different categories; one that finds problematic modules, and one that finds problematic data products. The framework needs to be able to run multiple instances of any particular module concurrently. Non-const static member data represents shared state between different instances of the same module. We can use our custom extensions to

Clang to look for modules that make use of non-const static data. The search has to include looking at all functions that the module calls, and each function they call, all the way down the stack.

Data products are shared between modules. They are the sole source of communication between the modules. Once a producer makes a data product and gives ownership of it to the Event, the framework can only allow const access to it. Problematic products contain non-const statics, mutable member data which is not std::atomic<>, member functions that cast away const on member data, pointer member data that is returned from a const function, and pointer member data that is passed as a non-const argument to a function. All of these are checked by our extensions to Clang. The checks have to be done recursively on all data members that are classes.

### 4.2. *Run time checking: Helgrind*

Helgrind is a tool within the Valgrind tool suite [4]. It can be used to search for possible race conditions between threads without requiring the race condition to happen. It does this by recording every memory read/write done in an executing program by each thread and flags if multiple threads use the same memory address where one of them is doing a write to that address. The tool understands posix synchronisation mechanisms that protect memory, i.e. mutex, semaphore, pthread_join, so that it can ignore these addresses. Unfortunately, Helgrind currently does not understand lock-free designs, e.g. designs that use std::atomics<>, generating lots of false positives in those cases. Even with this deficiency, we found Helgrind to be an invaluable tool in debugging both the implementation of the threaded framework, externals, and CMS software.

## 5. Performance

### 5.1. *Amdahl's Law*

Amdahl's law is a model for the relationship between the expected speedup of parallelized implementations of an algorithm relative to the serial algorithm, under the assumption that the problem size remains the same when parallelized. This means that with small enough overheads within the framework execution engine , it should be possible to approximate how much of the code is executing in parallel given the observed utilization and the number of cores the framework has been configured to allow. If P is the parallel fraction and n is the number of cores then the theoretical average utilization is = $[n(1-P)+P]^{-1}$ See [5] for a derivation and further discussion of Amdahl's Law.

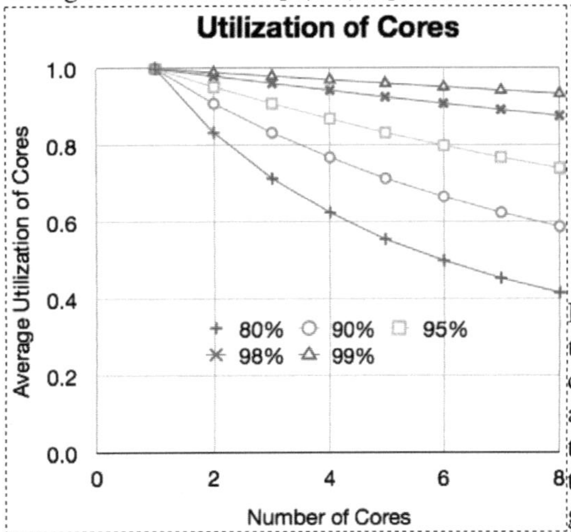


**Figure 2.** This figure plots the relationship between the theoretical average utilization of processing cores as a function of 2 variables: the number of allowed parallel tasks, and the fraction of the code that can be executed in parallel. The blue curve traces out this dependency when that faction is 80%, green 90%, yellow 95%, red 98% and purple 99%.

In a multi-threaded CMS reconstruction application run on a simulation of the most challenging data taking conditions we expect in Run 2, we observe a 90% average utilitization at 16 allowed parallel streams. This corresponds to a time weighted fraction of code which can be run in parallel, used in this application, of 99.3%. This highlights the high penalty for strictly serial code. Even if your profiler tells you that a function takes very little of the total time of your application, it can cause serious performance penalties if it requires serialization. A lot of effort went into finding serialization bottlenecks. We developed a log scanning tool that runs on the output of a tracer service that reports begin and end times of a modules execution as well as the stream that it was executing on. In this way it was possible to find gaps in time where the framework was forced to wait for some bit of serial code. A heuristic that worked well was to fix the module that was waiting to be executed. Where fix means make thread friendly enough to operate that module as a stream module instead of a legacy module.

### *5.1 Throughput measurements*

At the time of ACAT we ran our benchmark, described above and obtained the following results presented in Firgure 3. Note that there is good scaling up to 8 allowed streams on this 16 core, 64 GB AMD machine. At 8 cores, this application is 95% CPU efficient. Therefor running 2 processes in which each allowed 8 streams, CMS could utilize the full machine with an acceptable efficient hit.

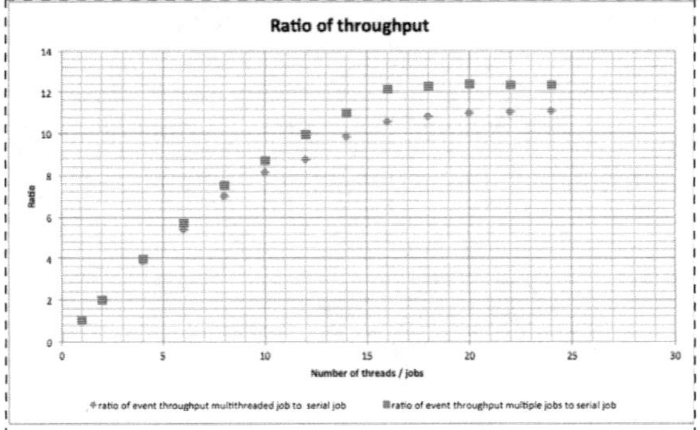


**Figure 3.** These two curves plot the ratio of throughput for multiple jobs to a serial job in red and a single multithreaded job to a serial job in blue

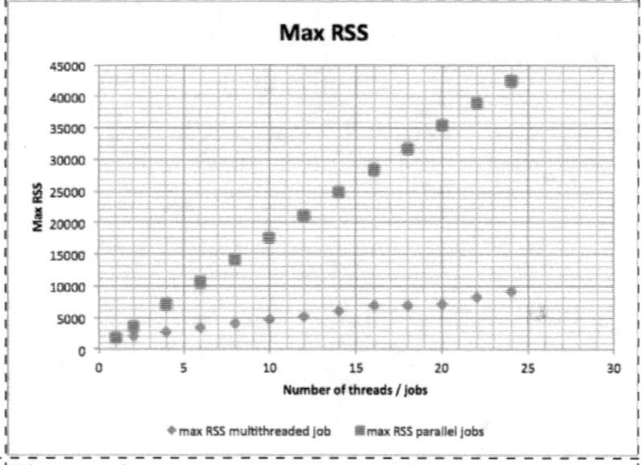


**Figure 4.** These two curves plot the maximum RSS used on the machine for the corresponding two cases in figure 3.

### *5.2. Memory, File Handle, and Network measurements*

Figure 4 demonstrates that the big win is in the memory use of the multi-threaded application. At 8 cores it requires 4 GB of memory where as running 8 simultaneous jobs requires 14 GB. The number of network connections also scales linearly with the number of jobs however in this case the network load of the multi-threaded application is the same as that for a single job of the serial application. A similar statement can also be made about open file handles.

## 6. Conclusions

The challenge of converting millions of lines of code, from serial to concurrent execution, is a huge task, and in some ways more difficult then the move from Fortran to C++ a decade ago. We have chosen a design that minimizes the work of the current framework users. By following well-defined rules about the subset of C++ that is allowed, their code can run in a concurrent framework that allows parallel execution of Events and modules. The rules are simple enough that with the help of automated tools we can find violations of these rules and get them fixed as they appear. Over the past year CMS has managed to make 99.3% of its reconstruction code thread friendly enough to work efficiently in our multi-threaded framework (in other words only 0.7% of the code must be serialized by the framework). At the target value of an 8 core multi-threaded job, our reconstruction application has a CPU efficiency of 95%. This is considered good enough for Run 2 of CMS. Future work in this project will be devoted to increasing the level of concurrency within the high level trigger and simulation applications.

## 7.References

[1] Sverre Jarp *et al* 2011 *J. Phys.: Conf. Ser.* **331** 052009, "Evaluating the scalability of HEP software and multi-core hardware" doi:10.1088/1742-6596/331/5/052009

**[2]** CMS Collaboration R. Adolphi et al., "The CMS experiment at the CERN LHC", JINST, 0803, S08004 (2008)

[3] C D Jones and E Sexton-Kennedy 2014 *J. Phys.: Conf. Ser.* **513** 022034, "Stitched Together: Transitioning CMS to a Hierarchical Threaded Framework"doi: 10.1088/1742-6596/513/2/022034

[4] http://valgrind.org, Jan. 25, 2014

[5] Amdah's Law http://en.wikipedia.org/wiki/Amdahl's_law

## Acknowledgments

Operated by Fermi Research Alliance, LLC under Contract No. DE-AC02-07CH11359 with the United States Department of Energy.

# WLCG Tier-2 site in Prague: a little bit of history, current status and future perspectives

**Dagmar Adamova[1], Jiri Chudoba[2], Marek Elias[2], Lukas Fiala[2], Tomas Kouba[2], Milos Lokajicek[2], Jan Svec[2]**

[1] Nuclear Physics Institute, Rez near Prague, Czech Republic
[2] Institute of Physics AS CR (FZU), Na Slovance 2, Prague, Czech Republic

E-mail: adamova@ujf.cas.cz

**Abstract.** High Energy Physics is one of the research areas where the accomplishment of scientific results is inconceivable without a complex computing infrastructure. This includes also the experiments at the Large Hadron Collider (LHC) at CERN where the production and analysis environment is provided by the Worldwide LHC Computing Grid (WLCG). A very important part of this system is represented by sites classified as Tier-2s: they deliver a half of the computing and disk storage capacity of the whole WLCG. In this contribution we present an overview of the Tier-2 site praguelcg2 in Prague, the largest site in the Czech Republic providing computing and storage services for particle physics experiments. A brief history flashback, current status report and future perspectives of the site will be presented.

## 1. Introducing WLCG Tier-2 site in Prague

Czech Republic is a member of the Worldwide LHC Computing Grid (WLCG) Collaboration [1] and provides computing and storage services and resources for the LHC experiments ATLAS [2] and ALICE [3]. These capacities are delivered by the Tier-2 site praguelcg2 [4] at the Institute of Physics (FZU [5]) in Prague.

In addition to ATLAS and ALICE, praguelcg2 provides computing capacities also to other High Energy Physics (HEP) projects including D0 [6], NOvA [7], STAR [8], astroparticle projects Pierre Auger Observatory (PAO) [9], Cherenkov Telescope Array (CTA) [10] and also to solid state physics.

It has an excellent network connectivity with the collaborating institutions local and worldwide and with other Grid projects, delivered by the Czech Republic National Research and Education Network (NREN) provider CESNET [11] including dedicated 1–10 Gb/s links. The site is also connected to LHCONE [12].

In this contribution, we present a short overview of the history of praguelcg2, the current status and activities and plans for meeting the demands of the upcoming Run2 and Run3 of the LHC.

## 2. praguelcg2 performance during last 10 years

The computer center at FZU officially started operation in 2004 when all hardware resources were installed in a new server room (UPS, air conditioning completed with water cooling in 2010), but was delivering computing services to HEP projects already in 1998. Since 2004,

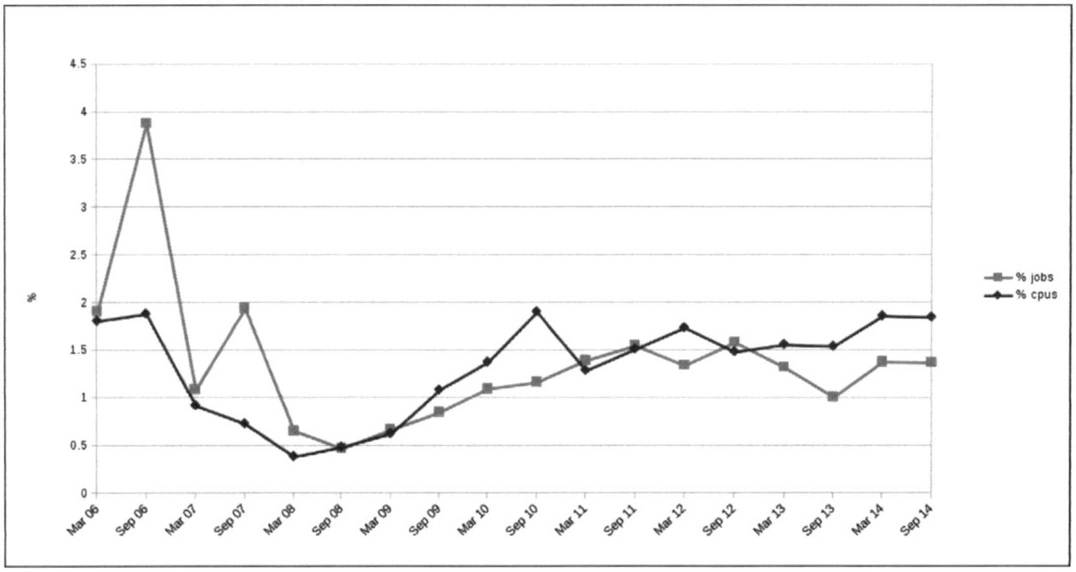


**Figure 1.** praguelcg2 contribution to WLCG ALICE and ATLAS computing resources since 2006 (in percents, http://accounting.egi.eu/). The blue line represents "Norm. Sum CPU (HEPSPEC06-hourS)", cf. the portal.

the site operated as a WLCG Tier-2 and participated already in the very first ALICE Data Challenges delivering up to 5 % of the total used resources. The site evolution details till 2009 are presented in [13] and [14]. praguelcg2 contribution to WLCG ALICE and ATLAS computing resources since 2006 as monitored by EGI [15] is shown in Fig. 1.

Important activities since 2004 include:

- in 2009 some of the first LHC data collected by the ATLAS experiment was processed in praguelcg2;
- start of IPv6 [16] implementation in 2011: the site router got an IPv6 address, IPv6 address configuration being done with DHCPv6 and an IPv6 testbed was set up. The first experience with IPv6 deployment on praguelcg2 was described in [17];
- an extensive tuning of the XrootD [18] storage cluster dedicated to ALICE in 2011/2012 which increased its performance and fixed the problems with occasional high load on the servers [19];
- since 2008 services management is done by CFengine version 2 [20]. An in-house developed cfagent Nagios sensor [21] enables to automatically fix problems that Nagios detects;
- the site was connected to LHCONE in 2013.

## 3. Current issues and activities

The current capacity of the site is about 4000 computer cores (CPU) and more than 3 PetaBytes of disk space. The running jobs profile for the last year is shown in Fig. 2. There is a continuous ongoing production of ALICE and ATLAS, the other projects use the site less intensively. The WLCG services provided by the site include: Apel publisher, Argus Authorization service, BDII, several UIs, ALICE VOBOX, Cream CEs and Storage Elements [22]. The use of virtualization [23] at the site is quite extensive.

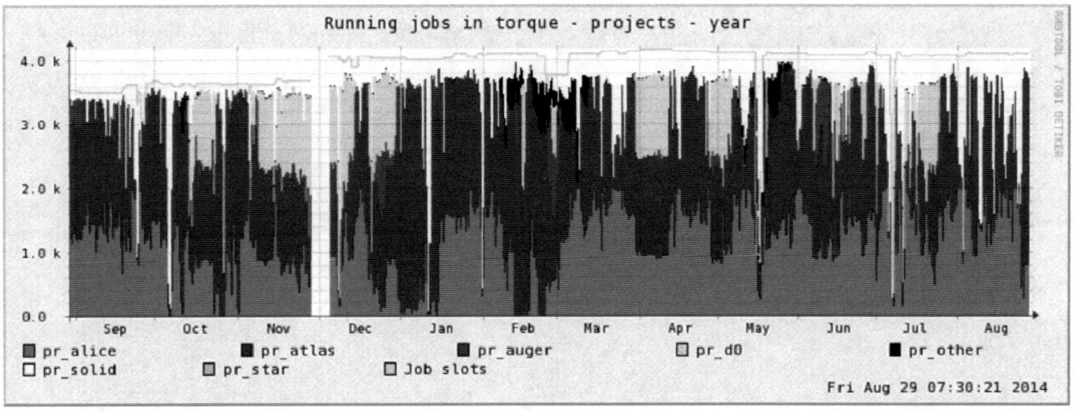


**Figure 2.** Running jobs profile at praguelcg2, 2013/2014.

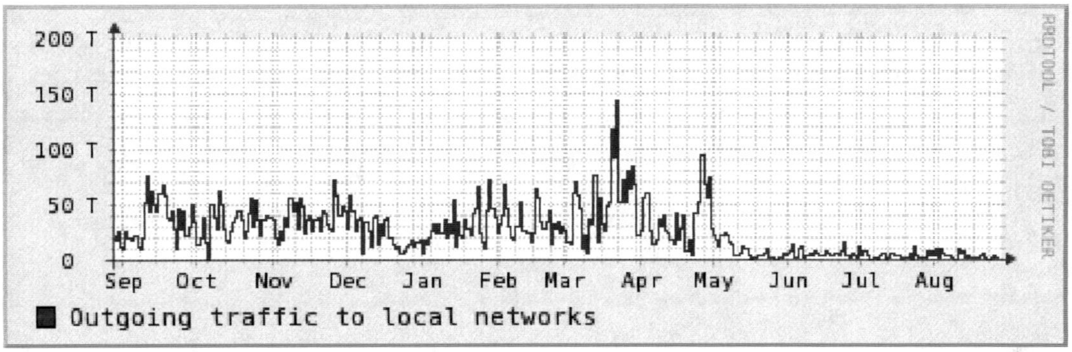


**Figure 3.** Outgoing IPv4 local traffic from DPM servers (in TeraBytes, monitoring by fzu_site_stat [29]).

As was already mentioned, the external connectivity of the site is provided by the CESNET state-of-the-art infrastructure. An extensive real-time monitoring of the utilization of all included connections can be found in [24].

There has been an intensive ongoing work on deployment of IPv6 on praguelcg2 [25]. The site runs an IPv6 testbed and also several production services are already available through both IPv4 and IPv6 including almost all WorkerNodes [22]. In addition, the site is actively involved in the activities of the HEPiX IPv6 working group [26]. Since May 2014 the GridFTP [27] data traffic inside the computing center runs via IPv6. The effect of this transition on the IPv4 and IPv6 traffic from our DPM [28] disk nodes is nicely demonstrated in Figs. 3 and 4.

A massive effort is given to building extensive monitoring of hardware elements and software services on the site. There are five products used on a daily basis: Nagios, Munin, RRD graphs created by local custom scripts, MRTG+Weathermap and Netflow [21]. The monitoring services are gradually migrated to IPv6 operation [25].

Another important issue is a relatively low efficiency of the ALICE jobs processed on the site. ALICE adopted storage solution is XrootD [18] combined with an optional remote data access.

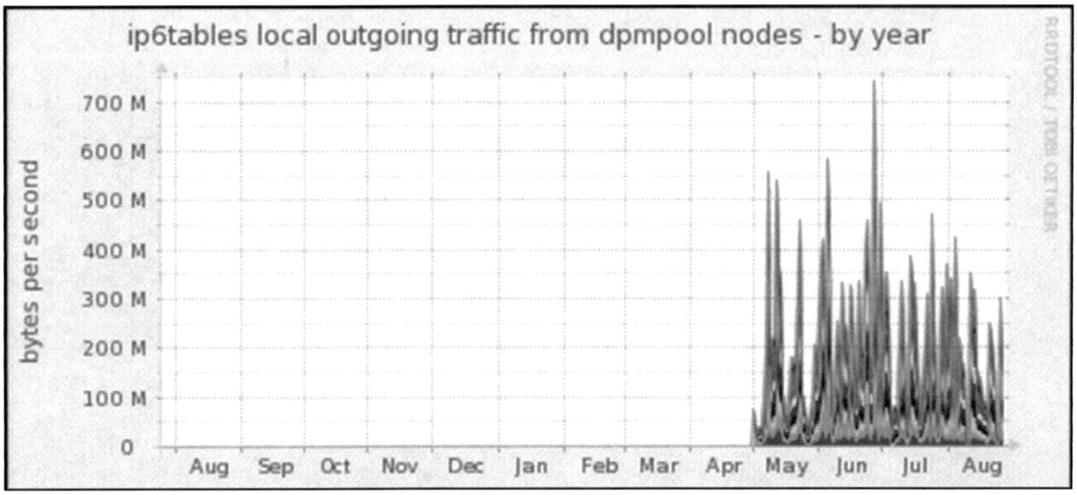


**Figure 4.** Outgoing IPv6 local traffic from DPM servers (monitoring by MUNIN [30]). Altogether 12 contributing disk servers identified by different colors.

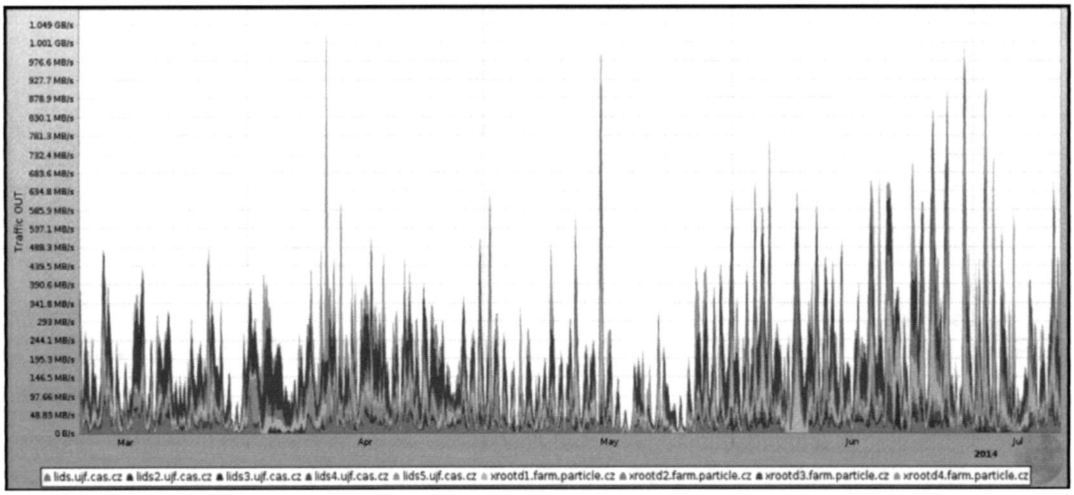


**Figure 5.** Outgoing network traffic on praguelcg2 XRootD cluster during 6 months in 2014. Altogether 9 servers, the peak aggregated traffic $\sim$ 1 GByte/s. Almost 2.5 PetaBytes of data were downloaded in total from the cluster.

Although the jobs preferably process local data it happens quite often that the data must be accessed remotely which can increase the time needed for input/output operations. praguelcg2 XrootD cluster is distributed: some of the servers are installed at Nuclear Physics Institute (NPI) about 15 km away from FZU. Although there is a dedicated 10 Gb/s connection and therefore when jobs running at FZU read data from the NPI servers, it is not a genuine remote access, it brings the efficiency down. We are working on identifying the problem with the data access from FZU to the NPI cluster. There again, a proper monitoring is the key issue. The aggregated traffic on our XRootD servers for a 6 months period is shown in Fig. 5.

## 4. Summary and outlook

praguelcg2 became a recognized Tier-2 site of WLCG providing practically 100 % accessibility, reliable resources and high-level services. Proper conditions have been set up for a full scale participation in the data processing during the LHC running phase. The site team members and collaborators have been actively involved in the study, use and development of advanced Grid technologies.

The upcoming Run2 and Run3 of the LHC will bring orders of magnitude higher particle production which translates into an increase in computing needs: a factor 10–100 is very likely, more is possible. praguelcg2 regularly upgrades its hardware resources on the basis of various financial sources, mainly academic grants. But the upgrade scale happening so far will not be enough for the future LHC demands. In this respect the situation of all WLCG sites is quite alike. Various ongoing activities to find the way to a sufficient resources stack include all kinds of re-engineering and optimization of the HEP software and an endeavor to pool resources between different communities with similar computing interests.

praguelcg2 team members are becoming involved in projects concerning the optimization of Grid software according to their participation in the LHC experiments. Concerning the search for non-HEP resources which might be used for the future processing of the LHC data, we accomplished so far only to get access to an external 35 TB disk storage provided by NGI_CZ [31] and installed about 100 km far from Prague. The storage was used as an opportunity to test the access to external resources [32] but the capacity was too small.

In the coming years, we will do our best to keep up the reliability and performance level of the services and engage fully into the data processing during the next LHC running periods.

### *4.1. Acknowledgments*

This work was supported by the Grant Agency of the Czech Republic (grants No. 202/97/0038, 104/97/S002 and 202/97/P042) and MSMT INGO LG13031 and LG13009.

## References

[1] Worldwide LHC Computing Grid, `http://wlcg.web.cern.ch/`
[2] A Toroidal LHC ApparatuS, `http://atlas.ch/`
[3] A Large Ion Collider Experiment, `http://aliceinfo.cern.ch/Public/Welcome.html`
[4] praguelcg2, `http://gstat2.grid.sinica.edu.tw/gstat/site/praguelcg2/`
[5] Institute of Physics of the AS CR, `http://www.fzu.cz/en/`
[6] The DZero Experiment, `http://www-d0.fnal.gov/`
[7] `http://www-nova.fnal.gov/`
[8] STAR Experiment, `https://www.star.bnl.gov/`
[9] Pierre Auger Observatory, `http://www.auger.org/`
[10] `https://portal.cta-observatory.org/Pages/Home.aspx`
[11] CESNET, `http://www.cesnet.cz/e-infrastruktura-3/?lang=en`
[12] HC Open Network Environment, `http://lhcone.net/`
[13] Lokajicek M et al. 2010 *Prague Tier2 computing centre evolution, Proc. of Conf. NEC'2009*, Dubna, p. 8
[14] Adamova D et al 2010 *Status of the Grid Computing for the ALICE Experiment in the Czech Republic, J. Phys.: Conf. Ser.* **219** 072023
[15] European Grid Infrastructure, `http://www.egi.eu/`
[16] World IPv6 Launch, `http://www.worldipv6launch.org/`
[17] Kouba T et al 2012 *IPv6 testing and deployment at Prague Tier-2, J. Phys.: Conf. Ser.* **396** 042034
[18] XRootD project webpage, `http://xrootd.org/`
[19] Adamova D and Horky J 2012 *An optimization of the ALICE XRootD storage cluster at the Tier-2 site in Czech Republic, J. Phys.: Conf. Ser.* **396** 042001
[20] CFEngine, `http://cfengine.com/learn/manage-processes-and-services-using-cfengine/`
[21] Chudoba J et al 2013 *JINR (Dubna) and Prague Tier2 Sites: Common experience in the WLCG Grid Infrastructure, Physics of Particles and Nuclei Letters* **10** (No.3) pp. 288-294.
[22] WLCG middleware, `http://wlcg.web.cern.ch/getting-started/software`
European Middleware Initiative, `http://www.eu-emi.eu/`

[23] virtualization, http://www.webopedia.com/TERM/V/virtualization.html
[24] Large Infrastructure CESNET, http://netreport.cesnet.cz/netreport/hep-cesnet-experimental-facility2/
[25] Kouba T et al 2014 *Enabling IPv6 at FZU - WLCG Tier2 in Prague, J. Phys.: Conf. Ser.* **513** 062027
[26] Campana S et al 2014 *WLCG and IPv6, the HEPiX IPv6 working group, J. Phys.: Conf. Ser.* **513** 062026
[27] GridFTP, http://toolkit.globus.org/toolkit/docs/latest-stable/gridftp/
[28] LCG Disk Pool Manager, https://twiki.cern.ch/twiki/bin/view/LCG/DpmAdminGuide
[29] Eliáš M et al 2011 *Network monitoring in the Tier2 site in Prague, J. Phys.: Conf. Ser.* **331** 052024
[30] MUNIN, http://munin-monitoring.org/
[31] NGI CZ, https://wiki.metacentrum.cz/metawiki/NGI_CZ
[32] Horky J et al 2013 *Influence of distributing a Tier-2 data storage on physics analysis, 15th Int. Workshop on Advanced Computing and Analysis Techniques in Physics Research (ACAT2013)*, Beijing

# STAR Online Framework: from Metadata Collection to Event Analysis and System Control

**D Arkhipkin, J Lauret**

Physics Department, Brookhaven National Laboratory, Upton, NY 11973-5000 USA

E-mail: `arkhipkin@bnl.gov, jlauret@bnl.gov`

**Abstract.** In preparation for the new era of RHIC running (RHIC-II upgrades and possibly, the eRHIC era), the STAR experiment is expanding its modular Message Interface and Reliable Architecture framework (MIRA). MIRA allowed STAR to integrate meta-data collection, monitoring, and online QA components in a very agile and efficient manner using a messaging infrastructure approach. In this paper, we briefly summarize our past achievements, provide an overview of the recent development activities focused on messaging patterns and describe our experience with the complex event processor (CEP) recently integrated into the MIRA framework. CEP was used in the recent RHIC Run 14, which provided practical use cases. Finally, we present our requirements and expectations for the planned expansion of our systems, which will allow our framework to acquire features typically associated with Detector Control Systems. Special attention is given to aspects related to latency, scalability and interoperability within heterogeneous set of services, various data and meta-data acquisition components coexisting in STAR online domain.

## 1. STAR Online Meta-Data Collection Framework

The STAR experiment [1] started in 1999, with just one Time-Projection Chamber and a few trigger detectors, but today it is equipped of 18 subsystems. Initially, STAR Slow Control system had 40,000 control variables, now it is expanded to over 60,000 variables and this list is still growing due to the RHIC II upgrade, beam energy scan program, and possible upgrade to eRHIC in future. STAR had just 120 structures to migrate to the calibrations database at the early days of the experiment, and we now migrate over 3,000 structures annually. STAR's Data Acquisition – physics data taking component – was upgraded three times, adding one order of magnitude to the rates each time.

As every experiment has slightly different grouping of internal components, and varying operational meanings for those, let us define a common terminology first. In this paper we define the following cornerstones of "a" physics experiment's backend: Data Acquisition (DAQ), Detector Control System (DCS), Meta-data Archiver, and Alarm system. In this paper, DAQ is assumed to operate on physics signal only, with all accompanied meta-data being handled by DCS, operating on detector control data, and meta-data archiver, which records and stores detector conditions for the data taking period.

STAR's Messaging Interface and Reliable Architecture framework (MIRA [2]), was created as an attempt to improve meta-data archiver operations in 2010. It relies on an advanced message-queuing middleware, which provides the asynchronous, payload-agnostic messaging. We have selected AMQP [3] as a messaging middleware standard. It allowed us to design a loosely coupled, modular framework architecture, resulting in a scalable service, suitable for a highly concurrent online environment.

During the deployment validation phase in 2010, just three subsystems used the MIRA framework. By 2014, all eighteen subsystems were completely integrated, with over sixty collector services deployed and continuously monitored by the framework. The total number of messages passing through the system reached three billion messages per year, with rates varying between one hundred and fifty messages per second to over two thousand messages per second. On average one message corresponds to structure of 24 variables, which gives a variable rate of three thousand to twenty five thousand channels readings processed per second. MIRA is using data compression technique to reduce archived record storage size. The overall disk usage is moderate: it progressed from thirty gigabytes (GB) to just over two hundred GB of recorded data per year. In addition to data archival feature, MIRA provides extensive data visualization capabilities. It allows detector experts to browse archived system state (variables) as structured hierarchical content, using user-friendly responsive web interface. Starting with around one hundred variables in 2010, MIRA was expanded to visualize 1,680 variables in 2014.

At its early stage of development, MIRA provided monitoring, storage and basic flow control over individual streams of data [4], covering for just the *Meta-Data Archiver* part of the fundamental building blocks. To allow further expansion, multi-stream real-time processing capabilities were requested and desirable. In the next paragraph, we describe the Complex Event Processing extension of the framework, implemented for Run 14.

## 2. Run 14 MIRA Extension: Complex Event Processing

While it is very convenient to have a framework which operates on individual data streams with automatic visualization and archival of the events, STAR experiment's needs required further development of MIRA to handle simultaneous multi-stream event processing. To accomplish this goal, we have integrated Complex Event Processor, bringing the possibility to merge streams, process them according to pre-assembled persistent queries, and produce output stream back to the system for processing via regular means. There are many open-source Event Processor toolkits, available at the market. We have settled on a WSO2 framework [5], opening the possibility for us to use the well-know Esper [6] and recently introduced Siddhi CEP engines interchangeably. Typical architecture of the CEP toolkit is presented at Fig. 1. It consists of a broker managing data format conversion, CEP processor instances handling event filtering, and a control component allowing to configure persistent queries for the processor. Complex Event Processor at STAR is configured to use existing AMQP topics and queues, provided by Apache *qpid* service [7] as input and output device, exactly in a way it was implemented for MIRA framework services.

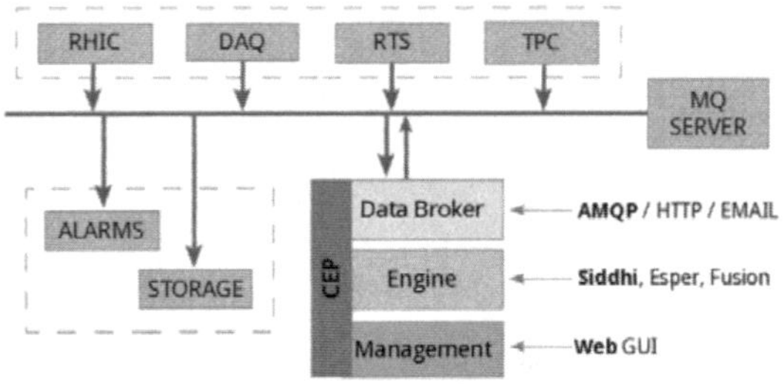


**Figure 1.** An overview of the Complex Event Processing components, attached to the messaging system: Broker, Engine and Management Console.

Let us describe a few practical use-cases, occurring in the STAR environment. The first example for CEP is to employ its stream processing power for a smart and flexible alarm system. In STAR, CEP provides the ability to reconfigure the existing alarm rules or add new sophisticated ones on the fly without stopping the existing monitoring services. This feature improves user experience and allows us to eliminate semi-permanent false or nuisance alarms (see Fig. 2). Second use-case is to perform CEP to detect anomalies in the detector conditions. By merging multiple streams and applying filters, which discard expected "normal" events, one

Journal of Physics: Conference Series **608** (2015) 012036 doi:10.1088/1742-6596/608/1/012036

can get a stream of "abnormal" events, indicating hidden problems with various components or subsystems of the experiment. For example, we can find voltage fluctuations, happening during stable beam conditions, data-taking flag set to "on" and non-fluctuating collision rate. Fig. 3 illustrates such behavior. In general, CEP increases modularity of online services by doing "external" stream aggregation. Dedicated Event Processor lifts the burden of manual stream manipulation from clients, simplifying their data handling code. Also, it allows to do pre-testing of services attached to message-queuing broker, by feeding simulated data stream and validating the response to the incoming events well before the production deployment. Another use-case, which is expected to be implemented for Run 15 is to perform triggered data migration from the online to the offline domain. Having CEP trigger stream, which combines readiness and completion states of many processes, allows to decreased resource consumption by eliminating unnecessary online database polling.

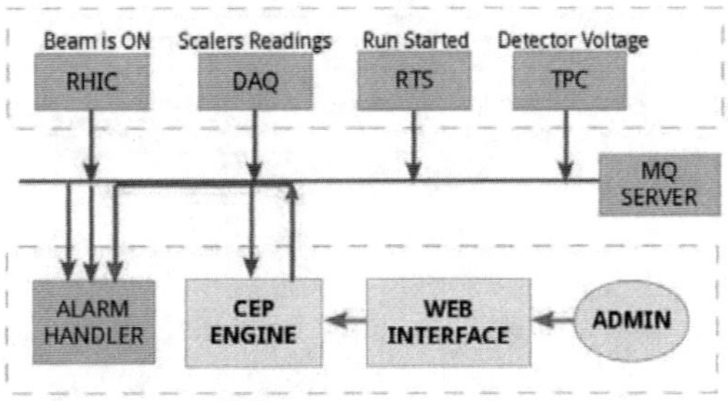


**Figure 2.** An overview of the Alarm handler use-case, describing the data flow and online configuration of persistent queries. Signals from many sources are combined to provide a clean and clear alarm one signal alone would not provide.

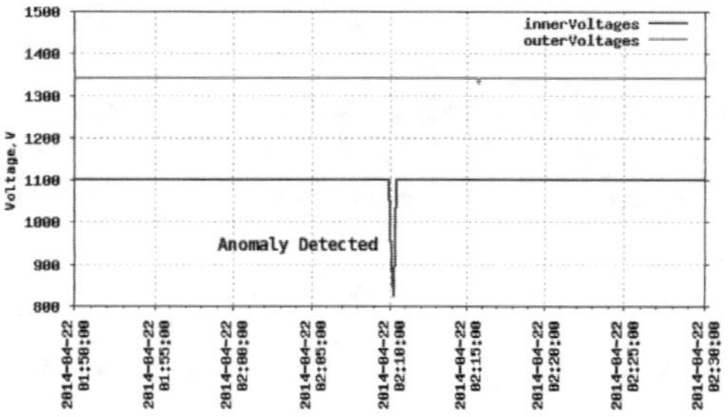


**Figure 3.** An example of the unexpected voltage drop, happening during data taking process, and registered by the event processing engine. Voltage drop is expected off physics running (and ramping up/down) but not during stable data taking (as in this example).

## 3. Framework Future: Control System Capabilities

The MIRA framework is still evolving. In the near future, we are planning to add features, commonly encountered in the Detector Control Systems domain: experiment workflow and hardware control. To implement MIRA's expansion to provide a *Detector Control* features, in addition to existing *Monitoring*, *Storage*, *Visualization* and *Alarms* capabilities, we have collected a list of requirements from our users and online software developers.

Requirements for the Detector Control system were collected during the initial preparation phase. We have identified the following key features: *Scalable Architecture*, *Low-overhead*, *Inter-operable*, *Messaging Protocol*, *Payload-agnostic Messaging*, *Quality of Service*, *Improved Finite State Machine*, *Real-time Remote Control*. Scalable Architecture is needed to accommodate for the growing number of detectors and channels. We expect STAR to double the number of channels in the next five years, hence, system scalability is our primary objective."A" low-overhead, inter-operable messaging protocol is desired to allow steady migration from wide variety of existing legacy hardware to the modern detector control equipment. It is very important to choose an easy-to-implement messaging protocol, as we need to support non-standard architectures from early days of the experiment. It implies that there are no publicly available, ready-to-use messaging libraries for this type of equipment. Payload-agnostic Messaging is needed to allow simultaneous handling of custom control protocols. The existing in-house developed protocols are predominantly text-based, and there is a strong desire to use and test legacy and upgraded components in parallel during the transition phase. Eventually, all archaic protocols will be upgraded to use JSON notation (text) or MsgPack (binary JSON equivalent) data serialization formats. Quality of Service regulation – there is a clear need to provide different service levels for three distinct groups of clients: (a) sensors, which provide continuous stream of data, are tolerant to sporadic loss of messages, but require maximum throughput, (b) data storage backend services, which need "receive message at most once" mode of operations, to avoid data duplication (c) control services, which operate in "receive at least once" mode to ensure persistent control behavior. Improved Finite State Machine – our preliminary survey indicates that existing Finite State Machine (FSM) implementations are lacking features like state stack, complex state handling and sub-states. We plan to investigate the possibility of implementing hierarchical, stack-based FSM for the expanded MIRA.Real-time Remote Control – we plan to study the possibility of making soft real-time, low-latency Human-Machine Interface (HMI) using newest web technologies like WebSockets [8] from the HTML5 technology stack, in combination with the messaging libraries. Our goal is to provide a consistent interface for all client types from desktops to smartphones and tables.

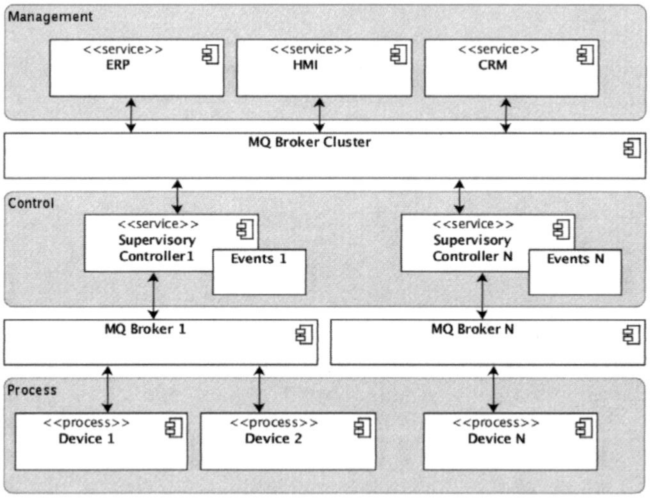


**Figure 4.** An overview of the preliminary version of tiered Control System architecture. Management, Control and Process tiers are communicating using messaging middleware services.

The development and transition to the fully upgraded MIRA is expected to happen in three years.

Important principle of our planned DCS implementation is the Observer Model methodology: clients are completely decoupled from data producers, any service attached to messaging broker may observe any process in the system. This allows us to embed rich monitoring, logging and reporting functionalities into the core of the framework. Also, we are going to implement this functionality as an extension of the MIRA framework, allowing us to reuse existing automatic archiving and visualization services.

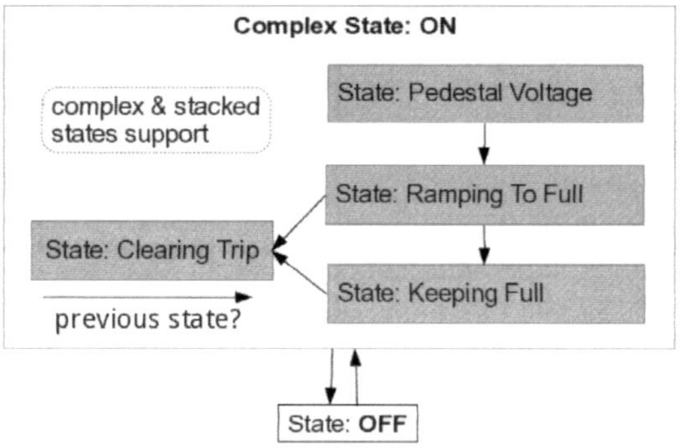


**Figure 5.** Illustration of the issue with the Finite State Machine having no stackable state support. Implementation of the exit state handlers using non-stackable state machine is possible, but is complicated compared to stack-enabled FSM.

The messaging part will be split into two distinct areas, corresponding to two distinct requirements: the *bottom layer* requires low-latency communications, while the *top layer* demands a rich transport protocol functionality and may not be satisfied by the same technology choice. We have selected MQTT protocol and AMQP protocol respectively. MQTT is a good choice for the sensor equipment due to its simplicity, while AMQP has larger overhead, but has a full stack of features required for reliable queue management. For the engine core we are going to implement brokered, hierarchical, stack-based finite state machine. In our future studies, we plan to do a detailed review the open-source implementations of FSMs used in HEP/NP experiments, with the objective to provide an implementation capable of resolving multi-exit scenarios, including complex-state and sub-state support (Fig. 5). Our project is going to use MQTT protocol over WebSockets to implement Human-Machine Interface of the Control System. This allows to utilize single technology stack for local, remote and web clients. For the messaging middleware we have selected Apache Apollo, an open-source multi-protocol broker, written in Scala. This broker handles MQTT, AMQP and WebSocket connections, allowing us to reduce the maintenance to the single node, or a single cluster of brokers, removing the necessity to maintain several independent brokers.

## 4. Summary and Outlook

We presented an update on the status of MIRA, the STAR Meta-Data Collection Framework, and its recent achievements, current state and plans for near-term development and extensions. Message-Queuing services became an instrumental part of STAR online infrastructure. MIRA framework, featuring flexible, loosely coupled system of components, was very well accepted by STAR's collaborators and detector experts, covering the meta-data collection and monitoring needs of all eighteen STAR subsystems, which gradually moved to the new framework. MIRA provided STAR with a solution to handle the growing number of channels (x15), and data structures (x25), allowing smooth operation during Runs 10-14.

In 2014 we have extended MIRA with the stream-based Complex Event Processing capability, which successfully passed our tests. A few alarms implemented for Run 14 saved months of work for the core team and collaborators. More use-cases will be implemented for Run 15 and beyond, expanding the usage of CEP to automatic data migration handling and extensive error conditions detection.

In addition to the Complex Event Processing features, we extensively described our future plans for extending MIRA to enable Control System capabilities. Extended framework will benefit from multi-tiered, scalable architecture, improved Finite State Machine functionality and a real-time remote control interface.

## Acknowledgments

This work was supported by the Office of Nuclear Physics within the U.S. Department of Energy's Office of Science.

## References

[1] The Solenoidal Tracker At Rhic (STAR) experiment [Online] URL `http://www.star.bnl.gov/`
[2] D Arkhipkin, J Lauret and W Betts 2011 *J. Phys.: Conf. Ser.* **331**
[3] Advanced Message Queuing Protocol [Online] URL `http://www.amqp.org/`
[4] D Arkhipkin, J Lauret, W Betts and G Van Buren 2012 *J. Phys.: Conf. Ser.* **396**
[5] WSO2 Complex Event Processor [online] URL `http://wso2.com/products/complex-event-processor/`
[6] ESPER: Event Stream and Complex Event Processing [online] URL `http://esper.codehaus.org/`
[7] RHEL6 High Performance Network with MRG - MRG Messaging: Throughput & Latency (*Preprint* `http://www.redhat.com/f/pdf/MRG-Messaging-1Gig-10Gig-IB-Xeon-v2.pdf`)
[8] RFC 6455: The WebSocket Protocol [online] URL `http://tools.ietf.org/html/rfc6455`

# Evolution of the ATLAS Software Framework towards Concurrency

**RWL Jones**[1]
Department of Physics, Lancaster University, Lancaster, UK

**GA Stewart**
School of Physics and Astronomy, University of Glasgow, Glasgow, UK

**C Leggett**
Physics Division, Lawrence Berkeley National Laboratory and University of California, Berkeley, CA, USA

**BM Wynne**
School of Physics and Astronomy, University of Edinburgh, Edinburgh, UK

E-mail: Roger.Jones@lancaster.ac.uk

**Abstract**. The ATLAS experiment has successfully used its Gaudi/Athena software framework for data taking and analysis during the first LHC run, with billions of events successfully processed. However, the design of Gaudi/Athena dates from early 2000 and the software and the physics code has been written using a single threaded, serial design. This programming model has increasing difficulty in exploiting the potential of current CPUs, which offer their best performance only through taking full advantage of multiple cores and wide vector registers. Future CPU evolution will intensify this trend, with core counts increasing and memory per core falling. Maximising performance per watt will be a key metric, so all of these cores must be used as efficiently as possible. In order to address the deficiencies of the current framework, ATLAS has embarked upon two projects: first, a practical demonstration of the use of multi-threading in our reconstruction software, using the GaudiHive framework; second, an exercise to gather requirements for an updated framework, going back to the first principles of how event processing occurs. In this paper we report on both these aspects of our work. For the hive based demonstrators, we discuss what changes were necessary in order to allow the serially designed ATLAS code to run, both to the framework and to the tools and algorithms used. We report on what general lessons were learned about the code patterns that had been employed in the software and which patterns were identified as particularly problematic for multi-threading. These lessons were fed into our considerations of a new framework and we present preliminary conclusions on this work. In particular we identify areas where the framework can be simplified in order to aid the implementation of a concurrent event

[1] To whom all correspondence should be addressed.

processing scheme. Finally, we discuss the practical difficulties involved in migrating a large established code base to a multi-threaded framework and how this can be achieved for LHC Run 3.

## 1. Introduction

The upcoming upgrades to the Large Hadron Collider both in energy and in luminosity, coupled with upgrades to various components of the detector, are in themselves enough to motivate changes to the ATLAS software framework. The changes bring with them increases in the pile-up of many collisions in one read-out of the detector, increased acceptance rates from the trigger to the offline and increased complexity in the events. These all present a challenging computing requirement. To illustrate this, the pile-up of events at the end of the run 1 data-taking was about 25; this will rise to well over 100 by run 4, presenting a particularly challenging pattern recognition problem for the tracking. This will imply a much larger processing load, which will be matched by increased simulation requirements. This also results in a larger raw event size of about 2MB (c.f. 1MB in run 1), which coupled with a planned factor 5-10 increase in the trigger output rate (to 5-10kHz in run 4) implies a much higher data rate, and a much bigger load on the storage, retrieval and archival systems.

The need for framework improvements is compounded by the changing hardware landscape. While the Moore's Law doubling of the number of transistors every two years still holds, the architectures are changing to sustain this. In particular, our existing code risks leaving transistors waiting for something to do, with the introduction of vector registers, out of order execution, hyperthreading etc. While such developments increase the theoretical power of the processor, achieving this with HEP applications is a challenge. The dominance of Xeon x86_64 as a platform for HEP is unlikely to last, with more inhomogeneous architectures becoming common: low power, low memory processors such as the ARM, Atom and Avaton; very high core count stronger processors such as the Xeon Phi; and even GPGPU architectures, such as NVidia CUDA. With computing budgets projected to be flat, at best, the experiments must make use of every available resource, often shared with other users; and this calls for performant and flexible code, run under a performant and flexible framework.

The offline processing in Run 1 was trivially coarse-grained parallel, with events handled by independent processes, each demanding about 2GB of physical memory per core. This was efficient in the usage of the cores, but becomes increasingly demanding in costly memory as the core count increases. ATLAS has made an important advance in addressing this for Run 2 by introducing a multi-process version of the Athena framework [1], Athena-MP. This makes use of the Linux kernel's copy-on-write feature. The framework is initialised only once by the initial mother process. Then worker processes are forked and each fork handles a distinct stream of events. All the memory pages corresponding to large static data structures (such as magnetic field and detector geometry) are only read from during event processing, so remain shared between the mother and workers, with only one physical page actually used. Memory pages that are written to by worker processes during event processing are unshared so the overall memory usage per processor is significantly reduced as the core count per processor increases. However, tests indicate that the memory saving of ~0.8GB/worker would be insufficient for post-Xeon architectures.

To go beyond this in terms of memory saving, multi-threading is required. This can provide huge savings, as all heap memory is shared. The cost is amore complicated programming model, as locks are required and data races and deadlocks need to be avoided. This is made doubly difficult, as a large (approximately 7 million line) successful code base exists that was written for serial mode and has been in use for over a decade. The evolution path needs to be considered very carefully, and adiabatic change made when possible.

There are three levels at which parallelism can be applied: the event, the algorithm and in-algorithm. While a less parallel option may be possible, with one event per thread, experience so far indicates that the requirement of multi-threading is sufficiently strong that the additional levels of parallelism are easily gained, and add to the flexibility of the code-base.

## 2. Prototypes and testbeds

### 2.1. CPU Crunchers

ATLAS has undertaken a series of experiments to determine the way forward. The first step, reported previously, was to use simulated algorithms (known as 'CPU-crunchers') to establish that multi-threading provides the gains expected. To do this, the GaudiHive [2] prototype was chosen (see other contributions to this meeting). This is a natural choice given ATLAS already shares the Gaudi framework, of which Athena is a derivative. The conclusion of the CPU-cruncher tests is that multi-threading events clearly works, with good use of the machine resources. In this regard, Athena-MP is already doing well. However, the Hive solution demonstrates that as well as scaling with multiple-events in flight, as with Athena-MP, significant gains can be made by having multiple copies of cpu-intensive algorithms (algorithm cloning).

### 2.2. Real reconstruction

The next step was to apply these ideas to real reconstruction. The current full reconstruction runs several hundred algorithms that produce hundreds of objects in the event store. A data flow and dependency analysis is sobering: there are tangled relations, and the simple analysis turns out to understate the issue. There are also hidden dependencies through the use of public tools, which would make scheduling for parallel running problematic. Rather than tackle the full problem at once, the approach was to take a more manageable part of the reconstruction, the calorimeter reconstruction, involving only seven algorithms and 16 data objects, and to implement it in GaudiHive. This had the additional benefit of being of interest to the software trigger activity.

Even so, many code modifications were required. The event selector was redesigned to work with Hive. There was also work on the event IO in the form of the convertors into the event storage, StoreGate. These were made thread safe by simple locking, with no attempt at parallelisation. Multiple events were bottlenecked by a single convertor algorithm. (There is clear scope for improvement here, as the IO step is about 12% of the processing time.) There was work on the AthAlgorithm and AthAlgTool base classes to enhance the thread and event slot information. A significant piece of the work was on the bypassing many uses of incidents, where execution jumps to plugins defined by the user. (These work well in serial processing, but in the parallel case even incidents such as BeginEvent and EndEvent become ambiguous with many events in flight). There was also work required in the user code to avoid the use of incidents to trigger actions, which would render the workflow unscheduleable. The user code was also changed such that data is only added to the event store when complete for use by other algorithms, not changed in-situ; again, without this scheduling becomes nearly impossible.

The code was benchmarked against regular Athena, serial Hive, and Hive with 1 event in flight and 30 threads. The number of events in flight with 5 algorithms and 30 threads was then compared, but without algorithmic cloning. The throughput improves using up to 3 events in flight, with a modest increase in memory usage and a factor 2x speed up. Beyond 3 events in parallel throughput is bottlenecked by the most time consuming algorithm and no further gain is achieved.

The exercise was then repeated with cloning of the most expensive algorithms and their tools, where this was possible. This is an intermediate solution on the way to full thread-safe algorithms, and helps alleviate the choke point of the most expensive algorithms. Even with this limited cloning, the bottlenecks were removed and the improvements continue to many events in flight. The best results, achieved with cloning just 4 algorithms, gave a factor 3.3 increase in speed (with the initialisation removed) for a 28% increase in memory usage.

### 2.3. Conclusions

The tests showed that multi-threading can speed up processing at a lower memory cost than serial or multi-process techniques, although not being able to clone certain algorithms is a handicap. Several approaches can be considered to avoid this issue, from making the tools fully thread safe to splitting-up very large algorithms into more tractable ones. It was also learned that incidents are a problem as these express workflows that are hidden from the scheduler. It became evident that incidents have often been used in the ATLAS code when alternatives exist. It is clear that access to the event store will need clearer patterns for concurrent processing, and that data dependencies need to be propagated to the lowest level where they are used; algorithms may know nothing of data they are to use via tools.

## 3. Requirements Capture

These lessons have been gathered and lead the formation of the Future Frameworks Requirement Group (FFREQ), which is charged with planning the evolution of the framework to Run 3. The ambition is to incorporate the requirements of the offline and of the high level trigger, to minimize or eliminate the differences between the two code bases. The initial emphasis has been on requirement capture and use cases. The event processing pattern, which has served well in particle physics, is still used, but now updated to incorporate concurrency. Events pass through a sequence of algorithms that read the event and other data, and generate new object. Much of the work is done through tools, which are now private to the algorithm. Incidents still exist and can be triggered by any event during the workflow, but are now handled by the scheduler in a controlled way.

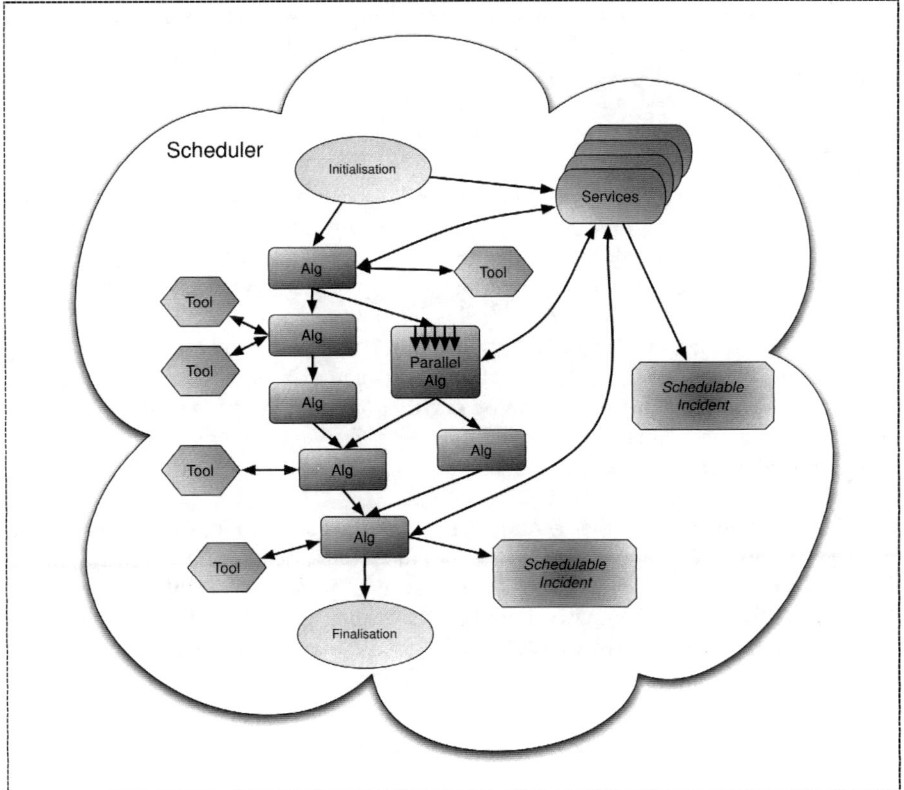


**Figure 1. The proposed new event flow, for a single event.**

Based on the experience from the prototype, public tools are avoided and replaced by thread-safe services. The reliance on incidents is minimised, and generally replaced by data flow and control flow dependencies (in a loose sense, case statements). Interactions in the event store will proceed via data handles, which provide a well-defined interface that declares the intent (read, write or update) of the algorithm with regards to an event store object. It is recognised that the event store data may be mutable – an algorithm can read and write to the same object(s) in the store. (An example might be in calibration or in augmenting an object.) This leads to the problem of multiple algorithms attempting to mutate the same data. The solution is to require that the order in which they mutate the object be declared at configuration time; if this is not done, there is a configuration error. Algorithms that only read the object will, by default, be scheduled after all mutations. The scheduling of the block of mutating algorithms is decided at run time. The block of algorithms should be treated as single schedulable unit, which will reduce the load on the scheduler during the event loop. E.g., it has been shown that such sequences can be implemented as Intel Threaded Building Blocks graphs, which are efficient even with very small computational units.

The High Level Trigger poses additional requirements on the framework. It takes multiple regions of interest in the detector and runs the reconstruction within these, and then runs multiple trigger chains in that region until a chain fails or the event is accepted. Different chains can configure the same algorithm, but for efficiency it should only be run once on the same data. Equally, data requests from the read out system and the time to clear the event build and read out buffers need to be minimised. In addition, the trigger can consider the multiple regions of interest in the same event and run the same algorithms in the trigger chains over this different input data in different detector regions. These considerations require that the event store must support multiple views. Algorithms and RoIs become mutually dependent in some contexts, which complicates the event scheduling. This concept of multiple views could be useful in the offline context, for instance allowing more expensive tracking to be run for candidate tau lepton decays.

## 4. Conclusions

Various important lessons have been learned concerning the required changes to the code base. Algorithms that are not heavy resource consumers that can be run in single instances without special constraints do not pose a problem; this generally means the physics analysis code can remain largely unchanged. One exception is the use of patterns hostile to threading, which seem to have been adopted by many coders, e.g., the use of public tools to cache data passed between algorithms without explicit use of the event store and the use of incidents, such as EndEvent to clear such caches. On the other hand, time consuming code must be made thread safe. This can mean either cloning in existing code or true thread safety in new code. This strategy allows work on thread safety to begin before the new framework is ready, and multithreading within single algorithms can already be implemented. Another key development is to ensure that services that can be called many times in different contexts must be made thread safe.

One inevitable overall conclusion is that good programming models and training must be provided to the developer community. With these lessons in mind, the requirements capture for the framework design in Runs 3 and 4 at very high luminosity is underway. There is reasonable hope that the trigger can be accommodated in this new framework, but the development of the scheduler and event store goes beyond the current Hive prototype. However, the offline may also benefit from these extensions. The plan is to design and implement the new framework during the next run, Run 2 and the code would then be adapted during the long shutdown that follows ready for Run 3.

**References**

[1] ATLAS Collabroation, *ATLAS Computing Techincal Design Report*, TDR-017. CERN-LHCC-2005-022

[2] B. Hegner et al, *Evolving LHC Data Processing Frameworks for Efficient Exploitation of New CPU Architectures*, Proceedings IEEE-NSS 2012

[3] D. Crooks et al, *Multi-core job submission and grid resource scheduling for ATLAS AthenaMP*, Computing in High Energy and Nuclear Physics 2012, New York, NY, USA

Journal of Physics: Conference Series **608** (2015) 012038 doi:10.1088/1742-6596/608/1/012038

# Modern Messaging for Distributed Sytems

**L Magnoni**[1]

[1] CERN, European Laboratory for Particle Physics (CERN), Geneva, Switzerland

E-mail: luca.magnoni@cern.ch

**Abstract.**
Modern software applications rarely live in isolation and nowadays it is common practice to rely on services or consume information provided by remote entities. In such a distributed architecture, integration is key. Messaging, for more than a decade, is the reference solution to tackle challenges of a distributed nature, such as network unreliability, strong-coupling of producers and consumers and the heterogeneity of applications. Thanks to a strong community and a common effort towards standards and consolidation, message brokers are today the transport layer building blocks in many projects and services, both within the physics community and outside. Moreover, in recent years, a new generation of messaging services has appeared, with a focus on low-latency and high-performance use cases, pushing the boundaries of messaging applications. This paper will present messaging solutions for distributed applications going through an overview of the main concepts, technologies and services.

## 1. Introduction
This paper presents an overview of messaging concepts, functionalities and modern technologies. It starts with an introduction of messaging for distributed communication and system integration. A review of the main messaging features is then provided, followed by an overview of the major technologies for messaging, from broker to broker-less systems. In conclusion, a list of successful stories concerning the use of messaging to solve the problem of communication for distributed applications is presented.

## 2. Messaging for loosely coupled communication
Modern distributed systems can be comprised of hundreds, if not thousands, of applications operating in multiple tiers and providing different services and functionality to each other. In such a distributed architecture, there are many challenges such as network unreliability, strong-coupling of producers and consumers and the heterogeneity of applications which need to be addressed to build a solid and reliable system.

### *2.1. Connection-oriented communication*
Connection-oriented communication is a simple solution to exchange information among remote entities. As presented in Figure 1, consider opening a socket over a connection-oriented protocol such as TCP/IP and transmit a raw stream of data through it. That would be a fast and inexpensive way to exchange information, but at the same time that tightly-coupled communication would be based on a number of assumptions which needs to be satisfied in order the communication to take place:

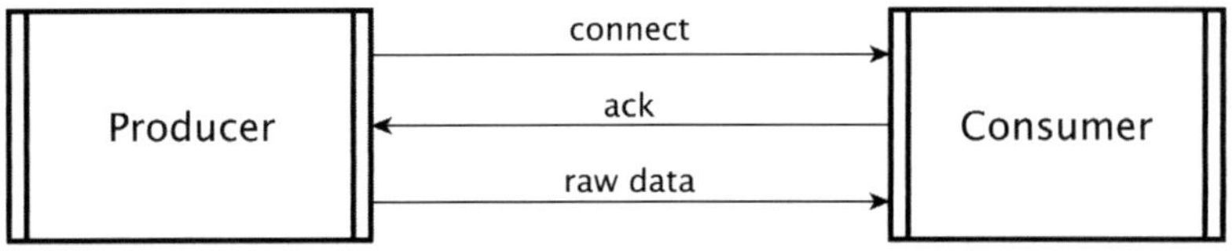


Figure 1: Tightly coupled communication.

- **Temporal dependency**: all components have to be available at the same time.
- **Location**: each component must know each other address.
- **Data structure and representation**: in its simplest implementation, all components have to agree on the data format and on the binary representation.

### *2.2. Messaging for loosely coupled communication*
Coupling can be measured as the number of assumptions parties make about each other when they communicate. Messaging is an example of loosely coupled communication solution where *message* is the information building block, which aims at minimizing those assumptions. Instead of sending information directly to a specific address, it can be sent to an addressable *channel*, to resolve the location dependency. In order to remove the temporal dependency, that channel can be enhanced to queue up the information until the remote components are ready to receive it. This way, the producer can now send requests into the channel and continue processing without having to worry about delivery. Messaging makes no assumption on data representation, so that standard data format, e.g. self-describing and platform independent as JSON or XML, can be used to remove the need to share data handling logic among all components.

### *2.3. Messaging scenario*
Typical messaging use cases are: *Information Publishing*: an entity publishes volatile information with no a-priori knowledge about who is interested (e.g. sensor); *Information Storing*: an entity collects information from multiple sources (e.g. log collector); *Remote Procedure Call*: an entity sends request to one or more remote entity and expect reply.

### *2.4. Messaging middleware*
Messaging is a loosely coupled communication solution which minimizes producer and consumer dependencies. Removing these dependencies makes the overall architecture more flexible and tolerant to changes, but it comes with additional complexity. Therefore, dedicated messaging middleware has been developed over the years to provide messaging functionality without having to deal with the inner complexity. The next section describes the main concepts and principles of messaging systems.

## 3. Messaging systems
A messaging system, as shown in Figure 2, acts as an indirection layer among entities that want to communicate. Usually referred as *message broker*, it is responsible for transferring data, as messages, from one application to another, so that producer and consumer can focus on what to share rather than on how to share it. Like many other technologies, messaging is based on some basic concepts and properties which are shared among all the different flavours and implementations.

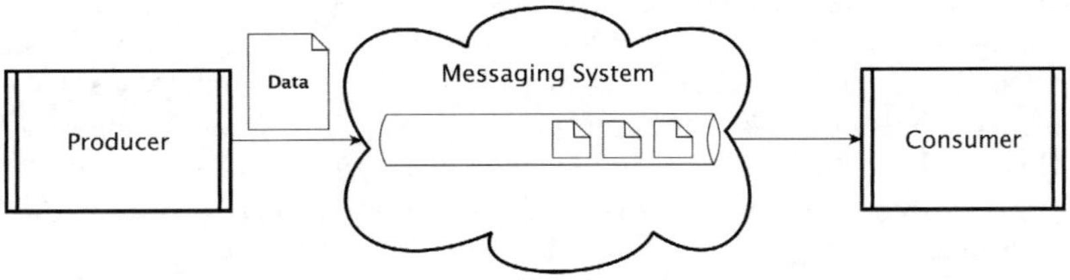


Figure 2: Messaging for loosely coupled communication.

### *3.1. Message*

A message is the information building block. It is composed by a *body*, which is immutable and contains the structured data (e.g. JSON, XML, serialization protocols) object of the communication and by a set of *header*, normally key value pairs which can be processed by the broker and used for routing.

### *3.2. Communication models: topic and queue*

Messaging systems support different communication models, each one defining how the information is exchanged among producer and consumer. The most common communication models are *queues* and *topics*. Queue are used to implement a point-to-point communication, where, if no consumer exists when the information is produced, the message is kept in the channel for later delivery, whilst if there are multiple consumers the message is delivered only once. Topic is for the classic publish/subscribe scenario, where if no consumer exists the message is discarded and in case of multiple consumers the message system delivers it to each of them. A part from queue and topic which are widely supported, more complex delivery semantics exist at protocol level (e.g. exchange/nodes from AMQP) and many others are middleware-specific.

### *3.3. Protocols*

The lack of a unique standard way to interact with message brokers has been a known issue for the messaging technology for many years. The AMQP protocol has been designed by the main messaging actors, companies and software producers, to overcome this limitation. Nevertheless, the inner complexity of a unified protocol which defines both wire communication and delivery semantics requires a major development effort for the messaging system to become fully compliant. This section presents an overview of the most common standard protocols which are supported today by the main messaging systems. The protocol choice is a crucial design decision for message-oriented architecture, in respect to the strong coupling it has within the application.

*3.3.1. AMQP* (Advanced Message Queuing Protocol) [1] is the result of a standardization effort by the major contributors in the messaging scene (e.g. Cisco, Microsoft, Red Hat, banks). It is designed for interoperability between different messaging systems. It provides the definition for a binary wire protocol and a complete delivery semantic, allowing, theoretically, for an AMQP messaging client to be able to interact seamlessly with different brokers implementations which are AMQP compliant. Nowadays, the adotion of the latest stable version (1.1) of the protocol is not yet extensive, but given that it is already supported by the major message brokers, a much wider implementation is expected in the upcoming years.

*3.3.2. STOMP* (Streaming Text Orientated Messaging Protocol) [2] is a text-based protocol meant to be simple and widely-interoperable. It is mainly a wire protocol, it comes with very basic messaging semantic built-in (e.g. no support for communication models, the destination is just a string message header) , requiring appropriate configuration at the message system level (e.g. the destination have to be appropriately mapped to a queue or a topic). Thanks to its simplicity, there is an extensive set of clients available in many languages and it is supported by the majority of brokers.

*3.3.3. MQTT* (Message Queue Telemetry Transport) [3] is lightweight protocol designed originally from IBM. It is meant for low bandwidth, high-latency networks. It defines a compact binary format with very limited overhead on the communication, few tens of Bytes, which makes it suitable for *Internet of Things* style applications (e.g. mobile phones, sensors) in a simple *produce-and-forget* scenario.

### *3.4. Capabilities*

As introduced in Section 2, a messaging system can be thought as a intermediate channel which is enhanced with additional features, such as queuing, to improve the communication experience of remote entities. Over the years, although no formal agreement exists, the different messaging systems converged over a common set of capabilities which become de-facto standard for messaging middleware. The list of features includes *Persistence*, which is the ability to save message on persistent storage, such as file-system or database; *Fail-over*, which allow clients to automatically reconnect in case of broker failure; *Guaranteed delivery*, which defines the policy for the message delivery (e.g. *at-least* once or *exactly-once*); *Ordering*, to deliver messages in the order they are produced; *Transaction*, the ability to consider multiple requests as part of a distributed transaction, with roll-back option and *Clustering*, which is the possibility to create network of message brokers for high-availability and load-balance. Nevertheless, each messaging system may provide different interpretation for the same features. Many other *unique* broker-specific features exist, but their usage imply hard coupling the application with a specific broker flavour.

## 4. Messaging technology

Message-oriented middleware has been developed for more than a decade to what today is a rich and solid ecosystem of services and libraries. Message brokers, as intermediate standalone services which offer messaging capabilities to distributed applications, are the most common type of messaging systems. Message brokers have been extensively used over the years for implementing communication and integration in distributed system [4], with the exception of data-intensive and high-performance use cases, where the existence of an intermediate entity is not a suitable option. In recent years, a new generation of messaging systems has appeared, with a focus on low-latency and high-performance use cases, pushing the boundaries of messaging applications. The next section is going to present an overview of the main messaging technology.

### *4.1. Message brokers*

Message brokers are the most common implementation of messaging system. A message broker is standalone intermediate entity which offers messaging functionality via standard or custom protocols. Many message brokers exist, different in capabilities, protocols, implementation languages, platform support. The focus of this review is on open-source solutions, but many exist as part of enterprise commercial software too.

Message brokers are the most feature-rich type of messaging system, in term of capabilities an protocol support as described in Section 2. Brokers can be *polyglot*, allowing producer and consumer to use different protocol (e.g. sender over AMQP, receiver over STOMP) and they can support *message transformation* (e.g. transforming message payload from XML to JSON) to further decouple applications.

*4.1.1. ActiveMQ* is one of the most widely adopted open-source message broker. It is an Apache project, it is written in Java and it is commercially supported by Red Hat. ActiveMQ has an extensive protocol support (e.g. AMQP, STOMP, MQTT, Openwire, HTTP and many others), it provides many cross language clients and it is fully JMS compliant. ActiveMQ offers many advanced capabilities, such as rich deliver semantic (e.g. virtual queues, composite destination, wildcards), JDBC message store (e.g. to persist messages in any JDBC compliant database) and advanced clustering configuration (e.g. master-slave, network of brokers). ActiveMQ is a feature-complete messaging solution, which can be used to implement many communication and integration patterns [4].

*4.1.2. RabbitMQ* is a lightweight open-source message broker written in Erlang which profits from the message passing capabilities from the language underneath. RabbitMQ architecture is deeply modular, it mainly supports AMQP and STOMP but additional protocols can be loaded as plug-in (e.g. MQTT, HTTP). It supports the main messaging capabilities, such as persistence, clustering, high-availability and federation. RabbitMQ remains a lightweight messaging solution which can be found embedded into several projects (e.g. Logstash) for its simplicity and reliability.

*4.1.3. Performance and Scalability* For messaging systems, the quantitative measure of *messages per second (msg/s)* has very little meaning without detailed contextualization. The protocol used (e.g. binary or textual), plays a major role, but many other latency factors exist: *persistent* messages can be orders of magnitude slower, the *amplification factor* (e.g. the number of topic consumers) can impact the system with multiple in-memory copy of the message, and the same is true for the *payload size*. Suboptimal clients can lead to high number of *open/close connections* and misbehaving consumers can lead to the *low-subscriber* problem, one of the most common issue for messaging infrastructure.

The comparison presented in [5], where several message brokers are evaluated via the STOMP protocol in several communication models, shows how, within realistic scenario, the performance may vary across 100000 msg/s and 1000 msg/s.

### *4.2. Apache Kafka*

Apache Kafka is an open-source project originally from LinkedIn, now part of the Apache foundation. It has been developed for *real-time* activity stream analytics, to solve the need for an effective way to move big amount of data (e.g. user metrics, computer farm monitoring) from the producers to many potential consumers. The scale and the data size (billions of messages and hundreds of gigabytes per day) and the time constraint makes the use case not suitable for standard brokers, as by the comparison in [6].

The innovative idea of Kafka is to be a stateless broker, so to not retain any information about consumers. A consumer has to retain its own state (e.g. the information about the last data read) and poll Kafka for new data when needed. This allows Kafka to persist a single message copy independently from the number of consumers (e.g. messages are not removed on consumption, but by retention period or other policy), with a resulting high-throughput for read and write operations. Kafka persistence is implemented as a distributed commit log, as shown in Figure 3, designed as distributed system easy to scale out (based on Zookeeper), which allows for automatic balancing of consumer/producer/broker.

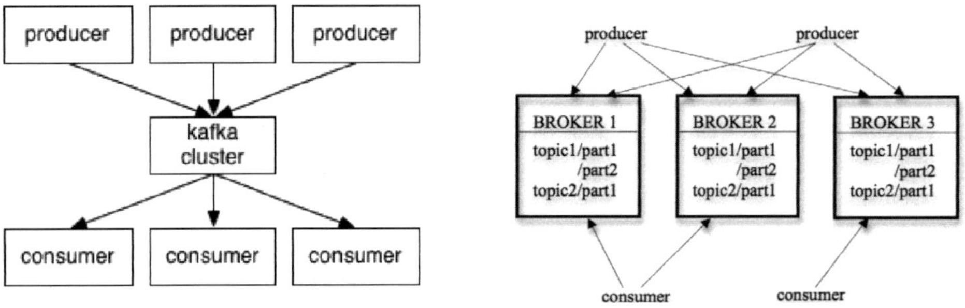


(a) Kafka cluster (b) Kafka topic partitioning

Figure 3: Kafka architecture.

In contrast with standard message brokers, Kafka provides limited messaging capabilities (e.g. mainly topic semantic, file-system as unique persistent storage, strict guaranteed ordering). Although many client libraries are available, it only supports its custom binary format over TCP. Kafka is an optimal solution for *data-movement*, frequently adopted as *pipe* to different processing systems (e.g. Hadoop, Storm).

### *4.3. ZeroMQ*

Despite the name, ZeroMQ (also known as 0MQ or ZMQ) [7] is not a standard message broker but a lightweight messaging library which provides messaging capabilities. Distributed applications can use ZeroMQ for high-throughput and low-latency communication, profiting from its ability to implement direct connection among producer and consumer, with no intermediate entities involved. Although this may seems in contradiction with one of the main assumption of messaging, ZeroMQ implements loosely coupled communication via an innovative approach, acting as a new layer on the networking stack. It expands the concept of socket with a similar API but enhaced with built-in messaging patterns: *Request/Reply*, *Publish/Subscribe*, *Pipeline* and *Exclusive pair*, as presented in Figure 4. In contrast with classic socket, each ZeroMQ socket comes with an internal queue to allow for asynchronous communication. The result is that, for example in a request/reply scenario used for point to point communication, if the data is produced when the consumers is not running, the ZeroMQ library will take care of deferred delivery with no additional load on the producer side.

The idea behind ZeroMQ is powerful, it allows for high-performance and low-latency communcation, but comes with additional complexity at the application level. ZeroMQ mainly supports its own binary protocol and provides limited messaging capabilities (e.g. failover, multicast support for 1-N topology). Although several features can be easily implemented using ZMQ API (e.g. acknowledgement), implementing advanced messaging capabilities (e.g. guaranteed delivery, persistence) may require considerable effort, making it suitable for data-intensive scenario where simple messaging semantic is needed.

## 5. Use cases

This section presents several use cases where messaging-based communication has been successfully adopted to solve the problem of exchange information in distributed system.

### *5.1. CERN Beam Control middleware*

The Beam Control department at the CERN laboratory is using messaging for highly reliable control/monitoring/alarm applications for the Large Hadron Collider (LHC). Since 2005, a

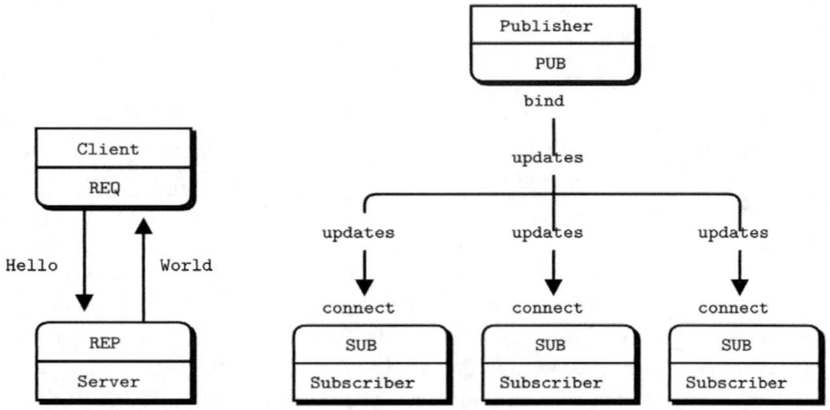


(a) Request/Reply (b) Publish/Subscribe

Figure 4: Examples of ZeroMQ sockets.

cluster of ActiveMQ brokers, in a store and forward configuration, is used to collect the critical data generated by the safety systems (e.g. 30 producers, 2MB/s, 4.5K msg/s) and to forward it to many consumers (e.g. monitoring tool, dashboards). Being safety data mission critical, the store and forward configuration allow to completely decouple data production from consumption, preventing misbehaving clients to affect data collection and archiving [8]. Moreover, the LHC Control framework has been recently migrated from CORBA to ZeroMQ as communication layer [9].

*5.2. DAQ Online Monitoring*

Messaging has been also extensively used in several monitoring tools for Data Acquisition (DAQ) systems, which are responsible to filter and collect data from detectors (e.g. high energy physic experiments) to storage facilities.

*5.2.1. The ATLAS TDAQ shifter assistant project* [10] relies on messaging to distribute operational alarms from private TDAQ network to GPN to a number of heterogeneous consumers. An ActiveMQ cluster is used in a master/slave configuration in order to minimize the impact on the required firewall configuration to a single outbound connection.

*5.2.2. The STAR Online framework* relies on an AMQP-based system for flexible, loosely-coupled distribution of detector metadata, using messaging as unified transport layer for processing, storage and monitoring. Moreover, investigation has been done to re-write the control framework over MQTT, profiting from the protocol flexibility and interoperability [11].

*5.3. WLCG Messaging Service*

Messaging has been also successfully used on large-scale geographically distributed infrastructure. The WLCG (Worldwide LHC Computing Grid) messaging service is the backbone transport layer used for monitoring WLCG sites and services around the world, with more than 50000 clients and an average message rate of 100 KHz. The monitoring infrastructure is based on STOMP with JSON payload. Thanks to the interoperability of the STOMP protocol across several broker flavours, heterogeneous message-broker clusters (ActiveMQ, Apollo or

RabbitMQ) are used in a scenario where client applications produce to any and consume to all [12].

## 6. Summary

*Messaging is fundamentally a pragmatic reaction to the problem of distributed systems* [4]. As presented in Section 2, it allows loosely coupled communication acting as intermediate layer between producer and consumer. It brings many benefits in distributed applications flexibility and scalability, with implications in application and infrastructure complexity. Messaging systems are still evolving technology, as shown in Section 3, with the AMQP standardization effort pointing in the good direction, but still with partial adoption. Message brokers are solid and reliable technology used as transport layer building blocks in many projects and services, both within the physics community and outside. In the recent years, a new generation of systems is promoting messaging for low-latency / high-throughput / data-intensive communication, as presented in Section 5, narrowing use cases and relaxing assumptions, but pushing the boundaries of messaging applications towards new domains.

## References

[1] AMQP (Advanced Message Queuing Protocol) http://www.amqp.org
[2] STOMP (Simple Text Oriented Messaging Protocol) http://stomp.github.io
[3] MQTT (MQ Telemetry Transport) http://mqtt.org
[4] G Hohpe and B Woolf 2003 *Enterprise Integration Patterns* Addison-Wesley Professional
[5] Chirino H *STOMP benchmark* http://hiramchirino.com/stomp-benchmark
[6] Kreps J, Narkhede N and Rao J *Kafka: A Distributed Messaging System for Log Processing.* NetDB Workshop (Athens, GREECE)
[7] Hintjens P *ZeroMQ: The Guide* http://zeromq.org
[8] Ehm F *Running a Reliable Messaging Infrastructure for CERN's Control System.* Proceedings of ICALEPCS2011 (Grenoble, FRANCE)
[9] Dworak A, Ehm F, Sliwinski W and Sobczak M 2011 *Middleware Trends and Market Leaders 2011.* Proceedings of ICALEPCS2011 (Grenoble, FRANCE)
[10] Kazarov A, Miotto G L and Magnoni L 2012 *The AAL project: automated monitoring and intelligent analysis for the ATLAS data taking infrastructure.* Journal of Physics: Conference Series, Volume 368
[11] Arkhipkin D, Lauret J and Betts W 2011 *A message-queuing framework for STARs online monitoring and metadata collection.* Journal of Physics: Conference Series, Volume 331
[12] Cons L and Paladin M 2011 *The WLCG Messaging Service and its Future.* Journal of Physics: Conference Series, Volume 396

# A Survey on Distributed File System Technology

**J Blomer**
CERN, CH-1211 Genève 23, Switzerland
E-mail: jblomer@cern.ch

**Abstract.** Distributed file systems provide a fundamental abstraction to location-transparent, permanent storage. They allow distributed processes to co-operate on hierarchically organized data beyond the life-time of each individual process. The great power of the file system interface lies in the fact that applications do not need to be modified in order to use distributed storage. On the other hand, the general and simple file system interface makes it notoriously difficult for a distributed file system to perform well under a variety of different workloads. This has lead to today's landscape with a number of popular distributed file systems, each tailored to a specific use case. Early distributed file systems merely execute file system calls on a remote server, which limits scalability and resilience to failures. Such limitations have been greatly reduced by modern techniques such as distributed hash tables, content-addressable storage, distributed consensus algorithms, or erasure codes. In the light of upcoming scientific data volumes at the exabyte scale, two trends are emerging. First, the previously monolithic design of distributed file systems is decomposed into services that independently provide a hierarchical namespace, data access, and distributed coordination. Secondly, the segregation of storage and computing resources yields to a storage architecture in which every compute node also participates in providing persistent storage.

## 1. Introduction

Distributed file systems provide persistent storage of unstructured data, which are organized in a hierarchical namespace of files that is shared among networked nodes. Files are explicitly created and they can survive the lifetime of processes and nodes until explicit deletion. As such they can be seen as the glue of a distributed computing infrastructure. Distributed file systems resemble the API of local file systems. To applications, it should be transparent whether data is stored on a local file system or on a distributed file system. This data model and the interface to applications distinguishes distributed file systems from other types of distributed storage such as databases.

Virtually all physics experiments store their data in distributed file systems. Large experiment collaborations, such as the experiment collaborations at the Large Hadron Collider (LHC), store data in a global federation of various cluster file systems rather than in a single, globally distributed file system. For LHC experiments, such globally federated and accessible storage sums up to more than 1 billion files and several hundred petabytes.

There is a variety of file systems available to choose from [1–14] and often it is not clear what are the particular strengths, weaknesses, and implications of using one distributed file system over the other. Several previous studies presented taxonomies, case studies, and performance comparisons on distributed file systems [15–20]. This survey is focused on the underlying building blocks of distributed file systems and what to expect from them with respect to physics applications.

The survey is mainly driven by the high-energy physics needs, which however cover many of the aspects relevant to distributed file systems in general.

## 2. How are Distributed File Systems Used?

Even though the file system interface is general and fits a broad spectrum of applications, most distributed file system implementations are optimized for a particular class of applications. For instance, the Andrew File System (AFS) is optimized for users' home directories [2], XrootD is optimized for high-throughput access to high-energy physics data sets [7], the Hadoop File System (HDFS) is designed as a storage layer for the MapReduce framework [10, 21], the CernVM File System (CVMFS) is optimized to distribute software binaries [12], and Lustre is optimized as a scratch space for cooperating applications on supercomputers [5]. These use cases differ both quantitatively and qualitatively. Consider a multi-dimensional vector describing different levels of properties or requirements for a particular class of data that consists of *data value*, *data confidentiality*, *redundancy*, *volume*, *median file size*, *change frequency*, and *request rate*. Every single use case above poses high requirements in only some of the dimensions. All of the use cases combined, however, would require a distributed file system with outstanding performance in every dimension. Moreover, some requirements contradict each other: a high level of redundancy (e. g. for recorded experiment data) inevitably reduces the write throughput in cases where redundancy is not needed (e. g. for a scratch area). The file system interface provides no standard way to specify quality of service properties for particular files or directories. Instead, we have to resort to using a number of distributed file systems, each with implicit quality of service guarantees and *mounted* at a well-known location (/afs, /eos, /cvmfs, /data, /scratch, ...). Quantitative file system studies, which are unfortunately rare, provide precise workload characterizations to guide file system implementers [22–24].

### *2.1. Application Integration*

From the point of view of applications there are different levels of *integration* a distributed file system can provide. Most file systems provide a library with an interface that closely resembles POSIX file systems. The advantage is that the interface can be adapted to the use case at hand. For instance, the Google File System (GFS) extends POSIX semantics by an atomic append [25], a feature particularly useful for the merging phase of MapReduce jobs. A library interface comes at the cost of transparency; applications need to be developed and compiled for a particular distributed file system.

So called *interposition* systems introduce a layer of indirection that transparently redirects file system calls of an application to a library or an external process. The Parrot system creates a sandbox around a user-space process and intercepts its system calls [26]. The Fuse kernel-level file system redirects file system calls to a special user-land process ("upcall") [27]. Interposition systems come with a performance penalty for the indirection layer. This penalty is smaller for Fuse than for a pure user-level interposition system, but Fuse requires co-operation from the kernel.

Some distributed file systems are implemented as an extension of the operating system kernel (e. g. NFS [1], AFS, Lustre). That can provide better performance compared to interposition systems but the deployment is difficult and implementation errors typically crash the operating system kernel.

Distributed file systems do not fully comply with the POSIX file system standard. Each distributed file system needs to be tested with real applications. Functionality that is often poorly supported in distributed file systems is file locking, atomic renaming of files and directories, multiple hardlinks, and deletion of open files. Sometimes, deviations from the POSIX standard are subtle. HDFS, for instance, writes file sizes asynchronously and thus it returns the real size of a file only some time after the file has been written.

## 3. Architecture Evolution

The simplest architecture for a distributed file system is a single server that exports a local directory tree to a number of clients (e. g. NFSv3). This architecture is obviously limited by the capabilities of the exporting server.

An approach to overcome some of these limitations is to delegate ownership and responsibility of certain file system subtrees to different servers, as done by AFS. In order to provide access to remote servers, AFS allows for lose coupling of multiple file system trees ("cells"). Across cells, this architecture is not network-transparent: moving a file from one cell to another requires a change of path. It also involves a copy through the node which triggers the move, e. g. move is not a namespace-only operation. Furthermore, the partitioning of a file system tree is static and changing it requires administrative intervention.

In *object-based* file systems, data management and meta-data management is separated (e. g. GFS). Files are spread over a number of servers that handle read and write operations. A meta-data server maintains the directory tree and takes care of data placement. As long as meta-data load is much smaller than data operations (i. e. files are large), this architecture allows for *incremental scaling*. As the load increases, data servers can be added one by one with minimal administrative overhead.

The architecture is refined by *parallel file systems* (e. g. Lustre) that cut every file in small blocks and distribute the blocks over many nodes. Thus read and write operations are executed in parallel on multiple servers for better maximum throughput.

A *distributed meta-data* architecture (as for instance in Ceph [9]) overcomes the bottleneck of the central meta-data server. Distributed meta-data handling is more complex than distributed data handling because all meta-data servers are closely coupled and need to agree on a single state of the file system tree. That involves either distributed consensus algorithms [28, 29] or distributed transaction protocols such as two-phase commit.

Object-based architectures involve two servers (meta-data server, data server) for read and write operations. Instead of asking a meta-data server, *decentralized* or *peer-to-peer* file systems (e. g. GlusterFS [13]) let clients *compute* the location of data and meta-data by means of a distributed hash table. Zero-hop routing in a distributed hash table is restricted to a local network, however, in which every node is aware of every other node. On a global scale, tree based routing as being done by XrootD is simpler to implement and it shows better lookup performance than globally distributed hash tables. Furthermore, high peer churn (servers that frequently join and leave the network) pose a hard challenge on distributed hash tables.

### *3.1. Decomposition*

There is a tendency of decomposition and modularization in distributed file systems. Examples are the offloading of authorization to Kerberos in AFS, the offloading of distributed consensus to Chubby [30] in GFS (resp. ZooKeeper [31] in HDFS), or the layered implementation of Ceph with the independent RADOS key-value store as building block beneath the file system. Another example is the separation of a distributed file system namespace and the data access. In the grid, for instance, the namespace is controlled by experiments' *file catalogs*, which, in combination with grid middleware, federates globally distributed cluster file systems.

For a globally distributed and administratively independent computing infrastructure used in high-energy physics, modularization is important because it allows for deployment of incremental improvements. A completely new file system takes many years to stabilize and roll-out throughout the grid.

## 4. Mechanisms and Techniques

A distributed file system should be fast and it should scale to many files, users, and nodes. At the same time, it should sustain hardware faults and recover gracefully from them and ensure the integrity of the file system over long storage periods and long-distance network links. This section highlights techniques that are used to achieve these goals and that are particularly relevant for distributed file systems in high-energy physics.

### *4.1. File System Integrity*

Global file systems often need to transfer data via untrusted connections and still ensure integrity and authenticity of the data. Cryptographic hashes of the content of files are often used to ensure data integrity. Cryptographic hashes provide a short, constant length, unique identifier for data of any size. Collisions are virtually impossible to occur neither by chance nor by clever crafting, which makes cryptographic hashes a means to protect against data tampering. Many globally distributed file systems use cryptographic hashes in the form of content-addressable storage [12, 32–34], where the name of a file is derived from its cryptographic content hash. This allows for verification of the data independently of the meta-data. It also results in immutable data, which eliminates the problem of detecting stale cache entries and keeping cache consistency. Furthermore, redundant data and duplicated files are automatically de-duplicated, which in some use cases (backups, scientific software binaries) reduces the actual storage space utilization by many factors [12, 35].

Cryptographic hashes are also used to protect the integrity of the file system tree when combined with a Merkle tree [36]. In a Merkle tree, nodes recursively hash their children's cryptographic hashes so that the root hash uniquely identifies the state of the entire file system. Copies of this root hash created at various points in time provide access to previous snapshots of file systems, which effectively allows for backups and for versioned file systems. The hashes in the tree can also be cryptographically signed in order to ensure data authenticity of a file system or a subtree (*who* created the content). An elegant way to solve the problem of key distribution inherent to digital signatures is the encoding of the public key as part of the path name [37].

To protect against *silent corruption*—the probabilistic decay of physical storage media over time—simple checksums such as CRC32 provide an easy means. Checksums can be faster verified than cryptographic hashes, fast enough to compute them on the fly on every read access [10].

### *4.2. Fault-Tolerance*

Fault-tolerance is an important property for a file system in high-energy physics because of the large scale of the storage systems and because mostly commodity hardware is used. Thus, hardware failures are not only the norm but they also tend to occur in a correlated manner. Power cuts are an example, or a failing controller with many connected drives.

*Replication* and *erasure codes* are the techniques used to avoid data loss and to continue operation in case of hardware failures. An engineering challenge is the placement of redundant data in such a way that the redundancy crosses multiple failure domains. The Ceph file system, for instance, can parse the physical layout of a data center together with policies to distribute redundant data on multiple racks or disk shelves. Another option is to keep a copy of all data at a remote data center. The second engineering challenge is to decide when to start a recovery, i. e. to distinguish between short-term glitches and permanent faults. Tight monitoring can help to distinguish between the two in some cases, but in general the only available technique are heartbeats between servers and properly tuned time-out values [38].

While replication is simple and fast, it also results in a large storage overhead. Most systems use a replication factor of three. Erasure codes, on the other hand, can be seen as more sophisticated, distributed RAID systems. Every file is chunked and additional redundancy blocks are computed. Typically the storage overhead of erasure codes is much smaller than a factor of two. Erasure codes, however, come at the cost of computational complexity. Modifications to a file as well as recovery from hardware faults is expensive because the redundancy blocks have to be recalculated and redistributed. Exploration of the trade-off between computational complexity and storage overhead in erasure codes are an active research area. Only a few of today's open-source distributed file systems implement erasure codes [9, 11, 14].

In order to determine the required level of redundancy and in order to balance the resources assigned to recovery and the resources assigned to normal usage, it is necessary to *predict* the failure rate of storage systems. This is a hard problem. In large scale storage systems, the mean time to failure of components can be measured, and simulations or a Markov chain model can be used to reason about the failure rate of the system as a whole. Although state of the art, the failure probabilities returned by these approaches are often off by large factors [38].

### *4.3. Efficiency*

Caching and file striping are standard techniques to improve the speed of distributed file systems. Caches can be located in memory, in flash memory, or on hard disks. They are transparently filled on access; opportunistic pre-placement of data sets is typically not part of distributed file systems but it is implemented in data management systems on top of distributed file systems. Cache sizes needs to be manually tuned according to the working set size of applications. Caches are most often managed per file system node. Co-operative caches between nodes in a local network have been discussed [33, 39, 40] but they are not implemented in today's production file systems. Some distributed file systems, however, support the notion of different *pools* with different performance characteristics (e. g. flash memory, hard drives, and tapes) and automatic migration of data sets between such pools [4, 8, 41].

Striping, reading and writing a file in small blocks on many servers in parallel, works only for large files. For small files, (memory) caches can improve the read performance but the write performance on hard drives can suffer from the `seek` operations necessary to write data and meta-data to disk. Log-structured file systems provide near-optimal write performance for small and large files because all changes, new data and meta-data are appended to the physical medium [42, 43]. While log-structured data organization is popular in distributed key-value stores, it is only occasionally used in distributed file systems, possibly because most of them assume large sequential writes. In high-energy physics, this assumption is often valid but not always. Counter examples are the modifications to a file system namespace or the merging phase of a parallelized physics analysis with many small distributed parts of the final result. Moreover, log-structured data organization is an efficient means to manage storage space throughout the memory hierarchy, for DRAM, flash memory, and hard disks [44].

*Dynamic workload adaptation* is a technique used in the Ceph file system to change the mapping of meta-data to meta-data servers based on the server load. The file system tree is dynamically re-partitioned and re-distributed and single hot directories are split across multiple servers. The implementation is challenging though; a recent study still revealed stability problems [20].

## 5. Current Developments and Future Challenges

In the upcoming years, the computing landscape will move towards the *exascale*. That means data sets that routinely sum up to exabytes and supercomputers that provide computing power in the exaflop range, which are expected by 2020. Table 1 shows the development of storage bandwidth and capacity in the last 20 years. The rapid increase in storage capacity allows for exabyte data sets already today. At the same time, however, storage bandwidth developed not nearly at the same pace. While the gap between capacity and bandwidth widened by one to two orders of magnitude in the last 20 years, the bandwidth of Ethernet networks scaled at a similar pace than the capacity of hard drives.

Raicu et al. predict the collapse of exaflop supercomputing applications due to the limited storage bandwidth and the architecture of today's distributed file systems [46, 47]. They suggest to break the segregation between storage networks and compute networks and to build distributed file systems with the following characteristics.

**Table 1.** Development of capacity and bandwidth in the last 20 years. Method and entries marked † from Patterson [45]. Other numbers refer to Seagate ST6000NM0034 hard drive (2014), DDR2-400 DRAM (2004) and DDR4-3200 DRAM (2014), and the 100 GbitE IEEE 802.3bj standard (2014).

| | **Hard Disk Drives** | | **DRAM** | | **Ethernet** |
|---|---|---|---|---|---|
| Year | Capacity | Bandwidth | Capacity | Bandwidth | Bandwidth |
| 1993 | | | 16 Mibit/chip† | 267 MiB/s† | |
| 1994 | 4.3 GB† | 9 MB/s† | | | |
| 1995 | | | | | 100 Mbit/s† |
| 2003 | 73.4 GB† | 86 MB/s† | | | 10 Gbit/s† |
| 2004 | | | 512 Mibit/chip | 3.2 GiB/s | |
| 2014 | 6 TB | 220 MB/s‡ | 8 Gibit/chip | 25.6 GiB/s | 100 Gbit/s |
| **Increase** | ×1395 | ×24 | ×512 | ×98 | ×1000 |

‡http://www.storagereview.com/seagate_enterprise_capacity_6tb_35_sas_hdd_review_v4

(i) Every compute node becomes a part of the storage network, counting on the fact that the bisection bandwidth of the network connecting compute nodes is sufficiently high to move data.

(ii) Flash memory is explicitly used as a layer in the storage hierarchy and data locality is exposed in the file system, so that in many cases data can stay in local fast storage.

Similarly, Seagate Kinetic technology suggests to replace storage management servers by network-attached hard drives that can be directly addressed from applications [48]. And in the BigData community, a move has taken place to coalesce storage and compute nodes using GFS and MapReduce, resp. its open source counterpart Hadoop and recent real-time and in-memory enhancements such as Spark [49]. In high-energy physics, workflows are often executed in multiple stages (e. g. simulation and reconstruction, filter and analysis and merge) and often the result of stage $n$ is explicitly written to a distributed file system only to be used as input for stage $n+1$. Investing in a computing framework and a distributed file system that exploits data locality and allows for keeping data at the node where they are produced would help to overcome the bandwidth limitations in today's and future storage hardware.

In the last few years, development efforts in distributed storage systems were mostly aimed at key-value stores, BLOB stores, and NoSQL databases rather than file systems. Due to their simpler interface, these technologies allow for very good scalability in local networks and fast provisioning of (mostly small) objects. As such, they become a building block of distributed file systems [9, 11] but they are not replacing them. They do not primarily address features necessary for data sharing, such as quotas, fine-grained access control, a standard interface to applications, or federation across multiple data centers.

Some recent distributed file system developments were aimed at storing and sharing personal data [50, 51]. Such systems explore extensions of the pure file system interface in order to support meta-data searches and instant dissemination of file system subtrees and document collections. These systems are decentralized and more flexible approaches to the problems addressed by Dropbox and AFS.

## 6. Conclusion

Distributed file systems provide a relatively well-defined and general purpose interface for applications to use large-scale persistent storage. The hierarchical namespace provides a natural and flexible way to store large amounts of unstructured data. The implementation of distributed file systems, however, is always tailored to a particular class of applications. Even though a large number of distributed file systems is available today, some practical and important use cases are still uncovered. For instance, a distributed file system for a cluster of commodity servers that provides at least half of the aggregated throughput of the hard drives, utilize at least 90 % of the available capacity, without central components and fault-tolerant to a small number of hardware failures would fit the storage needs of many small and mid-sized physics experiments [52].

For large-scale distributed file systems targeted at anticipated physics data sets in the exabyte range, the relatively small bandwidth of storage hardware poses a hard challenge. Approaches towards decentralized file systems that coalesce storage nodes and compute nodes and that make explicit use of locality and flash memory offer a possible solution to overcome such limits. Quantitative studies on the file system usage of physics applications should guide the selection and development of building blocks for future distributed file systems in physics experiments.

## References


[1] Sandberg R, Goldberg D, Kleiman S, Walsh D and Lyon B 1985 *Proc. of the Summer USENIX conference* pp 119–130

[2] Morris J H, Satyanarayanan M, Conner M H, Howard J H, Rosenthal D S H and Smith F D 1986 *Communications of the ACM* **29** 184–201

[3] Carns P H, III W B L, Ross R B and Thakur R 2000 *Proc. 4th Annual Linux Showcase and Conference (ALS'00)* pp 317–328

[4] Schmuck F and Haskin R 2002 *Proc. 1st USENIX conf. on File Storage and Technologies (FAST'02)* pp 231–244

[5] Schwan P 2003 *Proc. of the 2003 Linux Symposium* pp 380–386

[6] Nagle D, Serenyi D and Matthews A 2004 *Proc. of the 2004 ACM/IEEE conf. on SuperComputing (SC'04)*

[7] Dorigo A, Elmer P, Furano F and Hanushevsky A 2005 *WSEAS Transactions on Computers* **4** 348–353

[8] Fuhrmann P and Gülzow V 2006 *dCache, Storage System for the Future* (Springer) pp 1106–1113 (*Lecture Notes in Computer Science* no 4128)

[9] Weil S A 2007 *Ceph: reliable, scalable, and high-performance distributed storage* Ph.D. thesis University of California Santa Cruz

[10] Shvachko K, Kuang H, Radia S and Chansler R 2010 *Proc. of the 26th IEEE Sympoisum on Mass Storage and Technologies (MSST'10)* pp 1–10

[11] Peters A J and Janyst L 2011 *Journal of Physics: Conference Series* **331**

[12] Blomer J, Aguado-Sanchez C, Buncic P and Harutyunyan A 2011 *Journal of Physics: Conference Series* **331**

[13] Davies A and Orsaria A 2013 *Linux Journal*

[14] Ovsiannikov M, Rus S, Reeves D, Sutter P, Rao S and Kelly J 2013 *Proc. of the VLDB Endowment* vol 6 pp 1092 – 1101

[15] Satyanarayanan M 1990 *Annual Review of Computer Science* **4** 73–104

[16] Guan P, Kuhl M, Li Z and Liu X 2000 A survey of distributed file systems University of California, San Diego

[17] Agarwal P and Li H C 2003 A survey of secure, fault-tolerant distributed file systems `http://www.cs.utexas.edu/users/browne/cs395f2003/projects/LiAgarwalReport.pdf`

[18] Thanh T D, Mohan S, Choi E, Kim S and Kim P 2008 *Proc. int. conf. on Networked Computing and Advanced Information Management (NCM'08)* pp 144 – 149

[19] Depardon B, Séguin C and Mahec G L 2013 Analysis of six distributed file systems Tech. Rep. hal-00789086 Université de Picardie Jules Verne

[20] Donvito G, Marzulli G and Diacono D 2014 *Journal of Physics: Conference Series* **513**

[21] Dean J and Ghemawa S 2008 *Communications of the ACM* **51** 107–114

[22] Doraimani S and Iamnitchi A 2008 *Proc. 17th int. symposium on High performance distributed computing* pp 153–164

[23] Leung A W, Pasupathy S, Goodson G and Miller E L 2008 *Proc. of the USENIX Annual Technical Conference* pp 213–226

[24] Dayal S 2008 Characterizing hec storage systems at rest Tech. Rep. CMU-PDL-08-109 Carnegie Mellon University

[25] Ghemawat S, Gobioff H and Leung S T 2003 *ACM SIGOPS Operating Systems Review* **37** 29–43
[26] Thain D and Livny M 2005 *Scalable Computing: Practice and Experience* **6** 9
[27] Henk C and Szeredi M Filesystem in Userspace (FUSE) `http://fuse.sourceforge.net` URL `http://fuse.sourceforge.net/`
[28] Lamport L 1998 *ACM Transactions on Computer Systems* **16** 133–169
[29] Ongaro D and Ousterhout J 2014 *Proc. of the 2014 USENIX Annual Technical Conference (USENIX ATC 14)* pp 305–319
[30] Burrows M 2006 *Proc. 7th symposium on Operating systems design and implementation* pp 335–350
[31] Hunt P, Konar M, Junqueira F P and Reed B 2010 *Proc. of the 2010 USENIX annual technical conference*
[32] Kubiatowicz J, Bindel D, Chen Y, Czerwinski S, Eaton P, Geels D, Gummadi R, Rhea S, Weatherspoon H, Weimer W, Wells C and Zhao B 2000 *ACM SIGPLAN Notices* **35** 190–201
[33] Dabek F, Kaashoek M F, Karger D, Morris R and Stoica I 2001 *ACM SIGOPS Operating Systems Review* **35** 202–215
[34] Kutzner K 2008 *The Decentralized File System Igor-FS as an Application for Overlay-Networks* Ph.D. thesis University of Karlsruhe
[35] Quinlan S and Dorward S 2002 *Proc. of the 1st USENIX Conf. on File and Storage Technologies (FAST'02)* pp 89–102
[36] Merkle R C 1988 *A Digital Signature Based on a Conventional Encryption Function* (*Lecture Notes in Computer Science* vol 293) (Springer) pp 369–378
[37] Mazières D, Kaminsky M, Kaashoek M F and Witchel E 1999 *ACM SIGOPS Operating Systems Review* **34** 124–139
[38] Ford D, Labell F, Popovici F I, Stockly M, Truong V A, Barroso L, Grimes C and Quinlan S 2010 *Proc. 9th symposium on Operating Systems Design and Implementation (OSDI'10)*
[39] Nelson M N, Welch B B and Ousterhout J K 1988 *ACM Transactions on Computer Systems* **6** 134–154 ISSN 07342071
[40] Annapureddy S, Freedman M J and Mazières D 2005 *Proc. of the 2nd Symposium on Networked Systems Design and Implementation (NSDI'05)* pp 129–142
[41] Lo Presti G, Barring O, Earl A, Rioja R M G, Ponce S, Taurelli G, Waldron D and Santos M C D 2007 *Proc. of the 24th IEEE Conference on Mass Storage Systems and Technologies* pp 275–280
[42] Rosenblum M and Osterhout J K 1991 *ACM SIGOPS Operating Systems Review* **25**
[43] Hartman J H and Osterhout J K 1995 *ACM Transactions on Computer Systems* **13** 274–310
[44] Rumble S M, Kejriwal A and Ousterhout J 2014 *Proc. 12th USENIX Conference on File and Storage Technologies (FAST'14)*
[45] Patterson D A 2004 *Communications of the ACM* **47**
[46] Raicu I, Foster I T and Beckman P 2011 *Proc. 3rd int. workshop on Large-scale system and application performance (LSAP'11)* pp 11–18
[47] Zhao D, Zhang Z, Zhou X, Li T, Wang K, Kimpe D, Carns P, Ross R and Raicu I 2014 *Proc. of the 3rd IEEE int. conf. on Big Data (BigData'14)*
[48] Seagate Kinetic open storage documentation wiki `https://developers.seagate.com/display/KV/Kinetic+Open+Storage+Documentation+Wiki`
[49] Engle C, Lupher A, Xin R, Zaharia M, Li H, Shenker S and Stoica I 2012 *Proc. of the 9th USENIX conference on Networked Systems Design and Implementation (NSDI'12)*
[50] Fitzpatrick B *et al.* Camlistore `http://camlistore.org`
[51] Mashtizadeh A J, Bittau A, Huang Y F and Mazières D 2013 *Proc. 24th ACM Symposium on Operating Systems Principles (SOSP'13)*
[52] Allen B 2009 A distributed file system wish-list Talk at Max Planck Institut für Informatik, Saarbrücken

# Next Generation Workload Management System For Big Data on Heterogeneous Distributed Computing

**A Klimentov[1], P Buncic[2], K De[3], S Jha[4], T Maeno[1], R Mount[5], P.Nilsson[1], D Oleynik[3], S Panitkin[1], A Petrosyan[3], R J Porter[6], K F Read[7], A Vaniachine[8], J C Wells[7] and T Wenaus[1]**

[1] Brookhaven National Laboratory, Upton, NY, USA
[2] CERN, Geneva, Switzerland
[3] University Texas at Arlington, Arlington, TX, USA
[4] Rutgers University, Piscataway, NJ, USA
[5] SLAC National Accelerator Laboratory, Menlo Park, CA, USA
[6] Lawrence Berkeley National Laboratory, CA, USA
[7] Oak Ridge National Laboratory, Oak Ridge, TN, USA
[8] Argonne National Laboratory Lemont, IL, USA

E-mail: aak@bnl.gov

**Abstract.** The Large Hadron Collider (LHC), operating at the international CERN Laboratory in Geneva, Switzerland, is leading Big Data driven scientific explorations. Experiments at the LHC explore the fundamental nature of matter and the basic forces that shape our universe, and were recently credited for the discovery of a Higgs boson. ATLAS and ALICE are the largest collaborations ever assembled in the sciences and are at the forefront of research at the LHC. To address an unprecedented multi-petabyte data processing challenge, both experiments rely on a heterogeneous distributed computational infrastructure. The ATLAS experiment uses PanDA (Production and Data Analysis) Workload Management System (WMS) for managing the workflow for all data processing on hundreds of data centers. Through PanDA, ATLAS physicists see a single computing facility that enables rapid scientific breakthroughs for the experiment, even though the data centers are physically scattered all over the world. The scale is demonstrated by the following numbers: PanDA manages $O(10^2)$ sites, $O(10^5)$ cores, $O(10^8)$ jobs per year, $O(10^3)$ users, and ATLAS data volume is $O(10^{17})$ bytes. In 2013 we started an ambitious program to expand PanDA to all available computing resources, including opportunistic use of commercial and academic clouds and Leadership Computing Facilities (LCF). The project titled 'Next Generation Workload Management and Analysis System for Big Data' (BigPanDA) is funded by DOE ASCR and HEP. Extending PanDA to clouds and LCF presents new challenges in managing heterogeneity and supporting workflow. The BigPanDA project is underway to setup and tailor PanDA at the Oak Ridge Leadership Computing Facility (OLCF) and at the National Research Center "Kurchatov Institute" together with ALICE distributed computing and ORNL computing professionals. Our approach to integration of HPC platforms at the OLCF and elsewhere is to reuse, as much as possible, existing components of the PanDA system. We will present our current accomplishments with running the PanDA WMS at OLCF and other supercomputers and demonstrate our ability to use PanDA as a portal independent of the computing facilities infrastructure for High Energy and Nuclear Physics as well as other data-intensive science applications.

## 1. Introduction

The largest scientific instrument in the world, the Large Hadron Collider, is operating at the CERN Laboratory in Geneva, Switzerland [1]. The ATLAS [2], ALICE [3] and other LHC experiments explore the fundamental nature of matter and the basic forces that shape our universe. To address an unprecedented multi-petabyte data processing challenge, the LHC experiments rely on the computational grids infrastructure deployed in the framework of the Worldwide LHC Computing Grid (WLCG) [4]. WLCG is by far the largest academic distributed computing environment in the world and the levels of use are ground-breaking. Thanks to the outstanding LHC performance in 2010-2013, ATLAS manages over 160 petabytes of data, submitting up to two million jobs per day on more than 100 sites (figure 1). Following the massive data processing on the Grid, more than 8000 scientists analyze LHC data in search of new discoveries. ATLAS leads the WLCG usage in the number of jobs, processed data volume, and in core-hours.

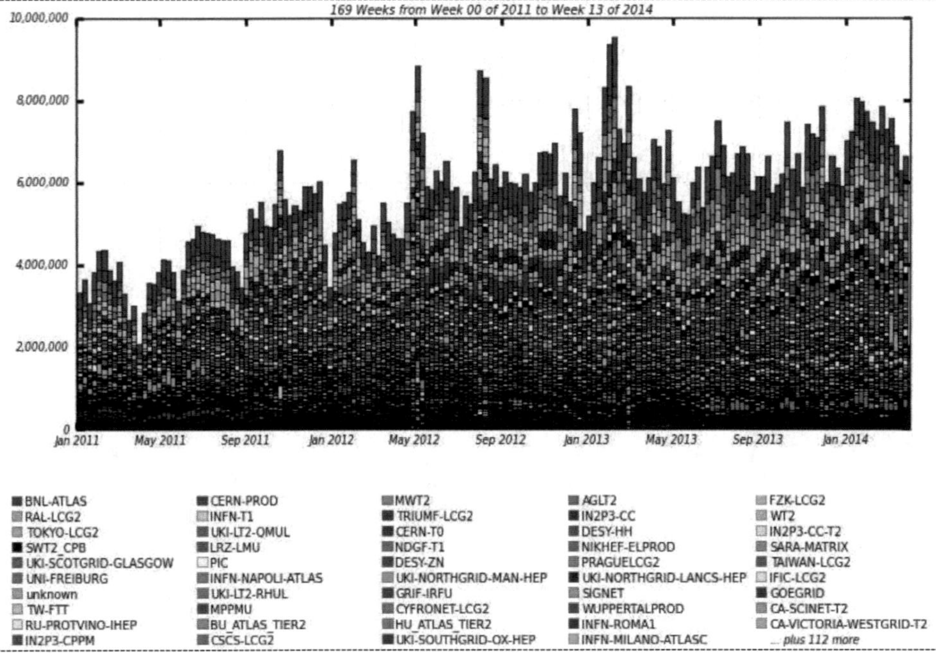


**Figure 1.** ATLAS: completed jobs (weekly average).

The challenges posed by such mega-science experiments are numerous and not limited to the unprecedented size of the data. Exascale data is often highly distributed and accessed by large international collaborations. A sophisticated Workload Management System (WMS) is needed to manage the distribution and processing of such data. One of the most successful WMS developed in the U.S. is PanDA (acronym for Production and Distributed Analysis System) [5], used by thousands of physicists in the ATLAS experiment at the Large Hadron Collider.

PanDA delivers transparency of data and processing in a distributed computing environment to ATLAS physicists. It provides execution environments for a wide range of experimental applications, automates centralized data production and processing, enables analysis activity of physics groups, supports custom workflow of individual physicists, provides a unified view of distributed worldwide resources, presents status and history of workflow through an integrated monitoring system, archives and curates all workflow, manages distribution of data as needed for processing or physicist access. The rich menu of features provided, coupled with support for heterogeneous computing environments, makes PanDA ideally suited for data intensive sciences.

PanDA has a highly scalable and flexible architecture. Scalability has been demonstrated in ATLAS through the rapid increase in usage over the past three years. PanDA was designed to have the flexibility to adapt to emerging computing technologies in processing, storage, networking as well as the underlying software stack (middleware and file management). This flexibility has also been successfully demonstrated through the past five years of evolving technologies adapted by computing centres in ATLAS, which span many continents and yet are seamlessly integrated into PanDA. This proven scalability and flexibility makes PanDA ideally suited for adoption by future megascience projects.

PanDA was conceived in 2005. At the time, a variety of workload management systems were deployed in ATLAS, separately for different applications, using distinct systems for the chaotic workflows from physicists and the organized workflows from central production. PanDA emerged as the best system and was adopted as the default and single system for ATLAS before the LHC started operating in 2009. Today, PanDA has grown to support all distributed workflows in ATLAS, and enjoys a huge user base with worldwide support.

PanDA was adapted and successfully used for molecular dynamics simulations of protein folding using the CHARMM [6] molecular modeling software, leading to a publication describing the implementation as general, flexible, easily modifiable for use with other molecular dynamics programs and other grids and automated in terms of job submission, monitoring, and resubmission. In 2012, the Alpha Magnetic Spectrometer (AMS) experiment [7] began using PanDA for Monte Carlo simulation and data processing. AMS set up the PanDA infrastructure at the largest academic computing center of Asia-Pacific region (ASGC). The Large Synoptic Survey Telescope collaboration [8] has conducted the first Monte Carlo simulation in 2013-2014 using a PanDA instance installed on the Amazon EC2 cloud.

Interest in PanDA by other big data sciences provided the primary motivation to make a proposal titled "Next Generation Workload Management and Analysis System for Big Data." The idea was to generalize PanDA as meta-application, providing location transparency of processing and data management, for High Energy Physics (HEP) community and other data-intensive sciences, and the wider exascale community. The DOE Advanced Scientific Computing Research office (ASCR) awarded a grant to expand PanDA beyond HEP, to add network awareness to PanDA, and to expand PanDA to Leadership Computing Facilities and cloud computing (BigPanDA) [9].

## 2. Next Generation Workload Management and Analysis System for Big Data

The project was started at the end of 2013 and is jointly funded by the DOE ASCR and HEP offices. The work proposed within the project will enable the use of PanDA by new scientific collaborations and communities as a means of leveraging extreme scale computing resources with a low barrier of entry. Particular attention is given to enabling the computing infrastructures supported by the DOE that were not initially supported by PanDA such as the Leadership Computing Facilities (LCF).

Support for Leadership Computing Facilities will expand the potential user community for PanDA and will also, even in the near term, benefit ATLAS and ALICE by running jobs on Leadership Computing Facilities with direct access to the data hosted by US National Laboratories, dynamically acquiring supplementary CPU resources when needed. Extending PanDA beyond the Grid and integrating distributed computing with cloud computing (or grids of clouds) will further expand the potential user community and the resources available to them, for example making it possible for non-LHC and non-HEP experiments such as AMS and LSST to use PanDA.

In the longer term, exascale facilities and cloud platforms could become a real alternative to very large data centers owned and managed by the scientific community. This work would create in PanDA a WMS system enabling users to bridge the near and longer terms smoothly, with minimal effort, and preserve the same global system view throughout, by leveraging PanDA support for both present and emerging computing infrastructures. ATLAS and ALICE would certainly lead the way in utilizing exascale platforms this way, and we would seek through this work to make it an easy path for others to follow as well.

There are three dimensions for the overall workload system evolution. First, we should make PanDA available beyond the LHC and HEP. Second, we should extend PanDA beyond the Grid. Third, we should integrate network services as a resource in workload management. The following four work packages have been identified:

- *Factorizing the core:* Factorizing the core components of PanDA to enable adoption by a wide range of exascale scientific communities.
- *Extending the scope:* Evolving PanDA to support extreme scale computing clouds and Leadership Computing Facilities.
- *Leveraging intelligent networks:* Integrating network services and real-time data access to the PanDA workflow.
- *Usability and monitoring:* Real time monitoring and visualization package for PanDA

We discuss below in more detail how to evolve PanDA to support computing clouds and HPC.

## 3. BigPanDA to support extreme scale computing clouds and Leadership Computing Facilities

### *3.1. Scale of Needs*

The ATLAS experiment uses a geographically distributed grid of approximately 130,000 cores continuously, (over 1000 million core-hours per year) to simulate and analyze its data. After the early success in discovering a new particle consistent with the long awaited Higgs boson [10], ATLAS is preparing for precision measurements and further discoveries that will be made possible by much higher LHC collision rates after early 2015. The need for simulation and analysis would overwhelm the expected capacity of WLCG computing facilities unless the range and precision of physics studies were to be curtailed.

From the usage information above, it becomes clear that HPC contributions of the order of 10 million or more core hours per year become important and valuable. ATLAS computing can also be a close-to-ideal "crack-filling" application. The ATLAS production management system is being upgraded to make it aware of dynamically changing resources, and thus able to exploit groups of processors that become available for relatively short times. Storage capacity on the petabyte scale will be an objective of the work package, adding extra resources such as computing clouds and Leadership Computing Facilities to those supported by PanDA. Extending PanDA beyond the Grid will further expand the potential user community and the resources available to them.

### *3.2. Challenges*

The major issues and challenges, which must be addressed to attain excellent scientific productivity on advanced computer architectures, are described below.

*3.2.1. Execution efficiency.* In common with most of HEP, ATLAS code typically executes around 0.8 instructions per clock cycle, even though the (Intel, AMD) hardware it executes on can perform more than 10 instructions/cycle and achieves between 2 and 3 instructions per cycle for much scientific code. The reasons for this "bad" performance are the many conditional branches and lack of repetitious activities, both due to the complex nature of the physics and devices being simulated or analyzed.

*3.2.2. Parallelism.* In its current form, ATLAS simulation and analysis is highly amenable to trivial high-level parallelism: events (i.e. an initial collision and the subsequent response of the detector to all the collision products) can be independently simulated or analyzed on thousands of cores. Current ATLAS simulation and analysis makes heavy use of this parallelism. ATLAS codes also have the potential for massive low-level parallelism that may be a good match for vector and GPU hardware. To expose this parallelism, the execution architecture must be transformed to present many similar (but unrelated) operations for simultaneous execution. The level of parallelism that can present "close-to-SIMD work packages" for efficient execution on many-core CPUs or GPUs will benefit from the high bandwidth communications network, typical of an HPC facility, to assemble work and distribute the results.

*3.2.3. Production environment.* Major HEP simulation and analysis, such as that of ATLAS, requires sophisticated systems to control the flow of data and tasks to computing resources, to monitor workflow and to assure automated recovery from most "environmental" errors. To make productive use of HPC, interfacing to such production environments is essential.

*3.2.4. Living code.* There are 6 million lines of active code in the ATLAS SVN repository. This is living code, being continuously improved as our understanding of the detector and of physics develops. The code must remain intelligible and maintainable. While small amounts of code can certainly be optimized for particular target architectures, the majority of the code must remain generic.

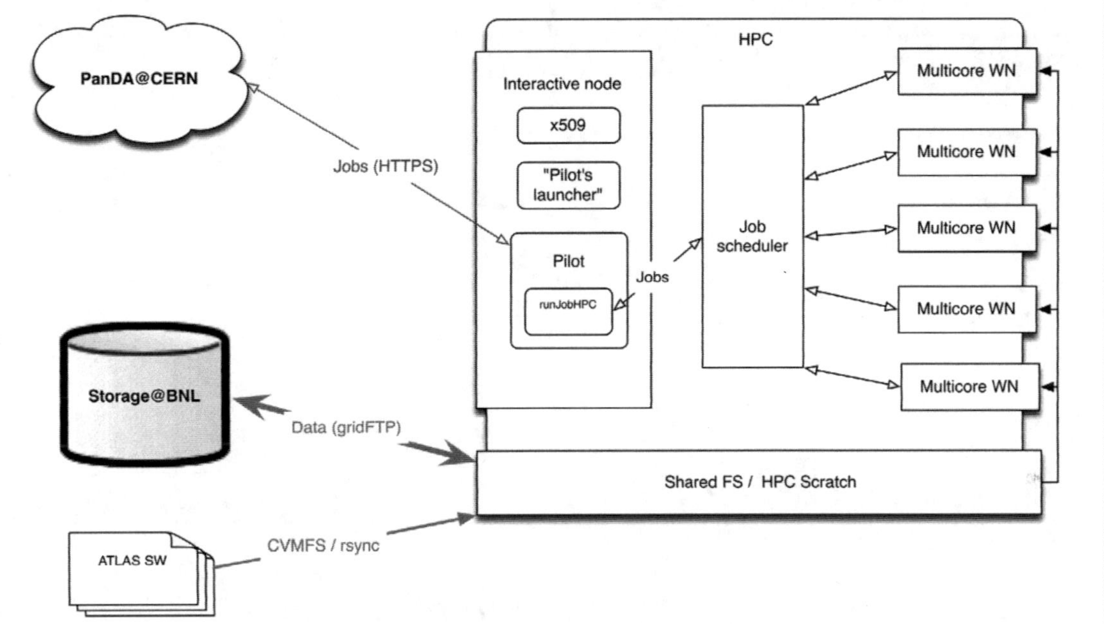


**Figure 2.** PanDA architecture on the Titan OLCF.

### *3.3. Extending PanDA to Leadership Computing Facilities*

Unlike the Grid, HPC centers are primarily built to execute parallel payloads and there are usually significant limitations: worker node set up is fixed, outbound network connections are not possible from the worker nodes, limited RAM per core, and a specialized Operating System. Standard Workload Management Systems (like PanDA) are not well-adapted to submit payloads to HPC resources. Together with colleagues at the OLCF, we adapted PanDA for the Titan LCF, trying to reuse existing PanDA components and workflow as much as possible. The PanDA connection layer runs on front-end nodes in user space. There is a predefined host to communicate with CERN from the OLCF, with connections initiated from the front-end nodes. We use the SAGA framework [10] (a Simple API for Grid Applications) as the local batch interface.

The PanDA architecture on Titan is shown in figure 2. The PanDA Pilot, responsible for payload submission, runs on a Titan interactive node, communicating with local batch scheduler to manage jobs on Titan. Outputs are transferred to the Brookhaven National Laboratory Tier-1 Grid center. The initial demonstration was done using Monte Carlo event generators as a payload. Event generators are mostly computational having stand-alone code with a small amount of stage-in/out data. For LHC experiments 10-15% of all CPU resources are utilized to run event generators.

It is important to define the mode of payload execution. One of possible scenarios would be to run HEP applications in backfill mode. Since HPC centers are geared toward large-scale jobs, about 10% of capacity of a typical HPC machine is unused due to a mismatch between job sizes and available resources. A rough estimation of a potential resource is 300 M CPU hours per year, primarily available via short jobs. PanDA can be a vehicle to harvest Titan resources opportunistically. Taking the above considerations into account, we have adapted the PanDA Pilot algorithm to use backfill information and to submit jobs in backfill mode as follows:

- Pilot periodically queries MOAB scheduler about available resources.
- Scheduler returns information about available (unscheduled) nodes and interval of availability.
- Pilot chooses the largest available block of nodes and submits jobs taken into account Titan's scheduling policy limitations.

The interest to run LHC experiment payloads using the scenario described above was expressed by several HPC centers: Czech IT4i, Swiss CSCS (Piz Daint), Taipei ASGC, and others.

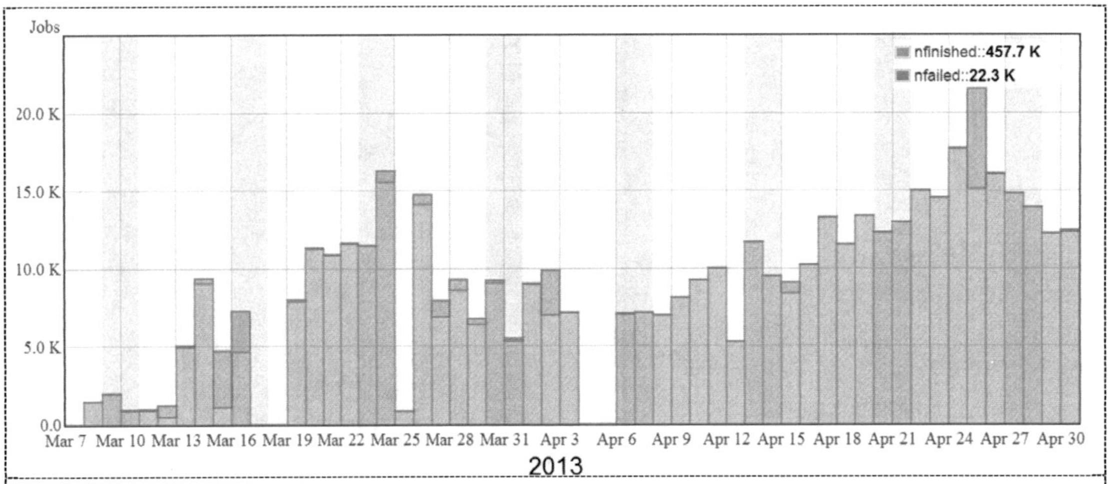


**Figure 3.** ATLAS jobs running on Google Compute Engine.

### *3.4. Extending PanDA to Clouds*

The demand on computing resources to accommodate physics calculation requirements increases inexorably. Commercial and research clouds are one of the venues we explored over the past two years. ATLAS was invited to participate in the Google Compute Cloud preview. After an initial period, Google allocated 5 M core hours (using 4 k cores) for two months. The idea was to run a realistic physics application to gain experience and to understand a long-term stability while running cloud facilities similar in size to a typical Tier-2 Grid center. We ran ATLAS Monte Carlo production for 10 weeks (figure 3). We reached a throughput of 15 k jobs per day, with 214 M events have been generated and processed. A similar test has been conducted on Amazon EC2.

In broad terms, commercial clouds are more expensive than Grid Computing Centers, but clouds resources may be used to advantage by scientific collaborations during peak periods. Figure 4 shows the variation in the number of Grid jobs waiting because of lack of resources, with peaks corresponding to the activities before major physics conferences.

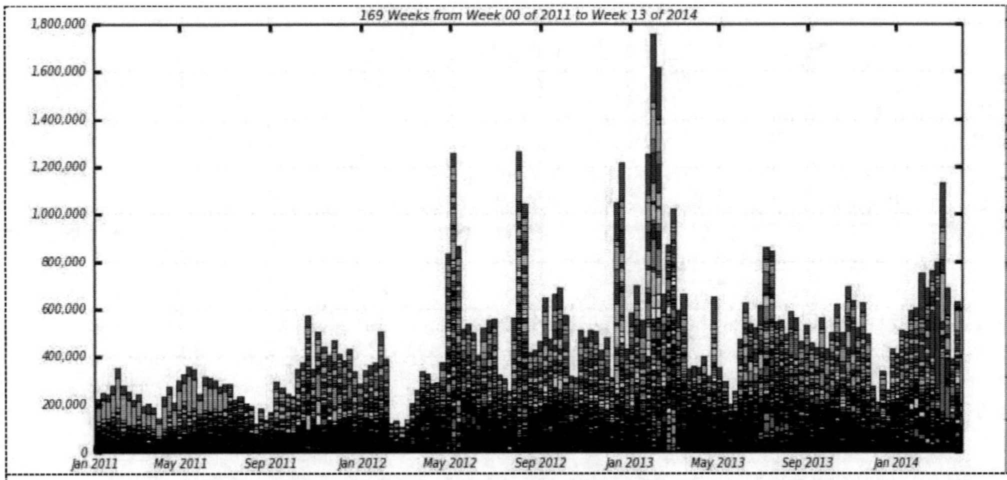


**Figure 4.** ATLAS: Waiting Jobs (weekly average).

## 4. Conclusions

It is now accepted that High Energy Physics and High Energy Nuclear Physics applications must undergo a transformation to be able to make good use of current and future computer architectures. Only by using hardware efficiently will the experiments in those (and other) communities be able to achieve the massive computing throughput that will be needed in the next decade. HPC facilities are a particular target because of the massive computing power they can bring to bear, and because they provide the lowest possible communications overhead for massively decomposed tasks.

Fully optimized use of HPC facilities is a long-term goal, certainly requiring work of five or more years. Early access to HPC facilities will provide an important stimulus to this work. Access to facilities coupled with collaborative help in the transformation of HEP code would be a major scientific contribution to the physics discoveries of the next ten years.

The PanDA system played a key role during LHC Run 1 (2009-2013) data reprocessing, simulation and analysis. The challenge of how to process and analyze the data and produce timely physics results was substantial, but at the end resulted in a great success.

The interest in PanDA by other big data sciences provided the primary motivation to generalize the PanDA system. DOE ASCR gave us a great opportunity to start the BigPanDA project.

- The work on extending PanDA to Leadership Computing Facilities has started. PanDA has been successfully ported to OLCF Titan.
- Large-scale PanDA deployments on commercial clouds are already producing valuable results.

The work reported here will enable the use of PanDA by new scientific collaborations and communities as a means of leveraging extreme scale computing resources with a low barrier of entry. The technology base provided by BigPanDA will enhance the usage of a variety of high-performance computing resources available to basic research.

**Acknowledgments**

This work was funded in part by the U. S. Department of Energy, Office of Science, High Energy Physics under Contracts No. DE-AC02-98CH10886 and DE-AC02-06CH11357. This research used resources of the Oak Ridge Leadership Computing Facility at the Oak Ridge National Laboratory, which is supported by the Office of Science of the U.S. Department of Energy under Contract No. DE-AC05-00OR22725.

## References

[1] LHC – The Large Hadron Collider, URL http://lhc.web.cern.ch

[2] The ATLAS Collaboration 2008 The ATLAS experiment at the CERN Large Hadron Collider *J. Inst.* **3** S08003

[3] The ALICE Collaboration 1995 *Technical proposal for A Large Ion Collider Experiment at the CERN LHC* LHCC, December 1995, ISBN 92-9083-077-8

[4] WLCG Worldwide LHC Computing Grid, URL http://lcg.web.cern.ch

[5] Maeno T *et al.* 2008 PanDA: Distributed Production and Distributed Analysis System for ATLAS *J. Phys.: Conf. Ser.* **119** (6) 062036

[6] Damjanović A, Miller B T, Wenaus T J, Maksimović P, García-Moreno E B, Brooks B R 2008 Open science grid study of the coupling between conformation and water content in the interior of a protein, *J Chem Inf Model* Oct; **48** (10) 2021-9 doi:10.1021/ci800263c

[7] Battiston R 2008 The antimatter spectrometer (AMS-02): A particle physics detector in space *Nucl. Instrum. Methods Phys. Res., Sect. A* **588**, 227-234

[8] Ivezic Z *et al.* 2008 LSST: from Science Drivers to Reference Design and Anticipated Data Products, *Preprint* astro-ph arXiv:0805.2366

[9] Next Generation Workload Management and Analysis System for Big Data, URL http://science.energy.gov/~/media/ascr/pdf/research/nextgen/KlimentovAlexei_BNL_abstract.pdf

[10] ATLAS Collaboration 2012 Observation of a new particle in the search for the Standard Model Higgs boson with the ATLAS detector at the LHC *Phys. Lett. B* **716** 1-29

[11] The SAGA Project, URL http://www.saga-project.org

*Notice:*

*This manuscript has been authored by employees of Brookhaven Science Associates, LLC under Contract No. DE-AC02-98CH10886 with the U.S. Department of Energy. The publisher by accepting the manuscript for publication acknowledges that the United States Government retains a non-exclusive, paid-up, irrevocable, world-wide license to publish or reproduce the published form of this manuscript, or allow others to do so, for United States Government purposes.*

# b-jet identification at High Level Trigger in CMS

**Eric Chabert**

Institue Pluridisplinaire Hubert Curien, Département Recherches Subatomiques, IPHC/DRS, 23 rue du Loess -BP 28- F67037 Strasbourg

E-mail: Eric.Chaber@cern.ch

**Abstract.** The CMS experiment at the CERN LHC has been designed with a 2-level trigger system. The Level 1 Trigger (L1) is implemented on custom-designed electronics. The High Level Trigger (HLT) is a streamlined version of the CMS offline reconstruction software running on a computer farm. Being able to identify b-quark jets (b-tagging) at trigger level will play a crucial role during the Run II data taking to ensure the top quark, beyond the Standard Model and Higgs boson physics program of the experiment. It will help to significantly reduce the trigger output rate which will increase due to the higher instantaneous luminosity and higher cross sections at 13 TeV. B-tagging algorithms based on the identification of tracks displaced from the primary proton-proton collision and/or on the reconstruction of secondary vertices have been successfully used during Run I. We will present their design and performance with an emphasis on the dedicated aspects of track and primary vertex reconstruction, as well as the improvements foreseen to meet the challenges of the Run II data taking.

## 1. The CMS experiment

The Compact Muon Solenoid (CMS) detector recorded proton-proton collisions occuring at the LHC during the first period of data-taking (2010-2013), called Run I. A description of the CMS detector can be found in [1]. CMS uses a right-handed coordinate system with the x-axis pointing towards the centre of the LHC, with the y-axis directed upwards along the vertical, and with the z-axis directed along the counterclock-wise beam direction. The azimutal angle $\phi$ is measured with respect to the x-axis in the $x-y$ plan and the polar angle $\theta$ is defined with respect to the z-axis. The pseudo-rapidity is defined as $\eta = -ln[\tan(\theta/2)]$.

## 2. b-jet identification: principles and algorithms

Many channels studied in the CMS physics program involve b-jets. Top quark[1], beyond the Standard Model[2] and Higgs boson[3] physics strongly rely on the capability to identify jets arising from the hadronization of b quarks.A good performance in terms of b-jet identification efficiency and rejection of non-b jets are essential to improve the purity and/or the significance of the selected events. Moreover, the use of b-tagging at trigger level is helpful if not mandory, to select efficiently the relevant events at trigger level while keeping the background sufficiently low to accommodate the online processing and bandwidth constraints.

[1] Top quarks decay almost exclusively into a W boson and a b-quark

[2] We can mention as example the searches for bottom or top quark partners, W' bosons, ... decaying to b quarks.

[3] H$\rightarrow b\bar{b}$ is the dominant decay mode for the Higgs boson

B-tagging algorithms exploit the properties of the B-hadrons such as their large decay lifetimes ($c\tau \approx 450\ \mu m$) or the presence of leptons in the final state to separate them from light hadrons (made of u,d,s-quarks). Several algorithms have been developed and optimized from the simplest ones which are cut-based, to more sophisticated ones using multivariate techniques. The Combined Secondary Vertex (CSV) algorithm [2] is currently one of the most frequently used algorithms within the CMS collaboration. It is based on a Likelihood Ratio technique to combine discriminating variables constructed from displaced track and secondary vertex information as well as jet kinematics. The offline b-jet identification efficiency is about 70% for a misidentification probability for light jets of about 1% for jets with a transverse momentum $p_T$ between 80 and 120 GeV/c. This algorithm has also been deployed in an online version at the High Level Trigger during Run I. Both online and offline versions use the same version of the algorithm. The difference comes from the tracks and primary vertices collections that are used as input. The measured performance shows a b-jet efficiency of about 60% for a misidentification probability of about 2%.

## 3. High Level Trigger

The CMS trigger system is based on two consecutive levels. The Level 1 Trigger (L1) is an hardware level implemented on FPGA and custom-designed ASICs. The goal is to select events at maximum rate of about 100 kHz from the 20 MHz incoming rate delivered by the LHC. L1 treats in parallel information coming from the electromagnetic and hadronic calorimeters as well as from the muons chambers. A global decision is taken based on the presence of energy deposits compatible with the presence of photons, electrons, muons, jets or hadronically decaying tau leptons.

The High Level Trigger (HLT) is a streamlined version of the CMS offline reconstruction software running on a cluster of commercial rack-mounted computers which consists of about 13000 processors. The goal is to reduce the rate to about 800 Hz for offline data storage. During Run I, only half of the selected events has been used for prompt reconstruction (within 48 hours) and reconstructed later on for physics purpose. The second half of the rate has been parked and the reconstruction has been delayed by several months. The parked data have been reconstructed during the first period of the LHC shut-down and are used in specific analyses where the gain of lowering the online conditions is substantial.

## 4. b-tagging sequence at HLT during Run I

HLT paths involving b-tagging have been used by many analyses during Run I. Many Higgs boson analyses[4] involving b-jets made the choice of using a trigger path based on the CSV algorithm like the analyses focusing on the Higgs boson decaying into two b-quarks [3][4] or on a Higgs boson produced in association with a Z boson [4]. Similarly the search of an invisible Higgs boson produced with a Z boson decaying into two b-quarks [5] used also b-tagging at HLT. Those examples show the interest for b-tagging at HLT to reduce the bandwidth when selecting hadronic events where the thresholds on the jet kinematics are not sufficient to reduce the background. The sequence used by the above mentioned analyses rely on the b-tagging sequence presented in the following:

- **Jets from the L1 trigger (L1 jets)**
  Before entering in the b-tagging sequence, events have to successfully pass a L1 trigger path requiring two central calorimeter jets with $p_T$ >40 GeV/c and $|\eta|$ <1.6. Jets using anti-$K_T$ algorithm with R = 0.4 are used.

[4] Higgs boson channels have been studied in the context of the SM or SUSY models like the MSSM.

- **Pixel clustering**
  A regional clustering is performed in the regions of interest determined by the direction of the L1 jets that have fired the L1 trigger.
- **Fast Primary Vertex reconstruction**
  The Fast Primary Vertex (FastPV) reconstruction is a procedure that allows to reconstruct a primary vertex based on pixel cluster information only. No tracks are involved in the determination of the vertex. The goal is to determine the position of the vertex along the beam axis (z-axis). Clusters in the pixel layers must have the same $\phi$ coordinate as the jet. Cluster sizes along the y-axis have to be compatible with the pseudo-rapidity $\eta$ of the jet, *i.e.*, the size increases for jet at high $\eta$. Cluster sizes along the x-axis have to be small to select only high $p_T$ tracks. The selected clusters are projected along the z-axis using the jet $\eta$ direction. The peak in the z projection direction is used to determine the position of the Fast Primary Vertex.
- **Pixel track reconstruction**
  The reconstruction of the pixel tracks is performed using the pixel clusters that are within a distance of 1.5 cm of the FastPV position obtained in the previous step.
- **Pixel Primary Vertex**
  The reconstruction of the primary vertex is performed using the pixel tracks reconstructed in the previous step.
- **Full track reconstruction**
  A regional track reconstruction using all the pixel and strip clusters is performed. The Combinatorial Track Finder (CTF) algorithm [6], based on Kalman filter method, is used. The regional track reconstruction is only performed when the distance between the pixel tracks and the pixel primary vertex is less than 1.5 cm.
- **Application of the b-tagging algorithm: Combined Secondary Vertex**
  A CSV discriminant is computed for each jet based on the information of the pixel primary vertex, and the regional CTF tracks. Then a cut on the discriminator is used to make a decision to select or reject the event.

## 5. Improvements of the b-tagging sequence at HLT for Run II

The b-tagging sequence has been revisited and optimized for the Run II. The main goal is to reduce the execution time of the sequence. This was achieved thanks to a more extensive use of regional reconstruction. The second goal is to improve the performance of the online version of the CSV b-tagging algorithm. This was also achieved thanks to the smart iterative tracking reconstruction and to a better resolution on the position of the primary vertex. The new sequence consists of the following steps:

- **Pixel clustering**
  This step is similar to the 2012 version.
- **Fast Primary Vertex**
  In the newest version, an improvement has been obtained by extending the pseudo-rapidity range for the considered jets ($|\eta|$ <2.4), and by using weights for the pixel clusters. The weight of a pixel cluster depends on its size in the x-direction, the cluster charge, $\rho_{cluster}$, the azimuthal angle between the jet and the cluster, $\Delta\Phi_{jet,cluster}$, the difference between the cluster size in the z-direction and the one expected from the knowledge of $\eta_{jet}$. The impact of the use of weights on the efficiency was estimated and is shown in the Figure 1. A gain of around 10% is observed on the efficiency to find the FastPV in a window of 2 cm around the true vertex. For jets with $p_T$ >80 GeV, the efficiency exceeds 90%. The use of the end-cap clusters also helps to significantly improve the efficiency for jets with

$|\eta|$ >1, leading to an improvement of about 20%. Moreover, the fraction of events where no FastPV is found decreased. Finally, the resolution on the z-coordinate of the primary vertex also improves as shown in Figure 2.

- **Pixel track reconstruction**
  The pixel track reconstruction is now regional only, based on the regions of interest around the jet axis. The constraint applied using the FastPV position remains the same. As one could expect, this speeds up the pixel track reconstruction.
- **Pixel Primary Vertex**
  A great feature of the use of regional tracking is the improved efficiency to find the real primary vertex. The constraint on the FastPV is loose due the relative poor resolution.Using regional track reconstruction suppresses the impact of tracks from pile-up collisions when reconstructing the primary vertex. The resolution on the position of the pixel primary vertex along the z-axis is about 100 $\mu m$.
- **Full track reconstruction**
  An iterative tracking procedure is performed based on pixel and strip clusters. The procedure is composed of three steps. First, high-$p_T$ tracks are reconstructed using pixel track seeds and applying a constraint on the pixel primary vertex. Then lower $p_T$ tracks are reconstructed using pixel triplets as seeds. The third step consists in recovering high $p_T$ tracks seeded by pixel pairs. The use of regional iterative tracking helps to reduce the execution time as it reduces the combinatorics.
- **Primary vertex reconstruction**
  In the current version, an additional step is performed in order to reconstruct a primary vertex using the reconstructed iterative tracks. The resolution obtained on the position of the pixel primary vertex along the z-axis is about 20-30 $\mu m$. The vertex finder algorithm is based on a method using deterministic annealing and referred to as "3D Adaptative Vertex Finder".
- **Application of the b-tagging algorithm: Combined Secondary Vertex**
  A CSV discriminant is computed for each jet based on the information of the primary vertex and the regional iterative tracks. Then a cut on the discriminator is used to make a decision to select or reject the event.

## 6. Comparison of the performance of b-tagging at HLT for Run 1 with the new sequence

The performance of the new sequence has been compared to the previous version. The studies have been performed with top-quark pair-produced events at a centre-of-mass energy of 13 TeV and with an average of 25 pile-up collisions. The Figure 3 presents the performance of the CSV algorithm in terms of b-jet identification efficiency and the misidentification probability for light jets. For an efficiency of 60%, the misidentification probability decreases from 2% to 1.3% with the new b-tagging sequence. The Figure 4 shows the processing time for the HLT path HLT_DiCentralPFJet30_PFMet80_BTagCSV07 using the old and the new b-tagging sequence. The mean processing time is reduced by about 15% and the tail at high values is strongly reduced.

## 7. Conclusion

B-jet identification algorithms have been successfully used at HLT during Run 1 for many analyses, in particular those involving $b\bar{b}$ resonances such as the Higgs boson decaying to $b\bar{b}$. Improvements with respect to the processing time and the performance have been developed during the last year leading to a promising use during Run II. The improvements are mainly due to a multi-step approach based on an intensive usage of regional reconstruction techniques.

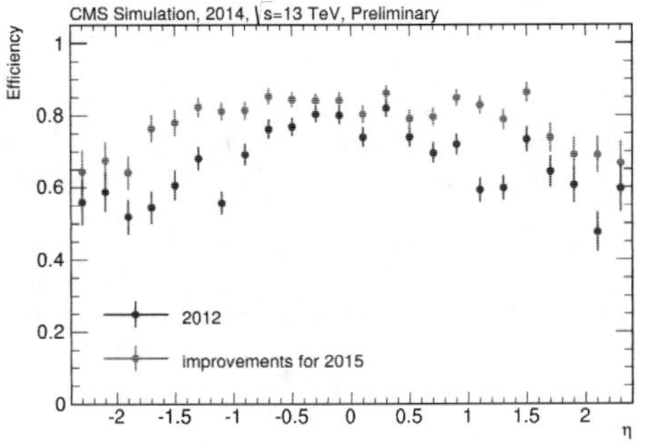


**Figure 1.** The efficiency to find a simulated primary vertex within 2 cm of the primary vertex reconstructed with the FastPV algorithm as a function of the pseudo-rapidity. The blue and red curves show respectively the performance before and after the improvements foreseen for 2015. This result is obtained using simulated $Z(\nu\nu)H(b\bar{b})$ events at $\sqrt{s} = 13$ TeV with on average 60 pile-up collisions.

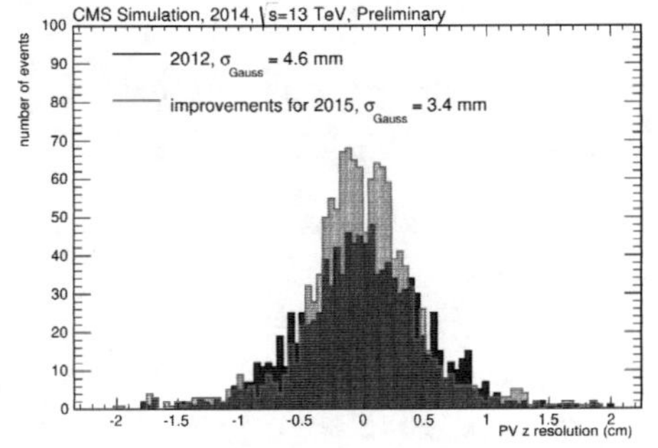


**Figure 2.** Resolution along the z axis of the primary vertex reconstructed with the FastPV algorithm, with respect to the simulated primary vertex (cm). The blue and red histograms show respectively the resolution distribution before and after the improvements foreseen for 2015. This result is obtained using simulated $Z(\nu\nu)H(b\bar{b})$ events at $\sqrt{s} = 13$ TeV with on average 60 pile-up collisions.

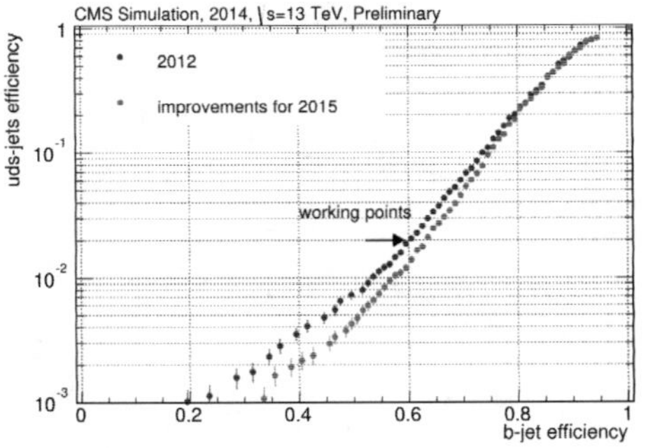


**Figure 3.** Efficiency of the Combined Secondary Vertex (CSV) b-tagging algorithm at HLT for u,d,s-jets vs b-jets. The blue and red curves show respectively the performance before and after the improvements foreseen for 2015. Jets from simulated $t\bar{t}$ events at $\sqrt{s}$=13TeV with on average 20 pile-up collisions and a bunch spacing of 25 ns are used.

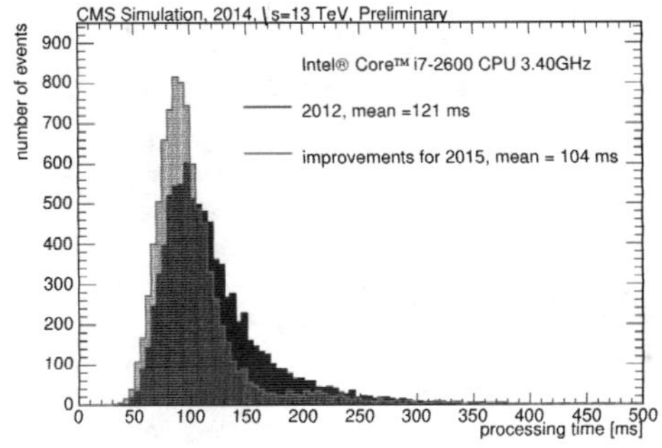


**Figure 4.** Timing of the HLT trigger path HLT_DiCentralPFJet30_PFMet80_BTagCSV07 designed for the 2012 $Z(\nu\nu)H(b\bar{b})$ analysis. The blue and red histograms show respectively the timing distribution before and after the improvements foreseen for 2015. Simulated $t\bar{t}$ events at $\sqrt{s} = 13$ TeV with on average 20 pile-up collisions and bunch spacing 25 ns are used. *Intel Core$^{TM}$* i7-2600 CPU 3.40 GHz processors are used.

## References

[1] CMS Collaboration 2008 *Journal of Instrumentation* **3** S08004

[2] CMS Collaboration 2013 Performance of b-tagging at $\sqrt{s}$=8 tev in multijet, $t\bar{t}$ and boosted topology events Tech. Rep. CMS-PAS-BTV-13-001
[3] Collaboration C 2010 *Physics Letters B* **722** 207 – 232 ISSN 0370-2693
[4] CMS Collaboration 2013 Higgs to $b\bar{b}$ in the vector boson fusion channel Tech. Rep. CMS-PAS-HIG-13-011
[5] CMS Collaboration 2013 Search for invisible higgs decays in z$h$ with z$\rightarrow$ $b\bar{b}$ Tech. Rep. CMS-PAS-HIG-13-028
[6] CMS Collaboration 2010 *The European Physical Journal C* **70** 1165–1192 ISSN 1434-6044

# GENFIT – a Generic Track-Fitting Toolkit

**Johannes Rauch[a] and Tobias Schlüter[b]**

[a] Technische Universität München, Physik-Department E18
James-Franck-Straße, 85748 Garching, GERMANY

[b] Ludwig-Maximilians-Universität München, Excellence Cluster Universe
Boltzmannstr. 2, 85748 Garching, GERMANY

E-mail: `j.rauch@tum.de`, `tobias.schlueter@physik.uni-muenchen.de`

**Abstract.** GENFIT is an experiment-independent track-fitting toolkit that combines fitting algorithms, track representations, and measurement geometries into a modular framework. We report on a significantly improved version of GENFIT, based on experience gained in the Belle II, $\overline{\text{P}}$ANDA, and FOPI experiments. Improvements concern the implementation of additional track-fitting algorithms, enhanced implementations of Kalman fitters, enhanced visualization capabilities, and additional implementations of measurement types suited for various kinds of tracking detectors. The data model has been revised, allowing for efficient track merging, smoothing, residual calculation, alignment, and storage.

## 1. Introduction

GENFIT provides an extensible modular open-source framework that performs track fitting and other related tasks and thus eliminates the redundancy of writing track-fitting programs for every experiment. Smaller experiments, which do not have the manpower to develop their own track fitters, and new experiments, which need working tools to research and develop, are especially encouraged to use GENFIT. GENFIT can also be a valuable teaching aid; the 3D display excellently illustrates track fitting.

All particle physics experiments need to identify and classify processes based on detector signals. Combining these signals to recover particle trajectories is the task called *tracking*: Suitable collections of measurements must be combined into track candidates. Tracks must be fitted and points of common origin or exit, vertices, must be found and fitted. Track finding and track fitting are not independent from each other. Steps of different levels of track finding and refining alternate with track-fitting steps. Fitted tracks from different tracking subdetectors can be matched and combined into larger tracks; and single measurements can be appended to existing tracks, which might come from another subdetector. Provided suitable collections of measurements, all of these tasks can be performed with GENFIT.

GENFIT was successfully used during a test of the Belle II high-level trigger architecture and the combined Belle II vertex-detector readout architecture [1]. There it served for online track reconstruction in the data-reduction stage of the high-level trigger, as well as for offline analysis. It was also used as the track fitter supplying input to the Millipede II alignment software [2].

## 2. The three pillars of track fitting

Track fitting in GENFIT is based on three pillars: Measurements, track representations, and fitting algorithms. Measurements serve as objects containing measured coordinates from a

Journal of Physics: Conference Series **608** (2015) 012042 doi:10.1088/1742-6596/608/1/012042

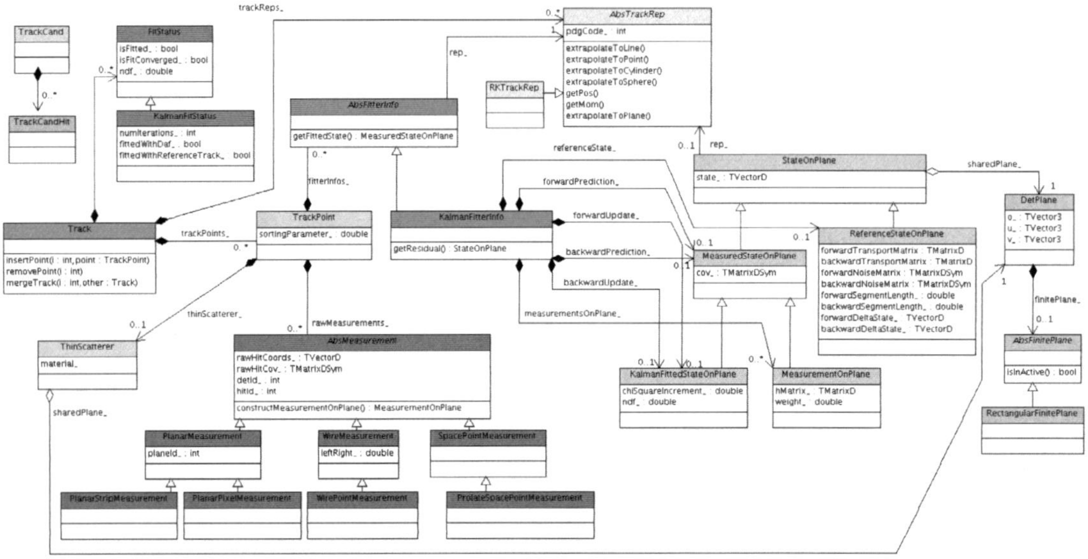


Figure 1: GENFIT Unified Modeling Language class diagram.

detector. They provide functions to construct a (virtual) detector plane and to provide measurement coordinates and covariance in that plane. The abstract base class `AbsMeasurement` defines the interface. GENFIT comes with predefined measurement classes for various detector types, including planar detectors, drift chambers, and time projection chambers.

For planar detectors, the detector plane is given by the detector geometry, whereas for wire and spacepoint measurements, so-called virtual detector planes are constructed. Further information can be used here, allowing one to compensate for detector deformations like plane bending, wire sag, and misalignment. Drift-time corrections and cluster fits (track dependent clustering) are possible.

Track representations combine track-parameterization and track-extrapolation code. GENFIT implements a track representation based on a Runge-Kutta extrapolator (`RKTrackRep`, see below).

Fitting algorithms use the measurements and track representations to calculate fit results, which are stored in corresponding objects in the `Track` object.

## 3. The track data structure

All per-track data is kept in the `Track` object (Fig. 1). It holds a sequence of `TrackPoint` objects, which can contain measurements, `FitterInfo` objects, which hold all fitter-specific information, and thin scatterers (currently only used in the GBL fitter).

A `Track` contains one or more track representations, representing the particle hypotheses that should be fitted. One of them must be selected as the cardinal track representation. This can be done by the user or by GENFIT, which selects the track representation that best fits the measurements (i.e. has the lowest $\chi^2$).

The `Track` also contains a `FitStatus` object, which stores general information (number of iterations, convergence, etc.) and fit properties ($\chi^2$, NDF, $p$ value, track length, etc.).

The track candidate (`TrackCand`) serves as a helper class, basically storing indices of raw detector hits in `TrackCandHit` objects, which can also be overloaded by the user to store additional information. `WireTrackCandHit` objects can store the left-right ambiguity. A use case is pattern recognition: algorithms can work on raw objects that are lighter weight than a

`Measurement` or implement other features needed by the pattern recognition but not necessarily by GENFIT. GENFIT's `MeasurementFactory` can build a `Track` from a `TrackCand`. This `Track` can then be processed by the various fitting algorithms.

After fitting, `Track` objects contain a lot of data: track representations, `TrackPoint` objects with measurements, `FitterInfo` objects, etc. Usually, not all of this information needs to be stored on disk. E.g. for physics analysis only parameters near the vertex and information on the track quality are necessary, whereas alignment needs residual information at all measurements. The user can decide which data to keep; e.g. only the fitted state of the first `TrackPoint`, and only for the cardinal track representation.

## 4. Runge-Kutta track representation

The Runge-Kutta track representation (`RKTrackRep`) is based on a Runge-Kutta extrapolator from GEANT 3 [3]. An abstract interface class interacts with the detector geometry. Implementations using ROOT's `TGeoManager` [4] and GEANT4's `G4Navigator` [5] are available. During fitting, material properties are used to calculate the following effects: energy loss and energy-loss straggling for charged particles according to the Bethe Bloch formula (code ported from GEANT 3); multiple scattering (according to Ref. [6] or using the Highland formula), where the full noise matrix is calculated; and soft Bremsstrahlung energy loss and energy-loss straggling for $e^-$ and $e^+$ (code ported from GEANT 3).

The step sizes used for the Runge-Kutta extrapolation should be as large as possible to save unnecessary computation, while still being small enough to keep errors reasonably small. An adaptive step-size calculation is done in the `RKTrackRep`, taking field inhomogeneities and curvature into account. To calculate material effects correctly, extrapolation stops at material boundaries and steps can only be so large that a maximum relative momentum loss in the material is not exceeded.

`RKTrackRep` provides different methods to find the point of closest approach (POCA) of the track to nonplanar measurements. These are used to construct *virtual detector planes*:

- *Extrapolate to point* finds the POCA of the track to a given spacepoint. The virtual detector plane contains the spacepoint and the POCA and is perpendicular to the track.
  The POCA can also be calculated with a weight matrix $G$ (i.e. the inverted three-dimensional covariance matrix of a spacepoint measurement), thus minimizing the weighted distance $\equiv \sqrt{\delta^T G \delta}$, where $\delta$ refers to the three-vector pointing from the measurement to the POCA.
- *Extrapolate to line* finds the POCA of the track to a given line or wire. The virtual detector plane contains the line and the POCA. This routine is used for fitting wire measurements.

The intersection of the virtual detector plane and the measurement covariance gives the covariance in the plane.

## 5. Fitter implementation details

Four different track-fitting algorithms are currently implemented: two smoothing Kalman filters, one which linearizes the transport around the state predictions and one which linearizes around a reference track; a deterministic annealing filter (DAF); and a general broken lines (GBL) fitter.

GENFIT provides the possibility to store several measurements of the same type in one `TrackPoint`, mainly for using the DAF to assign weights to them. Wire measurements also produce two `MeasurementOnPlane` objects, representing the passage of the particle on either side of the wire. These tracks can also be fitted with the Kalman fitters. GENFIT provides several ways to handle multiple measurements:

- The weighted average of the individual measurements is calculated. This is used by the DAF.

- The measurement closest to the state prediction or reference state is selected.
- If the `TrackPoint` has a wire measurement, the side that is closest to the prediction or reference is select; otherwise the average is selected.

For wire measurements, it turned out that selecting the side closest to the state prediction is the best option for the Kalman fitters.

As convergence criteria, a minimum and maximum number of iterations can be set, which are 2 and 4 by default. After the minimum number of iterations, GENFIT checks if the $p$ value has changed less than a certain amount since the previous iteration; the default is $10^{-3}$. However, tracks with a $p$ value close to zero are often considered as converged with this criterion, even though the $\chi^2$, albeit big, is still changing significantly, indicating that the fit is still improving. This occurs often for tracks that are given bad start parameters. To cure this issue, a nonconvergence criterion has been introduced: If the relative change in $\chi^2$ from one iteration to the next is larger than 20 %, the fit will continue. Again, this number is user-adjustable.

### *5.1. Kalman fitter*
The Kalman fitter is adapted from the formulas given in Ref. [7], with linearization around the state predictions. Optionally, a square-root formalism, adapted from Ref. [8] to the usecase relevant to track fitting, provides greater numerical stability at the expense of execution time.

### *5.2. Kalman fitter with reference track*
State predictions may stray very far from actual trajectories. Consequently, linearizing around them is not optimal; material and magnetic field lookups are also not done at the proper places. It is therefore common to linearize around reference states [7], which are calculated by extrapolating the start parameters to all `TrackPoint` objects. At later iterations, the smoothed states from the previous iterations are used as linearization points. However, if the change would be very small, reference states are not updated, saving computing time. It is also possible to let the fitter sort the measurements along the reference track, which can improve fitting accuracy.

In addition to the convergence criteria detailed above, the fit is regarded as converged if none of the reference states has been updated since the previous iteration.

### *5.3. Deterministic annealing filter*
The deterministic annealing filter [9] is a powerful tool for the rejection of outlying measurements. It is a Kalman filter that uses a weighting procedure between iterations based on the measurement residuals to determine the proper weights. The user can select which of GENFIT's two Kalman-fitter implementations should be used and specify the annealing scheme.

The DAF is also perfectly suited to resolving the left-right ambiguities of wire measurements. However, a problem can occur: The weights of the `MeasurementOnPlane` objects must be initialized. The basic solution is to initialize both left and right measurements with a weight of 0.5. Effectively the wire positions are taken as measurements in the first iteration, and their covariance is twice the mean of the individual covariances. So all the wire positions have the same covariance, no matter how far from the actual trajectory they are. This systematically false estimate of the covariances biases the fit. GENFIT implements a novel technique to initialize the weights that improves the fitting efficiency: Measurements with larger drift radii are assigned smaller weights, leading to larger covariances since the wire position is expected to be farther away from the trajectory. In contrast, measurements with smaller drift radii, which are closer to the trajectory, get larger weights.

After the annealing scheme, convergence is checked: If the absolute change of all weights is less than $10^{-3}$ (user configurable), the fit is regarded as converged. Otherwise another iteration

Table 1: Execution time of the GenFitter module in the Belle II software framework.

| Fitter | w/o matFX | w/ matFX | ∅ iterations |
|---|---|---|---|
| Kalman | 3.4 ms | 10 ms | 3 |
| Reference Kalman | 4.0 ms | 8.2 ms | 2.13 |
| DAF | 9.4 ms | 17 ms | 6 |

with the last temperature is done, until the fit converges or a maximum number of iterations is reached.

### *5.4. Generalized broken lines fitter*

The generalized broken lines method of track fitting [10] was implemented especially for the purpose of alignment with Millipede II. It is mathematically equivalent to the Kalman fitter (with thin scatterers instead of continuous materials), but fits tracks in their entirety in one step, providing a natural interface to the Millipede II software. For alignment purposes, GENFIT also provides a set of interfaces for alignment parameters and derivatives which can be implemented by the detector classes.

## 6. Vertex reconstruction with GFRave

GFRave, an interface to the vertex-fitting framework RAVE[1], has been implemented. RAVE [11] is a detector-independent toolkit for vertex reconstruction originally developed for the CMS experiment [12]. GFRave takes full advantage of the GENFIT material model and the sophisticated algorithms of RAVE, allowing for precise and fast vertex reconstruction.

## 7. Performance

GENFIT's performance will depend on many parameters like hardware and compiler, number and types of measurements, momentum, geometry, magnetic field, number of iterations, convergence settings, and so on. A small study was conducted in the Belle II analysis framework (basf2), to give a ballpark figure of the performance you can expect from GENFIT. The execution time of the `GenFitter` module in the Belle II software framework [13] was measured on a 3.4 GHz office PC in single-threaded operation. All code was compiled with `-O3` optimization settings. Besides fitting all tracks in an event, this module builds data structures for Belle II's data storage format, incurring an overhead of slightly less than 1 ms. Single track events were generated with a particle gun, with $\theta = 100°$ and a momentum of 0.9 GeV in a constant magnetic field. The resulting `Track` objects found by MC-based (perfect) track finding contain $72 \pm 2$ `TrackPoint` objects. The Kalman fitters have been configured to do 3 to 10 iterations with default convergence criteria, while the DAF uses its default annealing scheme with 5 temperatures.

From the results shown in Tab. 1, one can see that the Kalman fitter with reference track needs fewer iterations to converge. Material lookup requires approximately 2.2 ms per iteration for the Kalman fitter. The Kalman fitter with reference track is also faster here, since the reference states are not recalculated if they are close to the smoothed states of the previous iteration. This is also the reason why the reference Kalman can often finish after 2 iterations.

[1] Reconstruction (of vertices) in Abstract, Versatile Environments

## 8. Visualisation

GENFIT features a sophisticated 3D event display, which allows one to visualize fitted tracks. Detector geometry, measurements, detector planes, reference tracks, forward and backward fits (predictions and updates), smoothed tracks, and covariances of measurements and tracks can be drawn. Tracks can be refitted with different algorithms and settings, and fit results can be viewed instantly.

Fig. 2 shows the fit of a set of measurements with the Kalman fitter with reference track. For demonstration purposes, the different measurement types supported by GENFIT are used (starting from the origin): planar pixel measurement, spacepoint measurement, prolate spacepoint measurement, two perpendicular planar strip measurements, double-sided planar strip measurement, wire measurement, and wire measurement with second coordinate measurement.

## Acknowledgments

This research was supported by the DFG cluster of excellence "Origin and Structure of the Universe". T.S. was supported under BMBF Contract 05H12WM8.

## References

[1] T. Bilka et al. Demonstrator of the Belle II Online Tracking and Pixel Data Reduction on the High Level Trigger System. 2014. subm. IEEE Trans. Nucl. Sc.

[2] V. Blobel. Software alignment for tracking detectors. *Nucl.Instrum.Meth.*, A566:5–13, 2006. doi: 10.1016/j.nima.2006.05.157.

[3] CERN. GEANT - Detector description and simulation tool. *CERN Program Library Long Write-up*, W5013, 1993.

[4] Rene Brun and Fons Rademakers. Root - an object oriented data analysis framework. In *AIHENP'96 Workshop, Lausane*, volume 389, pages 81–86, 1996.

[5] GEANT 4 Collaboration. GEANT 4 – a simulation toolkit. *Nucl. Inst. Methods Phys. Res.*, A506:250–303, 2003.

[6] A. Fontana, P. Genova, L. Lavezzi, and A. Rotondi. Track Following in Dense Media and Inhomogeneous Magnetic Fields. *PANDA Report PV/01-07*, 2007.

[7] R. Frühwirth, M. Regler, R.K. Bock, H. Grote, and D. Notz. *Data Analysis Techniques for High-Energy Physics.* Cambridge University Press, 2000.

[8] Brian D.O. Anderson and John B. Moore. *Optimal Filtering.* Dover Publications, 1979.

[9] R. Frühwirth and A. Strandlie. Track Fitting with Ambiguities and Noise: A Study of Elastic Tracking and Nonlinear Filters. *Computer Physics Communications*, 120(2-3):197 – 214, 1999. ISSN 0010-4655. doi: 10.1016/S0010-4655(99)00231-3.

[10] C. Kleinwort. General Broken Lines as advanced track fitting method. *Nucl.Instrum.Meth.*, A673:107–110, 2012. doi: 10.1016/j.nima.2012.01.024.

[11] Wolfgang Waltenberger. RAVE: A Detector-independent Toolkit to Reconstruct Vertices. *IEEE Trans. Nucl. Sci.*, 58:434–444, 2011. doi: 10.1109/TNS.2011.2119492.

[12] CMS Collaboration. The CMS experiment at the CERN LHC. *JINST*, 0803:S08004, 2008. doi: 10.1088/1748-0221/3/08/S08004.

[13] Andreas Moll. The software framework of the Belle II experiment. *J.Phys.Conf.Ser.*, 331: 032024, 2011. doi: 10.1088/1742-6596/331/3/032024.

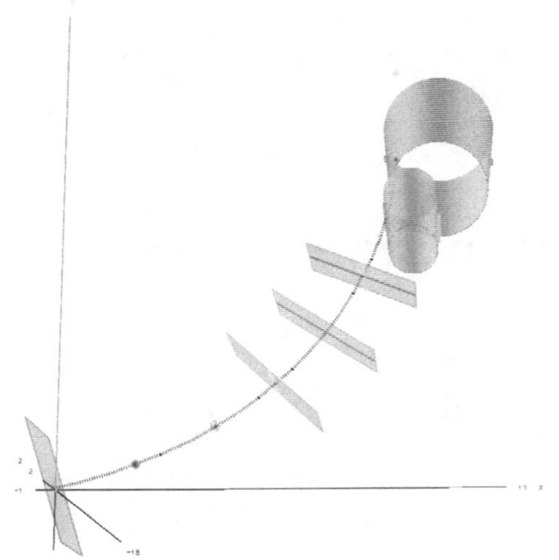

(a) Measurements with covariance (yellow), planar detectors and drift isochrones (cyan), and reference track (blue).

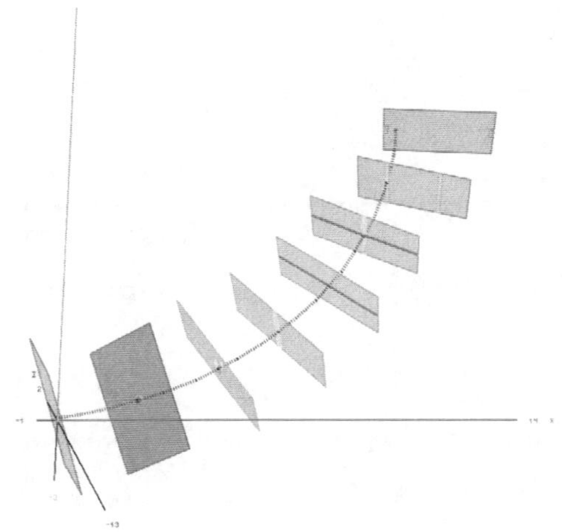

(b) Detector planes (gray). For the spacepoint and wire measurements virtual detector planes have been constructed.

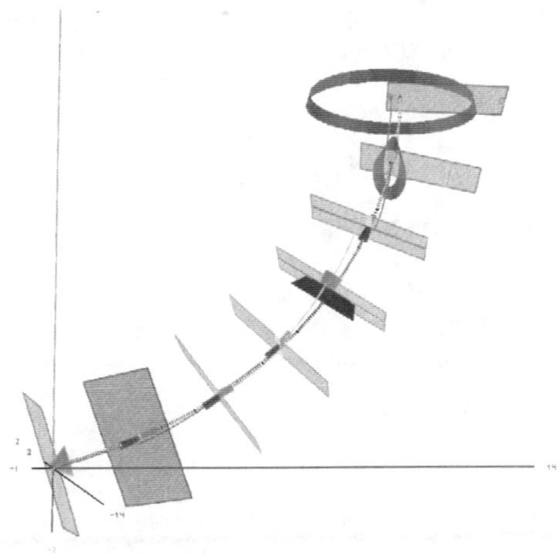

(c) Forward (cyan) and backward (magenta) fit with covariances of the state updates.

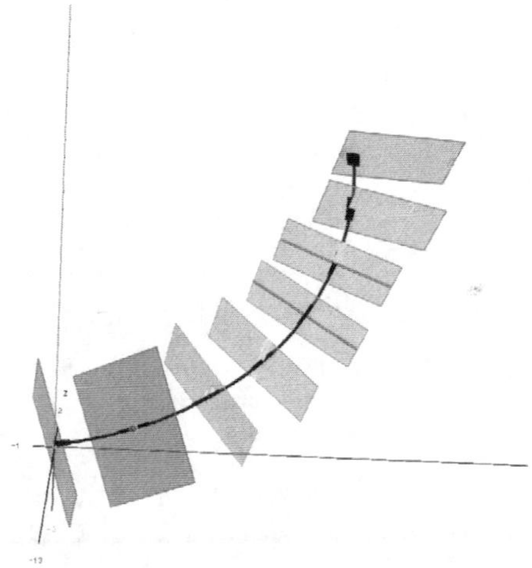

(d) Smoothed track with covariance (blue).

Figure 2: GENFIT event display screenshots.

# The TileCal Online Energy Estimation for the Next LHC Operation Period

**B Sotto-Maior Peralva on behalf of the ATLAS Collaboration**
Federal University of Juiz de Fora, Brazil
E-mail: bernardo@cern.ch

**Abstract.**
The ATLAS Tile Calorimeter (TileCal) is the detector used in the reconstruction of hadrons, jets and missing transverse energy from the proton-proton collisions at the Large Hadron Collider (LHC). It covers the central part of the ATLAS detector ($|\eta| < 1.6$). The energy deposited by the particles is read out by approximately 5,000 cells, with double readout channels. The signal provided by the readout electronics for each channel is digitized at 40 MHz and its amplitude is estimated by an optimal filtering algorithm, which expects a single signal with a well-defined shape. However, the LHC luminosity is expected to increase leading to pile-up that deforms the signal of interest. Due to limited resources, the current hardware setup, which is based on Digital Signal Processors (DSP), does not allow the implementation of sophisticated energy estimation methods that deal with the pile-up. Therefore, the technique to be employed for online energy estimation in TileCal for next LHC operation period must be based on fast filters such as the Optimal Filter (OF) and the Matched Filter (MF). Both the OF and MF methods envisage the use of the background second order statistics in its design, more precisely the covariance matrix. However, the identity matrix has been used to describe this quantity. Although this approximation can be valid for low luminosity LHC, it leads to biased estimators under pile-up conditions. Since most of the TileCal cell present low occupancy, the pile-up, which is often modeled by a non-Gaussian distribution, can be seen as outlier events. Consequently, the classical covariance matrix estimation does not describe correctly the second order statistics of the background for the majority of the events, as this approach is very sensitive to outliers. As a result, the OF (or MF) coefficients are miscalculated leading to a larger variance and biased energy estimator. This work evaluates the usage of a robust covariance estimator, namely the Minimum Covariance Determinant (MCD) algorithm, to be applied in the OF design. The goal of the MCD estimator is to find a number of observations whose classical covariance matrix has the lowest determinant. Hence, this procedure avoids taking into account low likelihood events to describe the background. It is worth mentioning that the background covariance matrix as well as the OF coefficients for each TileCal channel are computed offline and stored for both online and offline use. In order to evaluate the impact of the MCD estimator on the performance of the OF, simulated data sets were used. Different average numbers of interactions per bunch crossing and bunch spacings were tested. The results show that the estimation of the background covariance matrix through MCD improves significantly the final energy resolution with respect to the identity matrix which is currently used. Particularly, for high occupancy cells, the final energy resolution is improved by more than 20%. Moreover, the use of the classical covariance matrix degrades the energy resolution for the majority of TileCal cells.

## 1. Introduction

The ATLAS [1] Tile Calorimeter (TileCal) [2] is the detector used in the reconstruction of hadrons, jets and missing transverse energy from the proton-proton collisions at the Large Hadron Collider (LHC) [3]. It is composed by one central and two extended barrels covering the most central part of the ATLAS detector ($|\eta| < 1.6$). The energy deposited by the particles is read out by approximately 5,000 cells, with double readout channels. Figure 1 shows the TileCal readout granularity which consists of three radial layers (A,BC and D) with $\Delta\eta \times \Delta\phi = 0.1 \times 0.1$ ($0.2 \times 0.1$ in outermost layer). There is also the E layer, which consists of four scintillator plates per module and are located in the extended barrels.

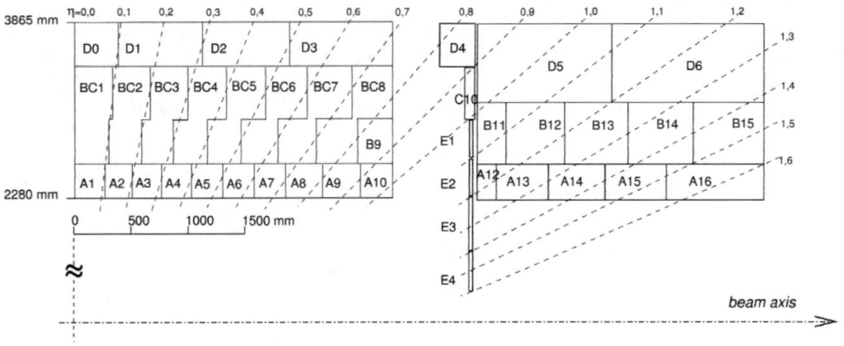


**Figure 1.** TileCal cell segmentation for half barrel and one extended partition.

The signal provided by the readout electronics [4] for each channel is digitized at 40 MHz and its amplitude is estimated by an optimal filtering algorithm, which expects a single signal with a well-defined shape. With the increase of the luminosity during Run2, the pile-up signal deforms the signal of interest. For illustration purposes, Figure 2 shows the readout window with the expected signal pulse shape (black line) and an out-of-time signal (red line). The resultant signal (magenta line) corresponds to the signal that is acquired by the front-end electronics.

Due to limited resources, the current Digital Signal Processors (DSP) based hardware setup does not allow the implementation of very sophisticated energy estimation methods, therefore, the online energy estimation must be based on fast filters such as the Optimal Filter (OF) [5]. The OF method uses the background second order statistics in its design, more precisely the covariance matrix. The identity matrix was used during Run1 to describe this quantity, although this approximation (that is valid for low luminosity LHC) can lead to larger variance and biased estimators under pile-up conditions, as it will be shown in Section 3.2.

This work evaluates the use of the background covariance matrix in the design of the OF weights to improve energy estimation performance under pile-up conditions. Next section describes the TileCal energy estimation algorithm and how we plan to improve it for LHC Run2. Section 3 shows the results using a simulated data set, where different versions of the OF method are evaluated. Finally, the conclusions are presented in Section 4.

## 2. TileCal Energy Estimation

The OF method was implemented to operate both online and offline in TileCal during Run1. The algorithm is based on a weighted sum of the received Analog to Digital Converter (ADC)

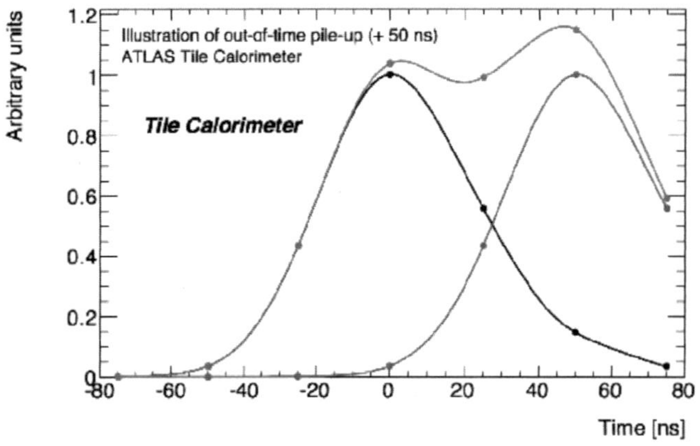


**Figure 2.** Illustration of the pile-up effect in TileCal.

samples aiming at minimizing the variance of the signal amplitude (see Equation1).

$$\hat{A} = \sum_{i=1}^{N} s_i w_i \tag{1}$$

The vectors **s** and **w** correspond to the received ADC samples and the OF weights, respectively. The parameter $N$ is the number of samples available.

The variance of the amplitude parameter to be minmized is given by:

$$var(\hat{A}) = \mathbf{w}'\mathbf{C}\mathbf{w} \tag{2}$$

where the matrix **C** corresponds to the background covariance matrix.

The current implementation of the OF in TileCal is called OF2, and it performs the optimization procedure subjected to the following constraints:

$$\sum_{i=1}^{N} g_i w_i = 1 \tag{3}$$

$$\sum_{i=1}^{N} g_i' w_i = 0 \tag{4}$$

$$\sum_{i=1}^{N} w_i = 0 \tag{5}$$

where the vectors **g** and **g**' are the TileCal reference pulse shape and its derivative, respectively.

The first constraint (Equation 3) regards the energy scale factor, while the additional second and third constraints (Equations 4 and 5) are added to make the estimation procedure immune against phase and baseline fluctuations, respectively.

The weights $\boldsymbol{w}$ can be found by solving the following matrix system using Lagrange multipliers:

$$\begin{pmatrix} C_{1,1} & \dots & C_{1,7} & -g_1 & -g'_2 & -1 \\ \vdots & \ddots & \vdots & \vdots & \vdots & \\ C_{7,1} & \dots & C_{7,7} & -g_7 & -g'_7 & -1 \\ g_1 & \dots & g_7 & 0 & 0 & 0 \\ g'_1 & \dots & g'_7 & 0 & 0 & 0 \\ 1 & \dots & 1 & 0 & 0 & 0 \end{pmatrix} \begin{pmatrix} w_1 \\ \vdots \\ w_7 \\ \lambda \\ \xi \\ v \end{pmatrix} = \begin{pmatrix} 0 \\ \vdots \\ 0 \\ 1 \\ 0 \\ 0 \end{pmatrix} \quad (6)$$

where $\lambda, \xi, v$ are the Lagrange multipliers.

Since the electronic noise can be modeled as an uncorrelated Gaussian process, the covariance matrix $\mathbf{C}$ was replaced by an identity matrix during LHC Run1 data taking.

If the third constraint (Equation 5) is removed from the optimization procedure, we call this method OF1. The difference is that OF2 computes the baseline value in an event-by-event basis, while OF1 relies on the stability of the baseline, and it subtracts a fixed value from each incoming ADC sample (see Equation 7).

$$\hat{A} = \sum_{i=1}^{N} (s_i - ped) w_i \quad (7)$$

The constant value *ped* is computed through a special run and stored in a data base.

It is worth mentioning that OF1 achieves similar performance as the Gaussian Matched Filter (MF) based approach, which was recently proposed for TileCal energy estimation [6]. The MF method is developed from the likelihood ratio test and, consequently, can be designed for any background model. However, for non-Gaussian background, its design leads to a nonlinear digital circuit, which is difficult to implement.

### *2.1. Online Energy Estimation for Run2*

As the pile-up introduces correlation between the samples, the use of the background covariance matrix is expected to improve the performance of the OF method under pile-up conditions. Moreover, in TileCal, the occupancy of most of the readout cells are low, and the pile-up signal is considered as outlier for the majority of the cells. Additionally, since the covariance matrix is very sensitive to outliers, alternative ways of computing this quantity must be considered.

The classical covariance matrix estimation [7] takes into account the whole dataset regarless the presence of outliers, leading to the miscalculation of the OF weights. As a result, the OF weights can be miscalculated leading to a larger variance and biased energy estimator.

Alternatively, the Minimum Covariance Determinant Estimator (MCDE) [8] provides a more careful way of computing the covariance matrix. The MCDE algorithm takes randomly a subset of the noise events and computes its classical covariance matrix and its determinant. The algorithm repeats this procedure several times and it selects the subset that resulted in the lowest determinant. This subset contains the events that has the lowest covariance between the samples (ADC digits) and therefore it consists of the most probable events, disgarding most of the outliers (high energy pile-up in this case).

## 3. Results

This section presents the data set used and the performance evaluation is carried out in terms of the estimation error for both OF1 and OF2 methods. The OF1 and OF2 weights were designed using the identity matrix as well as the covariance matrix, computed using the MCDE algorithm.

### *3.1. Data set*

The data set consists of 100,000 events for TileCal E3 ($\eta = 1.3$) and E4 ($\eta = 1.5$) cells, which are the highest occupancy cells in TileCal. The events are full ATLAS Monte Carlo simulation of Minimum Bias (MB) events with 25 ns of bunch spacing and an averaged number of $p-p$ interactions per bunch crossing ($< \mu >$) of 40 [9]. Only MB pile-up signals (both in-time and out-of-time) and electronic noise are present in the events.

### *3.2. Performance evaluation*

The designed OF1 and OF2 were applied to the data set described previously. Since the events considered above comprise only background, the energy estimation is expected to be centered at zero, with the smallest dispersion as possible. These distributions can be seen as the error introduced by each method to the final energy estimate, under this condition of pile-up. Figure 3 shows the energy estimation for the E3 cell.

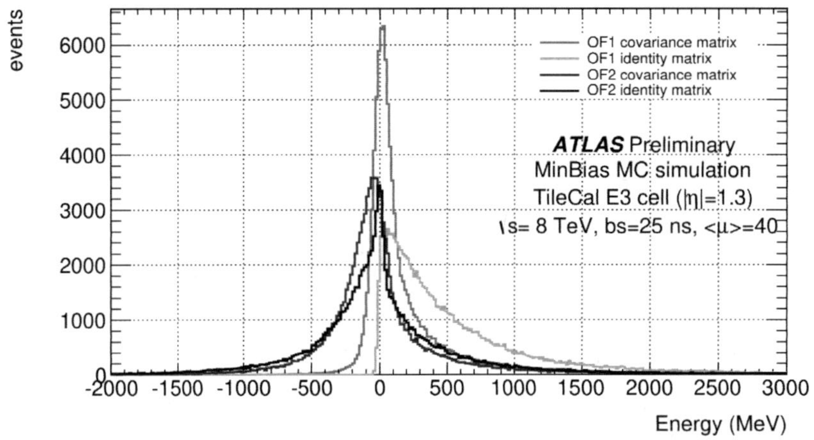


**Figure 3.** Energy distribution for cell E3 using both OF1 and OF2.

It can be noticed that the use of the covariance matrix in the computation of the OF1 weights shows the best performance in terms of estimation error (smaller dispersion). Additionally, the positive tail in the distribution is due to the high order statistics from the in-time pile-up signals that can not be accessed through the covariance matrix alone.

Figure 4 shows the energy distribution for the E4 cell which is the highest occupancy channel in TileCal. In order to summarize the parameters of each distribution, Table 1 shows the mean and RMS values for both E3 and E4 cells.

**Table 1.** Summary of the mean and RMS values of the energy distributions from E3 and E4 cells (in MeV).

| | **OF1 covariance** | | **OF1 identity** | | **OF2 covariance** | | **OF2 identity** | |
|---|---|---|---|---|---|---|---|---|
| | mean | RMS | mean | RMS | mean | RMS | mean | RMS |
| **E3** | 18.28 | 113.84 | 74.04 | 282.69 | -55.23 | 201.06 | -43.97 | 299.55 |
| **E4** | 145.79 | 311.26 | 971.25 | 687.82 | -163.13 | 424.67 | -98.75 | 674.40 |

As expected, the OF method becomes biased and it also increases its variance under pile-up, and the best performance is again achieved by OF1 using the covariance matrix (smaller

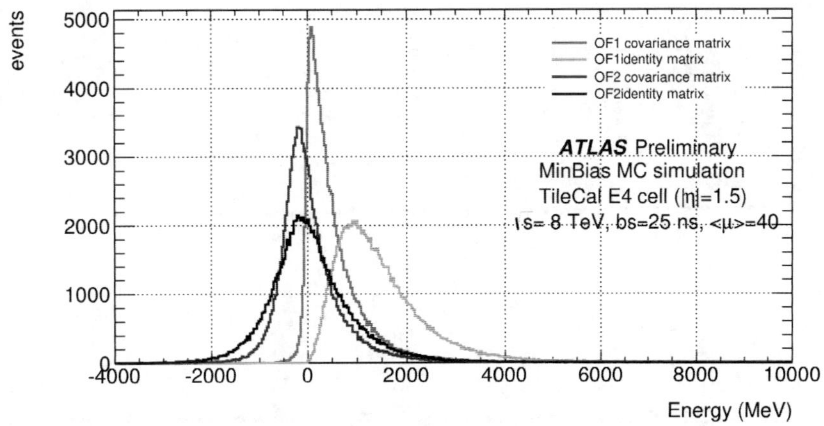


**Figure 4.** Energy distribution for cell E4 using both OF1 and OF2.

dispersion). In the case of these two highest occupancy cells, the bias (mean value) could be stored in a data base and used to be subtracted from each estimate for correction. This bias is not present for the other TileCal cells. It is worth mentioning that the mean and RMS were computed similarly as the covariance matrix, using a robust approach [8].

Figure 5 summarizes the improvement in the energy resolution (RMS of the energy distribution) using the covariance matrix with respect to the identity matrix for all cells in the central and extended barrels. The improvement is computed according to the following equation:

$$RI(\%) = 100 - \frac{RMS_{covariance}}{RMS_{identity}} \times 100 \tag{8}$$

For cells located in the layers BC and D, which are not highly affected by the pile-up, the improvement is minimal for TileCal central barrel ($|\eta| < 1$). In those cells, the identity approximation remains a reasonable description of the background for the majority of the events considered in this pile-up condition. While for the cells located in layers A and E in the extended barrel regions, the improvement becomes substantial, especially for the E cells which are closer to the beam and more exposed to the pile-up.

## 4. Conclusions

We present a study on the use of the background covariance matrix in the algorithm that will be used for TileCal online energy estimation during the next LHC run. The background covariance matrix can be used to reduce the uncertanties and bias due to the pile-up under high luminosity conditions. The results show that the OF using a fixed baseline value and the correct background covariance matrix presented the best performance in terms of estimation error. Most of the cells located in the central barrel did not show improvements due to their relatively low occupancy in the considered scenario. However, in some E cells in the extended barrel, the improvement is large as 60% with respect to the identity matrix.

Both the baseline value and the background covariance matrix are computed offline and stored in database. The OF weights are also computed offline and loaded in the DSPs for the online energy estimation.

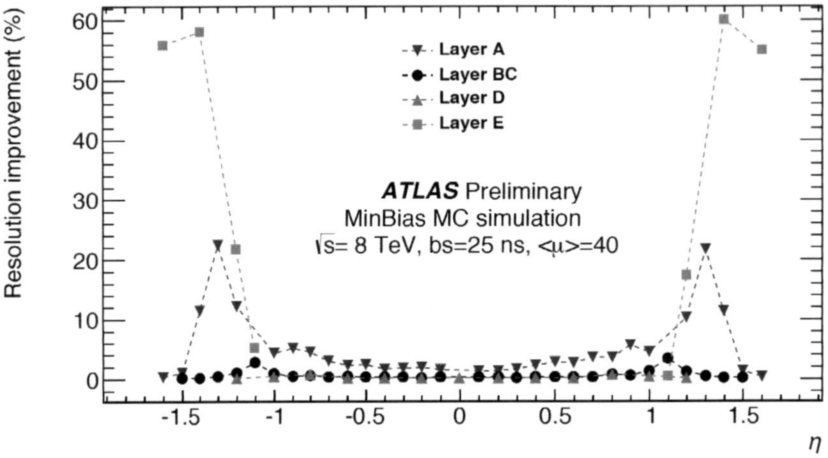


**Figure 5.** Improvement in function of $\eta$ by using OF1 designed with the covariance matrix with respect to OF1 designed with an identity matrix.

## Acknowledgments
We would like to thank the brazilian agencies CAPES, CNPq, FAPEMIG, FAPERJ and RENAVAE for the financial support. Also, we are very thankful to our colleagues from the ATLAS collaboration for the fruitful discussions concerning this work.

## References
[1] ATLAS Collaboration 2008 *The ATLAS Experiment at the CERN Large Hadron Collider* JINST **3** S08003
[2] ATLAS Collaboration 2010 *Readiness of the ATLAS Tile Calorimeter for LHC collisions* EPJC **70** pp 1193-1236
[3] Evans L and Bryant P (editors) 2008 *LHC Machine* JINST **3** S08001
[4] Anderson K *et al* 1998 *Front-end Electronics for the ATLAS Tile Calorimeter* Proceedings of Fourth Workshop on Electronics for LHC Experiments
[5] Fullana E *et al* 2006 *Digital Signal Reconstruction in the ATLAS Hadronic Tile Calorimeter* IEEE Transaction On Nuclear Science **53** 4
[6] Peralva B *et al* 2013 *The TileCal Energy Estimation for Collision Data Using the Matched Filter* IEEE Nuclear Science Symposium and Medical Imaging Conference pp 1-6
[7] Peebles P 2000 *Probability, Random Variables, and Random Signal Principles* McGraw-Hill
[8] Rousseeuw P and Dreissen K 1999 *A Fast Algorithm for the Minimum Covariance Determinant Estimator* Technometrics **41** 3
[9] Chapman J 2011 *ATLAS Simulation Computing Performance and Pile-Up Simulation in ATLAS* LPCC Detector Simulation Workshop

Journal of Physics: Conference Series **608** (2015) 012044 doi:10.1088/1742-6596/608/1/012044

# Quality Factor for the Hadronic Calorimeter in High Luminosity Conditions

**J M Seixas on behalf of the ATLAS Tile Calorimeter System**

Signal Processing Laboratory / UFRJ
Rio de Janeiro, RJ, Brasil

E-mail: seixas@lps.ufrj.br

**Abstract.** The Tile Calorimeter (TileCal) is the central section of the hadronic calorimeter of ATLAS experiment of the Large Hadron Collider (LHC) and has about 10,000 eletronic channels. An Optimal Filter (OF) has been used to estimate the energy sampled by the calorimeter and applies a Quality Factor (QF) for signal acceptance. An approach using Matched Filter (MF) has also been pursued. In order to cope with the luminosity rising foreseen for LHC operation upgrade, different algorithms have been developed. Currently, the OF measurement for signal acceptance is implemented through a chi-square test. At a low luminosity scenario, such QF measurement has been used as a way to describe how the acquired signal is compatible to the pulse shape pattern. However, at high-luminosity conditions, due to pile up, this QF acceptance is no longer possible when OF is employed, and the QF becomes a measurement to indicate whether the reconstructed signal suffers or not from pile up. Methods are being developed in order to recover the superimposed information, and the QF may be used again as signal acceptance criterion. In this work, a new QF measurement is introduced. It is based on divergence statistics, which measures the similarity of probability density functions.

## 1. Introduction

The Large Hadron Collider (LHC) collided protons with 8 TeV center-of-mass in 2012. The collision spacing was 50 ns and the mean number of interactions per crossing was 20 with the maximum value reaching 35. The planning for the next 10 years is to increase the number of colliding bunches. The center of mass energy will also increase the number of interactions [1]. The ATLAS detector [2] was designed to exploit the full potential of the LHC discoveries. As a general-purpose detector, ATLAS must be sensitive to a wide range of signatures left by particles.

In ATLAS, the calorimeter system plays a major role due to its excellent energy resolution and fast response. The system is split into electromagnetic [3] and hadronic [4] sections. The Tile Hadronic Calorimeter (Tilecal) is the central hadronic calorimeter in ATLAS. It is designed to identify and measure the jets, tau's, and contribute to the missing $E_T$. It is based on a sampling technique with scintillating tiles embedded in an absorber structure. The tiles are placed perpendicular to the beam axis for simplifying the mechanical construction and routing of the wavelength-shift (WLS) fibers for readout. It is mechanically divided into a central barrel and two extended barrels and each cylinder comprises 64 modules. Concerning the readout granularity, TileCal is divided into 3 radial layers with $\Delta\eta$ x $\Delta\phi$ = 0.1 x 0.1 (0.2 x 0.1 in outermost layer) resulting in approximately 5,000 double readout cells.

A particle traversing the calorimeter has part of its energy deposited in the scintillating tiles. This energy is converted into light that is transmitted by WLS fibers to photomultiplier tubes (PMT). The photomultipliers convert the light pulses into fast current signals, which are shaped into signals with full width at half maximum of 50 ns. The signals are sampled by two ADC converters operating at a frequency of 40 MHz, which is the nominal bunch crossing (BC) frequency of the LHC. The number of samples is programmable, and seven samples are being used for readout (see Figure 1). Such signal can be defined as

$$s(t) = Ag(t_i - \tau) + ped \tag{1}$$

where $A$ is the signal amplitude, $t_i$ is the time at which the sample $i$ was acquired, $ped$ is the pedestal level, $\tau$ is the phase of the pulse and $g$ is the expected pulse shape.

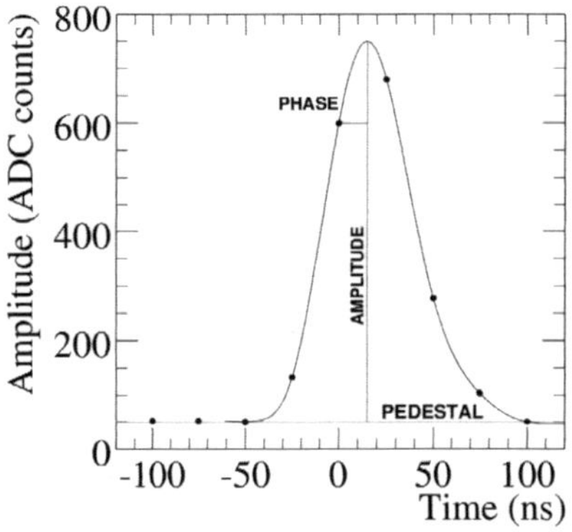


**Figure 1.** TileCal pulse shape generated at the front-end [7].

Most of energy reconstruction methods in calorimetry uses the knowledge of the pulse shape generated by the readout eletronics and assume Gaussian noise characteristics for each channel [5]. Due to the increase in luminosity, the signal pile-up will deform the signal of interest. Such effect is highly dependent on the conditions of the LHC operation and reduce the efficiency of current energy estimation techniques, as the background becomes highly non-gaussian [6].

## 2. Signal Reconstruction

The baseline energy reconstruction method for TileCal is the Optimal Filter (OF) [7], but other methods have also been developed for offline data processing such as the Matched Filter (MF) [8]. However, both methods suffer from the increase of pile-up events. It has to be mentioned that the planned LHC luminosity increase in the future will require pile-up mitigation in order to preserve unbiased signal reconstruction performance. Therefore, a more sophisticated approach to deal with the pile-up is under development. Figure 2 exemplifies the considerable deterioration of the pulse shape for a signal under the presence of an out-of-time (neighbouring bunch crossings) signal at 50 ns.

### *2.1. Energy Estimation*

The OF is based on a linear combination of the signal samples to obtain the pulse amplitude. The coefficients are chosen such that the noise impact on the resolution of the calorimeter is minimized [7]. The following equation describes how the OF reconstructs the signal:

$$A = \sum_{i=1}^{7} a_i s_i, \qquad \tau = \frac{1}{A}\sum_{i=1}^{7} b_i s_i \quad and \quad ped = \sum_{i=1}^{7} c_i s_i. \tag{2}$$

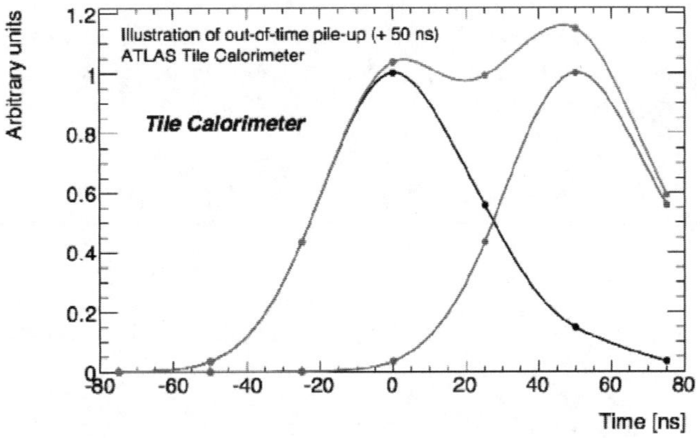


**Figure 2.** Pile-up effect in TileCal.

Here, $s_i$ is the sample in the bunch-crossing $i$ and $a$, $b$, and $c$ are the filter coefficients.

The MF also uses the reference signal to develop a linear filter for TileCal. The coefficients are calculated using a set of training data to estimate the covariance of the noise and the baseline of the input signal [8]:

$$\hat{A} = \frac{\sum_{i=1}^{7}(s_i - ped) \cdot g_i}{\sum_{i=1}^{7}(g_i \cdot g_i)} \tag{3}$$

where $ped$ corresponds to the pedestal.

Unlike OF and MF, signal deconvolution techniques estimate the amplitude for each BC simultaneously [9].

### 2.2. Quality Factor

These energy reconstruction methods use a $\chi^2$ to evaluate the goodness of the reconstructed pulse, but they have shown to be highly sensitive to the effect of pile-up. The $\chi^2$ is calculated according to

$$QF = \sqrt{\sum_{i=1}^{7}(s_i - \hat{a}_0 h_{0,i} - \sum_{j=1}^{N}\hat{a}_j h_{j,i})^2} \tag{4}$$

where $\hat{a}$ corresponds to the amplitude of each signal present in a given readout window, and $h$ is a matrix that contains the expected signals.

### 2.3. Divergence as Quality Factor

A new way to measure the quality factor is the use of divergence statistic, which is considered as a measurement of similarity between two probability distributions [10]. For this, the shaped TileCal pulse is considered as a distribution and its divergence is measured with respect to the one obtained from the signal sampled by the electronics chain. In this way, the quality factor qualifies a given signal if the measured divergence is low. The simplest way to use diverge as quality factor is by defining cuts in order to validate signals (physics of interest) against background (minimum bias or corrupted events). As the statistical fluctuations in the standard Tilecal shaped pulse are small, higher values of the divergence measurement indicate signal deterioration. These small statistical fluctuations of the Tilecal pulse are used to calibrate the divergence measurement, when a valid pulse is present.

Journal of Physics: Conference Series **608** (2015) 012044 doi:10.1088/1742-6596/608/1/012044

*2.3.1. KL Divergence* The Kullback-Leibler divergence, also known as KL divergence, uses mutual information for the calculation of this quasi-distance [11]. Considering two probability density functions $\hat{p}_A$ (as observed) and $\hat{p}_B$ (theoretical expectation), we can define the Kullback-Leibler divergence as shown by Equation (5).

$$D_{KL}(\hat{p_A}, \hat{p_B}) = \int_{-\infty}^{+\infty} \hat{p_A}(\xi) \log \frac{\hat{p_A}(\xi)}{\hat{p_B}(\xi)} d\xi \qquad (5)$$

*2.3.2. JS Divergence* The Jensen-Shannon divergence (JS) as defined by Equation (6), is a symmetric and smoother version of the Kullback-Leibler divergence [12]. Moreover, the square root of JS divergence is a metric:

$$D_{JS}(\hat{p_A}, \hat{p_B}) = \frac{D_{KL}(\hat{p_A}, \hat{p_M}) + D_{KL}(\hat{p_B}, \hat{p_M})}{2} \qquad (6)$$

where $\hat{p_M} = \frac{\hat{p_A + p_B}}{2}$ .

## 3. Results

The methods described above are applied for signal reconstruction using ATLAS Monte Carlo simulation [13] under high luminosity conditions, where the averaged number of interactions per bunch crossing reaches up to 40. The only information needed to compute the proposed quality factor is the standard TileCal pulse shape and the signal baseline (computed through special calibration runs).

If the reconstructed amplitude values are evaluated through their respective quality factors, a large difference between the quality factor values (for different reconstruction methods, OF and deconvolution techniques) is expected, even if both methods use the $\chi^2$ computation. Due to the out-of-time pile-up, large QF values are expected when the central bunch crossing is estimated with OF, so that QF becomes an useful measurement for pile-up detection, but loses efficiency as a measurement of signal reconstruction quality. For a simulation of a high occupancy TileCal cell (located in the A layer with $|\eta| = 1.2$ [4]) and LHC operating at 25 ns of bunch spacing with $< \mu >= 40$, the results show that the QF through $\chi^2$ is not able to validate signals that are affected by high pile-up. It discards approximately 15% of the events in the data set that have $\chi^2 > 15$. When the original signal samples are better recovered, signal reconstruction produces low QF values. In this case, QF recovers its capacity as an efficient measurement of signal reconstruction quality in such high pile-up conditions.

Because the divergence is measurement of similarity, it can not be compared in terms of absolute values with the $\chi^2$, which is currently used in ATLAS calorimeter system. However, divergence measurements allow the distinction between signals and background and, as a result, it can be applied in harsh pile-up environments. The divergence measurement is more sensitive to deviations from a valid pulse, as it considers a density estimation. For the same simulation mentioned previously (also the same TileCal cell), divergence measurement presents lower values than usual $\chi^2$ for the whole dinamic range, and it is able to validate approximately 95% of the events (JS divergence $< 0.4$).

## 4. Conclusions

Currently, the Optimal Filter is used for reconstruction of sampled signals in TileCal. However, due to the increase of luminosity in the near future, the number of pile-up events will increase substantially. In TileCal, the quality factor was conceived as a measurement of the adherence of an acquired signal to the reference shaped pulse. However, for high luminosity conditions this measurement becomes more of a pile-up indicator. New reconstruction methods are being developed for optimal reconstruction under severe pile-up conditions, which allow signal reconstruction quality evaluation.

The quality factor based on the $\chi^2$ from independent amplitude estimates is supposed to present lower values than those obtained by the OF. On the other hand, a quality factor through divergence statistics shows to be a a more appropriate measurement to distinguish the physics of interest from background for pile-up under non-Gaussian background that is introduced by the pile-up. The application of the proposed quality factor is under evaluation in TileCal, and studies considering higher pile-up condictions are on the way.

## 5. Acknowledgments
This work has been partially supported by CNPq, FAPERJ, CAPES and the National Network for High Energy Physics (RENAFAE) in Brazil.

## References
[1] The ATLAS Collaboration 2014 Plans for the Phase II upgrade to the ATLAS detector *13th Topical Seminar on Innovative Particle and Radiation Detectors* JINST 9 C02033
[2] The ATLAS Collaboration 2008 *The ATLAS Experiment at the CERN Large Hadron Collider* JINST **3** S08003
[3] The ATLAS Collaboration 2010 *R*eadiness of the ATLAS Liquid Argon Calorimeter for LHC Collisions Eur.Phys.J. C70:723-753
[4] The ATLAS Collaboration 2010 *Readiness of the ATLAS Tile Calorimeter for LHC collisions* EPJC **70** pp 1193-1236
[5] Valero A et al 2007 *DSP Online Algorithms for the ATLAS TileCal Read-Out Drivers* IEEE Real-Time Conference pp. 1-7
[6] The ATLAS Collaboration 2014 *Simulation and validation of the ATLAS Tile Calorimeter response* International Conference on Instrumentation for Colliding Beam Physics JINST 9 C09018
[7] Fullana E *et al* 2006 *Digital Signal Reconstruction in the ATLAS Hadronic Tile Calorimeter* IEEE Transaction On Nuclear Science **53** 4
[8] Peralva B et al 2013 *The TileCal Energy Reconstruction for Collision Data Using the Matched Filter* IEEE Nuclear Science Symposium and Medical Imaging Conference
[9] Andrade L et al 2012 *Calorimeter Signal Response Deconvolution for Online Energy Estimation in Presence of Pile-up* (Geneva: CERN CDS)
[10] Hyvarinen A, Karhunen J and Oja E 2009 *Independent Component Analysis* (United Kingdom: Wiley)
[11] Cichocki A, Zdunek R and Amari S 2009 *Nonnegative Matrix and Tensor Factorization* (United Kingdom: Wiley)
[12] Burbea J and Rao C 1991 *IEEE Transactions on Information Theory* **37** 145–151
[13] Chapman J 2011 *ATLAS Simulation Computing Performance and Pile-Up Simulation in ATLAS* LPCC Detector Simulation Workshop

# New features in Delphes 3

**Alexandre Mertens**

Centre for Cosmology, Particle Physics and Phenomenology (CP$^3$),
Université Catholique de Louvain,
Chemin du Cyclotron 2, B-1348 Louvain-la-Neuve, Belgium

E-mail: alexandre.mertens@uclouvain.be

**Abstract.** Delphes is an open source C++ framework to perform the fast simulation of the response of a multipurpose detector. The simulation includes a tracking system, embedded into a magnetic field, calorimeters and a muon system. The framework is interfaced to standard file formats and outputs observables such as isolated leptons, missing transverse energy and collection of jets that can be used for dedicated analyses. The simulation of the detector response takes into account the effect of magnetic field, the granularity of the calorimeters and subdetector resolutions. The program contains parmetrizations for the CMS and ATLAS detectors, based on published performances. Basic parametrizations for the LHCb and FCC detectors are also available. The Delphes framework also includes a simple event display.

Several new features are discussed, such as an emulation of the particle-flow algorithm, pile-up simulation, N-subjettiness and a simple b-tagging algorithm based on counting tracks with large impact parameter.

## 1. Introduction

High energy particle collisions can produce a large variety of final states. Highly sophisticated detectors are designed in order to detect and precisely measure particles originating from such collisions. Experimental collaborations often rely on Monte-Carlo event generation for designing and optimizing specific analysis strategies. Whenever such studies require a high level of accuracy, the interactions of long-lived particles with the detector matter content are fully simulated with the GEANT package [1], electronic response is emulated by dedicated routines, and final observables are reconstructed by means of complex algorithms. To face the limited computing resources and still allow the use of large samples (for example when scanning parameter spaces), the LHC collaborations have developed fast-simulation techniques [2, 3, 4, 5, 6] which are two to three orders of magnitude faster than the fully GEANT-based simulations.

These procedures require expertise and the deployment of large scale computing resources that can be handled only by large collaborations. For most phenomenological studies, such a level of complexity is not needed and a simplified approach based on the parameterization of the detector response is in general good enough. In 2009, the DELPHES framework [7] was designed to achieve such goal. The DELPHES framework takes as input the most common event generator output and performs a fast and realistic simulation of a general purpose collider detector. To do so, long-lived particles emerging from the hard scattering are propagated to the calorimeters within a uniform magnetic field parallel to the beam direction. The particle energies are computed by smearing the initial momenta according to the detector resolution. As a result, jets, missing energy, isolated electrons, muons and photons, and taus can be reconstructed.

With respect to its previous incarnation [7], the present version of DELPHES [8] includes an attempt to roughly emulate the particle-flow reconstruction philosophy used in ALEPH [20] and CMS [14], based on the optimally-combined use of the information from all the subdetectors to reconstruct and identify all particles indvidually. While the aim is not to re-implement the particle-flow algorithm in all its complexity (for example, electrons, muons, and photons, are assumed to be perfectly identified and have no fake rate in DELPHES), the simplified approach adopted here is particularly suitable for the treatment of pile-up, as well as for the emulation of b and $\tau$ tagging, and is able to reproduce the jet and missing energy resolutions observed in CMS with their complete reconstruction. From a technical perspective, the code structure is now fully modular, providing greater flexibility to the user and allowing the integration of DELPHES routines in other projects.

Among the new features of the DELPHES realease, the particle flow emulation is presented in section 2 and the pile-up implementation in section 3. Other developments such as B-tagging and jet substrure are described in section 4 and section 5.

## 2. Particle Flow emulation

The implementation of the Particle Flow emulation in DELPHES is described in detail in [8]. To illustrate the method, here are a few simple examples:

- a single charged pion is reconstructed as a track with energy $E_{HCAL,trk}$ and deposits some energy $E_{HCAL}$ in the HCAL. If $E_{HCAL} \leq E_{HCAL,trk}$ only a particle-flow track with energy $E_{HCAL,trk}$ is produced. If $E_{HCAL} > E_{HCAL,trk}$, a particle-flow track with energy $E_{HCAL,trk}$ and a particle-flow tower with energy $E_{HCAL} - E_{HCAL,trk}$ are produced.
- a single photon deposits its energy $E_{ECAL}$ in an ECAL cell. No tracks pointing to the cell is reconstructed. A particle-flow tower is created with energy $E_{ECAL}$.
- a photon and a charged pion reach the same calorimeter tower, the former deposits some energy $E_{ECAL}$ in ECAL and the latter $E_{HCAL}$ in HCAL. Furthermore, the charged pion is reconstructed as a track, with an energy $E_{HCAL,trk}$. If $E_{HCAL} \leq E_{HCAL,trk}$, a particle-flow track with energy $E_{HCAL,trk}$ and a particle-flow tower with energy $E_{ECAL}$ are created. If $E_{HCAL} > E_{HCAL,trk}$, a particle-flow track with energy $E_{HCAL,trk}$ and a particle-flow tower with energy $E_{ECAL} + E_{HCAL} - E_{HCAL,trk}$ are created.

Defined that way, the particle-flow tracks contain charged particles estimated with a good resolution, while the particle-flow towers contain in general a combination of neutral particles, charged particles with no corresponding reconstructed track and additional excess deposits induced by the positive smearing of the calorimeters, and are characterized by a lower resolution. Besides producing high-resolution inputs for jets and missing transverse energy, the particle-flow approach can be rather useful for addressing pile-up subtraction. While very simple when compared to what is actually required in real experiments, the algorithm illustrated above is shown to reproduce well the performance achieved at LHC [8].

*Physics application* In modern collider experiments at high energy, $t\bar{t}$ events are among the most copious signatures observed in the detectors. When one top quark decays leptonically and the other hadronically, the signature is characterized by one lepton, missing transverse energy and four jets, two of them originating from the fragmentation of b quarks. Moreover, at the LHC, about 50% of events have extra hard jets coming from initial or final state radiation. Following the semi-leptonic $t\bar{t}$ analysis described in ref. [21] we focus on the mass of the hadronically-decaying top quark.

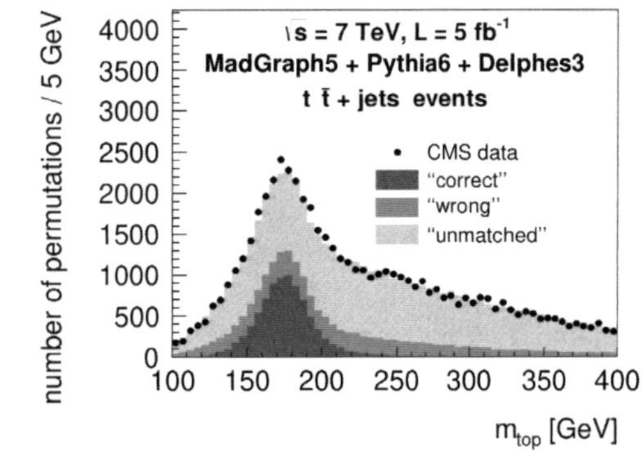


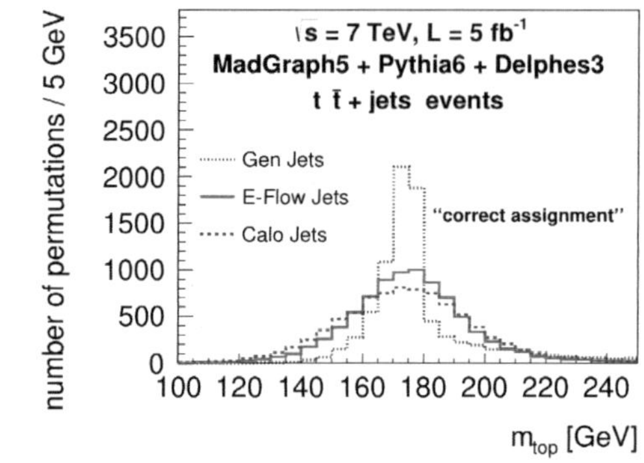


**Figure 1.** Left: Reconstructed hadronic top mass for all possible assignment categories. The DELPHES distribution is normalized to the CMS yield. The CMS contributions are taken from Ref. [21]. Right: Reconstructed hadronic top mass distribution for the *correct assignments* only. The distribution is shown for *Generated Jets*, *Particle-Flow Jets* and *Calorimeter Jets*.

The $t\bar{t}$+jets sample has been generated with MADGRAPH5 at a centre of mass energy $\sqrt{s} = 7$ TeV and PYTHIA6 was used for parton shower and hadronization. Backgrounds are not considered here. The reconstruction has been performed via DELPHES using the detector configuration designed to mimic the performance of the CMS detector.

Following the CMS approach, we select events with exactly one isolated lepton (electron or muon) with $p_T > 30$ GeV/c and $|\eta| < 2.1$. In addition, we require at least four particle-flow jets with $p_T > 30$ GeV/c and $|\eta| < 2.4$. The anti-$k_T$ [15] algorithm with a parameter $R = 0.5$ was used for jet clustering. Among the selected jets, at least two must be tagged as originating from the hadronization of a b quark (b-tagged) and at least two must be identified as *light jets* (i.e. fail the b-tagging criterion). The b-tagging efficiency parameterization has been extracted from [25]. The signal efficiency for this selection is 2.8%, compared to 2.3% in the CMS analysis, showing a reasonable agreement between DELPHES and CMS. Given the high jet multiplicity, the signal selection is extremely sensitive to changes in requirements that can affect the jet selection.

The top quark mass distributions obtained with DELPHES and CMS for the three permutation categories are shown in figure 1, left. The DELPHES distributions are normalized to the CMS total number of events. Overall the shapes and relative contributions corresponding to the three categories are well reproduced by DELPHES.

For the sake of illustration, the reconstructed hadronic top mass using correct permutations only is shown in figure 1 (right) using three different jet collections: *Generated Jets*, *Calorimeter Jets* and *Particle-Flow Jets*. We observe, as expected, a narrow peak when using *Generated Jets* and wider peaks when using *Particle-Flow Jets* or *Calorimeter Jets*. This illustrates the need for using realistically reconstructed objects rather that hadron-level quantities in prospective phenomenological studies.

## 3. Pile Up implementation

At the LHC, several collisions per bunch-crossing occur in high luminosity conditions, most of them resulting in a small amount of activity in the detector. Due to the elongated shape of the proton bunches constituting the beams, such additional *pile-up events*, take place in a similarly elongated region (called beam spot) around the nominal interaction point. In DELPHES, pile-up interactions are extracted from a pre-generated low-$Q^2$ QCD sample. These minimum-bias interactions are randomly placed along the beam axis according to some longitudinal spread that can be set by the user. The actual number of pile-up interactions per bunch-crossing is randomly extracted from a Poisson distribution.

Pile-up directly affects the performance of jets, $E_T^{miss}$ and isolation. Pile-up interactions are usually identified by means of vertex reconstruction. If such interactions occur far enough from the hard interaction, a precise vertexing algorithm is able to detect them. Combining vertexing and tracking information allows the identification of contaminating charged particles from pile-up. On the other hand, since neutral particles do not produce tracks, neutral pile-up contamination can only be estimated *on average.*

*Charged pile-up subtraction* In DELPHES we assume that vertices corresponding to pile-up interactions occurring at a distance z from the hard scattering, such that $|z| > \delta Z_{vtx}$ can be reconstructed. The parameter $\delta Z_{vtx}$ is the spatial vertex resolution of the detector. We assume that pile-up interactions occurring at a coordinate z, such that $|z| < \delta Z_{vtx}$ cannot be disentangled from those originating from the high-$Q^2$ process. Therefore every charged particle originating from such vertices cannot be subtracted from the event, while every charged particle originating from a vertex positioned at $|z| > \delta Z_{vtx}$ can be identified as originating from pile-up, provided that the corresponding track has been reconstructed. If the particle-flow algorithm is being used, the particle-flow tracks identified as originating from pile-up are removed from the list of 4-vector entering the jet clustering and the isolation procedures.

*Residual pile-up subtraction* Other techniques are needed in order to extract and remove residual contributions: these include particles that are too close to the hard interaction vertex to be identified as pile-up products with tracking information, charged particles that failed track reconstruction (or outside the tracker volume) and neutral particles. In DELPHES we have opted for the Jet Area method [17, 18]. This approach, widely used in present collider experiments, allows the extraction of an average contamination density $\rho$ on an event-by-event basis. In practice, this is performed in DELPHES with the help of the FASTJET package.

The pile-up density $\rho$, can then be used to correct observables that are sensitive to the residual contamination, the jet energies and the isolation variable. In the presence of residual pile-up contamination, these two quantities are corrected in the following way:

$$p_{jet} \rightarrow p_{jet} - \rho \cdot A_{jet}, \quad (1)$$

$$I(P) \rightarrow I(P) - \frac{\rho \cdot \pi R^2}{p_T(P)}, \quad (2)$$

where $A_{jet}$ is the jet area estimated via the FASTJET package, and $R$ is the diameter of the isolation cone.

As a validation of the Pile Up implementation, the resolution in missing transverse energy has been computed as a function of the number of reconstructed primary vertices in the left plot of the Figure 2.

## 4. B-tagging possibilities

The identification of jets that result from the hadronization of heavy flavour quarks — typically b or c quarks — is important in high energy collider experiments. In DELPHES a purely parametric approach based on Monte-Carlo generator is available as well as a track counting implementation.

*Parametric* The algorithm for b and $\tau$ jet identification proceeds as follows: the jet becomes a potential b($\tau$) jet if a generated b($\tau$) is found within some distance $\Delta R = \sqrt{(\eta^{jet} - \eta^{b,\tau})^2 + (\phi^{jet} - \phi^{b,\tau})^2}$ of the jet axis. The probability to be identified as b($\tau$) depends

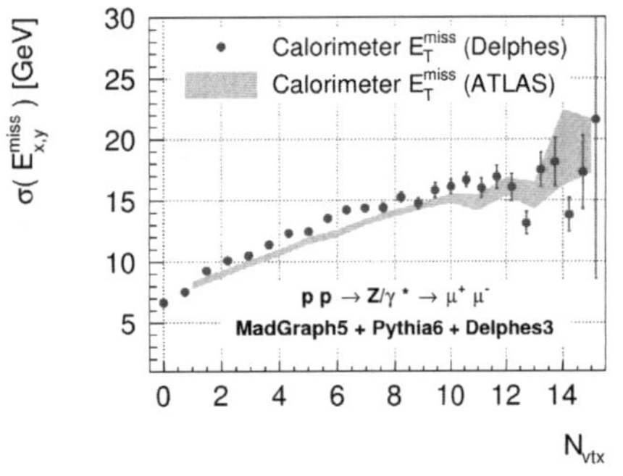


**Figure 2.** $E^{miss}_{x,y}$ resolution in DELPHES and ATLAS as a function of the number of reconstructed primary vertices [26]. The grey band represents the discrepancy between the ATLAS simulation and data.

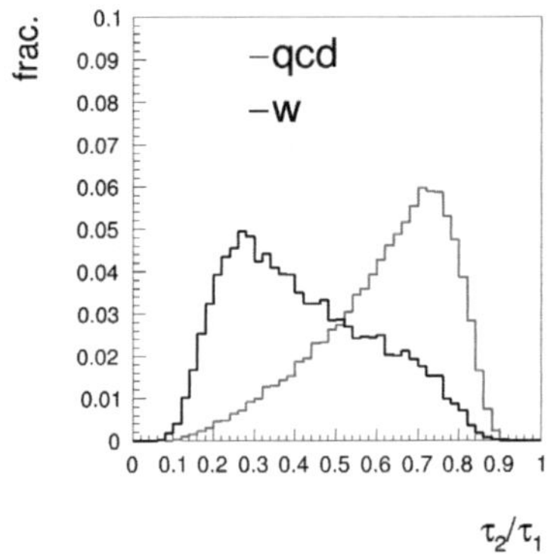


**Figure 3.** Ratio $\tau_2/\tau_1$ computed with N-subjettiness in DELPHES. We observe the discrimination between prompt jets and W-jets.

on user-defined parameterizations of the efficiency. The user can also specify a mis-tagging probability parameterization, that is, the probability that a particle other than b($\tau$) be wrongly identified. Modularity allows the user to use several b($\tau$)-tagging algorithms for the same jet collection and to easily implement other tagging algorithms, eventually involving an analysis of the jet constituents.

*Track Counting* When the Track counting b-tagging is selected, the impact parameter of each track inside all reconstructed jet is smeared according to a gaussian distribution. The width of the gaussian distribution is defined by the user and can be dependent on the $p_T$ and $\eta$ of the corresponding track. The *Track Counting* is then applied on each jet and require a minimum number of tracks (typically 2 or 3) with a high impact parameter significance for the jet to be tagged. While less precise than the parametric approach, this method provides predictive power needed for future detector design when a full simulation is not yet available.

## 5. Jet substructure techniques

At high energy, W, Z and Higgs bosons, as well as the top will produce only one jet when they are highly boosted, making the distinction between those objects more difficult. New techniques studying the substructure of the jets have been developped to disentangle boosted topologies from QCD backgrounds. Amongh these methods, the *N-Subjettiness* algorithm [22, 23, 24] from the FastJet-Contribs [13] has been implemented in DELPHES. The method returns the variables $\tau_1$ (1-subjettiness) to $\tau_5$ (5-subjettiness) which are saved as jet members in the DELPHES output. From those variables, it is then possible to build the ratios $\tau_N/\tau_M$ to discriminate jets originating from N or M-prongs.

An example of application of this method is shown in the Figure 3 where it is possible to discriminate jets comming from a W-boson decay from prompt jets originating from a QCD sample.

## Conclusion

Delphes is an advanced parametric simulation program allowing for a gain of four orders of magnitude in CPU time and widely tested by the community. A new version of the framework,

DELPHES 3 has been out since 2013 with major improvements: modularity, particle flow emulation, pile up implementation. Recently other new features such as the implementation of the Track Counting algorithm and jet substructure techniques were added to the framework. Thanks to those new implementations, we have shown that DELPHES is able to reproduce LHC analysis and produce qualitative predictive results.

The aim of DELPHES is to provide a simulation tool that goes beyond the run I for ATLAS and CMS. Thanks to its speed, DELPHES is able to handle the 200 PU scenario for the HL-LHC. Other detectors are also considered. Currently, Future Circular Collider (FCC), LHCb and Alice detector simulations are under preparation.

## Acknowledgments

First, I would like to thank the ACAT2014 workshop organisors for giving the opportunity to present this work. I wish to thank Christophe Delaere and Michele Selvaggi for their help and feedback. We acknowledge the support from the FNRS and the IAP Program, BELSPO VII/37. This work is partly supported by the IISN convention 4.4503.12.

## References

[1] J. Allison, K. Amako, J. Apostolakis, H. Araujo, P. A. Dubois, M. Asai, G. Barrand and R. Capra et al., *Geant4 developments and applications*, IEEE Trans. Nucl. Sci. **53**, 270 (2006).
[2] S. Hamilton, E. Kneringer, W. Lukas, E. Ritsch, A. Salzburger, K. Sliwa, S. Todorova and J. Wetter et al., *The ATLAS Fast Track Simulation Project*, ATL-SOFT-PROC-2011-038.
[3] W. Lukas [ATLAS Collaboration], *Fast Simulation for ATLAS: Atlfast-II and ISF*, J. Phys. Conf. Ser. **396**, 022031 (2012).
[4] R. Rahmat, R. Kroeger and A. Giammanco, *The fast simulation of the CMS experiment*, J. Phys. Conf. Ser. **396**, 062016 (2012).
[5] S. Abdullin *et al.* [CMS Collaboration], *The fast simulation of the CMS detector at LHC*, J. Phys. Conf. Ser. **331**, 032049 (2011).
[6] [CMS Collaboration], *Comparison of the Fast Simulation of CMS with the first LHC data*, CMS-DP-2010-039.
[7] Ovyn S, Rouby X and Lemaitre V 2009 DELPHES, a framework for fast simulation of a generic collider experiment *Preprint* hep-ph/09032225
[8] de Favereau J *et al*, 2013, arXiv:1307.6346 [hep-ex].
[9] Antcheva I *et al* 2009 A C++ framework for petabyte data storage, statistical analysis and visualization *Comput. Phys. Commun.* **180** 2499
[10] Dobbs M *et al* 2001 *Comput. Phys. Commun.* **134** 41
[11] Garren L and Lebrun P 2006 StdHep User Manual `http://cepa.fnal.gov/psm/stdhep/`
[12] Alwall J *et al* 2007 A Standard format for Les Houches event files *Comput. Phys. Commun.* **176** 300
[13] Cacciari M, Salam G P and Soyez G 2012 FastJet User Manual *Eur. Phys. J.* C **72** 1896
[14] CMS Collaboration 2009 Particle-Flow Event Reconstruction in CMS and Performance for Jets, Taus, and MET CMS-PAS-PFT-09-001
[15] Cacciari M, Salam G P and Soyez G 2008 The Anti-k(t) jet clustering algorithm *JHEP* **0804** 063
[16] Alwall J, Herquet M, Maltoni F, Mattelaer O and Stelzer T 2011 MadGraph 5 : Going Beyond *JHEP* **1106** 128
[17] Cacciari M, Salam G P and Soyez G 2008 The Catchment Area of Jets *JHEP* **0804** 005
[18] Cacciari M, Salam G P and Soyez G 2008 Pileup subtraction using jet areas *Phys. Lett.* B **659** 119
[19] Tadel M 2008 Raw-data display and visual reconstruction validation in ALICE *J. Phys. Conf. Ser.* **119** 032036
[20] D. Buskulic *et al.* [ALEPH Collaboration], *Performance of the ALEPH detector at LEP*, Nucl. Instrum. Meth. A **360**, 481 (1995).
[21] CMS Collaboration, *Measurement of the top-quark mass in* $t\bar{t}$ *events with lepton+jets final states in* pp *collisions at* $\sqrt{s} = 7$ *TeV*, JHEP **1212**, 105 (2012) [arXiv:1209.2319 [hep-ex]].
[22] J. Thaler and K. Van Tilburg, JHEP **1103**, 015 (2011) [arXiv:1011.2268 [hep-ph]].
[23] J. Thaler and K. Van Tilburg, JHEP **1202** (2012) 093 [arXiv:1108.2701 [hep-ph]].
[24] A. J. Larkoski, D. Neill and J. Thaler, JHEP **1404** (2014) 017 [arXiv:1401.2158 [hep-ph]].
[25] CMS Collaboration, *b-Jet Identification in the CMS Experiment*, CMS-PAS-BTV-11-004.
[26] ATLAS Collaboration, *Performance of Missing Transverse Momentum Reconstruction in ATLAS with 2011 Proton-Proton Collisions at sqrts = 7 TeV*, ATLAS-CONF-2012-101.

# Clustering analysis for muon tomography data elaboration in the Muon Portal project

**M. Bandieramonte*[1,2], V. Antonuccio-Delogu[2], U. Becciani[2], A. Costa[2], P. La Rocca[1,3], P. Massimino[2], C. Petta[1,3], C. Pistagna[2], F. Riggi[1,3], S. Riggi[2], E. Sciacca[2], F. Vitello[2]**

[1] Dept. of Physics and Astronomy, University of Catania, Italy
[2] INAF, Astrophysics Observatory of Catania, Italy
[3] INFN, Section of Catania, Italy

E-mail: *marilena.bandieramonte@oact.inaf.it

**Abstract.** Clustering analysis is one of multivariate data analysis techniques which allows to gather statistical data units into groups, in order to minimize the *logical distance* within each group and to maximize the one between different groups. In these proceedings, the authors present a novel approach to the muon tomography data analysis based on clustering algorithms. As a case study we present the *Muon Portal* project that aims to build and operate a dedicated particle detector for the inspection of harbor containers to hinder the smuggling of nuclear materials. Clustering techniques, working directly on scattering points, help to detect the presence of suspicious items inside the container, acting, as it will be shown, as a filter for a preliminary analysis of the data.

## 1. Introduction

Clustering or group analysis is a set of multivariate data analysis techniques that aims to select and to group homogeneous items in a data set. Since its appearance in the work of R. C. Tyron in 1939 [1], it was experienced with extended applications, starting from the '60s. Clustering techniques involve the division of the data into homogeneous subgroups. Through this multivariate data analysis technique, statistical units can be grouped together in order to minimize the *logical distance* within each group and to maximize the one between different groups. The *logical distance* is quantified by measures of similarity/dissimilarity between the statistical units. Informally, the goal of this partitioning is twofold. Data elements within a cluster must be similar to each other, while those within different clusters should be dissimilar, in such a way that the observations are the most possible homogeneous within classes and the most possible uneven between the different classes. The concept of homogeneity is specified in terms of distance and several criteria can be used, as it will be clarified later. Each of these classes is known as a *cluster*. It defines also a region where the objects density is locally higher than in other regions. Clustering techniques are successfully applied to problems as classification, pattern recognition and multivariate analysis. In these proceedings the authors present a density-based clustering algorithm and its application to the analysis of muon tomography data. Muon tomography is a technique that exploits the secondary particles of cosmic radiation and their interaction properties with matter. Recently it has been successfully applied to the problems of scanning and detection of radioactive material. The *Muon Portal* project, presented as a case study in this report, is aimed at creating a large prototype of a scanner for containers control at the borders, in order to combat the illicit transport of fissile elements and face terrorist attacks. To achieve this goal, a large-scale scanner that is composed of four logical planes made of plastic scintillators strips is being realized. The idea of using clustering algorithms to process data in the *Muon Portal* project arises from the need to make the tracks reconstruction and the visualization of the container content, to be independent from the grid and the 3D-voxels[1]. The clustering algorithms, working with scattering points, could be useful to detect the presence of suspicious items inside the container, and could act as a filter for a preliminary analysis of the data. Several algorithms have been tested so far. In these proceedings, a modified Friends-of-Friends (FOF) algorithm is presented and results of simulated scenarios are also showed. The FOF algorithm is a domain-specific clustering algorithm; it is a percolation algorithm often used to identify dark matter halos in N-body simulations [2]. We implemented a multiphase unsupervised clustering version of the algorithm which performs the clustering in two different steps, using at the first step the scattering angle and the euclidean distance at the second step. The algorithm has been optimized by using space-partitioning data structures for organizing points in a k-dimensional space (kd-trees [3]) and further optimization strategies to achieve a $\mathcal{O}(N\log(N))$ complexity which makes it usable in real time analysis. The report is organized as follows: in Section 2 a literature review is presented. Section 3 is dedicated to the description of the FOF algorithm and its modification. Section 4 shows some simulation results applied in the context of the muon tomography imaging. Conclusions are drawn in Section 5.

## 2. Related Works

The clustering problem has been addressed in many contexts and by researchers in many disciplines; it has been effectively applied in a variety of engineering and scientific disciplines such as psychology, biology, medicine, computer vision, communications and remote sensing (e.g. see Refs. [4, 5, 6]). Cluster analysis organizes data (a set of patterns where each pattern could be represented by a vector) by abstracting the underlying structure, playing therefore a vital role in the field of data mining. The clustering method implemented should be very fast, efficient, and robust. In literature, there are several clustering algorithms described which can be roughly grouped into four categories: partitioning methods, hierarchical methods, grid-based and density-based clustering (see e.g. the review in Ref. [7]). They differ not only in their algorithm principles (which determine the behavior at runtime and the scalability of the algorithm itself), but also in many of their most basic properties such as the type of data processed, the assumptions on the cluster shape, the final form of the partitioning or the parameters that must be provided in the input. Partitioning clustering algorithms generate a single partition with a specified or estimated number of non-overlapping clusters of the data in an attempt to recover natural groups present in the data. Depending on the kind of prototypes, it is possible to distinguish k-means [8], k-modes [9] and k-medoid [10] algorithms. In the k-means algorithm, the similarity between clusters is measured with respect to the mean value of the objects belonging to a cluster. The k-modes extends the k-means paradigm to categorical domains. The k-medoid algorithms use a prototype, called the medoid, that is one of the objects located near the center of a cluster, the so-called centroid. The algorithm *Clarans* introduced in Ref. [11] is an improved k-medoid type algorithm restricting the large search space by using two additional user-supplied parameters. It is significantly more efficient than the well-known k-medoid algorithms *Pam* and *Clara* presented in Ref. [10], nonetheless producing a result of nearly the same quality. Hierarchical clustering algorithms construct a hierarchy of partitions,

[1] A voxel (volumetric pixels, or more precisely volumetric picture element) is a volume element, representing a value of signal intensity or color in a three dimensional space, similar to the pixel that represents a given two-dimensional image. The voxels are often used as a basis for the visualization and analysis of medical and scientific data.

represented as a dendrogram in which each partition is nested within the partition at the next level in the hierarchy [12]. A common way to find regions of high-density in the data space is based on grid cell densities [13]. A histogram is constructed by partitioning the data space into a number of non-overlapping regions or cells. Cells containing a relatively large number of objects are potential cluster centers and the boundaries between clusters fall in the "valleys" of the histogram. The success of this method depends on the size of the cells which must be specified by the user. The limit of this algorithm is that cells of small volume give a very "noisy" estimate of the density, whereas large cells tend to excessively smooth the density estimate. Density-based approaches apply a local cluster criterion and are very popular for the purpose of database mining. Clusters are regarded as regions in the data space in which the objects are dense, and which are separated by regions of low object density (noise). These regions may have an arbitrary shape and the points inside a region may be arbitrarily distributed.

In Ref. [14], a density-based clustering algorithm not grid-based, called *DbScan*, is presented. The basic idea of the *Dbscan* algorithm is that for each point of a cluster the neighborhood of a given radius, $\epsilon$, has to contain at least a minimum number of points, $N_{min}$, where $\epsilon$ and $N_{min}$ are input parameters. In Ref. [15] the density-based algorithm, named *DenClue*, is proposed. This algorithm uses a grid but is very efficient because it only keeps information about grid cells that do actually contain data points and manages these cells in a tree-based access structure. This algorithm generalizes some other clustering approaches which, however, result in a large number of input parameters. Another recent approach to clustering is the *Birch* method [16]. It cannot entirely be classified as a hierarchical or partitioning method since it constructs a CF-tree [17]. CF-tree is a hierarchical data structure designed for a multiphase clustering method. First, the database is scanned to build an initial in-memory CF-tree which can be seen as a multi-level compression of the data that tries to preserve the inherent clustering structure of the data. Then, an arbitrary clustering algorithm can be used to cluster the leaf nodes of the CF-tree. Because *Birch* is reasonably fast, it can be used as a more intelligent alternative to data sampling in order to improve the scalability of clustering algorithms.

## 3. Multivariate Data Analysis Techniques: a multi-phase unsupervised FOF algorithm

Clustering analysis provides a very sensible tool to detect "objects" within a spatial domain. It is a customary tool adopted e.g. in Astrophysics to quantify the hierarchical evolution of gravitationally bound systems like clusters and superclusters of galaxies. Clustering analysis has a significant advantage over other tools, such as classification algorithms [18, 19], statistical learning algorithms [20, 21] and association analysis algorithms [22, 23]. In fact, because of its sensitivity it can detect objects (clusters) composed of a small number of units.

The reason for this sensitivity is in its scaling with the number of composing units $N_c$. Therefore the algorithm has a computational complexity $\mathcal{O}(N_c^2)$. This arises because all clustering algorithms are based on pair counting and the number of pairs in a given set of $N_c$ objects is $N_c(N_c - 1)/2$. This scaling however is also computationally very expensive, in general. For this reason all the numerical implementations of the clustering algorithms have to be thoroughly tested to determine the minimum/optimal value(s) of $N_c$, as a result of a trade off between the statistical significance and the numerical complexity.

Common to all clustering algorithms is the comparison between the number of pairs in the sample with the number of cells in a random comparison sample, prepared under the controlled conditions. Two questions arise naturally:

- How much "random" should the control sample be?
- How large should $N_c$ be to guarantee a reasonable accuracy?

There is no universal answer to these two questions. Instead what one can often do is to test many samples carrying a given signal against the null hypothesis, and to use this statistics to determine the level of significance of a given problem. This is also what was done in the presented work.

Probably the most well known clustering algorithm is the *Friends-of-Friends* algorithm (FOF) [24, 2]. It depends on two parameters, the *linking length* $r_{ll}$ and the minimum number of points $N_{min}$. Its principle is very simple. Starting from an arbitrary initial element in the dataset it looks for all the elements contained within a distance $r_{ll}$. If $r_i$ and $r_j$ are two independent elements and their relative distance is $< r_{ll}$, then they are considered as members of a new group. The algorithm defines uniquely groups that contain all the particles separated by a distance smaller than the given linking length $r_{ll}$. Once the length is defined, the algorithm identifies all pairs of particles which have a mutual distance smaller than the linking one. These pairs are designated friends and clusters are defined as sets of particles that are connected by one or more of the friendly relations, so that they are friends of friends. Another parameter in FOF algorithm is the minimum number of particles $N_{min}$, in the input list. This is particularly important to reject spurious clusters, i.e. transient objects which have a low statistical significance. Since it is much more likely that a spurious cluster (noise) involves a small number of events then viceversa, choosing $N_{min}$ sufficiently large allows to eliminate these spurious clusters.

To summarize, the FOF requires: $r_{ll}$ and $N_{min}$ as input parameters and it is based on the following steps:

1. Choose an arbitrary unvisited data point as starting point.
2. Find the neighborhood of this point, e.g. all points within the radius $r_{ll}$. If more than $N_{min}$ neighborhoods are found around this point, a cluster is started and the point is marked as visited, otherwise the point is labelled as noise.
3. If a point is found to be a part of the cluster, then its $r_{ll}$ neighborhood is considered as a part of the cluster and the above procedure from step 2 is repeated for all $r_{ll}$ neighborhood points. This is repeated until all points in the cluster are determined.
4. A new unvisited point is retrieved and processed, leading to the definition of a new cluster or noise.
5. This process continues until all points are marked as visited.

The complexity of this algorithm grows at most as $N_c(N_c - 1) \sim \mathcal{O}(N_c^2)$. In practice, as elements are added to the list of clusters they are subtracted from the "active" list so that, at each step, the actual complexity grows as $N_{act}(N_{act} - 1)$, where $N_{act} \leq N_c$.

The only freedom left lies in the choice of $r_{ll}$. A common choice is to take it as the average of the distances between elements for a representative subset of data. This proves to be a very effective choice, as it has also been demonstrated that both the final list of objects and their average properties are largely independent from the choice of $r_{ll}$, provided that the input data catalogue is an unbiased, statistically representative sample. FOF is thus stable to its only parameter.

### *3.1. A modified multivariate unsupervised FOF algorithm*

An important property of many real datasets is that their intrinsic cluster structure can not be characterized by a single global density parameter. Very different local densities may be needed to reveal clusters in different regions of the data space. Furthermore, one of the main weaknesses of clustering algorithms is that most of them are supervised and require the inclusion of a set of input parameters, and the result is dependent on the input values. In order to perform an optimized exploratory data analysis of the tomographic data from the *Muon Portal* detector, we implemented a modified version of the FOF an unsupervised and multi-phase FOF clustering version. To make the algorithm independent from the settings of input parameters and in order to identify the density-based clustering structure, we implemented an augmented sorting technique of the dataset, based on the method called *OPTICS* [25]. In this method, the points in the dataset are ordered linearly according to the criteria of spatial distances. The points that are spatially close, are placed next to each other in the ordered database. An object in the dataset is defined a *core object* if it has at least $N_{min}$ points in its $\epsilon$ neighborhood. The algorithm define two metrics. The Core Distance is the minimum value of $\epsilon$ that would allow an object to be a core object ($\epsilon'$). The Reachability Distance between two objects, p and q, is the maximum between the Euclidean distance and the Core Distance $\max(d(p,q), \epsilon')$. For every object which has not been processed, the algorithm finds the points close to the $\epsilon$ neighborhood of the object, calculates its core distance $\epsilon'$ and save the object in the new database. During the data sorting operation, this distance is used as a criterion for two points to be accepted as a part of the same cluster. The algorithm proceeds ordering the points in ascending order to the nearest core object from which they can be reached. This process can be visualized by a dendrogram (see Fig. 1). It is a versatile base both for the automatic and interactive analysis of the cluster, useful to extract the intrinsic structure of clusters.

The grouping operation is performed in two different steps. First, the scattering angle $\theta_{scatt}$ is used as a criterion for grouping, performing an initial screening of the dataset and filtering out the noise from the data of interest. In this way, the scattering points are grouped according to the scattering angle. Then, the Euclidean distance is used in the second step, in order to distribute the clusters in the spatial domain and to distinguish different items inside the scanned volume.

The FOF requires the computation of the nearest neighbour of each point in the volume, so it has, in its original implementation, a computational complexity of $\mathcal{O}(N_c^2)$. The modified version of the algorithm has been optimized by using kd-trees to achieve a $\mathcal{O}(N_c \log(N_c))$ complexity which makes the algorithm usable for real time analysis.

## 4. Simulation results in the context of muon tomography imaging: the *Muon Portal* project

As mentioned in the *Introduction*, the *Muon Portal* project, used as a case study in these proceedings, has as its target the construction of a prototype of a large area detector for cargo containers inspection in ports in order to counteract the illicit carrying of fissile material. The originality and strength of the project lie in the use of the muon tomography technique to identify the presence of suspicious radioactive materials and to reconstruct the container content, discriminating the type of materials without the need to open it for inspection. The logical principle of the technique is the following: two detection planes are placed above the container that has to be inspected and two below it. The planes are made up of modules segmented in plastic scintillators strips, sensitive to muons, which emit light as they cross the strip. This light is then collected by Wavelength Shifting fibers, which re-emit the light in the green wavelength and which convey it towards Silicon Photomultiplier devices, which are able to read the light signal, then processed by the front-end electronics which provides as output the coordinates of the particle impact point on the plane. More details concerning the detector geometry, the electronic readout and the mechanical structure simulation study can be found in Refs. [26, 27, 28]. Using the coordinates, it is then possible to derive the trajectories of the incident ray and of the outgoing one, to determine the position and the amplitude of the scattering. This process is accomplished through the application of tracks reconstruction algorithms and 3D visualization techniques which allow to realize the tomography of the container. The track reconstruction is a challenging task and consists in the elaboration of data from the planes in order to obtain information on the deflection occurred by the muons within the scanned volume. Different statistical algorithms have been applied within the project for tomographic image reconstruction, as for example the simple POCA (*Point of Closest Approach*), the EM-

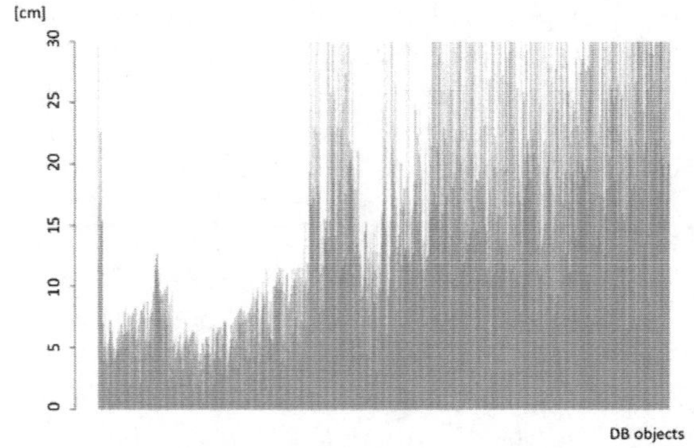


**Figure 1.** Reachability-plot for a simulation dataset with two blocks of high-Z material.

LM (*Expectation-Maximization Likelihood*), and the modified multivariate unsupervised FOF algorithm. They have been developed in C++ using links to GEANT4 [29] and ROOT [30] frameworks for detector geometry building, navigation and mathematical routines. See Ref. [31] for further details. In the following we show the results of the application of the modified FOF algorithm to a simulated dataset.

The simulated dataset refers to a scenario in which there is a container at the center of which blocks of different materials have been placed. They create a MUON shape built of voxels of size 10 cm×10 cm×10 cm. Each letter is made of a different material: M = Uranium, U = Iron, Lead = O, N = Aluminum. The container load related to this scenario is $\simeq$ 480 kg. In the first step the algorithm OPTICS [25] is run. Figure 1 shows the reachability plot relative to our dataset, through a dendrogram. It shows on the $x$-axis data of the set reordered according to a criterion of "distance" defined initially as a spatial distance. Each line represents a single point. The $y$-axis shows the distance values (according to the metric set) that give rise to the presence of clusters. The presence of dips in the plot indicates the existence of a cluster. The analysis made using OPTICS shows that the value of the distance, which is useful for the identification of two clusters, is in the range between 5 and10 cm. It can be seen also graphically, indeed if we draw a horizontal line and move it along the dendrogram, the height of the line that corresponds to the identification of two valleys, it is found around that range.

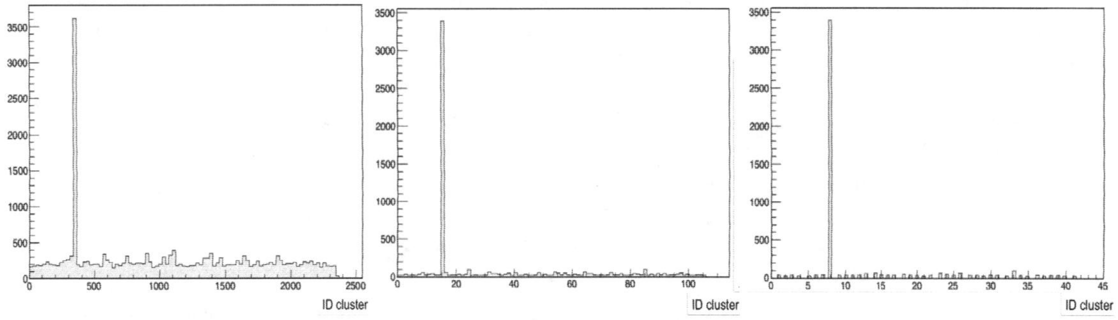


**Figure 2.** Results of the modified FOF algorithm applied on a dataset with high-Z material using a $r_{ll} = 5$ cm and $N_{min}$ equals to 5, 20, 30 points respectively.

The parameter $r_{ll}$ can be chosen in the identified range and then it can be passed as input parameter to the modified multiphase FOF algorithm. Figure 2 shows several histograms representing the results obtained from FOF algorithm by setting the parameter $r_{ll} = 5$ cm, and assigning to $N_{min}$ different values, 5, 20 and 30, respectively. On the $y$-axis we have the number of elements for each cluster, while in the $x$-axis we find the Id of the identified clusters. As it can be seen from the image, the parameter $N_{min}$, affects only the total number of clusters identified, but in all three cases, the presence of suspicious material is revealed and this can be seen from the high number of elements part of the same cluster. The algorithm calculates the mean value of the scattering angles for each cluster, giving consequently the alarm. In figure 3 is shown the word *MUON* simulated scenario with the results of the application of the clustering method. We can see a 3D visualization of the points filtered by the clustering and the volume rendering of the scenario. As it can be seen the method is successfully applied, and the two high-Z material letters (the M and the O) are identified.

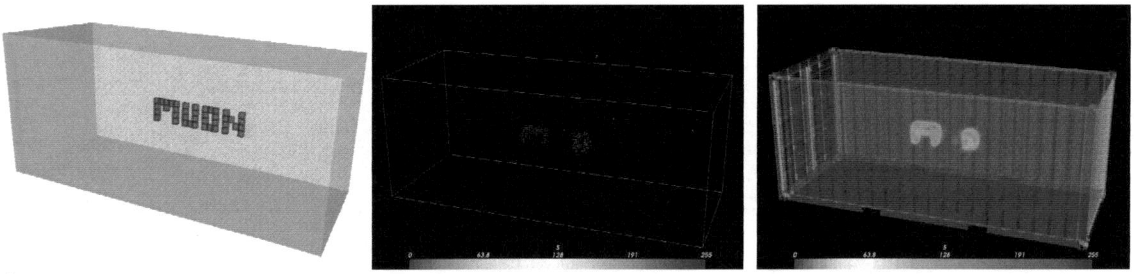


(a) (b) (c)

**Figure 3.** A simulated scenario of a container with a M= Uranium, U= Iron, O= Lead, N= Aluminum writing inside (a). A 3D visualization of the points filtered by the clustering (b). Volume rendering of the scenario realized by the clustering technique (c).

## 5. Conclusions

In these proceedings a study of the application of density based clustering algorithms to a muon tomography dataset in the context of the *Muon Portal* project has been shown. A modified version of the FOF algorithm, multi-phase and unsupervised, has been implemented and the simulation results were shown. The clustering method proposed can be used both as a stand-alone tool to analyze and characterize a large dataset, for example for data mining applications and data processing, or it is suitable to be used as a filter, in pre-processing phases for other algorithms that operate, subsequently, on the identified clusters. The implemented method has many advantages. Firstly, there is a linear scaling behavior, which makes it interesting for the application on big datasets. Secondly, the nature of the algorithm makes it quite robust to the effects of outliers within the data. Also, the proposed clustering method has the ability to work with all types of datasets that can be described by a metric and does not impose any assumptions about the shape of the clusters within which it works, being free from the limits of three-dimensional grids. Finally, an important aspect of the algorithm is its ability to automatically determine the number of clusters without a supervision. Statistical analysis of different scenarios are currently in progress, in order to verify the validity of the proposed method even in realistic and highly dense scenarios.

## References

[1] Tryon R 1939 *Cluster Analysis: Correlation Profile and Orthometric (factor) Analysis for the Isolation of Unities in Mind and Personality* (Edwards brother, Incorporated, lithoprinters and publishers)

[2] Lacey C G and Cole S 1994 *Mon.Not.Roy.Astron.Soc.* **271** 676 (*Preprint* astro-ph/9402069)
[3] Bentley J L 1975 *Communications of the ACM* **18** 509–517
[4] Sciacca E, Spinella S, Ienco D and Giannini P 2011 Annotated stochastic context free grammars for analysis and synthesis of proteins. *Evolutionary Computation, Machine Learning and Data Mining in Bioinformatics* (*Lecture Notes in Computer Science* vol 6623) ed Pizzuti C, Ritchie M and Giacobini M (Springer Berlin Heidelberg) pp 77–88 ISBN 978-3-642-20388-6
[5] Frigui H and Krishnapuram R 1999 *Pattern Analysis and Machine Intelligence, IEEE Transactions on* **21** 450–465 ISSN 0162-8828
[6] Bandyopadhyay S, Maulik U and Mukhopadhyay A 2007 *Geoscience and Remote Sensing, IEEE Transactions on* **45** 1506–1511 ISSN 0196-2892
[7] Jain A K, Murty M N and Flynn P J 1999 *ACM Comput. Surv.* **31** 264–323 ISSN 0360-0300
[8] MacQueen J B 1967 Some methods for classification and analysis of multivariate observations. *Proc. of the fifth Berkeley Symposium on Mathematical Statistics and Probability* vol 1 ed Cam L M L and Neyman J (University of California Press) pp 281–297
[9] Huang Z 1997 A fast clustering algorithm to cluster very large categorical data sets in data mining. *In Research Issues on Data Mining and Knowledge Discovery* pp 1–8
[10] Kaufman L and Rousseeuw P J 2008 *An Introduction to Cluster Analysis* (John Wiley Sons, Inc.) pp 320–331 ISBN 9780470316801
[11] Ng R T and Han J 1994 Efficient and effective clustering methods for spatial data mining. *Proceedings of the 20th International Conference on Very Large Data Bases* VLDB '94 (San Francisco, CA, USA: Morgan Kaufmann Publishers Inc.) pp 144–155 ISBN 1-55860-153-8
[12] Johnson S 1967 *Psychometrika* **32** 241–254 ISSN 0033-3123
[13] Jain A K and Dubes R C 1988 *Algorithms for Clustering Data* (Upper Saddle River, NJ, USA: Prentice-Hall, Inc.) ISBN 0-13-022278-X
[14] Ester M, Kriegel H P, Sander J and Xu X 1996 A density-based algorithm for discovering clusters in large spatial databases with noise. *Kdd* vol 96 (AAAI Press) pp 226–231
[15] Hinneburg A, Hinneburg E and Keim D A 1998 An efficient approach to clustering in large multimedia databases with noise. (AAAI Press) pp 58–65
[16] Zhang T, Ramakrishnan R and Livny M 1996 Birch: An efficient data clustering method for very large databases.
[17] Barbará D 2002 *SIGKDD Explor. Newsl.* **3** 23–27 ISSN 1931-0145
[18] Kukar M 2006 *Knowledge and information systems* **9** 364–384
[19] Li T, Zhu S and Ogihara M 2006 *Knowledge and information systems* **10** 453–472
[20] Fung G and Stoeckel J 2007 *Knowledge and Information Systems* **11** 243–258
[21] Tao D, Li X, Wu X, Hu W and Maybank S 2007 *Knowledge and Information Systems* **13** 1–42 ISSN 0219-1377
[22] Ahmed S, Coenen F and Leng P 2006 *Knowledge and information systems* **10** 315–331
[23] Bonchi F and Lucchese C 2006 *Knowledge and information systems* **9** 180–201
[24] Davis M, Efstathiou G, Frenk C S and White S D 1985 *Astrophys.J.* **292** 371–394
[25] Ankerst M, Breunig M M, Kriegel H P and Sander J 1999 Optics: Ordering points to identify the clustering structure. (ACM Press) pp 49–60
[26] Pugliatti C, Antonuccio V, Bandieramonte M, Becciani U, Belluomo F, Belluso M, Billotta S, Blancato A A, Bonanno D L, Bonanno G, Costa A, Fallica G, Garozzo S, Indelicato V, Rocca P L, Leonora E, Longhitano F, Longo S, Presti D L, Massimino P, Petta C, Pistagna C, Puglisi M, Randazzo N, Riggi F, Riggi S, Romeo G, Russo G V, Santagati G, Valvo G, Vitello F, Zaia A and Zappal G 2014 *Journal of Instrumentation* **9** C05029
[27] Antonuccio V, Bandieramonte M, Becciani U, Belluomo F, Belluso M, Billotta S, Blancato A, Bonanno D, Bonanno G, Carbone B *et al.* 2012 Design of a large area tomograph to search for high-z materials inside containers by cosmic muons. *Nuclear Science Symposium and Medical Imaging Conference (NSS/MIC), 2012 IEEE* (IEEE) pp 5–8
[28] Rocca P L, Antonuccio V, Bandieramonte M, Becciani U, Belluomo F, Belluso M, Billotta S, Blancato A A, Bonanno D, Bonanno G, Costa A, Fallica G, Garozzo S, Indelicato V, Leonora E, Longhitano F, Longo S, Presti D L, Massimino P, Petta C, Pistagna C, Pugliatti C, Puglisi M, Randazzo N, Riggi F, Riggi S, Romeo G, Russo G V, Santagati G, Valvo G, Vitello F, Zaia A and Zappala' G 2014 *Journal of Instrumentation* **9** C01056
[29] Geant4 C 2003 *Nuclear Instruments and Methods in Physics Research Section A: Accelerators, Spectrometers, Detectors and Associated Equipment* **506** 250 – 303 ISSN 0168-9002
[30] Brun R and Rademakers F 1997 *Nuclear Instruments and Methods in Physics Research Section A: Accelerators, Spectrometers, Detectors and Associated Equipment* **389** 81 – 86 ISSN 0168-9002 new Computing Techniques in Physics Research V

[31] Riggi S, Antonuccio-Delogu V, Bandieramonte M, Becciani U, Costa A, La Rocca P, Massimino P, Petta C, Pistagna C, Riggi F *et al.* 2013 *Nuclear Instruments and Methods in Physics Research Section A: Accelerators, Spectrometers, Detectors and Associated Equipment* **728** 59–68

# Developments in the ATLAS Tracking Software ahead of LHC Run 2

**Nicholas Styles[1], Massimiliano Bellomo[2], Andreas Salzburger[3], on behalf of the ATLAS collaboration**

[1]Deutsches Elektronen-Synchtron (DESY), [2]University of Massachusetts, [3]CERN

E-mail: nicholas.styles@desy.de

**Abstract.** After a hugely successful first run, the Large Hadron Collider (LHC) is currently in a shut-down period, during which essential maintenance and upgrades are being performed on the accelerator. The ATLAS experiment, one of the four large LHC experiments has also used this period for consolidation and further developments of the detector and of its software framework, ahead of the new challenges that will be brought by the increased centre-of-mass energy and instantaneous luminosity in the next run period. This is of particular relevance for the ATLAS Tracking software, responsible for reconstructing the trajectory of charged particles through the detector, which faces a steep increase in CPU consumption due to the additional combinatorics of the high-multiplicity environment. The steps taken to mitigate this increase and stay within the available computing resources while maintaining the excellent performance of the tracking software in terms of the information provided to the physics analyses will be presented. Particular focus will be given to changes to the Event Data Model, replacement of the maths library, and adoption of a new persistent output format. The resulting CPU profiling results will be discussed, as well as the performance of the algorithms for physics processes under the expected conditions for the next LHC run.

## 1. Introduction

The first data-taking run of the Large Hadron Collider (LHC), which ended in 2013, was an enormous success. It provided almost 30 $\text{fb}^{-1}$ of proton-proton collision data at centre-of-mass energies of $\sqrt{s}$ =7-8 TeV to the detectors. This data allowed a vast range of groundbreaking physics measurements to be made, including but not limited to the discovery of the Higgs Boson by the ATLAS and CMS experiments [1],[2]. It is planned that for the second run which will begin in 2015, the delivered luminosity will increase even further, with around 45 $\text{fb}^{-1}$ per year at $\sqrt{s}$=13 TeV anticipated to be delivered to the experiments. This will be coupled with a move from a bunch spacing of 50 ns to 25 ns. During Run 2, the detectors will see an increase in the average number of interactions per bunch crossing, $\langle\mu\rangle$, starting at around 40 and potentially increasing up to 80 by the end of the run, compared to an average of around 25 for 2012 data, as shown in Figure 1. These multiple interactions per bunch crossing are referred to as 'pile-up'.

The ATLAS experiment [4] has used the long shutdown between Run 1 and Run 2 to perform various detector upgrades and consolidation. The ATLAS Inner Detector (ID) [5], as the subsystem closest to the interaction point, responsible for reconstructing the trajectory of charged particles, is particularly affected by increases in pile-up. The ID comprises three detector technologies; the high-granularity silicon Pixel detector forms the innermost part of the

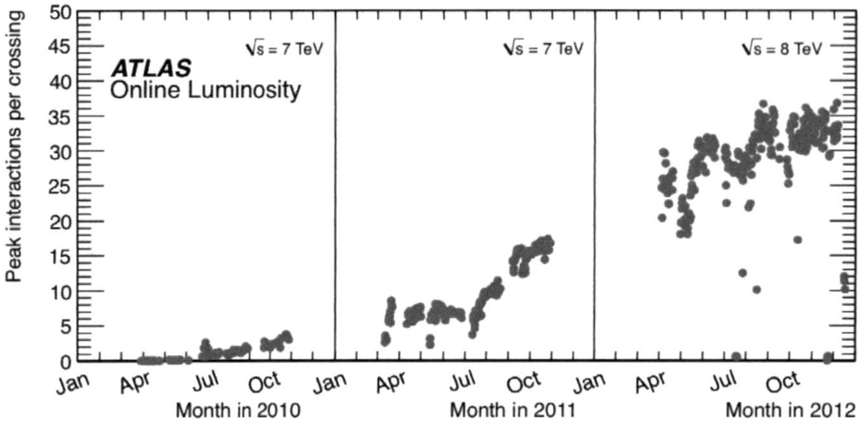


**Figure 1.** Evolution of the peak number interactions per bunch crossing observed by the ATLAS detector with time during the LHC Run 1 [3].

ID, followed by the silicon microstrip Semiconductor Tracker (SCT), and finally the Transition Radiation Tracker (TRT), based on drift tubes. During the long shutdown, a new innermost Pixel layer has been added to the detector, the Insertable B-Layer (IBL) [6], but this is far from the only activity going on during the shutdown period. Extensive work has also been undertaken to make significant improvements to the ATLAS event reconstruction software, part of the Athena [7] software framework. The ID track reconstruction algorithms [8] already take up a significant fraction of the total event reconstruction time, and the required CPU rises quickly as the multiplicity increases, exacerbating the situation further. Significant improvements to the ID tracking software were therefore essential to allow the goals for Run 2 data processing to be met; in particular, the requirement of 1 kHz first-pass event processing at the CERN Tier-0 computing centre [9].

## 2. Tracking Software Developments

In the following sections, the various developments made to the ATLAS ID Tracking software in order to help meet the 1 kHz Tier-0 processing target will be discussed.

### *2.1. Event Data Model Developments*

The Event Data Model (EDM) describes the various elements of the track data in the ATLAS software framework. It is principally based upon the concept of `TrackParameters` objects, which describe the helix parameters of a track in a given reference frame. The previous EDM implementation [10] was analysed, and among the observations was that a large number of `dynamic_cast` operations were being used, necessitated by large number of virtual functions used in the EDM, which taken together constituted a non-negligible contribution to the overall CPU usage in track reconstruction. The EDM was redesigned using templated classes, and the concept of `MeasuredParameters` which inherited from TrackParameters, was removed, with measurements being distinguished from other types of TrackParameters by the presence of an associated covariance matrix. Through these changes, in addition to reducing the number of `dynamic_cast` operations, the EDM code base was also massively reduced in size, making it much easier to maintain in the future.

These changes were all made within the context of the development of a new persistent output format for ATLAS reconstruction, which necessitated a re-design of the persistent output object of the track reconstruction, the `TrackParticle`. In redesigning this output object, priority was given to simplifying its inheritance structure and user interface. This was necessary in view of its future role as an object for use in data analyses, where previously there would have been further processing steps before reaching the objects used in analyses.

### *2.2. Maths Library Changes*

Track reconstruction makes extensive use of linear algebra, and so there are potentially large savings available if the CPU spent performing these calculations can be reduced. The CLHEP maths library [11] was previously used throughout the ATLAS software, but after investigating potential alternatives (including SMatrix [12]), the decision was made to move to Eigen [13], a vectorised C++ maths library, for linear algebra within the ATLAS reconstruction software. Eigen was chosen since it offered the largest performance improvements for ATLAS use cases of the options investigated.

Macros, with optimisations for fixed size and symmetric matrices (which make up a large fraction of the matrices used in track reconstruction) were also developed to be used within the ATLAS track reconstruction, further improving the speed and efficiency of linear algebra calculations.

### *2.3. Track Seeding Optimisations*

The first stage in reconstructing a track is the formation of track seeds from the various ID measurements. This was the focus of optimisation ahead of Run 2, leading to a number of changes in the strategy applied. Table 1 shows the percentage of seeds that result in a 'good' (i.e. satisfying various track quality criteria) track for various 3-measurement seed types (Pixel only, Pixel and SCT, and SCT only) at pile-up values of $\langle\mu\rangle = 0$ and 40. The presence of the IBL for run 2 allows more stringent requirements for seeds to be applied, which helps to mitigate the effect of increasing pile-up, and in the case of the best-performing seed type (3 SCT plus 1 extra measurement) the performance is actually improved with respect to the previous configuration at $\langle\mu\rangle = 0$.

**Table 1.** Percentage of seeds resulting in a good track being reconstructed, for the Run 1 (top row) and Run 2 (bottom row) configurations under different pile-up conditions. $P$ represents a Pixel measurement being used in the seed, $S$ an SCT measurement, and $I$ an additional 4th measurement of either type.

| $\langle\mu\rangle$ | PPP | PPS | PSS | SSS |
|---|---|---|---|---|
| 0 | 57% | 26% | 29% | 66% |
| 40 | 17% | 6% | 5% | 35% |
| | PPP+I | PPS+I | PSS+I | SSS+I |
| 0 | 79% | 53% | 52% | 86% |
| 40 | 39% | 8% | 16% | 70% |

During Run 1, the search region for seeds was defined by the maximum/minimum longitudinal impact parameter, while for Run 2 the region will be defined by the maximum/minimum vertex position in the longitudinal direction. This change was accompanied by tightening of track quality requirements.

A further development was made for the so-called 'TRT seeded tracks', which are reconstructed using measurement points not used during the initial seeding from Pixel and SCT measurements. TRT seeded track finding will only be processed for Run 2 in the case that a compatible seed cluster matching the track is found in the Electromagnetic calorimeter.

Together, these changes have resulted in a significant speed-up in the software, with no apparent loss of tracks useful for physics analyses.

### *2.4. Results of Developments*

In implementing the changes described above (in addition to others not mentioned here,such as improved access to the magnetic field information), a huge number of software packages needed to be carefully modified, and in some cases substantially re-written. The result of this large effort by many people can be seen in Figure 2, which shows the average CPU time per event for reconstructing simulated $\sqrt{s} = 14$ TeV $t\bar{t}$ events with $\langle\mu\rangle = 40$, for different releases of the ATLAS reconstruction software.

Between releases 17.2.7.9 and 19.0.3.3, the changes to the EDM and Maths libraries were applied, in addition to other changes including moving from 32-bit to 64-bit, and applying compiler optimisations. Between 19.0.3.3 and 19.1.1.1, the optimisations of the track seeding strategy were applied (along with other changes). A significant, more than two-fold reduction in the overall event reconstruction time, and a corresponding reduction in the Inner Detector reconstruction time can be observed when going from 17.2.7.9 to 19.1.1.1, and the $\sim 20$ s average event reconstruction time is compatible with a 1 kHz Tier 0 event processing rate.

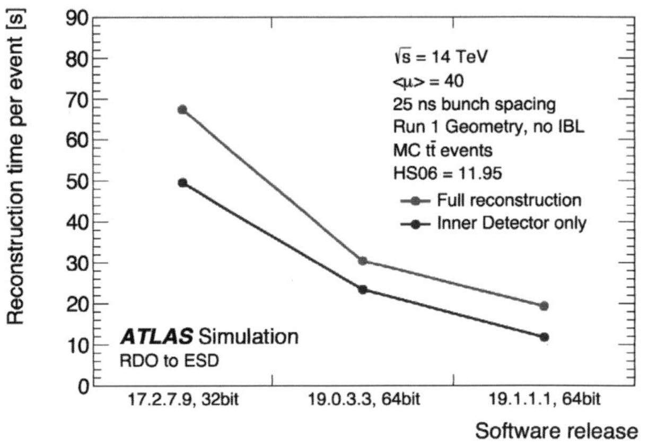


**Figure 2.** Average CPU time per event for reconstructing simulated $\sqrt{s} = 14$ TeV $t\bar{t}$ events with $\langle\mu\rangle = 40$ in various ATLAS software releases [14].

### *2.5. Developments required beyond Run 2*

The High Luminosity LHC (HL-LHC) planned for the next decade, is foreseen to provide even higher integrated luminosities (a projected total dataset of 3000 fb$^{-1}$), which implies further increases in pile-up, with anticipated $\langle\mu\rangle$ of 140 or beyond.

A completely new tracking detector will replace the ID during the so-called 'Phase 2' upgrade [15]. In order to reconstruct events within whatever computing budget is available at the time, it is likely that dramatic changes will be required to the track reconstruction, to mitigate the 'combinatorial explosion' under such a high-multiplicity environment.

The current approach to track reconstruction is highly sequential, and so not ideally suited to make use of the increased parallelism offered by Many-Core/GPU technologies, and so this may imply that a fundamentally different approach is necessary in the future to make best use of available resources. Increased use of Vectorisation/SIMD approaches (i.e. within certain parts of the ATLAS code base itself) may also be a possible avenue of exploration.

## 3. Conclusions

An extensive program of software development has been undertaken within the ATLAS reconstruction framework during the LHC's long shutdown between Run 1 and Run 2, necessitated by the anticipated increase in pile-up for Run 2 LHC collisions, and the corresponding increases in CPU time for event reconstruction, in particular for the Inner Detector track reconstruction.

A variety of developments were made to the track reconstruction software, including moving to a redesigned Event Data Model, replacement of the Maths library for linear algebra with a vectorised library (Eigen), and optimisation of the track seeding strategy.

Together with changes elsewhere in the reconstruction software, this has resulted in a reduction by a factor $> 2$ of the CPU time required for event reconstruction. This will allow first-pass event reconstruction to run at the CERN Tier 0 site at a rate of 1 kHz.

## References

[1] ATLAS Collaboration, G. Aad et al., *Observation of a new particle in the search for the Standard Model Higgs boson with the ATLAS detector at the LHC*, Phys.Lett. **B716** (2012) 1–29, `arXiv:1207.7214 [hep-ex]`.

[2] CMS Collaboration, S. Chatrchyan et al., *Observation of a new boson at a mass of 125 GeV with the CMS experiment at the LHC*, Phys.Lett. **B716** (2012) 30–61, `arXiv:1207.7235 [hep-ex]`.

[3] "ATLAS Experiment Luminosity - Public Results." https://twiki.cern.ch/twiki/bin/view/AtlasPublic/LuminosityPublicResults.

[4] ATLAS Collaboration, G. Aad et al., *The ATLAS Experiment at the CERN Large Hadron Collider*, JINST **3** (2008) S08003.

[5] ATLAS Collaboration, G. Aad et al., *The ATLAS Inner Detector commissioning and calibration*, Eur.Phys.J. **C70** (2010) 787–821, `arXiv:1004.5293 [physics.ins-det]`.

[6] ATLAS Collaboration, G. Aad et al., *ATLAS Insertable B-Layer Technical Design Report*, Tech. Rep. CERN-LHCC-2010-013. ATLAS-TDR-19, CERN, Geneva, Sep, 2010. https://cds.cern.ch/record/1291633.

[7] ATLAS Collaboration, G. Duckeck et al., *ATLAS Computing: technical design report*, Tech. Rep. CERN-LHCC-2005-022. ATLAS TDR-017, CERN, Geneva, 2005. https://cds.cern.ch/record/837738.

[8] T. Cornelissen, M. Elsing, I. Gavrilenko, W. Liebig, E. Moyse, and others on behalf of the ATLAS Inner Detector Software Group, *The new ATLAS track reconstruction (NEWT)*, J.Phys.Conf.Ser. **119** (2008) 032014.

[9] M. Elsing, L. Goossens, A. Nairz, and G. Negri, *The ATLAS Tier-0: Overview and operational experience*, J.Phys.Conf.Ser. **219** (2010) 072011.

[10] T. Cornelissen, N. Van Eldik, M. Elsing, W. Liebig, E. Moyse, et al., *Updates of the ATLAS Tracking Event Data Model (Release 13)*, ATL-SOFT-PUB-2007-003 (2007).

[11] "CLHEP - a class library for high energy physics." http://proj-clhep.web.cern.ch/proj-clhep/.

[12] "ROOT Mathematical Libraries - SMatrix Package." http://project-mathlibs.web.cern.ch/project-mathlibs/sw/html/SMatrix.html.

[13] "Eigen - a C++ template library for linear algebra: matrices, vectors, numerical solvers, and related algorithms." http://eigen.tuxfamily.org.

[14] "ATLAS Experiment Computing and Software - Public Results." https://twiki.cern.ch/twiki/bin/view/AtlasPublic/ComputingandSoftwarePublicResults.

[15] ATLAS Collaboration, G. Aad et al., *Letter of Intent for the Phase-II Upgrade of the ATLAS Experiment*, Tech. Rep. CERN-LHCC-2012-022. LHCC-I-023, CERN, Geneva, Dec, 2012. https://cds.cern.ch/record/1502664.

ACAT2014 IOP Publishing
Journal of Physics: Conference Series **608** (2015) 012048 doi:10.1088/1742-6596/608/1/012048

# An automated framework for hierarchical reconstruction of $B$ mesons at the Belle II experiment

**C Pulvermacher, T Keck, M Feindt, M Heck and T Kuhr**

Institut für Experimentelle Kernphysik, Karlsruhe Institute of Technology (KIT), Wolfgang-Gaede-Str. 1, 76131 Karlsruhe, DE

E-mail: `christian.pulvermacher@kit.edu`

**Abstract.** We present a software framework for Belle II that reconstructs $B$ mesons in many decay modes with minimal user intervention. It does so by reconstructing particles in user-supplied decay channels, and then in turn using these reconstructed particles in higher-level decays. This hierarchical reconstruction allows one to cover a relatively high fraction of all $B$ decays by specifying a limited number of particle decays. Multivariate classification methods are used to achieve a high signal-to-background ratio in each individual channel. The entire reconstruction, including the application of pre-cuts and classifier trainings, is automated to a high degree and will allow users to retrain to account for analysis-specific signal-side selections.

## 1. Introduction

Belle II is an experiment being built at the $e^+e^-$ SuperKEKB $B$ factory in Tsukuba, Japan, and is planned to have a luminosity 40 times higher than its predecessor Belle [1]. Like other $B$ factories the collider will operate mainly on the energy of the $\Upsilon(4S)$ resonance, which practically always decays into pairs of $B$ mesons. Since these events consist of only $B^+B^-$ or $B^0\bar{B}^0$ pairs and their decay products—with only minor contamination from beam or electronics background—they enable the use of reconstruction techniques that would not be available at a hadron collider.

Specifically, reconstructing one $B$ meson (called $B_{\text{tag}}$) allows one to infer information about the other $B$, including its four-momentum, without explicitly reconstructing it. Combined with a signal selection, the additional information about $B_{\text{sig}}$ allows to improve the reconstruction, e.g. in channels with neutrinos. Additionally, one can also make use of the fact that in a correctly reconstructed event every track must come from one of the $B$ mesons, so requiring the absence of additional particles improves the purity of the selection. This is equivalent to reconstructing the $\Upsilon(4S)$, but of course requires a high efficiency in the $B_{\text{tag}}$ reconstruction to be of use.

## 2. Hierarchical reconstruction

To this end, a neural-network-assisted reconstruction of $B$ mesons in many decay channels was pioneered at the Belle experiment which reconstructs decays of particles into their immediate daughters and combines the output until reaching the $B$ meson level [2]. This principle is illustrated in Figure 1. To reduce the computational burden imposed by the combinatorics of the reconstruction, soft cuts based on the network output remove background at each stage without overly compromising efficiency. The network output can be interpreted as a probability

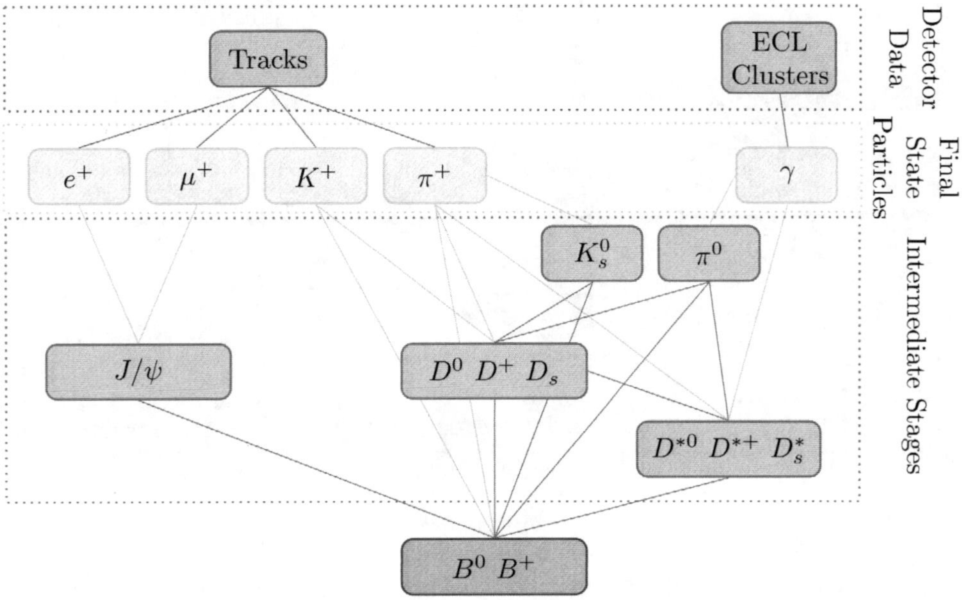


Figure 1: Illustration of the hierarchical reconstruction from final state particles to $B$ mesons.

for the candidate to be correctly reconstructed, which improves its usefulness as an input for subsequent networks even though candidates can originate from a number of different decay channels.

Compared to an explicit reconstruction of each exclusive decay channel, this hierarchical approach greatly reduces the number of multivariate classifiers needed and consequently increases the statistics in each of these channels. For the Full Reconstruction method used at Belle, about 1000 exclusive decay channels are handled by $\mathcal{O}(100)$ artificial neural networks. The efficiency of this method is about 0.18 % for $B^0$ mesons and 0.28 % for $B^{\pm}$ mesons, i. e. for the final Belle $\Upsilon(4S)$ data sample containing $771.6 \times 10^6$ $B\bar{B}$ pairs 3.5 million tag-side $B$ mesons are correctly reconstructed.

For Belle II, we want to further increase the efficiency, by being able to easily add additional decay channels and by optimising cuts and multivariate methods. To gain the most benefit from this, the process should be automatic to a significant degree, so that after a user adds or modifies a decay channel, the training of classifiers and determination of applied cuts should proceed without further intervention. This also extends to evaluation of the training and the resulting performance, where the necessary plots and statistics should be produced automatically.

Another advantage of this highly automatic training process is that it becomes possible for users running physics analyses to retrain the tag-side reconstruction for their own analysis. Normally, if the software is trained on a generic $B$-decay Monte Carlo and reused for a certain signal-side selection, changes in the prior distribution may decrease classifier performance and introduce biases; i. e. the trained methods assume combinatorial background from generic decays on the signal side, but receive selected data with much higher purity. These differences should be especially notable for signal channels with only few tracks, e.g. $B \to \tau\nu$. Retraining for a specific signal selection helps avoid these problems.

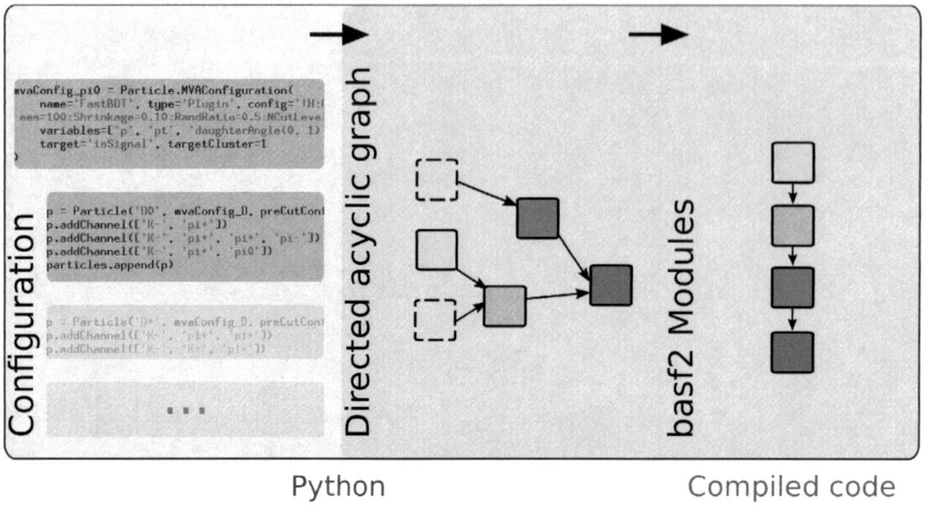


Figure 2: Architecture of the reconstruction framework's implementation, illustrating the configuration, dependency resolution between resulting reconstruction tasks, and the final data-intensive processing handled by basf2 modules.

### 3. Implementation in the Belle II software

The Belle II analysis software framework (`basf2`) consists of independent modules mostly written in C++, which can be combined by the user using Python-based steering files [3]. For physics analyses, all common data-intensive tasks are performed by such modules and are provided via a common high-level syntax. The following example demonstrates the usage of these tools in reconstructing $\pi^0 \to \gamma\gamma$ decays.

```
selectParticle('gamma')

reconstructDecay('pi0 -> gamma gamma', '0.11 < M < 0.15')
matchMCTruth('pi0')
```

This selects final-state particles with a photon hypothesis, saves all possible combinations of two photons satisfying the given cut in a list of particle objects, and relates those objects to their Monte Carlo counterparts.

To fully use these existing analysis tools, the hierarchical $B$ meson reconstruction framework itself is implemented in Python. The main interface for the user is the configuration of particles and channels to reconstruct, as well as of the multivariate classifiers. This high-level description is converted into separate reconstruction tasks, such as combining particles according to a specific decay channel, vertex fitting, or the training/application of a multivariate classifier. Each of these tasks is represented by a functor object whose inputs define dependencies that need to be fulfilled by the output of other functors, or (user-supplied) configuration items. Inputs and outputs thus produce a directed acyclic graph, which naturally defines an ordering in which functors should be executed to fulfill their dependencies. Most functors do not directly process data, but only add modules to the execution path using the aforementioned reconstruction tools. After the dependency resolution, the modules do data-intensive processing in the correct order. This overall scheme is illustrated in Figure 2.

For some tasks, outputs cannot be provided immediately. For example, the multivariate classifiers provide a probability for a given candidate to be correctly reconstructed, but need to be trained first. This is accomplished by checking for the existence of the training file and to only provide the output if it is detected. If the file is not found, the training is done instead and execution of dependent tasks is deferred. All file names and lists of particles include SHA1 hashes of the inputs used, so they will change once their configuration changes. This in turn affects all subsequent functors, so the changes are propagated through the entire dependency graph.

## 4. Technical details

### *4.1. Automatic cut determination*

To reduce the combinatorial background in the reconstruction, cuts need to be applied in between stages, before saving candidates. Depending on the decay channel, we might cut on the invariant mass of a candidate, on the product of classifier output probabilities of the candidate's daughters, or any other variable. Since the optimal cut often depends on the particular decay channel, requiring users to manually configure these cuts would discourage retraining or the addition of channels. We address this by creating histograms of the signal and background distributions of the candidates in the cut variable and use them to automatically determine cuts that satisfy common requirements on the purity and efficiency. Figure 3 shows pre-cuts determined for the $D^{*+} \to D^0\pi^+$ channel as an example. Channels in which the selection does not accept enough signal events to warrant the training of a multivariate classifier are excluded from further processing.

As a result, including potentially less clean channels poses no disadvantage to the process, since they will either be assigned a cut that offers a useful compromise between the number of signal events and necessary CPU time, or the channels will be discarded entirely.

### *4.2. TMVA interface*

Multivariate classification to discard incorrectly reconstructed particles is performed using TMVA, a multivariate analysis package for ROOT containing a number of different classification methods, pre-processing and evaluation tools [4]. To make full use of the package inside basf2 two interface modules, `TMVATeacher` and `TMVAExpert`, are provided, which are meant to train or run classifiers on a list of particles. Users can specify input variables, the classification target, and a list of methods that should be trained using module parameters, where variables are functions that return floating-point values for a candidate particle, e.g. the invariant mass of a particle or the $\chi^2$ of a vertex fit. Some variables also take additional arguments to, e.g. return the angle between the candidate's momentum and that of its $i^{\text{th}}$ daughter, or to compute the product of a given variable for all daughter particles. All variables are written in C++ and made available to all analysis modules through a string identifier.

### *4.3. Automatic reporting*

To control the quality of the final $B_{\text{tag}}$ reconstruction, plots are generated automatically for each decay channel and its associated network. This includes the pre-cut histograms shown in Figure 3, plots provided by TMVA for the classifiers (receiver operating characteristic curves, overtraining checks, etc.), the ranking of input variables, and CPU usage statistics differentiated by channel and reconstruction task. A table of the efficiencies and purities for each particle is also produced and allows inspection of the performance of the applied cuts.

## 5. Summary

Belle II, as a $B$ factory, allows combining $B$ mesons selected as signal with a more generically reconstructed $B_{\text{tag}}$ to the original $\Upsilon(4S)$ resonance to improve many analyses. As the usefulness of this technique depends strongly on the total efficiency of the $B_{\text{tag}}$ reconstruction, our

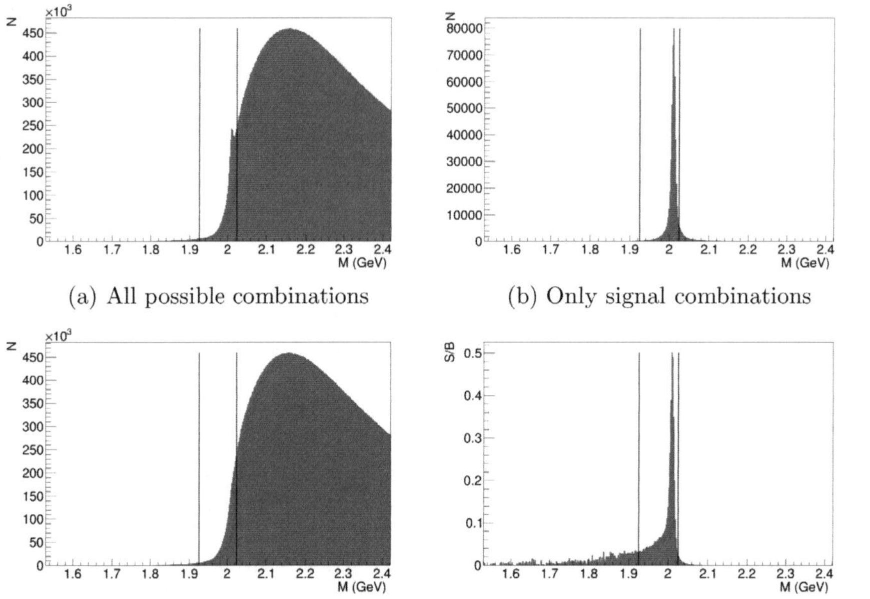


(a) All possible combinations

(b) Only signal combinations

(c) Only background combinations

(d) Ratio $S/B$ of signal and background

Figure 3: Plots illustrating the combinatorial background in the $D^{*+} \to D^0\pi^+$ channel and its distribution in the $D^{*+}$ invariant mass $M$. The lower signal-to-background ratio $S/B$ for higher masses automatically results in asymmetric pre-cuts (vertical lines).

framework makes it easy to add additional channels or optimize parts of the reconstruction. This is achieved by automating many tasks, including the determination of intermediate cuts, the decision of which classifiers need to be retrained, and the generation of evaluation plots and tables. The reconstruction framework itself is written in Python, the language used for basf2 steering files, and is built on top of high-level tools that can also be used by physics analyses.

The framework can be configured by defining particles and the channels in which to reconstruct them, plus related attributes like the configuration of multivariate methods or cuts. It is generic enough to be reused for similar tasks like $B_{\rm tag}$ reconstruction with semileptonic channels, or reconstructing $B_s$ mesons on $\Upsilon(5S)$ data samples.

## References

[1] Abe T *et al.* (Belle II Collaboration) 2010 Belle II Technical Design Report Tech. rep. (*Preprint* `1011.0352`)
[2] Feindt M, Keller F, Kreps M, Kuhr T, Neubauer S *et al.* 2011 *Nucl.Instrum.Meth.* **A654** 432–440 (*Preprint* `1102.3876`)
[3] Moll A 2011 *Journal of Physics: Conference Series* **331** 032024
[4] Hoecker A, Speckmayer P, Stelzer J, Therhaag J, von Toerne E and Voss H 2007 *PoS* **ACAT** 040 (*Preprint* `physics/0703039`)

# HistFitter - A flexible framework for statistical data analysis

**J M Lorenz[1,2], M Baak[3], G J Besjes[4,5], D Côté[6], A Koutsman[7] and D Short[8]**

[1] Fakultät füer Physik, LMU München, Am Coulombwall 1, 85748 Garching, Germany
[2] Excellence Cluster Universe, Boltzmannstr. 2, 85748 Garching, Germany
[3] CERN, Route de Meyrin 385, 1217 Meyrin, Switzerland
[4] IMAPP, Radboud University Nijmegen, P.O. Box 9010, 6500 GL Nijmegen, The Netherlands
[5] Nihkef, Science Park 105, 1098 XG Amsterdam, The Netherlands
[6] The University of Texas at Arlington, Department of Physics, Box 19059, Arlington, Texas 76019
[7] TRIUMF, 4004 Wesbrook Mall, Vancouver, BC V6T 2A3, Canada
[8] Department of Physics, Denys Wilkinson Building, University of Oxford, Keble Road, Oxford, OX1 3RH, Great Britain

E-mail: Jeanette.Lorenz@physik.uni-muenchen.de

**Abstract.** We present a software framework for statistical data analysis, called *HistFitter*, that has extensively been used in the ATLAS Collaboration to analyze data of proton-proton collisions produced by the Large Hadron Collider at CERN. Most notably, HistFitter has become a de-facto standard in searches for supersymmetric particles since 2012, with some usage for Exotic and Higgs boson physics. HistFitter coherently combines several statistics tools in a programmable and flexible framework that is capable of bookkeeping hundreds of data models under study using thousands of generated input histograms. The key innovations of HistFitter are to weave the concepts of control, validation and signal regions into its very fabric, and to treat them with rigorous statistical methods, while providing multiple tools to visualize and interpret the results through a simple configuration interface.

## 1. Introduction

HistFitter [1] is a software framework for statistical data analysis extensively used in the ATLAS Collaboration [2] in (mainly) searches for supersymmetric particles, analyzing data of proton-proton collisions produced by the Large Hadron Collider at CERN. HistFitter consists of a C++ part for CPU-intensive calculations and of a Python part for configuration purposes. The tool is built on top of the packages ROOT [6, 7], RooFit [5], HistFactory [3] and RooStats [4], where RooFit and HistFactory are used to build parametric models, RooFit to fit those models and RooStats to perform statistical tests.

HistFitter extends the functionalities of RooFit, HistFactory and RooStats in four key areas:

- HistFitter offers a programmable framework to perform complete statistical analyses starting from a user-defined configuration file.
- Typical analysis strategies in particle physics using control, validation and signal regions are deeply woven into the design of HistFitter.

- HistFitter keeps track of numerous data models including the construction and statistical tests in an organized way.
- HistFitter provides a collection of tools for interpreting results of statistical analyses such as determining the statistical significance of signal hypotheses, estimating the quality of likelihood fits and for producing plots and tables presenting the results.

## 2. Data analysis strategy with HistFitter

Particle physics experiments analyze large samples of data to measure properties of fundamental particles or to discover new physical processes. The data is usually interpreted using external predications for background and signal processes which are summarized in statistical models aiming to describe the observed data. HistFitter configures and builds such parametric models and provides various tools to help in the interpretation of the data.

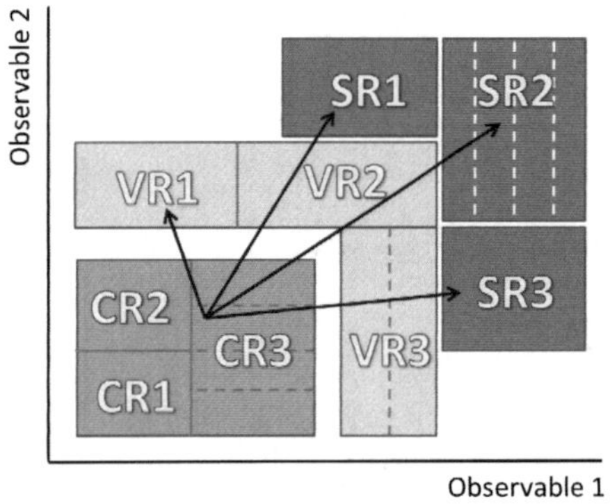


**Figure 1.** Schematic analysis strategy with control (CR1 - CR3), validation (VR1 - VR3) and signal regions (SR1 - SR3).

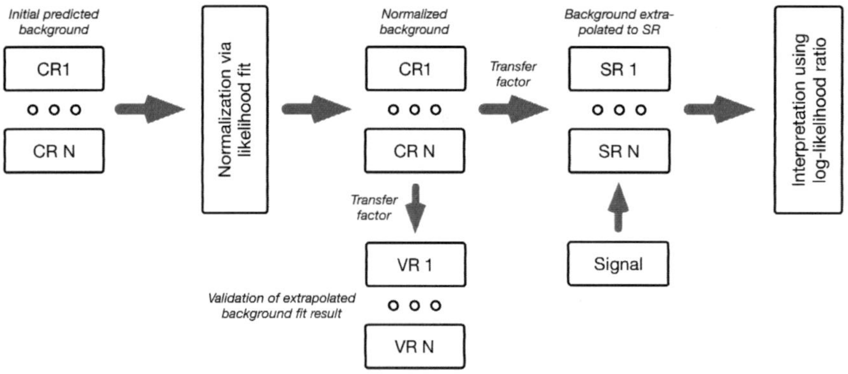


**Figure 2.** A typical analysis strategy flow with HistFitter involving the control regions CR 1 - CR N, validation regions VR 1 - VR N and signal regions SR 1 - SR N.

In the construction and handling of the models HistFitter deeply relies on the concept of statistically independent control, validation and signal regions as illustrated in Figures 1 and 2. Signal regions are regions in a phase space, defined by cutting on kinematic variables, that are enriched in potential signal in comparison to the predicted background level. To estimate the background in signal regions in a semi-data-driven way, control regions enriched in background are defined. The predicted background is normalized to data in the control regions using a likelihood fit to data. The normalized background model is extrapolated to a signal region using a transfer factor, which is the ratio of expected event counts between each control and signal region. This extrapolation is verified in validation regions located between control and signal regions.

The parametric model describing the data is represented by a Probability Density Function (PDF). PDFs are built for every control, validation and signal region using HistFactory [3]. The regions being statistically independent, the PDFs can be fitted simultaneously to data, adjusting the parameters of the PDFs. An important point in the analysis strategy of HistFitter is the possibility to share parameters of the PDFs in different regions. This allows the consistent use of information on signal and background processes and systematic uncertainties in all regions.

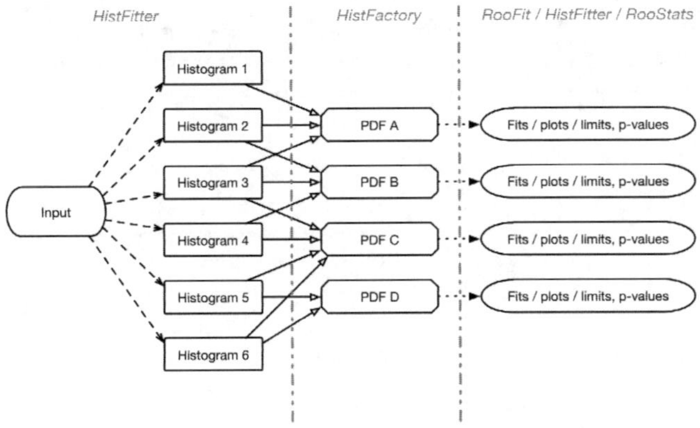


**Figure 3.** The HistFitter software framework.

The HistFitter software framework, illustrated in Figure 3, is designed based on the analysis strategy in Figure 2. Starting from a user-defined configuration and input data, binned histograms are created by HistFitter in a first processing step. In a second step, PDFs are constructed based on the histograms using HistFactory tools. In subsequent steps, the model is analyzed by performing fits, calculating limits and presenting the results in plots and tables. These steps use tools both from RooFit and RooStats, but also HistFitter specific tools as discussed in Section 4 and 5.

## 3. Configuration and construction of PDFs

The analysis flow presented requires substantial bookkeeping and an extended configuration machinery, in particular if working with hundreds of different models e.g. for different hypotheses for new physics beyond the Standard Model.

The configuration consists of a user-defined configuration file which interacts with a configuration manager within HistFitter. The configuration manager is realized as two singleton objects, one in Python with which the user interacts, the other one in C++. It organizes and creates `fitConfig` objects. A `fitConfig` object contains the PDF of a certain model along with meta-data providing information about fitting, visualization and interpretation of the model. The `fitConfig` object thus represents one row in Figure 3 and acts as factory of a model. The configuration manager thus functions as 'factory of factories'.

PDFs are constructed using HistFactory. The resulting likelihood has the general form of

$$L(\boldsymbol{n}, \boldsymbol{\theta}^0 | \mu_{\text{sig}}, \boldsymbol{b}, \boldsymbol{\theta}) = P_{\text{SR}} \times P_{\text{CR}} \times C_{\text{syst}}, \tag{1}$$

being the product of Poisson distributions of event counts in signal and control regions ($P_{\text{SR}}$ and $P_{\text{CR}}$ ) and of additional constraint terms for systematic uncertainties ($C_{\text{syst}}$). The likelihood depends on number of observed events in all regions ($\boldsymbol{n}$), nuisance parameters parameterizing the impact of systematic uncertainties ($\boldsymbol{\theta}$) with their central values $\boldsymbol{\theta}^0$, signal strength $\mu_{\mathrm{sig}}$ and predictions $\boldsymbol{b}$ for various background sources.

The likelihood has three different building blocks: `channels`, `samples` and `systematics`. `Channels` include all control, validation and signal regions. Signal and background processes compose the `samples` and `systematic` uncertainties summarize statistical, experimental and theoretical uncertainties.

HistFitter extends and mirrors the classes in HistFactory for these building blocks as illustrated in the following.

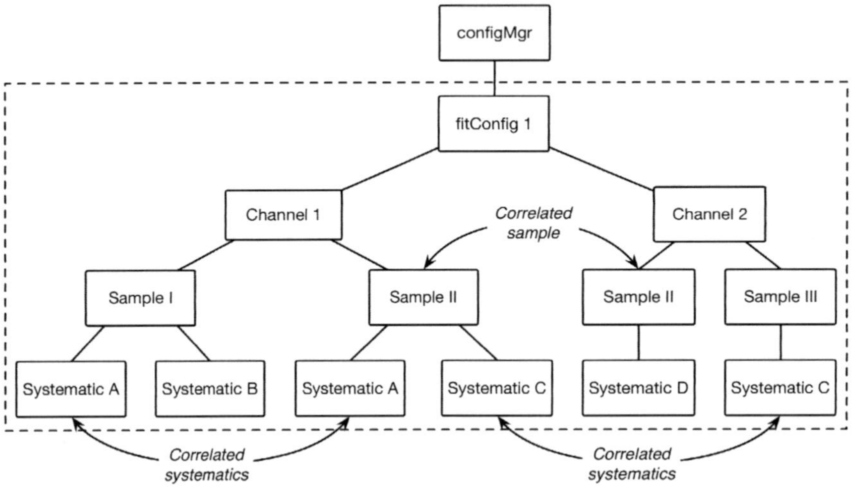


**Figure 4.** Illustration of an example fit strategy in HistFitter.

The three building blocks are put together by a `fitConfig` object in the construction of each PDF, as illustrated in Figure 4, and links them to the input data. A `fitConfig` object can have multiple `channels` that may be one- or multi-bin and are either control, validation or signal regions. `Samples`, corresponding to components of the PDF decorated with meta-data, are attached to a `channel`, and may also be correlated between multiple `channels`. The data input for the sample may be provided as ROOT `TTree` or `TH1` or also as float numbers by the user. `Systematics` attached to a `sample` and possibly correlated between `samples` are typically provided by $1\sigma$ up and down histograms of the nominal histogram of the sample(s). `Systematics` can be implemented using different methods of interpolation between up and down histograms and extrapolations. HistFitter allows the description of a complicated PDF by few lines of code through the 'trickle-down mechanism': this mechanism takes care of adding `samples` that were added to a `fitConfig` object also to all depending `channels`. Similarly, all `systematics` added to a `fitConfig` or `channel` are also propagated to the depending `channels` and `samples`.

The trickle-down mechanism together with the concepts of the `fitConfig` class and the configuration manager eases the construction of complex analyses setups considerably.

## 4. Fit strategies and presentation of fit results

HistFitter provides different fit strategies to fit the PDF as illustrated in Table 1. The *background-only* fit using only control regions and background samples aims for estimating the

Journal of Physics: Conference Series **608** (2015) 012049 doi:10.1088/1742-6596/608/1/012049

background yields in validation and signal regions by extrapolating from the control regions. The *model-dependent signal fit* is used to derive exclusion limits on a specific model or to measure the properties of an excess. This fit strategy uses background and signal samples in control and signal regions. The simultaneous fitting of multiple signal regions is possible and often used to increase the exclusion power ('shape fit'). The *model-independent signal fit* is used to derive model-independent limits on the number of events beyond the expected number of events in a specific non-binned signal region (other signal regions are not allowed in this fit). No assumption is made on the signal which may only appear in the signal region but not in the control regions.

**Table 1.** Summary of the different fit strategies possible with HistFitter.

| **Fit setup** | *Background-only fit* | *Model-dependent signal fit* | *Model-independent signal fit* |
|---|---|---|---|
| **Samples used** | backgrounds | backgrounds + signal | backgrounds + dummy signal |
| **Fit regions** | CR(s) | CR(s) + SR(s) | CR(s) + SR |

The results of the fits can be presented in e.g. tables or plots indicating the before and after fit event yields in the regions or in pull plots illustrating the agreement of the fitted background estimates with the observed data. Examples are given in Figures 5 and 6.

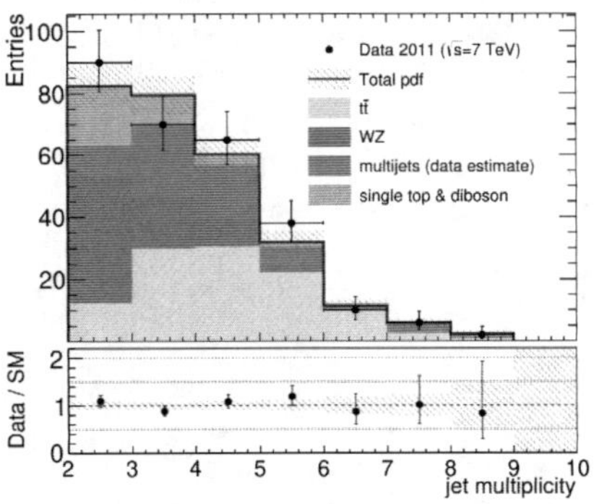


**Figure 5.** The jet multiplicity distribution after fitting it to data.

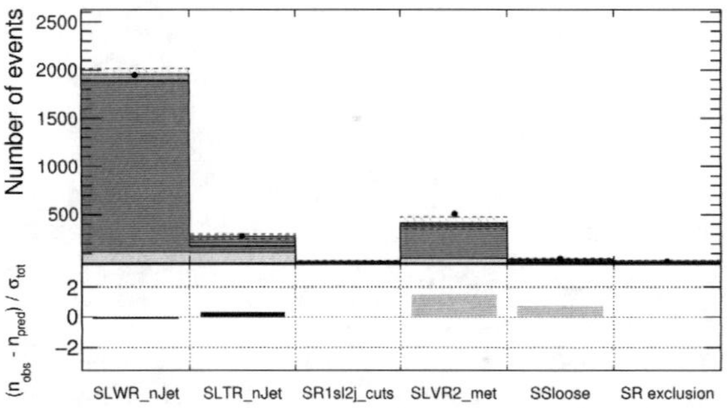


**Figure 6.** The agreement between estimated background yields and observed data (top) can be illustrated in a pull plot (bottom).

## 5. Interpretation of results

The observed data can be interpreted in hypothesis tests in HistFitter through calls to the appropriate functions and classes of RooStats [4]. HistFitter also provides macros for plots and tables to display the results.

The interpretations are based on either the model-dependent or model-independent signal fit. Using e.g. the model-dependent fit strategy, limits on a certain signal model can be placed by either deriving exclusion limits on this specific model in comparison to a background assumption in a hypothesis test or by evaluating an upper limit on the allowed signal strength given the data. Figure 7 shows an example giving the exclusion limits as function of parameters important for a specific model class.

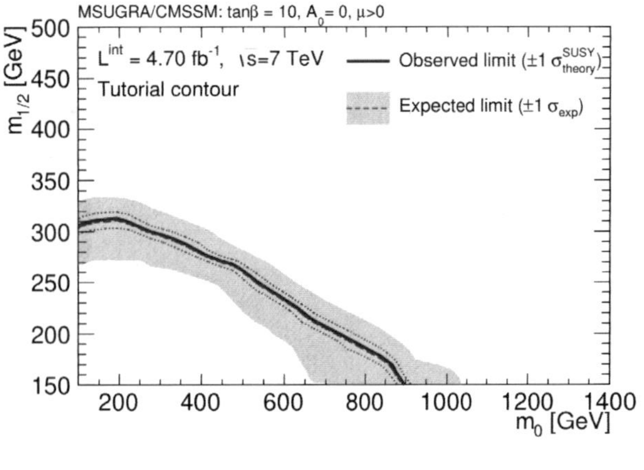


**Figure 7.** Exclusion limits in a class of supersymmetric models depending on the parameters on the $x$- and $y$-axes. Models below the red observed limit are excluded by hypothesis tests comparing a background-only with background+signal models.

## 6. Summary

We have presented a software framework for statistical analyses, called HistFitter. This programmable framework is used to build, fit and test data models of nearly arbitrary complexity. Starting from a user-defined configuration file and using HistFactory and RooFit functionalities PDFs are configured, constructed and fit. Interpretations of the models are possible through statistical tests using RooStats functionalities.

Among the innovative features of HistFitter is the modular configuration interface with a trickle-down mechanism which eases the construction of complicated PDFs. HistFitter is designed to provide the bookkeeping to work with hundreds of different signal models at once and thus provides an additional level of abstraction over existing tools. Built-in concepts of control, validation and signal regions allow a particular rigorous statistical treatment of extrapolations from control to signal regions as used in numerous background estimation techniques in particle physics. In addition, HistFitter offers a sizable collection of tools for presenting final results in a publication-style quality.

## References

[1] Baak M, Besjes G J, Côté D, Koutsman A, Lorenz J and Short D 2014 HistFitter software framework for statistical data analysis *Preprint* arXiv:1410.1280 (`http://cern.ch/histfitter`)
[2] ATLAS Collaboration 2008 it JINST **3** S08003 (`http://atlas.web.cern.ch/Atlas/Collaboration`)
[3] ROOT Collaboration, Cranmer K, Lewis G, Moneta L, Shibata A and Verkerke W 2012 *Preprint* CERN-OPEN-2012-016
[4] Moneta L et al. 2010 *PoS* **ACAT2010** 057 *Preprint* arXiv:1009.1003
[5] Verkerke W and Kirkby D 2003 *eConf* **C0303241** (*Preprint* arXiv:physics/0306116)
[6] Brun R and Rademakers F 1997 *Nucl.Instrum.Meth.* **A389** 81
[7] Antcheva I et al. 2011 *Comput.Phys.Commun.* **182** 1384

# New exclusive CHIPS-TPT algorithms for simulation of neutron-nuclear reactions

**M Kosov and D Savin**
All-Russia Research Institute of Automatics, Moscow, Russia
E-mail: Kosov@vniia.ru, sd57@inbox.ru

**Abstract.**
The CHIPS-TPT physics library for simulation of neutron-nuclear reactions on the new exclusive level is being developed in CFAR VNIIA. The exclusive modeling conserves energy, momentum and quantum numbers in each neutron-nuclear interaction. The CHIPS-TPT algorithms are based on the exclusive CHIPS library, which is compatible with Geant4. Special CHIPS-TPT physics lists in the Geant4 format are provided. The calculation time for an exclusive CHIPS-TPT simulation is comparable to the time of the corresponding Geant4-HP simulation. In addition to the reduction of the deposited energy fluctuations, which is a consequence of the energy conservation, the CHIPS-TPT libraries provide a possibility of simulation of the secondary particles correlation, e.g. secondary gammas, and of the Doppler broadening of gamma lines in the spectrum, which can be measured by germanium detectors.

## 1. Introduction
CHIPS-TPT is the successor of CHIPS model formerly included in Geant4 package. Starting from the Geant-4.10 release, CHIPS is no longer supported by the Geant collaboration. The main goals of the CHIPS-TPT project are further maintenance of the universal CHIPS algorithms and the development of exclusive TPT (Toolkit for Particle Transport) algorithms of nuclear reactions at low energies for neutrons and light charged nuclear fragments. The exclusive TPT model can be a replacement for the inclusive NeutronHP model [1] currently used in the Geant4 framework for neutrons below 20 MeV. Exclusive modeling also reproduces kinematic effects such as secondary particle correlations and Doppler broadening of spectral lines and allows to avoid the fluctuations in the simulated response due to energy non-conservation in an inclusive approach. Therefore it is essential for the development of detectors measuring correlation spectra applied to identification of nuclear isotopes.

The main features of CHIPS-TPT are:

- Charge and baryon number conservation.
- Energy and momentum conservation in each nuclear interaction.
- Full isotope coverage.
- All neutron-nuclear reactions below 20 MeV.

## 2. Reactions
The following reactions have been implemented:

- Elastic neutron scattering.
  It is easy to comply conservation laws since elastic scattering is a binary reaction. However, at incident neutron energies near the material temperature scattering is sensitive to the thermal motion of atoms leading to effective cross section

$$\sigma_{eff} = \sigma_0 \cdot \left[ \left(1 + \frac{T}{mV^2}\right) \cdot erf\left(\sqrt{\frac{mV^2}{2T}}\right) + \sqrt{\frac{2T}{\pi m V^2}} e^{-\frac{mV^2}{2T}} \right]$$

  where $V$ is the incident neutron velocity, $m$ is the mass of the target nucleus, $T$ is the temperature, $\sigma_0$ - the cross section at zero temperature. The $\sigma_0$ value is a constant at incident energies below the first resonance, so $|\bar{v} + \bar{V}|\sigma_0$ can be averaged over Maxwell distribution to obtain the reaction rate $R = < |\bar{v} + \bar{V}|\sigma_0 > = V\sigma_{eff}$. The same result can be found in [2].
- Inelastic neutron scattering.
  Inelastic neutron scattering is divided into scattering with excitation of the recoil nucleus and its following step-by-step de-excitation cascade. De-excitation of moving nuclei leads to Doppler broadening.
- Radiative capture.
  Radiative capture mainly takes place at thermal energies, so the compound nucleus is at rest, but the first gamma radiation accelerates the compound nucleus and broadens the following gamma-lines of the cascade, whilst the first gamma line remains narrow.
- Production of light nuclear fragments.
  In reactions with multiple secondaries the compound nucleus is formed and then fragmented emitting nuclear fragments according to the phase space distribution. The inclusive double-differential spectra available in ENDF-6 databases [3] are not taken into account yet, instead the phase space distributions are used for all nuclear fragments except for the leading fragment (mainly a neutron), which is generated in accordance with the spectrum provided by the database.
- Neutron induced fission.
  Fission is implemented by forming the compound nucleus, emitting prompt neutrons and dividing the residual nucleus into two fragments, conserving charge and baryon number in the reaction.

## 3. Results

We ran a few simulations for testing and verification. Either a simple geometry or scattering on a thin target were used.

Figure 1 shows the energy-momentum conservation up to millielectronvolts for the most complicated CHIPS-TPT process of neutron-induced fission, while an inclusive fission simulation does not conserve neither energy and momentum nor charge and baryon number.

Figure 2 shows the simulated neutron fluence induced by a 14 MeV neutron source located in the center of a water ball of 50 cm radius. The benchmark is similar to the one used for comparison of several other Monte-Carlo neutron codes [4]. Fluence was computed as the sum of the lengths of neutron tracks divided by the integration volume. A good agreement can be seen between CHIPS-TPT and NeutronHP results. The low energy maximum is due to thermal motion of the material atoms.

Figure 3 shows an example of correlated gamma spectra obtained by the scattering of 14 MeV neutrons on nitrogen and oxygen. The peaks correspond to pairs of photons emitted in a single nuclear de-excitation cascade. Similar distribution can be provided by Geant4 in a form of narrow spikes without Doppler broadening if the excited nucleus is stopped in matter and then decayed in gamma and the residual nucleus.

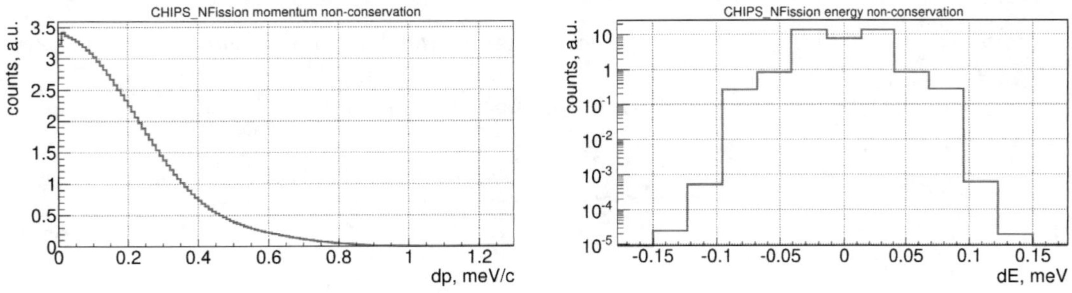


(a) (b)

Figure 1: Energy and momentum conservation. Horizontal axis shows the difference of the momentum (left) or energy (right) between initial and final particles in each event.

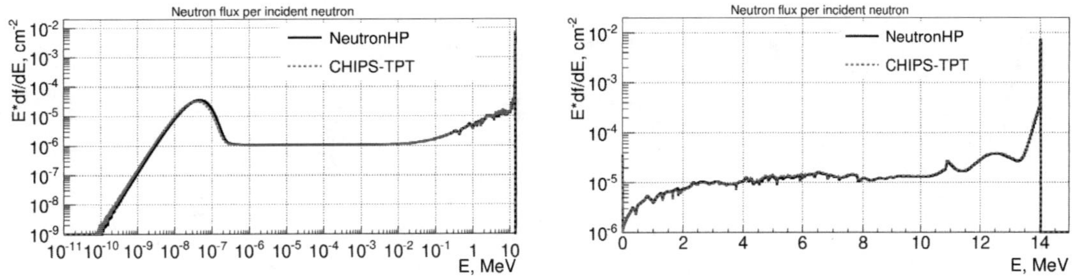


(a) (b)

Figure 2: Neutron fluence in water. Horizontal axes show neutron energy, vertical axes show fluence multiplied by the energy. Left panel shows the horizontal axis in the logarithmic scale, while the right panel shows the horizontal axis in the linear scale.

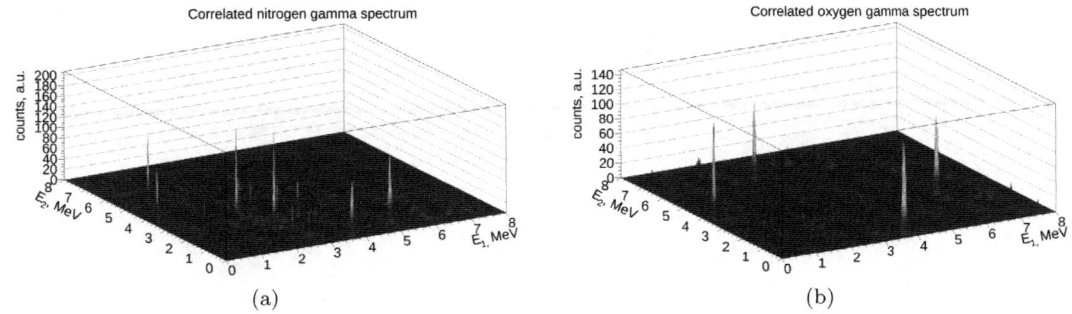


(a) (b)

Figure 3: Energy spectra of pairs of photons emitted in a single nuclear de-excitation cascade. Left panel shows spectra for nitrogen, right panel shows spectra for oxygen.

## 4. Summary and Outlook

The exclusive CHIPS-TPT code reproduces the results of simulations with inclusive Monte-Carlo codes, being capable to simulate quantities obtainable with exclusive codes only. CHIPS-TPT also strictly complies the energy and momentum conservation requirements inherited from

CHIPS. The plans for future development are:

- Improvement of multiple secondary particle production by comparison with experimental data.
- Development of fragmentation models for inelastic interactions of soft charged nuclear fragments.
- Calculation speed-up by caching cross section for each material used in a simulation run.

Updated versions of CHIPS will be publicly available online [5]. For the full CHIPS-TPT library contact the authors via e-mail.

## References

[1] Wellisch J P 2005 *The Monte Carlo Method: Versatility Unbounded in a Dynamic Computing World CD-ROM* (American Nuclear Society)

[2] Larson N M, Moxon M C, Leal L C and Derrien H 1998 Doppler broadening revisited Tech. Rep. ORNL/TM-13525 Oak Ridge National Laboratory

[3] Chadwick M B *et al.* 2011 *Nuclear Data Sheets* **112**(12) 2887–2996

[4] Cullen D E *et al.* 2006 How accurately can we calculate neutrons slowing down in water? Tech. Rep. UCRL-TR-220605 Lawrence Livermore National Laboratory

[5] `http://vniia.ru/cfpi/eng/wptpt.html`

Journal of Physics: Conference Series **608** (2015) 012051 doi:10.1088/1742-6596/608/1/012051

# HERAFitter - an open source QCD fit framework

**Andrey Sapronov**
**on behalf of the `HERAFitter` team**

Joint Institute for Nuclear Research, 141980 Russia, Dubna, Moscow Region, Joliot-Curie 6.

E-mail: `Andrey.Sapronov@cern.ch`

**Abstract.** The open source project **`HERAFitter`** was established to increase applicability of the QCD analysis in the hadron collider experiments. The framework may be used to extract parton density functions from a variety of experimental measurements and to assess the impact of different data on the parton density determination. It may also be employed to perform data consistency checks and to test theoretical models.
This short article covers a parallel session contribution of the ACAT2014 conference.

## 1. Introduction

In the era of hadron colliders the proton parton density functions (PDF) have been gaining increasing attention. While the parton interactions within a hadron cannot be duly described using perturbative approximation, the factorisation theorem helps isolating the hard scattering cross section components which can be determined empirically. Improving precision of PDFs is important to reduce the uncertainty of theoretical predictions for both Standard Model (SM) and new physics processes. The **`HERAFitter`** [1] is an open source project created to facilitate QCD analysis of experimental data and to improve the current knowledge of PDF.

The PDF extraction procedure in QCD analysis implies fitting the experimental data with theoretical predictions based on parametrised partonic distributions at a starting scale which are then evolved to the scale of the measurements using DGLAP [2, 3, 4, 5, 6] evolution equations. The treatment of the uncertainties and accounting for possible correlations of the precision data is of high importance in order to improve precision of theoretical predictions for high energy processes.

Several groups are currently involved in the PDF fitting analysis: MSTW [7], CT(EQ) [8], NNPDF [9], AB(K)M [10], GJR [11] and HERAPDF [12]. Their methods and distinguishing features are reviewed in [13, 14]. In short, all groups perform global fits of the deep inelastic scattering (DIS), fixed target and $pp(\bar{p})$ collider data, except HERAPDF which relies purely on the HERA data and hence justified usage of the $\chi^2$ tolerance of $\Delta\chi^2 = 1$. The NNPDF approach is notable for their analysis being performed via neural network training in association with a Monte Carlo method to estimate their uncertainties and minimise the input parametric form dependence on PDFs. There are other various differences among PDF groups: higher orders counting, heavy flavour corrections, treatment of $\alpha_S$, etc.

The interest of achieving a better level of agreement among these PDF groups, as well as an overall better understanding of the proton's PDF is driven by the recent results from the LHC.

## 2. Overview of the HERAFitter program

HERAFitter is an open source QCD fitting framework aimed to provide means of experimental data QCD analysis and extraction of PDFs. The program is designed with a user friendly but highly configurable interface and includes a large number of theoretical models. It allows extensive study of the impact of experimental data. A detailed information about the project status and available releases with documentation can be found in http://herafitter.org.

The analysis is based on the $\chi^2$ estimate of how well the theoretical prediction describe experimental data for a given PDF parametrisation. Various ansatz on PDF parametric form can be tested within this framework. The framework is designed such that a user can plug in his own sets of data with corresponding theory prediction, adjust input PDF parametrisation form, select a specific desired scheme for matching the heavy quark thresholds, a calculation order, cuts, etc. The QCD fit output will provide a quantitative evaluation of level of agreement between data and theory, and if desired, the resulting PDF grids in LHAPDF format can be obtained. The package contains elaborate plotting tools for prompt inspection of the results.

The overall analysis approach is schematically shown in the figure 1. The procedure starts with reading the input parameters and definition of initial parametrisation at the starting scale $Q_0^2$. Initial scale is chosen to be below the charm mass threshold, as evolution is performed starting from a three active quarks as required by the QCDNUM [15] program, which performs the QCD evolution. The minimisation of the $\chi^2$ is performed via the MINUIT [16] package.

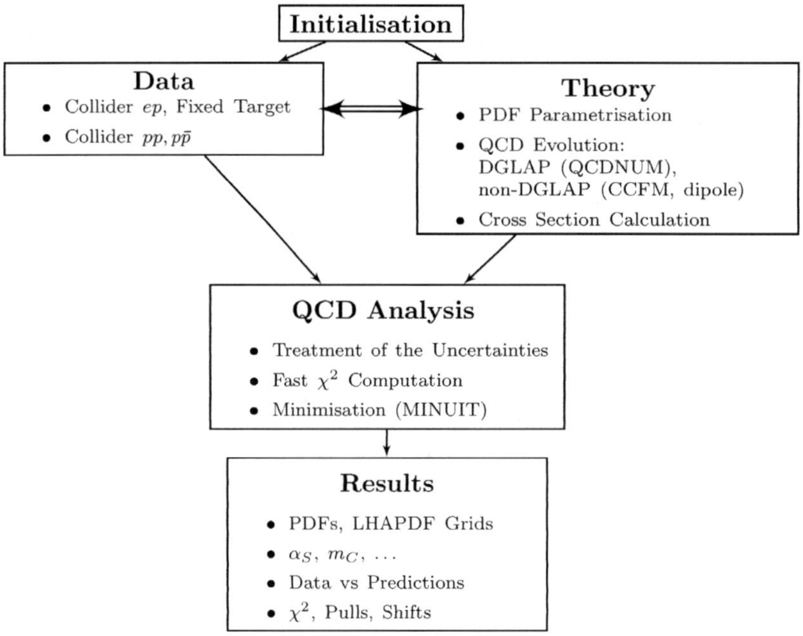


**Figure 1.** Schematic structure of the HERAFitter program.

The experimental data are introduced as a text table which format is defined in a header needed to be specified by the user in the same file. The header includes e.g. reaction name, data table format and theory parameters. The HERAFitter program features elaborate correlated data uncertainty treatment which can be presented in various forms: as a covariance matrix, as a table with nuisance parameters or a combination of both. Depending on the choice, corresponding $\chi^2$ definition will be used for minimisation procedure. For a single data set the $\chi^2$ function is defined in a form

$$\chi^2 = \sum_i \frac{\left[m^i - \sum_j \gamma^i_j m^i b_j - \mu^i\right]^2}{\left(\delta_{i,\mathrm{stat}} m^i\right)^2 + \left(\delta_{i,\mathrm{uncor}}\, m^i\right)^2} + \sum_j b_j^2, \tag{1}$$

Here $\mu^i$ is the measured central value at a point $i$ with relative statistical $\delta_{i,stat}$ and relative uncorrelated systematic uncertainty $\delta_{i,unc}$, the quantity $m^i$ is the theoretical prediction, $\gamma^i_j$ represents relative correlated systematic uncertainties and $b_j$ their shifts. In the case of the covariance matrix, the $\chi^2$ function takes the form

$$\chi^2 = \sum_{ij} (m^i - \mu^i) C^{-1}_{\mathrm{tot}\ ij} (m^j - \mu^j), \tag{2}$$

where the $C^{-1}_{\mathrm{tot}\ ij}$ is the total covariance matrix given by the sum of the statistical and systematic covariance matrices. Due to flexible implementation in the framework, different approaches can be combined together, e.g. some dataset's systematic uncertainties can be treated using the matrix method, others can be treated using nuisance parameters.

Various types of uncertainties can be handled using following implemented methods: hessian [17], Monte Carlo [18, 19] or offset [20]. The hessian method is a standard way of computing the eigenvectors corresponding to the uncertainty sources by diagonalizing a matrix of second derivatives of the $\chi^2$ distribution. In Monte Carlo method the PDF uncertainties are estimated using pseudo-data replicas randomly generated from the measurement's central values and their systematic and statistical uncertainties. The offset method consists in constraining the PDFs in a flexible parametrisation style.

The current version of the `HERAFitter` framework allows fitting the following types of data: inclusive cross section measurements from HERA DIS and fixed target experiments and Drell–Yan, jet and single-top quark production data from *ep*, *pp* and $p\bar{p}$ colliders.

The DIS structure functions may be computed directly in a variety of heavy quark schemes including the fixed-flavour (FFN) and variable flavour number (VFN) schemes. VFN schemes with various treatments for the heavy quark thresholds include the Thorne Roberts (TR) scheme at LO, NLO and NNLO [21, 22] as provided by the MSTW group, the ACOT scheme at LO and NLO as provided by the CTEQ group. The QCDNUM also provides the calculations of the DIS structure functions in the zero-mass VFN and FFN schemes. The FFN scheme is alternatively available via the OPENQCDRAD [23] interface with running mass definition [24].

Theoretical predictions for other processes are provided via the interfaces to fast cross section estimation tools, such as APPLGrid [25] or FastNLO [26]. Evaluation of the differential (or total) cross section is performed by convoluting the parametrised PDFs with pregenerated tables of higher order coefficients (NLO or NNLO). Further corrections may be applied via K-factor tables, for example NNLO QCD computed with FEWZ [27] or NLO EW obtained from SANC [28, 29]. Drell–Yan, jet production in *pp*, $p\bar{p}$ collisions are the processes with fast cross section estimates available. The HATHOR [30] program interfaced in the `HERAFitter` can be used to calculate complete NNLO total cross section of $t\bar{t}$ production in *pp*, $p\bar{p}$ collisions while differential NLO cross section for this process is available from MCFM [31, 32] via APPLGrid or FastNLO.

## 3. Application in QCD analysis of experimental data

The `HERAFitter` program has been used in a number of experimental and theoretical analyses. This list includes several LHC analyses of SM processes, namely inclusive Drell-Yan and $W$and $Z$ production [33, 34, 35, 36, 37], inclusive jet production [38], and inclusive photon production [39]. One of the notable results is the strange quark density measurement from 2010 data on single gauge boson production performed by the ATLAS experiment [33]. The analysis had the

HERA-I *ep* data as a basis with additional 30 data points of the ATLAS Drell–Yan measurement fitted with QCD NNLO predictions. The partial $\chi^2$ of the ATLAS data shows significant improvement from 44.5 when the $\bar{s}$ density was tied with the $\bar{d}$-quark to 33.9 when it was released. Figure 2 shows example plots for gluon density measured with ATLAS inclusive jets data and $r_s = \bar{s}/\bar{d}$ value measured from muon charge asymmetry at the CMS experiment. The results of QCD analyses using `HERAFitter` were also published by HERA experiments for inclusive [12, 40] and heavy flavour production measurements [41, 42]. The following phenomenological studies have been performed with `HERAFitter`: a determination of the transverse momentum dependent gluon density using precision HERA data [43], an analysis of HERA data within a dipole model [44], the study of the low-x uncertainties in PDFs determined from the HERA data using different parametrisations [45] and the impact of QED radiative corrections on PDFs [46]. A recent study based on a set of PDFs determined with the `HERAFitter` and addressing the correlated uncertainties between different orders has been published in [47].

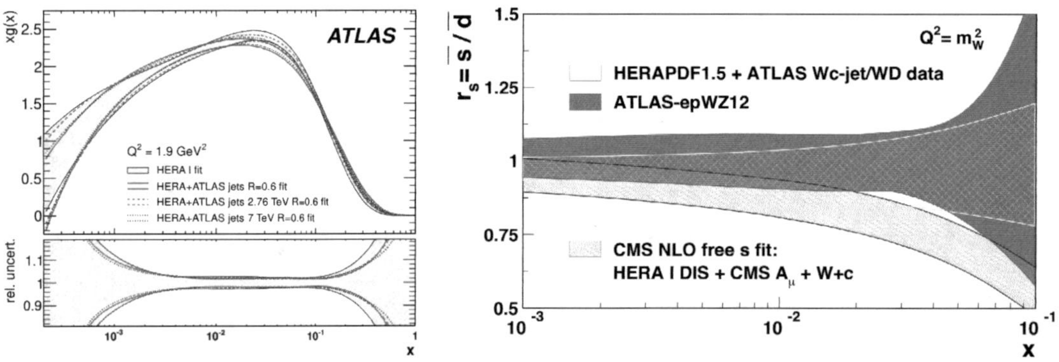


**Figure 2.** LHC analysis results, obtained with `HERAFitter` program: (*left*) gluon momentum distribution, measured in the ATLAS inclusive jets analysis for $\sqrt{s} = 2.76$ TeV and $\sqrt{s} = 7$ TeV data [38] and (*right*) $r_s = \bar{s}/\bar{d}$ value measurement by the CMS experiment [34].

The `HERAFitter` framework has been used to produce PDF grids from QCD analyses performed at HERA [12, 48] and at the LHC [49], using measurements from ATLAS [33, 38]. These PDFs can be used to calculate predictions for SM or beyond SM processes. Furthermore, `HERAFitter` provides the possibility to perform various benchmarking exercises [50] and impact studies for possible future colliders as demonstrated by QCD studies at the LHeC [51].

## 4. Summary

The `HERAFitter` project provides diverse means for QCD interpretation of the experimental measurements at hadron colliders. The source code of the package is freely available under GPL v3 license. The framework design allows robust extraction of proton PDFs and assessment of the data impact on the current PDF determination. The package contains necessary ingredients to study the PDFs incorporating a variety of different data processes and theory calculations and is an optimal platform for QCD related benchmarks.

## References

[1] HERAFitter https://www.herafitter.org
[2] Gribov V N and Lipatov L N 1972 *Sov. J. Nucl. Phys.* **15** 438
[3] Gribov V N and Lipatov L N 1972 *Sov. J. Nucl. Phys.* **15** 675
[4] Lipatov L N 1975 *Sov. J. Nucl. Phys.* **20** 94
[5] Dokshitzer Y L 1977 *Sov. Phys. JETP* **46** 641

[6] Altarelli G and Parisi G 1977 *Nucl. Phys. B* **126** 298
[7] Martin A, Stirling W, Thorne R and Watt G 2009 *Eur.Phys.J.* **C63** 189–285 (*Preprint* 0901.0002)
[8] Gao J, Guzzi M, Huston J, Lai H L, Li Z *et al.* 2014 *Phys.Rev.* **D89** 033009 (*Preprint* 1302.6246) URL http://hep.pa.msu.edu/cteq/public/
[9] Ball R D, Bertone V, Carrazza S, Deans C S, Del Debbio L *et al.* 2013 *Nucl.Phys.* **B867** 244–289 (*Preprint* 1207.1303) URL https://nnpdf.hepforge.org/
[10] Alekhin S, Blümlein J and Moch S 2013 (*Preprint* 1310.3059)
[11] Jimenez-Delgado P and Reya E 2009 *Phys.Rev.* **D80** 114011 (*Preprint* 0909.1711) URL http://www.het.physik.tu-dortmund.de/pdfserver/index.html
[12] Aaron F *et al.* (H1 and ZEUS Collaborations) 2010 *JHEP* **1001** 109 (*Preprint* arXiv:0911.0884)
[13] De Roeck A and Thorne R 2011 *Prog.Part.Nucl.Phys.* **66** 727–781 (*Preprint* 1103.0555)
[14] Watt G 2011 *JHEP* **1109** 069 (*Preprint* 1106.5788)
[15] Botje M 2011 *Comput.Phys.Commun.* **182** 490–532 (*Preprint* 1005.1481)
[16] James F and Roos M 1975 *Comput.Phys.Commun.* **10** 343–367
[17] Pumplin J, Stump D, Brock R, Casey D, Huston J *et al.* 2001 *Phys.Rev.* **D65** 014013 (*Preprint* hep-ph/0101032)
[18] Giele W T and Keller S 1998 *Phys.Rev.* **D58** 094023 (*Preprint* hep-ph/9803393)
[19] Giele W T, Keller S A and Kosower D A 2001 (*Preprint* hep-ph/0104052)
[20] Botje M 2002 *J.Phys.* **G28** 779–790 (*Preprint* hep-ph/0110123)
[21] Thorne R S and Roberts R G 1998 *Phys. Rev. D* **57** 6871 (*Preprint* hep-ph/9709442)
[22] Thorne R S 2006 *Phys. Rev.* **D73** 054019 (*Preprint* hep-ph/0601245)
[23] Alekhin S, Blümlein J and Moch S OPENQCDRAD a program description and the code are available via: http://www-zeuthen.desy.de/~alekhin/OPENQCDRAD
[24] Alekhin S and Moch S 2011 *Phys. Lett.* **B699** 345 (*Preprint* arXiv:1011.5790)
[25] Carli T *et al.* 2010 *Eur. Phys. J.* **C66** 503–524 (*Preprint* arXiv:0911.2985)
[26] Kluge T, Rabbertz K and Wobisch M 2006 483–486 (*Preprint* hep-ph/0609285)
[27] Li Y and Petriello F 2012 *Phys.Rev.* **D86** 094034 (*Preprint* arXiv:1208.5967)
[28] Andonov A, Arbuzov A, Bardin D, Bondarenko S, Christova P *et al.* 2006 *Comput.Phys.Commun.* **174** 481–517 (*Preprint* hep-ph/0411186)
[29] Bondarenko S G and Sapronov A A 2013 *Comput.Phys.Commun.* **184** 2343–2350 (*Preprint* 1301.3687)
[30] Aliev M, Lacker H, Langenfeld U, Moch S, Uwer P *et al.* 2011 *Comput.Phys.Commun.* **182** 1034–1046 (*Preprint* arXiv:1007.1327)
[31] Campbell J M and Ellis R K 2010 *Nucl. Phys. Proc. Suppl.* **205-206** 10–15 (*Preprint* arXiv:1007.3492)
[32] Campbell J M and Ellis R K 2012 Top-quark processes at NLO in production and decay report FERMILAB-PUB-12-078-T (*Preprint* 1204.1513)
[33] Aad G *et al.* (ATLAS Collaboration) 2012 *Phys.Rev.Lett.* **109** 012001 (*Preprint* 1203.4051)
[34] Chatrchyan S *et al.* (CMS Collaboration) 2014 *Phys.Rev.* **D90** 032004 (*Preprint* 1312.6283)
[35] Aad G *et al.* (ATLAS Collaboration) 2013 *Phys. Lett.* **B725** 223 (*Preprint* arXiv::1305.4192)
[36] Aad G *et al.* (ATLAS Collaboration) 2014 *JHEP* **1406** 112 (*Preprint* 1404.1212)
[37] Aad G *et al.* (ATLAS Collaboration) 2014 *JHEP* **1405** 068 (*Preprint* 1402.6263)
[38] Aad G *et al.* (ATLAS Collaboration) 2013 *Eur.Phys.J.* **73** 2509 (*Preprint* arXiv:1304:4739)
[39] Aad G *et al.* (ATLAS Collaboration) 2013 A study of the sensitivity to the proton parton distributions of the inclusive photon production cross section in *pp* collisions at 7 TeV measured by the ATLAS experiment at the LHC Tech. Rep. ATL-PHYS-PUB-2013-018 CERN Geneva
[40] Aaron F *et al.* (H1 Collaboration) 2012 *JHEP* **1209** 061 (*Preprint* arXiv:1206.7007)
[41] Abramowicz H *et al.* (H1 and ZEUS Collaborations) 2013 *Eur. Phys. J.* **C73** 2311 (*Preprint* arXiv:1211.1182)
[42] Abramowicz H *et al.* (ZEUS Collaboration) 2014 (*Preprint* 1405.6915)
[43] Hautmann F and Jung H 2014 *Nuclear Physics B* **883** 1 – 19 (*Preprint* 1312.7875)
[44] Luszczak A and Kowalski H 2014 *Phys.Rev.* **D89** 074051 (*Preprint* 1312.4060)
[45] Glazov A, Moch S and Radescu V 2011 *Phys. Lett. B* **695** 238–241 (*Preprint* arXiv:1009.6170)
[46] Sadykov R 2014 (*Preprint* 1401.1133)
[47] HERAFitter Developers Team and Lisovyi M 2014 (*Preprint* arXiv:1404.4234)
[48] H1prelim-13-141 and ZEUS-prel-13-003, H1prelim-10-142 and ZEUS-prel-10-018, H1prelim-11-042 and ZEUS-prel-11-002 HERAPDF1.5LO, NLO and NNLO available via: http://lhapdf.hepforge.org/pdfsets
[49] ATLAS NNLO epWZ12 available via: http://lhapdf.hepforge.org/pdfsets
[50] Butterworth J, Dissertori G, Dittmaier S, de Florian D, Glover N *et al.* 2014 (*Preprint* 1405.1067)
[51] Abelleira Fernandez J L *et al.* (LHeC Study Group) 2012 *Journal of Phys.* **G** 075001 (*Preprint* arXiv:1206.2913)

# A neural network $z$-vertex trigger for Belle II

**S Neuhaus[1], S Skambraks[2], F Abudinen[3], Y Chen[1], M Feindt[4], R Frühwirth[5], M Heck[4], C Kiesling[3], A Knoll[1], S Paul[2] and J Schieck[5]**

[1] Institute for Robotics and Embedded Systems, Technische Universität München, Germany
[2] Physik Department, Technische Universität München, Germany
[3] Max-Planck-Institut für Physik, München, Germany
[4] Institute for Experimental Nuclear Physics, Karlsruher Institut für Technologie, Germany
[5] Institute for High Energy Physics, Österreichische Akademie der Wissenschaften, Austria

E-mail: sara.neuhaus@in.tum.de

**Abstract.** We present the concept of a track trigger for the Belle II experiment, based on a neural network approach, that is able to reconstruct the $z$ (longitudinal) position of the event vertex within the latency of the first level trigger. The trigger will thus be able to suppress a large fraction of the dominating background from events outside of the interaction region. The trigger uses the drift time information of the hits from the Central Drift Chamber (CDC) of Belle II within narrow cones in polar and azimuthal angle as well as in transverse momentum (sectors), and estimates the $z$-vertex without explicit track reconstruction. The preprocessing for the track trigger is based on the track information provided by the standard CDC trigger. It takes input from the 2D ($r - \varphi$) track finder, adds information from the stereo wires of the CDC, and finds the appropriate sectors in the CDC for each track in a given event. Within each sector, the $z$-vertex of the associated track is estimated by a specialized neural network, with a continuous output corresponding to the scaled $z$-vertex. The input values for the neural network are calculated from the wire hits of the CDC.

## 1. Introduction to the Belle II track trigger

The Belle II experiment [1] will go into operation at the upgraded KEKB collider (SuperKEKB [2]) in 2016. SuperKEKB is designed to deliver an instantaneous luminosity $\mathcal{L} = 8 \cdot 10^{35}\,\mathrm{cm}^{-2}\mathrm{s}^{-1}$, a factor of 40 larger than the previous KEKB world record. The Belle II experiment will therefore have to cope with a much larger machine background than its predecessor Belle, dominated by Touschek scattering [3, 4]. This background produces a high rate of background events with the longitudinal position of the vertex ($z$-vertex) outside of the nominal interaction region, where the physically interesting reactions from the $e^+e^-$ collisions are produced. By reconstructing the $z$-vertex within the latency of the first level (L1) trigger, a large fraction of this background can be suppressed. The total latency of $5\,\mu\mathrm{s}$ for the full L1 trigger allows for only $\approx 1\,\mu\mathrm{s}$ for the $z$-vertex reconstruction. Neural networks of the Multi Layer Perceptron (MLP) type are well suited to meet these constraints due to their inherent parallelism and deterministic runtime. The approach presented here uses MLPs with only the hits in the Central Drift Chamber (CDC) as input to make an estimation of the $z$-vertex without explicit track reconstruction, based on the concepts presented in [5, 6, 7].

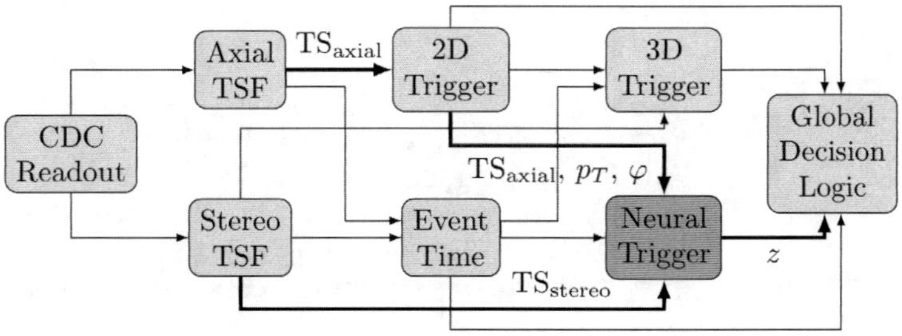


**Figure 1.** Schematic of the CDC trigger. The proposed neural $z$-vertex trigger takes input from the TS finders and from the 2D trigger.

The CDC contains 56 layers of sense wires of either "axial" orientation (aligned with the solenoidal magnetic field of 1.5 T along the $z$-axis) or "stereo" (skewed with respect to the axial wires). 6 or 8 adjacent layers have the same orientation and number of wires per layer and are combined in a so-called SuperLayer (SL). In total there are 9 SL alternating between axial and stereo SL. By combining the information of axial and stereo wires it is possible to reconstruct a full 3D helix track, which can be parametrized by $(p_T, \varphi, \theta, z, d)$, where $p_T$ is the transverse momentum, $\varphi$ and $\theta$ are the azimuthal and polar angle at the vertex, $z$ is the vertex position along the beamline and $d$ is the displacement of the vertex from the beamline in the $r-\varphi$ plane. The latter is expected to be small and is therefore assumed to be $d = 0$ cm in our studies.

A schematic view of the track trigger is shown in figure 1. For each of the 9 SL the wire hits are combined in so-called Track Segments (TS) by the Track Segment Finder (TSF) to reduce noise and the amount of data for the trigger. A TS is a predefined arrangement of several wires from 5 layers within a SL which produces a hit if at least 4 wires in different layers are active. For each TS the TSF transmits an id, the drift time of a reference wire in the TS with a resolution of 2 ns and the position of the track relative to this wire. The latter is obtained from a lookup table based on the hit pattern within the TS and can be *left*, *right* or *undecided*[1]. The event time that is needed to calculate the absolute drift times is estimated from the fastest wire hits in an event.

The 2D trigger combines TS from the axial SL to provide 2D tracks in $r-\varphi$ space [8]. The 3D trigger then adds the information from the stereo TS to determine the $z$-vertex. The proposed neural trigger will run in parallel to the conventional analytical 3D trigger and employ MLPs trained on sample tracks to predict the $z$-vertex. The outputs of the 2D/3D triggers are fed into the Global Decision Logic, which combines the results of all subtriggers and makes the final decision. More details about the Belle II detector and the trigger system can be found in [1, 8].

## 2. Structure of the neural network $z$-vertex trigger

The core concept of the neural trigger are networks of the Multi Layer Perceptron (MLP) type. The MLP is a universal function approximator [9] with the structure of an acyclic graph (see figure 2). Each neuron calculates a weighted sum of the input values and evaluates it with a non-linear bounded activation function, in our case the tanh function:

$$y_j = \tanh\left(\sum_{i=0}^{N} w_{ij} \cdot x_i\right) \tag{1}$$

where $x_i$ are the input values and $w_{ij}$ are the weights, which contain the information of the calculated function. The first input is $x_0 = 1$ such that the corresponding weight $w_{j0}$ acts as a constant bias. During training the weights of the MLP are adjusted with a back propagation

[1] The left/right information was not included in earlier studies [5, 6, 7], which were based on an older version of the Belle II simulation.

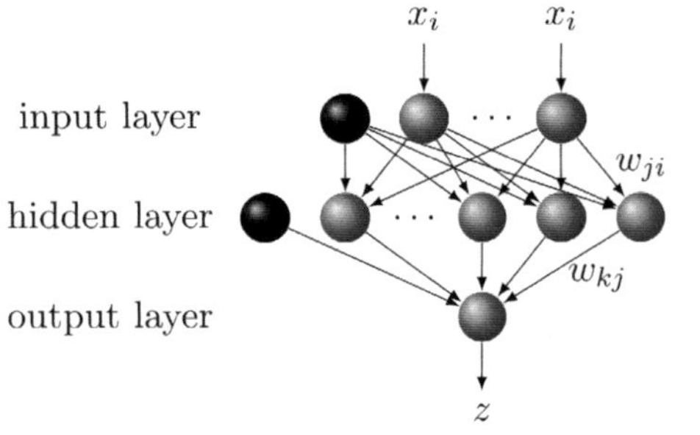


**Figure 2.** Structure of an MLP with one hidden layer. In the nodes the weighted sum of the input values $x_i$ is evaluated by the activation function to produce one output value. The black nodes denote constant bias nodes. Input and target values are always scaled to the co-domain of the activation function $[-1, 1]$.

algorithm such that for given input values the output values converge to certain target values. These target values are the $z$-vertex or other track parameters, scaled to the co-domain of the activation function $[-1, 1]$. The input values are calculated from the TS hits and scaled to the same range, following different input representations that are explained in detail below.

We use an ensemble of "expert" MLPs that are specialized to small sectors in phase space [5]. Only tracks with helix parameters from limited intervals are selected as training samples for a given sector. For each sector a number of "relevant" TS are identified based on a histogram of active TS generated from tracks within the sector. Only hits from the relevant TS are considered for the MLP, limiting the amount of input data for each sector. While the sector in $p_T$ and $\varphi$ can be selected using the estimation of the 2D trigger, the polar angle $\theta$ is unknown. As the stereo hits depend on both $z$ and $\theta$, one of them cannot be reconstructed precisely without knowing the other. Therefore we use a multi-step cascade prediction [5, 7], where a primary MLP specialized to a small sector in $p_T$ and $\varphi$ is used to predict $z$ and $\theta$. This prediction is used to select another MLP restricted in all track parameters, which can in the ideal case predict $z$ with the desired resolution. Otherwise the concept can be extended by adding another step, as was successfully demonstrated in [7].

To find an input representation for the MLP of a given sector, the relevant TS must be mapped to the input nodes. The most straightforward way to do this is to use one input node per relevant TS, with the drift time scaled to $[-1, 1]$ as input value and the left/right information encoded in the sign ("rep1"). For undecided left/right information the input value is 0, which is also the default value for inactive TS. While this mapping produces good results for small sectors, it is not an efficient representation for larger sectors, as the number of relevant TS grows with the sector size while the typical number of hits per track is just one hit per SL. In a second approach we therefore fix the number of inputs by using always one hit per SL, choosing the hit with the shortest drift time in case of several neighboring hits. In addition to the drift time we also use the wire position as input for the MLP, conveniently encoded in the TS id within a SL, which is proportional to a discrete azimuthal angle. This input representation ("rep2") results

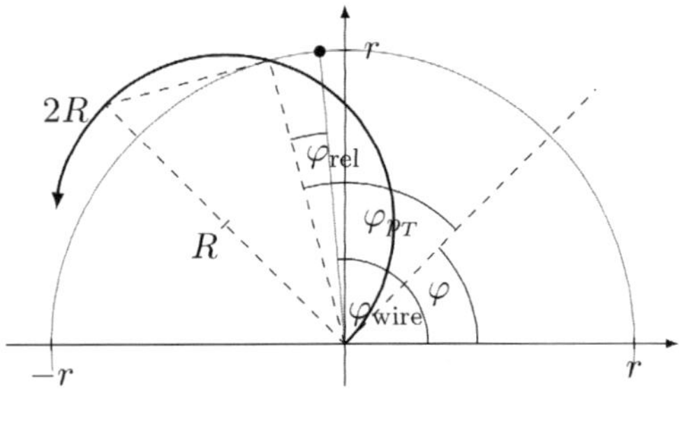


**Figure 3.** Using the estimation for the curvature and the azimuthal angle provided by the 2D track finder, the relative azimuthal angle between the track and a given wire can be calculated.

**Table 1.** Estimates for the precision of the $z$ prediction in small sectors for different input representations. $\Delta z$ is the RMS of the difference between true and predicted $z$-vertex.

| | input values | high $p_T$ $\Delta z$ | inputs | low $p_T$ $\Delta z$ | inputs |
|---|---|---|---|---|---|
| rep1 | 1 per TS ($t$) | 1.40 cm | 17 | 1.89 cm | 31 |
| rep2 | 2 per SL ($t, \varphi_{\text{wire}}$) | 1.10 cm | 18 | 1.50 cm | 18 |
| rep3 | 2 per SL ($t, \varphi_{\text{rel}}$) | 1.18 cm | 18 | 1.32 cm | 18 |
| rep4 | 3 per SL ($t, \varphi_{\text{rel}}, \mu$) | 1.15 cm | 27 | 1.18 cm | 27 |

in a constant number of 18 inputs, independent of the sector size.

Making use of the $\varphi$ and $p_T$ estimation of the 2D trigger it is possible to reduce the number of sectors by calculating the azimuthal angle *relative* to the reconstructed 2D track and use this as input for the MLP together with the drift times ("rep3"). The input representation is then independent of $\varphi$ and a single sector[2] is enough for this variable. The relative wire position can be obtained from the position of the wire and the estimated 2D values (see figure 2):

$$\varphi_{\text{rel}} = \varphi_{\text{wire}} - \varphi - \arcsin\left(\frac{r}{2R}\right) \equiv \varphi_{\text{wire}} - \varphi - \varphi_{p_T} \tag{2}$$

where $\varphi$ is the estimated track direction at the vertex, $r$ is the radial position of the wire, $R \propto p_T$ is the radius of the helix and $\varphi_{\text{wire}}$ is the azimuthal coordinate of a reference position on the wire. $\varphi_{p_T} = \arcsin\left(\frac{r}{2R}\right)$ is the change in the azimuthal angle due to the curling of the track in the magnetic field and is also proportional to the arc length $\mu$ of the track in the $r$-$\varphi$-plane:

$$\mu(r, R) = 2\varphi_{p_T} R \tag{3}$$

In the $\mu$-$z$-projection the track is linear, facilitating the extrapolation of the track to the vertex:

$$z(r) = \mu(r, R)\cot(\theta) + z_0 \tag{4}$$

where $z_0$ is the $z$-coordinate of the vertex and $\theta$ is the polar angle at the vertex. Therefore the scaled arc length $\mu(r, R)$ for each SL is used as an additional input for the MLP, leading to a total of 27 inputs ("rep4").

## 3. Experimental comparison of input representations

To evaluate the different input representations we have used simulated single muon tracks generated within limited ranges of $p_T$, $\varphi$, $\theta$ and $z$. The generated distribution within these ranges is uniform, except for $p_T$ where we use a uniform distribution in $p_T^{-1}$, i.e. in the curvature. When sectorizing $p_T$ we also choose bins of equal size in $p_T^{-1}$. Ranges in $\varphi$ apply only to the simpler input representations rep1 and rep2, otherwise we use a whole quarter of the detector ($\varphi \in [135°, 225°]$). The Geant4 based detector simulation of the Belle II CDC includes material interaction with the inner detector components (beampipe and vertex detectors), non-linear drift velocity due to inhomogeneities in the electric field and the gravitational wire sag effect.

The presented input representations are first studied with very small sectors to compare their performance under optimal conditions. We train one sector for low $p_T$ tracks

[2] The hardware implementation for a fully pipe-lined neural trigger will consist of 4 FPGA boards, each covering a quarter of the detector. This hardware sectorization is used also by the 2D trigger to cope with the high bandwidth requirements.

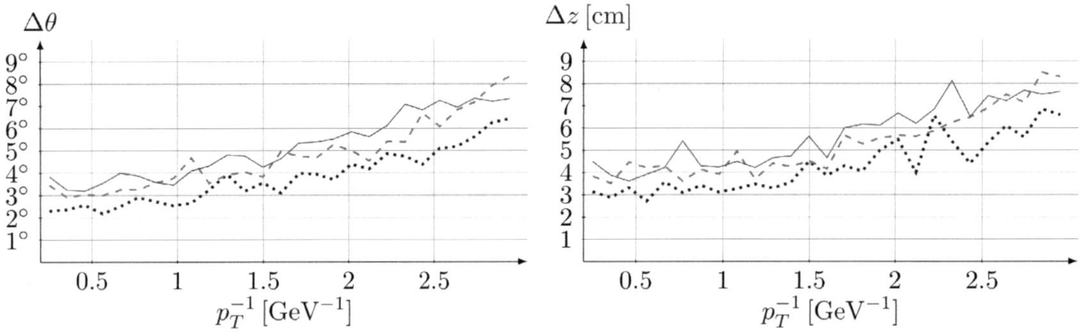


**Figure 4.** $\theta$ and $z$ prediction without a priori 3D knowledge for different input representations depending on $p_T^{-1}$. Blue solid line: rep2, red dashed line: rep3, black dotted line: rep4.

($p_T^{-1} \in [1.95, 2]\,\text{GeV}^{-1}$) and one for high $p_T$ tracks ($p_T^{-1} \in [0.2, 0.25]\,\text{GeV}^{-1}$). The remaining track parameters are constrained to $\varphi \in [180°, 181°]$, $\theta \in [50°, 60°]$ and $z \in [-10, 10]\,\text{cm}$. The results are summarized in table 1. Based on the $z$-vertex distribution in Belle [1] we estimate that a resolution of $\approx 2\,\text{cm}$ will be required to efficiently suppress the background, which is reached in both tested sectors for all input representations. Fixing the network size (rep2, rep3 and rep4) shows an advantage not only for the hardware implementation but also for the resolution. For the low $p_T$ case the resolution improves when correcting the curvature (rep3) and even more so when explicitly using the arc length $\mu$ (rep4).

Next we compare the prediction capability for the first step of the cascade prediction, when no information about $\theta$ and $z$ is available. In $\varphi$ and $p_T^{-1}$ we use a gaussian smearing with $\sigma_\varphi = 1°$, $\sigma_{p_T^{-1}} = 0.1\,\text{GeV}^{-1}$ on the simulated values before using them to simulate the effect of errors in the 2D trigger prediction. The 2D trigger is still under construction and not yet fully tested, but will likely exceed this resolution. The track parameters are constrained to $p_T^{-1} \in [0.2, 3]\,\text{GeV}^{-1}$, $\varphi \in [179°, 181°]$, $\theta \in [35°, 123°]$ and $z \in [-50, 50]\,\text{cm}$. The $p_T$ range is divided into 9 sectors of width $\Delta p_T^{-1} = 0.31\,\text{GeV}^{-1}$. In figure 4 the resolution is shown depending on $p_T^{-1}$. rep1 is not included in the test, as it was least effective in the previous test and the MLPs are too big to train efficiently. The uncertainty of both the $z$ prediction and the $\theta$ prediction increases with rising curvature for all tested input representations. The best results are again achieved for rep4. Although the target resolution is not reached, the prediction of the primary MLP can be used to significantly decrease the sector size for the next step.

Even in this first step the MLP prediction exceeds the estimate for the 3D trigger of $\Delta z = 9.3\,\text{cm}$ given in [1], where no drift times were used. Recent progress in the development of the 3D trigger including the drift time information suggests that a better $z$ resolution can be achieved. A comparison of the neural trigger and the 3D trigger will be subject of future studies.

## 4. Conclusion

We have presented a setup for a first level $z$-vertex trigger based on an ensemble of "expert" MLPs, specialized to sectors in phase space, and demonstrated the influence of different input representations on the final $z$-vertex resolution. Efficiently using the information provided by earlier trigger components and knowledge about the detector geometry itself, we can reduce not only the size of each expert network, but also the total number of sectors.

In future investigations we will try to improve the resolution for low momentum tracks,

Journal of Physics: Conference Series **608** (2015) 012052 doi:10.1088/1742-6596/608/1/012052

combining different training algorithms for the neural networks with the presented preprocessing. The fully pipe-lined neural network trigger will be implemented in hardware on Virtex 7 FPGA boards and installed in the Belle II detector.

## References


[1] Abe T, Adachi I, Adamczyk K, Ahn S, Aihara H, Akai K, Aloi M, Andricek L, Aoki K, Arai Y *et al.* 2010 Belle II Technical Design Report (*Preprint* `1011.0352`)

[2] Aushev T, Bartel W, Bondar A, Brodzicka J, Browder T *et al.* 2010 Physics at Super B Factory (*Preprint* `1002.5012`)

[3] Bernardini C, Corazza G F, Di Giugno G, Ghigo G, Haissinski J, Marin P, Querzoli R and Touschek B 1963 *Phys. Rev. Lett.* **10**(9) 407–409 URL `http://link.aps.org/doi/10.1103/PhysRevLett.10.407`

[4] Piwinski A 1998 The Touschek effect in strong focusing storage rings (*Preprint* `physics/9903034`)

[5] Skambraks S 2013 *Use of Neural Networks for Triggering in the Belle II Experiment* Master's thesis Ludwig-Maximilians-Universität München MPP-2013-377

[6] Abudinén F 2014 *Studies on the neural z-Vertex-Trigger for the Belle II Particle Detector* Master's thesis Ludwig-Maximilians-Universität München MPP-2014-42

[7] Skambraks S, Abudinén F, Chen Y, Feindt M, Frühwirth R, Heck M, Kiesling C, Knoll A, Neuhaus S, Paul S and Schieck J 2014 A $z$-Vertex Trigger for Belle II (*Preprint* `1406.3319`)

[8] Iwasaki Y, Cheon B, Won E, Gao X, Macchiarulo L, Nishimura K and Varner G 2011 *Nuclear Science, IEEE Transactions on* **58** 1807–1815 ISSN 0018-9499

[9] Hecht-Nielsen R 1989 Theory of the backpropagation neural network *Neural Networks, 1989. IJCNN., International Joint Conference on* pp 593–605 vol.1

# A novel robust and efficient algorithm for charge particle tracking in high background flux

**C Fanelli[1,2], E Cisbani[2], A Del Dotto[3]**

[1] Dipartimento di Fisica, Università "La Sapienza", I-00185 Rome, Italy
[2] INFN, Sezione di Roma, gruppo Sanità and Istituto Superiore di Sanità, I-00161 Rome, Italy
[3] Dept. of Phys. and Math. Univ. of Rome "Roma Tre" and INFN, Sezione di "Roma Tre", I-00146 Rome, Italy

E-mail: cfanelli@iss.infn.it

**Abstract.** The high luminosity that will be reached in the new generation of High Energy Particle and Nuclear physics experiments implies large high background rate and large tracker occupancy, representing therefore a new challenge for particle tracking algorithms. For instance, at Jefferson Laboratory (JLab) (VA,USA), one of the most demanding experiment in this respect, performed with a 12 GeV electron beam, is characterized by a luminosity up to $10^{39}\mathrm{cm}^{-2}\mathrm{s}^{-1}$. To this scope, Gaseous Electron Multiplier (GEM) based trackers are under development for a new spectrometer that will operate at these high rates in the Hall A of JLab. Within this context, we developed a new tracking algorithm, based on a multistep approach: (i) all hardware - time and charge - information are exploited to minimize the number of hits to associate; (ii) a dedicated Neural Network (NN) has been designed for a fast and efficient association of the hits measured by the GEM detector; (iii) the measurements of the associated hits are further improved in resolution through the application of Kalman filter and Rauch-Tung-Striebel smoother. The algorithm is shortly presented along with a discussion of the promising first results.

## 1. The Tracking Challenge

We developed a new charged particle tracking algorithm, able to operate in high luminosity experiments such as those at Jefferson Laboratory (JLab) (VA,USA). In particular, the algorithm is designed for the experimental Hall A in JLab, where an hybrid tracker has been developed to optimally exploit the new 12 GeV, high intensity, electron beam. The tracker consists of six large GEM chambers [1] and two small planes of Silicon microstrip Detectors (SIDs). The GEM tracker will be placed in the Super Big Bite Spectrometer (SBS) (see, e.g., Fig. 1), after a momentum analyzing dipole, while the silicon tracker will be sitting very close to the scattering chamber, in order to increase the tracked flying path and the lever arm for better tracking. Each GEM chamber is composed (Fig. 2) by three $40 \times 50$ cm$^2$ GEM individual modules, having three GEM foils, and with two-dimensional strip readout, with expected spatial resolution of about 80 $\mu$m.

In the most demanding experiment (the measurement of the proton form factors at high momentum transfer [2]), we expect almost 20 kHz (the coincidence trigger rate of the elastically scattered electron and proton) of signal and 400 kHz/cm$^2$ of background hits on each chamber ($\gamma$, e, $\pi^{+,-}$); the hits rates in a cone of 10 cm$^2$ in the first GEM plane are of 1 signal hit plus

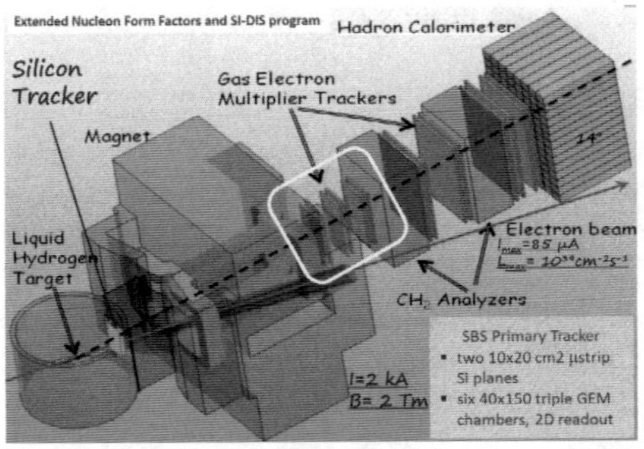


**Figure 1.** Layout of the SBS spectrometer in Jefferson Lab Hall A.

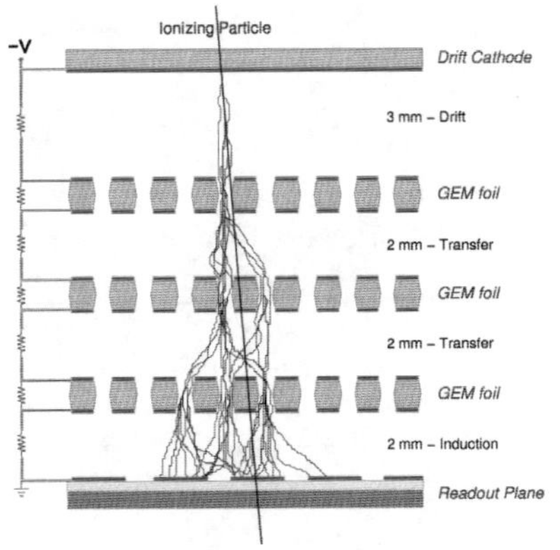


**Figure 2.** Schematic diagram of the 3-GEM foils detector.

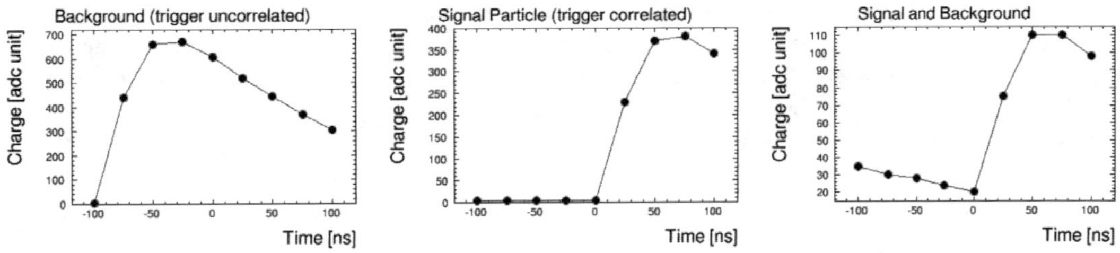


**Figure 3.** Time evolution of the charge coming from the strips of the GEM chamber for three different events: left a background particle uncorrelated to the trigger (starting at time = 0 ns); center a signal particle correlated to the trigger; right a combination of two particles (background and signal).

roughly 100 hits of background. To cope with these high rates of hits, we developed a new multistep tracking algorithm, based on (i) "hardware reduction", (ii) a Neural Network (NN) for a fast "association of hits", and (iii) Kalman filter for "precise tracking", as explained in what follows.

***Hardware Reduction*** We simulated the response of the electronics to charge deposits of GEM chambers. In particular, the protons represent the signal, whereas photons or charged particles (electrons or pions) account for background. In Fig. 3, the time evolution of the signals registered by the electronics is shown: the time evolution of proton signals develops around 0 ns (25 ns maximum jitter distribution is included in the simulation), having been synchronized with the trigger starting at 0 ns by definition. Background hits are uncorrelated with the trigger. This simulation has been performed with Geant4 [3], along with a custom, realistic digitization algorithm transforming the energy deposited by the particle into an electric discharge in the GEM chamber and then into an electronic signal. This temporal correlation, as well as x/y charge correlation, are used to significantly reduce the number of hits passed to the second step of the procedure.

***Association of hits*** Lots of successful applications of artificial NN's in High Energy and Nuclear Physics have been realized in the last two decades. The possibility to use an Hopfield-like network in track finding problems has been firstly demonstrated in [4, 5]. In this proceeding, we present the preliminary results obtained with a NN defined within the framework of Mean Field Theory (MFT). Here we recall the essential features of a NN in MFT: neurons $S_{ij}$ correspond to connections between two hits, $i \to j$, belonging to two adjacent planes. $S_{ij}$ assume binary values (0 - connection off, 1 - connection on), but, in MFT, their values are permitted to range continuously $\in$[0,1], and to make a clear distinction with the digital case, we use the notation $V_{ij}$. In order to connect a neuron to another, namely $i \to j \to l$, it is necessary to introduce the so called synaptic strengths, formally quadri-dimensional $T_{ijkl}$, with the constraint $j = k$. To this respect, two essential ingredients are the lengths of two adjacent segments $r_{ij}$, $r_{jl}$ and the angle between them $\theta_{ijl}$. The general features of the MFT approach remain the same, but, in a bi-dimensional case, we can write

$$V_{ij} = \frac{1}{2}\left[1 + \tanh\left(-\frac{\partial E}{\partial V_{ij}}\frac{1}{T}\right)\right]. \tag{1}$$

It is important to point out that the particular definition of the energy function $E$ is a matter of choice and it depends strictly on the particular features of the problem. In our case, we have 6 subsequent tracking planes, spanning a certain angular region, which is crossed by a heavy number of charged particles, whose trajectories are approximately straight lines (magnetic field is negligible in the tracker region). We define an energy function able to describe and discriminate the trajectories even when they occur in a very small angular region (high background conditions). Given a hit $i$ in a certain plane $p$ (from first to penultimate, $p = 1, ..., n-1$), we consider connections $i \to j$ to hits belonging to the next plane, where $j \in p+1$, and from this to the subsequent plane ($j \to l$, where $l \in p+2$). These requirements are encoded in the energy function as:

$$E = \underbrace{-\sum_{ijl}\frac{d_G\cos\theta_{ijl}}{r_{ij}+r_{jl}}\delta_{kj}V_{ij}V_{kl} + \sum_{ijl}\frac{d_G\cos\theta_{ijl}\delta_{li}V_{ij}}{r_{ij}+r_{jl}}}_{cost} + \underbrace{\alpha(\sum_{ijl\neq j}V_{ij}V_{il} + \sum_{ijk\neq j}V_{ij}V_{kj})}_{constraint}, \tag{2}$$

where $d_G$ is the distance between two tracking planes, whereas $r_{ij}$, $r_{jl}$ and $\theta_{ijl}$ are the segment lengths between the $ij$, $jl$ hits and the angle between them, respectively; the second term of the 'cost' part, comprises the so called 'firing thresholds' $I_{ij} = (d_G\cos\theta_{ijl}\delta_{li}/(r_{ij}+r_{jl}))$, to correctly weight the connections (neurons) at the edge planes. In the 'constraint' part, we assume a bifurcation inhibitor by using a Lagrange multiplier $\alpha$.

The energy function (2), and, in general, any *cost* term based on trigonometric functions, works well if one has to consider tracks separated enough in space and cases in which the occupancy of the planes (namely the density of hits) is not too high. On the other hand, these two conditions are not encountered in the high luminosity experiments. To recover the full discrimination power of the energy function even in the case of small spatial region, we propose to use a novel approach, based on an affine scaling transformation, that preserves the linearity of trajectories. This has been done by setting $r'_{ij} \equiv \mathcal{S}_{xy}(\xi)r_{ij} = \sqrt{\xi^2\Delta x_{ij}^2 + \xi^2\Delta y_{ij}^2 + \Delta z_{ij}^2}$, where $\xi$ is a continuous parameter $\in [1, +\infty)$. One has to apply this transformation to the cost term $\cos\theta_{ijl}/(r_{ij}+r_{jl})$ present in Eq. (2). In our case, the optimal value that distinguishes more effectively the 'on' and 'off' connections in the energy cost term, was found to be $\xi \approx 100$. The introduced affine scaling has revealed very effective, and guides the NN to fast convergence. As illustrated in Fig. 4, the energy decreases to a plateau value (and it behaves similarly to a Lyapunov function) as well as the convergence $\sum_{ij}|V_{ij}(t+1) - V_{ij}(t)|$, as a function of the number of sweep cycles $t$[1]. Correspondingly, all the neurons saturate to values close to 0 (if connections are 'off') or

[1] After each sweep cycle, a full asynchronous updating of all the neurons is completed

close to 1 (if 'on'), as shown in Fig. 5. Common difficulties in practical applications, arise from the arbitrary choice of the NN parameters (Lagrange multipliers, temperature, scaling variable). We provided general criteria which allow the convergence of the algorithm and an effective and robust association of the hits: for instance, we used a 'Network digitalization' (see, e.g., [6]) for setting the Lagrange multiplier, and a technique similar to an 'a posteriori annealing' (as suggested in [7]) to investigate the dependence on temperature of the system. We then applied

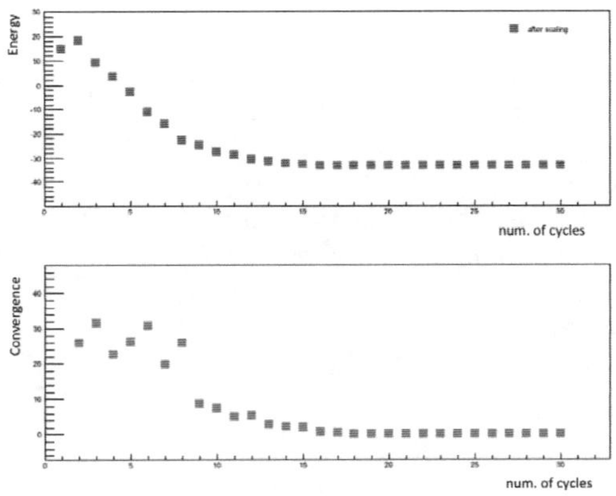


**Figure 4.** (top) energy and (bottom) variation of the average variation of neuron activities as a function of the sweep cycles.

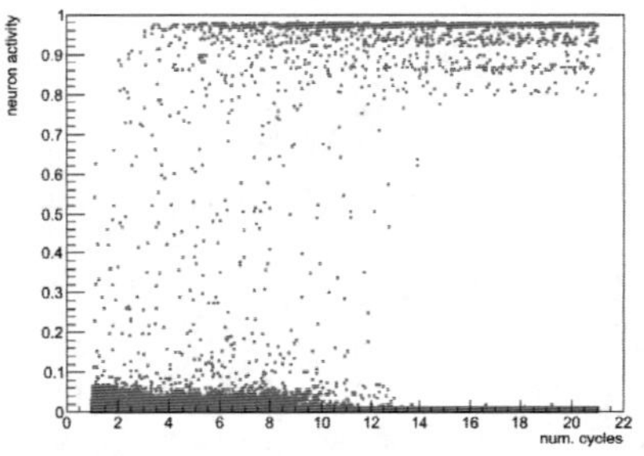


**Figure 5.** Neuron activity. In MFT, values are continuous and range $\in [0,1]$. After a certain number of sweep cycles, all the neurons stabilize. A threshold of 0.5 is chosen, and the neurons are transformed to binary values, 0 or 1, depending on whether smaller or larger than the threshold.

our NN based tracking algorithm to the case of JLab12 experiments. As mentioned earlier, the typical rates expected in an area of 10 $cm^2$ in the first tracking plane are 1 signal (proton) track from the interaction vertex, 1 charged background particle, few background photons per plane, and about 100 ghost hits per plane, which arise from combinatorial x/y strips association. Notably, as shown in Fig. 6, a 97% tracking efficiency is obtained with 100 ghost hits per plane.

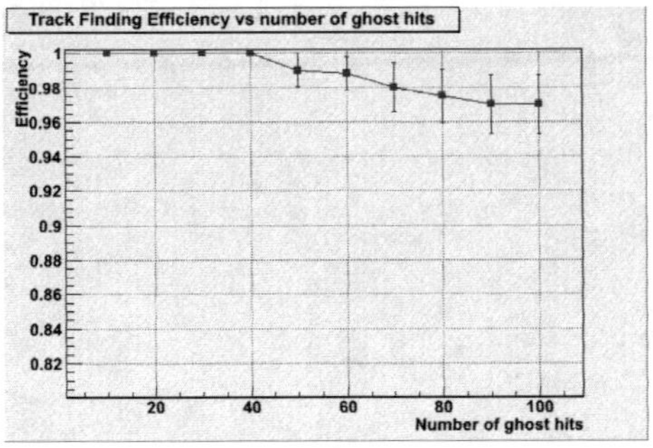


**Figure 6.** Track association efficiency as a function of the number of ghost hits. The NN parameter settings were determined by optimizing the track finding efficiency for the case of 100 ghost hits (97%). In all the other cases, efficiency always exceeds 97% and it is larger than 99% below 50 ghost hits.

***Precise Tracking*** Once the association is accomplished, one can provide even more accurate measurements of the associated hits through filtering techniques. We used both Kalman filter (KF) [8], to filter the tracks in the forward direction, and Rauch Tung Striebel (RTS) smoother [9], for the reverse backward filtering. It is worth mentioning here that these are basically

Bayesian algorithms, that combine both experimental and theoretical information, starting from a measured hit, and according to a certain evolution model, one can infer the predicted hit in the subsequent plane, and compare it to its effective measurement in this plane. In this way, one can easily obtain a posterior distribution, from which the most likely 'true' position of the hit is estimated. We simulated a large number of correctly associated tracks, in order to calculate the accuracy of the filters (KF+RTS) in rather realistic experimental conditions. As reported in Fig. 7, the high accuracy obtained by combining the filters method is $\sim 10\ \mu$m, to be compared to the hit spatial resolution of about 80 $\mu$m.

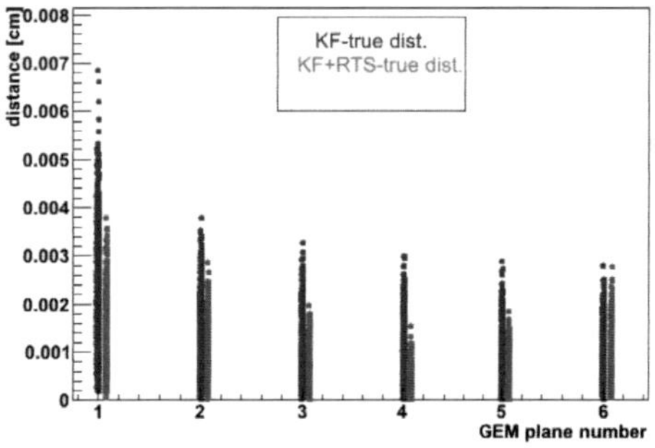


**Figure 7.** Distance between KF filtered and true hits (blue) and distance between KF+RTS filtered and true hits (gray) for 6000 hits belonging to associated tracks. The accuracy of the combined KF+RTS filters method is $\sim$10 $\mu$m.

## 2. Conclusions

The overall procedure (NN+filtering) looks terrific promising, with high performances both in terms of association efficiency ($\geq$ 97%), provided by the NN, and of reconstruction accuracy ($\sim$ 10 $\mu$m, a factor 8 smaller than GEM designed spatial resolution), this latter provided by the combined filters. Next development will concentrate on the comparison of NN with other "standard" algorithms. Furthermore, a study and optimization of the computational time is also compulsory. Multiple scattering is under implementation for a deeper testing of the KF+RTS filtering technique, which has proved to be effective on the associated hits of the NN algorithm. Study of the refined and optimized procedure on real data will be the ultimate goal.

## References

[1] Bellini V, Cisbani E, Capogni M, Colilli S, De Leo R, De Oliveira R, De Smet V, Fratoni R, Frullani S, Giuliani F *et al.* 2012 *Journal of Instrumentation* **7** C05013
[2] Perdrisat C F *et al.* 2008 *JLab approved experiment E12-07-109*
[3] Agostinelli S, Allison J, Amako K a, Apostolakis J, Araujo H, Arce P, Asai M, Axen D, Banerjee S, Barrand G *et al.* 2003 *Nuclear instruments and methods in physics research section A: Accelerators, Spectrometers, Detectors and Associated Equipment* **506** 250–303
[4] Denby B 1988 *Computer Physics Communications* **49** 429–448
[5] Peterson C 1989 *Nuclear Instruments and Methods in Physics Research Section A: Accelerators, Spectrometers, Detectors and Associated Equipment* **279** 537–545
[6] Kamgar-Parsi B 1992 *IJCNN, International Joint Conference on Neural Networks* vol 4 (IEEE) pp 566–571
[7] Peterson C and Söderberg B 1989 *International Journal of Neural Systems* **1** 3–22
[8] Kalman R E 1960 *Journal of Fluids Engineering* **82** 35–45
[9] Rauch H E, Striebel C and Tung F 1965 *AIAA journal* **3** 1445–1450

ACAT2014 IOP Publishing
Journal of Physics: Conference Series **608** (2015) 012054 doi:10.1088/1742-6596/608/1/012054

# New features of MadAnalysis 5 for analysis design and reinterpretation

**Eric Conte**[1], **Béranger Dumont**[2,3], **Benjamin Fuks**[4,5] **and Thibaut Schmitt**[5]

(1) Groupe de Recherche de Physique des Hautes Énergies (GRPHE), Université de Haute-Alsace, IUT Colmar, 34 rue du Grillenbreit BP 50568, 68008 Colmar Cedex, France
(2) Center for Theoretical Physics of the Universe, Institute for Basic Science (IBS), Daejeon 305-811, Korea
(3) LPSC, Université Grenoble-Alpes, CNRS/IN2P3, 53 Avenue des Martyrs, F-38026 Grenoble, France
(4) CERN, PH-TH, CH-1211 Geneva 23, Switzerland
(5) Institut Pluridisciplinaire Hubert Curien/Département Recherches Subatomiques, Université de Strasbourg/CNRS-IN2P3, 23 Rue du Loess, F-67037 Strasbourg, France

E-mail: eric.conte@iphc.cnrs.fr, beranger.dumont@lpsc.in2p3.fr, fuks@cern.ch, thibaut.schmitt@iphc.cnrs.fr

**Abstract.** We present MadAnalysis 5, an analysis package dedicated to phenomenological studies of simulated collisions occurring in high-energy physics experiments. Within this framework, users are invited, through a user-friendly Python interpreter, to implement physics analyses in a very simple manner. A C++ code is then automatically generated, compiled and executed. Very recently, the expert mode of the program has been extended so that analyses with multiple signal/control regions can be handled. Additional observables have also been included, and an interface to several fast detector simulation packages has been developed, one of them being a tune of the Delphes 3 software. As a result, a recasting of existing ATLAS and CMS analyses can be achieved straightforwardly.

## 1. MadAnalysis 5 in a nutshell

While both LHC experiments are currently pushing limits on particles beyond the Standard Model to higher and higher scales, tremendous progress has been made in the development of Monte Carlo event generators and satellite programs, in particular with respect to precision predictions, new physics implementations and event analysis. The public MadAnalysis 5 package [1, 2] addresses this last aspect and provides a framework for analyzing events generated either at the parton-level, after hadronization or after detector simulation. The program hence allows one to efficiently design and recast LHC analyses. The user can in this way investigate any physics model and determine the LHC sensitivity to its signatures by either conceiving a novel analysis or recasting existing ATLAS and CMS studies. MadAnalysis 5 starts by reading event samples as generated by any Monte Carlo event generator that satisfies community-endorsed output formats. Next, it applies selection cuts and computes differential distributions as requested by the user. In its normal mode of running, the results are represented under the form of histograms and cut-flow tables that are collected within Html and LaTeX reports. In the expert mode of the program, they are presented in text files compliant with the Saf format,

Journal of Physics: Conference Series **608** (2015) 012054 doi:10.1088/1742-6596/608/1/012054

an XML-inspired format internally used by MADANALYSIS 5 that has been described in Ref. [2]. Moreover, events can be preprocessed. For instance, the application of a jet-clustering algorithm or the simulation of a detector response can be achieved effortlessly before the execution of the analysis itself, and preprocessed events can be possibly saved in output files.

## 2. Main concepts

We recall in this section the main concepts of the MADANALYSIS 5 package. We refer to Refs. [1, 2] for more details.

- **Universal**: One can in the same way design new phenomenological analyses and recast existing LHC studies, and a given analysis can be equally applied on parton-level, hadron-level or reconstructed-level events stored under the STDHEP [3], HEPMC [4], LHE [5, 6], LHCO [7] or ROOT [8] format.
- **User friendly**: In the normal mode of running, the user designs her/his analysis by interacting with a PYTHON console through a metalanguage designed to be intuitive. The syntax is moreover easy to handle thanks to tab completion and in-line help.
- **Efficient**: The PYTHON module of the program takes care of exporting the analysis encoded with the MADANALYSIS 5 metalanguage to an optimized C++ program readily to be compiled and executed.
- **Flexible**: MADANALYSIS 5 is shipped with a series of basic (transverse momentum, *etc.*) and more sophisticated ($\alpha_T$, $M_{T2}$, *etc.*) built-in observables that can be used in analysis design. Operations on particle four-momenta are also available. For more complicated selections, the user can directly write his/her analysis in C++ (the so-called expert mode of the tool) and implement the desired observables.
- **Interfacing**: MADANALYSIS 5 is interfaced to ZLIB [9], ROOT [8], FASTJET [10] (and FASTJET-CONTRIB) and DELPHES [11]. The installation of these packages can be performed automatically from the PYTHON console. The program also contains the so-called MA5TUNE of DELPHES [12] (see Section 4).

## 3. Analysis design with the MADANALYSIS 5 metalanguage

From the PYTHON console, the user is invited to describe its selection and configure MADANALYSIS 5 through intuitive and human-understandable commands that inherit from the PYTHON syntax. As an exhaustive review of this metalanguage is beyond the scope of this paper, we focus on a detailed illustrative example and refer to Ref. [1] for more information. We consider a simple analysis defined by the following commands:

```
[0] ma5> import DrellYan*.hep as dy
[1] ma5> import ttbar*.hepmc.gz as tt
[2] ma5> set detector.fastsim.package = fastjet
[3] ma5> set detector.fastsim.algo = kt
[4] ma5> plot MET
[5] ma5> define mu = mu+ mu-
[6] ma5> select (mu) PT > 25
[7] ma5> plot M(mu+ mu-)
[8] ma5> set main.outputfile = output.lhco
[9] ma5> submit
```

The first two commands (lines 0 and 1) enable the import of the event samples to be considered. We recall that MADANALYSIS 5 can read samples provided in the STDHEP, HEPMC, LHE, LHCO or ROOT format, while compressed files can be used if the ZLIB package is available. The next two lines (lines 2 and 3) indicate how to use the interface to the FASTJET program and select the $k_T$ algorithm for jet reconstruction (see also Section 4). The computation of two differential distributions, namely the missing transverse energy and the muon-pair invariant mass spectra, are then requested after enforcing the selection of muons with a transverse momentum of at least 25 GeV (lines 4 to 7). The next command (line 8) requires MADANALYSIS 5 to save the reconstructed events in the LHCO format, the ROOT and LHE formats being also available. The final command (line 9) requests the program to operate. A C++ code corresponding to the analysis is generated and executed. The results (histograms and selection efficiencies) are collected by a report generator based on ROOT (other graphical components will be available in the next release of MADANALYSIS 5) and further exported into HTML and LaTeX reports.

## 4. Detector fast-simulation packages

The MADANALYSIS 5 package offers the option to install, configure and launch the fast simulation of a detector response on simulated events given as input in order to mimic instrumental and experimental effects and increase the realism of the simulation. Three possible choice are available, a very simple (but often sufficient for, *e.g.*, generator-level based phenomenology) and fast simulation based on FASTJET, and two more sophisticated and slower simulations based on DELPHES 3.

- MADANALYSIS 5 is shipped with its own detector simulation based on FASTJET. All the jet-clustering algorithms included in FASTJET can be applied to hadron-level events, and the user can further switch on several detector effects like reconstruction efficiencies, resolution, misidentification rates, *etc.*
- MADANALYSIS 5 is interfaced to DELPHES 3, which offers a realistic description of the ATLAS and CMS detectors. Default configuration can be used, but the user is also invited to tune the DELPHES setup according to his/her needs. Pile-up effects can be included.
- A modified version of DELPHES 3, the so-called DELPHES-MA5TUNE, can be employed. Based on DELPHES, it includes the calculation of new observables and optimizes the size of the output file. The main improvements address lepton and photon isolation. Unlike the official package, isolation variables are saved in the output file so that isolation requirements can be implemented at the analysis level instead of at the detector simulation level.

## 5. Recasting existing LHC analyses with the MADANALYSIS 5 expert mode

The *expert mode* of the program offers the possibility to write an analysis directly in C++, using the MADANALYSIS 5 framework. It is developer-friendly, and, in addition, a large collection of methods dedicated to common high-energy physics issues are available, as well as various services allowing one, for instance, to produce cut-flow charts and/or histograms. A detailed manual can be found in Ref. [2]. With version 1.1.10 onwards, one can also recast any cut-based ATLAS and CMS analysis possibly containing multiple subanalyses or regions. This allows reinterpretation of experimental results under the approximation (compared to a real experimental setup) of using a fast, simple, detector simulation software. Our framework offers hence a way to work out the implications of the LHC for any new physics model, derive limits in a realistic-enough manner, point out possible loopholes in the current searches and help to design future analyses. It could also be used to improve the content and realism of fast detector simulation packages.

The adopted strategy consists in first applying the DELPHES-MA5TUNE fast detector simulation on signal events stored under the STDHEP or HEPMC format. Secondly, a reimplementation of a public ATLAS or CMS analysis is executed to derive the number of selected signal events from the input samples. A PYTHON script finally extracts limits by confronting the numbers associated with the more sensitive signal region to the Standard Model expectation provided in the experimental publication. In the case where a user would need a more sophisticated procedure, he/she is welcome to extract directly the limits from

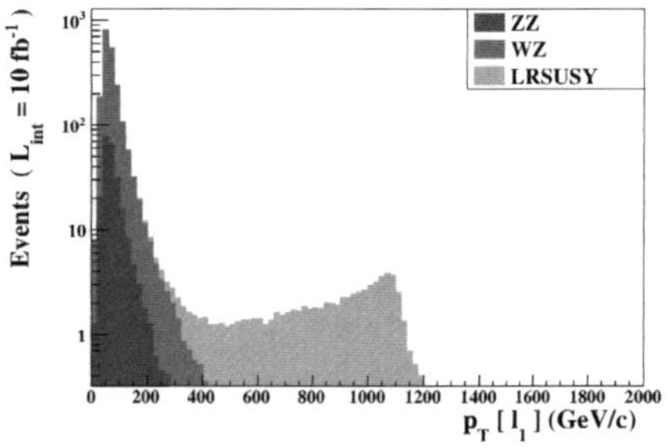


**Figure 1.** Transverse-momentum distribution of the hardest lepton for the LRSUSY signal and the diboson background.

**Cut-flow chart**

- How to compare signal (S) and background (B): S/B.
- Associated uncertainty: 1./(B**2)*sqrt(B**2*ES**2+S**2*EB**2).

| Cuts | Signal (S) | Background (B) | S vs B |
|---|---|---|---|
| Initial (no cut) | 387.033 +/- 0.584 | 2029.72 +/- 1.44 | 0.190683 +/- 0.000318 |
| Cut 1 | 84.95 +/- 8.14 | 2029.72 +/- 1.44 | 0.04185 +/- 0.00401 |
| Cut 2 | 81.56 +/- 8.02 | 26.09 +/- 5.08 | 3.126 +/- 0.681 |
| Cut 3 | 76.19 +/- 7.82 | 19.45 +/- 4.39 | 3.92 +/- 0.97 |

Signal and Background comparison

**Figure 2.** Snippet of the MadAnalysis 5 report with the cut-flow chart associated with the selection described in the text. The figure of merit consists of the ratio of the number of signal to the number of background events.

the MadAnalysis output independently. Since the validation of reimplemented analyses can sometimes be complicated (see, *e.g.*, Refs. [12, 13]), validated codes can be shared through our public analysis database [12] and submitted to Inspire [14], that additionally assigns each submission a DOI [15].

## 6. Selection of results

In order to demonstrate the performances of the program, we focus on two examples. A first analysis is related to the search for a multileptonic signature of left-right symmetric supersymmetry (LRSUSY) in simulated LHC collisions at a center-of-mass energy of 13 TeV [16]. A second example is devoted to the recast of the CMS-SUS-13-012 analysis focusing on squark and gluino searches in multijet events in collisions at a center-of-mass energy of 8 TeV [17, 18].

### *6.1. Prospective phenomenology in the left-right symmetric supersymmetric context*

LRSUSY predicts additional gaugino and higgsino states that can potentially lead to signals in multileptonic events to be produced at the LHC. In the present example, we perform a simple selection in order to extract the LRSUSY signal from the Standard Model $WZ$ and $ZZ$ background. In the MadAnalysis 5 metalanguage, the analysis reads

```
ma5> define l = e+ e- mu+ mu-
ma5> select N(l) >= 3
ma5> plot PT(l[1]) 200 0 2000 [ logY ]
ma5> select PT(l) > 200
ma5> select MET > 50
```

Events containing at least three charged leptons are selected. The transverse momentum ($p_T$) distribution of the hardest lepton is then calculated, the results being shown in Figure 1. The signal exhibiting a harder spectrum, the $p_T$ of the hardest lepton is constrained to be larger than 200 GeV, while at least 50 GeV of missing transverse energy (assumed to be carried by the lightest supersymmetric state that is stable and escapes detection) is demanded. The corresponding cut-flow table is presented on Figure 2, that shows that the selection yields a figure of merit of about $4\sigma$.

### *6.2. Example of recast analysis*

A recasting of the CMS-SUS-12-012 analysis [17] has been recently performed, the validated MadAnalysis 5 code being available in Ref. [18]. This validation has been achieved by

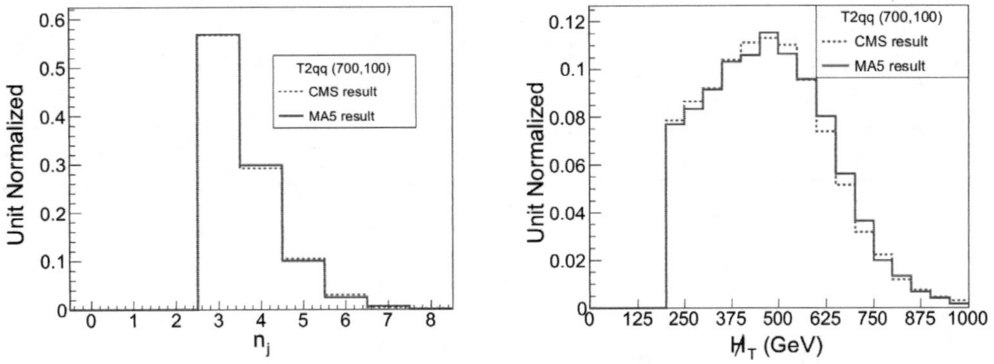


**Figure 3.** Comparison between CMS and MA5 for the distribution in the jet multiplicity (left) and in the $\not{H}_T$ variable (right) for a specific signal scenario and after the baseline selection described in Ref. [17].

comparing CMS and MadAnalysis 5 (MA5) results, both at the cut-flow level and for various observables. Two examples are shown in Figure 3, where we present distributions for the jet multiplicity and the missing transverse hadronic energy $\not{H}_T$ in the case of the *T2qq* simplified model [19]. We confront MA5 predictions to CMS official results and observe a good agreement.

## 7. Summary

The public package MadAnalysis 5 offers a simple and efficient way to design and recast LHC analyses. The user can either benefit from its Python console and dedicated metalanguage to implement his/her analysis, or make use of the expert mode of the program which provides a flexible developer-friendly C++ environment. We have also discussed detector simulation within the MadAnalysis 5 framework, presenting the Delphes-MA5tune simulation of a detector response, and introduced our recently initiated database of reimplementations of LHC analyses.

## Acknowledgments

We thank the organizers for setting up this nice conference. This work has been partially supported by the French ANR projects 12-JS05-002-01 BATS@LHC and 12-BS05-0006 DMAstroLHC, by the Theory-LHC France initiative of the CNRS/IN2P3 and by the *Investissement d'avenir, Labex ENIGMASS.*

## References

[1] Conte E, Fuks B and Serret G 2013 *Comput.Phys.Commun.* **184** 222–256 (*Preprint* 1206.1599)
[2] Conte E, Dumont B, Fuks B and Wymant C 2014 (*Preprint* 1405.3982)
[3] URL http://cepa.fnal.gov/psm/stdhep/c++/
[4] Dobbs M and Hansen J B 2001 *Comput.Phys.Commun.* **134** 41–46
[5] Boos E, Dobbs M, Giele W, Hinchliffe I, Huston J *et al.*
[6] Alwall J, Ballestrero A, Bartalini P, Belov S, Boos E *et al.* 2007 *Comput.Phys.Commun.* **176** 300–304 (*Preprint* hep-ph/0609017)
[7] URL http://www.jthaler.net/olympicswiki/
[8] Brun R and Rademakers F 1997 *Nucl.Instrum.Meth.* **A389** 81–86
[9] http://www.zlib.net
[10] Cacciari M, Salam G P and Soyez G 2012 *Eur.Phys.J.* **C72** 1896 (*Preprint* 1111.6097)
[11] de Favereau J *et al.* (DELPHES 3) 2014 *JHEP* **1402** 057 (*Preprint* 1307.6346)
[12] Dumont B, Fuks B, Kraml S, Bein S, Chalons G *et al.* 2014 (*Preprint* 1407.3278)
[13] Kraml S, Allanach B, Mangano M, Prosper H, Sekmen S *et al.* 2012 *Eur.Phys.J.* **C72** 1976 (*Preprint* 1203.2489)

[14] http://inspirehep.net, see also http://www.projecthepinspire.net
[15] http://www.datacite.org
[16] Alloul A, Frank M, Fuks B and Rausch de Traubenberg M 2013 *JHEP* **1310** 033 (*Preprint* 1307.5073)
[17] Chatrchyan S *et al.* (CMS Collaboration) 2014 *JHEP* **1406** 055 (*Preprint* 1402.4770)
[18] Bein S and Sengupta D 10.7484/INSPIREHEP.DATA.83GG.U5BW
[19] Chatrchyan S *et al.* (CMS Collaboration) 2013 *Phys.Rev.* **D88** 052017 (*Preprint* 1301.2175)

# Clad – Automatic Differentiation Using Clang and LLVM

**V Vassilev[1,2], M Vassilev[2], A Penev[2], L Moneta[1], and V Ilieva[3]**
[1] CERN, PH-SFT, Geneva, Switzerland
[2] FMI, University of Plovdiv Paisii Hilendarski, Plovdiv, Bulgaria
[3] Princeton University, Princeton, NJ, USA

E-mail: vvasilev@cern.ch

**Abstract.** Differentiation is ubiquitous in high energy physics, for instance in minimization algorithms and statistical analysis, in detector alignment and calibration, and in theory. Automatic differentiation (AD) avoids well-known limitations in round-offs and speed, which symbolic and numerical differentiation suffer from, by transforming the source code of functions. We will present how AD can be used to compute the gradient of multi-variate functions and functor objects. We will explain approaches to implement an AD tool. We will show how LLVM, Clang and Cling (ROOT's C++11 interpreter) simplifies creation of such a tool. We describe how the tool could be integrated within any framework. We will demonstrate a simple proof-of-concept prototype, called Clad, which is able to generate n-th order derivatives of C++ functions and other language constructs. We also demonstrate how Clad can offload laborious computations from the CPU using OpenCL.

## 1. Introduction

Both industry and science often use the mathematical apparatus of differential calculus. Modeling financial markets, climatic changes or searching for the Higgs boson use function optimization and thus derivatives. The numerical calculation of the derivative values yields precision losses. They come from machine's floating point representation and the stability of the used numerical method. The computation fragility becomes even worse when computing higher order derivatives. In practice the user (a programmer) must consider very carefully the input values and the stepping delta, which sometimes can be far from trivial. Moreover, the derivative is hard-coded and becomes a maintenance issue. The developer has to differentiate the function mentally or using an external tool and translate it to the implementation language.

An alternative approach is the so called symbolic differentiation which overcomes the above-described issues, but sometimes it is slow [1]. It does not offer straight forward framework integration. For example, the function to be differentiated is hardcoded the programming language (e.g. C++) and it has to be translated to the symbolic language, differentiated and the result needs to be translated back to the framework's programming language. Moreover, both numerical and symbolic differentiation methods suffer when computing gradients and higher-order derivatives.

Despite sometimes overlooked, there is a hybrid approach which lays in the middle of the both extremes, resolving the mentioned issues. It is the automatic/algorithmic differentiation (AD). A widely accepted definition of AD is "a set of techniques to numerically evaluate the derivative of a function specified by a computer program. AD exploits the fact that every computer program, no matter how complicated, executes a sequence of elementary arithmetic operations (addition, subtraction, multiplication, division, etc.) and elementary functions (exp, log, sin, cos, etc.). By applying the chain rule repeatedly to these operations, derivatives of arbitrary order can be computed automatically, accurately to working precision, and using at most a small constant factor more arithmetic operations than the original program.".

This paper is divided into sections as follows: Section 2, Related Concepts, discusses in brief some of the existing tools in the field and demonstrates their advantages and disadvantages. Section 3, Concepts of Clad, lays down the key concepts of our prototype – Clad. Section 4, Implementation, uncovers some of the important technical details of the concrete realization. Section 5, Applications, shows a few use cases in the context of the computer graphics and high-energy physics. Section 6, Computation Offload, describes a conceptual implementation employing parallel derivative computations. Section 7, Conclusion and Plans.

## 2. Related Concepts

Informally, the variety of incarnations of the AD can be classified in three major classes:

- Implemented via operator overloading – quick to implement at the cost of excessive amount of memory; most compilers evaluate expressions containing overloaded operators exactly as they are written, without performing any compile-time optimizations [2].
- Implemented via source-to-source transformations – results in faster derivative generation than the operator overloading. The concept of the idea is to rewrite an expression in a particular computer language, such as C++ or Fortran [3]. In order for the previously mentioned operations to be executed, the AD tool have to perform compiler-like processes like code parsing, code analyses, intermediate representations. The source transform approach is considered harder to implement but on the other hand it has advantages such as: better compiler optimizations and relatively lower amount of used memory [4].
- Implemented via compiler modules – an ideal realization of the source-to-source transformations is implementing the AD compiler modules. Integrating automatic differentiation capabilities in a compiler combines the advantages of both the operator overloading and source transform approaches [5].

ADOL-C is an AD tool, based on operator overloading. It can produce first and higher order derivatives of vector functions written in C/C++. One of its advantages is that the derived functions are valid C/C++ routines. ADOL-C can handle codes based on classes, templates and other C++ features. The tool supports the computation of standard objects required for optimization purposes such as gradients, Jacobians and Hessians [6]. ADOL-C uses the concept of active variables for denoting the possible differentiation independent variables. All of those variables have to be declared with a special variable type. The process of derivation is based on an ADOL-C specific internal representation and its start and finish are denoted with calls to special service routines. Every calculation incorporating active variables within the derivation process are recorded in a special data type. Once this is done, ADOL-C proceeds with executing its internal algorithms for computing the derivatives.

ADIC2 is a project following the source-to-source transform for Fortran, C and C++ based programs. It is built on top of the OpenAD project, which incorporates multiple, independent software components [7]. ADIC2 uses the ROSE compiler framework for parsing the input source code programs and for generating corresponding abstract syntax trees (ASTs). Unfortunately ADIC2 is closely coupled with ROSE thus changes in ROSE may result in failures in ADIC2. The tool uses configuration files specifying parameters and settings related to the differentiation process. The AST is passed to a dedicated analyzer which is reducing the amount of code passed to the differentiation algorithms. A XML based data structure is used in order to denote which parts of the source code should be differentiated and which parts should be marked as statements and thus ignored by the differentiation algorithms. This XML based representation is used by another tool producing the differentiation and a differentiated AST. ADIC2 uses again the ROSE compiler for creating the output source code from the resulted AST.

The Fortran 95 compiler AD incorporates both the operator overloading and source transform techniques thus it is considered as a hybrid compiler. The mathematics support is handled by a compiler integrated module which provides overloaded versions of the arithmetic operators. The user is allowed to select the dependent and independent variables. The compiler uses a special active data type that is used to hold the function value as well as a vector for the direction derivatives. The overloaded operators as well as the active data types are contained in a compiler specific module. The independent variables as well as sections of the code that needs to be differentiated need to be marked using directive-like statements, in order for the compiler to recognize them. Unfortunately, no static data flow analyses of the code is conducted therefore the achieved efficiency is not optimal [8].

Source-to-source transformation technique is considered as the ideal approach for building large scale and run time crucial applications [9]. Furthermore, if the source-to-source transformers are implemented as a part of an existing compiler, it makes the implementation extremely efficient and easy to maintain at the cost of limited portability.

## 3. Concepts of Clad

An automatic differentiation algorithm takes a function ($F$) written in a programming language ($L$), translates it and yields another function ($F'$) written in an another programming language ($R$), where the translation between ($F$) $\rightarrow$ ($F'$) follows the rules of the differential calculus and turning $F'$ into a derivative. For many tools ($L$) matches ($R$), i.e. the implementation language of the input function is the same as the programming language of the differentiated function. Usually it is so, because of design or technical limitations of the implementation. Most of the uses of ($F'$) tend to be in the same framework and programming environment. An interesting domain of research is when ($L$) is different from ($R$) but ($L$) compatible with ($R$). For example, ($F$) is written in C/C++ and ($F'$) is written in OpenCL [10] or CUDA [11].

A derivative is produced by transforming the body of $F$ operation by operation. Every operation is transformed following the well-known differentiation rules. Every statement in a C++ function is treated as a standalone transformation entity. The translation employs the chain rule of differentiation.

The chain rule in differential calculus provide mathematically proved simplification of the translation process. It reduces the implementation complexity of the algorithm responsible for the differentiation. There are two general flavors of implementing the differentiation: top-down or bottom-up. In the top-down approach (also called forward mode or tangent mode) the computations are done on every step and the byproducts are thrown away. In the bottom-up approach (also called reverse mode or adjoint mode) the computations of the byproducts are stored and reused. This is particularly useful when computing derivatives of the same function with respect to different independent variables, for example when calculating function's gradient.

$$\frac{\partial z}{\partial t} = \frac{\partial z}{\partial x}\frac{\partial x}{\partial t} + \frac{\partial z}{\partial y}\frac{\partial y}{\partial t}, \tag{1}$$

where $z = f(x, y), x = g(t), y = h(t)$.

AD works with the assumption that a derivative exists, i.e. it does not check whether the differentiated function is continuous at any interval. If the logic of the original function is mathematically incorrect, there is no way how AD can produce a correct derivative.

## 4. Implementation

Clad is an AD tool, implemented on top of the LLVM compiler infrastructure [12]. The underlying compiler platform (Clang) builds an internal representation of the source code in the form of an AST. Parts of AST are used to communicate with third-party libraries, which can further specialize the compilation process. Because of the internal design and the communication through well-defined representations, Clad can be packed in different ways. It can be easily transformed from a plugin into a standalone tool or as an extension to third-party tools such as the interactive C++ interpreter Cling [13].

Once Clad receives the necessary information, it can decide whether a derivative is requested and produce it. It transforms the body of the candidate for differentiation and clones its AST nodes while applying the differentiation rules. If the body is not present it tries to find user provided directions how to proceed.

### *4.1. Usage*

Currently, Clad is shipped as a plugin library for the Clang compiler. One can attach the library to the Clang compiler and it will produce derivatives at compilation time as a part of the current object file. However, this approach has some limitations: it is bound to a particular compiler and compiler version. It is also required to compile the project, using derivatives with the same compiler. For this reason, Clad can operate in three conceptually different modes (Figure 1), generating derivatives in different representations. Derivatives can be a part of, either:

- an object file – if a derivative is requested Clad would put it as if it was present in the source file;
- a source file – Clad can write out valid source code of the derivative into a source file. This is handy when the user wants to produce the list of the requested derivatives and compile them with another compiler;
- a shared library – Clad can write out the derivatives into a shared object (dynamic-link library). This is useful when the user prefers another compiler, which is binary compatible with Clang, For instance, the user wants to compile the application with GCC/ICC but wants the derivatives to be still used.

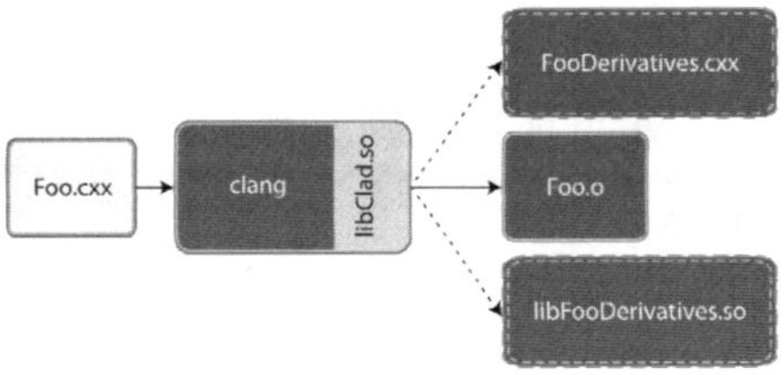


Figure 1: Clad usage scenarios.

Listing 1 shows how simple is to use Clad. The example demonstrates how to produce the first derivative of power of two (*pow2*). One needs to include a small header file, introducing Clad's runtime and use *clad::differentiate* function to specify which function needs to be differentiated. It takes two arguments: function to differentiate and the position of the independent variable. In the example below, Clad will fill in the body of *pow2_darg0* statement-by-statement following the derivation rules.

Listing 2 and 3 illustrate the textual mode in Clad. Clad can use the produced derivative straight away or to write its source code into a file. In the Listing 2 there is the user code and in the Listing 3 there are the automatically generated derivatives.

```
#include "clad/Differentiator/Differentiator.h"
double pow2(double x) { return x * x; }
double pow2_darg0(double); // Body will be filled by Clad.
int main() {
  clad::differentiate(pow2, 0);
  printf("Result is %f\n", pow2_darg0(4.2)); // Prints out 8.4
  return 0;
}
```

Listing 1: Clad will produce the body of the forward declared function.

```
#include "clad/Differentiator/Differentiator.h"

float example1(float x, float y) {
  return x * x * y * y;
}
void diffExamples() {
  clad::differentiate(example1, 0);
  clad::differentiate(example1, 1);
}
```

Listing 2: Differentiation of function *example1.*

```
float example1_darg0(float x, float y) {
  return (((x + x) * y) * y);
}

float example1_darg1(float x, float y) {
  return ((x * x) * y + x * x * y);
}
```

Listing 3: Derivatives of the function *example1.*

Valid C++ functions are generated with a system-defined name inferred from original function name, the independent variable of differentiation, and the derivative order. By construction, the signature of the newly-generated function is the same as the derivation template function.

Differentiation may be applied not only on simple functions but also on more complex language constructs such as:

- Templated C++ constructs;
- Classes and structs (eg. functors);
- Virtual functions.

*4.1.1. Builtin Derivatives* Some functions don't have bodies, because they are only forward declared and their implementation is in a library. The differentiation of some functions could be steered for performance improvements. Clad has a mechanism allowing to override the default differentiation policy and provide user-directed substitutions. Listing 4 specializes the default behavior of the differentiation, by replacing the default cosine (*cos*) with a user-specified one.

The implementation of the user-specific substitutions relies on a namespace with overloaded semantics. All substitutions need to be inside a special namespace called *custom_derivatives* and to follow specific naming rules. Before Clad builds a derivative, it checks if there is already a predefined derivative candidate. If this is the case it simply uses the one available.

All built-in derivatives rely on this mechanism. Differentiation of trigonometric functions and the derivatives of other special functions is done using user substitutions.

*4.1.2. Higher Order Derivatives and Mixed Partial Derivatives* Clad provides a convenient way of obtaining n-th order derivative out of a specific function. This is possible by invoking the templated version of the *clad::differentiate* function for instance see Listing 4.

Mixed derivatives are produced on Listing 5.

Journal of Physics: Conference Series **608** (2015) 012055 doi:10.1088/1742-6596/608/1/012055

```
//...
namespace custom_derivatives { double sin_darg0(double x) { return my_better_impl::cos(x); } }
void secondDerivative() {
  clad::differentiate<2>(sin, 0);
}
```

Listing 4: Generation of the second derivative of sin, using substitutions.

```
float example2(float x, float y) {
  return x * x * y;
}
float example2_darg0(float x, float y);
auto example_darg0 = clad::differentiate(example2, 0);
auto example_darg0_darg1 = clad::differentiate(example2_darg0, 1);
```

Listing 5: Generation of mixed partial derivatives.

### *4.2. Performance*
Clad works in synergy with the Clang compiler. Table 1 shows the times to compile the test cases with and without Clad. We test the overhead in two extreme cases – one, when it differentiates a function body with many statements (Listing 6), and another, where the body contains a very large expression (Listing 7). In more realistic scenarios Clad's overhead is negligible. It is so because the functions to be differentiated are much shorter, and the volume of these functions compared to the rest of the code is much smaller.

Table 1: Compilation times of Listing 6 and Listing 7.

| Test | With Clad | No Clad | Clad overhead |
|---|---|---|---|
| Large body | 1.007s | 0.993s | 0.014s ( 1.39%) |
| Large expression | 3.258s | 2.432s | 0.826s (25.35%) |

```
double f1(double x) {
  x = x + 2; x = x * x; x = x + x;
  // ... repeated 1020 times.
  return x;
}
```

Listing 6: Performance – Large function body.

```
double f2(double x) {
  return
    x+2 + x*x + x+x +
    // ... 1020 repetitions.
}
```

Listing 7: Performance – Large expression.

## 5. Applications
We explored two major application domains for the automatic differentiator Clad – computer graphics and high-energy physics.

### *5.1. Computer Graphics*

We embedded Clad in a demonstration path tracer, called SmallPT [14]. We investigated how difficult and laborious would the integration be. SmallPT shoots rays of light from the viewer towards the scene and computes lighting properties at the intersection points. In order to do that, it computes three partial derivatives of the function describing the surface. They form the normal vector at the intersection point between a ray and a surface. We replaced the hand-written implementation with invocations of Clad. The performance is comparable to the performance of the hand-written calculations. Calculating the normal vectors using numerical approximation is about three times slower. The timing was collected from $10^{10}$ calculations of the gradient at random points on a sphere and a hyperbolic solid. Another advantage of embedding Clad allows even to improve the flexibility of our modification of SmallPT implementation, because it allows to find derivatives of arbitrary implicit surfaces (not only spheres as in the original code). The addition of a new surface is now easier, because the hand-written derivative will not be needed.

### *5.2. High-Energy Physics*

The ROOT Framework [15] is widely adopted in high-energy physics for data analysis. It offers a wide range of mathematical tools for fitting and minimization. These tools use extensively derivatives and some of them are hand-written while others are numerically calculated. We plan on adopting Clad in ROOT6 through its C++ interpreter – Cling [13]. We expect performance improvement and Clad is expected to become a gateway for derivative computation on General-Purpose Computing on Graphics Processing Units (GPGPU). This work is still to be done soon.

## 6. Computation Offload

Derivative calculation is very computing-intense and time-consuming process. Clad's immediate goal is to increase the computational performance and to make use of all computing power of the environment. The natural evolution towards execution on GPGPUs is prominent. Computation can be accelerated by using the offloading of calculations from CPU towards the available GPGPUs. There are mainly two conceptually-different approaches:

- Using built-in approaches in the compiler to guide compilation (usually by 'pragma' directives) to turn on the parallel execution, the automatic offload of specific computation-intense parts of the code to selected accelerators. This is the approach in OpenMP 4.0 [16]. The advantage is that the code does not change significantly. If the compiler does not support these options, it ignores the parallelisation directives gracefully. The disadvantages are: very hard to make full use of the underlying computing architectures, because the pragma directives cannot specialize the algorithm for all architectures; requires some effort and re-engineering to get some parallelism.
- Using a specific programming language (such as OpenCL and CUDA) to talk to the hardware. This approach supposes rewriting the algorithms, or parts of them. The main advantage of this approach is that the algorithm can fully comply with the architecture of the target hardware. The disadvantage is that it requires a lot of effort and expertise to port it for every architecture and hardware.

We target scalability by changing semantics of the language syntax constructs and based on them we enhance the compiler actions. A key goal for this transformations is to be as transparent as possible. Most actions are performed automatically – at compile time. The goal is to use most of the advantages of the above-mentioned approaches, without forcing major changes in the user code.

Listing 8 presents an example of CPU-calculated gradient. In the implementation we find a sum of the partial derivatives of the Rosenbrock function.

```
float rosenbrock(float x[], int size) {
  auto rosenbrockX = clad::differentiate(rosenbrock_func, 0);
  auto rosenbrockY = clad::differentiate(rosenbrock_func, 1);
  float sum = 0;
  for (int i = 0; i < size-1; i++) {
    float one = rosenbrockX.execute(x[i], x[i + 1]);
    float two = rosenbrockY.execute(x[i], x[i + 1]);
    sum += one + two;
  }
  return sum;
}
```

Listing 8: Rosenbrock function implementation using Clad.

Listing 9 shows how the user should specify that he/she wants the compilation to be offloaded. The original implementation needs to be transformed into a lambda function, which is very straight-forward and almost transparent conversion. It is not much harder than putting a pragma directive in the source code. The code is not trivial to offload, because it has a for-loop, which should be made parallel to be able to exploit the properties of the GPGPU architecture. Furthermore, we need to calculate the sum of the results of the calculations, that requires reduction in the highly parallel hardware.

```
float rosenbrock_offloaded(float x[], int size) {
  return clad::experimental_offload([=] {
    auto rosenbrockX = clad::differentiate(rosenbrock_func, 0);
    auto rosenbrockY = clad::differentiate(rosenbrock_func, 1);
    float sum = 0;
    for (int i = 0; i < size-1; i++) {
      float one = rosenbrockX.execute(x[i], x[i + 1]);
      float two = rosenbrockY.execute(x[i], x[i + 1]);
      sum += one + two;
    }
    return sum;
  });
}
```

Listing 9: Computation offloading – Conceptual implementation using C++11 lambda function.

After the minimalistic transformation is done, Clad should take over and perform the rest automatically. Every call to *clad::experimental_offload* is replaced by a call to a function generated by the Clad plugin. This is done by an AST transform of the lambda function, allowing to transmit the parameters and to invoke one or more kernels.

In order to achieve optimal performance, Clad has to keep the computing units busy, which is far from a trivial task. Computation offload to GPGPU has to be very well planned and should happen when processing large data sets. One of the reasons is that all data needs to be copied through the system bus to a peripheral device, which introduces big overheads. Clad-generated code which passes data and executes kernels, should provide optimal parallel load as well as minimum data transfer between different types of memory (as it is possible to get too much latency). The use of read only, write only parameters and other similar techniques are recommended in order to achieve better performance.

The given example was tested for performance with 100 calls to Rosenbrock's function over 1024*1024*48 float numbers. The resulting benchmarks are shown in Table 2.

Table 2: OpenCL parallel execution results.

| Test | Device | Clock | Compute Units | Work Group | Global Memory | Local Memory | Time |
|---|---|---|---|---|---|---|---|
| Original | Intel i7-2635QM | 2GHz | 1 | – | – | – | 12.466s |
| Multicore | Intel i7-2635QM | 2GHz | 4 | 1024 | 8192MB | 32KB | 11.128s |
| ATI | Radeon 6490M | 150MHz | 2 | 256 | 256MB | 32KB | 18.183s |
| AMD | Devastator | 844MHz | 6 | 256 | 2047MB | 32KB | 15.479s |
| NVIDIA | Tesla K20m | 705MHz | 13 | 1024 | 4800MB | 47KB | 10.615s |

The benchmarking showed that the CPU multicore is slightly faster than the original single core computation. This is because of the copying of large amounts of data. Overall, the GPGPU offload is very sensitive to the particular hardware. This is mainly due to the copy of large data sets sequentially between the host and the device. There is also latency because of the synchronous manner of execution. The next data transfer waits for the previous computation to be executed. In the case of NVIDIA Tesla, the CPU offload reached 50%, which is very good result for a prototype implementation.

## 7. Conclusion and Plans

Derivative production is important not only in high-energy physics but in many other domains. The automatic differentiation is often an overlooked approach to compute derivatives. It eliminates the precision losses in the numerical differentiation and it is faster than symbolic differentiation. It can also simplify the complexity of gradient computations. The AD tool mainly focuses on C and partially C++, because of the complexity of the language. We presented an innovative proof-of-concept prototype facilitating automatic differentiation, called Clad. It is based on the industrial-strength compiler technologies Clang and LLVM. Clad can differentiate non-trivial C++ routines and it is getting closer to production grade quality. Adding C support is trivial, the only work that needs to be done is writing the runtime environment to be C compliant.

We explained how it can be used to produce derivatives of various orders; how to produce mixed derivatives; how to perform user-based substitutions, steering the differentiation process; and how to offload computations in heterogeneous environments. There is still a lot of room for improvements. We have a conceptual implementation in OpenCL, providing a way how to offload computations in heterogeneous environments (making extensive use of GPGPUs). However, this work is still experimental and it requires a lot of efforts for Clad to be made robust.

We plan to generalize the computation of gradient and Jacobian of functions. We plan to reduce the computational complexity of these computations by making use of the reverse automatic differentiation mode in cases of many seeds.

Another immediate plan is to integrate Clad into Cling – the C++ interpreter of ROOT6, which would make Clad available to the entire high-energy physics community. Then, we can proceed using Clad in ROOT's minimization and fitting algorithms of ROOT.

## Acknowledgments
The work was partially facilitated by Google Summer of Code Program 2013 and 2014, through CERN-SFT mentoring organization. The heterogeneous environment benchmarking are supported by the University of Plovdiv "Paisii Hilendarski" through Fund "Research" under contract NIS14-FMIIT-002/26.03.2014.

## References
[1] Castro M, Vieira R and Biscaia Jr E 2000 Automatic differentiation tools in the dynamic simulation of chemical engineering processes *Brazilian Journal of Chemical Engineering* **17** 373–382 ISSN 0104-6632 URL http://www.scielo.br/scielo.php?script=sci_arttext&pid=S0104-66322000000400002&nrm=iso

[2] Bartholomew-Biggs M, Brown S, Christianson B and Dixon L 2000 Automatic differentiation of algorithms *Journal of Computational and Applied Mathematics* **124** 171–190 ISSN 0377-0427 numerical Analysis 2000. Vol. IV: Optimization and Nonlinear Equations URL http://www.sciencedirect.com/science/article/pii/S0377042700004222

[3] Gay D 2006 Semiautomatic differentiation for efficient gradient computations *Automatic Differentiation: Applications, Theory, and Implementations* (*Lecture Notes in Computational Science and Engineering* vol 50) ed Bücker M, Corliss G, Naumann U, Hovland P and Norris B (Springer Berlin Heidelberg) pp 147–158 ISBN 978-3-540-28403-1 URL http://dx.doi.org/10.1007/3-540-28438-9_13

[4] Bischof C H, Hovland P D and Norris B 2002 Implementation of automatic differentiation tools. *PEPM* ed Thiemann P (ACM) pp 98–107 ISBN 1-58113-455-X URL http://dblp.uni-trier.de/db/conf/pepm/pepm2002.html#BischofHN02

[5] Cohen M, Naumann U and Riehme J 2003 Towards differentiation-enabled Fortran 95 compiler technology *Proceedings of the 2003 ACM Symposium on Applied Computing* SAC '03 (New York, NY, USA: ACM) pp 143–147 ISBN 1-58113-624-2 URL http://doi.acm.org/10.1145/952532.952564

[6] Walther A and Griewank A 2012 Getting started with ADOL-C *Combinatorial Scientific Computing* ed Naumann U and Schenk O (Chapman-Hall CRC Computational Science) chap 7, pp 181–202

[7] Narayanan S H K, Norris B and Winnicka B 2010 ADIC2: Development of a component source transformation system for differentiating C and C++ *Procedia Computer Science* **1** 1845–1853 ISSN 1877-0509 iCCS 2010 URL http://www.sciencedirect.com/science/article/pii/S1877050910002073

[8] Naumann U and Riehme J 2005 A differentiation-enabled Fortran 95 compiler *ACM Trans. Math. Softw.* **31** 458–474 ISSN 0098-3500 URL http://doi.acm.org/10.1145/1114268.1114270

[9] Voßbeck M, Giering R and Kaminski T 2008 Development and first applications of TAC++ *Advances in Automatic Differentiation* (*Lecture Notes in Computational Science and Engineering* vol 64) ed Bischof C H, Bücker H M, Hovland P D, Naumann U and Utke J (Springer Berlin Heidelberg) pp 187–197 ISBN 978-3-540-68935-5 URL http://dx.doi.org/10.1007/978-3-540-68942-3_17

[10] Gaster B, Howes L, Kaeli D R, Mistry P and Schaa D 2013 *Heterogeneous Computing with OpenCL: Revised OpenCL 1.2 Edition* 2nd ed (San Francisco, CA, USA: Morgan Kaufmann Publishers Inc.) ISBN 9780124055209

[11] Cook S 2013 *CUDA Programming: A Developer's Guide to Parallel Computing with GPUs* 1st ed (San Francisco, CA, USA: Morgan Kaufmann Publishers Inc.) ISBN 9780124159334

[12] Lattner C and Adve V 2004 LLVM: A compilation framework for lifelong program analysis & transformation *Proceedings of the 2004 International Symposium on Code Generation and Optimization (CGO'04)* (Palo Alto, California)

[13] Vasilev V, Canal P, Naumann A and Russo P 2012 Cling – the new interactive interpreter for ROOT 6 *Journal of Physics: Conference Series* **396** 052071 URL http://stacks.iop.org/1742-6596/396/i=5/a=052071

[14] Beason K 2014 SmallPT: Global illumination in 99 lines of C++ URL http://www.kevinbeason.com/smallpt/

[15] Brun R and Rademakers F 1997 ROOT – an object oriented data analysis framework *Nuclear Instruments and Methods in Physics Research Section A: Accelerators, Spectrometers, Detectors and Associated Equipment* **389** 81–86 ISSN 0168-9002 new Computing Techniques in Physics Research V URL http://www.sciencedirect.com/science/article/pii/S016890029700048X

[16] OpenMP A R B 2013 OpenMP application program interface URL http://www.openmp.org/mp-documents/OpenMP4.0.0.pdf

Journal of Physics: Conference Series **608** (2015) 012056 doi:10.1088/1742-6596/608/1/012056

# Upgrades for the CMS simulation

**D J Lange[1], M Hildreth[2], V N Ivantchenko[3], and I Osborne[4] for the CMS Collaboration**

[1]Lawrence Livermore National Laboratory, Livermore, Ca USA 94550.
[2]University of Notre Dame, Notre Dame, In, 64556, USA
[3]CERN, Geneva, Switzerland.
[4]Fermi National Laboratory, Batavia, Il USA 60510.

**Abstract**. Over the past several years, the CMS experiment has made significant changes to its detector simulation application. The geometry has been generalized to include modifications being made to the CMS detector for 2015 operations, as well as model improvements to the simulation geometry of the current CMS detector and the implementation of a number of approved and possible future detector configurations. These include both completely new tracker and calorimetry systems. We have completed the transition to Geant4 version 10, we have made significant progress in reducing the CPU resources required to run our Geant4 simulation. These have been achieved through both technical improvements and through numerical techniques. Substantial speed improvements have been achieved without changing the physics validation benchmarks that the experiment uses to validate our simulation application for use in production. In this presentation, we will discuss the methods that we implemented and the corresponding demonstrated performance improvements deployed for our 2015 simulation application.

## 1. Introduction

The increasing beam energy and pileup conditions of the LHC will bring new challenges to the simulation program of the CMS experiment for the next decade. Given both short and long term upgrades to the accelerator and experimental apparatus, there are significant challenges to be faced to ensure good physics performance of the experiment for both today and tomorrow. In particular, higher LHC beam energy means more complex interactions, a larger number of interactions per bunch crossing (higher pileup) which are more time consuming to simulate and reconstruct. At the same time, CMS plans a much higher output rate of trigger (~1 kHz, or 2.5x higher than in Run 1), which means a demand for correspondingly larger samples of simulated events. In the longer term, simulating a potential CMS detector for the high-luminosity LHC program (HL-LHC) has additional challenges including a still to be determined CMS detector configuration, a higher output rate of trigger (potentially 10 kHz), and further increases in luminosity and pileup conditions (i.e., more than 140 interactions per crossing).

These proceedings discuss a number of recent changes made to the CMS simulation application to allow CMS to continue to rely on a high-fidelity detector simulation as the basis for its analysis and detector optimization work. The detector response for the primary CMS simulation application is performed using Geant4 [1]. Our recent work has brought substantial technical performance improvements without changing the physics accuracy of our simulation.

## 2. CMS Detector upgrades

The CMS experiment is working on instrumentation changes on three different timescales. The most immediate detector upgrades are being implemented during the on-going LHC shutdown period and will be fully completed before the start of the Run 2 data collection. The most significant changes are to replace the central region of the beam pipe, to complete the muon detector coverage as it was originally designed, and to update the electronics in part of the hadronic calorimeter.

The "Phase-I" upgrades begin in 2017 and continue through the next long LHC downtime in 2018 [2,3,4]. The Phase-I upgrade will include a new pixel detector installed during the annual shutdown between 2016 and 2017, and more extensive changes during the LHC technical stop. These include substantial changes to the hadronic calorimeter electronics and Level-1 trigger.

Finally, the proposed "Phase-II" upgrades of CMS are major detector improvements for operation in the high luminosity LHC (HL-LHC) conditions. HL-LHC presents increased challenges for triggering, tracking and calorimetry, in particular for low to medium $P_T$ objects. As summarized in Figure 1, CMS is presently working on a Technical Proposal including a tracker replacement and extensive forward calorimetry and muon improvements. These extensive changes require a flexible simulation program. While physics results can be derived using a parameterized simulation, a full detector simulation is invaluable for doing validation of any fast simulation, in particular for a beam intensity regime that is very different from that of the LHC Run 1.

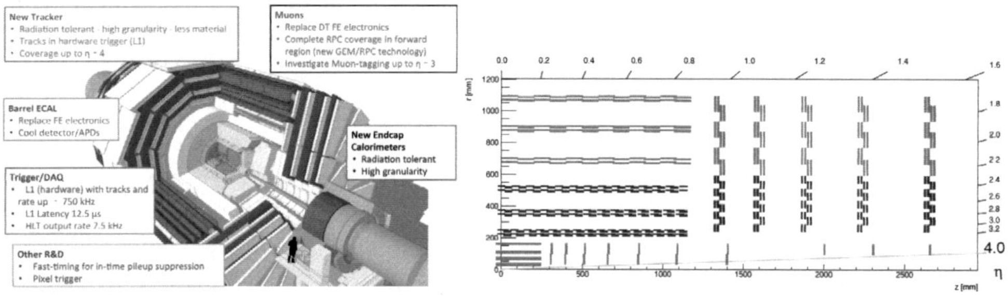


**Figure 1: Summary of CMS detector upgrade concepts for Phase II (left), and cross section of upgraded tracker detector showing simulated hits (right).**

## 3. Changes to CMS simulation for 2015

CMS has made a number of changes in our simulation application and infrastructure to facilitate both Run 2 and long-term upgrade work within the experiment. Users for both applications rely on a full Geant4-based detector simulation for reliable results.

An important aspect of this work is an underlying geometry model that is very flexible [5]. We are able to simultaneously support numerous geometries and sub-detector developers can easily update their XML-based geometry definitions as detector requirements are updated.

Goals of our recent work include achieving significantly better throughput (reduced time to simulate each event), easier configuration management in order to support additional detectors, and to enable very high pileup simulation within the CMS computing resource constraints. A number of projects have been undertaken over the course of several years in anticipation of the Run 2 requirements for larger simulation samples and for more complex events.

This Section describes our approach and the results of our work in these areas. We focus primarily on changes made to achieve large gains in throughput, and the significant changes we have implemented in our pileup simulation.

### *3.1. Monte Carlo sampling techniques: Russian Roulette*

A significant portion of the CMS simulation time is due the tracking of low-energy particles, particularly in the electromagnetic and hadronic calorimeter systems. We investigated and have now

Journal of Physics: Conference Series **608** (2015) 012056 doi:10.1088/1742-6596/608/1/012056

implemented a sampling technique used widely in Monte Carlo applications such as neutron shielding calculations ("Russian Roulette") that significantly reduces the time spent tracking these particles [6]. Specifically, we changed our simulation to track only a small fraction of gamma and neutron particles below a threshold energy (5 and 10 MeV, respectively) where the fraction and threshold were tuned such that the final physics output of our simulation was not affected. We found that it was necessarily to have sampling factors and thresholds that depend on both detector region and particle type.

Figure 2 illustrates the technique for a particle interaction. In this example, only one of the six low-energy particles in the shower resulting from the original particle is tracked using Geant. The energy deposits from this particle and its daughters are given an additional weight, $W$, corresponding to the inverse of the fraction of particles kept. We found that $W$=0.1 for neutrons was an adequate compromise between loss of physics performance and improvement in CPU performance.

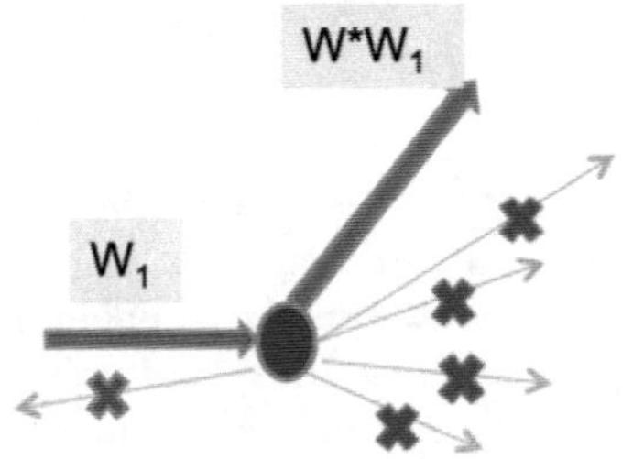


**Figure 2: Russian Roulette sampling technique.**

The performance improvement we observe from the application of this technique depends on the physics channel being simulated, but is typically between 25% and 40%. Our validation process showed that the energy and shower shape response in our high-resolution barrel electromagnetic calorimeter were the most sensitive to parameter tuning of the Russian Roulette technique.

### *3.2. Library packaging*

One purely technical improvement to the CMS simulation is that we have repackaged all shared libraries within our software that depend on Geant4 into a single static library. We also packaged all of the Geant4 libraries that we use into a single static library. These changes allow us to hide Geant4 from the rest of the CMSSW code base, and subsequently to evaluate the benefit of more aggressive linker options than used for the rest of CMSSW. The result of these investigations was that adding "-*flto -Wl,--exclude-libs,ALL*" to our linker options gave the best performance improvement, providing approximately an 10% reduction in the CPU needed per event with no detrimental impact on memory or other constrained resources.

This technique does impose a number of constraints. While these are not a large burden for simulation developments, they do need to be considered before extending this technique to more of the CMSSW codebase.

First, we must control dependencies to use Geant4 only within this single library. This is relatively straightforward for CMSSW code that depends on Geant4, as only a small fraction of our libraries (<2%) depend on it. Second, there is some impact on simulation developers due to the creation of this single large library. We have minimized this impact by keeping the original shared libraries cached in release. Therefore, any developer needing to rebuild one of the affected libraries just incurs an extra step of rebuilding the static library rather than all of the packages that depend on Geant4.

In limited experience so far, we have had no problems controlling dependencies that would impact the effectiveness of this single static library. More experience will allow us to better understand any operational overhead that we have introduced.

Journal of Physics: Conference Series **608** (2015) 012056 doi:10.1088/1742-6596/608/1/012056

*3.3. Summary of improvements*

The use of Monte Carlo sampling techniques and other purely technical changes in code optimization, together with the integration of Geant4.10, have provided a total reduction of about 50% in the CPU per event needed for the CMS simulation at 13 TeV within only the past one year of development. This improvement has been achieved without introducing a reduction in physics accuracy in our detector simulation. The largest components are the Russian Roulette technique and other CMS simulation code optimizations. We continue to work on other optimizations as we anticipate the need to further reduce the computing resources needed for our simulation application in the future.

## 4. Pileup Simulation

An import component of the CMS Monte Carlo simulation is the modeling of the minimum bias pileup events, both in the bunch crossing of the hard-scatter event simulated and in the nearby out-of-time crossings. Here we describe our original approach and recent developments undertaken to improve performance at high pileup.

*4.1. Run 1 approach*

CMS has always simulated pileup interactions by separately generating simulated samples of hard-scatter interactions and minimum bias events for pileup. We have employed this approach both to save processing time (by using events in the minimum bias samples more than one time) and to increase our flexibility to change the distribution of the number of pileup events much after the simulation samples themselves are generated. For example, this approach allows us to change the pileup distribution during a re-digitiziation and re-reconstruciton pass through the simulation samples.

Figure 3 shows the pileup simulation approach. The combination is done at the "SimHit" level as the input to the electronics simulation step. These SimHits corresponds to the full information saved from the Geant4 detector simulation.

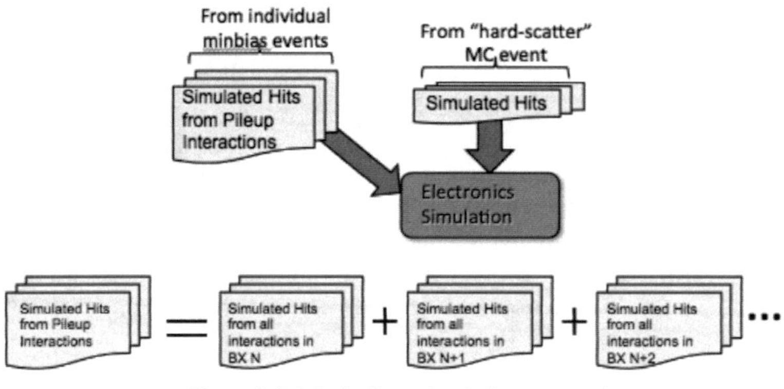


Figure 3 Original pileup simulation approach.

The figure shows the critical limitation of this approach. All SimHits from all pileup interactions (from all 16 bunch crossings that we model) are loaded into memory simultaneously. Given practical limits on the memory footprint of the simulation application, this limits our pileup simulation essentially the Run 1 operating conditions.

*4.2. Addressing the memory footprint problem*

Given expectations for higher pileup in both Run 2 and upgrade simulations, we refactored the pileup simulation to process each interaction sequentially, as shown in Figure 4. This work required a substantial rewrite of the digitization code and the re-organization of internal event processing framework.

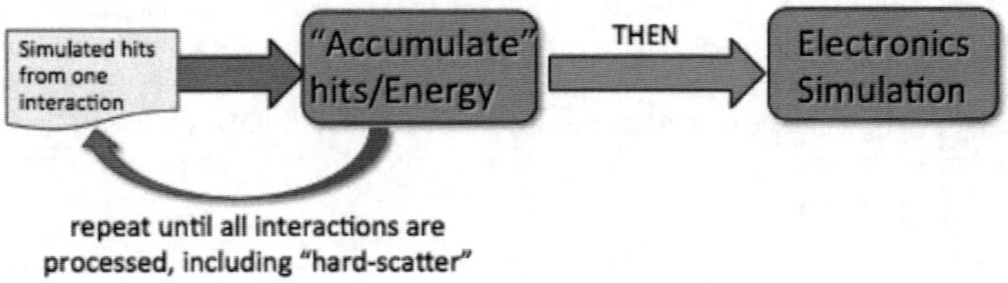


Figure 4: New pileup simulation approach.

The benefit of this new approach is that the content of individual interaction events is dropped once they have been processed so that only one event is in memory at any given time. This means that essentially an arbitrarily number of pileup events can be included in the digitization within the memory footprint limits of our computing resources. This work has been successfully deployed and used in a number of simulation production campaigns (primarily for the upgrade program).

### *4.3. Addressing the I/O problem*

While our current pileup-mixing scheme has reduced the memory requirement for this simulation to fit within resource constraints for the foreseeable future, we continue to read a large number of minimum bias event one-by-one to simulate pileup. In particular, we still read more than 2000 minimum bias events to produce a single event with appropriate pileup at HL-LHC luminosities in order to simulate the interactions in all of the bunch crossings needed to accurately model signals processed by the CMS detector.

We have developed a potential solution to the I/O component of the pileup simulation, which we call "pre-mixing". The idea, illustrated in Figure 5, is to create a library of events containing only pileup contributions, following pre-determined luminosity profile to calculate how many interactions to include. The result is saved in a raw data format. The hard-scatter sample is created and processed through the digitization step with no pileup, convert to our raw data format. The final step is then that the two streams are merged

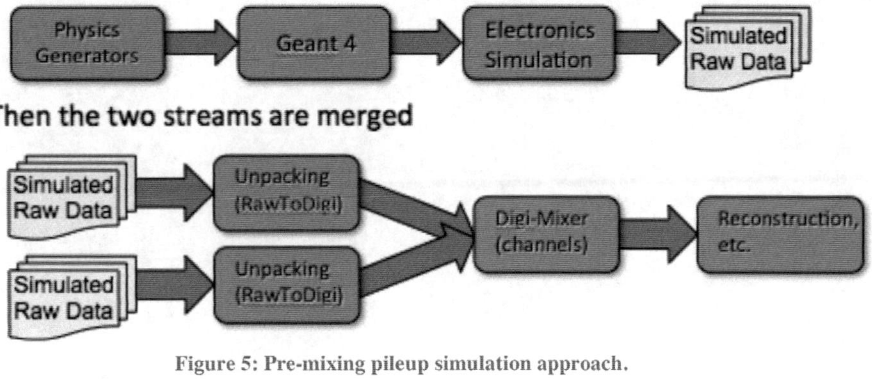


Figure 5: Pre-mixing pileup simulation approach.

The advantage of this approach is that only one pileup event is needed for the pileup simulation of each hard-scatter event. Thus it is much easier to process the pre-mixing approach through computing infrastructure once the initial premixed sample is created.

The initial version of pre-mixing has been implemented and deployed for part of our large 2015 preparation samples. Production tests went smoothly and demonstrated gains in I/O and CPU efficiency at high pileup. We have done an extensive validation against our normal mixing simulation approach. At this point, there are only a few known physics issues left to solve (saturation effects).

Another remaining issue is that we need to take additional care with the re-use of the individual minimum bias events. While our current mixing scheme reuses individual minimum bias events, the pre-mixing scheme can potentially reuse an entire set of minimum bias events, thus enhancing the effect of any important correlations between minimum bias events. We could avoid this entirely by generating one pre-mixed event for each hard-scatter event (e.g., 10 billion), but this would significantly reduce the potential benefit of this approach. We are evaluating the statistical issues with reuse of pileup events, but in any case, care is needed in the Monte Carlo production system to avoid excessive reuse of pre-mixed events.

## 5. Conclusion

We have summarized a number of areas of recent development work in CMS that have brought a big reduction in simulation resource needs for 2015 even in the face of higher event complexity and trigger rates. CMS detector upgrades push us to use today's software/computing infrastructure to simulate tomorrow's event complexity. In particular, the detector upgrade developments have proven to be an excellent platform for the quick deployment of new simulation features.

## References

[1] Agostinelli, S, *et. al.*, "Geant4 – a simulation toolkit", *Nucl. Inst. Meth.* **A506** (2003) 250.
[2] CMS Collaboration, "CMS Technical Design Report for the Pixel Detector Upgrade", CERN-LHCC-2012-016, (2012).
[3] CMS Collaboration, "CMS Technical Design Report for the Phase 1 Upgrade of the Hadron Calorimeter", CERN-LHCC-2012-015, (2012).
[4] CMS Collaboration, "CMS Technical Design Report for the Level-1 Trigger Upgrade", CERN-LHCC-2013-011, (2013).
[5] Osborne I, et al., "CMS geometry through 2020," CHEP 2013.
[6] Kalos, M H, "Importance sampling in Monte Carlo shielding calculations, *Nuci. Sci. Eng.*, **16**, 227, 1963.

ACAT2014 IOP Publishing
Journal of Physics: Conference Series **608** (2015) 012057 doi:10.1088/1742-6596/608/1/012057

# Traditional Tracking with Kalman Filter on Parallel Architectures

**Giuseppe Cerati[1], Peter Elmer[2], Steven Lantz[3], Ian MacNeill[1], Kevin McDermott[4], Dan Riley[4], Matevž Tadel[1], Peter Wittich[4], Frank Würthwein[1], Avi Yagil[1]**

[1] University of California - San Diego , La Jolla, CA, 92093, USA
[2] Department of Physics, Princeton University, Princeton, NJ 08540, USA
[3] Center for Advanced Computing, Cornell University, Ithaca NY 14853, USA
[4] Laboratory of Elementary Particle Physics, Cornell University, Ithaca NY 14853, USA

E-mail: Peter.Elmer@cern.ch

**Abstract.** Power density constraints are limiting the performance improvements of modern CPUs. To address this, we have seen the introduction of lower-power, multi-core processors, but the future will be even more exciting. In order to stay within the power density limits but still obtain Moore's Law performance/price gains, it will be necessary to parallelize algorithms to exploit larger numbers of lightweight cores and specialized functions like large vector units. Example technologies today include Intel's Xeon Phi and GPGPUs. Track finding and fitting is one of the most computationally challenging problems for event reconstruction in particle physics. At the High Luminosity LHC, for example, this will be by far the dominant problem. The most common track finding techniques in use today are however those based on the Kalman Filter. Significant experience has been accumulated with these techniques on real tracking detector systems, both in the trigger and offline. We report the results of our investigations into the potential and limitations of these algorithms on the new parallel hardware.

## 1. Introduction

Track finding and fitting is one of the most computationally challenging problems for event reconstruction in particle physics. At the High Luminosity LHC (HL-LHC), for example, this will be by far the dominant problem. The need for greater parallelism has driven investigations (e.g. [1, 2, 3]) of very different track finding techniques including Cellular Automata or returning to Hough Transform techniques originating in the days of bubble chambers. The most common track finding techniques in use today are based on the Kalman Filter [4]. Significant experience has been accumulated with these techniques on real tracking detector systems, both in the trigger and offline. They are known to provide high physics performance, are robust and are exactly those being used today for the design of the tracking system for HL-LHC. We report the results of our investigations into the potential and limitations of these algorithms on the new parallel hardware.

**Kalman Filter tracking:** We choose to attack charged particle tracking as it is one of the most CPU-time intensive tasks in the hadron collider event reconstruction chain. The CPU processing time is also acutely sensitive to increases in instantaneous luminosity and, therefore, pileup. Figure 1 shows the CPU time for a typical event reconstruction job. The growth term,

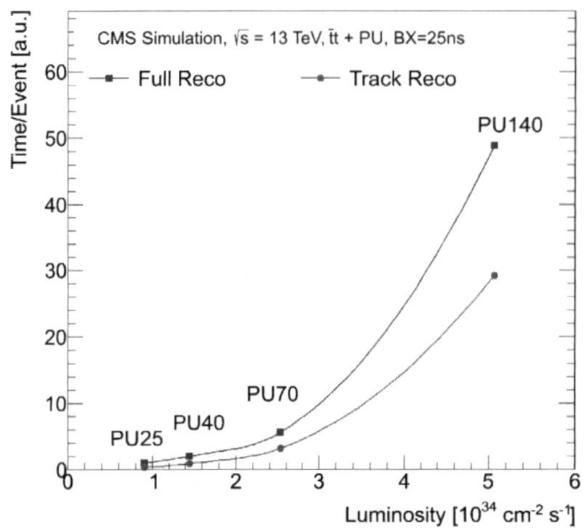


**Figure 1.** Timing vs instantaneous luminosity for $pp \to t\bar{t}$ simulated events, with overlaid pile-up events, with 25 ns bunch crossing. Samples with different average pile-up are used as reported on the plot. We show the timing of the full CMS reconstruction, and of the iterative tracking sequence by itself. As can be seen, the time required for tracking dominates the total time and increases sharply at higher values of the instantaneous luminosity.

*i.e.*, the increase in CPU time as a function of instantaneous luminosity, is exponential and the dominant contribution clearly comes from track reconstruction [5].

We choose a Kalman Filter-based algorithm as this high-performance algorithm is the standard in modern HEP experiments. For instance, the CMS experiment uses a Kalman Filter in both its high-level trigger as well as its offline reconstruction application [6]. The algorithm has well-understood timing performance characteristics in serial computing environments. The physics performance is also well understood in terms of efficiencies and fake rates.

Additionally, the Kalman Filter algorithm is adaptable for the wide vector units of modern CPUs and Xeon Phi. The algorithm can be implemented via matrix manipulations, which are well-suited for these hardware platforms.

To be able to harness new computing platforms, event-level parallelization across cores is the key to gaining more speed. Due to limitations in memory and I/O bandwidth, it is insufficient to run the serial algorithm for many events at once in an 'embarrassingly parallel' approach.

**Many-core:** There has been a sea change in the progess of computing in the past decade. Previously, one could expect that a given piece of code would run progressively faster with each generation of new processors with exponential gains consistent with Moore's-law-like growth. These gains were largely achieved by increases in the processor clock frequency. However, these gains have stalled in the last decade as processors have started to hit scaling limits, as has been widely recognized [7, 8]. The limits are largely driven by constraints on power consumption.

Moore's Law itself, which states that the number of transistors in an integrated circuit doubles approximately every two years, is still alive and kicking, but exists now in new forms. The first new form is the advent of so-called 'multi-core' processors, with more than one functional processor on a chip, however without increases in the main processor frequency. The next anticipated change is the advent of so-called 'small core' processors, such as ARM, GPUs, or MIC (Intel Xeon Phi). These small cores individually have less processing power than the big cores they replace, but in aggregate offer an immense amount of computing power. Sequential applications, such as those favored by HEP experiments, do not run out of the box on these platforms, however, and must be adapted.

As an example of deploying a complex algorithm on a many-core system, we explore the implementation of the CMS Kalman Filter tracking code on an Intel Xeon Phi.

## 2. Challenges

New architectures are "easy" to deploy for simple problems and tailored algorithms. This is not the problem we are facing; we need to migrate tools from a running experiment such as CMS at the CERN LHC. The serial algorithms currently in use have been optimized by many person-years of effort and set a high bar for the physics performance that we must meet before being able to deploy a many-core system. Existing algorithms require new thinking to refactor and adapt them to parallel architectures. Additional constraints come from the heterogeneous environment that we expect to use. Software for modern HEP experiments must run on a diverse array of platforms (basically, whatever is available across countries and continents.) Our code must be usable on big cores, little cores, and GPUs; we cannot use algorithms that only work well on one architecture. We are exploring a family of algorithms consistent with these constraints. Our current attempt focuses on the Intel Xeon Phi.

To exploit the Intel Xeon Phi, we need to vectorize and multithread a Kalman Filter algorithm for track building and fitting. Some of the expected difficulties include the fact that the algorithm uses small matrices, and has multiple decision points along the building of each track. Details of cache occupancy, pipelining, and latencies must be understood to achieve maximum possible performance. An expected benefit is that vectorization and multithreading should improve the serial (big-core) performance as well, which would immediately be applicable to a running experiment without the need for new hardware purchases.

## 3. Detector, Geometry, Simulation

For this study, we used a simplified detector geometry consisting of equally spaced, fixed length concentric cylinders, with radii at fixed steps (ten layers, 2 m long, spaced every 4 cm in radius). This was implemented using USOLIDS, a light-weight geometry library with interfaces based on ROOT and GEANT4 libraries which was created as part of the AIDA project [9]. The primary advantage of using an existing geometry library is that more complex geometries (polygonal geometry, z-segmentation, etc.) will be straightforward to implement as algorithm development progresses. We are also investigating eventual use of VECGEOM from GEANTV for better vectorization when generating our simulated data sets. To provide input data for the Kalman Filter tests, a standalone simulation was written which uses this full geometry for track propagation and simulates the detector resolution via Gaussian smearing of the hit positions.

We generate sets of tracks with transverse momentum in the range $0.5 < p_T < 10.0$ GeV. The tracks are limited to be within the central region of a barrel detector ($|\eta| < 1.0$) and generated uniformly in $\phi$. In this configuration, every track crosses every layer of our simplified detector. At the intersection of the generated tracks with the detector, we create a hit whose position is smeared in $x - y$ by $\sigma_{xy} = 0.01$ cm and in $z$ by $\sigma_z = 0.1$ cm.

## 4. Tracking

Track reconstruction can be divided into three steps: seeding, building, and fitting. While seeding is based on a different kind of algorithm, both track building and fitting are often based on a Kalman Filter. The Kalman Filter is an iterative procedure where a basic logic unit is applied repeatedly. The unit consists of the propagation of parameters and uncertainties (*track state*) from one layer to the next. At each layer, the track state is updated with the hit measurement information of the hits on that layer. (See Fig. 2 left.)

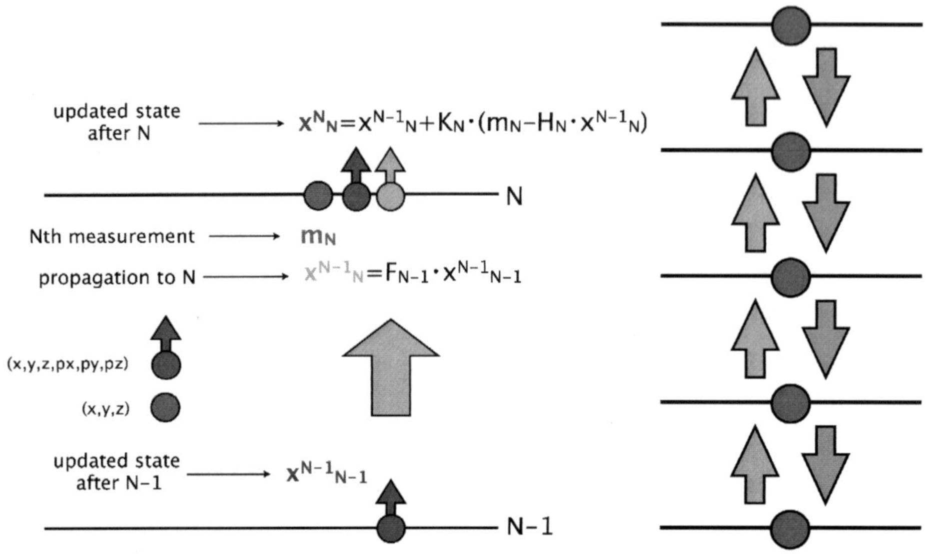


**Figure 2.** Left: Basic unit of the Kalman Filter algorithm. At each step, position information from hits is used to estimate the track parameters and their uncertainties. The red circle represents the measurement (a hit). The blue point on layer $N$ represents the estimated state (position and direction) at layer $N$ before taking into account information from hits on that layer. The yellow point is the updated state at layer $N$, taking into account all hits from up to and including layer $N$. Right: Cartoon representing the two stages of fitting: forward fit and backward smoothing. For this test we do not perform a smoothing step.

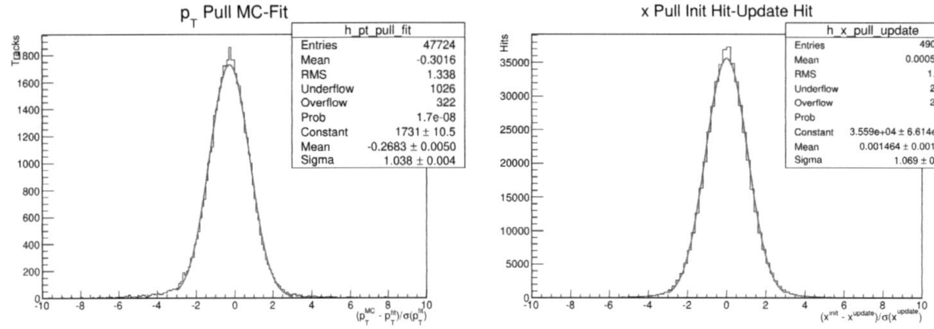


**Figure 3.** $p_T$ and position pull plots for the fit. The performance of the fit is under control.

As can be seen in Fig. 2 (right), the track fit consists of the simple repetition of the basic logic unit for all the pre-determined track hits and therefore it is the easiest case to test. It is divided in two steps: a forward fit and a backward smoothing stage for optimal performance. For this test, only forward fit is run. The Kalman Filter requires initial estimation of track parameters to get started. In our test, the starting state is taken directly from simulation with 100% uncertainty. As a more realistic option, we also implemented a parabolic fit in the conformal space to define initial parameters. For the fitting test, the hits are attached to a track "by name" (*i.e.*, no pattern recognition is performed) and a Kalman Filter fitting stage is performed. Figure 3 shows the $p_T$ and position pulls for the resulting fit with Gaussian distributions consistent with

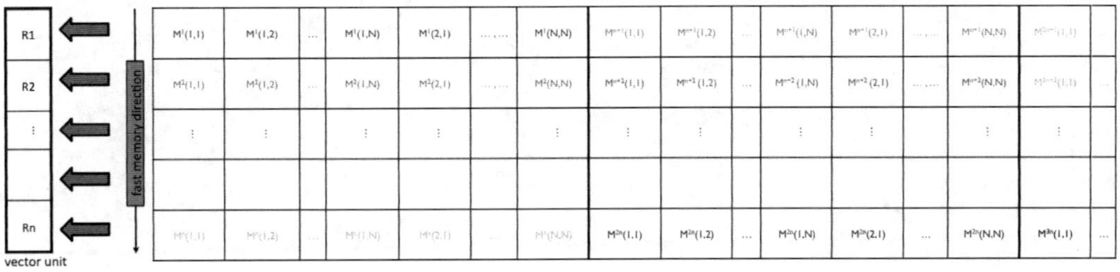

**Figure 4.** Memory layout for the new matrix library MATRIPLEX. The layout is optimized for our problem, which consists of matrix manipulations of low-dimensional matrices. The memory layout is matrix-major. In the Figure, the matrix dimension is $N \times N$ and the vector unit size is $n$.

unit width and demonstrates that the fit results are reasonable. The achieved $p_T$ resolution is roughly $\sigma_{p_T}/p_T = 0.005 \times p_T$.

## 5. Optimized Matrix Library Matriplex

The computational problem of Kalman Filter-based tracking consists of a sequence of matrix operations on matrices of sizes from $N \times N = 3 \times 3$ up to $N \times N = 6 \times 6$. To allow maximum flexibility for exploring SIMD operations on small-dimensional matrices, and to decouple the specific computations from the high level algorithm, we have developed a new matrix library, MATRIPLEX. The MATRIPLEX memory layout is optimized for the loading of vector registers for SIMD operations on a set of matrices as shown in Fig. 4. MATRIPLEX includes a code generator for generation of optimized matrix operations supporting symmetric matrices and on-the-fly matrix transposition. Patterns of elements which are known by construction to be zero or one can be specified, and the resulting generated code will be optimized accordingly to reduce unnecessary register loads and arithmetic operations. The generated code can be either standard C++ or simple intrinsic macros that can be easily mapped to architecture-specific intrinsic functions.

## 6. Results

We present the results of this study in two stages: vectorization and parallelization. In the first step we restructure the code to allow use of the vector units in Xeon[1] and Xeon Phi[2] processors. In the second step we use OPENMP to parallelize the vectorized fitting procedure across the cores on the large-core and small-core devices.

Figure 5 shows the timing for fitting 1M tracks as a function of the vector size, using a single thread. Results are compared to scaling of serial processing time ("ideal scaling"), defined as the time with vector unit size=1 divided by the vector unit size. Both for Xeon and Xeon Phi, a significant vectorization speedup is achieved, with an effective utilization of the vector units of $\sim 50\%$ .

Figure 6 shows the timing for fitting the same set of tracks as a function of the number of threads, in case all vector units are used. We test two approaches for distributing threads on the cores: filling every core with one thread or adding a second thread on the same core before moving to a different one. We compare to ideal parallelization performance ("ideal scaling"), assuming no hyperthreading (*i.e.* a maximum of 12 threads on Xeon, maximum 120 threads on Xeon Phi). Performance for one thread/core approach follows the ideal curve, with a small

[1] CentOS 6.5, $2 \times 6$ core Xeon E5-2620 @ 2GHz, 64 GB RAM, turbo off, hyperthreading enabled
[2] Xeon Phi 7150, 16 GB RAM, 61 cores @ 1.24GHz

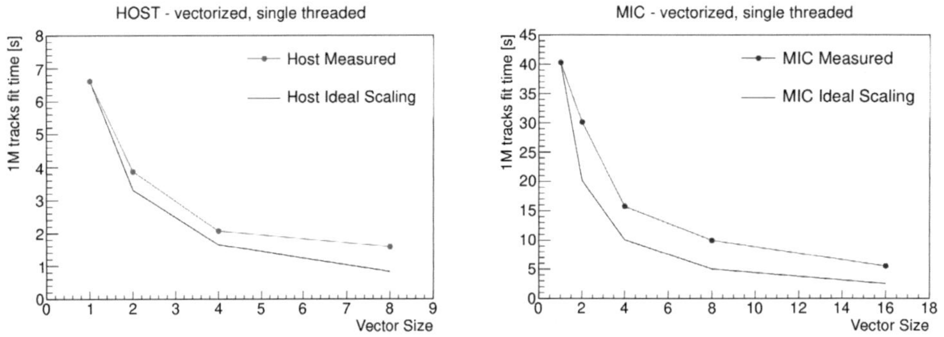


**Figure 5.** Timing results for vectorized code, as a function of the vector size, for host (left) and MIC (right). The results are compared to an ideal scaling described in the text. Significant speedups compared to serial code are observed.

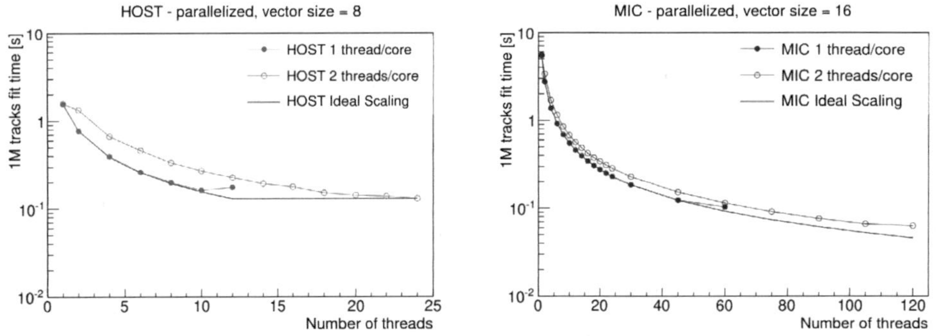


**Figure 6.** Timing results for parallelization tests, as a function of the number of threads, for host (left) and MIC (right). Two different methods of distributing threads across cores is shown, and compared to an ideal scaling. We observe ideal scaling when distributing one thread/core.

overhead only when all cores are filled. The two threads per core approach shows deviation from ideal behavior immediately: on Xeon, this is due to the use of hyperthreading slots [10], while on Xeon Phi we surmise it is related to L1 cache contention issues.

## 7. Future Work: Track Building

Track building is by far the most time-consuming step in the whole event reconstruction and thus it is the most important next step of development. With respect to track fitting, track building adds significantly more complexity to the problem as hits are searched for within a compatibility window when moving to the next layer. The track candidate needs to branch in case of multiple matches, and the algorithm needs to be robust against missing (due to detector inefficiency) or outlier hits (Fig. 7). We implemented a preliminary serial version and tested it on events with 500 simulated tracks/event, finding a hit finding efficiency close to 100%. The inherent branching of the algorithm, along with the variable size of the combinatorial problem, make it difficult to vectorize and parallelize it efficiently; thus, specific design choices have to be made to boost its computing performance on the coprocessor.

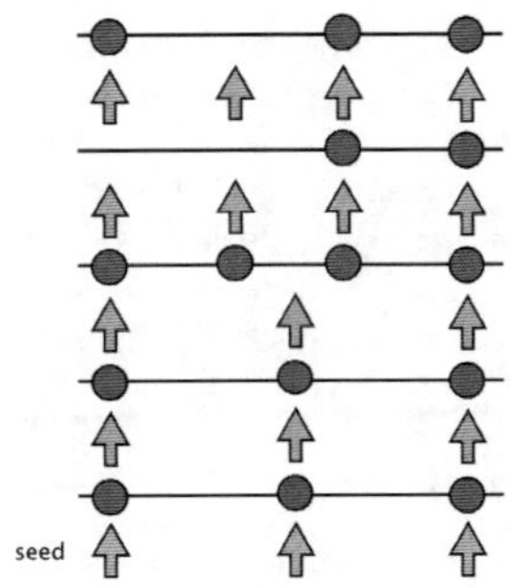


**Figure 7.** Schematic representation of track building. Unlike in track fitting, the algorithm has many branch points, *e.g.* when missing hits on layers or when multiple hit candidates are encountered on a layer. This is among the challenges for achieving efficient vectorization of the track building code.

## 8. Progress and Plans

To summarize, we presented the start of a task of adapting a complex algorithm from running HEP experiments for use in new parallel hardware. We showed a preliminary implementation of Kalman Filter-based track fitting, vectorized and parallelized on Intel Xeon and Intel Xeon Phi. In both cases, we achieve $\sim 50\%$ vectorization efficiency and $\sim 100\%$ parallelization efficiency when running one thread per core.

We are currently working towards the implementation of more realistic geometries, particularly polygonal/planar geometries typical of pixel detectors, and of material effects, both in the simulation and reconstruction models, to complete the study of track fitting. The next major step is the implementation of a fully vectorized and parallelized track building algorithm. At the same time, we will continue work on optimizing fitting and matrix operations so that overheads are further reduced and computing efficiency is increased.

At this point the results shown encourage us to continue pursuing implementation of the Kalman Filter algorithm for charged particle tracking on parallel hardware, and demonstrate successful first steps towards our goal of exploiting parallel hardware for HEP experiments.

## Acknowledgements

This work was partially supported by the National Science Foundation, partially under Cooperative Agreement PHY-1120138, and by the U.S. Department of Energy.

## References

[1] Daniel Funke, Thomas Hauth, V. Innocente, G. Quast, P. Sanders, et al. Parallel track reconstruction in CMS using the cellular automaton approach. *J.Phys.Conf.Ser.*, 513:052010, 2014.

[2] V. Halyo, P. LeGresley, P. Lujan, V. Karpusenko, and A. Vladimirov. First Evaluation of the CPU, GPGPU and MIC Architectures for Real Time Particle Tracking based on Hough Transform at the LHC. *JINST*, 9:P04005, 2014.

[3] David Rohr et al. ALICE HLT TPC tracking of Pb-Pb Events on GPUs. *J.Phys.Conf.Ser.*, 396:012044, 2012.

[4] R Frühwirth. Application of Kalman filtering to track and vertex fitting. *Nuclear Instruments and Methods in Physics Research Section A: Accelerators, Spectrometers, Detectors and Associated Equipment*, 262(2-3):444–450, December 1987.

[5] G Cerati. Vertexing and tracking algoritms at high pile-up (CMS). *Proceedings of VERTEX2014 (Doksy, Czech Republic)*, 2014.

[6] S. Chatrchyan et al. Description and performance of track and primary-vertex reconstruction with the CMS tracker. Submitted to JINST, May 2014.

[7] Samuel H. Fuller and Editors; Committee on Sustaining Growth in Computing Performance; National Research Council Lynette I. Millett. *The Future of Computing Performance: Game Over or Next Level?* The National Academies Press, 2011.

[8] Peter Elmer, Salvatore Rappoccio, Kevin Stenson, and Peter Wittich. The Need for an R&D and Upgrade Program for CMS Software and Computing. Contributed paper to SNOWMASS 2013, 2013.

[9] AIDA Collaboration. Advanced European Infrastructures for Detectors at Accelerators (AIDA). `http://aida.web.cern.ch/aida/index.html`. Accessed September 2014.

[10] Antonio Valles. Performance Insights to Intel Hyper-Threading Technology. `https://software.intel.com/en-us/articles/performance-insights-to-intel-hyper-threading-technology`. Accessed January 2015. See the Section, "Understanding Limiations...".

# Successes, Challenges and Future Outlook of Multivariate Analysis In HEP

**Helge Voss**

H. Voss, Max Plank Institute for Nuclear Physics, Heidelberg, Germany

E-mail: helge.voss@cern.ch

**Abstract.** Multivariate techniques using machine learning algorithms have become an integral part in many High Energy Physics data analyses. This article is intended to sketch how this development took place by pointing out a few analyses that pushed forward the exploitation of these powerful analysis techniques. This article does not focus on controversial issues like for example how systematic uncertainties can be dealt with when using such techniques, which have been widely discussed previously by other authors. The main purpose here is to point to the gain in physics reach of the physics experiments due to the adaptation of machine learning techniques and to the challenges the HEP community faces in the light a rapid development in the field of machine learning if we want to make successful use of these powerful selection and reconstruction techniques.

## 1. Introduction

Multivariate techniques have been used successfully since the very beginning of High Energy Physics data selection and reconstruction. The widespread use of TMVA [1], a software package which gives easy access to a variety of machine learning algorithms and which is integrated in the popular ROOT [2] analysis framework, helped to increase the knowledge about these techniques and to overcome prejudice against using such - at first glance "black box" - selection algorithms. While important issus like possible complications in addressing systematic uncertainties, apparent lack of control about the actual important figure of merit that a selection should optimise and other challenges are well known and have been reported previously by a number of authors (see for example [3]), this report focuses on the challenges of staying up to date with the rapid development in the field of machine learning in order to get the most out the very expensive accelerators and detectors.

## 2. Past Successes with Multivariate Analyses in HEP

This section does not intend to give a full historic overview about the use of multivariate techniques in high energy physics, but merely tries to point out a few examples which are meant to illustrate the importance of venturing into new analysis techniques in order to boost the performance of an analysis. In fact, one of the most powerful multivariate pattern recognition tools for image recognition, the human visual cortex, was used in the very early days when photos of bubble chambers were analysed. Only with the advent of computer based analysis of large data sets, many physicists reverted to simple one-dimensional, rectangular cuts to select their events of interest. These cuts are easy to formulate in a program code, their operation is easy to understand and interpret and last but not least, they are easy to communicate.

Nonetheless, there were permanent successful attempts to venture into more sophisticated selection algorithms, as for example the naïve Bayesian (Likelihood) $\tau$-particle identification algorithm ("TAUPID") in ALEPH [4]. In the literature one can also find even earlier attempts of promoting even more complicated techniques like neural networks at dedicated HEP statistics conferences. However, it was also already then pointed out that the adaptation of these powerful classification techniques in high energy physics is 'somewhat slow' as expressed for example in the summary of [5]: *The progress of exploiting ANN in high energy physics has been somewhat slow. Partly this conservatism is due to the a misconception that ANN approaches contain an element of "black box magic" as compared to conventional approaches. I hope I have convinced the reader that this is not the case. Statistical interpretation of the answers makes the ANN approach as well-defined to use as the discriminant ones.*

As far as the author is aware, the use of multivariate pattern recognition algorithms was basically taboo in new particle searches until the LEP2 aera Higgs searches [6] when people started to apply Likelihood or Neural Network algorithms and overcame widespread scepticism by demonstrating the superior results obtained compard to classical rectangular cuts.

With the pioneering analysis of MiniBooNE that used for the first time a Boosted Decision Tree in a HEP analysis as outlined in [7], a lot of interest in has been raised in the community to exploit this "new" selection algorithm. Subsequently, boosted decision trees for sure became the favourite algorithm within the TMVA framework used in many analysis, including the first evidence for the Higgs Boson as reported by CMS [8].

Boosted Decision Trees are of course popular for a good reason. They are very robust, meaning they do not require very careful tuning in order to get close to their optimal performance, and powerful. They had for this reason sometimes been called the best 'out of the box' classifiers. Simple Decision Tree are easy to interpret and very similar to standard rectangular cuts. Each branch of the tree simply represents a cut sequence. Boosted decision trees however loose basically all the advantage of easy interpretation or understanding of the selection during the boosting sequence. They are probably as difficult to understand as Neural Networks in terms of how they place the decision boundaries in the feature space of the discriminating variables that are fed into it. Moreover, if one carefully examines the decision boundary that boosted decision trees parameterise if a certain threshold is imposed on the response variable, this is typically a very irregular, non smooth hyperplane. This is easily understandable from the fact that the Boosted Decision Trees' response is a weighted sum of rectangular volumes that are given by each of the individual trees as signal or background like areas. Even for very large number of boosting steps, this feature remains[1] as typically the boosting algorithm intrinsically doesn't give a sizeable contribution to those classifiers in the ensemble that are derived at late boosting steps. Therefore no matter how many boosting steps there are, they will not really smooth out the decision boundaries.

On the other hand, the decision boundaries from neural networks, as they are a superposition of smooth activation functions of the various nodes, are generally much smoother and for sure, after looking at a comparison of the decision boundaries from a Neural Network and a Boosted

[1] AdaBoost, the most popular boosting algorithm, for example keeps reweighting events of the original training data sample such that event, which are mis-classified by the previously trained classifier (decision tree) in the ensemble of boosted classifiers, keep getting higher weights compared to events out of the training set that we correctly classified. This results in variable distributions of the reweighted training set that keep getting more and more equal between signal and background. Decision trees trained on these reweighted data samples then obviously have a larger error rate on the training set and hence get a smaller weight in the ensemble of classification trees. Eventually the error rate approaches 50% which translates to a contribution of to the boosted ensemble classifier of zero. This behaviour is actually a very nice feature of the boosting algorithm, as it removes the necessity of actively carefully choosing the number of boosting steps as configuration parameter.

Journal of Physics: Conference Series **608** (2015) 012058 doi:10.1088/1742-6596/608/1/012058

Decision Tree, most physicist would rather evaluate possible systematic on a smooth Neural Network output rather than on a very irregular one Boosted Decision Tree. Unfortunately, standard Neural Networks are much harder to train properly, particularly with limited amount of labelled training data and hence they are in reality often less performant than Boosted Decision Trees and therefore less popular.

## 3. New Developments in Machine Learning and Multivariate Analysis

At this point it is worth noting that at the time Boosted Decision Trees became popular in the HEP community, the hype about Boosted Decision Tree in the machine learning community had been already over and new developments in the field of neural network training have all but totally revolutionised the main areas of pattern recognition like image and speech recognition. In 2006/2007, a major breakthrough in neural network research finally allowed successful training of so called 'deep neural networks'[2]. This was achieved using a meaningful initialisation of the weights[3] connecting the neurons in each layer before starting the usual training via back-propagation. This initialisation is done by pre-training each individual layer as Restricted Boltzmann Machines [9] or as auto-encoders [10].

This next generation of neural networks proved extremely powerful and it seems that all major players in the 'industry' that apply machine learning at a large scale (e.g. Google, Facebook, Microsoft and IBM) are replacing their hand crafted and fine tuned speech and image recognition algorithms by deep neural networks which often substantially outperform their predecessors. This development can be followed in the news where major acquisitions of deep learning companies or hiring of deep learning specialist from academia by one of these companies are reported.

Also in physics, first attempts have been made to apply these new techniques and some promising results have been reported [11]. In order to understand how profound the changes are that other fields have experienced due to the advances of deep learning technologies, let's imagine we discard all of our carefully, with physics knowledge crafted high level features like invariant masses, jets, secondary vertices or impact parameters and let a large deep network learn these features by itself from simple 4-vectors as input.[4] While this might not be the best approach in HEP, the technology might well be useful for much more changes to our analyses than a simple replacement of a Boosted Decision Tree by a Deep Neural Network.

Given the sweeping success the paradigm change in terms of which type of features are used as input to the machine learning together with the deep learning network technologies, it would certainly be unwise not to study such and other possibilities of deep network applications in detail for our HEP data. Rather than lagging behind in the application of modern machine learning techniques, high energy physicist should rather be using state of the art technologies if not trying to drive the development as it has been in so many other fields.

An interesting additional aspect of these deep neural networks is that the layer-by-layer pre-training uses unlabled data, meaning it doesn't need to be known whether an event is of type "signal" or "background". This of course gives the potential to use recorded data from the experiment rather than extensive amounts of generated Monte Carlo events for this step.

A completely unsupervised training of a large deep layer neural network on HEP data might eventually perhaps not 'discover cats' as it was done by Google [12], but could be used for

[2] Neural Networks with more than 2 hidden layers are typically referred to as 'deep networks'

[3] Previously, the weights of a neural network were initialized randomly with small values, such that the network acted in the beginning much like a random linear classifier which through backpropagation was intended to quickly learn the best linear separating boundary which would then gradually be refined as the training process progresses.

[4] In speech recognition for example, over the years carefully tuned high level features (phonemes) which were previously used as input to the learning algorithms have apparently been simply replaced by a large deep neural network which learns these features.

completely model independent searches, or perhaps help in understanding the detector and hence be a vital tool for systematic studies. Obviously, the requirements in HEP are not the same as for speech or image recognition and it will probably require a substantial effort to evaluate the possibilities of these new Deep Learning strategies. Research on these Deep Neural Networks is developing at a fast pace[5] Understanding such complex algorithms and their behaviour on such data samples will certainly exceed the scope of a typical PhD thesis, let alone the ever more complex code base required for building, training and monitoring such advanced deep networks. It should be noted that particular effort is necessary to monitor and understand the working of these algorithms in order to understand all systematic aspects [6]

## 4. Summary

It is hard to quantify the impact of the increased use of sophisticated MVA techniques on the physics output of our current HEP experiments, but one could perhaps say that an improvement of selection performance w.r.t a simple cut based analysis of somewhere between 5% and 20% are certainly not unheard of, but rather common. This leads to a very substantial increase of the physics reach for our experiments! Compared to how much an equivalent gain in physics reach by upgrading an accelerator or detector would cost, funding of machine learning reasearch and development for physicists should be regarded as an at least equally good investment.

Simply reaching out to the other groups that do either plain machine learning research or within the context of other industrial or business applications as it is has been done so far with the funding of interdisciplinary activities which lead to initiatives like the Kaggle Higgs Challenge (https://www.kaggle.com/c/higgs-boson) is probably not enough to really make proper use of modern machine learning in HEP.

In order to deal properly with the ever increasing complexity of machine learning it will be difficult for individuals to really grasp the full potential and manage the pitfalls of those analysis techniques. While it was still possible to simply program your own likelihood selection algorithm, this was already much more challenging with standard neural networks or boosted decision trees and will be even more so with the new algorithms. But rather than simply adopting tools developed outside of (high energy) physics, the community would certainly profit considerably from an engagement into an 'in house' machine learning development and support group for HEP usage, similar as it was done with ROOT [2], which has a CERN funded core development and support team for analysis software dedicated to physicists. This would help to ensure that the data we have collected in our experiments is exploited in the best way using state of the art analysis technologies while making sure that we understand all the results to the level that we feel comfortable with as physicists.

Perhaps our data selection is less difficult than pattern recognition in other fields and we do not need such sophisticated analysis tools. However it would certainly be a loss if the possible gains would not be thoroughly studied. These new machine learning techniques are thorougly studied in the context of self driving cars for example, where 'errors' are much more severe than an under/over estimated systematic error on a physics parameter. This clearly shows that potential criticism about the use of such machine learning algorithms in High Energy Physics because we need to understand our systematic uncertainties much more thoroughly than it is the habit in other field, does not seem very compelling. Moreover it simply means that more effort is needed to be put into a deeper understanding of these state of the art techniques within

[5] Recently it has been reported for example that using so called 'rectified linear units' [13] instead of the typical sigmodial activation functions used in traditional neural networks or 'dropout' [14] again may lead to major improvement of (deep) neural network performances

[6] Note: Systemiatic uncertainties or errors never are a result of a particular training of an algorithm. The come into effect once we do not exactly understand the efficiency and background rejection of the trained algorithm, globally and for individual regions in phase space.

Journal of Physics: Conference Series **608** (2015) 012058 doi:10.1088/1742-6596/608/1/012058

the context of high energy physics.

## References


[1] A. Hoecker, P. Speckmayer, J. Stelzer, J. Therhaag, E. von Toerne, and H. Voss, "TMVA: Toolkit for Multivariate Data Analysis," PoS A CAT 040 (2007) [physics/0703039].

[2] http://root.cern.ch

[3] L. Lyons, *"Open statistical issues in particle physics"* The Annals of Applied Statistics, Vol. 2, No. 3, (2008) 887.

[4] D. Decamp *et al.*, ALEPH Collaboration, *"Measurement of tau branching ratios"*, Z. Phys. C **54** (1992) 211.

[5] C. Peterson, *"Neural Networks and High Energy Physics"*, Proc. of International Workshop on Soft- ware Engineering, Artificial Intelligence and Expert Systems for High Energy and Nuclear Physics, Lyon Villeurbanne, France, March , 1990 , eds. D. Perret-Gallix and W. Woijcik, Editions du CRNS (Paris 1990); B. Denby, *"Neural Networks and Cellular Automata in Experimental High Energy Physics"*, Computer Physics Communications 49 (1988) 429.

[6] ALEPH, DELPHI, L3 and OPAL Collaboration, Phys. Lett. B565 (2003) 61.

[7] B.P. Roe, H-J. Yang, J. Zhu, Y. Liu, I. Stancu, G. McGregor, *"Boosted decision trees as an alternative to artificial neural networks for particle identification"*, NIMA, Vol543, (2005), 577.

[8] CMS Collaboration, *"Observation of a new boson at a mass of 125 GeV with the {CMS} experiment at the {LHC}"*, Phys Lett. B, 716 (2012) 30. bibitemAdaBoost Y. Freund and R.E. Schapire, J. of Computer and System Science 55, (1997) 119.

[9] Hinton, G. E., Osindero, S. and Teh, Y. W. *"A Fast Learning Algorithm For Deep Belief Networks"*, Neural Computation, Vol 18 (2006) 1527.

[10] Bengio, P. Lamblin, D. Popovici and H. Larochelle, *"Greedy Layer-Wise Training of Deep Networks"*, Advances in Neural Information Processing Systems 19 (NIPS06), MIT Press (2007) 153.

[11] P. Baldi, P. Sadowski, and D. Whiteson *"Searching for Exotic Particles in High-Energy Physics with Deep Learning"*, arXiv hep-ph/1402.4735v2; de Oliveira, L.P., Golling, T., *"Optimizing Deep Learning Neural Network Architectures for Jet Identification at $\sqrt{s} = 8\,TeV$ using the ATLAS detector"*, (In preparation).

[12] Q. V. Le, M. Ranzato, R. Monga, M. Devin, G. Corrado, K. Chen, J. Dean and A.Y. Ng, *"Building high-level features using large scale unsupervised learning"*, In International Conference on Machine Learning (ICML), Omnipress, (2012) 103; (arXiv:1112.6209v5).

[13] V. Nair and G.E. Hinton, *"Rectified linear units improve restricted boltzmann machines"*. In Proc. 27th International Conference on Machine Learning (ICML), Omnipress Madison, (2010) 807.

[14] G.E. Hinton, N. Srivastava, A. Krizhevsky, I. Sutskever and R.R. Salakhutdinov, *"Improving neural networks by preventing co-adaptation of feature detectors"*, CoRR (2012), arXiv:1207.0580v1.

# Data Analysis – Algorithms and Tools

**Martin Spousta**

Charles University in Prague,
Institute of Particle and Nuclear Physics,
V Holešovičkách 2, 180 00 Prague 8,
Czech republic

E-mail: martin.spousta@cern.ch

**Abstract.**
Modeling of detector response, modeling of physics, and software tools for modeling and analyzing are three fields among others that were discussed during $16^{\text{th}}$ International workshop on Advanced Computing and Analysis Techniques in physics research – ACAT 2014. This short report represents a summary of track two where the current status and progress in these fields were reported and discussed.

## 1. Introduction

There are three fields connected with the data analysis: modeling of the detector response, modeling of the physics and extracting physics from data, and software tools for modeling and analyzing. Each effort in any of these fields requires a knowledge of the status and progress in other fields. Many efforts also require a feed-back to be given to make the work successful and useful. In this short report, we summarize the presented methods and tools along with some of the feed-back given during the discussions. Each section below is dedicated to one of these three fields.

## 2. Modeling of detector response

The standard candle in the field of modeling of detectors is Geant4 [1]. Geant4 is a toolkit for the simulation of passage of particles through the matter. It is used in most of the high energy physics experiments including all experiments at the LHC for a reliable full simulation of the detector response. It has also found users in the field of medical science (for simulation of interactions of radiations used for treatment), in the field of space science (to study interactions between the natural space radiation environment and space hardware or astronauts), nuclear physics, and other fields where ionizing effects have to be taken into account. Recent updates of Geant4 consist of updates in the design and updates in the modeling of physics. Updates in the design are namely the use of multi-threading and a strong reproducibility of events (that is the independence of simulation of one event on the history of previous running). Updates in the physics modeling are namely fixing the energy response in QGSP_BERT physics list and improving lateral shower shape, improving the FTF physics list, and the use of a new unique solid library (USolid) [2].

Successful Geant4 project is being followed by GeantV [3] that will have several important new features. Main motivation for the new Geant is the need to speed up the process of particle transport simulation. Indeed, most of the time is spent on few percents of the volume that is simulated. The main new features of GeantV are following: the vectorization and locality (not a single track but a group of tracks are transported), multi-threading, adding new entities to control the work flow, optimizing simulation geometry, and a possibility to include some fast simulation.

Fast simulation itself will certainly be needed during the LHC run II. Robustness of Geant4 simulation was a part of the success of LHC run I and the Higgs discovery [4, 5]. However, at the same time, many important physics analyses suffered from the lack of Monte-Carlo (MC) statistics. Thus, fast simulation will be indispensable during the run II. The core of the fast simulation is the use parameterizations of the response or pre-generated samples. Each LHC experiment has its own fast simulation of the detector (e.g. Atlfast of ATLAS or FastSim of CMS). A typical enhancement by about a factor of twenty in the computation speed was reported by ATLAS when using the fast simulation compared to the full simulation. An alternative to the fast simulation only is to combine the fast simulation with the full simulation (e.g. to use the fast tracking in tandem with the full simulation of the rest of the detector) or to replace costly physics objects by pre-clustered ones (e.g. frozen showers used by ATLAS).

Besides the fast simulation tools that are specific for a given experiment, there are tools on the market that can be used to simulate any detector or its part. One of such tools is Delphes [6] which is being developed since 2007. Delphes has several attractive features: it contains the simulation of propagation in the magnetic field and the calorimetry; it can simulate multiple simultaneous proton-proton collisions (pileup); it can be used to simulate realistic detector response to high-luminosity LHC (HL-LHC) environment; it has particle-flow algorithm (a strategy for jet reconstruction by CMS). Delphes is community based, modular tool that has many interfaces (e.g. to MadGraph, Hep-MC, FastJet) and a standard root output.

To simulate precisely the tracking system e.g. for a new detector to be used at ILC or other future collider facility one can use Genfit [7]. Genfit is a generic track-fitting toolkit that provides a modular track fitting framework and simulation of a tracker. Genfit can use fitting based on standard Kalman Filter (which does a linearization around predictions) [8] or it can use an improved fitting algorithm, which utilizes a Deterministic Annealing Filter [9]. This improved algorithm allows namely a better handling of out-layer hits.

Tracking and the vertexing may be improved by a use of multivariate analysis techniques (MVA). This was demonstrated by a project of robust tracking with neural network for JLab. JLab experiments investigate the hadron structure using polarized $e^-$ beams at the energy up to 12 GeV. In the new setup, experiments will have to deal with lager pile-up and backgrounds. It was proposed that the new tracking with Gas Electron Multiplier (GEM) may use the neural network implemented within a Mean Field theory framework [10] to get the association of hits into tracks. Energy which is being minimized is defined in terms of a natural distance measure (reciprocal value of distances between hits times the cosine of an angle formed by line segments connecting the hits) and matrices quantifying connections between two points in subsequent GEM planes. An improved resolution by a factor of eight was reported, however it was also concluded during the discussion that a careful evaluation of computation time requirements is needed.

The use of neural network is planned also to improve the ability of vertex triggering at Belle II at SuperKEKB accelerator. Future SuperKEKB is expected to provide a luminosity of $8 \cdot 10^{35}$ cm$^{-2}$s$^{-1}$ which is forty times larger than the world record by KEKB accelerator. Topological and time information will be provided at Belle II by a Central Drift Chamber. This information is the input to a simple and fast neural network with one hidden layer, one neuron, and fully forward connected multi-layer percepton. The training can be done with back-propagation algorithm and the output is scaled $z$-vertex. This trigger is being implemented in firmware on Virtex 7 FPGA board. Estimated vertex resolution is $1.3 - 2.3$ cm.

Besides track and vertex reconstruction, the particle identification is also important for some of the physics analyses. A promising tool for particle identification without modeling the response, called Pridix [11], was introduced and discussed. Pridix uses minimization of generalized Kullback-Leibler divergence. Kullback-Leibler divergence (also known as information gain or relative entropy) is a non-symmetric measure of the difference between two probability distributions $P$ and $R$. Specifically, the Kullback-Leibler divergence of $R$ from $P$ is a measure of the information lost when $R$ is used to approximate $P$. The technique utilizing this tool from probability theory was shown to be useful and to provide a very good performance of PID at PHENIX detector at RHIC collider.

An interesting application of the knowledge of particle interaction with matter was reported by members of Muon Portal Project [12], an innovative non-destructive scanning based on muon tomography to be used to scan containers in docks to search for weapons and radioactive material. The deflection of muon trajectory in the presence of materials with high proton number was demonstrated to be sufficient for detecting the problematic materials. The basic approach for reconstructing the tracks uses the point of closest approach (POCA) between incoming and outgoing tracks. The scanning algorithm with POCA neglects multiple scattering and it has generally a poor resolution. A performance of an alternative approach which uses the density based clustering with friends-of-friends percolation algorithm (also used in cosmology) was reported. This clustering algorithm acts also as a filter and may be combined with other algorithms to further boost the performance. The estimated start-up of the muon-based scanning is March 2015.

## 3. Software tools for modeling and analyzing

The standard candle software tool used in high-energy physics is ROOT [13]. ROOT is an object-oriented program and library tailored for particle physics data analysis but it is used in many other fields, e.g. in astronomy, nuclear physics, and data mining. Current ROOT 5 is versatile tool but sometimes it is mentioned that has also some drawbacks such as difficulty for beginners, heavy use of global variables, or limitations coming from CINT interpreter. Some of these issues are addressed in the new ROOT 6 which contains many great improvements. The main new feature is the use of new interpreter called CLING which use CLANG frontend for LLVM compiler which are standards being used for C++. This allows e.g. for interpreting from includes without dictionaries, improved compilation diagnostics, or C++14 support. Besides the new interpreter, there are many other news such as a new TTreeReader which simplifies reading from trees, a new TFormula which profits from the intime compiler, a new possibility to export graphics to LaTeX, or improved GUI with better axis zooming or guides for object placements. The ROOT 6.00 was released in May 2014. The ROOT 6.02 which is targeted to LHC run II was released in September 2014.

A new tool for automatic differentiation (CLAD) [14] was discussed that may also be available in later releases of ROOT 6. CLAD extends the symbolic differentiation to the compiler as a module. This allows for automatic defining the function used for differentiation which is only declared by the user. Such versatile tool can be used also to boost the performance of minimization and fitting.

More complex tools connected with the ROOT environment are RooFit, HistFactory, RooStat, and RooWorkspace [15]. RooFit [16] is a toolkit for modeling of expected distribution of events. In RooFit, each mathematical concept (such as variable, probability density function (PDF) or its integral) is implemented as a RooFit class. This allows for an abstract manipulation with the mathematical objects within a simple and transparent code. HistFactory [17] allows structured statistical model building for binned likelihood template models. RooStats [18] provides a wide set of statistical tests that can be performed on RooFit models. RooWorkspace [19] is a concept that allows to persist complete likelihood models. These tools were used during the Higgs discovery. The Higgs discovery and subsequent analysis of its properties are truly a result of collaborative statistical analysis of many signal and control samples which is allowed by the use of these tools.

A framework which is built on top of RooFit, HistFractory, and RooStat is HistFitter [20]. HistFitter is a programmable framework to build and test complex data models. It allows to construct and fit PDFs and provide a statistical interpretation. It has a built-in concepts of control, validation and signal regions with rigorous treatment of extrapolations. It allows simple book-keeping of multiple models and provides tools for graphical representation of the results of complex statistical analysis. HistFitter was used within ATLAS since 2012 and now the framework is publicly available [21].

Reconsideration of data model is being done by all the experiments at LHC to anticipate the upcoming LHC run II. In particular, ATLAS has reported an improvement in the data handling by replacing the Analysis Object Data (AOD). The AOD is a data structure that contains information about reconstructed object such as four-momenta of jets or charged particles. The AOD is legible only in a complicated software of ATLAS called ATHENA which made it practically unusable for the end user and it had to be translated to a legible structure called Derived Physics Data. Now, the AOD is replaced by so called xAOD which is a structure legible both in ATHENA and ROOT. This should generally save PB of memory and lot of computation time. The goal of the new data model is also to provide a simple access to events and kinematics of objects used in the data analysis by a light weighted code structure. CMS has reported the upgrade in the simulation which, among other issues, targets to solve the problem of demanding simulation of pileup.

## 4. Modeling the physics and extracting physics from data

MVA techniques boosted the ability of high energy physics (HEP) to analyze complex event topologies. The important question raised in the conference was if the field of HEP is still the state of the art in MVA, in particular whether some novel techniques used in other fields of science should be adapted to further enhance the performance of HEP analyses. A partial answer to that question was provided by a report discussing a comparison of different MVA techniques within inclusive top pair production cross-section measurement in D0 at Tevatron. In that study, several MVA techniques have been compared:

- Generalized Linear Models (GLM) that are used namely in econometrics
- Model based clustering (MBC) that is used often in acoustics
- Neural networks with cooperative switching units (NNSU) – an enhanced model of neural network designed originally for classification tasks
- Boosted Decision Tree (BDT) from standard TMVA package [22]
- Multilayer Perceptrons (MLP) from TMVA package

Last two above mentioned techniques are golden standards used for years in HEP. It was demonstrated that if the techniques are used in an appropriate way and with a good training, then the standard HEP methods exhibit similar performance as alternative techniques used in other fields of science.

The most complex events at LHC come from heavy ion collisions. The question of the parton radiation and knowledge of properties of the quark-gluon plasma created in those collisions may shed light on fundamentals of physics [23, 24]. A new statistical method of Markov chain Monte Carlo was proposed for extracting the viscosity and equation of state of the quark-gluon plasma [25].

The knowledge of what fraction of momentum of hadron is carried by partons within protons, which is commonly characterized by parton distribution functions (PDF), is essential for all physics studies at hadron colliders. Many physics observables strongly depend on PDFs. Relatively large uncertainties in the knowledge of PDFs thus translate into a large uncertainty in final state observables of interest. PDFs cannot be calculated within the theory of strong interaction, quantum chromodynamics, but they need to be extracted from the data. Parametrized PDFs at a certain momentum scale can then be evolved to a different scale which is relevant for a particular measurement of interest. HERAFitter [26] is a framework which allows the extraction of PDFs from data as well as comparing theory predictions with the data. PDFs can be represented in standard LHAPDF grids [27] and griding tools like APPLgrid [28] allowing a fast use of beyond-leading order pQCD observables. HERAFitter was used in more than fifteen LHC publications.

Typical problem in LHC analyses is a reconstruction of particles from decay chain. A typical example is the decay of top quark to $b$-quark and $W$ boson which further decays to lepton and neutrino or a quark anti-quark pair. Cut based analysis or an analysis utilizing MVA techniques are often used to reconstruct the kinematics of original particle (e.g. the top quark). CMS presented a performance of a different technique called Matrix Elements method (ME). ME method establishes a direct link between theory and event reconstruction. It builds a discriminant in terms of probability that the event with a given reconstructed kinematics matches a given hypothesis about the original particle. This probability can be written as a convolution of PDFs, perturbative QCD matrix element, and a transfer function which is obtained from simulation. Advantages of the ME method are: its versatility (it can be used in many different physics analyses); no need for complicated training (such as in the case of MVA methods); maximizing the amount of theoretical information in the discrimination. The disadvantage is the fact that it can easily use only the information from leading order pQCD calculation and that it may be relatively slow.

An example of the use of MVA techniques in a similar problem of the decay of $\Upsilon(4S)$ to two $B$ mesons was reported by members of Belle II Collaboration. Hierarchical $B$ meson reconstruction takes the advantage of events in which one $B$ meson is accurately reconstructed which allows to estimate the four-momentum of the other $B$ meson without a need to explicitly reconstruct the full decay chain. This is done by combining the particles into different decays, each with its own multivariate classifier. The order of 100 classifiers can handle order of 1000 exclusive $B$ decay modes. Hierarchical reconstruction is implemented in an automated framework which needs only minimal interventions by users. The MVA techniques are in use also for the Higgs boson reconstruction. An alternative technique for the Higgs boson reconstruction in $\gamma\gamma$ decay channel was presented which formulates the Quadratic Unconstrained Binary Optimization problem in speech of a Hamiltonian and use the Adiabatic Quantum Annealing to find a optimal classifier [29] for identifying the Higgs decay.

One of important problems present in different fields of physics is the deconvolution of a signal. Generally, a signal information can be deteriorated by a presence of noise or by a finite detector resolution. There is a lot of techniques present on the market, e.g. RooUnfold package [30] where several unfolding techniques are implemented. One new promising unfolding technique was presented here. It uses a Mixture Density Model for representation of unknown true distribution and a cross-validation approach to define optimal parameters of the unfolding. Despite to the fact that method seems to be promising it was generally concluded that any new method is useful only when being followed by a publicly available implementation which allows for an independent testing and comparison with other existing techniques. An overview comparing different deconvolution methods used in particular for $\gamma$-ray spectroscopy was provided. Besides other techniques, it discussed a comparison between Gold deconvolution algorithm, Richardson-Lucy deconvolution, Maximum A Posteriori Deconvolution Algorithm, and other algorithms implemented with TSpectra class in root.

**Acknowledgment:** This work was supported by Charles University in Prague, projects PRVOUK P45 and UNCE 204020/2012.

## References

[1] S. Agostinelli et al. GEANT4: A Simulation toolkit. *Nucl.Instrum.Meth.*, A506:250–303, 2003.
[2] http://aidasoft.web.cern.ch/USolids.
[3] http://geant.cern.ch.
[4] ATLAS Collaboration. Observation of a new particle in the search for the Standard Model Higgs boson with the ATLAS detector at the LHC. *Phys.Lett.*, B716:1–29, 2012.
[5] CMS Collaboration. Observation of a new boson at a mass of 125 GeV with the CMS experiment at the LHC. *Phys.Lett.*, B716:30–61, 2012.
[6] J. de Favereau et al. DELPHES 3, A modular framework for fast simulation of a generic collider experiment. *JHEP*, 1402:057, 2014.
[7] C. Hoppner, S. Neubert, B. Ketzer, and S. Paul. A Novel Generic Framework for Track Fitting in Complex Detector Systems. *Nucl.Instrum.Meth.*, A620:518–525, 2010.
[8] R. E. Kalman. A new approach to linear filtering and prediction problems. *J. Fluids Eng.*, 82:35, 1960.
[9] R. Frhwirth and A. Strandlie. Track fitting with ambiguities and noise: A study of elastic tracking and nonlinear filters. *Comput.Phys.Commun.*, 120(2-3):197–214, 1999.
[10] N. Kurita and K. Funahashi. *Neural Networks*, 9:1531–1540, 1999.
[11] http://github.com/sangaline/pidrix.
[12] S. Riggi, V. Antonuccio-Delogu, M. Bandieramonte, U. Becciani, A. Costa, et al. Muon tomography imaging algorithms for nuclear threat detection inside large volume containers with the Muon Portal detector. *Nucl.Instrum.Meth.*, A728:59–68, 2013.
[13] http://root.cern.ch.
[14] http://github.com/vgvassilev/clad.
[15] http://root.cern.ch/root/html/ROOFIT_Index.html.
[16] http://root.cern.ch/drupal/content/roofit.
[17] https://cdsweb.cern.ch/record/1456844.
[18] https://twiki.cern.ch/twiki/bin/view/RooStats/WebHome.
[19] http://root.cern.ch/root/html/RooWorkspace.html.
[20] M. Baak, G.J. Besjes, D. Cote, A. Koutsman, J. Lorenz, et al. HistFitter software framework for statistical data analysis. *arXiv:1410.1280*.
[21] http://cern.ch/histfitter.
[22] http://tmva.sourceforge.net.
[23] Edward V. Shuryak. Quark-Gluon Plasma and Hadronic Production of Leptons, Photons and Psions. *Phys.Lett.*, B78:150, 1978.
[24] Martin Spousta. Jet Quenching at LHC. *Mod.Phys.Lett.*, A28:1330017, 2013.
[25] Scott Pratt, Claudia Ratti, and William Patrick McCormack. Chemical properties of super-hadronic matter created in relativistic heavy ion collisions. *arXiv:1409.2164*.
[26] P. Belov et al. Parton distribution functions at LO, NLO and NNLO with correlated uncertainties between orders. *arXiv:1404.4234*.
[27] J. Butterworth, G. Dissertori, S. Dittmaier, D. de Florian, N. Glover, et al. Les Houches 2013: Physics at TeV Colliders: Standard Model Working Group Report. *arXiv:1405.1067*.
[28] Tancredi Carli, Dan Clements, Amanda Cooper-Sarkar, Claire Gwenlan, Gavin P. Salam, et al. A posteriori inclusion of parton density functions in NLO QCD final-state calculations at hadron colliders: The APPLGRID Project. *Eur.Phys.J.*, C66:503–524, 2010.
[29] K Pudenz and D Lidar. Quantum adiabatic machine learning. *Quant. Inf. Proc.*, 12:2027–2070, 2013.
[30] Andreas Hocker and Vakhtang Kartvelishvili. SVD approach to data unfolding. *Nucl.Instrum.Meth.*, A372:469–481, 1996.

Journal of Physics: Conference Series **608** (2015) 012060 doi:10.1088/1742-6596/608/1/012060

# Redberry: a computer algebra system designed for tensor manipulation

**Stanislav Poslavsky**
Institute for High Energy Physics, Protvino, Russia
SRRC RF ITEP of NRC Kurchatov Institute, Moscow, Russia
E-mail: stvlpos@mail.ru

**Dmitry Bolotin**
Institute of Bioorganic Chemistry of RAS, Moscow, Russia
E-mail: bolotin.dmitriy@gmail.com

**Abstract.** In this paper we focus on the main aspects of computer-aided calculations with tensors and present a new computer algebra system Redberry which was specifically designed for algebraic tensor manipulation. We touch upon distinctive features of tensor software in comparison with pure scalar systems, discuss the main approaches used to handle tensorial expressions and present the comparison of Redberry performance with other relevant tools.

## 1. Introduction

General-purpose computer algebra systems (CASs) have become an essential part of many scientific calculations. Focusing on the area of theoretical physics and particularly high energy physics, one can note that there is a wide area of problems that deal with tensors (or more generally — objects with indices). This work is devoted to the algebraic manipulations with abstract indexed expressions which forms a substantial part of computer aided calculations with tensors in this field of science. Today, there are many packages both on top of general-purpose systems (Maple Physics [1], xAct [2], Tensorial etc.) and standalone tools (Cadabra [3, 4], SymPy [5], Reduce [6] etc.) that cover different topics in symbolic tensor calculus. However, it cannot be said that current demand on such a software is fully satisfied [7].

The main difference of tensorial expressions (in comparison with ordinary indexless) lies in the presence of contractions between indices. This additional structure must be reflected in computer representation of tensorial expressions. It makes the existing pure list-based CASs not a good basis for implementing algorithms for tensor manipulation [4]. Because of contractions, implementation of even such a fundamental atomic operation like expression comparison becomes much more complicated, which in turn complicates implementation of general operations such as substitutions, reduction of similar terms etc. Since solution of a real-world problem comprises a long sequence of such common operations, their performance is critical for obtaining of result in a reasonable time.

Here we present Redberry — a computer algebra system for algebraic manipulations with tensorial expressions. A comprehensive documentation accompanied by lots of examples can be found on the Redberry website http://redberry.cc and in Ref. [8]. This paper gives a very brief overview of features specific for tensorial software, approaches used in the field and describes Redberry in comparison with other systems putting a particular stress on performance.

## 2. Specific features of tensorial CAS

The presence of indices brings an additional complexity to tensorial expressions. It makes the algorithms used in pure scalar CASs inapplicable and thus requires fundamentally different approach. Let's consider few noteworthy aspects which set tensorial CAS apart from scalar ones.

One of the most frequent operations arising in majority of computer-aided calcualtions with tensors is automatic relabeling of dummy indices. Consider a trivial substitution:

$$R_{\alpha\beta} = R^{\mu}{}_{\alpha\mu\beta} \quad \text{into} \quad R_{\rho\tau}\, R^{\tau\mu}$$

$$R^{\mu}{}_{\rho\mu\tau}\, R^{\mu\rho}{}_{\mu}{}^{\mu} \quad (\text{incorrect ✗}) \quad , \quad R^{\alpha}{}_{\rho\alpha\tau}\, R^{\beta\rho}{}_{\beta}{}^{\mu} \quad (\text{correct ✓})$$

One can see that substitution of r.h.s. without appropriate relabeling of dummy index $\mu$ leads to an indices clash. Thus correctness of CAS operations depends on its ability to automatically resolve such conflicts (as shown).

Another nontrivial atomic operation is comparison or matching of tensorial expressions (this operation is straightforward for scalar expressions). Consider a common simplification routine — reduction of similar terms:

$$(1) \quad A_{\mu}\, A^{\mu} + A_{\nu}\, A^{\nu} = 2\, A_{\mu}\, A^{\mu} \qquad (2) \quad F_{abcd}\, F^{acef}\, F_{ef}{}^{bd} - F_{bfed}\, F^{cafd}\, F^{be}{}_{ca} = 0$$

In both cases (and second one is really non trivial) CAS has to match dummy indices in order to check that the summands are equal. The problem of comparison of indexed expressions is very similar to a well known *pattern matching problem* from scalar CASs, while here it arises even in simplest calculations.

Additional aspect of comparison emerges in substitutions where both free and dummy indices must be matched. Consider for example a substitution:

$$F^{\alpha\beta}\, F_{\alpha\gamma} = T^{\alpha}{}_{\gamma} \quad \rightarrow \quad F^{\beta\gamma}\, F^{\mu}{}_{\gamma}\, F_{\beta\mu}, \quad \text{which gives} \quad T^{\gamma}{}_{\mu}\, F^{\mu}{}_{\gamma}.$$

The ability to define arbitrary permutational symmetries and antisymmetries of tensors is required for nearly all calculations. The presence of symmetries significantly complicates issues mentioned above. For example, supposing that $R_{abcd} = R_{cdab} = -R_{bacd}$ one can check that

$$R^{abcd}\, R_{efdc}\, R^{ef}{}_{ab} + R_{rc}{}^{df}\, R_{ab}{}^{rc}\, R_{fd}{}^{ba} = 0.$$

Checking of this equality requires both matching of indices and considering information about symmetries. Here is a more complicated example (suppose $W_{abcde} = W_{ebcda} = W_{cbade}$):

$$\begin{aligned}(W_{bde}{}^{ij} + W_{bde}{}^{ji} + W_{bed}{}^{ij} + W_{dbe}{}^{ji} + W_{de}{}^{i}{}_{b}{}^{j})(W_{cfhji} + W_{chfji} + W_{cjhfi} + W_{fchij} + W_{fchji}) - \\ (W_{bde}{}^{ij} + W_{bde}{}^{ji} + W_{bed}{}^{ji} + W_{dbe}{}^{ij} + W_{de}{}^{i}{}_{b}{}^{j})(W_{cfhij} + W_{chfij} + W_{cihfj} + W_{fchij} + W_{fchji}) = 0\end{aligned} \tag{3}$$

This identity can be proved by using information about contractions of indices and symmetries of tensor $W_{abide}$ without expansion of brackets.

Journal of Physics: Conference Series **608** (2015) 012060 doi:10.1088/1742-6596/608/1/012060

## 3. A brief overview of Redberry

Today there are a lot of packages written both on top of general CASs and standalone systems that cover different topics in computer computations with tensors. The most relevant packages that are actively supported are `xAct` [2] (open-source, written on top of Mathematica), `Maple Physics` [1] (commercial, not open-source) and `Cadabra` [3] (open-source, standalone). These packages well cover most of the features described above.

On the other hand, many modern calculations in physics involve extremely huge expressions, thus performance of CAS becomes critical. This aspect is not well addressed in the existing software. Redberry is aimed at solution of large scale problems with millions of tensorial terms in a reasonable time. For example, calculation of one-loop counterterms of gravitational field[1] requires processing of about 700 000 tensorial terms with about 8 multipliers and about 10 indices per term; Redberry solves this problem in less than 8 minutes (on a standard laptop).

Another example that one can easily check using mentioned packages is the identity (3). Time used by different systems to prove it is the following: Redberry — 2 ms, Cadabra — 180 ms, xAct — 180 ms, Maple Physics — 200 ms. In this example Redberry shows nearly 100 fold higher performance. Moreover, Redberry does not perform expansion of brackets while other systems do.

Along with high performance Redberry provides a comprehensive set of both general-purpose and HEP-specific tools for tensor handling which include: native dummy indices handling (particularly automatic resolving of clashes), support of arbitrary permutational symmetries, extensive tools for simplification of tensorial expressions, multiple index types, LaTeX-style input/output and a wide range of other tensor-specific transformations. Out of the box Redberry provides instruments for computations in high energy physics including tools for Feynman diagrams calculation (Dirac & SU(N) traces, simplification of Levi-Civita tensors etc.) and for calculation of one-loop counterterms in quantum field theory in curved space-time.

Redberry allows to use general purpose programming language equipped with a dozen of computer algebra and particularly tensor-specific syntax constructions which facilitates common usage as well as implementation of custom functionality for particular problems.

In order to demonstrate Redberry syntax let us consider a "toy" example. The following example proves that $\left(2\, R_{abcd}\, R^{acbd} - R_{abcd}\, R^{abcd}\right) = 0$ where $R_{abcd}$ is a Riemann tensor by using the identity $R_{abcd} = \left(2\, R_{abcd} + R_{acbd} - R_{adbc}\right)/3$:

```
1 // Setup symmetries
2 addSymmetries 'R_abcd', [[0, 2], [1, 3]].p, -[[1, 0]].p
3 // Define substitution
4 subs = 'R_abcd = (2*R_abcd + R_acbd - R_adbc)/3'.t
5 // Input expression
6 expr = '2*R_abcd*R^acbd - R_pqrs*R^pqrs'.t
7 // Apply substitution and expand
8 expr = (subs & Expand) >> expr
9 println expr //prints zero

  ▷ 0
```

In line 2 we define symmetries by providing two generators of corresponding permutation group written in disjoint cycle notation (`.p` converts list of integers into a `Permutation` object). In line 4 and 6 we input substitution and target expression (`.t` converts string into a computer object). Finally in line 8 we apply substitution and expand out brackets. The result is zero since Redberry automatically reduces similar terms taking into account all symmetries.

[1] for details on one-loop calculations in Redberry see [8] and online documentation at http://redberry.cc

Journal of Physics: Conference Series **608** (2015) 012060 doi:10.1088/1742-6596/608/1/012060

One can find lots of examples (including physical examples such as Feynman diagrams and one-loop counterterms calculation) along with comprehensive documentation on the Redberry website (http://redberry.cc).

## 4. Algorithms

Since the most frequently invoked atomic operation during any calculation is comparison of expressions, its performance makes the main contribution to the overall performance of CAS. This operation arises in such common simplification as reduction of similar terms. The algorithms of tensor comparison (or matching) is the main topic in the field of tensorial CAS.

The existing systems (known to the authors) are based on the so-called *indices canonicalization approach* [9, 10]. This procedure brings indices of products into unique canonical order using the information about symmetries of its multipliers; when indices of tensors are brought into canonical form, it becomes trivial to test whether two expressions are equal. As it is shown in [9, 10], this problem is equivalent to the problem of *double coset enumeration* which is known to be $\mathcal{NP}$-hard (see Sec. 4.6.8 in [11]). Unfortunately, no satisfactory algorithm to solve this problem has been found to date (see Sec. 4.6.8 in [11]). Importantly, indices canonicalization works only with products of simple tensors, e.g. if product contains a sum one need to expand out brackets first.

Redberry uses another approach to the problem which is based on the idea that contractions between indices in product constitutes a graph (see for example Fig. 1). From such point of view the problem of testing whether two expressions are equal is equivalent to problem of testing whether corresponding graphs are isomorphic (this approach was discussed in [9], but was not implemented in any CAS). While no worst-case polynomial-time algorithms are known for the general Graph Isomorphism (GI) problem, there are several algorithms which solve it very effectively in all practical cases (see e.g. [12] and overview given in [13]). The GI algorithm implemented in Redberry is specifically optimized for typical problems arising in physics (i.e. when one needs to compare huge number of small graphs; see below).

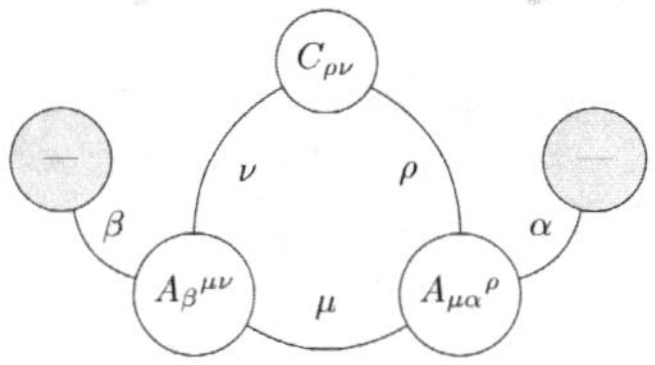


Figure 1: Graph representation of tensor $A_{\beta}{}^{\mu\nu}\,C_{\rho\nu}\,A_{\mu\alpha}{}^{\rho}$

In order to find out how Redberry performance compares to existing approaches we carried out performance evaluation of Redberry, Cadabra (which is open source and uses the same canonicalisation engine as Mathematica xAct ) and Maple Physics (commercial, implementation is unknown). We performed two types of benchmarks.

In our first benchmark we measured performance of comparison operation. For this purpose, we randomly generated sums of products of tensors[2] and measured time needed by each system to obtain zero when subtracting it from itself rewritten in equivalent form by renaming dummies, rearranging terms and shuffling indices according to symmetries. For example:

$$\Big(F_{\mu\nu\alpha}T^{\nu}{}_{\beta}W^{\alpha\beta} + \ldots\,(99\ terms)\Big) - \Big(T^{\nu}{}_{\beta}T^{\alpha}{}_{\rho}W^{\nu\rho}F_{\mu\alpha\nu} + \ldots\,(99\ terms)\Big) = 0$$

The results of our benchmarks are shown on Fig. 2. We analysed cases without and with symmetries (symmetric or antisymmetric with respect to all indices of particular simple tensor; in case with symmetries Maple failed to obtain zero on majority of examples, so it was excluded from the benchmark). We found that in case without symmetries, both Redberry and Maple have polynomial dependence on size of products (Fig. 2 left), while Cadabra has exponential; in case with symmetries, both Cadabra and Redberry have exponential dependence in worst case, but Redberry has better performance on larger problems. The dependence on the number of

[2] all expressions used for benchmarks can be found at http://redberry.cc/documentation:benchmarks

indices (Fig. 2 right) in case without symmetries is also polynomial for Redberry and Maple and exponential for Cadabra; on the other hand, the presence of symmetries almost does not change the behaviour of Cadabra, while Redberry fails to simplify large expressions in a reasonable time.

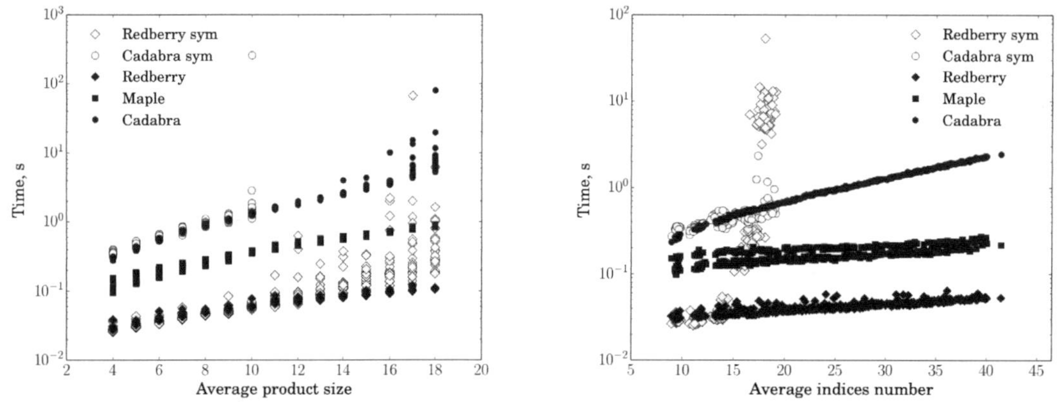


Figure 2: **Dependence of time spent in collect routine on product size (left) and on average number of indices in products (right)**. For these benchmarks we generated sums (100 summands) of products of simple tensors, then subtracted it from itself (in equivalently rewritten form: shuffled summands and multipliers, renamed dummy indices and interchanged indices of simple tensors according to their symmetries) and measured time needed to obtain zero. For Redberry and Cadabra we performed benchmarks both with (marked as "sym" in the legend) and without symmetries of simple tensors, while Maple was unable to simplify such expressions in case with symmetries.

In the second benchmark we measured performance of the most common simplification procedure — expand out product of sums and reduce similar terms (*expand and collect*). For this benchmark we used sums of products of sums of products[2]. We applied *expand and collect*, subtracted result from the initial expression and measured time needed by each system to obtain zero. For example:

$$\Big(F_{\mu\nu}\Big(T^{\mu\alpha}R^{\nu\beta}+\dots\Big)+\dots\Big)\Big(R_{\alpha\rho}\Big(F^{\rho}{}_{\beta}R_{\tau\gamma}+\dots\Big)+\dots\Big)\times\Big(\dots\Big) \\ -\Big(F_{\nu\mu}T^{\nu\beta}R^{\mu\alpha}R_{\beta\rho}F^{\rho}{}_{\alpha}R_{\tau\gamma}+\dots\Big)=0$$

Fig. 3 shows the comparison of time spent in *expand and collect* operation for the same input expressions for Redberry vs. Maple and for Redberry vs. Cadabra. On average in these benchmarks Redberry is 38 times faster than Maple and 62 times faster than Cadabra.

## 5. Conclusions

In this paper we considered the main features of algebraic computer-aided manipulations with tensors. We discussed two main approaches used for implementation of symbolic tensor algebra and presented a new computer algebra system Redberry based on the alternative to common approach for tensor handling. We demonstrated that Redberry outperforms other tools (Maple Physics and Cadabra) in typical operations on large tensorial expressions. Further information about Redberry including comprehensive documentation accompanied by lots of examples can be found on the Redberry website http://redberry.cc and in Ref. [8]

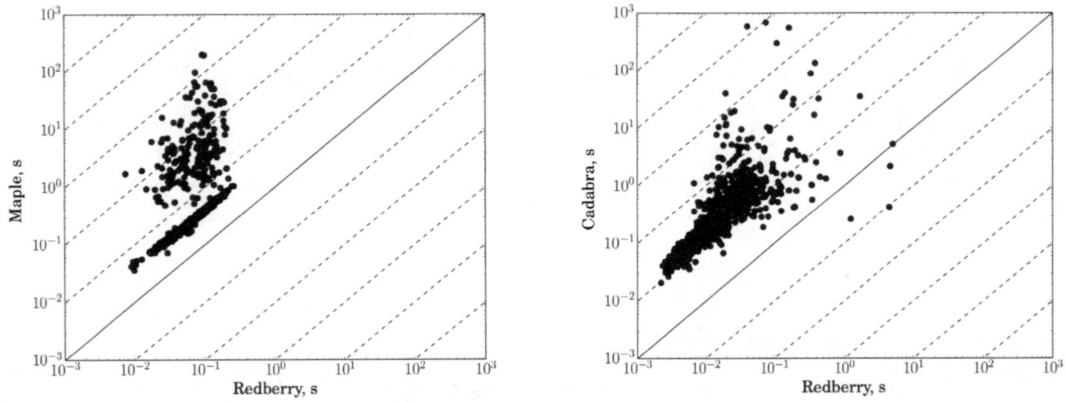


Figure 3: **Comparison of time spent in expand and collect routine for randomly generated expressions.** Each input problem plotted as filled circle. Each plot represents comparison of execution times needed to simplify input expression to zero between two systems (left for Maple versus Redberry, right for Cadabra versus Redberry). Solid line corresponds to equal execution times, dashed lines shows 10x, 100x, 1000x, etc. execution time ratio. For benchmark presented here, Maple failed to obtain zero for more then 16% of input expressions, this points are not presented on the left plot. On average in these benchmarks Redberry is 38 times faster than Maple and 62 times faster then Cadabra.

## Acknowledgements

The authors would like to thank A. Kataev and V. Ilyin for valuable and stimulating discussions. The work of S.P. was supported by the RFBR (grant #14-02-00096 A), grant of SAEC "Rosatom" and Helmholtz Association. The work of D.B. was financially supported by RSCF (grant #14-14-00533).

## References

[1] Maple Inc *Maple Physics homepage: http://www.maplesoft.com*
[2] J Martin-Garcia *xAct homepage: http://www.xact.es/*
[3] Peeters K *arXiv:hep-th/0701238*
[4] Peeters K 2007 *Comput.Phys.Commun.* **176** 550–558 arXiv:cs/0608005 [cs.SC]
[5] SymPy Development Team *SymPy homepage http://www.sympy.org*
[6] Anthony Hearn *REDUCE homepage: http://www.reduce-algebra.com*
[7] Korol'kova A, Kulyabov D and Sevast'yanov L 2013 *Programming and Computer Software* **39** 135–142 ISSN 0361-7688
[8] Bolotin D A and Poslavsky S V *arXiv:1302.1219 [cs.SC]*
[9] Rodionov A and Taranov A 1989 *Eurocal '87* (*Lecture Notes in Computer Science* vol 378) ed Davenport J (Springer Berlin Heidelberg) pp 192–201 ISBN 978-3-540-51517-3 URL `http://dx.doi.org/10.1007/3-540-51517-8_113`
[10] Manssur L R U, Portugal R and Svaiter B F 2002 *International Journal of Modern Physics C* **13** 859–879
[11] Holt D, Eick B and O'Brien E 2005 *Handbook of Computational Group Theory* Discrete Mathematics and Its Applications (Taylor & Francis) ISBN 9781420035216
[12] McKay B D 1981 *Congressus Numerantium* **30** 4587 10th. Manitoba Conference on Numerical Mathematics and Computing (Winnipeg, 1980)
[13] McKay B D and Piperno A *arXiv:1301.1493 [cs.DM]*

Journal of Physics: Conference Series **608** (2015) 012061 doi:10.1088/1742-6596/608/1/012061

# GoSam 2.0: a tool for automated one-loop calculations

**J F von Soden-Fraunhofen**
Max Planck Institute for Physics, Föhringer Ring 6, 80805 Munich, Germany
E-mail: `jfsoden@mpp.mpg.de`

**Abstract.** We present the version 2.0 of the program package GoSam, which is a public program package to compute one-loop QCD and/or electroweak corrections to multi-particle processes within and beyond the Standard Model. The extended version of the Binoth-Les-Houches-Accord interface to Monte Carlo programs is also implemented. This allows a large flexibility regarding the combination of the code with various Monte Carlo programs to produce fully differential NLO results, including the possibility of parton showering and hadronisation. We illustrate the wide range of applicability of the code by showing phenomenological results for multi-particle processes at NLO, both within and beyond the Standard Model.

## 1. Introduction

In the last years, there has been a remarkable progress in the development of automated NLO tools for multi-particle processes. This leads to the possibility of having NLO corrections matched to parton-showers as the new standard at the LHC.

Here we want to summarize the new features of the GoSam 2.0 release [1], which is able to automatically compute one-loop QCD and/or electroweak corrections to user-defined processes, and show the results of two applications.

The developments since the former release of GoSam described in [2,3] has allowed various phenomenological studies within and beyond the Standard model [4–15]. The new features are now publicly available in the GoSam 2.0 release.

## 2. GoSam 2.0 workflow

GoSam can be used standalone to generate code for one-loop amplitudes[1] or in combination with recent Monte Carlo programs via the Binoth-Les-Houches-Accord (BLHA) described in Sec. 3.1.

In the standalone mode, which is shown schematically in Fig. 1, the user provides the needed information to GoSam via an input card. GoSam calls QGraf [16] to generate all Feynman diagrams and the related algebraic expressions. These are further processed by FORM [17,18]. The spinor algebra is handled by the Spinney library [19]. The code is then transformed to compilable expressions using the new FORM 4 features [18] (see Sec. 3).

At runtime, the generated code can choose between different integrand-level reduction methods [20,21], implemented in Ninja [22–24] and Samurai [25,26], and improved tensor reduction methods implemented in Golem95 [27–30].

[1] The generation of colour/spin correlated tree level amplitudes is also possible.

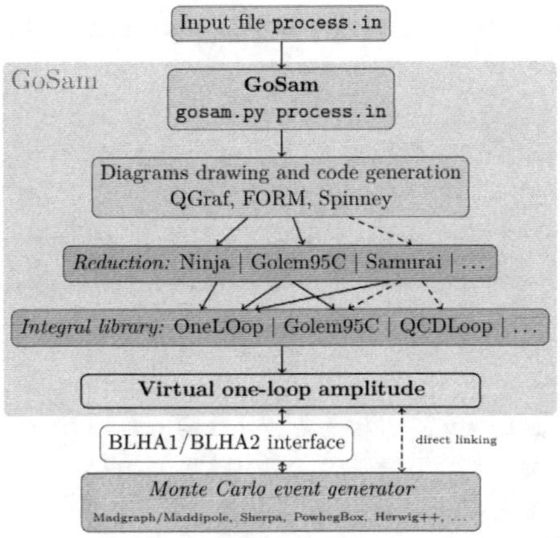


**Figure 1.** Basic workflow of GoSam

## 3. New features of GoSam 2.0

In this section we present some highlights of the version 2.0 of GoSam.

*Higher rank integrals* While in calculations in Feynman gauge within in the Standard Model, the tensor rank is always smaller or equal to the number of propagators in a loop integral, this is not necessarily the case in other gauges, effective theories or theories with spin-2 particles. For the GoSam 2.0 release, all of the supported reduction libraries (Ninja, Samurai, Golem95) have been extended to the case where the rank of the tensor integrals exceeds the number of the propagators [22, 26, 30].

*New reduction method* In GoSam 2.0, the new default reduction library is Ninja [22–24]. Ninja is a new developed integrand-reduction method and exploits the fact that the integrands are provided by GoSam in analytic expressions. It uses the Laurent expansion of the integrands performed by semi-numerical polynomial divisions.

*Improved code generation* The generation of optimized code is now mainly handled by the new FORM 4.0 optimization features [18, 31], which enables faster and more compact code.

*Grouping and summing of diagrams with common substructure* GoSam 2.0 has a new option `diagsum` which automatically groups or combines diagrams which share a common loop-structure and/or which can be distinguished only by external tree parts. This reduces not only the code generation time, but also the number of calls to the reduction method and, therefore, increases the overall performance. This extends the option of grouping diagrams already present in GoSam 1.0.

*Numerical polarisation vectors* By default, GoSam 2.0 creates code with generic polarisation vectors for massless gauge bosons and evaluates it for the various helicity configurations numerically. This saves the generation of separate code for each helicity configuration and therefore reduces the code generation time and the disk and memory footprint.

*Improved tensorial reconstruction* The new GoSam release enables by default the `derive` extension, which allows to isolate the tensor coefficients entering tensorial reconstruction [32] in

Journal of Physics: Conference Series **608** (2015) 012061 doi:10.1088/1742-6596/608/1/012061

a more efficient way. It is based on the fact that the numerator can be written as a (tensorial) Taylor series in the loop momenta $q^\mu$ around $q = 0$.

*Electroweak scheme choices* GoSAM 2.0 supports various schemes defining which of the electroweak parameters are input parameters and which ones are derived parameters. From a certain set of input parameters defining an electroweak scheme, GoSam will automatically determine the derived parameters. If not otherwise specified, GoSAM uses the set ($M_W$, $M_Z$, $\alpha$) by default.

The settings can be provided to GoSAM using the `model.options` keyword, which also needs to contain `ewchoice` if the choice should be changeable at runtime. If GoSAM is accessed via the Binoth Les Houches interface (cf. Sec. 3.1), the electroweak scheme is determined automatically according to the `OLP_SetParameter` calls.

### *3.1. BLHA2 Interface*

GoSAM 2.0 supports the Binoth Les Houches Accord 2 (BLHA2), an updated standard interface between Monte Carlo event generators and one-loop programs [33]. The first version of the BLHA standard [34] is still supported.

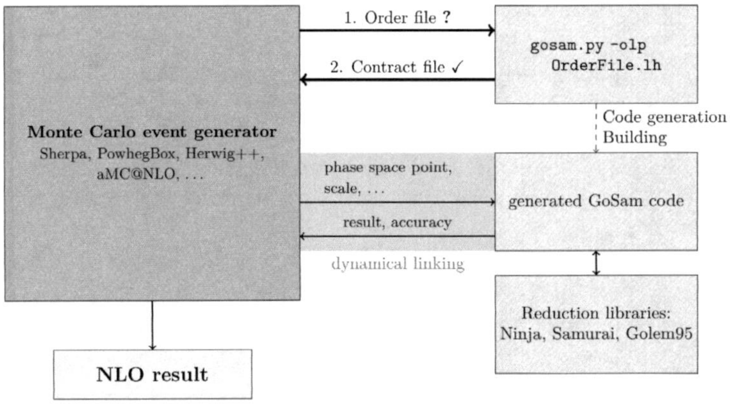


**Figure 2.** Workflow of the BLHA2 interface with GoSAM

The workflow of this interface is described in Fig. 2. It can be split into two phases. The first phase is the "order/contract phase", where the Monte Carlo programs "asks" the one-loop-provider (e.g. GoSAM) for the needed processes in an *order file*. It can then "reply" with an *contract file* and generate the code. As soon as the contract is "signed", i. e. all the items placed in the order file can be provided, the second phase can start. The Monte Carlo program is able to dynamically link the generated code for the virtual amplitude and can evaluate the amplitude at phase space points, thus performing the integration and event generation steps.

The updated interface allows to specify processes with different multiplicities and settings in one order file. GoSAM can also provide colour- and spin-correlated LO matrix elements. Both features allows the Monte Carlo program to built subtraction terms of NLO real radiation in addition to the virtual part.

The accuracy can be assessed by various tests: The first test compares the infrared pole coefficients to those obtained from the general structure of infrared singularities. The second, more time-consuming test is a rotation test, where the scattering amplitude is re-evaluated after an azimuthal rotation around the beam axis, which should keep the amplitude invariant. The thresholds triggering the various tests can be adjusted by the user via the input card.

Journal of Physics: Conference Series **608** (2015) 012061 doi:10.1088/1742-6596/608/1/012061

These tests are also used to trigger the rescue system (switching to GOLEM95 as reduction library) in the case of unstable results and are, if necessary, repeated to obtain the estimated accuracy of the final (rescued) result.

This procedure has been shown to be a good compromise between speed and quality of the assessed accuracy.

## 4. Installation

GOSAM 2.0 is distributed with a new install script, which installs the code automatically. It also helps to update GOSAM and its components.

The install script can be downloaded by calling

```
wget http://gosam.hepforge.org/gosam-installer/gosam_installer.py
```

The next step is to run the installer with the following commands

```
chmod +x gosam_installer.py
./gosam_installer.py --prefix=path/to/install
```

or alternatively:

```
python gosam_installer.py --prefix=path/to/install
```

If the `--prefix` option is left out, the script installs GOSAM into a subdirectory `./local` of the current working directory. Upon installation, the installer asks the user if it should use existing programs on the system (FORM 4, QGRAF, reduction libraries) or should install new versions of them which are tested with GOSAM. For a default installation, the questions can be safely answered by pressing the `ENTER` key. After this, the installer downloads, builds and installs all components. Finally, a script `[prefix]/bin/gosam_setup_env.sh` is created which sets up (temporarily) all environment variables if sourced in a shell.

It is suggested to keep the `installer-log.in`, which may be needed to update or uninstall GOSAM later. The update process can be initiated by calling the install script again without any arguments in the same folder where this file is located.

## 5. Using GoSam

By calling `gosam.py --template process.in`, the user is able to generate an input file where all possible settings are explained and default settings are shown. After filling all settings, `gosam.py process.in` generates a subdirectory specified by the `process_path` setting in the input card. The remaining steps for code generation and building can be performed by calling `make source` and `make compile` in this process directory. In the `doc/` subfolder, `make doc` generates documentation containing process specific informations as well as Feynman diagrams. The `matrix/` subdirectory contains a file `test.f90` which contains an example how to interface the compiled code. It can be built by calling `make test.exe` and outputs the NLO result for one random phase space point by default.

An example input-file for QCD corrections to $e^+e^- \to t\bar{t}$ is shown in Fig. 3.

```
process_name=eett
process_path=eett_code
in=e+,e-
out=t,t
order=QCD,0,2
one=gs,e
zero=me,mU,mD,mS,mC,mB
```

**Figure 3.** Example of an input card for QCD corrections to $e^+e^- \to t\bar{t}$

## 6. Applications

### *6.1. First example of* GoSam *plus* Herwig++/Matchbox

As a first example demonstrating the features of the BLHA2 interface implemented in GoSam 2.0 and an upcoming version of Herwig++/Matchbox [35–37], the NLO QCD corrections to the process $pp \to Z/\gamma^* + \text{jet} \to e^+e^- + \text{jet}$ were calculated.

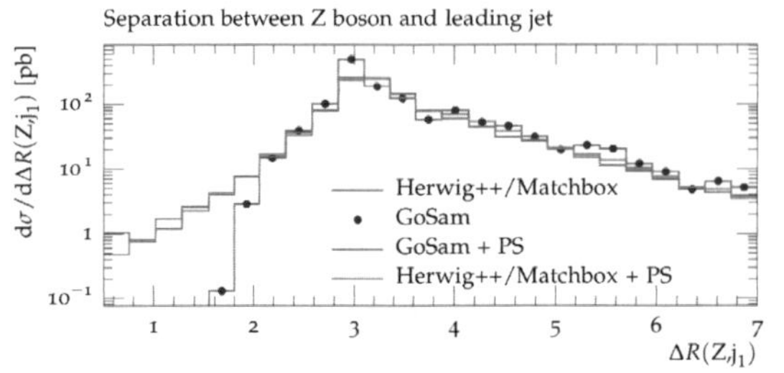


**Figure 4.** Z+jet production: R separation between the Z-boson and the leading jet. The blue dots and the green line show the result of Herwig++/Matchbox using matrix elements generated by GoSam without and with parton shower; the red and pink lines show the corresponding results with built-in Herwig/Matchbox matrix elements.

Besides the virtual matrix elements, GoSam provides various tree-level matrix elements (normal Born matrix elements and spin- and colour-correlated Born matrix elements) via the BLHA2 interface, which supports now order files with subprocesses of different amplitude types and multiplicities. This allows Herwig++ to calculate the full NLO real radiation part.

As this specific process is also directly built-in in Herwig++/Matchbox, it was easily possible to validate the results. Fig. 4 shows the matching results. Both ways are largely automated, i.e. only small changes in the Herwig++ input card (and optionally in an additional GoSam setting card) are needed.

Furthermore, Herwig++ can match a parton shower to the fixed-order calculation, which gives access to new kinematical regions. The result is also plotted in Fig. 4.

Full details of this calculation and information about the implementations can be found in [38].

### *6.2. Diphoton plus jet production through graviton exchange*

The NLO QCD corrections to the production of a photon pair through graviton exchange in association with a jet has been calculated with GoSam within the ADD model [39, 40], which assumes flat, large extra dimensions. The model file was generated by FeynRules [41] and imported into GoSam using the UFO file format [42]. Four to six extra dimensions were assumed and an ADD scale $M_S = 4\,\text{TeV}$ was chosen. According to this scale, the invariant mass of the photon pair was restricted to $140\,\text{GeV} \leq m_{\gamma\gamma} \leq 3.99\,\text{TeV}$. A variation of this upper cut-off showed that the results are not considerably affected by the unknown UV completion of the ADD model.

Since rank-5 boxes and rank-4 triangles appear, the higher rank extensions of GoSam and Golem95C (cf. Sec. 3) were used. Additionally, the sum over all discrete Kaluza-Klein mode propagators was approximated by a density function, which required a new `custompropagator` extension in GoSam, and a Lorentz structure where on-shell conditions were assumed. All of the needed features are available in the Golem95 1.3 and GoSam 2.0 release.

For this process, large enhancements are expected in the tail of the diphoton mass spectrum which is shown in Fig. 5. In particular, we observe a non-constant K-factor for this observable, which existing experimental searches have not considered.

More details and results can be found in [15], including the case of five and six extra dimensions.

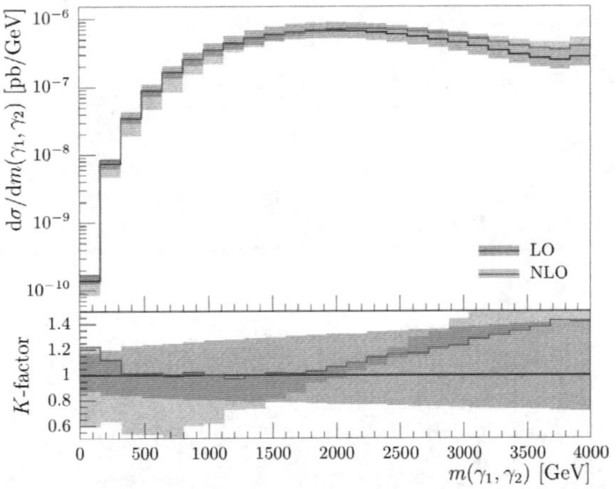


**Figure 5.** NLO corrections to the invariant di-photon mass distribution stemming from the graviton decay including a scale variation by a factor two around the central scale in proton-proton collisions at $\sqrt{s} = 8\,\text{TeV}$

## 7. Conclusion

GoSAM 2.0 provides many new features, leading to faster and more compact code, increased usability and a broader range of possible applications.

In particular, GoSAM supports now higher-rank integrals, where the tensor rank exceeds the number of propagators. This is needed in effective field theories and in calculations with spin-2 particles. Furthermore, complex masses and couplings, and various electroweak schemes are supported.

A new installation script simplifies the installation of GoSAM and related libraries and tools.

The overall performance was greatly increased by introducing a new integrand-level reduction method, implemented in the NINJA library, by advanced techniques of combining and grouping Feynman diagrams and by using the new FORM 4 features for code optimization.

Many of the new features were already successfully used and tested in various phenomenological studies [4–15].

The implementation of the new standardized interface to Monte Carlo programs (BLHA2) allows to use parton showered events at NLO level and to use different Monte Carlo generators for comparisons.

## Acknowledgements

The author thanks all the current and former members of the GoSAM collaboration for the joint effort in developing GoSAM 2.0. Furthermore, he is grateful to the members of the HERWIG++ collaboration for the common work on combining both programs via the BLHA2 interface, and Joscha Reichel for his contribution to the diphoton plus jet through graviton exchange calculation.

## References

[1] Cullen G, van Deurzen H, Greiner N, Heinrich G, Luisoni G, Mastrolia P, Mirabella E, Ossola G, Peraro T, Schlenk J, von Soden-Fraunhofen J F and Tramontano F 2014 *Eur.Phys.J.* **C74** 3001 (*Preprint* 1404.7096)

[2] Reiter T 2009 *Automated Evaluation of One-Loop Six-Point Processes for the LHC* Ph.D. thesis (*Preprint* 0903.0947)

[3] Cullen G, Greiner N, Heinrich G, Luisoni G, Mastrolia P *et al.* 2012 *Eur.Phys.J.* **C72** 1889 (*Preprint* 1111.2034)

[4] Greiner N, Heinrich G, Mastrolia P, Ossola G, Reiter T *et al.* 2012 *Phys.Lett.* **B713** 277–283 (*Preprint* 1202.6004)
[5] van Deurzen H, Greiner N, Luisoni G, Mastrolia P, Mirabella E *et al.* 2013 *Phys.Lett.* **B721** 74–81 (*Preprint* 1301.0493)
[6] Gehrmann T, Greiner N and Heinrich G 2013 *JHEP* **1306** 058 (*Preprint* 1303.0824)
[7] Luisoni G, Nason P, Oleari C and Tramontano F 2013 *JHEP* **1310** 083 (*Preprint* 1306.2542)
[8] Hoeche S, Huang J, Luisoni G, Schoenherr M and Winter J 2013 *Phys.Rev.* **D88** 014040 (*Preprint* 1306.2703)
[9] Cullen G, van Deurzen H, Greiner N, Luisoni G, Mastrolia P *et al.* 2013 *Phys.Rev.Lett.* **111** 131801 (*Preprint* 1307.4737)
[10] van Deurzen H, Luisoni G, Mastrolia P, Mirabella E, Ossola G *et al.* 2013 *Phys.Rev.Lett.* **111** 171801 (*Preprint* 1307.8437)
[11] Gehrmann T, Greiner N and Heinrich G 2013 *Phys.Rev.Lett.* **111** 222002 (*Preprint* 1308.3660)
[12] Dolan M J, Englert C, Greiner N and Spannowsky M 2014 *Phys.Rev.Lett.* **112** 101802 (*Preprint* 1310.1084)
[13] Heinrich G, Maier A, Nisius R, Schlenk J and Winter J 2014 *JHEP* **1406** 158 (*Preprint* 1312.6659)
[14] Cullen G, Greiner N and Heinrich G 2013 *Eur.Phys.J.* **C73** 2388 (*Preprint* 1212.5154)
[15] Greiner N, Heinrich G, Reichel J and von Soden-Fraunhofen J F 2013 *JHEP* **1311** 028 (*Preprint* 1308.2194)
[16] Nogueira P 1993 *J.Comput.Phys.* **105** 279–289
[17] Vermaseren J 2000 (*Preprint* math-ph/0010025)
[18] Kuipers J, Ueda T, Vermaseren J and Vollinga J 2013 *Comput.Phys.Commun.* **184** 1453–1467 (*Preprint* 1203.6543)
[19] Cullen G, Koch-Janusz M and Reiter T 2011 *Comput.Phys.Commun.* **182** 2368–2387 (*Preprint* 1008.0803)
[20] Ossola G, Papadopoulos C G and Pittau R 2007 *Nucl.Phys.* **B763** 147–169 (*Preprint* hep-ph/0609007)
[21] Giele W T, Kunszt Z and Melnikov K 2008 *JHEP* **0804** 049 (*Preprint* 0801.2237)
[22] Mastrolia P, Mirabella E and Peraro T 2012 *JHEP* **1206** 095 (*Preprint* 1203.0291)
[23] van Deurzen H, Luisoni G, Mastrolia P, Mirabella E, Ossola G *et al.* 2014 *JHEP* **1403** 115 (*Preprint* 1312.6678)
[24] Peraro T 2014 *Comput.Phys.Commun.* **185** 2771–2797 (*Preprint* 1403.1229)
[25] Mastrolia P, Ossola G, Reiter T and Tramontano F 2010 *JHEP* **1008** 080 (*Preprint* 1006.0710)
[26] van Deurzen H 2013 *Acta Phys.Polon.* **B44** 2223–2230
[27] Binoth T, Guillet J P, Heinrich G, Pilon E and Schubert C 2005 *JHEP* **0510** 015 (*Preprint* hep-ph/0504267)
[28] Binoth T, Guillet J P, Heinrich G, Pilon E and Reiter T 2009 *Comput.Phys.Commun.* **180** 2317–2330 (*Preprint* 0810.0992)
[29] Cullen G, Guillet J P, Heinrich G, Kleinschmidt T, Pilon E *et al.* 2011 *Comput.Phys.Commun.* **182** 2276–2284 (*Preprint* 1101.5595)
[30] Guillet J P, Heinrich G and von Soden-Fraunhofen J 2014 *Comput.Phys.Commun.* **185** 1828–1834 (*Preprint* 1312.3887)
[31] Kuipers J, Ueda T and Vermaseren J 2013 (*Preprint* 1310.7007)
[32] Heinrich G, Ossola G, Reiter T and Tramontano F 2010 *JHEP* **1010** 105 (*Preprint* 1008.2441)
[33] Alioli S, Badger S, Bellm J, Biedermann B, Boudjema F *et al.* 2014 *Comput.Phys.Commun.* **185** 560–571 (*Preprint* 1308.3462)
[34] Binoth T, Boudjema F, Dissertori G, Lazopoulos A, Denner A *et al.* 2010 *Comput.Phys.Commun.* **181** 1612–1622 (*Preprint* 1001.1307)
[35] Bellm J, Gieseke S, Grellscheid D, Papaefstathiou A, Plätzer S *et al.* 2013 (*Preprint* 1310.6877)
[36] Plätzer S and Gieseke S 2012 *Eur.Phys.J.* **C72** 2187 (*Preprint* 1109.6256)
[37] Bahr M, Gieseke S, Gigg M, Grellscheid D, Hamilton K *et al.* 2008 *Eur.Phys.J.* **C58** 639–707 (*Preprint* 0803.0883)
[38] Bellm J, Gieseke S, Greiner N, Heinrich G, Plätzer S, Reuschle C and von Soden-Fraunhofen J 2014 Gosam plus herwig++/matchbox *Les Houches 2013: Physics at TeV Colliders: Standard Model Working Group Report* (*Preprint* 1405.1067)
[39] Arkani-Hamed N, Dimopoulos S and Dvali G 1998 *Phys.Lett.* **B429** 263–272 (*Preprint* hep-ph/9803315)
[40] Antoniadis I, Arkani-Hamed N, Dimopoulos S and Dvali G 1998 *Phys.Lett.* **B436** 257–263 (*Preprint* hep-ph/9804398)
[41] Christensen N D and Duhr C 2009 *Comput.Phys.Commun.* **180** 1614–1641 (*Preprint* 0806.4194)
[42] Degrande C, Duhr C, Fuks B, Grellscheid D, Mattelaer O *et al.* 2012 *Comput.Phys.Commun.* **183** 1201–1214 (*Preprint* 1108.2040)

Journal of Physics: Conference Series **608** (2015) 012062 doi:10.1088/1742-6596/608/1/012062

# Numerical multi-loop calculations with the program SecDec

**Sophia Borowka**[1], **Gudrun Heinrich**[2]

[1] Institute for Physics, University of Zurich, Winterthurerstr.190, 8057 Zurich, Switzerland
[2] Max Planck Institute for Physics, Föhringer Ring 6, 80805 Munich, Germany

E-mail: `sborowka@physik.uzh.ch`, `gudrun@mpp.mpg.de`

**Abstract.** SecDec is a program which can be used for the evaluation of parametric integrals, in particular multi-loop integrals. For a given set of propagators defining the graph, the program constructs the graph polynomials, factorizes the endpoint singularities, and finally produces a Laurent series in the dimensional regularization parameter, whose coefficients are evaluated numerically. In this talk we discuss various features of the program, which extend the range of applicability. We also present a recent phenomenological example of an application entering the momentum dependent two-loop corrections to neutral Higgs boson masses in the MSSM.

## 1. Introduction

In view of the absence of "smoking gun" signals of new physics at the LHC so far, the importance of precision calculations cannot be over-emphasized. This means that higher order corrections in both the QCD and the electroweak sector need to be evaluated, ideally without neglecting mass effects, and including matched parton showers as far as possible.

These calculations have many facets, however most of them have in common that they involve multi-dimensional integrations over some parameters, for example Feynman parameters in the case of (multi-)loop integrals, or parameters related to the integration over a factorized phase space of subtraction terms for infrared-divergent real radiation. Usually, these calculations are performed within the framework of dimensional regularization, and one of the challenges is to factorize the poles in the regulator $\varepsilon$.

The program SecDec [1, 2, 3] is designed to perform this task, and to integrate the coefficients of the resulting Laurent series in $\varepsilon$ numerically, based on the sector decomposition algorithm described in [4, 5]. Other public implementations of sector decomposition can be found in [6, 7, 8, 9].

## 2. The program SecDec

### 2.1. *Basic structure*

The program consists of two main parts, one being designed for loop integrals, to be found in a directory called `loop`, the other one for more general parametric integrals, located in a directory called `general`. The procedure to isolate the poles in the regulator $\varepsilon$ and to do the subtractions and integrations is very similar in the two branches. However, only the `loop` part contains the possibility of contour deformation, because only for loop integrals the analytic continuation can be performed in an automated way, following Feynman's "$i\delta$" prescription. The basic flowchart

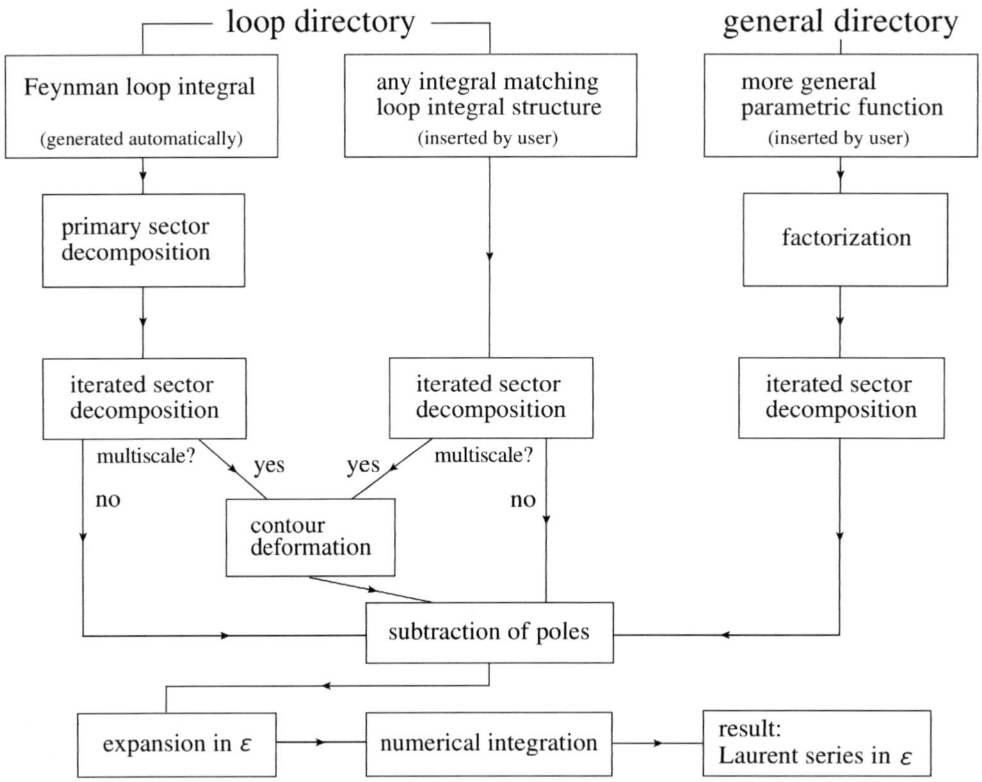


**Figure 1.** Flowchart showing the main steps the program performs to produce the numerical result as a Laurent series in $\varepsilon$.

of the program is shown in Fig. 1. More details about the central column will be given in section 2.5.

### *2.2. Installation and usage*

*Installation*

The program can be downloaded from `http://secdec.hepforge.org`. Unpacking the tar archive via *tar xzvf SecDec.tar.gz* will create a directory called `SecDec`. Running *./install* in the `SecDec` directory will install the package. Prerequisites are Mathematica, version 6 or above, perl (installed by default on most Unix/Linux systems), a C++ compiler, and a Fortran compiler if the Fortran option is used. Contour deformation is only available in C++. The libraries CUBA [10, 11] and BASES [12], which are used for the numerical integration, come with the package SecDec and will be compiled at the installation stage.

*Usage*

The user should edit two files (where templates are provided): a text file where parameters like the name of the graph, the order to be expanded in $\varepsilon$ or numerical integration parameters are set, and a file in Mathematica syntax, where the graph is defined, optionally either in terms of propagators or in terms of lines connecting numbered vertices. The template to define the parameters for standard loop integrals is called `paramloop.input`, and the one in Mathematica syntax is called `templateloop.m`. It is recommended that the user copies these template files and renames them before editing them. We will call the edited files `myparamfile.input` and `mytemplatefile.m` in the following. Examples for a number of specific graphs or input functions can be found in the subdirectories `demos` of both the `loop` and the `general` directory. After having edited the input files, SecDec is called as follows:

(i) from the `loop` (or `general`) directory, execute the following command in the shell:
*./launch -p myparamfile.input -t mytemplatefile.m*
If your files `myparamfile.input, mytemplatefile.m` are in a different directory, say, *myworkingdir*, use the option *-d myworkingdir*, i.e. the full command then looks like *./launch -d myworkingdir -p myparamfile.input -t mytemplatefile.m*, executed from the directory `SecDec/loop` or `SecDec/general`.
If the option *-p myparamfile.input* is omitted, the file `paramloop.input` will be taken as default. Likewise, if the option *-t mytemplatefile.m* is omitted, the file `templateloop.m` will be taken as default.

(ii) Collect the results. If the calculations are done sequentially on a single machine, the results will be collected automatically. If the jobs have been submitted to a cluster, when all jobs have finished, use the command *perl results.pl [-d myworkingdir -p myparamfile]*. In both cases, the files containing the final results will be located in the `graph` subdirectory specified in the input file.

### *2.3. Topology definition*

In the Mathematica input file `mytemplatefile.m`, the graph can be defined in two ways: Either by labeling the vertices and listing the connections between vertices (corresponding to the flag `cutconstruct=1`), or by specifying the momentum flow explicitly (corresponding to `cutconstruct=0`). The program will then construct the integrand in terms of Feynman parameters automatically. For tensor integrals, i.e. integrals with loop momenta in the numerator, the propagators have to be given in terms of momenta, i.e. `cutconstruct=0` must be used, as tensor integrals are not shift invariant.

An example for the construction of the graph polynomials based on labelled vertices is given in subsection 3.1.

### *2.4. Controlling the different stages of the calculation*

As the program consists of a purely algebraic part and a numerical part, it can be useful to perform the calculation only up to a certain level, instead of launching immediately the full program chain including the numerical integration. For example, to get an idea about the pole structure of an integral, one can first perform only the iterated sector decomposition to factor out the parameters exhibiting the poles in $\varepsilon$ to expect. This can be achieved by setting `exeflag=0` in `myparamfile.input`. The following stages can be selected with the `exeflag`:

- 0: The iterated sector decomposition is done. The scripts to do the subtractions, the expansion in epsilon and to launch the numerical part are created (scripts batch* in the subdirectory *graph*), but not run.
- 1: The subtraction and epsilon expansion is performed and the resulting functions are written to Fortran/C++ files.
- 2: All the files needed for the numerical integration are created.
- 3: The compilation of the Fortran/C++ files is launched to make the executables.
- 4: The executables are run to perform the numerical integration.

Journal of Physics: Conference Series **608** (2015) 012062 doi:10.1088/1742-6596/608/1/012062

### *2.5. Evaluation of user-defined functions with arbitrary kinematics*

In the standard setup, the program will construct the graph polynomials $\mathcal{F}$ and $\mathcal{U}$ and then proceed directly to the so-called *primary sector decomposition* [4, 5], which serves to integrate out the constraint $\delta(1 - \sum_i x_i)$ in a way which preserves the property that all the endpoint singularities are located at the origin of parameter space, rather than creating singularities at $x_i = 1$.

However, there are certain situations where it is useful to start the iterated decomposition at a later stage, where the delta-function already has been integrated out analytically, or for an integrand which does not contain such a delta-function at all. If one can find a convenient parametrisation and integrate out one Feynman parameter analytically, this can be beneficial for complicated integrals, because it reduces the number of integration variables for the subsequent Monte Carlo integration and therefore will improve the numerical efficiency.

In such cases, the user can skip the primary sector decomposition step and insert the functions to be factorized directly into the Mathematica input file. The purpose of this option is to be more flexible with regard to the functions to be integrated, such that expressions for loop integrals which are not in the "standard form" can be dealt with as well. This includes the possibility to perform a deformation of the integration contour into the complex plane, taking the user-defined functions as a starting point. Oriented at the functions $\mathcal{F}$ and $\mathcal{U}$ for the "standard" loop case, the user-defined functions can encompass the product of two arbitrary polynomial functions with different exponents, and an additional numerator.

The command to launch the calculation is the same as for the standard setup, except that the Mathematica input file looks different (for an example see the file `templateuserdefined.m`), and that the extension `-u` should be appended:

*./launch -p myparamfile.input -t mytemplateuserdefined.m -u*

The "-u" stands for "user defined" and skips the primary sector decomposition step.

### *2.6. Scanning over ranges of numerical parameters*

The algebraic part of SecDec can deal with symbolic expressions for the kinematic invariants or other parameters contained in the integrand. Therefore, the decomposition and subtraction parts only need to be done once and for all. The produced functions then can be employed for the calculation of many different numerical points. The program comes with scripts which facilitate the scanning over ranges of numerical values for the kinematic invariants. There is a perl script `helpmulti.pl` which can be used to produce an input file `multiparam.input` containing ranges of numerical values for the invariants/symbolic parameters contained in the integrand. This way the user does not have to type in all the numerical values by hand. Each line in `multiparam.input` defines a new run. An example of a `multiparam.input` file is contained in both the `loop` and the `general` directories. A detailed description of this option is also given in [1].

In order to launch the calculations for the values specified in `multiparam.input`, the command

*perl multinumerics.pl [-d myworkingdir] -p multiparamfile*

should be issued.

Please note that before executing the script `multinumerics.pl`, the Mathematica-generated functions must already be existent. The simplest way to do this is to make one run with *exeflag=1* in the single-run parameter file.

To collect the results: in single-machine mode (*clusterflag=0*), the results will be collated automatically and written to the `graph` directory specified in the parameter file. In *cluster* mode the *multinumerics.pl* script has to be run again with the argument '1' appended i.e. *perl multinumerics.pl [-d myworkingdir] -p multiparamfile 1* to collect the results. In both cases files with the extension `[pointname][i].gpdat` will be created for each coefficient of $\varepsilon^i$ in the Laurent

expansion, which can be plotted easily. For example, if the parameter $s$ has been scanned over, and $s$ is the first element in the list of invariants, setting `xplot=1` in `multiparam.input` will define $s$ to be on the x-axis, and therefore $s$ will be printed into the first column of the `.gpdat` file. Thus, the columns in these result files will be
`(s) (real part) (error_real_part) (imaginary part) (error_imaginary_part) (timings)`.
Result files for each individual point (extension `.res`) will also be created.

## 3. Applications

### *3.1. Two-loop four-point integrals with two mass scales*

Here we give an example for the definition of a non-planar two-loop four-point function, containing both internal masses and massive external legs with a different mass.

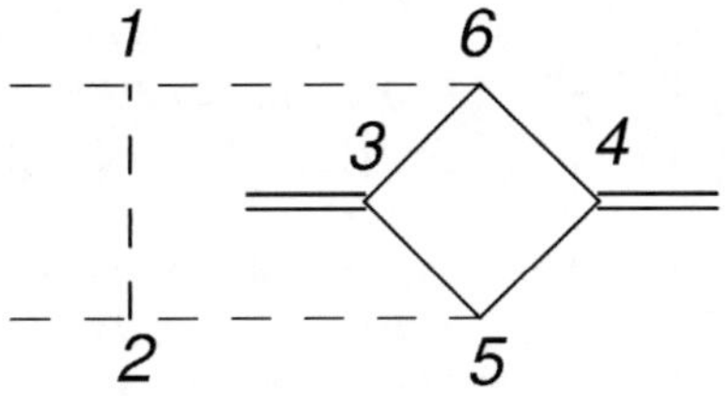


**Figure 2.** Example of a non-planar two-loop box diagram with two mass scales. Dashed lines denote massless propagators, solid lines massive propagators. The double lines denote massive legs with a mass different from the internal mass.

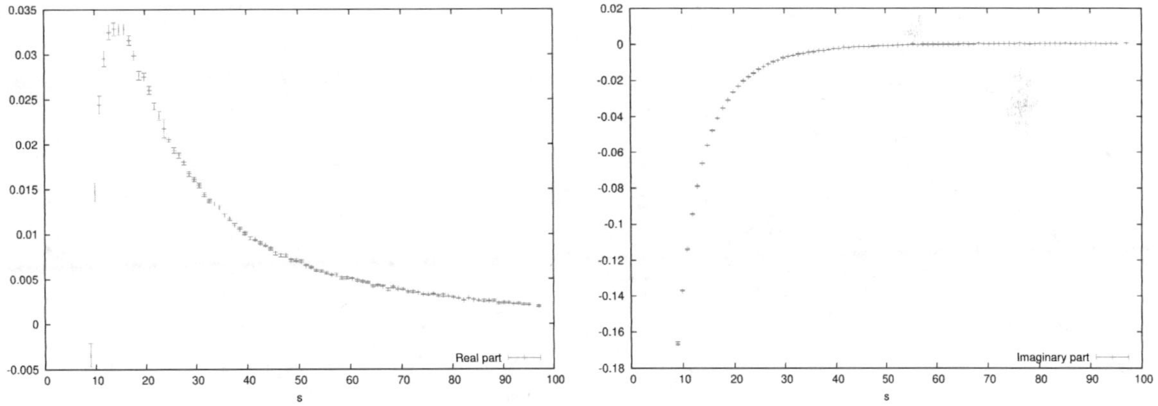


**Figure 3.** Real and imaginary parts of the finite part of the graph shown in Fig. 2, for the values $m_1^2 = 1, m_2^2 = 0.522, t = -3.978, u = -s - t + 2m_2^2$.

To define the scalar graph depicted in Fig. 2, the user can use the option `cutconstruct=1` as explained above. To this end, the vertices containing external legs need to be labelled with the same number as the corresponding external leg. Then the vertices containing only internal lines are labelled (in arbitrary order). The graph is defined in `mytemplate.m` by a list `proplist` containing the two vertex labels $i_1, i_2$ a certain propagator is connecting, and the mass $m$ of that propagator, as $\{m, \{i_1, i_2\}\}$. For example, for the graph of Fig. 2

```
proplist={{0,{1,2}},{0,{1,6}},{0,{2,5}},{ms[1],{3,6}},{ms[1],{3,5}},
```

```
{ms[1],{4,5}},{ms[1],{4,6}}}; .
```

The kinematic conditions have to be given as

```
onshell={ssp[1]->0,ssp[2]->0,ssp[3]->ms[2],ssp[4]->ms[2]}; .
```

The expression `ssp[i]` is the standard name for $p_i^2$. Squared masses $m_i^2$ are denoted by `ms[i]`. The restriction to "standard names" will be lifted in the next release of SecDec, where the user will be able to define arbitrary symbols for the invariants.
Numerical results for this graph are shown in Fig. 3.

### *3.2. Momentum dependent two-loop corrections to neutral Higgs boson masses in the MSSM*

In this section we briefly describe an application of SecDec where two-loop two-point integrals with up to four different mass scales are entering. Until recently, the $\mathcal{O}(\alpha_t\alpha_s)$ corrections to the neutral $\mathcal{CP}$-even Higgs boson masses in the MSSM have been known adopting a full $\overline{\text{DR}}$ scheme only [13, 14, 15], or neglecting the momentum dependence [16, 17, 18]. For a consistent inclusion of the corrections in the program FeynHiggs [16, 19, 20, 21, 22, 23], the top-quarks and top-squarks need to be renormalized on-shell. These momentum dependent $\mathcal{O}(\alpha_t\alpha_s)$ corrections missing so far in FeynHiggs were calculated in [24], where the 34 mass configurations of the analytically unknown two-loop two-point integrals are calculated numerically using SecDec. For further details we refer to [24, 25]. Very recently, an independent calculation of these corrections appeared [26], where the unknown integrals have been calculated numerically using the program `TSIL` [15], which is a dedicated program for the evaluation of two-loop two-point functions, based on differential equations.

At tree level, the mass matrix of the neutral $\mathcal{CP}$-even Higgs bosons in the $(\phi_1^0, \phi_2^0)$ basis can be written as

$$M^{2,\text{tree}}_{\text{Higgs}} = \begin{pmatrix} M_A^2 \sin^2\beta + M_Z^2 \cos^2\beta & -(M_A^2 + M_Z^2)\sin\beta\cos\beta \\ -(M_A^2 + M_Z^2)\sin\beta\cos\beta & M_A^2\cos^2\beta + M_Z^2\sin^2\beta \end{pmatrix}, \tag{1}$$

where $M_A$ is the mass of the $\mathcal{CP}$-odd neutral Higgs boson $A$. The rotation to the basis formed by the mass eigenstates $H^0, h^0$ is given by

$$\begin{pmatrix} H^0 \\ h^0 \end{pmatrix} = \begin{pmatrix} \cos\alpha & \sin\alpha \\ -\sin\alpha & \cos\alpha \end{pmatrix} \begin{pmatrix} \phi_1^0 \\ \phi_2^0 \end{pmatrix}. \tag{2}$$

The higher-order corrected $\mathcal{CP}$-even Higgs boson masses in the MSSM are obtained from the corresponding propagators dressed by their self-energies. The inverse propagator matrix in the $(\phi_1^0, \phi_2^0)$ basis is given by

$$(\Delta_{\text{Higgs}})^{-1} = -\mathrm{i}\begin{pmatrix} p^2 - m_{\phi_1}^2 + \hat{\Sigma}_{\phi_1}(p^2) & -m_{\phi_1\phi_2}^2 + \hat{\Sigma}_{\phi_1\phi_2}(p^2) \\ -m_{\phi_1\phi_2}^2 + \hat{\Sigma}_{\phi_1\phi_2}(p^2) & p^2 - m_{\phi_2}^2 + \hat{\Sigma}_{\phi_2}(p^2) \end{pmatrix}, \tag{3}$$

where the $\hat{\Sigma}(p^2)$ denote the momentum-dependent renormalized Higgs-boson self-energies, $p$ being the external momentum. The latter have been calculated at the two-loop level, at order $\alpha_s\alpha_t$.

Our calculation is performed in the Feynman-diagrammatic approach. We adopt a hybrid on-shell/$\overline{\text{DR}}$ scheme, in line with the renormalization of previous higher-order contributions included in the program FeynHiggs[1], see Ref. [24] for more details. To obtain expressions for the unrenormalized self-energies and tadpoles at $\mathcal{O}(\alpha_t\alpha_s)$, the evaluation of genuine two-loop diagrams and one-loop graphs with counterterm insertions is required. Example diagrams for the neutral Higgs-boson self-energies are shown in Fig. 4.

[1] We neglect a numerically insignificant shift in the value of $\tan\beta$.

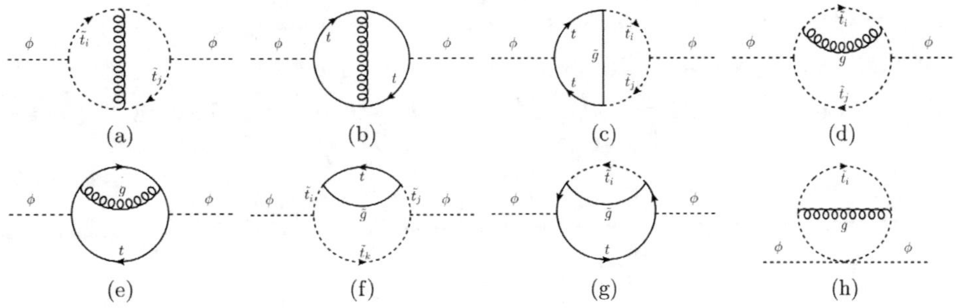

**Figure 4.** Examples of two-loop diagrams enetring the Higgs-boson self-energies ($\phi = h, H, A$).

For the counterterm-insertions, one-loop diagrams with external top quarks/squarks have to be evaluated as well. The complete set of contributing Feynman diagrams has been generated with the program `FeynArts` [27, 28] (using the model file including counterterms from Ref. [29]). Tensor reduction and the evaluation of traces was done with support from the programs `FormCalc` [30] and `TwoCalc` [31]. The resulting two-loop integrals which depend on the external momentum contain four topologies for which only partial analytical results are available. These topologies are shown in Fig. 5. They occur in 34 different mass configurations, and have been evaluated with SecDec.

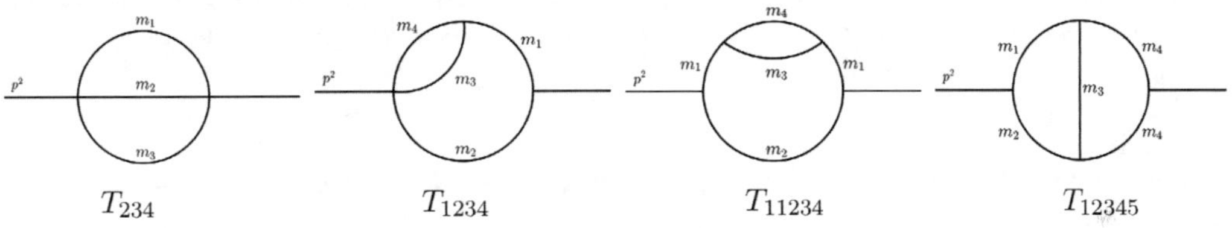

**Figure 5.** Topologies which have been calculated numerically using SecDec.

*Computation of mass shifts*

The calculation of the self-energies is performed in the $(\phi_1^0, \phi_2^0)$ basis. To be consistent with all other higher-order contributions to the Higgs-boson masses incorporated in the program FeynHiggs, the renormalized self-energies in the $(\phi_1^0, \phi_2^0)$ basis is rotated into the physical $(h^0, H^0)$ basis, where the tree-level propagator matrix is diagonal, via

$$\hat{\Sigma}^{(2)}_{H^0H^0} = \cos^2\alpha\, \hat{\Sigma}^{(2)}_{\phi_1^0\phi_1^0} + \sin^2\alpha\, \hat{\Sigma}^{(2)}_{\phi_2^0\phi_2^0} + \sin(2\alpha)\, \hat{\Sigma}^{(2)}_{\phi_1^0\phi_2^0}\,, \tag{4a}$$

$$\hat{\Sigma}^{(2)}_{h^0h^0} = \sin^2\alpha\, \hat{\Sigma}^{(2)}_{\phi_1^0\phi_1^0} + \cos^2\alpha\, \hat{\Sigma}^{(2)}_{\phi_2^0\phi_2^0} - \sin(2\alpha)\, \hat{\Sigma}^{(2)}_{\phi_1^0\phi_2^0}\,, \tag{4b}$$

$$\hat{\Sigma}^{(2)}_{h^0H^0} = \sin\alpha\cos\alpha\, (\hat{\Sigma}^{(2)}_{\phi_2^0\phi_2^0} - \hat{\Sigma}^{(2)}_{\phi_1^0\phi_1^0}) + \cos(2\alpha)\, \hat{\Sigma}^{(2)}_{\phi_1^0\phi_2^0}\,, \tag{4c}$$

and $\alpha$ is the tree-level mixing angle.

The resulting new contributions to the neutral $\mathcal{CP}$-even Higgs-boson self-energies, containing all momentum-dependent and additional constant terms, are assigned to the differences

$$\Delta\hat{\Sigma}^{(2)}_{ab}(p^2) = \hat{\Sigma}^{(2)}_{ab}(p^2) - \tilde{\Sigma}^{(2)}_{ab}(0)\,, \qquad ab = \{H^0H^0, h^0H^0, h^0h^0\}\,. \tag{5}$$

Note the tilde (not hat) on $\tilde{\Sigma}^{(2)}(0)$, which signifies that not only the self-energies are evaluated at zero external momentum, but also the corresponding counter-terms, following Refs. [32, 33, 16].

A finite shift $\Delta\hat{\Sigma}^{(2)}(0)$ therefore remains in the limit $p^2 \to 0$ due to $\delta m_{A^0}^{2(2)} = \mathrm{Re}\,\Sigma_{A^0A^0}^{(2)}(m_{A^0}^2)$ being computed at $p^2 = m_{A^0}^2$ in $\hat{\Sigma}^{(2)}$, but at $p^2 = 0$ in $\tilde{\Sigma}^{(2)}$.

According to Eq. (3), the $\mathcal{CP}$-even Higgs boson masses are determined by the poles of the $h^0$-$H^0$-propagator matrix. This is equivalent to solving the equation

$$\left[p^2 - m_{h^0}^2 + \hat{\Sigma}_{h^0h^0}(p^2)\right]\left[p^2 - m_{H^0}^2 + \hat{\Sigma}_{H^0H^0}(p^2)\right] - \left[\hat{\Sigma}_{h^0H^0}(p^2)\right]^2 = 0 \ , \qquad (6)$$

yielding the loop-corrected pole masses, $M_h$ and $M_H$.

*Numerical results*
In our numerical analysis we find that the effects on the light $\mathcal{CP}$-even Higgs boson mass, $M_h$, are sensitive to the value of the gluino mass, $m_{\tilde{g}}$. For values of $m_{\tilde{g}} \sim 1.5$ TeV corrections to $M_h$ of about $-50$ MeV are found, at the level of the anticipated future ILC accuracy. For very large gluino masses, $m_{\tilde{g}} \gtrsim 4$ TeV, on the other hand, substantially larger corrections are found, at the level of the current experimental accuracy of $\sim$ 500 MeV. The new results of $\mathcal{O}(\alpha_t\alpha_s)$ including momentum dependence have been implemented into the program FeynHiggs.

## 4. Conclusions
We have described the program SecDec, with emphasis on useful features and new applications. SecDec is a flexible tool which can be used to factorize poles from dimensionally regulated parameter integrals, and to evaluate the pole coefficients numerically. As there is no restriction on the kinematic invariants, the program is particularly useful for two-loop integrals with several mass scales, where analytic approaches have difficulties.

The program further contains scripts to facilitate scans over large ranges of numerical parameters, based on functions for the Laurent series expansion of the integral which can be produced once and for all.

The forthcoming version 3.0 of the program will contain further improvements, in particular offer new decomposition strategies.

## Acknowledgments
We would like to thank the organizers of ACAT2014 for the nice conference. We also thank Thomas Hahn, Sven Heinemeyer and Wolfgang Hollik for collaboration on the MSSM Higgs project, and Johannes Schlenk for his contributions towards SecDec version 3.0.

## References
[1] Carter J and Heinrich G 2011 *Comput.Phys.Commun.* **182** 1566–1581 (*Preprint* 1011.5493)
[2] Borowka S, Carter J and Heinrich G 2013 *Comput.Phys.Commun.* **184** 396–408 (*Preprint* 1204.4152)
[3] Borowka S and Heinrich G 2013 *Comput.Phys.Commun.* **184** 2552–2561 (*Preprint* 1303.1157)
[4] Binoth T and Heinrich G 2000 *Nucl. Phys.* **B585** 741–759 (*Preprint* hep-ph/0004013)
[5] Heinrich G 2008 *Int. J. Mod. Phys.* **A23** 1457–1486 (*Preprint* 0803.4177)
[6] Bogner C and Weinzierl S 2008 *Comput. Phys. Commun.* **178** 596–610 (*Preprint* 0709.4092)
[7] Gluza J, Kajda K, Riemann T and Yundin V 2011 *Eur.Phys.J.* **C71** 1516 (*Preprint* 1010.1667)
[8] Smirnov A and Tentyukov M 2009 *Comput.Phys.Commun.* **180** 735–746 (*Preprint* 0807.4129)
[9] Smirnov A V 2014 *Comput.Phys.Commun.* **185** 2090–2100 (*Preprint* 1312.3186)
[10] Hahn T 2005 *Comput. Phys. Commun.* **168** 78–95 (*Preprint* hep-ph/0404043)
[11] Agrawal S, Hahn T and Mirabella E 2011 (*Preprint* 1112.0124)
[12] Kawabata S 1995 *Comp. Phys. Commun.* **88** 309–326
[13] Martin S P 2004 *Phys.Rev.* **D70** 016005 (*Preprint* hep-ph/0312092)
[14] Martin S P 2005 *Phys.Rev.* **D71** 016012 (*Preprint* hep-ph/0405022)
[15] Martin S P and Robertson D G 2006 *Comput.Phys.Commun.* **174** 133–151 (*Preprint* hep-ph/0501132)
[16] Heinemeyer S, Hollik W and Weiglein G 1999 *Eur.Phys.J.* **C9** 343–366 (*Preprint* hep-ph/9812472)

[17] Heinemeyer S, Hollik W and Weiglein G 1999 *Phys.Lett.* **B455** 179–191 (*Preprint* hep-ph/9903404)
[18] Degrassi G, Slavich P and Zwirner F 2001 *Nucl.Phys.* **B611** 403–422 (*Preprint* hep-ph/0105096)
[19] Heinemeyer S, Hollik W and Weiglein G 2000 *Comput.Phys.Commun.* **124** 76–89 (*Preprint* hep-ph/9812320)
[20] Hahn T, Heinemeyer S, Hollik W, Rzehak H and Weiglein G 2009 *Comput.Phys.Commun.* **180** 1426–1427
[21] Hahn T, Heinemeyer S, Hollik W, Rzehak H and Weiglein G 2014 *Phys.Rev.Lett.* **112** 141801 (*Preprint* 1312.4937)
[22] Degrassi G, Heinemeyer S, Hollik W, Slavich P and Weiglein G 2003 *Eur.Phys.J.* **C28** 133–143 (*Preprint* hep-ph/0212020)
[23] Frank M, Hahn T, Heinemeyer S, Hollik W, Rzehak H *et al.* 2007 *JHEP* **0702** 047 (*Preprint* hep-ph/0611326)
[24] Borowka S, Hahn T, Heinemeyer S, Heinrich G and Hollik W 2014 *Eur.Phys.J.* **C74** 2994 (*Preprint* 1404.7074)
[25] Borowka S 2014 *Ph.D. thesis, Max Planck Institute for Physics/Technical University Munich* (*Preprint* 1410.7939)
[26] Degrassi G, Di Vita S and Slavich P 2014 (*Preprint* 1410.3432)
[27] Hahn T 2001 *Comput.Phys.Commun.* **140** 418–431 (*Preprint* hep-ph/0012260)
[28] Hahn T and Schappacher C 2002 *Comput.Phys.Commun.* **143** 54–68 (*Preprint* hep-ph/0105349)
[29] Fritzsche T, Hahn T, Heinemeyer S, von der Pahlen F, Rzehak H *et al.* 2014 *Comput.Phys.Commun.* **185** 1529–1545 (*Preprint* 1309.1692)
[30] Hahn T and Perez-Victoria M 1999 *Comput.Phys.Commun.* **118** 153–165 (*Preprint* hep-ph/9807565)
[31] Weiglein G, Scharf R and Böhm M 1994 *Nucl.Phys.* **B416** 606–644 (*Preprint* hep-ph/9310358)
[32] Heinemeyer S, Hollik W and Weiglein G 1998 *Phys.Rev.* **D58** 091701 (*Preprint* hep-ph/9803277)
[33] Heinemeyer S, Hollik W and Weiglein G 1998 *Phys.Lett.* **B440** 296–304 (*Preprint* hep-ph/9807423)

# Modern Particle Physics Event Generation with WHIZARD

**J Reuter[1], F Bach[1], B Chokoufé[1], W Kilian[2], T Ohl[3], M Sekulla[2] and C Weiss[1,2]**

[1]DESY Theory Group, Notkestr. 85, D–22607 Hamburg, Germany
[2]University of Siegen, Department of Physics, Walter-Flex-Str. 3, D–57068 Siegen, Germany
[3]University of Würzburg, Department of Physics and Astronomy, Emil-Hilb-Weg 22, D–97074 Würzburg

E-mail: juergen.reuter@desy.de, fabian.bach@desy.de, bijan.chokoufe@desy.de, kilian@hep.physik.uni-siegen.de, ohl@physik.uni-wuerzburg.de, sekulla@hep.physik.uni-siegen.de, christian.weiss@desy.de

**Abstract.** We describe the multi-purpose Monte-Carlo event generator WHIZARD for the simulation of high-energy particle physics experiments. Besides the presentation of the general features of the program like SM physics, BSM physics, and QCD effects, special emphasis will be given to the support of the most accurate simulation of the collider environments at hadron colliders and especially at future linear lepton colliders. On the more technical side, the very recent code refactoring towards a completely object-oriented software package to improve maintainability, flexibility and code development will be discussed. Finally, we present ongoing work and future plans regarding higher-order corrections, more general model support including the setup to search for new physics in vector boson scattering at the LHC, as well as several lines of performance improvements.

## 1. Introduction

WHIZARD [1] is a general event generator for all kinds of scattering and decay processes at high-energy hadron and lepton colliders.

The default matrix element generator of WHIZARD is O'Mega [2]. This latter subpackage provides matrix elements for multi-leg tree-level processes, using the helicity formalism. The high-dimensional phase-space integrations are performed by the multi-channel Monte-Carlo integrator VAMP [3]. Its algorithm is adaptive both between and within channels, and thus computes accurate phase-space integrals and efficiently generates weighted and unweighted event samples.

The WHIZARD core acts as a connector of these different components. The core contains the algorithm for multi-channel phase-space parameterization and mapping, provides the user interface, and also interfaces external programs (e.g., parton distributions, event formats, hadronization), the routines for writing and reading event files, and modules for parton shower and jet physics. Further modules allow for the for numerical analyses and visualization of event samples.

In order to be able to describe realistic ILC and CLIC environments, WHIZARD contains a dedicated package for beam-spectrum simulation, CIRCE [4]. As an alternative, GuineaPig beam-event samples can be fed into `WHIZARD` using `CIRCE2`.

## 2. Program Overview

`WHIZARD` – initiated along the lines of the TESLA design study [5] to provide improved simulations for electroweak processes at lepton colliders – has developed into a universal generator for partonic events at all types of present and future (lepton and hadron) colliders.

With the new millenium, the implementation of QCD color flows finalized the support for the full Standard Model (SM), and parton distributions, different event samples and an ever enlarging list of models beyond the SM (BSM) came up. Here, we should mention particularly the implementation of the MSSM and several variations thereof, like the NMSSM [6]. For this purpose, the SUSY Les Houches Accord (SLHA) [7, 8, 9] has been provided in order to interface SUSY spectrum generators. `WHIZARD` has been used for a plethora of both theoretical and experimental studies for TESLA and the ILC as well as the LHC. The SLAC event database for Linear Collider events has been generated with the package.

To meet the enormous technical demands of event simulations in the LHC and ILC era, the core of the program has been thoroughly rewritten in 2007-2010, which gave rise to the `WHIZARD` 2 release series.

The rigid input files of the early versions have been replaced in `WHIZARD` 2 by input commands within the domain-specific language SINDARIN. Besides a simplification and unification of the input inside the user interface, the user gets the full power of a programming language at hand. With the help of SINDARIN, cuts, analyses, interfaces, process collection, parameter scans and many more things can be defined.

By means of external packages like SARAH [10] or `FeynRules` [11], interfaced to `WHIZARD` [12], new physics models can easily be added. Inside `WHIZARD`, complete quantum-mechanical correlations are kept using a generic density-matrix formalism. In addition, the program is endowed with its own parton-shower module [13] as an alternative to showering being performed externally. Exact matrix elements can be matched to the shower by means of the MLM scheme, while other merging and matching schemes are planned for the near future.

`WHIZARD` 2 is written in Fortran2003 that is supported in modern Fortran compilers. Backwards compatibility is kept until version 4.7.4 of `gfortran`. The matrix element generator `O'Mega` is written in the functional programming language `OCaml`, for which complete packages exist for all major platforms. The current `WHIZARD` production version is 2.2.2 (as of October 2014), while the next release, 2.2.3, is planned for end of November. The complete package is thoroughly tested and can be installed on all recent Linux and MAC OS systems.

## 3. Technical Details

The standard installation of `WHIZARD` is foreseen centrally on a machine, but it can of course also be installed locally in a user directory. It uses the standard toolchain of `automake`, `autoconf`, and `libtool`. The program is located at the HepForge server [14]. Downloads are available either as tagged versions (`.tar.gz` format), or as a development version from the public `svn` repository. The package conforms to the standard installation procedure, using the usual `configure` – `make` – `make install` chain and the optional `make check` and `make installcheck` steps for additional safety checks.

Event generation and simulation projects of the users can then be set up in arbitrary directories without any predefined structure. For this purpose, a single command `whizard` exists, calling a single SINDARIN command file as input. Alternative modes of using `WHIZARD` exist, namely an interactive mode, or linking it as a subroutine library that is interoperable with C, C++ or any other C-compatible language (e.g., Python). The package generates and processes matrix element codes on-the-fly as dynamically linked libraries. There is also a statically linked mode for the work e.g. on batch clusters.

Journal of Physics: Conference Series **608** (2015) 012063 doi:10.1088/1742-6596/608/1/012063

According to the modern object-oriented programming paradigm, there should be a clear separation between abstract type declarations and specific implementations. This is reflected in the modern Fortran standard, and hence WHIZARD 2 is broken down into these building blocks. There has been a major refactoring of the code between versions 2.1 and 2.2 (which is still partially ongoing) along these lines. The consequent separation of interface from implementation greatly improves the maintainability and enhances the possibilities for future module replacements, reimplementations, and extensions enormously. This is now realized in most parts of the core code like the process structure, matrix element calculation methods, beam structure, integration methods, decays, the phase space and the shower. For parallelization, WHIZARD 2 uses OpenMP. This is especially used in a new high-performance virtual machine and will be made available in the upcoming version 2.2.3. An MPI implementation is foreseen for the nearer future.

Quality assignment for WHIZARD is provided at first by using the svn version control at HepForge. Secondly, a continuous integration system runs all commits automatically through a chain of several hundreds of unit and function tests. Bugs and feature requests are steered by means of HepForge's tracking system.

## 4. Physics

The main method for matrix elements inside WHIZARD is provided by the matrix element generator O'Mega [2] which is able to generate complete tree-level matrix elements with multiple external legs (successful tests on standard hardware have involved up to 15 external particles). O'Mega is based on a recursive algorithm that reuses *all* common subexpressions and replaces the forest of all tree diagrams by the equivalent directed acyclical graph (DAG). It is able to provide the complete color correlations of QCD matrix elements, using the color-flow formalism with phantom $U(1)$ particles [15] as an efficient way to generate colorized DAGs that can be evaluated exactly or be projected onto color-flow amplitudes.

WHIZARD now supports processes to consist of several different components, e.g. for inclusive production samples using process containers or to combine NLO subtraction terms with real emission and virtual matrix elements. Also available are flavor sums which are however technically different, as masses have to be equal to commonly use the same phase space. Within the SINDARIN steering language this feature is easily available for the user with an appropriate syntax for inclusive processes and a detailed specification of decay chains. The latter will be refined in upcoming versions. Exact spin and color correlations are kept using the internal density-matrix formalism inside WHIZARD.

WHIZARD 2 can not only process complete matrix elements, but also generate decay chains and cascades. They consist of arbitrarily chosen elementary processes to be integrated each separately, but concatenated for the event generation. The default option is to take full spin correlation among intermediate states into account. There is also an option to restrict to classical spin correlations (only diagonal entries in the spin-density matrix), or to even switch off spin correlations completely in order to test the importance of spin correlations. WHIZARD can also set up decays and branching fractions for chosen physics models automatically.

Arbitrary factorization and renormalization scale settings (affecting QCD) can be used in WHIZARD 2 using the same syntax expressions as the one for cuts or analyses. Furthermore, since version WHIZARD 2.2 there is the possibility to reweight existing event samples, generated either internally or read from file, when changing the setup of the original process, e.g. parameters in the hard matrix elements, the event scale, the chosen structure functions, or the QCD parton shower.

## 5. Linear Collider Simulation

WHIZARD is an event generator for all kinds of high-energy colliders, but a particular focus has always been its vast support for a realistic lepton collider environment. A description as accurate as possible of the beam properties is manadatory for studies and analyses, given the required level of precision at ILC and CLIC.

WHIZARD incorporates the CIRCE1 [4] package that parameterizes the beams of an $e^+e^-$ collider and provides an event generator for factorized beam spectra. Adapting this generator (or, alternatively, the parameterized spectrum directly), WHIZARD can integrate and simulate any $e^+e^-$ process with a realistic beam description. As an upgrade to previous versions, WHIZARD 2.2 ships with beam spectra that correspond to the current ILC design parameters. To account for cases where such a factorized form is insufficient, WHIZARD can alternatively read beam-event files as they are produced by GuineaPig. From version 2.2.3, the CIRCE2 package inside WHIZARD now directly interfaces GuineaPig(++) and allows for the use of correlated lepton collider beam spectra with their steeply rising peaks where a power-law parameterization is insufficient.

Lepton-collider processes are not only influenced by beamstrahlung, but also strongly affected by electromagnetic initial-state radiation (ISR). WHIZARD uses a standard structure-function formalism that resums the corrections from infrared (leading) and collinear (3rd order) radiation to implement ISR. It takes ISR into account both in kinematics and dynamics, if requested.

WHIZARD enables the user to specify arbitrary beam polarizations, ranging from unpolarized, left- or right-handed circular or transversal polarization to arbitrary spin-density matrices. The polarization and polarization fractions are specified for both beams independently. The user can also define asymmetric beam setups and a crossing angle, which will be taken into account in the kinematics setup.

Photons as initial particles are available in various incarnations: on-shell, radiated from $e^\pm$ (effective photon approximation), or beamstrahlung photons generated by CIRCE1. A photon-collider option that uses the CIRCE2 beam description is also available but no longer maintained, due to the lack of current ILC or CLIC photon-collider mode beam parameters and simulations of the corresponding beam-beam interactions.

For cuts, reweighting and internal analysis, WHIZARD employs its dedicated language SINDARIN that allows for computing a wide range of event-specific and generic observables.

WHIZARD supports various event output formats, including the traditional HEPEVT format and its derivatives, StdHEP, LHEF, HepMC, and others. A direct interface to LCIO is planned.

## 6. QCD

For a precise calculation of exclusive processes, a program needs fine control over QCD corrections. Regarding real radiation, the multi-leg capability of WHIZARD allows to include high orders of the QCD coupling. Collinear and soft radiation in exclusive events is affected by large logarithms, which are conveniently resummed in the semi-classical approach of a parton shower algorithm.

Using standard event formats and suitable cuts, WHIZARD allows for attaching an external parton-shower generator. WHIZARD 2 furthermore contains an internal showering module in two different incarnations: a $k_T$ ordered shower along the lines of the Pythia shower [16], and an analytic parton shower, which keeps the complete shower history and allows to reweight it [13]. There is support for combining exact matrix elements and QCD radiation from the parton shower using the MLM matching prescription. These modules are foreseen to receive a more detailed validation, tuning and further improvements after the 2.2 release.

Beyond the parton shower, hadronization and hadronic decays are not performed by WHIZARD internally, but can be applied to the generated partonic event samples via the Pythia 6 package [16] which is attached to the WHIZARD distribution, or using e.g. LHE event samples that are then fed into a external (shower and) hadronization package.

Journal of Physics: Conference Series **608** (2015) 012063 doi:10.1088/1742-6596/608/1/012063

A high-luminosity linear collider will be capable of a high-precision scan of the top-quark pair-production threshold [17]. To match this on the theoretical side, one needs to resum logarithms of the top velocity $\sim \alpha_s \ln v$ as well as gluon Coulomb potential terms $\sim \alpha_s/v$ in a non-relativistic approach and to relate this to the relativistic matrix elements in the continuum. There is an ongoing project for including these effects in `WHIZARD` 2 which will make the theoretical calculation available in the simulation of exclusive events. As a first step, the next-to-leading-logarithmic approximation matched to next-to-leading-order matrix elements has already been implemented and will be included in the upcoming release 2.2.3 [18].

## 7. Status of next-to-leading order calculations

`WHIZARD` 2 with `O'Mega` matrix elements is an event generator of tree-level processes. There have been several projects that extended it to next-to-leading order, including loop corrections and proper infrared-collinear subtraction. Ref. [19, 20] describes the extension of `WHIZARD` 1 to a positive-definite NLO event generator for the electroweak pair production of charginos in the MSSM, including full electroweak SUSY corrections matched to the photon initial and final state radiation. Independently, Ref. [21, 22] implemented the QCD NLO correction with subtraction for a particular LHC process. Along these lines, the Binoth Les Houches Accord (BLHA) interface [23, 24] has been implemented for reading and writing contract files with one-loop programs (OLP), and has been validated.

Building upon the new data structures of `WHIZARD` 2.2, an implementation of automatic NLO QCD corrections is currently being developed.

`WHIZARD` is being extended to calculate cross sections at next-to-leading order in $\alpha_s$. For this purpose, we have implemented the FKS subtraction scheme [25], which relies on a partition of the phase space into disjoint regions. Using this scheme, `WHIZARD` computes the real-subtracted and virtual-subtracted part of the cross section.

The virtual amplitude is calculated using `GoSam` [26], which can be interfaced very easily with `WHIZARD` using the BLHA conventions [23].

For the real amplitude, standard `O'Mega` matrix elements are used. The subtraction terms require color- and spin-correlated Born matrix elements. However, the treatment of the latter has not been adressed yet, because spin-correlated matrix elements are only non-zero if at least one external particle is a gluon. For the color-correlated Born matrix element we use the usual Born matrix element multiplied with $C_F$, which is suited for 2-jet cross sections. These parts of the calculation will be fully generalized soon, also using `GoSam`.

The implementation has been tested using the analytically known R-ratio for 2-jet production in lepton collisions. `WHIZARD` correctly reproduces the result $\sigma_{\text{NLO}} = (1 + \alpha_s/\pi)\,\sigma_{\text{LO}}$. It has also been tested for the processes $e\nu_e \to q\bar{q}'Z$ and $e^+e^- \to q\bar{q}'l\nu_l$ and has been validated for those processes.

The calculation of NLO-QCD cross sections will be an experimental feature of Release 2.2.3 and at least supports processes including two colored particles in the final state.

## 8. Physics models

As a generator of hard matrix elements, `WHIZARD` has to support various particle species and interactions. The allowed spin representations for particles are 0, 1, 2 (bosons) and 1/2, 3/2 (fermions), all massive or massless, both Dirac and Majorana spinors, optionally colored (triplet or octet). The `WHIZARD` libraries support all Lorentz structures for interactions in the models described below. A completely general framework supporting all possible Lorentz structures is under construction.

*BSM Models* Beyond the SM and its QCD and QED subsets, `WHIZARD` supports the minimal supersymmetric Standard Model (MSSM) with different variants and extensions. These include models with gravitinos [27, 28, 29] and the NMSSM [6] (see also [30]).

Among models with strongly interacting sectors WHIZARD includes Little Higgs models in different incarnations, with and without discrete symmetries, cf. [31, 32, 33, 34, 35, 36].

WHIZARD has also been used for studying more exotic models such as the noncommutative SM [37, 38, 39] (not included in the official release). It further supports the completely general two-Higgs doublet model (2HDM), as well as generic models containing a $Z'$ state, and extra-dimensional models like Universal Extra Dimensions (UED). A more detailed list can be found in the WHIZARD documentation [14].

*Effective Theories* As an alternative tool for studying deviations from the SM, WHIZARD contains SM extensions with anomalous couplings, expressible as coefficients of higher-dimensional operators in an effective theory. Several models in WHIZARD's library define either anomalous triple and quartic gauge boson couplings, which have been used for studies at LCs [40, 41] or at LHC [42, 43]. Anomalous top-quark couplings are also supported [44, 45].

Recent interest in the physics of high-energy vector-boson scattering has triggered the development and addition of simplified models for strong interactions and compositeness (SSC) [43]. In addition to generic new degrees of freedom, they implement a unitarization procedure that is required for extrapolating into the energy range that will become accessible at ILC and, in particular, at CLIC.

## 9. Conclusion and Outlook

WHIZARD is a versatile and user-friendly tool for both SM and BSM physics at all possible high-energy colliders. In these proceedings, we report the recent progress in the technical development of the program as well as new physics features. Two main points to be mentioned are the special focus on lepton-collider beam spectra, which involves a detailed account of correlated beam spectra, arbitrary polarization and interfaces to the machine simulation. Future development will be devoted to a better description of QED radiation from both the initial and final states at lepton colliders and their proper matching to hard matrix elements. The second topic is the progress in the automation of QCD higher-order processes by interfacing external one-loop programs through the BLHA interface, the automatic generation of subtraction terms for soft-collinear singularities in the real emission as well as the virtual part and an automatic phase-space integration of the different components by using the method of FKS regions.

Further plans for new features include the support for more general Lorentz and color structures in models, a more convenient model interface, power-counting of coupling constants in the matrix element, further refinements in the beam description, and many technical improvements.

## Acknowledgments

JRR wants to thank the organizers of ACAT 2014 for a great conference in the wonderful and lovely city of Prague.

## References

[1] Kilian W, Ohl T and Reuter J 2011 *Eur.Phys.J.* **C71** 1742 (*Preprint* 0708.4233)
[2] Moretti M, Ohl T and Reuter J 2001 (*Preprint* hep-ph/0102195)
[3] Ohl T 1999 *Comput.Phys.Commun.* **120** 13–19 (*Preprint* hep-ph/9806432)
[4] Ohl T 1997 *Comput.Phys.Commun.* **101** 269–288 (*Preprint* hep-ph/9607454)
[5] Kilian W 2001 (*Preprint* LC-TOOL-2001-039)
[6] Reuter J and Braam F 2010 *AIP Conf.Proc.* **1200** 470–473 (*Preprint* 0909.3059)
[7] Skands P Z, Allanach B, Baer H, Balazs C, Belanger G *et al.* 2004 *JHEP* **0407** 036 (*Preprint* hep-ph/0311123)
[8] Allanach B, Balazs C, Belanger G, Bernhardt M, Boudjema F *et al.* 2009 *Comput.Phys.Commun.* **180** 8–25 (*Preprint* 0801.0045)

[9] Aguilar-Saavedra J A, Ali A, Allanach B C, Arnowitt R L, Baer H A *et al.* 2006 *Eur.Phys.J.* **C46** 43–60 (*Preprint* hep-ph/0511344)
[10] Staub F 2014 *Comput.Phys.Commun.* **185** 1773–1790 (*Preprint* 1309.7223)
[11] Christensen N D and Duhr C 2009 *Comput.Phys.Commun.* **180** 1614–1641 (*Preprint* 0806.4194)
[12] Christensen N D, Duhr C, Fuks B, Reuter J and Speckner C 2012 *Eur.Phys.J.* **C72** 1990 (*Preprint* 1010.3251)
[13] Kilian W, Reuter J, Schmidt S and Wiesler D 2012 *JHEP* **1204** 013 (*Preprint* 1112.1039)
[14] WHIZARD-Manual URL http://whizard.hepforge.org/
[15] Kilian W, Ohl T, Reuter J and Speckner C 2012 *JHEP* **1210** 022 (*Preprint* 1206.3700)
[16] Sjostrand T, Mrenna S and Skands P Z 2006 *JHEP* **0605** 026 (*Preprint* hep-ph/0603175)
[17] Baer H, Barklow T, Fujii K, Gao Y, Hoang A *et al.* 2013 (*Preprint* 1306.6352)
[18] Bach F, Hoang A, Reuter J, Stahlhofen M and Teubner T *in preparation*
[19] Robens T, Kalinowski J, Rolbiecki K, Kilian W and Reuter J 2008 *Acta Phys.Polon.* **B39** 1705–1714 (*Preprint* 0803.4161)
[20] Kilian W, Reuter J and Robens T 2006 *Eur.Phys.J.* **C48** 389–400 (*Preprint* hep-ph/0607127)
[21] Binoth T, Greiner N, Guffanti A, Reuter J, Guillet J P *et al.* 2010 *Phys.Lett.* **B685** 293–296 (*Preprint* 0910.4379)
[22] Greiner N, Guffanti A, Reiter T and Reuter J 2011 *Phys.Rev.Lett.* **107** 102002 (*Preprint* 1105.3624)
[23] Binoth T, Boudjema F, Dissertori G, Lazopoulos A, Denner A *et al.* 2010 *Comput.Phys.Commun.* **181** 1612–1622 (*Preprint* 1001.1307)
[24] Alioli S, Badger S, Bellm J, Biedermann B, Boudjema F *et al.* 2014 *Comput.Phys.Commun.* **185** 560–571 (*Preprint* 1308.3462)
[25] Frederix R, Frixione S, Maltoni F and Stelzer T 2009 *JHEP* **0910** 003 (*Preprint* 0908.4272)
[26] Cullen G, van Deurzen H, Greiner N, Heinrich G, Luisoni G *et al.* 2014 (*Preprint* 1404.7096)
[27] Hagiwara K, Kilian W, Krauss F, Ohl T, Plehn T *et al.* 2006 *Phys.Rev.* **D73** 055005 (*Preprint* hep-ph/0512260)
[28] Kalinowski J, Kilian W, Reuter J, Robens T and Rolbiecki K 2008 *JHEP* **0810** 090 (*Preprint* 0809.3997)
[29] Pietsch N, Reuter J, Sakurai K and Wiesler D 2012 *JHEP* **1207** 148 (*Preprint* 1206.2146)
[30] Reuter J and Wiesler D 2011 *Phys.Rev.* **D84** 015012 (*Preprint* 1010.4215)
[31] Kilian W and Reuter J 2004 *Phys.Rev.* **D70** 015004 (*Preprint* hep-ph/0311095)
[32] Kilian W, Rainwater D and Reuter J 2005 *Phys.Rev.* **D71** 015008 (*Preprint* hep-ph/0411213)
[33] Kilian W, Rainwater D and Reuter J 2006 *Phys.Rev.* **D74** 095003 (*Preprint* hep-ph/0609119)
[34] Reuter J and Tonini M 2013 *JHEP* **1302** 077 (*Preprint* 1212.5930)
[35] Reuter J, Tonini M and de Vries M 2014 *JHEP* **1402** 053 (*Preprint* 1310.2918)
[36] Reuter J and Tonini M 2014 (*Preprint* 1409.6962)
[37] Ohl T and Reuter J 2004 *Phys.Rev.* **D70** 076007 (*Preprint* hep-ph/0406098)
[38] Alboteanu A, Ohl T and Ruckl R 2006 *Phys.Rev.* **D74** 096004 (*Preprint* hep-ph/0608155)
[39] Alboteanu A, Ohl T and Ruckl R 2007 *eConf* **C0705302** TEV05 (*Preprint* 0709.2359)
[40] Beyer M, Kilian W, Krstonosic P, Monig K, Reuter J *et al.* 2006 *Eur.Phys.J.* **C48** 353–388 (*Preprint* hep-ph/0604048)
[41] Boos E, He H, Kilian W, Pukhov A, Yuan C *et al.* 1998 *Phys.Rev.* **D57** 1553 (*Preprint* hep-ph/9708310)
[42] Alboteanu A, Kilian W and Reuter J 2008 *JHEP* **0811** 010 (*Preprint* 0806.4145)
[43] Kilian W, Ohl T, Reuter J and Sekulla M 2014 (*Preprint* 1408.6207)
[44] Bach F and Ohl T 2012 *Phys.Rev.* **D86** 114026 (*Preprint* 1209.4564)
[45] Bach F and Ohl T 2014 *Phys.Rev.* **D** in print (*Preprint* 1407.2546)

# Mathematica and Fortran programs for various analytic QCD couplings

**César Ayala and Gorazd Cvetič**

Department of Physics, Universidad Técnica Federico Santa María, Casilla 110-V, Valparaíso, Chile

E-mail: gorazd.cvetic@gmail.com

**Abstract.** We outline here the motivation for the existence of analytic QCD models, i.e., QCD frameworks in which the running coupling $\mathcal{A}(Q^2)$ has no Landau singularities. The analytic (holomorphic) coupling $\mathcal{A}(Q^2)$ is the analog of the underlying pQCD coupling $a(Q^2) \equiv \alpha_s(Q^2)/\pi$, and any such $\mathcal{A}(Q^2)$ defines an analytic QCD model. We present the general construction procedure for the couplings $\mathcal{A}_\nu(Q^2)$ which are analytic analogs of the powers $a(Q^2)^\nu$. Three analytic QCD models are presented. Applications of our program (in Mathematica) for calculation of $\mathcal{A}_\nu(Q^2)$ in such models are presented. Programs in both Mathematica and Fortran can be downloaded from the web page: gcvetic.usm.cl.

## 1. Why analytic QCD?

Perturbative QCD (pQCD) running coupling $a(Q^2)$ [$\equiv \alpha_s(Q^2)/\pi$, where $Q^2 \equiv -q^2$] has unphysical (Landau) singularities at low spacelike momenta $0 < Q^2 \lesssim 1\ \text{GeV}^2$.

For example, the one-loop pQCD running coupling

$$a(Q^2)^{(1-\ell.)} = \frac{1}{\beta_0 \ln(Q^2/\Lambda^2_{\text{Lan.}})} \tag{1}$$

has a Landau singularity (pole) at $Q^2 = \Lambda^2_{\text{Lan.}}$ ($\sim 0.1\ \text{GeV}^2$). The 2-loop pQCD coupling $a(Q^2)^{(2-\ell.)}$ has a Landau pole at $Q^2 = \Lambda^2_{\text{Lan.}}$ and Landau cut at $0 < Q^2 < \Lambda^2_{\text{Lan.}}$.

It is expected that the true QCD coupling $\mathcal{A}(Q^2)$ has no such singularities. Why?

General principles of QFT dictate that any spacelike observable $\mathcal{D}(Q^2)$ (correlators of currents, structure functions, etc.) is an analytic (holomorphic) function of $Q^2$ in the entire $Q^2$ complex plane with the exception of the timelike axis: $Q^2 \in \mathbb{C}\backslash(-\infty, -M^2_{\text{thr.}}]$, where $M_{\text{thr.}} \sim 0.1$ GeV is a threshold mass ($\sim M_\pi$). If $\mathcal{D}(Q^2)$ can be evaluated as a leading-twist term, then it is a function of the running coupling $a(\kappa Q^2)$ where $\kappa \sim 1$: $\mathcal{D}(Q^2) = \mathcal{F}(a(\kappa Q^2))$. Then the argument $a(\kappa Q^2)$ is expected to have the same analyticity properties as $\mathcal{D}$, which is not the case with the pQCD coupling in the usual renormalization schemes ($\overline{\text{MS}}$, 't Hooft, etc.).

A QCD coupling $\mathcal{A}(Q^2)$ with holomorphic behavior for $Q^2 \in \mathbb{C}\backslash(-\infty, -M^2_{\text{thr.}}]$, represents an analytic QCD model (anQCD).

Such holomorphic behavior comes usually together with (IR-fixed-point) behavior [$\mathcal{A}(0) < \infty$]. The IR-fixed-point behavior of $\mathcal{A}(Q^2)$ is suggested by:

- lattice calculations [1, 2, 3]; calculations based on Dyson-Schwinger equations (DSE) [4, 5]; Gribov-Zwanziger approach [6, 7];

- The holomorphic $\mathcal{A}(Q^2)$ with IR-fixed-point behavior was proposed in various analytic QCD models, among them:
  (i) Analytic Perturbation Theory (APT) of Shirkov, Solovtsov et al. [8, 9, 10, 11, 12];
  (ii) its extension Fractional APT (FAPT) [13, 14, 15];
  (iii) analytic models with $\mathcal{A}(Q^2)$ very close to $a(Q^2)$ at high $|Q^2| > \Lambda^2_{\rm Lan.}$: $\mathcal{A}(Q^2) - a(Q^2) \sim (\Lambda^2_{\rm Lan.}/Q^2)^N$ with $N = 3, 4$ or 5, [16, 17, 18, 19];
  (iv) Massive Perturbation Theory (MPT), [20, 21, 22, 23].

Perturbative QCD (pQCD) can give analytic coupling $a(Q^2)$ in specific schemes with IR fixed point; the condition of reproduction of the correct value of the (strangeless and massless) semihadronic $\tau$ lepton $V + A$ decay ratio $r_\tau \approx 0.20$ strongly restricts such schemes [24, 25, 26].

If the analytic coupling $\mathcal{A}(Q^2)$ is not perturbative, $\mathcal{A}(Q^2)$ differs from the pQCD couplings $a(Q^2)$ at $|Q| \gtrsim 1$ GeV by nonperturbative (NP) terms, typically by some power-suppressed terms $\sim 1/Q^{2N}$ or $1/[Q^{2N} \ln^K(Q^2/\Lambda^2_{\rm Lan.})]$.

An analytic QCD model which gives $\mathcal{A}(0) = \infty$ was constructed in [27, 28, 29].

## 2. The formalism of constructing $\mathcal{A}_\nu$ in general anQCD

Having $\mathcal{A}(Q^2)$ [the analytic analog of $a(Q^2)$] specified, we want to evaluate the physical QCD quantities $\mathcal{D}(Q^2)$ in terms of such $\mathcal{A}(\kappa Q^2)$.

Usually $\mathcal{D}(Q^2)$ is known as a (truncated) power series in terms of the pQCD coupling $a(\kappa Q^2)$:

$$\mathcal{D}(Q^2)^{[N]}_{\rm pt} = a(\kappa Q^2)^{\nu_0} + d_1(\kappa) a(\kappa Q^2)^{\nu_0+1} + \ldots + d_{N-1}(\kappa) a(\kappa Q^2)^{\nu_0+N-1}. \tag{2}$$

In anQCD, the simple replacement $a(\kappa Q^2)^{\nu_0+m} \mapsto \mathcal{A}(Q^2)^{\nu_0+m}$ is not correct, it leads to a strongly diverging series when $N$ increases, as argued in [30]; a different formalism was needed, and was developed for general anQCD, first for the case of integer $\nu_0$ [31, 32], and then for the case of general $\nu_0$ [33]. It results in the replacements

$$a(\kappa Q^2)^{\nu_0+m} \mapsto \mathcal{A}_{\nu_0+m}(Q^2) \quad [\neq \mathcal{A}(Q^2)^{\nu_0+m}] \ , \tag{3}$$

where the construction of the analytic power analogs $\mathcal{A}_{\nu_0+m}(Q^2)$ from $\mathcal{A}(Q^2)$ was obtained.

The construction starts with logarithmic derivatives of $\mathcal{A}(Q^2)$ [where $\beta_0 = (11 - 2N_f/3)/4$]:

$$\widetilde{\mathcal{A}}_{n+1}(Q^2) \equiv \frac{(-1)^n}{\beta_0^n n!} \left( \frac{\partial}{\partial \ln Q^2} \right)^n \mathcal{A}(Q^2) \ , \qquad (n = 0, 1, 2, \ldots) \ , \tag{4}$$

and $\widetilde{\mathcal{A}}_1 \equiv \mathcal{A}$. Using the Cauchy theorem, these quantities can be expressed in terms of the discontinuity function of anQCD coupling $\widetilde{\mathcal{A}}$ along its cut, $\rho(\sigma) \equiv {\rm Im}\mathcal{A}(-\sigma - i\epsilon)$

$$\widetilde{\mathcal{A}}_{n+1}(Q^2) = \frac{1}{\pi} \frac{(-1)}{\beta_0^n \Gamma(n+1)} \int_0^\infty \frac{d\sigma}{\sigma} \rho(\sigma) {\rm Li}_{-n}(-\sigma/Q^2) \ . \tag{5}$$

This construction can be extended to a general noninteger $n \mapsto \nu$

$$\widetilde{\mathcal{A}}_{\nu+1}(Q^2) = \frac{1}{\pi} \frac{(-1)}{\beta_0^\nu \Gamma(\nu+1)} \int_0^\infty \frac{d\sigma}{\sigma} \rho(\sigma) {\rm Li}_{-\nu}\left( -\frac{\sigma}{Q^2} \right) \quad (-1 < \nu) \ . \tag{6}$$

This can be recast into an alternative form, involving $\mathcal{A}$ ($\equiv \widetilde{\mathcal{A}}_1$) instead of $\rho$

$$\widetilde{\mathcal{A}}_{\delta+m}(Q^2) \ = \ K_{\delta,m} \left( \frac{d}{d \ln Q^2} \right)^m \int_0^1 \frac{d\xi}{\xi} \mathcal{A}(Q^2/\xi) \ln^{-\delta}\left( \frac{1}{\xi} \right) \ , \tag{7}$$

where: $0 \leq \delta < 1$ and $m = 0, 1, 2, \ldots$; $K_{\delta,m} = (-1)^m \beta_0^{-\delta-m+1}/[\Gamma(\delta+m)\Gamma(1-\delta)]$. This expression was obtained from Eq. (6) by the use of the following expression for the $\mathrm{Li}_{-\nu}(z)$ function [34]:

$$\mathrm{Li}_{-n-\delta}(z) = \left(\frac{d}{d\ln z}\right)^{n+1} \left[\frac{z}{\Gamma(1-\delta)} \int_0^1 \frac{d\xi}{1-z\xi} \ln^{-\delta}\left(\frac{1}{\xi}\right)\right] \quad (n = -1, 0, 1, \ldots; 0 < \delta < 1)\,. \tag{8}$$

The analytic analogs $\mathcal{A}_\nu$ of powers $a^\nu$ are then obtained by combining various generalized logarithmic derivatives (with the coefficients $\widetilde{k}_m(\nu)$ obtained in [33])

$$\mathcal{A}_\nu = \widetilde{\mathcal{A}}_\nu + \sum_{m\geq 1} \widetilde{k}_m(\nu) \widetilde{\mathcal{A}}_{\nu+m}\,. \tag{9}$$

## 3. The considered anQCD models

We constructed Mathematica and Fortran programs for three anQCD models: 1.) Fractional Analytic Perturbation Theory (FAPT) [13, 14, 15]; 2.) $2\delta$ analytic QCD ($2\delta$anQCD) [19]; 3.) Massive Perturbation Theory (MPT) [20, 21, 22, 23]. These three models are described below.

### 3.1. anQCD models: FAPT

Application of the Cauchy theorem to the function $a(Q'^2)^\nu/(Q'^2 - Q^2)$ gives

$$a(Q^2)^\nu = \frac{1}{\pi} \int_{\sigma=-\Lambda^2_{\mathrm{Lan.}}-\eta}^{\infty} \frac{d\sigma \mathrm{Im}(a(-\sigma - i\epsilon)^\nu)}{(\sigma + Q^2)}, \quad (\eta \to +0). \tag{10}$$

In FAPT, the integration over the Landau part of the cut in the above integral is eliminated; since $\sigma \equiv -Q^2$, the Landau cut is $-\Lambda^2_{\mathrm{Lan.}} < \sigma < 0$. This leads to the FAPT coupling

$$\mathcal{A}_\nu^{(\mathrm{FAPT})}(Q^2) = \frac{1}{\pi} \int_{\sigma=0}^{\infty} \frac{d\sigma \mathrm{Im}(a(-\sigma - i\epsilon)^\nu)}{(\sigma + Q^2)}\,. \tag{11}$$

### 3.2. anQCD models: $2\delta$QCD

Here, $\rho(\sigma) \equiv \mathrm{Im}\mathcal{A}(-\sigma - i\epsilon)$ is approximated at high momenta $\sigma \geq M_0^2$ by $\rho^{(\mathrm{pt})}(\sigma)$ [$\equiv \mathrm{Im}\, a(-\sigma - i\epsilon)$], and in the unknown low-momentum regime by two deltas:

$$\rho^{(2\delta)}(\sigma) = \pi F_1^2 \delta(\sigma - M_1^2) + \pi F_2^2 \delta(\sigma - M_2^2) + \Theta(\sigma - M_0^2)\rho^{(\mathrm{pt})}(\sigma) \quad \Rightarrow \tag{12}$$

$$\widetilde{\mathcal{A}}_\nu^{(2\delta)}(Q^2) = \frac{(-1)}{\beta_0^\nu \Gamma(\nu+1)} \left\{ \sum_{j=1}^{2} \frac{F_j^2}{M_j^2} \mathrm{Li}_{-\nu}\left(-\frac{M_j^2}{Q^2}\right) + \frac{1}{\pi} \int_{M_0^2}^{\infty} \frac{d\sigma}{\sigma} \mathrm{Im}a(-\sigma - i\epsilon) \mathrm{Li}_{-\nu}\left(-\frac{\sigma}{Q^2}\right) \right\}. \tag{13}$$

The parameters $F_j^2$ and $M_j$ ($j = 1, 2$) are fixed in such a way that the resulting deviation from the underlying pQCD at high $|Q^2| > \Lambda^2$ is: $\mathcal{A}_\nu^{(2\delta)}(Q^2) - a(Q^2)^\nu \sim (\Lambda^2/Q^2)^5$. The pQCD-onset scale $M_0$ is determined so that the model reproduces the measured (strangeless and massless) $V + A$ tau lepton semihadronic decay ratio $r_\tau \approx 0.20$.

The underlying pQCD coupling $a$ is chosen in $2\delta$anQCD, for calculational convenience, in the Lambert-scheme form

$$a(Q^2) = -\frac{1}{c_1} \frac{1}{1 - c_2/c_1^2 + W_{\mp 1}(z_\pm)}\,, \tag{14}$$

where: $c_1 = \beta_1/\beta_0$; $Q^2 = |Q^2| e^{i\phi}$, the upper (lower) sign when $\phi \geq 0$ ($\phi < 0$), and

$$z_\pm = (c_1 e)^{-1} (|Q^2|/\Lambda^2)^{-\beta_0/c_1} \exp\left[i(\pm\pi - \beta_0\phi/c_1)\right]. \tag{15}$$

### *3.3. anQCD models: MPT*

Nonperturbative physics suggests that the gluon acquires at low momenta an effective (dynamical) mass $m_{\rm gl} \sim 1$ GeV, and that the coupling then has the form

$$\mathcal{A}^{\rm (MPT)}(Q^2) = a(Q^2 + {m_{\rm gl}}^2) \ . \tag{16}$$

Since $m_{\rm gl} > \Lambda_{\rm Lan.}$, the new coupling has no Landau singularities.

The (generalized) logarithmic derivatives $\widetilde{\mathcal{A}}^{\rm (MPT)}_{\delta+m}(Q^2)$ are then uniquely determined

$$\widetilde{\mathcal{A}}_{\delta+m}(Q^2) = K_{\delta,m} \left( \frac{d}{d \ln Q^2} \right)^m \int_0^1 \frac{d\xi}{\xi} \mathcal{A}^{\rm (MPT)}(Q^2/\xi) \ln^{-\delta}\left(\frac{1}{\xi}\right) . \tag{17}$$

## 4. Numerical implementation and results

Programs of numerical implementation in anQCD models:

- for integer power analogs $\mathcal{A}_n(Q^2)$ in APT and in "massive QCD" [35, 36]: Nesterenko and Simolo, 2010 (in Maple) [37], and 2011 (in Fortran) [38];
- for general power analogs $\mathcal{A}_\nu(Q^2)$ in FAPT: Bakulev and Khandramai, 2013 (in Mathematica) [39];
- for general power analogs $\mathcal{A}_\nu(Q^2)$ in $2\delta$anQCD, MPT and FAPT: the presented work in Mathematica [40] and Fortran (programs in both languages can be downloaded from the web page: **gcvetic.usm.cl**).

The basic relations for the numerical implementation of $\mathcal{A}_\nu$ are: in FAPT Eq. (11); in $2\delta$anQCD Eqs. (13) and (9); in MPT Eqs. (17) and (9).

In Mathematica, $\mathrm{Li}_{-\nu}(z)$ is implemented as PolyLog$[-\nu, z]$. In Mathematica 9.0.1 it is unstable for $|z| \gg 1$. Therefore, we provide a subroutine Li_nu.m (which is called by the main Mathematica program anQCD.m) and gives a stable version under the name polylog$[-\nu, z]$. This problem does not exist in Mathematica 10.0.1.

In Fortran, program Vegas [41] is used for integrations. However, in Fortran, $\mathrm{Li}_{-\nu}(z)$ function is not implemented for general (complex) $z$, and is evaluated as an integral Eq. (8). Therefore, the evaluation of $\widetilde{\mathcal{A}}_\nu$'s in $2\delta$anQCD is somewhat more time consuming in Fortran than in Mathematica. Further, more care has to be taken in Fortran to deal correctly with singularities of the integrands.

## 5. Main procedures in Mathematica for three analytic QCD models

1.) AFAPT$N$l$[N_f, \nu, 0, |Q^2|, \Lambda^2, \phi]$ gives $N$-loop ($N = 1,2,3,4$) analytic FAPT coupling $\mathcal{A}_\nu^{\rm (FAPT,\mathit{N})}(Q^2, N_f)$ with real power index $\nu$, with fixed number of active quark flavors $N_f$, in the Euclidean domain [$Q^2 = |Q^2| \exp(i\phi) \in \mathcal{C}$ and $Q^2 \not< 0$]

$$\begin{aligned}\mathrm{AFAPT}N\mathrm{l}[Nf,\nu,0,Q2,L2,\phi] &= \mathcal{A}_\nu^{\rm (FAPT,\mathit{N})}[Q2 = |Q^2|, \phi = \arg(Q^2); Nf = N_f; L2 = \overline{\Lambda}^2_{N_f}] \\ &\quad (N = 1,2,3,4;\ Nf = 3,4,5,6).\end{aligned}$$

2.) A2d$N$l$[N_f, M, \nu, |Q^2|, \phi]$ gives "$N$-loop" $2\delta$anQCD coupling $\mathcal{A}_{\nu+M}^{(2\delta)}(Q^2, N_f)$, with power index $\nu + M$ ($\nu > -1$ and real; $M = 0, 1, \ldots, N-1$), with number of active quark flavors $N_f$, in the Euclidean domain. It is used in the N$^{N-1}$LO truncation approach [where in (9): $\nu_0 \mapsto \nu$ and $n \mapsto M$, and we truncate at $\widetilde{\mathcal{A}}_{\nu+N-1}$]

$$\begin{aligned}\mathrm{A2d}N\mathrm{l}[Nf,M,\nu,Q2,\phi] &= \mathcal{A}_{\nu+M}^{(2\delta)}[Q2 = |Q^2|, \phi = \arg(Q^2); Nf = N_f], \\ &\quad (N = 1,2,3,4,5;\ Nf = 3,4,5,6;\ M = 0,1,\ldots,N-1).\end{aligned}$$

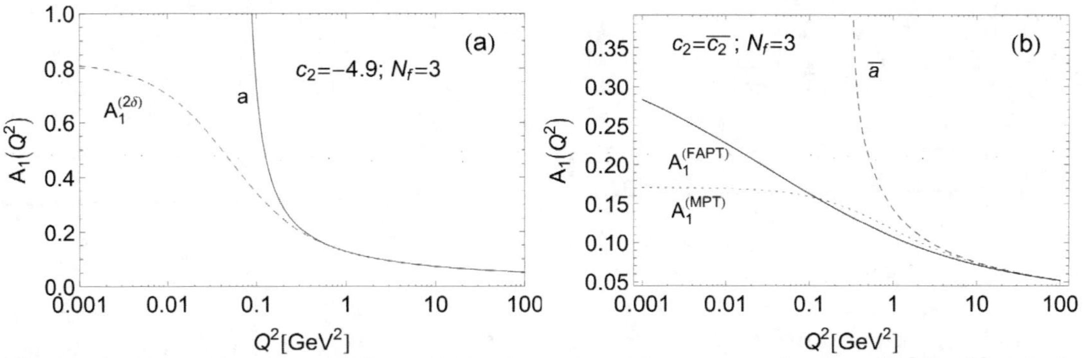


**Figure 1.** $\mathcal{A}_1 \equiv \mathcal{A}$ in three anQCD models with $\nu = 1$ and $N_f = 3$, as a function of $Q^2$ for $Q^2 > 0$; the underlying pQCD coupling $a$ is included for comparison: (a) 2$\delta$anQCD coupling and pQCD coupling, in the Lambert scheme with $c_2 = -4.9$ (and $c_j = c_2^{j-1}/c_1^{j-2}$ for $j \geq 3$); (b) FAPT and MPT in 4-loop $\overline{\rm MS}$ scheme and with $\overline{\Lambda}_3^2 = 0.1$ GeV$^2$; MPT is with $m_{\rm gl}^2 = 0.7$ GeV$^2$.

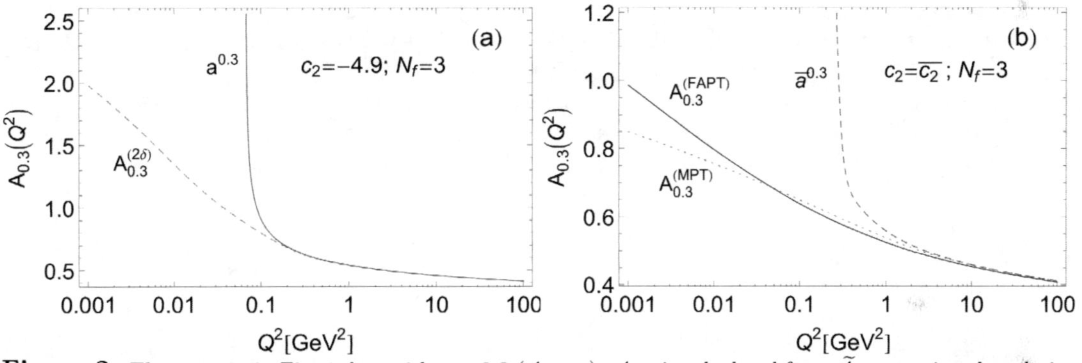


**Figure 2.** The same as in Fig. 1, but with $\nu = 0.3$ ($\mathcal{A}_{\nu=0.3}$). $\mathcal{A}_{0.3}$ is calculated from $\widetilde{\mathcal{A}}_{0.3+m}$ using the relation (9) with $\nu_0 = 0.3$, and truncation at $\widetilde{\mathcal{A}}_{0.3+4}$ in 2$\delta$anQCD, and at $\widetilde{\mathcal{A}}_{0.3+3}$ in MPT; and in FAPT using Eq. (11). Figs. 1 and 2 are taken from [40].

3.) AMPT$N$l$[N_f, \nu, Q^2, {m_{\rm gl}}^2, \overline{\Lambda}_{N_f}^2]$ gives $N$-loop ($N = 1,2,3,4$) analytic MPT coupling $\mathcal{A}_\nu^{(\rm MPT,\mathit{N})}(Q^2, {m_{\rm gl}}^2, N_f)$, with real power index $\nu$ ($0 < \nu < 5$) and with number of active quark flavors $N_f$, in the Euclidean domain ($Q^2 \in \mathcal{C}$ and $Q^2 \not< 0$)

$$\text{AMPT}N\text{l}[Nf, \nu, Q2, M2, L2] \;=\; \mathcal{A}_\nu^{(\rm MPT,N)}[Q2{=}Q^2{\in}\mathcal{C}; Nf{=}N_f; M2{=}{m_{\rm gl}}^2; L2{=}\overline{\Lambda}_{N_f}^2]$$
$$(N = 1,2,3,4;\; Nf = 3,4,5,6);\; 0 < \nu < 5). \qquad (18)$$

Examples:
Input scale of the underlying $\overline{\rm MS}$ pQCD for FAPT and MPT is $\overline{\Lambda}_3^2 = 0.1$ GeV$^2$. The times are for a typical laptop, using Mathematica 9.0.1; the first entry in the results is the time of calculation, in $s$.

```
In[1]:= <<anQCD.m;
In[2]:= AFAPT3l[5, 1, 0, 10^2, 0.1, 0] // Timing
Out[2]= {0.382942, 0.0624843}
In[3]:= AMPT3l[5, 1, 10^2, 0.7, 0.1] // Timing
Out[3]= {0.108983, 0.0627726}
```

```
In[4]:= A2d3l[5, 0, 1, 10^2, 0] // Timing
Out[4]={0.768884, 0.0559182}
In[5]:= AFAPT3l[3, 1, 0, 0.5, 0.1, 0] // Timing
Out[5]= {0.378943, 0.121853}
In[6]:= AMPT3l[3, 1, 0.5, 0.7, 0.1] // Timing
Out[6]= {0.106984, 0.132199}
In[7]:= A2d3l[3, 0, 1, 0.5, 0] // Timing
Out[7]= {0.775882, 0.163402}
In[8]:= AFAPT3l[3, 0.3, 0, 0.5, 0.1, 0] // Timing
Out[8]= {0.456930, 0.556644}
In[9]:= AMPT3l[3, 0.3, 0.5, 0.7, 0.1] // Timing
Out[9]={0.110983, 0.569473}
In[10]:= A2d3l[3, 0, 0.3, 0.5, 0] // Timing
Out[10]= {3.125525, 0.576005}
```

## 6. Conclusions

We constructed programs, in Mathematica and Fortran, which evaluate couplings $\mathcal{A}_\nu(Q^2)$ in three models of analytic QCD (FAPT, $2\delta$anQCD, and MPT). These couplings are holomorphic functions (free of Landau singularities) in the complex $Q^2$ plane with the exception of the negative semiaxis, and are analogs of powers $a(Q^2)^\nu \equiv (\alpha_s(Q^2)/\pi)^\nu$ of the underlying perturbative QCD. We checked that our results in FAPT model agree with those of Mathematica program [39].

## Acknowledgments

This work was supported by FONDECYT (Chile) Grant No. 1130599 and DGIP (UTFSM) internal project USM No. 11.13.12 (C.A and G.C).

## References

[1] Cucchieri A and Mendes T 2008 *Phys. Rev. Lett.* **100** 241601 (arXiv:0712.3517 [hep-lat])
[2] Bogolubsky I L, Ilgenfritz E M, Müller-Preussker M and Sternbeck A 2009 *Phys. Lett.* B **676** 69 (arXiv:0901.0736 [hep-lat])
[3] Furui S 2009 *PoS LAT* **2009** 227 (arXiv:0908.2768 [hep-lat])
[4] Lerche C and von Smekal L 2002 *Phys. Rev.* D **65** 125006 (hep-ph/0202194)
[5] Aguilar A C and Papavassiliou J 2008 *Eur. Phys. J.* A **35** 189 (arXiv:0708.4320 [hep-ph])
[6] Zwanziger D 2004 *Phys. Rev.* D **69** 016002 (hep-ph/0303028)
[7] Dudal D, Gracey J A, Sorella S P, Vandersickel N and Verschelde H 2008 *Phys. Rev.* D **78** 065047 (arXiv:0806.4348 [hep-th])
[8] Shirkov D V and Solovtsov I L 1996 *JINR Rapid Commun.* 2[76] 5–10 (arXiv:hep-ph/9604363)
[9] Shirkov D V and Solovtsov I L 1997 *Phys. Rev. Lett.* 79 1209–1212 (arXiv:hep-ph/9704333)
[10] Milton K A and Solovtsov I L 1997 *Phys. Rev.* D **55** 5295–98 (arXiv:hep-ph/9611438)
[11] Shirkov D V 2001 *Eur. Phys. J.* C **22** 331 (hep-ph/0107282)
[12] Karanikas A I and Stefanis N G 2001 *Phys. Lett.* B **504** 225 (hep-ph/0101031)
[13] Bakulev A P, Mikhailov S V and Stefanis N G 2005 *Phys. Rev.* D **72** 074014 (arXiv:hep-ph/0506311)
[14] Bakulev A P, Mikhailov S V and Stefanis N G 2007 *Phys. Rev.* D **75** 056005 (arXiv:hep-ph/0607040)
[15] Bakulev A P, Mikhailov S V and Stefanis N G 2010 *JHEP* **1006** 085 (arXiv:1004.4125 [hep-ph])
[16] Webber B R 1998 *JHEP* **9810** 012 (hep-ph/9805484)
[17] Alekseev A I 2006 *Few Body Syst.* **40** 57 (arXiv:hep-ph/0503242)
[18] Contreras C, Cvetič G, Espinosa O and Martínez H E 2010 *Phys. Rev.* D **82** 074005 (arXiv:1006.5050)
[19] Ayala C, Contreras C and Cvetič G 2012 *Phys. Rev.* D **85** 114043 (arXiv:1203.6897 [hep-ph])
[20] Simonov Yu A 1995 *Phys. Atom. Nucl.* **58** 107 (hep-ph/9311247)
[21] Simonov Yu A 2010 arXiv:1011.5386 [hep-ph]

[22] Badalian A M and Kuzmenko D S 2001 *Phys. Rev.* D **65** 016004 (hep-ph/0104097)
[23] Shirkov D V 2013 *Phys. Part. Nucl. Lett.* **10** 186 (arXiv:1208.2103 [hep-th])
[24] Cvetič G, Kögerler R and Valenzuela C 2010 *J. Phys.* G **37** 075001 (arXiv:0912.2466 [hep-ph])
[25] Cvetič G, Kögerler R and Valenzuela C 2010 *Phys. Rev.* D **82** 114004 (arXiv:1006.4199 [hep-ph])
[26] Contreras C, Cvetič G, Kögerler R, Kröger P and Orellana O 2014 arXiv:1405.5815 [hep-ph].
[27] Nesterenko A V 2001 *Phys. Rev.* D **64** 116009 (arXiv:hep-ph/0102124)
[28] Nesterenko A V 2003 *Int. J. Mod. Phys.* A **18** 5475 (arXiv:hep-ph/0308288)
[29] Aguilar A C, Nesterenko A V and Papavassiliou J 2005 *J. Phys.* G **31** 997 (hep-ph/0504195)
[30] Cvetič G 2014 *Phys. Rev.* D **89** 036003 (arXiv:1309.1696 [hep-ph]).
[31] Cvetič G and Valenzuela C 2006 *J. Phys.* G **32** L27 (arXiv:hep-ph/0601050)
[32] Cvetič G and Valenzuela C 2006 *Phys. Rev.* D **74** 114030 (arXiv:hep-ph/0608256)
[33] Cvetič G and Kotikov A V 2012 *J. Phys.* G **39** 065005 (arXiv:1106.4275 [hep-ph])
[34] Kotikov A V, Krivokhizhin V G and Shaikhatdenov B G 2012 *Phys. Atom. Nucl.* **75** 507
[35] Nesterenko A V and Papavassiliou J 2005 *Phys. Rev.* D **71** 016009 (hep-ph/0410406)
[36] Nesterenko A V 2009 *Nucl. Phys. Proc. Suppl.* **186** 207 (arXiv:0808.2043 [hep-ph])
[37] Nesterenko A V and Simolo C 2010 *Comput. Phys. Commun.* **181** 1769 (arXiv:1001.0901 [hep-ph])
[38] Nesterenko A V and Simolo C 2011 *Comput. Phys. Commun.* **182** 2303 (arXiv:1107.1045 [hep-ph])
[39] Bakulev A P and Khandramai V L 2013 *Comput. Phys. Commun.* **184** 183 (arXiv:1204.2679 [hep-ph])
[40] Ayala C and Cvetič G 2014 arXiv:1408.6868 [hep-ph].
[41] Lepage G P 1980 CLNS-80/447.

# Computations and generation of elements on the Hopf algebra of Feynman graphs

**Michael Borinsky**
Humboldt-University Berlin
Departments of Physics and Mathematics
Unter den Linden 6
10099 Berlin

E-mail: `borinsky@physik.hu-berlin.de`

**Abstract.** Two programs, feyngen and feyncop, were developed. feyngen is designed to generate high loop order Feynman graphs for Yang-Mills, QED and $\phi^k$ theories. feyncop can compute the coproduct of these graphs on the underlying Hopf algebra of Feynman graphs. The programs can be validated by exploiting zero dimensional field theory combinatorics and identities on the Hopf algebra which follow from the renormalizability of the theories. A benchmark for both programs was made.

## 1. Introduction

The Hopf algebra structure of Feynman graphs has been explored extensively in the last years. It proved to be valuable for the analytic computation of Feynman amplitudes by means of systematic parametric integration techniques and could lead to new non-perturbative results in the scope of Dyson-Schwinger equations. **feyngen** and **feyncop** were developed to provide input for the powerful new techniques. **feyngen** is a tool for the fast generation of higher loop Feynman diagrams. **feyncop** can be used to calculate the coproduct on the Hopf algebra of Feynman graphs. This coproduct encodes the BPHZ algorithm necessary to evaluate the finite amplitude of a Feynman diagram and fits well into the world of Dyson-Schwinger equations. In this framework certain identities can be obtained which were used to validate the two programs.

## 2. Feynman diagram generation with feyngen

The **python** program **feyngen** can generate $\phi^k$ for $k \geq 3$, QED, QED with Furry's theorem, Yang-Mills and $\phi^3 + \phi^4$ diagrams ready to be used in green's function calculations. Developing **feyngen**, the focus was on the generation of Feynman diagrams with comparatively large loop orders. Additionally to the generation of non-isomorphic diagrams, **feyngen** calculates the symmetry factors of the resulting graphs. Handling of graphs with fixed external legs and without is supported. Furthermore, options are available to filter for connected, one-particle-irreducible (1PI), vertex-2-connected and snail free graphs. To achieve the high speed for the computation **feyngen** relies on the established **nauty** package [1, 2]. The output of **feyngen** is designed to be readable by a **maple** program.

Details to the implementation, theoretical background and handling are given in [3].

### *2.1. Examples*

Consider the sum of all two loop, photon propagator residue type, 1PI, QED diagrams. For convenience the vertices of the graph in the illustration are labeled as in the output of **feyngen**. The labels do not have further meaning. Note that, also the external source vertices are labeled, because they also appear in the output of **feyngen**.

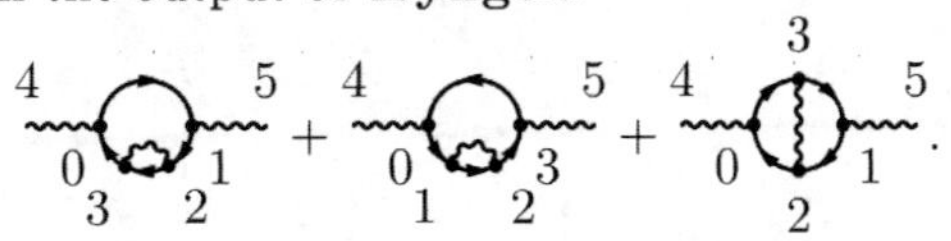


**feyngen** generates them if it is called with the command line

```
$ ./feyngen --qed 2 -b2 -p
qed_f0_b2_h2 :=
+G[[0,1,f],[1,2,f],[2,3,f],[3,0,f],[3,2,A],[4,0,A],[5,1,A]]/1
+G[[0,1,f],[1,2,f],[2,3,f],[3,0,f],[2,1,A],[4,0,A],[5,3,A]]/1
+G[[0,3,f],[1,2,f],[2,0,f],[3,1,f],[3,2,A],[4,0,A],[5,1,A]]/1
;
```

**--qed** indicates QED graph generation, **2** stands for 2-loop diagrams ($\hbar^2$), **-b2** makes **feyngen** generate graphs with 2 photon legs and the **-p** option filters out non 1PI graphs.

For the sum of all one loop, gauge boson propagator residue type, 1PI, Yang-Mills diagrams,

the call,

```
$ ./feyngen --ym 1 -tp -b2
```

where the generation of Yang-Mills graphs is triggered with the **--ym** option, gives the desired result:

```
ym_f0_g0_b2_h1 :=
+G[[0,1,c],[1,0,c],[2,0,A],[3,1,A]]/1
+G[[0,1,f],[1,0,f],[2,0,A],[3,1,A]]/1
+G[[1,0,A],[1,0,A],[2,0,A],[3,1,A]]/2
;
```

### *2.2. Validation*

To validate the Feynman graph generation with the program **feyngen**, the perturbation expansion of a zero dimensional quantum field theory was used. Given for instance the generating function for $\phi^k$ theory in zero dimensions:

$$Z_{\phi^k}(\lambda, j) := \int_{\mathbb{R}} \frac{d\phi}{\sqrt{2\pi}} e^{-\frac{\phi^2}{2}+\lambda\frac{\phi^k}{k!}+j\phi}, \tag{1}$$

a powerseries expansion in terms of the coupling $\lambda$ can be readily obtained,

$$\widetilde{Z}_{\phi^k}(\lambda, j) = \sum_{l\geq 0} \sum_{\substack{n,m\geq 0 \\ nk+m=2l}} \frac{(2l-1)!!}{n!m!(k!)^n} \lambda^n j^m. \tag{2}$$

With this multivariate powerseries the sum of the symmetry factors of disconnected $\phi^k$ with a fixed number of vertices and external legs can be obtained. Here, $\lambda$ counts the number of vertices

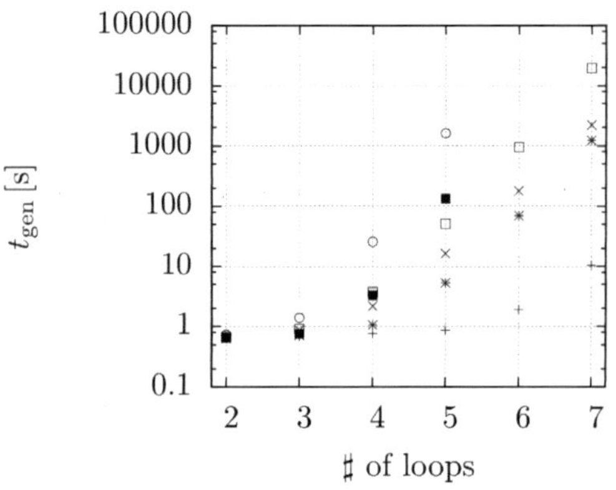


(a) Time to generate all 1PI diagrams of given loop order.

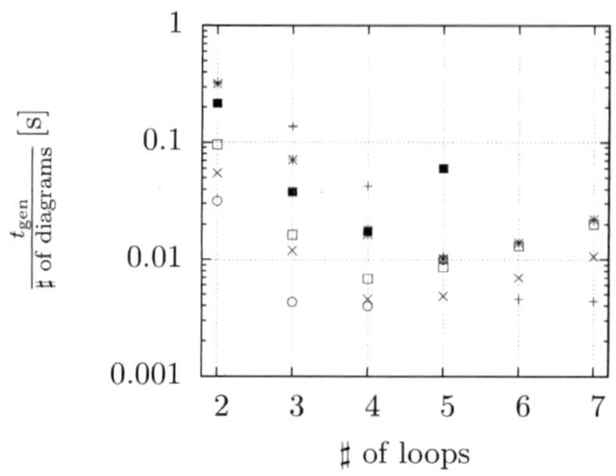


(b) Average generation time for one diagram for the given loop order.

**Figure 1.** Plot of the results of the benchmark for **feyncop**. **Legend:** $+$ : $\phi^4$ proper propagator, $\times$ : $\phi^4$ proper vertex, $*$ : $\phi^3$ proper propagator, $\square$ : $\phi^3$ proper vertex, $\blacksquare$ : QED proper photon propagator, $\circ$ : QCD proper gluon propagator.

and $j$ the number of external legs. The corresponding powerseries for the connected diagrams can be calculated by taking the logarithm. The reason for this is that Feynman diagrams are a labeled combinatorial class for which the exponential theorem holds [4]:

$$W(\lambda, j) = \log(Z(\lambda, j)). \tag{3}$$

For the computation of the numbers for 1PI diagrams the classical field,

$$\phi_c(\lambda, j) := \frac{\partial W}{\partial j}, \tag{4}$$

is needed. The source variable $j \to j' + j_0$ is redefined such that $\phi_c(j')$ vanishes at $j' = 0$. Using the definition of the effective action as Legendre transformation of $W$, changing $j'$ for $\phi_c$,

$$\Gamma = W - j'\phi_c, \tag{5}$$

the sum of the symmetry factors of the 1PI diagrams can be calculated using the Lagrange inversion theorem [4]:

$$[\phi_c^m]\,\Gamma(\lambda, \phi_c) = -\frac{1}{m}\left[j'^{(m-2)}\right]\frac{\partial^2 W(\lambda, j')}{\partial j'^2}\left(\frac{j'}{\phi_c(\lambda, j')}\right)^m, \tag{6}$$

where $[\cdot]$ is the coefficient extraction operator. $\Gamma(\lambda, \phi_c)$ generates the proper green's functions in zero dimension.

### *2.3. Benchmarks*

Figure 1 depicts an example for the computation time for the 1PI diagrams generation of a given loop order. Additionally to the non-isomorphic diagrams the corresponding symmetry factors was computed. The benchmark was performed on a `Intel(R) Core(TM) i7-3770 CPU @ 3.40GHz`. Although **feyngen** does not explicitly use parallelization, a speedup was gained because the generation of graphs using **nauty** runs in parallel to the refinement of the graphs to Feynman diagrams.

The benchmark clearly shows that the $\phi^4$-diagram generation is the fastest. The generation of these diagrams was the main purpose for the development of **feyngen**. Therefore, the highest loop orders can be achieved in this theory. The exponential rise in the computation time for the QED diagram generation can be explained by the very naive diagram refinement algorithm used. The same explosion in computation time can be expected for higher loop order QCD diagrams, where the same simple algorithm was applied.

## 3. The Hopf algebra of Feynman graphs

As was shown by Kreimer et al. [5, 6, 7] a Hopf algebra $\mathcal{H}_D$ can be used to describe the self-similar structure of Feynman graphs and their renormalization. The index $D$ stands for the dimension of spacetime. The coproduct $\Delta_D$ on this Hopf algebra corresponds to the forest formula in BPHZ renormalization [8, 9, 10]. For 1PI graphs $\Gamma$ the coproduct is defined as,

$$\Delta_D \Gamma := \sum_{\gamma \trianglelefteq \Gamma} \gamma \otimes \Gamma/\gamma \qquad : \qquad \mathcal{T} \to \mathcal{H}_D \otimes \mathcal{H}_D \tag{7}$$

where $\mathcal{T}$ is the set of all 1PI graphs and

$$\gamma \trianglelefteq \Gamma \Leftrightarrow \gamma \in \left\{ \delta \subseteq \Gamma \middle| \delta = \bigcup_i \delta_i, \text{ such that } \delta_i \in \mathcal{T} \text{ and } \omega_D(\delta_i) \leq 0 \right\} \tag{8}$$

denotes the membership of $\gamma$ in the set of subgraphs of $\Gamma$, whose connected components are superficially divergent 1PI graphs. Disconnected graphs $\gamma = \bigcup_i \gamma_i$ are identified with the product $\left(\prod_i \gamma_i\right) \in \mathcal{H}_D$. $\Gamma/\gamma$ denotes the contraction of the subgraph $\gamma$ in $\Gamma$. The cograph $\Gamma/\Gamma$ and the empty graph $\gamma = \emptyset$ in (7) are identified with the unit $\mathbb{I} \in \mathcal{H}_D$.

The function $\omega_D : \mathcal{T} \to \mathbb{Z}$ has an important role. It assigns the superficial degree of divergence in $D$ dimensions to a 1PI Feynman graph. $\omega_D$ performs power counting on a graph in the sense of Weinberg's theorem [11].

Additionally, the reduced coproduct $\widetilde{\Delta}_D$ is defined as

$$\widetilde{\Delta}_D := \Delta_D - \mathrm{id} \otimes \mathbb{I} - \mathbb{I} \otimes \mathrm{id} \qquad : \qquad \mathcal{H}_D \to \mathcal{H}_D \otimes \mathcal{H}_D, \tag{9}$$

giving rise to the space of primitive elements of $\mathcal{H}_D$:

$$\mathrm{Prim}\,(\mathcal{H}_D) := \ker \widetilde{\Delta}. \tag{10}$$

The primitive 1PI graphs are also called skeleton graphs.

Details to the Hopf algebra of Feynman graphs in the scope of the coproduct calculation with **feyncop** are given in [3].

## 4. Coproduct computation with feyncop

The **python** program **feyncop** can be used to compute the reduced coproduct $\widetilde{\Delta}_D$ of given 1PI graphs as defined in (9). The output of **feyngen** can be piped into **feyncop** to calculate the reduced coproduct of all 1PI graphs of a given loop order and residue type.

By default, the subgraphs composed of superficially divergent, 1PI graphs of the input graphs are computed and given as output. These correspond to the left-hand factor of the tensor product originating from the coproduct. Optionally, the complementary cographs, giving account to the right-hand factor of the tensor product, can be computed. Furthermore, there is the option to identify the sub- and cographs with unlabeled 1PI graphs i.e. elements of $\mathcal{H}_D$. Additionally, the input graphs can be filtered for primitive graphs.

The coproduct calculation does only take the degree of divergence obtained by power counting, formulated by the map $\omega_D$ into account. Further information, as gained by Furry's theorem in the case of QED, is not used.

### *4.1. Examples*

The graph  is represented as an edge list using an auxiliary vertex labeling, as in the output of **feyngen**:

```
G[[1,0],[2,0],[2,0],[3,1],[3,1],[3,2],[4,0],[5,1],[6,2],[7,3]].
```

This can be used as input for **feyncop**:

```
$ echo "G[[1,0],[2,0],[2,0],[3,1],[3,1],[3,2],[4,0],[5,1],[6,2],[7,3]]"
  | ./feyncop -D4
```

This will yield the output:

```
+ D[G[[1,0],[2,0],[2,0],[3,1],[3,1],[3,2],[4,0],[5,1],[6,2],[7,3]],
[{{1,2}}, {{3,4}}, {{1,2},{3,4}}]]
;
```

The output line

```
[{{1,2}}, {{3,4}}, {{1,2},{3,4}}]
```

corresponds to the subgraphs which are composed of superficially divergent, 1PI graphs, represented a by their edge sets. The edges are indexed by their order of appearance in the edge list.

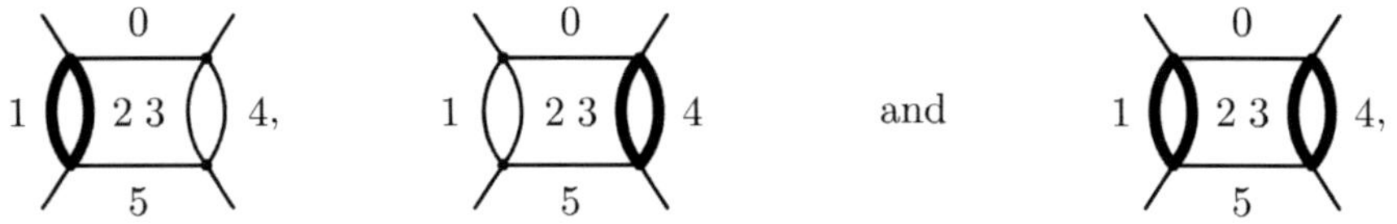


represented as the sets of sets,

{{1,2}}, {{3,4}} and {{1,2},{3,4}}.

**feyncop** can also be used to identify the subgraphs with elements of the Hopf algebra of Feynman graphs. Giving again 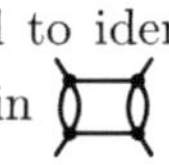 as input:

```
$ echo "G[[1,0],[2,0],[2,0],[3,1],[3,1],[3,2],[4,0],[5,1],[6,2],[7,3]]"
  | ./feyncop -D4 -u
```

The output will be:

```
+ 2/1 * T[ G[[1,0],[1,0],[2,0],[3,0],[4,1],[5,1]],
  G[[1,0],[1,0],[2,0],[2,1],[3,2],[4,2],[5,0],[6,1]] ]
+ T[ (G[[1,0],[1,0],[2,0],[3,0],[4,1],[5,1]])^2,
  G[[1,0],[1,0],[2,0],[3,0],[4,1],[5,1]] ]
;
```

This output corresponds to the tensor products on the right-hand side of

where the graphs are again represented using an auxiliary vertex labeling.

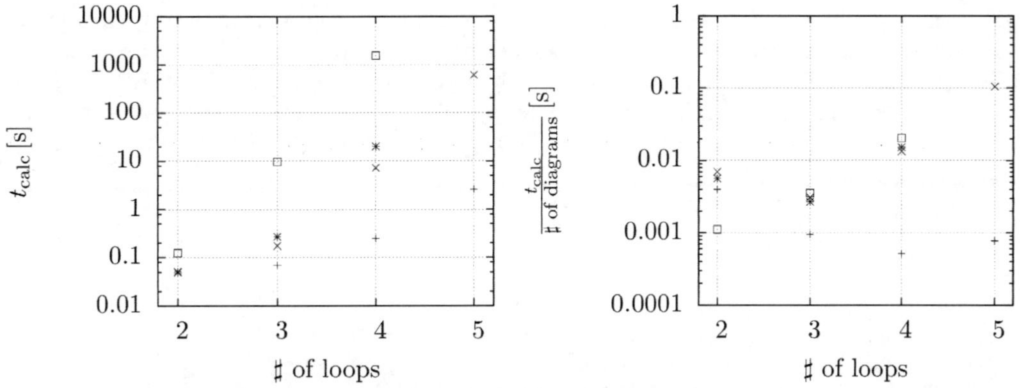


(a) Time to calculate the coproduct of all 1PI diagrams of given loop order.

(b) Average calculation time for one diagram for the given loop order.

**Figure 2.** Plot of the results of the benchmark for **feyncop**. **Legend:** $+$ : $\phi^4$ vertex type diagrams, $\times$ : $\phi^3$ vertex type diagrams, $*$ : QED vertex type diagrams, $\square$ : QCD 3-gluon vertex type diagrams.

### *4.2. Benchmarks*

Figure 2 depicts an example of the computation time of the coproducts of certain classes of 1PI diagrams of a given loop order. The benchmark was performed on a `Intel(R) Core(TM) i7-3770 CPU @ 3.40GHz`. No parallelization was used.

Because $\phi^4$-diagrams have less edges per loop in comparison to the other diagram classes, this coproduct computation is by far the fastest. There are no significant differences in the coproduct computation times of the other classes. This difference in the performance could be made much smaller by implementing a more elaborate 1PI subdiagram detection algorithm. The fast handling $\phi^4$-theory was the main priority during the development of **feyngen** and **feyncop**, so this optimization was not implemented.

### *4.3. Validation*

The output of **feyncop** can be checked by using an identity from [12] on sums of Feynman graphs:

$$\sum_{\Gamma \in \mathcal{T}} \frac{\Delta_D \Gamma}{|\mathrm{Aut}(\Gamma)|} = \sum_{\substack{\gamma = \left(\prod_i \gamma_i\right) \in \mathcal{F} \\ \omega_D(\gamma_i) \le 0}} \sum_{\widetilde{\Gamma} \in \mathcal{T}} \frac{\left|\mathcal{I}(\widetilde{\Gamma}|\gamma)\right|}{|\mathrm{Aut}(\gamma)| \left|\mathrm{Aut}(\widetilde{\Gamma})\right|} \gamma \otimes \widetilde{\Gamma}, \tag{11}$$

where $\left|\mathcal{I}(\widetilde{\Gamma}|\gamma)\right|$ is the number of insertions of $\gamma$ into $\widetilde{\Gamma}$, $\mathcal{T}$ the set of all 1PI graphs, $\mathcal{F}$ the set of all products of 1PI graphs and $|\mathrm{Aut}(\Gamma)|$ is the number of automorphisms of the graph $\Gamma$.

For the validation the coproduct is calculated for the left hand side of equation (11) and is compared to the right hand side which just depends on simple topological properties of the underlying Feynman graphs.

## 5. Conclusions

Two programs were presented. **feyngen** can generate Feynman diagrams of various theories and with certain optional properties and **feyncop** calculates the coproduct on the Hopf algebra of Feynman graphs. The method of validation and a benchmark were presented for both programs.

Both programs are publicly available at `http://people.physik.hu-berlin.de/~borinsky/`.

## Acknowledgements

I wish to thank Dmitrii Batkovich for fruitful discussions and comparisons of my results with the ones obtained in the work [13]. Also, I wish to thank the organizers for the nice conference.

## References

[1] McKay B D and Piperno A 2014 *Journal of Symbolic Computation* **60** 94
[2] McKay B D 1998 *Journal of Algorithms* **26** 306
[3] Borinsky M 2014 *Computer Physics Communications* **185** 3317
[4] Flajolet P and Sedgewick R 2009 *Analytic Combinatorics* (Cambridge University Press)
[5] Kreimer D 1998 *Adv. Theor. Math. Phys* **2** 303
[6] Connes A and Kreimer D 2000 *Communications in Mathematical Physics* **210** 249
[7] Kreimer D 1999 *Communications in Mathematical Physics* **204** 669
[8] Bogoliubov N and Parasiuk O 1957 *Acta Math* **97** 227
[9] Hepp K 1966 *Communications in Mathematical Physics* **2** 301
[10] Zimmermann W 1969 *Communications in Mathematical Physics* **15** 208
[11] Weinberg S 1960 *Phys. Rev.* **118**(3) 838
[12] van Suijlekom W D 2007 *Communications in Mathematical Physics* **276** 773
[13] Batkovich D, Kirienko Y, Kompaniets M and Novikov S 2014 *preprint hep-ph/arXiv:1409.8227*

# Concurrent Cuba

**Thomas Hahn**
Max-Planck-Institut für Physik, Föhringer Ring 6, D–80805 München
E-mail: `hahn@feynarts.de`

**Abstract.** Parallel features of the multidimensional numerical integration package Cuba are presented and achievable speed-ups discussed.

## 1. Introduction

Cuba is a C99 library for multidimensional numerical integration with interfaces for Fortran, C/C++, and Mathematica [1, 2, 3]. It includes four independent algorithms: Vegas, Suave, Divonne, and Cuhre, with very different characteristics (Table 1). Numerical integration is particularly suited for parallelization, which can significantly speed up the computation as it generally incurs only a small overhead. The following design considerations were relevant for the parallel version of Cuba.

1. No additional software shall be needed: Parallelization thus proceeds through operating-system functions only; no external message-passing interface is used. This of course fills the internal cores (typically 4–8) only, i.e. parallelization across the network is not possible. On the other hand, as the performance figures will show, the speed-ups cannot be expected to increase linearly, so utilizing many more cores is more of a theoretical option anyway.

2. Parallelization shall work for any integrand: The programmer shall not need to rewrite the integrand function for parallel execution, which means in particular that reentrancy cannot be taken for granted. Cuba uses `fork/wait` rather than the `pthread*` functions since the former create a completely independent copy of the running process and thus work for any integrand.

3. Parallelization shall work 'automatically': The number of cores to use is determined through an environment variable and hence requires no re-compile. More importantly, its default value is set automatically to the number of idle cores on the present machine, so a program using Cuba will automatically take advantage of the system's capabilities.

4. Parallelization shall be available on all platforms: The POSIX functions `fork` and `wait` are unavailable on native Windows, but Cygwin provides a suitably licensed (GPL) emulation. The emulated `fork` is rather slow, however, so Cuba keeps the `fork` calls at a minimum.

The present note describes the concurrent features of Cuba 4 and discusses their performance. Usage details will be rather brief here but can be found in Ref. [4] and in the Cuba manual.

## 2. Parallelization model

Cuba operates in a master–worker model. The master process orchestrates the parallelization but does not count towards the number of cores, e.g. `CUBACORES = 4` means four workers and one master. Very importantly, the samples are generated by the master process only and distributed to the workers, such that random numbers are never used more than once.

**Table 1.** Overview of the Cuba routines

| Routine | Basic method | Type | Variance reduction |
|---|---|---|---|
| Vegas | Sobol sample<br>*or* Mersenne Twister sample | quasi Monte Carlo<br>pseudo Monte Carlo | importance sampling |
| Suave | Sobol sample<br>*or* Mersenne Twister sample | quasi Monte Carlo<br>pseudo Monte Carlo | globally adaptive subdivision<br>+ importance sampling |
| Divonne | Korobov sample<br>*or* Sobol sample<br>*or* Mersenne Twister sample<br>*or* cubature rules | lattice Monte Carlo<br>quasi Monte Carlo<br>pseudo Monte Carlo<br>deterministic | stratified sampling,<br>aided by methods from<br>numerical optimization |
| Cuhre | cubature rules | deterministic | globally adaptive subdivision |

### *2.1. 'Simple' parallelization*

The parallelization of Cuba centers on the main sampling routine `DoSample`, which in principle parallelizes straightforwardly on $N$ cores:

| *Serial version:* | *Parallel version:* |
|---|---|
| sample $n$ points. | sample $n/N$ points on core 1,<br>⋮<br>sample $n/N$ points on core $N$. |

The actual distribution strategy is somewhat more involved and is described in Sect. 4.2 below.

### *2.2. Parallelization in Divonne*

Speed-ups measured with Divonne by parallelizing `DoSample` alone were generally unsatisfactory and significantly below those of the other integrators, e.g. $\lesssim 1.5$ on four cores.

The Divonne algorithm works in three phases:

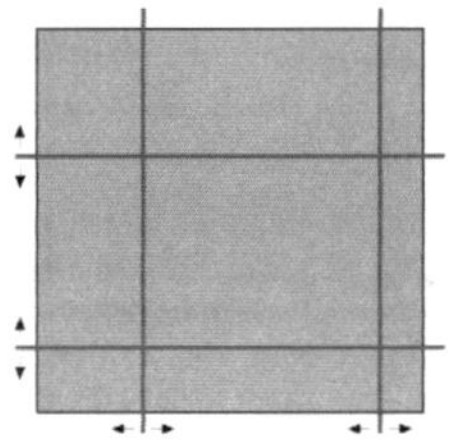

- Partitioning Phase: Split the integration region into subregions.
  a) For each subregion $r$, 'actively' determine $\sup f$ and $\inf f$ using methods from numerical optimization.
  b) Move 'dividers' around until all $r$ have approximately equal
  $$\text{Spread}(r) = \frac{1}{2}\,\text{Vol}(r)\Big(\sup_{\vec{x}\in r} f(\vec{x}\,) - \inf_{\vec{x}\in r} f(\vec{x}\,)\Big).$$
- Sampling Phase: Sample the subregions independently with the same number of points each. That number is extrapolated from the results of the Partitioning Phase.
- Refinement Phase: Further subdivide or sample again if results from the Partitioning and Sampling Phase do not agree within their estimated errors.

It turned out that the Partitioning Phase was crucial for attaining reasonable speed-ups and needed special treatment: Firstly, the original partitioning algorithm divided the regions recursively (and with a minimum recursion depth, too) and had to be 'un-recursed', mainly by better bookkeeping of the subregions.

Secondly, the Partitioning Phase was modified such that each core receives an entire region to subdivide, not just a list of points (as `DoSample` does). In particular the minimum/maximum search, during which only one point at a time is sampled, is distributed much more efficiently this way. The other two phases were not so critical precisely because they sample more points per region.

## 3. Parallelization in Mathematica

The parallelization procedure is rather different in Fortran/C/C++ and in Mathematica, and much of this has to do with the way the MathLink protocol works: As all MathLink programs, Cuba's run independently of Mathematica, so all parallelization must be done by Mathematica means. (A clever way, too, of preventing the user from acquiring more licenses than paid for.)

Cuba's Mathematica executables evaluate the integrand in Mathematica only and can thus integrate arbitrary Mathematica functions, even ones without a C equivalent:

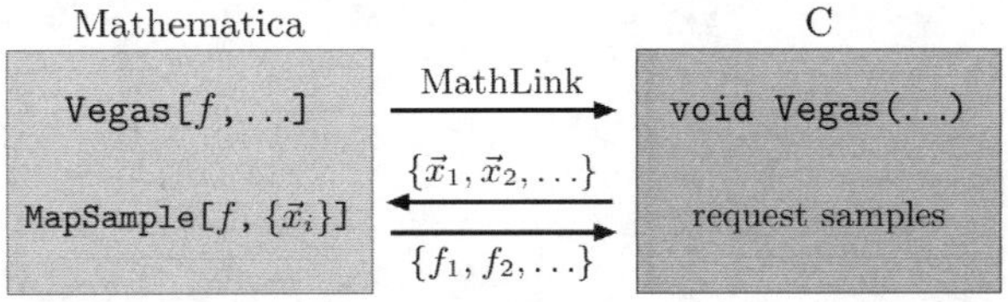


Sampling is handled through a function `MapSample` which by default is identical to `Map`, i.e. the serial version. To parallelize sampling one merely needs to redefine `MapSample = ParallelMap` (after loading Cuba). If the integrand depends on user-defined symbols or functions, their definitions must be distributed to the workers beforehand using `DistributeDefinitions` and likewise required packages must be loaded with `ParallelNeeds` instead of `Needs`. This is standard procedure, however, and is explained in detail in the Mathematica manual.

## 4. Parallelization in Fortran and C/C++

In Fortran and C/C++ the Cuba library can (and usually does) automatically parallelize the sampling of the integrand. It parallelizes through `fork` and `wait` which, though slightly less performant than pthreads, do not require reentrant code. (Reentrancy may not even be under full control of the programmer, for example Fortran's I/O is usually non-reentrant.) Worker processes are started and shut down only as few times as possible, however, so the performance penalty is really quite minor even for non-native fork implementations such as Cygwin's. Parallelization is not available on native Windows for lack of the `fork` function.

The communication of samples to and from the workers is done through IPC shared memory (`shmget` etc.), or if that is not available, through a `socketpair` (two-way pipe). The control information is always transferred via `socketpair` since this elegantly and portably solves the synchronization issues. Remarkably, shared memory's anticipated performance advantage turned out to be hardly perceptible. Possibly there are cache-coherence issues introduced by several workers writing simultaneously to the same shared-memory area.

### *4.1. Starting and stopping the workers*

The workers are usually started and stopped automatically by Cuba's integration routines, but the user may choose to start them manually or keep them running after one integration and shut them down later, e.g. at the end of the program, which can be slightly more efficient. The latter mode is referred to as 'Spinning Cores' and must be employed with certain care, for running workers will not 'see' subsequent changes in the main program's data (e.g. global variables, common blocks) or code (e.g. via `dlsym`) unless special arrangements are made (e.g. shared memory).

### *4.2. Accelerators and Cores*

Based on the strategy used to distribute samples, Cuba distinguishes two kinds of workers.

Workers of the first kind are referred to as 'Accelerators' even though Cuba does not actually send anything to a GPU or Accelerator in the system by itself – this can only be done by the integrand routine. The assumption behind this strategy is that the integrand evaluation is running on a device so highly parallel that the sampling time is more or less independent of the number of points, up to the number of threads $p_{\text{accel}}$ available in hardware. Cuba tries to send exactly $p_{\text{accel}}$ points to each core – never more, less only for the last batch. To sample e.g. 2400 points on three accelerators with $p_{\text{accel}} = 1000$, Cuba sends batches of 1000/1000/400 and not, for example, 800/800/800 or 1200/1200.

CPU-bound workers are just called 'Cores'. Their distribution strategy is different in that all available cores are used and points are distributed evenly. In the example above, the batches would be 800/800/800 thus. Each core receives at least 10 points, or else fewer cores are used. If no more than 10 points are requested in total, Cuba uses no workers at all but lets the master sample those few points. This happens during the partitioning phase of Divonne, for instance, where only single points are evaluated in the minimum/maximum search. Conversely, if the division of points by cores does not come out even, the remaining few points ($< n_{\text{cores}}$) are simply added to the existing batches, to avoid an extra batch because of rounding. Sampling 2001 points on two cores with $p_{\text{cores}} = 1000$ will hence give two batches 1001/1000 and not three batches 1000/1000/1.

Unless the integrand is known to evaluate equally fast at all points, a moderate number for $p_{\text{cores}}$ (10000, say) may actually increase performance because it effectively load-levels the sampling. For, a batch always goes to the next free core so it doesn't matter much if one core is tied up with a batch that takes longer.

The integrand can tell by an additional argument '`core`' [4] whether it is running on an Accelerator (`core` $< 0$) or a Core (`core` $\geqslant 0$). The special value `core` $= 32768$ indicates that the master itself is doing the sampling.

The user may register subroutines for (de)initialization. On worker processes, the functions are respectively executed after `fork` and before `wait`, independently of whether the worker actually receives any samples. The master executes them only when actual sampling is done. For Accelerators, the init and exit functions are typically used to set up the device for the integrand evaluations, which for many devices must be done per process, i.e. after the `fork`.

### *4.3. Concurrency issues*

By creating a new process image, `fork` circumvents all memory concurrency, to wit: each worker modifies only its own copy of the parent's memory and never overwrites any other's data. The programmer should be aware of a few potential problems nevertheless:

- Communicating back results other than the intended output from the integrand to the main program or exchanging data between workers is not straightforward because, by the same token, each worker modifies only its own memory space.
- `fork` does not guard against competing use of other common resources. For example, if the integrand function writes to a file (debug output, say), there is no telling in which order the lines will end up in the file, or even if they will end up as complete lines at all.
- Fortran users should flush (or close) open files before calling Cuba, i.e. `call flush(`*unit*`)`, for the workers inherit all file buffers, and *each* will write out the buffer content at exit.

### *4.4. Vectorization*

Vectorization means evaluating the integrand function for several points at once. This is different from ordinary parallelization where independent threads are executed concurrently. Vector instructions are commonly available in hardware and it is usually possible to employ vectorization on top of parallelization.

Cuba cannot automatically vectorize the integrand function, of course, but it does pass (up to) `nvec` points per integrand call. This value need not correspond to the hardware vector length – computing several points in one call can also make sense e.g. if the computations have significant intermediate results in common. The actual number of points passed is indicated through the corresponding `nvec` argument of the integrand.

## 5. Performance

### 5.1. Test setup

Parallelization entails a certain overhead as usual, so the efficiency will depend on the 'cost' of an integrand evaluation, i.e. the more 'expensive' (time-consuming) it is to sample the integrand, the better the speed-up will be.

The timing measurements in the following figures were made with the 11 integrands of the demo code included in the Cuba distribution, which were originally chosen to highlight different aspects of the integrators. These are simple one-liners and for timing purposes 'infinitely' fast. To tune the cost of the integrands, a calibrated delay loop was inserted into the integrand functions. All measurements were done on the same i7-2600K Linux machine, idle except for the timing program. The i7 has four real and eight virtual (hyperthreaded) cores, i.e. eight register sets but only four arithmetic units. Beyond four cores, indicated by a line in the plots, CPU effects are thus expected on top of Cuba's 'pure' scaling behavior.

The plotted quantity is the speed-up, defined as the quotient of the serial and parallel running times, i.e. speed-up $= t_{\text{serial}}/t_{N\text{-cores}}$. The speed-up ideally reaches $N$.

### 5.2. Timing measurements

Fig. 1 shows the speed-ups for an 'easy' and a 'hard' one of the 11 integrands of the demo program included in the Cuba package for two different integrand delays. Also in the one-core case the parallel version was deployed (one master, one worker), which explains why the timings normalized to the serial version are below 1, in the top row visibly so.

The first, expected, observation is that parallelization is worthwhile only for not-too-fast integrands. This is not a major showstopper, however, as many interesting integrands fall into this category anyway. What appears to be a drastic underperformance of Cuhre in the 'easy' case can in fact be attributed to Cuhre's outstanding efficiency: it delivers a result correct to almost all digits with around 300 samples. In such a case, Cuba may for efficiency choose not to fill all available cores and relative to the full number of cores this shows up as a degradation.

The second observation is that parallelization works best for 'simple-minded' integrators, e.g. Vegas. This is showcased even better in Fig. 2. The 'intelligent' algorithms are generally much harder to parallelize because they don't just do mechanical sampling but take into account intermediate results, make extra checks on the integrand (e.g. try to find extrema), etc. This is particularly true for Divonne, see Sect. 2.2. Then again, the 'intelligent' algorithms are usually faster to start with, i.e. converge with fewer points sampled, which compensates for the lack of parallelizability.

## 6. Summary

The Cuba library for multidimensional numerical integration now features concurrent sampling and thereby achieves significant speed-ups. No extra software needs to be installed since only operating-system functions are used. No reentrancy is required for the integrand function since `fork`/`wait` is applied. Parallelization is switched on automatically in Fortran and C/C++ but can be controlled through API calls or the environment.

Cuba is available from `http://feynarts.de/cuba` and licensed under the LGPL. The download contains a manual which gives full details on installation and usage.

[1] Hahn T, 2005 Comput. Phys. Commun. **168** 78 [hep-ph/0404043].
[2] Agrawal S, Hahn T, Mirabella E, 2012 J. Phys. Conf. Ser. **368** 012054 [arXiv:1112.0124].
[3] Chokoufe Nejad B, Hahn T, Lang J-N, Mirabella E, 2014 J. Phys. Conf. Ser. **523** 012050 [arXiv:1310.0274].
[4] Hahn T, arXiv:1408.6373.

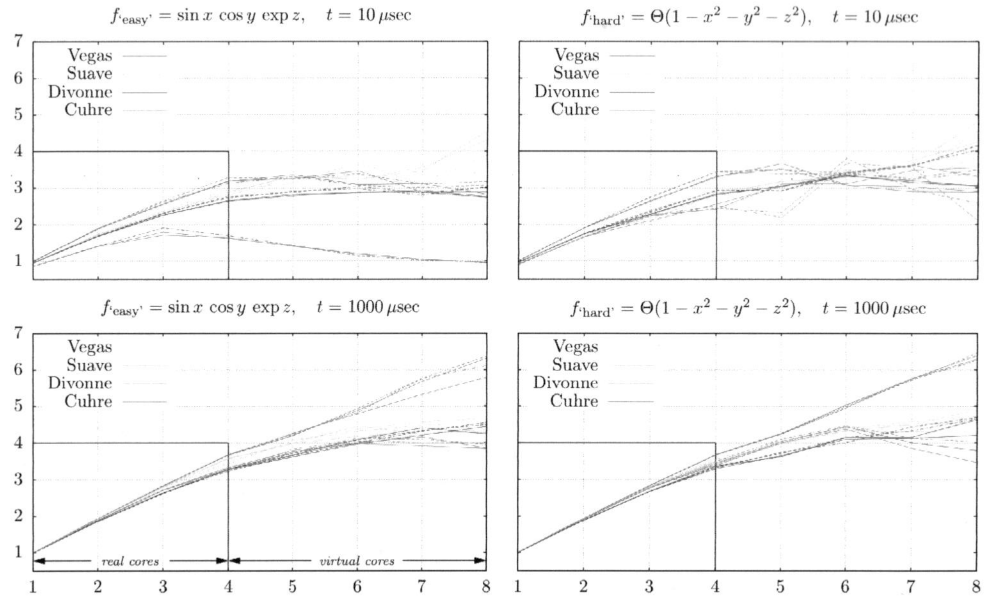


**Figure 1.** Cuba speed-ups for three-dimensional integrals with $10^{-4}$ requested accuracy. *Left:* 'easy' integrand, *right:* 'hard' integrand. *Top:* 'fast' integrand ($10\,\mu$sec), *bottom:* 'slow' integrand ($1000\,\mu$sec per evaluation). *Solid:* shared memory, *dashed:* socketpair communication (two curves each to show fluctuations in timing).

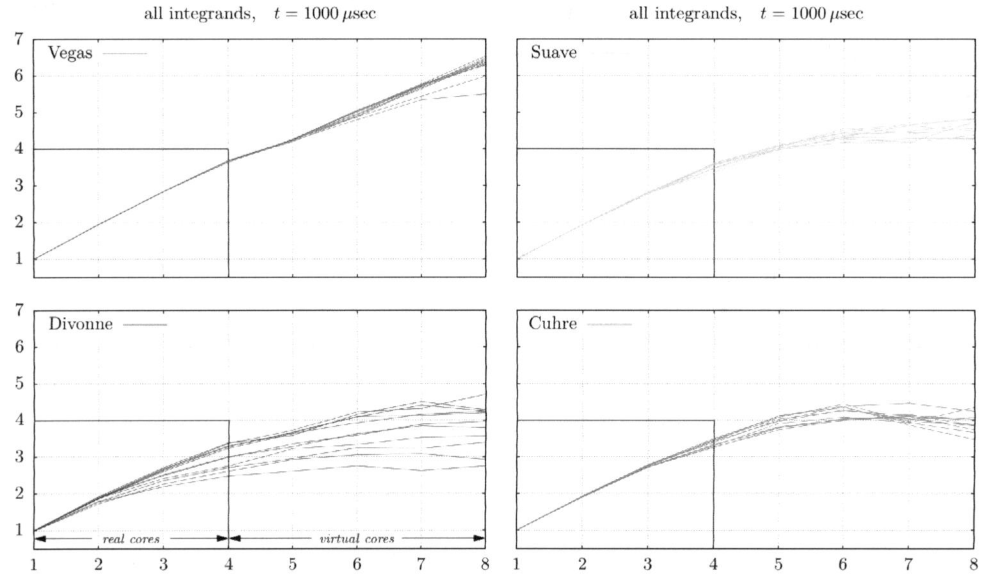


**Figure 2.** Comparison of the parallelization efficiency for all 11 integrands.

Journal of Physics: Conference Series **608** (2015) 012067 doi:10.1088/1742-6596/608/1/012067

# Statistical methods for cosmic ray composition analysis at the Telescope Array Observatory

**Grigory I. Rubtsov, Sergey V. Troitsky for the Telescope Array collaboration**
Institute for Nuclear Research of the Russian Academy of Sciences, Moscow 117312, Russia
E-mail: grisha@ms2.inr.ac.ru

**Abstract.** The Telescope Array surface detector stations record the temporal development of the signal from the extensive air shower front which carries information about the type of the primary particle. We develop the method to study the primary mass composition of the ultra-high-energy cosmic rays based on multivariate analysis (MVA). We report the preliminary mass composition results based on the Telescope array 5 years data.

## 1. Introduction

The Telescope Array (TA) experiment [1, 2] is an ultra-high-energy cosmic ray (UHECR) detector operating in Millard County, Utah, USA since May 2008. The TA is a hybrid experiment with a surface detector array of 507 plastic scintillator stations with 1.2 km spacing covering 700 $km^2$ area overlooked by 38 fluorescence telescopes located at the three sites. This *Talk* is dedicated to the study of the primary mass composition of UHECRs based on the time-resolved data of the TA surface detector (SD).

The primary composition of the UHECRs is one of the most intriguing questions. It is a key component in the studies of the potential sources of the cosmic rays and for predicting the flux of secondary photons and neutrino [3]. Moreover, one needs to know the primary composition in order to probe the interaction cross-section at the highest energies [4, 5] and to perform the precision tests of the Lorentz-invariance [6, 7]. The most accurate composition measurements up to date are based on the observation of the shower maximum $X_{\mathrm{MAX}}$ with the fluorescence telescopes (FD) working in stereo or in hybrid mode with the surface detectors (SD). A couple of alternative techniques are developed based on the data of the surface and muon detectors. The list of the available methods is given in Table 1.

It is well established that the photons and neutrino produce a minor fraction of the extensive air showers (EAS) in the atmosphere and therefore the most of the EAS are produced by hadrons [3]. Nevertheless, there is no agreement on the interpretation of the hadronic composition results between the experiments. The data of HiRes and TA are consistent with the pure protonic composition [8, 9], while the Pierre Auger observatory (PAO) indicate the presence of heavier nuclei at the energies higher than 10 EeV [10, 11]. The composition puzzle may be resolved by increasing the statistics of the Telescope Array experiment and increasing the control over systematic uncertainties in both TA and PAO.

In the present *Talk* the technique of the composition analysis based on the data of the TA surface detector is developed. The surface detector is capable of measuring chemical composition

**Table 1.** The techniques used for the composition measurements. $X_{\rm MAX}$ – depth of the shower maximum, $X^{\mu}_{\rm MAX}$ – muon production depth, $\rho_{\mu}$ – muon density, risetime – time interval between 10% and 50% of the total integral signal.

| Experiment | detector | observable | ref. |
|---|---|---|---|
| HiRes | FD stereo | $X_{\rm MAX}$ | [8] |
| Pierre Auger | FD+SD (hybrid) | $X_{\rm MAX}$ | [10] |
| Telescope Array | FD stereo | $X_{\rm MAX}$ | |
| Telescope Array | FD hybrid | $X_{\rm MAX}$ | [9] |
| Yakutsk | SD + muon detector | $\rho_{\mu}$ | [12] |
| Pierre Auger | SD | $X^{\mu}_{\rm MAX}$ | [11] |
| Pierre Auger | SD | risetime asymmetry | [11] |

independently of the fluorescence telescopes. It also provides better statistics at the highest energy due to the 95% duty cycle. The multivariate analysis (MVA) method is implemented using the two observables related to the area-over-peak of the signals at the surface detector stations. The data registered for the five years of observations are confronted with the proton and iron Monte-Carlo leading to the preliminary results on the mass-composition.

## 2. Method

Each of 507 TA surface detectors records time-resolved signals with a time step of 20 ns. For each detector we define Area-over-Peak ($AoP$) as the ratio of the integral of the FADC trace to it's peak value. The $AoP$ therefore is measured in time units (ns). The muons arrive close in time to the shower front making $AoP$ smaller, while electron-photon cascade shifts $AoP$ towards larger values. The $AoP$ observable was previously introduced by Pierre Auger collaboration for neutrino identification [13]. We fit $AoP$ as a linear function of the core distance:

$$AoP(r) = \alpha - \beta(r - 1200\,\mathrm{m})\,,$$

where $\alpha$ has a meaning of $AoP$ value at 1200 meters and $\beta$ is the $AoP$ slope.

Following [14, 15] we define percentile ranks of $\alpha$ and $\beta$ parameters relative to the proton showers $\mathcal{C}_{\alpha}$,$\mathcal{C}_{\beta}$:

$$\mathcal{C}^i_{\alpha} = \int\limits_{-\infty}^{\alpha^i} f^i_{MC,p}(\alpha)d\alpha\,,$$

$$\mathcal{C}^i_{\beta} = \int\limits_{-\infty}^{\beta^i} f^i_{MC,p}(\beta)d\beta\,,$$

where $f^i_{MC,p}(\alpha)$ is an $\alpha$ distribution function for proton QGSJETII-03 Monte-Carlo events compatible by zenith angle with the real event "i".

The composition analysis is based on the proton-iron classification procedure using the method of boosted decision trees (BDT). The *TMVA* package [16] for ROOT is used as an implementation of the method. The decision forest is constructed using the three observables: $\mathcal{C}_{\alpha}$, $\mathcal{C}_{\beta}$ and zenith angle. We construct independent BDT classifier for each energy bin spaced by 0.2 in $\log_{10}(E)$. The BDT is trained using the proton Monte-Carlo set as a background and the iron Monte-Carlo as a signal. The result of the BDT classifier is a single parameter $\xi^i$ for

Journal of Physics: Conference Series **608** (2015) 012067 doi:10.1088/1742-6596/608/1/012067

each event "i" which has a limited range by definition $-1 < \xi^i < 1$. The $\xi$-parameter is finally used for traditional one-parametric composition analysis.

We note that the reduction of the composition problem to proton-iron classification is based on an approximation the all relevant shower properties scale linearly with the logarithm of the atomic mass - $\log A$. The latter is supported by the results of simulations with nitrogen, silicon and helium primaries.

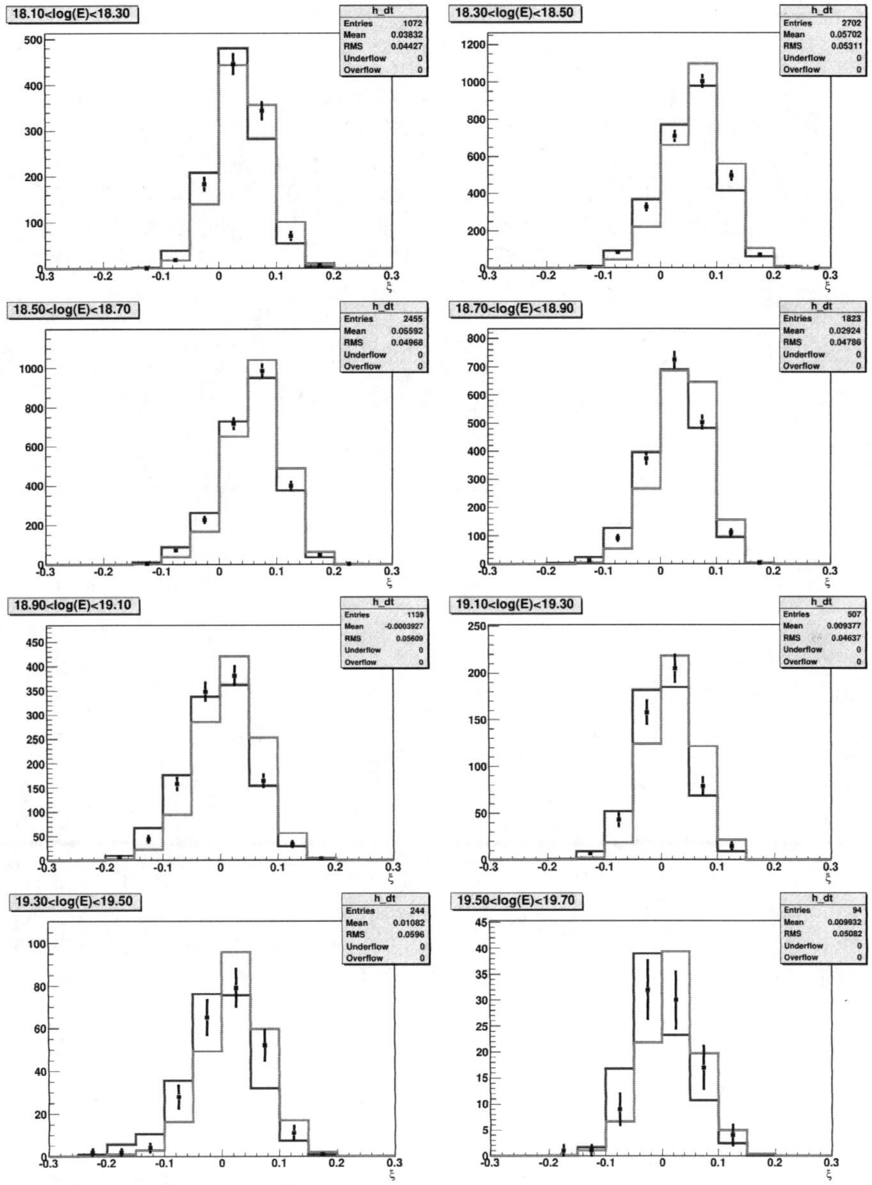


**Figure 1.** Distributions of the $\xi$ parameter for data and QGSJET-II Monte-Carlo (blue - protons, magenta - iron). /PRELIMINARY/

## 3. Data set, Monte-Carlo set and results

We use Telescope Array surface detector data set covering five years of observation from 2008-05-11 to 2013-07-13. Surface detector has been collecting data for more than 95% of time during that period [17]. The following cuts are applied to both data and MC events:

(i) Quality cuts used for spectral analysis [17];
(ii) Zenith angle cut: $0^\circ < \theta < 45^\circ$;
(iii) The number of detectors triggered is 7 or more;
(iv) The reconstructed energy is greater than $10^{18}$ eV.

The dataset contains 10242 events after the cuts. For $AoP(r)$ fit we require that the detector is not saturated and has a core distance larger than 600 m.

We use CORSIKA [18] with QGSJET II-03 [19] model to generate hadronic showers induced by primary protons and iron. The showers are simulated with thinning and the dethinning procedure is adopted [20] to simulate realistic shower fluctuations. The detector response is accounted for by using look-up tables generated by GEANT4 [21] simulations. Real-time array status and detector calibration information are used for each Monte Carlo (MC) simulated event. The Monte-Carlo events are produced in the same format as real events and the analysis procedures are applied in uniformly to both [22]. Each Monte-Carlo set is split into two halves - first is used for training the decision trees and the second for the final result.

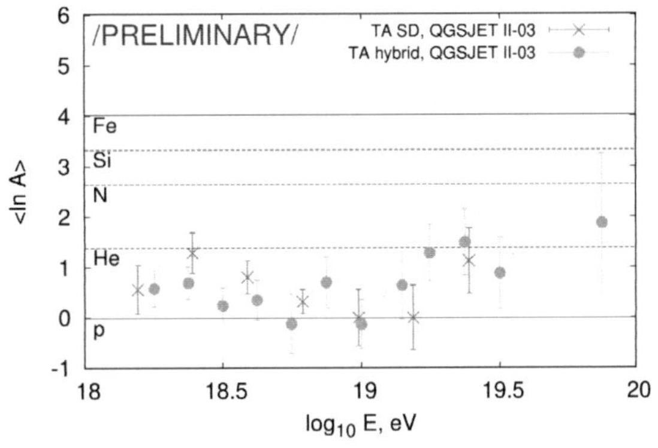


**Figure 2.** The preliminary result on the primary atomic mass based on the TA SD data compared to the TA hybrid result [9]

The histograms of $\xi$-parameter for data and Monte-Carlo are and QGSJET-II-03 and SIBYLL Monte-Carlo are shown in Figures 1. The data histograms are then compared to the mixture of proton and iron Monte-Carlo with the Kolmogorov-Smirnov probability. The preliminary result on the primary atomic mass from TA SD data is shown in Figure 2 in comparison with TA hybrid result. One may see the agreement of the results within the statistical errors. The advancemen of the method is required to reach sensitivity needed to resolve the composition puzzle. The systematic uncertainties and dependence on the hadronic model will be explained elsewhere.

## 4. Conclusion

We designed the new method of composition study based on the data of the array of scintillators. The preliminary composition based on the 5 years of TA SD is presented. Further development of the method involving more observable parameters is required to reach the sensitivity to discriminate between mono and mixed composition based on the TA surface detector data alone.

## Acknowledgments
We thank the organizers of the ACAT-2014 for warm hospitality and inspiring atmosphere of the meeting. The work is supported by the Russian Science Foundation grant 14-12-01340. The numerical part of the work was done at the cluster of the Theoretical Division of INR RAS.

## References
[1] Abu-Zayyad T et al. for the Telescope Array Collaboration 2012 *Nucl. Instrum. Meth. A* **689** 87
[2] Tokuno H et al. for the Telescope Array Collaboration 2012 *Nucl. Instrum. Meth. A* **676** 54
[3] Alvarez-Muniz J, Risse M, Rubtsov G and Stokes B 2013 *EPJ Web Conf.* **53** 01009
[4] Belov K et al. for the HiRes collaboration 2006 *Nucl. Phys. Proc. Suppl.* **151** 197
[5] Abreu P et al. for the Pierre Auger Collaboration 2012 *Phys.Rev.Lett.* **109** 062002
[6] Galaverni M and Sigl G 2008 *Phys. Rev. Lett.* **100** 021102
[7] Rubtsov G, Satunin P and Sibiryakov S 2014 *Phys. Rev.* D **89** 123011
[8] Abbasi R et al. for the HiRes collaboration 2010 *Phys.Rev.Lett.* **104** 161101
[9] Abbasi R et al. for the Telescope Array collaboration 2014 *Astropart.Phys.* **64** 49
[10] Abraham J et al. for the Pierre Auger collaboration 2010 *Phys.Rev.Lett.* **104** 091101
[11] Abreu P et al. for the Pierre Auger Collaboration 2011 *Studies of Cosmic Ray Composition and Hadronic Interaction models* (Proceeding of ICRC'11, Beijing) *preprint* arXiv:1107.4804
[12] Glushkov A, Makarov I, Pravdin M, Sleptsov I, Gorbunov D, Rubtsov G and Troitsky S 2008 *JETP Lett.* **87** 190
[13] Abreu P et al. for the Pierre Auger Collaboration 2013 *JCAP* **1305** 009
[14] Gorbunov D, Rubtsov G, Troitsky S 2007 *Astropart. Phys.* **28** 28
[15] Abu-Zayyad T et al. for the Telescope Array Collaboration 2013 *Phys. Rev. D* **88** 112005
[16] Hoecker A, Speckmayer P, Stelzer J, Therhaag J, von Toerne E and Voss H 2007 PoS ACAT 040,2007 *preprint* arXiv:physics/0703039
[17] Abu-Zayyad T et al. for the Telescope Array Collaboration 2013 *Astrophys. J.* **768** L1
[18] Heck D et al., 1998, *Forschungszentrum Karlsruhe* Report FZKA-6019.
[19] Ostapchenko S 2006 *Nucl. Phys. Proc. Suppl.* **151** 143
[20] Stokes B T et al. 2012 *Astropart. Phys.* **35**, 759
[21] Agostinelli S et al. for the GEANT4 Collaboration 2003 *Nucl. Instrum. Meth. A* **506** 250
[22] Abu-Zayyad T et al. for the Telescope Array Collaboration 2014 CORSIKA Simulation of the Telescope Array Surface Detector *Preprint* arXiv:1403.0644

Journal of Physics: Conference Series **608** (2015) 012068 doi:10.1088/1742-6596/608/1/012068

# Toolbox for multiloop Feynman diagrams calculations using $R^*$ operation

**D.V. Batkovich**[1], **M.V. Kompaniets**

Department of Theoretical Physics, St. Petersburg State University, Uljanovskaja 1, St. Petersburg, Petrodvorez, 198504 Russia

E-mail: `batya239@gmail.com`, `mkompan@gmail.com`

**Abstract.** We present the toolbox for analytical calculation of $UV$-counterterm of Feynman diagrams. It combines the power of $R^{*\prime}$-operation with modern analytical methods. Written in pure `Python` our toolbox can be easily used and extended.

## 1. Introduction

In this paper we present the toolbox that allows one to calculate $UV$-counterterm of Feynman diagrams in the framework of Bogoliubov–Parasiuk $R'$-operation [1] within $MS$-scheme (see e.g. [2, 3]). It uses a number of methods that include infrared rearrangement ($IRR$) [4, 5], $R^*$-operation [6–9], $G$-functions method [5] and integration-by-parts ($IBP$) reduction [10–17].

Infrared rearrangement is one of the oldest tricks for calculating $UV$-counterterm of diagrams, it is based on the fact that $UV$-counterterm of a logarithmically divergent diagram is independent on external momenta and masses of this diagram [18]. This fact allows one to set some masses or external momenta to zero or to introduce new masses or external momenta, the only restriction is that such a rearrangement should not produce nonphysical infrared (IR) divergences. It is possible to overcome this restriction by $R^{*\prime}$-operation and significantly increase the number of counterterms that can be calculated using $G$-functions and $IBP$-reduction.

Toolbox presented in this paper can be applied to a wide range of models. Additionally, it can be easily extended to calculate graphs not only using $G$-functions and $IBP$-reduction, but also with some other analytical or numerical methods (graph calculators) using API (application program interface) provided.

The paper is organized as follows: in the next section we shortly discuss $R^*$-, $R^{*\prime}$- operations, $IR$-counterterms and present an example that shows benefits of $R^{*\prime}$-operation. The third section describes architecture of the toolbox and packages it contains. In the fourth section we present examples on how to calculate $UV$-counterterms using `RStar` package. Fifth section is devoted to the implementation of the $IBP$-reduction and the API for the graph calculator extensions. The last section shortly describes toolbox installation procedure.

## 2. Short introduction to $R^{*\prime}$ operation

$R^*$-operation was introduced in [8, 9] as a method that allows to overcome restrictions of the infrared rearrangement: if one calculates $UV$-countertem of the diagram with improper $IRR$

[1] Speaker

using $R'$-operation there are some nonphysical $IR$-divergences arising in the integrals calculated. This $IR$-divergences as well as $UV$-divergences are poles in $\varepsilon$ (parameter of the dimensional regularization) and can not be distinguished from $UV$-divergences, this results in the incorrect $UV$-counterterm. $R^{*\prime}$-operation allows to separate this nonphysical $IR$-divergences and subtract them to get correct value for counterterm. Thus counterterm $\Delta_{UV}(\gamma)$ for diagram $\gamma$ is defined as a combination of standard $R'$ operation and infrared $\widetilde{R}'$ operation:

$$\Delta_{UV}(\gamma) = -KR^{*\prime}(\gamma) = -K\widetilde{R}'R'(\gamma) \tag{1}$$

In the case of the proper infrared rearrangement, i.e. no IR divergences, $\widetilde{R}' \equiv 1$, but in the case of improper IRR it becomes nontrivial. $\widetilde{R}'$-operation can be defined by introducing $IR$-counterterm ($\Delta_{IR}$) as follows:

$$\widetilde{R}'(\gamma) = \begin{cases} \Delta_{IR}(\gamma) & \text{, where } \gamma \text{ is 0-tadpole,} \\ \gamma + \sum\limits_{\delta \in W_u(\gamma)} \Delta_{IR}(\gamma/\delta)\delta & \text{, other,} \end{cases} \tag{2}$$

here $W_u(\gamma)$ - set of all $UV$-divergent subgraphs of $\gamma$ through which external momentum of $\gamma$ can be completely passed (these subgraphs contain all boundary vertices of initial graph) and which also don't contain tadpoles. By 0-tadpole we denote massless tadpole. If one treats 0-tadpole as integral it is definitely 0, but in case of $R^{*\prime}$-operation combinatorics it is important to take these graphs into account. $IR$-counterterm in case of 0-tadpole $\gamma$ can be given by the following relation:

$$\Delta_{UV}(\gamma) + \Delta_{IR}R(\gamma) = 0. \tag{3}$$

Advantages of using $R^{*\prime}$-operation can be demonstrated by 6-loop diagram example:

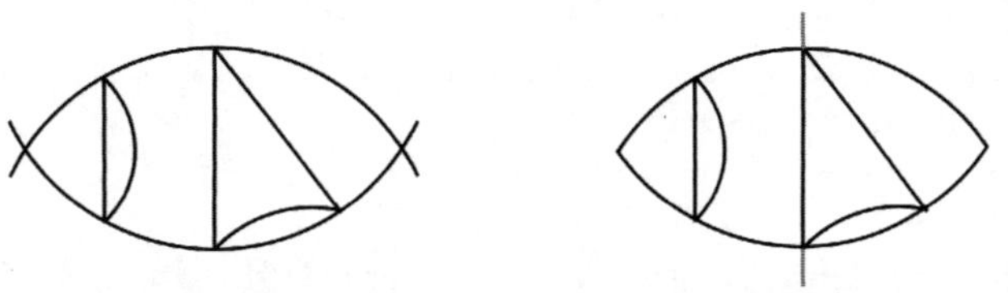

$UV$-counterterm of the left diagram can't be calculated with proper $IRR$ by using $G$-functions an 4-loop $IBP$-reduction, but if we rearrange external momenta(and set all masses to zero) as shown on the right diagram $UV$-counterterm can be calculated by using $R^{*\prime}$ with use of $G$-functions method only.

Further information about $R^*$- and $R^{*\prime}$-operations can be found in [6–9].

## 3. Architectural overview

Toolbox is built on top of `GraphState` and `Graphine` [19] libraries, all graph manipulations and serialization are provided by these libraries. All input graphs must be given in form readable by `GraphState` [19] which is the extension of graph labeling algorithm introduced by [20].

`RStar` package plays the central role in the toolbox and is responsible for all combinatorial operations to evaluate $R^{*\prime}$-operation. Additionally this module can calculate components of $R^{*\prime}$, namely $\Delta_{IR}$ and $\widetilde{R}$ of some given graph.

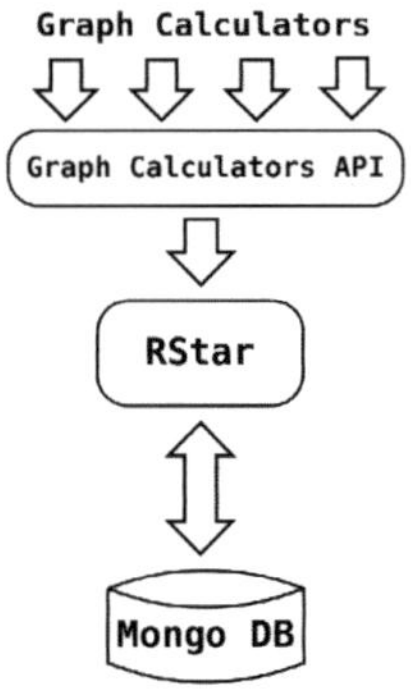


*3.1. Brief packages overview*

- `RStar` - main package serves to calculate $\Delta_{UV}$, $\Delta_{IR}$ and $\widetilde{R}'$ on diagrams. Package contains graph calculators API to calculate required diagram values. Default list of calculators can be extended. Calculated values of diagrams and counterterms are placed on `Mongo DB` [21] -based storage. This storage saves all values which can be reused for another diagrams and has separate collections for different methods and models. Environment of this package can be simply configured as shown below.
- `RgGraphEnv` and `RgGraphUtil` - some set of utils containing connector to `Mongo DB`, `swiginac` CAS extensions and $G$-functions calculator.
- `Reduction` package contains implementation of $IBP$-reduction and wrappers of this procedure for graph calculators API.

*3.2. Symbolic calculations*

As backend for computer algebra we use `swiginac` library. It's a python wrapper of the most known `GiNaC` computer algebra written in `C++`. It gives us fast analytical computations and provides better performance than pure `Python` libraries.

## 4. Getting started with RStar

Different models require different environmental parameters. We provide a simple API to configure user own environment including user own graph calculators, $UV$-relevance conditions to determine $UV$-divergent subgraphs (note that condition of 1-irreducibility is automatically applied), dimensions as well as `Mongo DB` storage settings. These parameters must be configured before starting any calculations as shown in the following program code example:

```
from rstar import Configure
from rggraphenv.symbolic_functions import cln, e
from rggraphenv import GLoopCalculator, StorageSettings
from graphine import Graph
from graphine.filters import graph_filter

d_phi4 = cln(4) - cln(2) * e

@graph_filter
```

```
def uv_condition(graph_edges, super_graph):
    g = Graph(graph_edges)
    uv_index = 4 * g.loops_count - 2 * g.internal_edges_count
    return uv_index >= 0

Configure()\
  .with_uv_filter(uv_condition)\
  .with_dimension(d_phi4)\
  .with_calculators(GLoopCalculator(d_phi4))\
  .with_storage_holder(StorageSettings("test_project", "localhost", 27017))\
  .configure()
```

When the program environment is configured one can use $R^{*\prime}$−operation to evaluate $UV$- and $IR$-counteritems. All the calculated data will be automatically saved to structured storage. In the following snippet we show how to calculate $\Delta_{IR}$ and $\Delta_{UV}$ on examples of:

$\Delta_{IR}($ 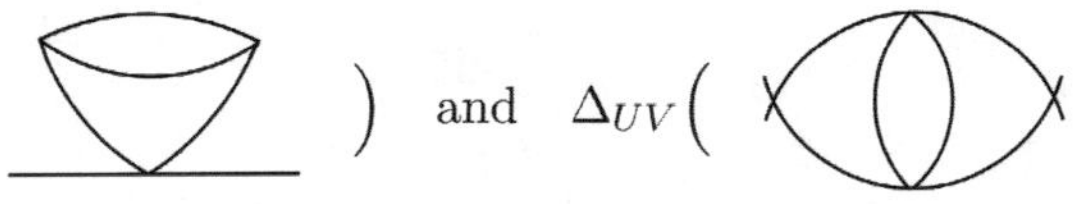 $)$ and $\Delta_{UV}($ $)$

**Figure 1.** Diagram examples.

To calculate these counterterms user needs to create `RStar` object and call correspondent methods (`delta_ir` and `delta_uv`):

```
from rstar import RStar, graph_util
operator = RStar()

eye_tadpole = graph_util.graph_from_str("ee12|22||")
print operator.delta_ir(eye_tadpole)
# 1/2*(1+e)*e**(-2)

three_loop_diagram = graph_util.graph_from_str("ee12|223|3|ee|")
print operator.delta_uv(three_loop_diagram)
# -1/3*e**(-1)+2/3*e**(-2)-1/3*e**(-3)
```

Note that both of graphs must be given in `GraphState` [19] graph representation based on Nickel index [19, 20]. In current example $ee12|22||$ is representation of left diagram and $ee12|223|3|ee|$ is representation of right diagram pictured on Fig. 1.

## 5. Calculator extensions

`RStar` graph calculator API allows to create custom calculators. These calculators can be written in any language and can use any software or hardware instruments. To create custom calculator one should implement `GraphCalculator` class

```
from rggraphenv.abstract_graph_calculator import AbstractGraphCalculator

class MyCustomGraphCalculator(AbstractGraphCalculator):
    def get_label(self):
        return "description of calculator"
```

```
    def init(self):
        # do initialization here
        pass

    def dispose(self):
        # clean resources here
        pass

    def is_applicable(self, graph):
        # check is calculator applicable to graph
        return can_calculate_graph

    def calculate(self, graph):
        # calculation logic...
        return calculated_value
```

In the `calculate` method the logic of the user graph calculation process must be incapsulated. Return value must be `None` if the calculator can't calculate a given graph or a pair where first component is value of `swiginac` expression type that depends on $\varepsilon$. Second component must have the type `rggraphutil.VariableAwareNumber` and is representing power of $p^{-2}$ where $p$ is the external momentum of the graph:

```
import rggraphutil
rggraphitul.VariableAwareNumber("l", a, b)
```

means $a + bl$ number where $l = d/2 - 1$ and will represent $p^{-2(a+bl)}$. For example:

```
def calculate(self, graph):
    # calculation logic...
    return 1/rggraphenv.symbolic_functions.e, VariableAwareNumber("l", 1, -1)
```

means that returned value equals to $\frac{1}{\varepsilon}p^{-2(1-l)}$.

### 5.1. `Reduction` *package*

`Reduction` provides reduction by *IBP* [15] and *DRR* [14, 17, 22] rules and implemented using rules of LiteRed [11, 12, 16] program. Values of master integrals for 4-loop reduction obtained from [10, 13]. In our implementation we use some cache for more effective utilization of already calculated integrals. Our cache is common for all run reduction processes at the same time and stored to `Mongo DB`. Additionally cache is separate for different reduction rules such as rules for 2, 3 and 4 loop reduction. `Reduction` can be used separately from other parts of the program.

To be calculated Feynman integral must be given in the following form:

$$J(n_1, n_2, n_3, \ldots, n_m) = \int \frac{\prod\limits_{l}^{L} dk_l}{D_1^{n_1} D_2^{n_2} \ldots D_m^{n_m}}, \qquad (4)$$

where $D_i$ - i-th propagator of L-loop basis. To define this integral from `Python` one need following code:

```
import reduction
integral = reduction.J(n1, n2, ..., n_m)
```

One can see the list of basis propagators by typing `reduction.BASIS_2`, `reduction.BASIS_3` or `reduction.BASIS_4`. Last number means count of loops in basis.

But sometimes it can be very uncomfortable to generate these `reduction.J` objects manually from diagrams. For this case we allow to set graph as the parameter and our package will find corresponding momentum layout. The only thing which are needed to specify in this case is how to obtain propagator weights. This procedure is carried out using additional function that has one parameter (edge) and returns weight of propagator corresponding to given edge. It must be specified by user.

Here is the example how to calculate following diagram with weights of propagators shown in picture:

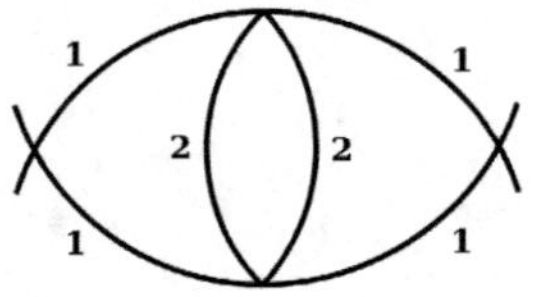


To calculate this diagram we use graph representation and integral representation. For graph representation we define function `weight_extractor` that returns `weight` attribute of given edge.

```
import reduction
from rggraphenv import symbolic_functions

reductor = reduction.THREE_LOOP_REDUCTOR.with_cache("localhost", 27017)

weight_extractor = lambda e: e.weight

def calculate_3loop_graph_using_reduction(graph):
    unsubstituted_result = reductor.calculate_diagram(graph, weight_extractor)
    # in variable unsubstituted_result:
    #     - master integral values not substituted
    #     - dimension not substituted
    #     - no expansion in series by epsilon
    return unsubstituted_result.evaluate(substitute_masters=True,
                                         _d=4-2*symbolic_functions.e,
                                         series_n=4,

                                         remove_o=True)


def calculate_3loop_integral(integral):
    unsubstituted_result = reductor.calculate_j(integral)
    return unsubstituted_result.evaluate(substitute_masters=True,
                                         _d=4-2*symbolic_functions.e,
                                         series_n=4,
                                         remove_o=True)


def main():
    three_loop_diagram =
        reduction.graph_util.graph_from_str("e12|223|3|e|:0_1_1|2_2_1|1|0|")
    print calculate_3loop_graph_using_reduction(three_loop_diagram)
    # 1/3*e**(-3)+1/3*e**(-2)+1/3*e**(-1)+...
```

Journal of Physics: Conference Series **608** (2015) 012068 doi:10.1088/1742-6596/608/1/012068

```
    three_loop_diagram_as_integral = reduction.J(1, 1, 0, 2, 2, 0, 1, 0, 1)
    print calculate_3loop_integral(three_loop_diagram_as_integral)

if __name__ == "__main__":
    main()
```

Result of `calculate_j` and `calculate_diagram` methods is linear combination of master integrals, where values of these master integrals and dimensions in coefficients are not substituted. To substitute them we call `evaluate` method. It should be noted that in bases included in `Reduction` package values of master integrals are given only for dimension $d = 4-2\varepsilon$. If one need to perform calculations in dimension that differs from $d = 4 - 2\varepsilon$, it is necessary to clone basis needed and set values of master integrals for required space dimension. Also it is possible to create own basis and reduction rules.

## 6. Installation

Our toolbox supports only Unix-like systems. As external dependencies one need `swiginac`, `pymongo`, `repoze.lru` and `inject` packages which one can find in `Python Package Index` `https://pypi.python.org/`. Additionally following packages are needed to install `GraphState`, `Graphine`, `RgGraphEnv`, `RgGraphUtil`, `Reduction`, `RStar` from `https://code.google.com/p/rg-graph/`.

## 7. Acknowledgments

Authors are grateful to L.Ts. Adzhemyan, K.G. Chetyrkin and R.N. Lee for helpful discussions and advices, to ACAT'14 Organizing Committee for support and hospitality. The work was supported in part by Saint-Petersburg State University (project 11.38.185.2014).

## References

[1] Bogoliubov N N and Parasiuk O S 1957 *Acta Math.* **97** 227

[2] Bogoliubov N N and Shirkov D V 1976 *Introduction to the theory of quantized fields* [in Russian] (Moscow: Nauka)
English transl.: 1980 *Introduction to the theory of quantized fields*, (New York: Interscience)

[3] Vasiliev A N 1998 *Quantum field renormalization group in critical behavior theory and stochastic dynamics* [in Russian] (St.Petersburg: Petersburg Inst. Nucl. Phys.)
English transl.: 2004 *The field theoretic renormalization group in critical behavior theory and stochastic dynamics*, (Boca Raton: Chapman & Hall/CRC)

[4] Vladimirov A A 1978 *Theor. Math. Phys.* **36** 732

[5] Chetyrkin K G, Kataev A L and Tkachov F V 1980 *Nucl. Phys. B* **174** 345

[6] Chetyrkin K G, Gorishny S G, Larin S A and Tkachov F V 1986 Analytical calculations 5-loops approximations RG functions for $g\phi^4$ model in MS-scheme (*Preprint* `INRP-0453`)

[7] Chetyrkin K G 1991 Combinatorics of $R$-, $R^{-1}$ and $R^*$ operations and asymptotic expansions of Feynman integrals in the limit of large momenta and masses (*Preprint* `MPI-Ph/PTh13/91`)

[8] Chetyrkin K G and Tkachov F V 1982 *Phys. Lett. B* **114** 340

[9] Chetyrkin K G and Smirnov V A 1984 *Phys. Lett. B* **144** 4194

[10] Baikov P A and Chetyrkin K G 2010 *Nucl. Phys. B* **837** 186

[11] Lee R N 2012 Presenting LiteRed: a tool for the loop integrals reduction (*Preprint* `hep-ph/1212.2685`)

[12] Lee R N 2014 *J. Phys.: Conf. Series* **523** 012059 (*Preprint* hep-ph/1310.1145)

[13] Lee R N, Smirnov A V and Smirnov V A 2012 *Nucl. Phys. B* **856** 95

[14] Lee R N 2012 *J. Phys.: Conf. Series* **368** 012050

[15] Chetyrkin K G and Tkachov F V 1981 *Nucl. Phys. B* **192** 159

[16] Lee R N LiteRed http://www.inp.nsk.su/~lee/programs/LiteRed/

[17] Baikov P 1996 *Phys. Lett. B* **385** 404 (*Preprint* hep-ph/9603267)

[18] Collins J C 1974 *Nucl. Phys. B* **80** 341

[19] Batkovich D, Kirienko Y, Kompaniets M and Novikov S 2014 Graphstate - a tool for graph identification and labelling (*Preprint* hep-ph/1409.8227)

[20] Nickel B G, Meiron D I and Baker Jr G A 1977 *Compilation of 2-pt and 4-pt graphs for continuous spin model* (University of Guelph report)

[21] Mongo DB http://www.mongodb.org/

[22] Tarasov O V 1996 *Phys. Rev. D* **54** 6479

# A new generator for the Drell-Yan process

**Yahor Dydyshka, Vitaly Yermolchyk**

National Scientific and Educational Center of Particles and High Energy Physics BSU, Pervomayskaya str. 18, 220088 Minsk, Belarus

E-mail: `Yahor.Dydyshka@cern.ch,vitaly.yermolchyk@cern.ch`

**Abstract.** We present the Monte Carlo event generator `LePaProGen` for lepton pair production at hadron colliders. `LePaProGen` focuses on the description of higher-order electroweak radiative corrections. The generator is implementing a new algorithm for the selection of the optimal variables for phase space parametrization.

## 1. Introduction

The processes of single electroweak boson production are very significant for physical studies at LHC. There are many groups working on them to provide adequate theoretical predictions. `FEWZ` [1], `POWHEG` [2–4], `SANC` [5], `HORACE` [6] et al. programs has been developed for this purpose. At energies 1-5 TeV charged and neutral current Drell-Yan processes are background to the processes of new physics. To make distinguish between different models it is important to have both fast and accurate tool for Monte-Carlo simulation. Designed for this purpose code `READY` [7] based on Sudakov and collinear logarithms approximation working only at extra large dilepton invariant mass. It is convenient to have a tool which works equally good at all kinematic regions and nevertheless is very fast. For this reason we introduce our event generator `LePaProGen`.

## 2. LePaProGen

LePaProGen (Lepton Pair Production Generator) is a Monte-Carlo generator for the neutral and charged current Drell-Yan process. It is intended for fast and accurate generation of events and uses independent generation algorithm. Now it implements one loop radiative corrections with the exact hard QED Bremsstrahlung contribution. But the higher order QCD or electroweak corrections could be easily added by design. LePaProGen is written in C++ using modular architecture. To overcome the C++ templates limitations we use a Python module which calculates all the constants and preprocesses the input settings. A glue layer between `Python` and C++ parts of the program is `Mako` template library which performs optimized code generation for the process chosen by user. LePaProGen can be used as a `Pythia8` plug-in. It has all needed interfaces for convenient work: it generates events in the Les Houches Accord (LHA) format, the `HepMC` output format; it uses the `LHAPDF` interface for parton density functions. It supports a variety of renormalization schemes and performs the `POWHEG`-like matching.

In the following sections we will describe the main algorithms implemented in LePaProGen in detail.

### 3. Generation of phase space points

First of all we start with phase space points generation. All the variables can be divided into two groups: the "slow" and the "fast". The "slow" variables can be generated uniformly but the "fast" ones require importance sampling. In tree level amplitude all "fast" variables are due to propagators. In order to generate a point in phase space we need to parametrize it by the invariant variables, which appear in the propagators. It can be done by effectively replacing all the propagators by $\delta$-functions

$$\int dR_n \left(\frac{1}{p^2 - m^2}\ldots\right) = \int ds' \frac{1}{s' - m^2}\left[\int dR_n \delta(p^2 - s')\ldots\right]. \tag{1}$$

Such integrals appear in generalized unitarity methods and can be expressed by Feynman graphs (Fig. 1) with the help of the optical theorem. In fact all intermediate and final particles in such

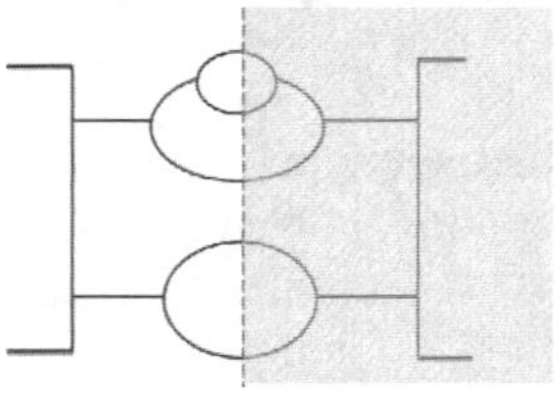

**Figure 1.** The Feynman graph for integral.

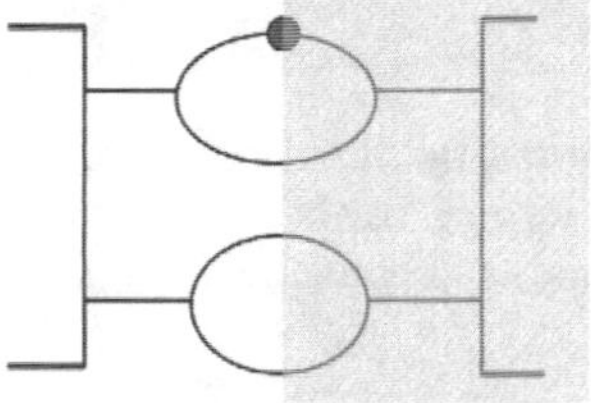

**Figure 2.** The contraction of the first loop.

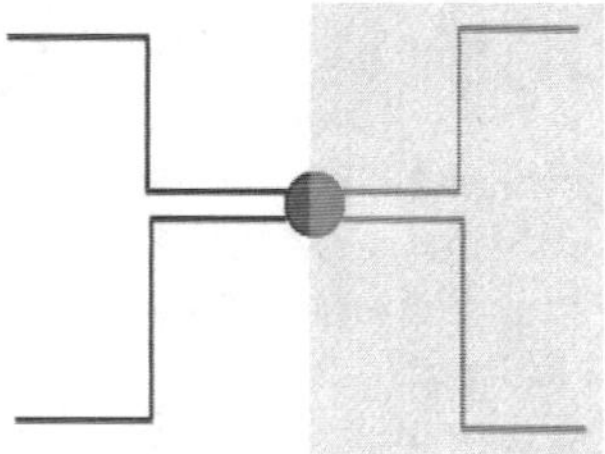

**Figure 3.** The result of the final contraction.

graph are on-shell. To calculate the integral we subsequently contract one loop subdiagrams starting with ones containing lesser number of propagators (Fig. 2-3). As a result we obtain a Jacobian determinant of transformation. This algorithm is a generalization of the recurrence relation for phase space volume from Byckling and Kajantie [8].

### 4. Particles Momenta Reconstruction

The cross-section for the process $a + b \to 1 + \ldots + n$ is the following phase space integral:

$$\sigma_n = \frac{(2\pi)^4}{2\lambda^{1/2}(s, m_a^2, m_b^2)} \frac{1}{(2\pi)^{3n}} \int \prod_{i=1}^{n} \frac{d^3 p_i}{2E_i} |M|^2 \delta(p_a + p_b - \sum_i p_i). \tag{2}$$

For now lets formulate the problem of phase space reconstruction.

- We select some set of invariant valiables sutable for generation.
- Then we sample these variables according to some importance density.
- We must evaluate the matrix element at the selected point of phase space. If the matrix element is in the invariant form then we need only the full set of scalar products $p_i \cdot p_j$.
- If the matrix element is expressed in terms of noninvariant components of vectors in some reference frame, then we must express them in our invariant set of variables. This task we call "reconstruction".

Consider reconstruction in the case of 3-propagators phase space integral.

Journal of Physics: Conference Series **608** (2015) 012069 doi:10.1088/1742-6596/608/1/012069

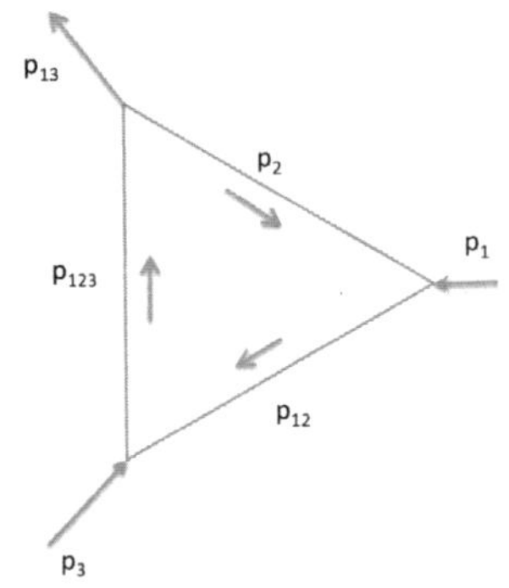


**Figure 4.** Momentum flow diagram.

### *4.1. Three-Point Reconstruction*

This case corresponds to the interference between the decay and the t-channel scattering, or to the subdiagram of interference between two 3-particle decay channels. Our reconstruction algorithm is universal and works for both interpretations. Differences appear only in the integration limits.

As input parameters we have two 4-momenta. These are the momenta of the particles scattered in t-channel $p_1$ and $p_{13}$ (Fig. 4). Let the momenta after scattering be $p_2$ and $p_{12}$. The masses of the decay products and the intermediate t-channel propagators are known ($m_1$, $m_{12}$ and $m_{123}$). We use only the invariant squared masses $s_1 = p_1^2$, $s_{12} = p_{12}^2$ and $s_{123} = p_{123}^2$ because a contracted one-loop subdiagram can appear at the place of a particle. It is evident, that due to the 4-momentum conservation it is enough to reconstruct only the 4-momentum of one of the particles.

The first necessary condition for the decay is time-likeness of the 2-plane spanned by external momenta:

$$\pi_{13} = p_1 \wedge p_3 = p_1 \wedge p_{13}, \qquad \|\pi_{13}\|^2 > 0. \tag{3}$$

This condition guarantees finiteness of the phase space volume and is weaker then the corresponding one from [8]. Consequently our approach is more general and can be applicable in cases when relations from [8] does not work.

The scalar products are fixed by the masses of decay products:

$$p_1 \cdot p_2 = \frac{s_{12} - s_1 - s_2}{2}, \qquad p_2 \cdot p_{13} = \frac{s_{123} - s_2 - s_{13}}{2}. \tag{4}$$

The next step is to construct reciprocal basis:

$$\mathfrak{p}_1 = p_{13}\pi_{13}^{-1}, \qquad \mathfrak{p}_{13} = -p_1\pi_{13}^{-1}. \tag{5}$$

Using this basis we can reconstruct the longitudinal component of $p_2$ and norm of the transverse component:

$$p_{2L} = (p_1 \cdot p_2)\mathfrak{p}_1 + (p_2 \cdot p_{13})\mathfrak{p}_{13}, \qquad p_{2T}^2 = s_2 - p_{2L}^2. \tag{6}$$

Here is the place to apply the second necessary condition on the input parameters: it is evident that the longitudinal part is time-like ($p_{2L}^2 > 0$) so the transverse part should be space-like:

$$p_{2T}^2 < 0. \tag{7}$$

The direction of the transverse part of $p_2$ cannot be determined from provided data and should be generated randomly and uniformly. Lets generate a unit 2-vector $n$ uniformly on circle.

In order to use this as the direction of $p_{2T}$ we should Lorentz-transform it to the plane orthogonal to the all external momenta. We are going to construct a Lorentz transformation which maps $\pi_{13}/\|\pi_{13}\|$ to the bivector $\gamma_0 \wedge \gamma_3$. In Clifford algebra this transformation is determined by the rotor $R$:

$$R = \sqrt{S}, \qquad S = \mathrm{sgn}\big(\langle \pi_{13}\gamma_{03}\rangle\big)\pi_{13}\gamma_{03}, \qquad \gamma_{03} = \gamma_0\gamma_3 = \gamma_0 \wedge \gamma_3. \tag{8}$$

$$\begin{aligned}\sqrt{S} &\propto \big(\|S\| + S\big)\Big(\langle S\rangle_+ + \sqrt{\langle S\rangle_+ \langle S\rangle_-}\Big)\langle S\rangle_-,\\ \langle S\rangle_\pm &= \|S\| + \langle S\rangle \pm \langle S\rangle_4.\end{aligned} \tag{9}$$

$$p_{2T} = \sqrt{-p_{2T}^2}\,RnR^{-1}, \qquad n = n_x\gamma_1 + n_y\gamma_2. \tag{10}$$

The above mentioned procedure is correct because Lorentz-transformations preserve orthogonality. Moreover, this 4-rotation has minimum norm in parametric space.

The Jacobian determinant we need is

$$J_3 = \frac{\pi}{2(2\pi)^6\sqrt{\lambda(s_1, s_3, s_{13})}}. \tag{11}$$

## 5. The Amplitude

In this section we turn our focus on the hard Bremsstrahlung amplitudes. They are generated using helicity amplitude technique by our MetaAmp package. We use the Kleiss-Stirling convention [9]. Lets choose two basis vectors $k_0$ and $k_1$, which satisfy relations:

$$k_0^2 = 0, \qquad k_0 \cdot k_1 = 0, \qquad k_1^2 = -1. \tag{12}$$

So then we construct basis spinors $u_+(k_0)$ and $u_-(k_0)$, such that

$$u_\lambda(k_0)\bar{u}_\lambda(k_0) = \omega_\lambda \hat{k_0}, \qquad u_\lambda(k_0) = \lambda \hat{k_1} u_{-\lambda}(k_0), \qquad \omega_\lambda = \frac{1+\lambda\gamma_5}{2}. \tag{13}$$

If we take $k_0 = \{1/2, 0, 0, 1/2\}$ and $k_1 = \{0, 1, 0, 0\}$ than in the Weyl representation (up to an arbitrary phase) the basis spinors look like $u_+(k_0) = \{1, 0, 0, 0\}^T$ and $u_-(k_0) = \{0, 0, 0, -1\}^T$.

Now we can express a spinor for arbitrary external particle with momentum $p$:

$$u_\lambda(p) = \frac{\hat{p} \pm m}{\sqrt{2p\cdot k_0}} u_{-\lambda}(k_0). \tag{14}$$

To generate amplitudes MetaAmp uses Fiertz identities and Dirac matrix algebra. In that way we reduce the task to calculation of the following expressions:

$$\begin{aligned}\langle pq\ldots k\rangle &= \bar{u}_+(k_0)\hat{p}\hat{q}\ldots\hat{k}\omega_+ u_+(k_0) = \mathrm{Tr}[\hat{p}\hat{q}\ldots\hat{k}\omega_+\hat{k_0}],\\ \langle pq\ldots k] &= \bar{u}_+(k_0)\hat{p}\hat{q}\ldots\hat{k}\omega_- u_-(k_0) = \mathrm{Tr}[\hat{p}\hat{q}\ldots\hat{k}\omega_-\hat{k_0}\hat{k_1}],\\ [pq\ldots k\rangle &= \bar{u}_-(k_0)\hat{p}\hat{q}\ldots\hat{k}\omega_+ u_+(k_0) = \mathrm{Tr}[\hat{p}\hat{q}\ldots\hat{k}\omega_+\hat{k_1}\hat{k_0}],\\ [pq\ldots k] &= \bar{u}_-(k_0)\hat{p}\hat{q}\ldots\hat{k}\omega_- u_-(k_0) = \mathrm{Tr}[\hat{p}\hat{q}\ldots\hat{k}\omega_-\hat{k_0}].\end{aligned} \tag{15}$$

Normalization factors from external wave-functions are factorized out from the whole amplitude so square roots are not presented. We have chosen $k_0 = \{1/2, 0, 0, 1/2\}$ and $k_1 = \{0, 1, 0, 0\}$, so operations (15) are reduced to taking the component of the resulting matrix making numerical calculation of generated amplitudes really fast.

This is an example of the single bremsstrahlung amplitudes generated by MetaAmp:

$$A^{\gamma}_{+--+-} = 4e^3\delta^{i_1}_{i_2}\Bigg[Q(l)D_{RR}(p^2_{12})\left(\frac{1}{p^2_{45}}[p_2p_3\rangle\langle p_4p_5]\langle p_1p_{45}] - \frac{1}{p^2_{35}}[p_3]\langle p_1p_4][p_2p_3p_5]\right)$$
$$+ Q(q)D_{RR}(p^2_{34})\left(\frac{1}{p^2_{25}}[p_2]\langle p_1p_4][p_3p_2p_5] - \frac{1}{p^2_{15}}[p_2p_3\rangle\langle p_1p_5]\langle p_4p_{15}]\right)\Bigg], \quad (16)$$

where $D_{LR}(s) = Q(q)Q(\mu) + g_L(q)g_R(l)\frac{s}{s-M_Z^2+i\Gamma_Z M_Z}$; $Q(q)(Q(\mu))$ – charge of quark (muon); $g_L(q)$, $g_R(l)$ – weak current coupling constants; $i_1$,$i_2$ – color indices.

## 6. Plots

In order to demonstrate work of our generator we present some plots Fig. 5-7 for the neutral-current Drell-Yan process generated using LePaProGen.

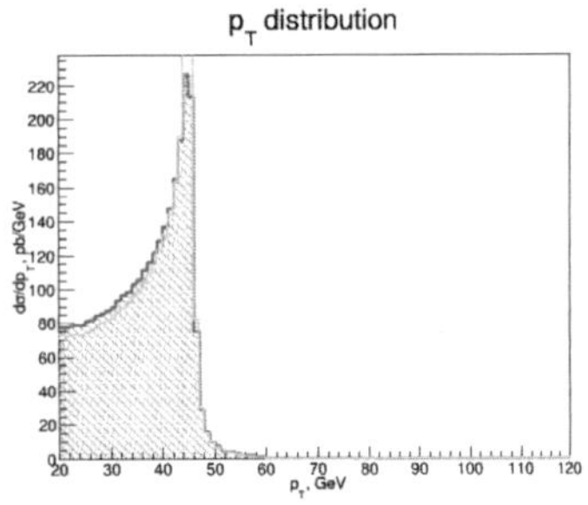


**Figure 5.** The transverse momentum distribution of the muon. Born – green, Born + EW –guaranty finteness red.

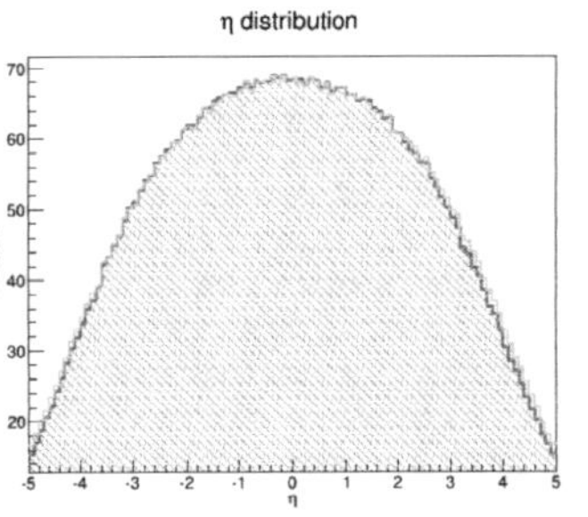


**Figure 6.** The pseudorapidity distribution of the muon. Born – green, Born + EW – red.

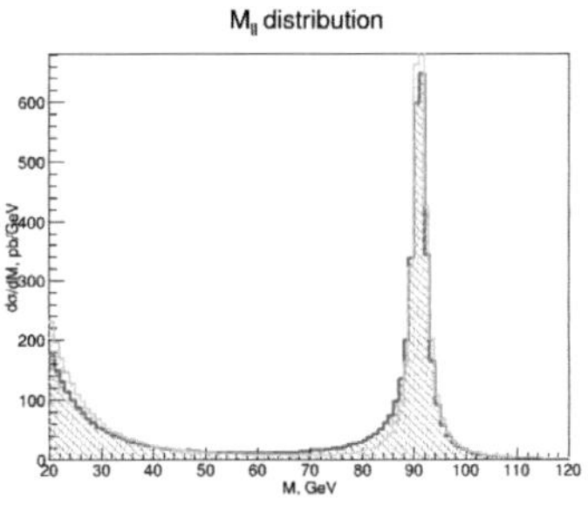


**Figure 7.** The muon pair invariant mass distribution. Born – green, Born + EW – red.

## 7. Conclusions

In this paper a new generator `LePaProGen` for Drell-Yan processes is presented. A new algorithm for the selection of the optimal variables for phase space parametrization is implemented. The resulting efficiency is under investigation. A comparison with existing Monte-Carlo algorithms, like `Vegas` and `FOAM`[10], will be performed. A fast modification of helicity amplitude method is used. The detailed numerical comparison between the predictions of `LePaProGen` and other MC event generators for charged and neutral current Drell-Yan processes will be fulfilled in the near future.

## Acknowledgments

We thank organizing committee of ACAT2014 and personally Prof. Milos Lokajicek, Prof. Andrei Kataev and Prof. Grigory Rubtsov for help and financial support.

Journal of Physics: Conference Series **608** (2015) 012069 doi:10.1088/1742-6596/608/1/012069

## References

[1] Li Y and Petriello F 2012 *Phys. Rev. D* **86**(9) 094034
[2] Barze L, Montagna G, Nason P, Nicrosini O, Piccinini F and Vicini A 2013 Neutral current drell-yan with combined qcd and electroweak corrections in the powheg box (*Preprint* `hep-ph/1302.4606`)
[3] Bernaciak C and Wackeroth D 2012 *Phys. Rev. D* **85**(9) 093003
[4] Barze L, Montagna G, Nason P, Nicrosini O and Piccinini F 2012 Implementation of electroweak corrections in the powheg box: single w production (*Preprint* `hep-ph/1202.0465`)
[5] Sadykov R, Arbuzov A, Bardin D, Bondarenko S, Christova P, Kalinovskaya L, Kolesnikov V, Sapronov A and Uglov E 2014 *Journal of Physics: Conference Series* **523** 012043
[6] Calame C, Montagna G, Nicrosini O and Vicini A 2007 *Journal of High Energy Physics* **2007** 109
[7] Zykunov V A 2014 *Journal of Physics: Conference Series* **523** 012053
[8] Byckling E and Kajantie K 1973 *Particle Kinematics* (Wiley)
[9] Kleiss R and Stirling W 1985 *Nuclear Physics B* **262** 235–62
[10] Jadach S 2003 *Comput.Phys.Commun.* **152** 55–100

# Non-planar Feynman diagrams and Mellin-Barnes representations with AMBRE 3.0

**Ievgen Dubovyk**[1], **Janusz Gluza**[2], **Tord Riemann**[3]

[1] Institute of Electrophysics and Radiation Technologies, Chernyshevsky st. 28, 61002 Kharkiv, Ukraine

[2] Institute of Physics, University of Silesia, Uniwersytecka 4, 40007 Katowice, Poland

[3] 15711 Königs Wusterhausen, Germany

E-mail: e.a.dubovyk@gmail.com, janusz.gluza@us.edu.pl, tordriemann@gmail.com

**Abstract.** We introduce the Mellin-Barnes representation of general Feynman integrals and discuss their evaluation. The Mathematica package AMBRE has been recently extended in order to cover consistently non-planar Feynman integrals with two loops. Prospects for the near future are outlined. This write-up is an introduction to new results which have also been presented elsewhere.

## 1. Introduction

The evaluation of Feynman integrals is a central numerical problem of perturbative quantum field theory and is not solved in generality beyond the one-loop case. One has to consider arbitrary $L$-loop integrals $G(X)$ with loop momenta $k_l$ , with $E$ external legs with momenta $p_e$, and with $N$ internal lines with masses $m_i$ and propagators $1/D_i$,

$$G(X) = \frac{1}{(i\pi^{d/2})^L} \int \frac{d^dk_1 \dots d^dk_L \, X(k_1, \dots, k_L)}{D_1^{n_1} \dots D_i^{n_i} \dots D_N^{n_N}}, \tag{1}$$

with $d = 4 - 2\epsilon$, $D_i = q_i^2 - m_i^2 = [\sum\limits_{l=1}^{L} c_i^l k_l + \sum\limits_{e=1}^{E} d_i^e p_e] - m_i^2$, and some tensors $X(k_1, \dots, k_L)$ in the loop momenta. For several reasons, one is interested in analytical solutions either:

- Analytical results are well suited for the exact cancellation of certain intermediate terms, arising either from the renormalization procedure, from regularization, or just from the organization of the calculation.
- Analytical results may stabilize the numerics.
- With analytical results, analytical continuations of representations of Feynman integrals into regions of physical interest may be performed. Typically, it is easier to determine a Feynman integral in the Euclidean region, but often it is needed in the Minkowskian.

There are quite different approaches to tackle multi-loop Feynman integrals. Without aiming at completeness, we like to mention:

- Direct evaluation of a Feynman parameter representation (see e.g. the seminal articles [1, 2], and for later improvements [3, 4, 5] and references therein)

- Solving systems of differential or difference equations (see e.g. [6, 7, 8] and refs. therein)
- Expansion by regions (see e.g. [9] and refs. therein)
- Sector decomposition (see e.g. [10, 11, 12, 13, 14, 15, 16, 17, 18, 19] and refs. therein)
- Mellin-Barnes (MB-) representations (see e.g. [20, 21, 22, 23, 24, 25, 26] and refs. therein)

For an overview of most recent developments we recommend [27]. We are concentrating here on the Mellin-Barnes approach. Quite some work has been devoted to an automation of it, see [28, 29, 30, 31, 32, 33, 34] and software at the webpage http://projects.hepforge.org/mbtools. Nevertheless, the formalism is much less developed than e.g. that with differential equations, where algorithms have been worked out for solving them in terms of certain classes of functions. An analogue idea would be here using Cauchy's theorem in order to derive from a Mellin-Barnes representation multiple sums of residues, and then to sum up these sums analytically [35]. How this might work with the software of the RISC (Linz) group [36, 37, 38, 39, 40, 41, 42, 43, 44] around the packages `SIGMA` [45, 46], `EvaluateMultiSums`, and `SumProduction` [47, 48, 49] is indicated in [50]. For not too involved classes of functions, one may try to apply the packages `SUMMER` (http://www.nikhef.nl/~t68/) or `XSUMMER` [51]. For automation, several problems have to be solved:

- Determine if the topology is planar or non-planar – Mathematica package `PlanarityTest` [34], see http://prac.us.edu.pl/~gluza/ambre/planarity/.
- Construct the appropriate Mellin-Barnes representations – Mathematica package `AMBRE` [29, 30, 31, 50], see http://prac.us.edu.pl/~gluza/ambre/. Here, also the Mathematica packages `MB` [28] and `MBresolve` [32] are integrated – for the analytic continuation of Mellin-Barnes integrals in $\epsilon$; as well as `MBasymptotics` (2005, http://projects.hepforge.org/mbtools/) – for the parametric expansions of Mellin-Barnes integrals and `barnesroutines` by D. Kosower (2008, http://projects.hepforge.org/mbtools/) – for the automatic application of the first and second Barnes lemmas.
- Change the MB-integrals into multiple nested sums – Mathematica package `MBsums`, under development by M. Ochman (DESY) et al., see [50].
- Finally, try to perform the multiple sums analytically, see [50, 52].
- Alternatively, a purely numerical evaluation of the multi-dimensional Mellin-Barnes integral may be envisaged. Here it is wishful to cover not only Euclidean cases, but also the Minkowskian kinematics; this has not been studied so far.

Of course, a solution of the general case is not to be expected. The limitations of the method have several reasons; we mention here:

- The number of loops;
- The number of different scales due to internal masses and the kinematics;
- The number of external legs;
- A planar or non-planar topology.

A Feynman parameter integral with $N$ internal legs has, essentially, a dimensionality $N-1$. For the corresponding MB-integral, the dimensionality will be different, and for complex problems it often will come out much higher.

## 2. Mellin-Barnes representations

For the Feynman integral (1) one may derive the following Feynman parameter integral:

$$G(X) = \frac{(-1)^{N_\nu}\Gamma\left(N_\nu - \frac{d}{2}L\right)}{\prod_{i=1}^{N}\Gamma(n_i)} \int \prod_{j=1}^{N} dx_j \; x_j^{n_j-1}\delta(1-\sum_{i=1}^{N}x_i)\frac{U(x)^{N_\nu-d(L+1)/2}}{F(x)^{N_\nu-dL/2}} \tag{2}$$

Journal of Physics: Conference Series **608** (2015) 012070 doi:10.1088/1742-6596/608/1/012070

The functions $U$ and $F$ are called graph or Symanzik polynomials.

The repeated application of the Mellin-Barnes representation [20]

$$\frac{1}{(A+B)^{\lambda}} = \frac{1}{\Gamma(\lambda)}\frac{1}{2\pi i}\int_{-i\infty}^{+i\infty} dz\Gamma(\lambda+z)\Gamma(-z)\frac{B^z}{A^{\lambda+z}} \quad (3)$$

to the Symanzik polynomials $U$ and $F$ allows to replace the sums of monomials in their definitions (here: $A(x)$ and $B(x)$) by products of these monomials and $\Gamma$-functions and, subsequently, to perform the $x$-integrations. The integration path separates poles of $\Gamma[\lambda+z]$ and $\Gamma[-z]$. `AMBRE 1.1, 1.2, 2.0` perform so-called loop-by-loop integrations with $U = 1$ and are efficient enough for many applications (figure 1, left). For the massless, on-shell non-planar diagram of figure 1, right, which was first solved in [24], one would have to treat:

$$\begin{aligned} U(x) &= + x[1]x[2] + x[1]x[4] + x[2]x[4] + x[1]x[5] + x[2]x[5] + x[2]x[6] \quad (4)\\ &\quad + x[4]x[6] + x[5]x[6] + x[1]x[7] + x[4]x[7] + x[5]x[7] + x[6]x[7], \\ F(x) &= -s\ x[1]x[2]x[5] - s\ x[1]x[3]x[5] - s\ x[2]x[3]x[5] - u\ x[2]x[4]x[6] \quad (5)\\ &\quad - s\ x[3]x[5]x[6] - t\ x[1]x[4]x[7] - s\ x[3]x[5]x[7] - s\ x[3]x[6]x[7]. \end{aligned}$$

Here it becomes operational to apply the Cheng-Wu theorem [53, 54] which states that (2) holds also with a modified delta function $\delta(1-\sum_{i\in\Omega} x_i)$ where $\Omega$ is an arbitrary subset of the lines $1,\ldots,N$, when the integration over the rest of the variables, i.e. for $i \notin \Omega$, is extended to the integration from 0 to $\infty$. With `AMBRE 3.0`, non-planar Feynman integrals with two loops may be efficiently represented by a direct approach with use of the Cheng-Wu theorem. Planar integrals are treated by the loop-by-loop ansatz with the earlier `AMBRE` versions. For the example of the *massless non-planar double box* mentioned, `AMBRE 3.0` derives without manual interaction the 4-dimensional Mellin-Barnes representation of [24]. It might look more promising to apply here instead the loop-by-loop approach, but experience shows that the dimensionality becomes, without further interventions, higher. At the other hand, for the *massive non-planar double box*, the loop-by-loop approach gives the MB-presentation derived first in [25]. We mention here shortly that 3-loop integrals may be treated in a hybrid way by a loop-by-loop approach, subsequently using `AMBRE 1.2` and `AMBRE 3.0` [50]; see the example files `MB_hybrid_3loopNP_massless.nb, MB_hybrid_3loopNP_massless.m, out_MB_hybrid_3loopNP_massless` at the webpage http://prac.us.edu.pl/~gluza/ambre/.

Finally we would like to mention that the mathematica package `AMBRE 3.0` is released for public use at the webpage http://prac.us.edu.pl/~gluza/ambre/. There is no source file made publicly available so far, but it may be made available on request. We consider this to be appropriate; for closer information we refer to the webpages http://fh.desy.de/projekte/-gfitter01/Gfitter01.htm (March 2013) and http://zfitter-gfitter.desy.de/ (April 2014). For a collection of experts' views on proper distribution of scientific software in basic research, we would like to refer to [55, 56], a summary of a round table discussion at ACAT 2014.

## 3. Summary

We scetched essential features of the Mellin-Barnes approach to Feynman integrals and its implementation in `AMBRE`. In the new version `AMBRE 3.0` the Cheng-Wu theorem is implemented and the replacement of the Symanzik polynomials by MB-integrals is performed globally. The alternative loop-by-loop approach is implemented in `AMBRE 2.0` or older versions. Which of the approaches is more appropriate for a given problem has to be investigated.

A treatment of non-planar three-loop integrals deserves the combined application of `AMBRE 1.2` and of `AMBRE 3.0` in a hybrid approach. This will be improved in the nearest future. The *analytical* summation of series of MB-residues is under study. Concerning the alternative to analytical summation, namely a *numerical* evaluation of the MB-integrals, we are restricted with `AMBRE` to the Euclidean kinematics so far. The Minkowskian numerics is on our to-do list.

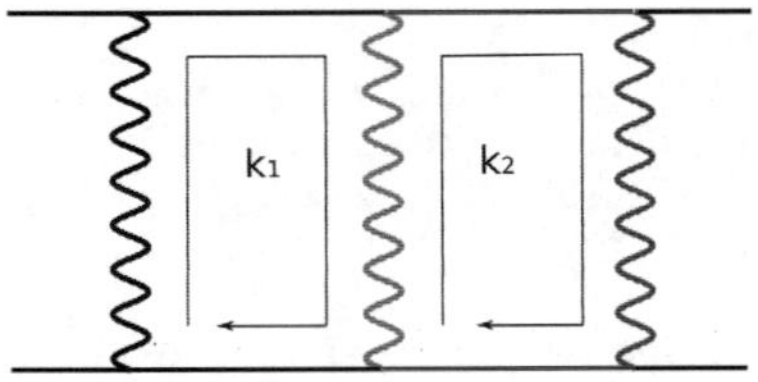


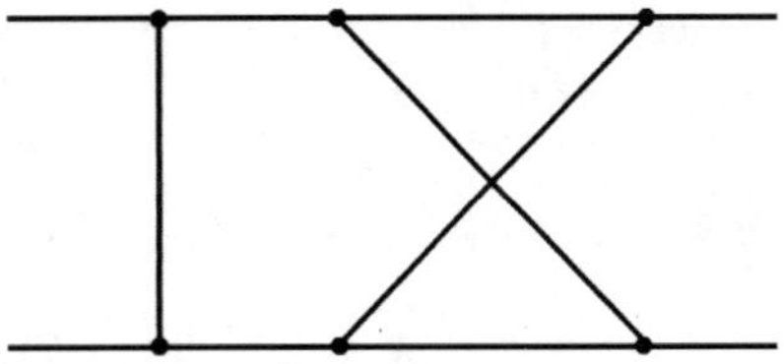

**Figure 1.** Left: The planar double box with loop momenta as used for the loop-by-loop representation. Right: The non-planar double box.

## Acknowledgements

Work is supported in part by Sonderforschungsbereich/Transregio SFB/TRR 9 "Computergestützte Theoretische Teilchenphysik" of Deutsche Forschungsgemeinschaft (DFG), by the Polish National Center of Science (NCN) under the Grant Agreement number DEC-2013/11/B/ST2/04023, and by the European Initial Training Network LHCPHENOnet PITN-GA-2010-264564.

## References

[1] 't Hooft G and Veltman M 1979 *Nucl. Phys.* **B153** 365–401
[2] Passarino G and Veltman M 1979 *Nucl. Phys.* **B160** 151
[3] Fleischer J and Riemann T 2011 *Phys. Rev.* **D83** 073004 (*Preprint* 1009.4436)
[4] Fleischer J and Riemann T 2011 *Phys. Lett.* **B701** 646–653 (*Preprint* 1104.4067)
[5] Fleischer J and Riemann T 2012 *Phys. Lett.* **B707** 375–380 (*Preprint* 1111.5821)
[6] Kotikov A 1991 *Phys. Lett.* **B254** 158–164
[7] Remiddi E 1997 *Nuovo Cim.* **A110** 1435–1452 (*Preprint* hep-th/9711188)
[8] Henn J M 2013 *Phys. Rev. Lett.* **110** 251601 (*Preprint* 1304.1806)
[9] Beneke M and Smirnov V A 1998 *Nucl. Phys.* **B522** 321–344 (*Preprint* hep-ph/9711391)
[10] Hepp K 1966 *Commun. Math. Phys.* **2** 301–326
[11] Roth M and Denner A 1996 *Nucl. Phys.* **B479** 495–514 (*Preprint* hep-ph/9605420)
[12] Binoth T and Heinrich G 2004 *Nucl. Phys.* **B680** 375–388 (*Preprint* hep-ph/0305234)
[13] Denner A and Pozzorini S 2005 *Nucl. Phys.* **B717** 48–85 (*Preprint* hep-ph/0408068)
[14] Bogner C and Weinzierl S 2008 *Comput. Phys. Commun.* **178** 596–610 (*Preprint* 0709.4092)
[15] Heinrich G 2008 *Int. J. Mod. Phys.* **A23** 1457–1486 (*Preprint* 0803.4177)
[16] Smirnov A and Tentyukov M 2009 *Comput. Phys. Commun.* **180** 735–746 (*Preprint* 0807.4129)
[17] Smirnov A and Smirnov V 2009 *JHEP* **0905** 004 (*Preprint* 0812.4700)
[18] Carter J and Heinrich G 2011 *Comput. Phys. Commun.* **182** 1566–1581 (*Preprint* 1011.5493)
[19] Borowka S and Heinrich G 2014 (*Preprint* 1411.0994)
[20] Barnes E W 1908 *Proc. Lond. Math. Soc. (2)* **6** 141–177
[21] Usyukina N 1975 *Teor. Mat. Fiz.* **22** 300–306
[22] Boos E and Davydychev A I 1991 *Theor. Math. Phys.* **89** 1052–1063
[23] Smirnov V A 1999 *Phys. Lett.* **B460** 397–404 (*Preprint* hep-ph/9905323)
[24] Tausk J 1999 *Phys. Lett.* **B469** 225–234 (*Preprint* hep-ph/9909506)
[25] Heinrich G and Smirnov V 2004 *Phys. Lett.* **B598** 55–66 (*Preprint* hep-ph/0406053)
[26] Czakon M, Gluza J and Riemann T 2006 *Nucl. Phys.* **B751** 1–17 (*Preprint* hep-ph/0604101)
[27] Blümlein J, Marquard P and Riemann T (eds) 2014 *PoS* **LL2014** Proceedings of the 12th DESY Workshop on Elementary Particle Theory, April 27 – May, 2, 2014, Weimar, Germany
[28] Czakon M 2006 *Comput. Phys. Commun.* **175** 559–571 (*Preprint* hep-ph/0511200)
[29] Gluza J, Kajda K and Riemann T 2007 *Comput. Phys. Commun.* **177** 879–893 (*Preprint* 0704.2423)
[30] Gluza J, Kajda K, Riemann T and Yundin V 2010 *Nucl. Phys. Proc. Suppl.* **205-206** 147–151 (*Preprint* 1006.4728)
[31] Gluza J, Kajda K, Riemann T and Yundin V 2011 *Eur. Phys. J.* **C71** 1516 (*Preprint* 1010.1667)
[32] Smirnov A and Smirnov V 2009 *Eur. Phys. J.* **C62** 445–449 (*Preprint* 0901.0386)
[33] Jantzen B 2013 *J. Math. Phys.* **54** 012304 (*Preprint* 1211.2637)
[34] Bielas K, Dubovyk I, Gluza J and Riemann T 2013 *Acta Phys. Polon.* **B44** 2249–2255 (*Preprint* 1312.5603)

Journal of Physics: Conference Series **608** (2015) 012070 doi:10.1088/1742-6596/608/1/012070


[35] Gluza J and Riemann T *Simple Feynman diagrams and simple sums* Talk held at RISC - DESY *Workshop on Advanced Summation Techniques and their Applications in Quantum Field Theory* on the occasion of the 5th year jubilee of the RISC-DESY cooperation, May 7-8, 2012, RISC Institute, Castle of Hagenberg, Linz, Austria. Conference link: http://www.risc.jku.at/conferences/RISCDESY12/, talk link: http://www-zeuthen.desy.de/~riemann/Talks/riemann-Linz-2012-05.pdf

[36] Karr M 1981 *J. ACM* **28** 305–350

[37] Schneider C 2001 *Symbolic Summation in Difference Fields* Tech. Rep. 01-17 RISC-Linz, J. Kepler University phD Thesis, http://www.risc.jku.at/publications/download/risc_3017/SymbSumTHESIS.pdf

[38] Schneider C 2005 *J. Differ. Equations Appl.* **11** 799–821

[39] Schneider C 2007 *J. Algebra Appl.* **6** 415–441

[40] Schneider C 2008 *J. Symbolic Comput.* **43** 611–644 (*Preprint* 0808.2543)

[41] Schneider C 2010 *Appl. Algebra Engrg. Comm. Comput.* **21** 1–32

[42] Schneider C 2010 *Motives, Quantum Field Theory, and Pseudodifferential Operators* (*Clay Mathematics Proceedings* vol 12) ed Carey A, Ellwood D, Paycha S and Rosenberg S (Amer. Math. Soc) pp 285–308, http://www.risc.jku.at/publications/download/risc_3806/DepthSequ.pdf

[43] Schneider C 2010 *Ann. Comb.* **14** 533–552 (*Preprint* 0808.2596)

[44] Schneider C 2014 *Computer Algebra and Polynomials* Lecture Notes in Computer Science (LNCS), to appear ed J Guitierrez J Schicho M W (Springer) (*Preprint* 1307.7887)

[45] Schneider C 2007 *Sém. Lothar. Combin.* **56** 1–36

[46] Schneider C 2013 *Computer Algebra in Quantum Field Theory: Integration, Summation and Special Functions* Texts and Monographs in Symbolic Computation ed Schneider C and Blümlein J (Springer) pp 325–360 (*Preprint* 1304.4134)

[47] Ablinger J, Blumlein J, Klein S and Schneider C 2010 *Nucl. Phys. Proc. Suppl.* **205-206** 110–115 (*Preprint* 1006.4797)

[48] Blumlein J, Hasselhuhn A and Schneider C 2011 *PoS* **RADCOR2011** 032 (*Preprint* 1202.4303)

[49] Schneider C 2014 *J. Phys. Conf. Ser.* **523** 012037 (*Preprint* 1310.0160)

[50] Blümlein J, Dubovyk I, Gluza J, Ochman M, Raab C G *et al.* 2014 *PoS* **LL2014** 052 (*Preprint* 1407.7832)

[51] Moch S and Uwer P 2006 *Comput. Phys. Commun.* **174** 759–770 (*Preprint* math-ph/0508008)

[52] Ablinger J, Blümlein J, Raab C G and Schneider C 2014 *PoS* **LL2014** 020 (*Preprint* 1407.4721)

[53] Nakanishi N 1961 *Prog. Theor. Phys. Supplement* **18** 1

[54] Nakanishi N 1971 *Graph Theory and Feynman Integrals* (Gordon and Breach)

[55] Carminati F, Perret-Gallix D and Riemann T 2014 *J. Phys. Conf. Ser.* **523** 012066

[56] Carminati F, Perret-Gallix D and Riemann T 2014 (*Preprint* 1407.0540)

# Automatic numerical integration methods for Feynman integrals through 3-loop

**E de Doncker[1], F Yuasa[2], K Kato[3], T Ishikawa[2] and O Olagbemi[1]**

[1] Department of Computer Science, Western Michigan University, Kalamazoo MI 49008, U. S. A.
[2] High Energy Accelerator Research Organization (KEK), Oho 1-1, Tsukuba, Ibaraki, 305-0801, Japan
[3] Department of Physics, Kogakuin University, Shinjuku, Tokyo 163-8677, Japan

E-mail:
elise.dedoncker@wmich.edu, fukuko.yuasa@kek.jp, kato@cc.kogakuin.ac.jp, tadashi.ishikawa@kek.jp, omofolakunmiel.olagbemi@wmich.edu

**Abstract.** We give numerical integration results for Feynman loop diagrams through 3-loop such as those covered by Laporta [1]. The methods are based on automatic adaptive integration, using iterated integration and extrapolation with programs from the QUADPACK package, or multivariate techniques from the ParInt package. The Dqags algorithm from QUADPACK accommodates boundary singularities of fairly general types. PARINT is a package for multivariate integration layered over MPI (Message Passing Interface), which runs on clusters and incorporates advanced parallel/distributed techniques such as load balancing among processes that may be distributed over a network of nodes. Results are included for 3-loop self-energy diagrams without IR (infra-red) or UV (ultra-violet) singularities. A procedure based on iterated integration and extrapolation yields a novel method of numerical regularization for integrals with UV terms, and is applied to a set of 2-loop self-energy diagrams with UV singularities.

## 1. Introduction

The techniques in this paper are based on *automatic integration*, which is a black-box approach to generate an approximation $Q(f)$ to an integral $If = \int_{\mathcal{D}} f(\vec{x})\, d\vec{x}$ and an absolute error estimate $\mathcal{E}f$, in order to satisfy a specified accuracy requirement for the actual error. We use an accuracy requirement of the form

$$|\,Qf - If\,| \;\leq\; \mathcal{E}f \;\leq\; \max\{\,t_a\,,\,t_r\,|\,If\,|\,\} \tag{1}$$

for a given integrand function $f$, region $\mathcal{D}$ and (absolute/relative) error tolerances $t_a$ and $t_r$, respectively. In order to achieve Eq (1), the actual error should not exceed the error estimate $\mathcal{E}f$, and the error estimate should not exceed the weakest of the absolute and relative error tolerances (indicated by the maximum taken on the right of Eq (1)). This type of accuracy requirement is based on [2] and used extensively in QUADPACK [3].

Known methods for parallelization of these procedures include:
(i) Parallelization on the rule or points level: typically in *non-adaptive* algorithms, e.g., for *Monte-Carlo (MC)* algorithms and composite rules using *grid* or *lattice* points. Then in $If = \int_{\mathcal{D}} f \approx \sum_k w_k f(\vec{x}_k)$ the function evaluations $f(\vec{x}_k)$ are performed in parallel.
(ii) Parallelization on the region level: in *adaptive* (region-partitioning) methods. These lead to task pool strategies, which may benefit from load balancing on distributed memory systems; or maintain a shared priority queue on shared memory systems.

(iii) We added multi-threading to iterated integration [4, 5, 6]: the inner integrals are independent and computed in parallel, e.g., over a subregion $\mathcal{S} = \mathcal{D}_1 \times \mathcal{D}_2$ (with inner region $\mathcal{D}_2$) consider $\int_{\mathcal{S}} F(\vec{x})dx \approx \sum_k w_k F(\vec{x}_k)$, with $F(\vec{x}_k) = \int_{\mathcal{D}_2} f(\vec{x}_k, \vec{y})d\vec{y}$. The inner integrations can be performed adaptively, which we applied in iterated versions of the 1D QUADPACK program DQAGS [3, 7].

We apply numerical extrapolation to integrals with an asymptotic expansion in the dimensional regularization parameter $\varepsilon$, of the form

$$S(\varepsilon) \sim \sum_{k \geq K} C_k \varphi_k(\varepsilon) \qquad \text{as } \varepsilon \to 0. \tag{2}$$

For example, the $\varphi_k(\varepsilon)$ functions may be integer powers of $\varepsilon$, $\varphi_k(\varepsilon) = \varepsilon^k$. Then for finite integrals $K = 0$ and the integral is represented by $C_0$. Linear extrapolation can be applied when the $\varphi_k(\varepsilon)$ functions are known. In that case, $S(\varepsilon)$ is approximated for decreasing values of $\varepsilon = \varepsilon_\ell$, and Eq (2) is truncated after $2, 3, \ldots, \nu$ terms to form linear systems of increasing size in the $C_k$ variables.

As the integral approximation generally becomes harder with smaller $\varepsilon$, we use slowly decreasing sequences $\{\varepsilon_\ell\}$, such as a geometric sequence with base $1/1.2$. Another sequence of interest is based on the Bulirsch sequence $1, 2, 3, 4, 6, 8, 12, 16, 24, \ldots$, (see [8]); we employ $\{1/b_j\}_{j \geq J}$, from a starting index $J$ in the Bulirsch sequence. We resort to non-linear extrapolation when the structure of the asymptotic expansion is not known. In previous work we have made ample use of the $\epsilon$-algorithm [9, 10], which can be applied with geometric sequences of $\varepsilon$.

Section 2 covers some background and notations for multi-loop integrals, and presents numerical results obtained with the PARINT package [11] for a set of 3-loop self-energy Feynman diagrams. These do not exhibit IR or UV singularities. Integrals with UV singularities are specified in Section 3, with their asymptotic expansions in the dimensional regularization parameter $\varepsilon$. In Section 3 we describe a novel method of numerical regularization for integrals with UV singularities, based on iterated integration and linear extrapolation, which is applied to a set of 1- and 2-loop self-energy diagrams. Section 4 shows numerical results of the regularization procedure.

## 2. Feynman loop integrals

Higher order corrections are required for accurate theoretical predictions of the cross-section for particle interactions. Loop diagrams are taken into account, leading to the evaluation of loop integrals. The derivation of a closed analytic form is sometimes hard for the higher order loop integrals with arbitrary internal masses and external momenta. Thus we resort to numerical calculations.

A $L$-loop integral with $N$ internal lines can be represented in Feynman parameter space by

$$\mathcal{I} = \frac{\Gamma\left(N - \frac{nL}{2}\right)}{(4\pi)^{nL/2}} (-1)^N \int_0^1 \prod_{r=1}^{N} dx_r \, \delta(1 - \sum x_r) \frac{C^{N-n(L+1)/2}}{(D - i\varrho C)^{N-nL/2}}, \tag{3}$$

where $C$ and $D$ are polynomials determined by the topology of the corresponding diagram and physical parameters ($C = 1$ for 1-loop ($L = 1$) integrals). The integration in Eq (3) is taken over the $N$-dimensional unit cube. However, as a result of the $\delta$-function one of the $x_r$ can be expressed in terms of the other ones, which reduces the integral dimension to $N-1$ and the domain to the $(N-1)$-dimensional unit simplex. In the absence of IR and UV singularities, $n = 4$. For dimensional regularization in case of IR singularities we set $n = 4 + 2\varepsilon$ (cf, [12]); and for UV singularities $n = 4 - 2\varepsilon$. We apply the regularization by a numerical extrapolation as $\varepsilon \to 0$.

The term $i\varrho C$ prevents the integral from diverging if the denominator vanishes in the interior of the domain. Results using iterated integration with QUADPACK programs and extrapolation were given in [13, 7]. However with the parameters of Laporta [1], which we use in this paper, the denominator does not vanish inside the integration domain. In this case we can take $\rho = 0$ in the integrand of (3).

Integral approximations obtained with PARINT for 2-loop double-triangle ($N = 5$), tetragon-triangle ($N = 6$), pentagon-triangle ($N = 7$), ladder and crossed ladder ($N = 7$) were presented in [6]. A set of

Journal of Physics: Conference Series **608** (2015) 012071 doi:10.1088/1742-6596/608/1/012071

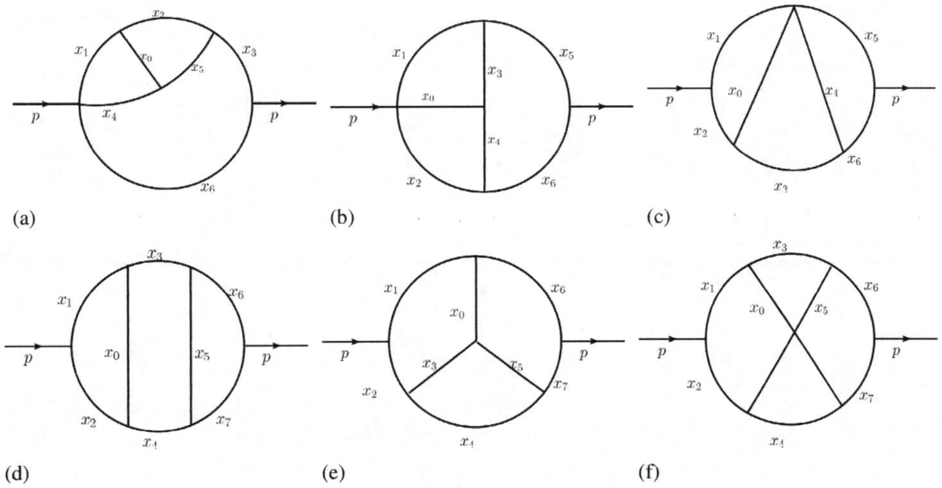


**Figure 1.** 3-loop diagrams (a) $N = 7$ (Laporta[1] Fig 2(q)), (b) $N = 7$ (Laporta[1] Fig 2(r)), (c) $N = 7$ (Laporta[1] Fig 2(s)), (d) $N = 8$ (Laporta[1] Fig 2(t)), (e) $N = 8$ (Laporta[1] Fig 2(u), (f) $N = 8$ (Laporta[1] Fig 2(v))

**Table 1.** Parallel performance of PARINT (on MPI) for 3-loop diagrams of Fig 1, abs. tolerance $t_a = 5 \times 10^{-10}$, and max. number of evaluations = 10B

| 3-loop diag. | $N$ | Result Laporta [1] | Result $p = 1$ | Result $p = 64$ | $T_1[s]$ | $T_{64}[s]$ | $S_{64}$ |
|---|---|---|---|---|---|---|---|
| Fig 1 (a) | 7 | 1.32644820827 | 1.326448206 | 1.32644819 | 902.7 | 15.8 | 57.1 |
| Fig 1 (b) | 7 | 1.34139924145 | 1.34139924147 | 1.3413992416 | 1026.2 | 14.4 | 71.3 |
| Fig 1 (c) | 7 | 2.00250004111 | 2.00250004113 | 2.0025000412 | 879.3 | 13.4 | 65.6 |
| Fig 1 (d) | 8 | 0.27960892328 | 0.2796089227 | 0.279608920 | 1019.7 | 15.9 | 64.1 |
| Fig 1 (e) | 8 | 0.18262723754 | 0.1826272372 | 0.1826272368 | 1018.3 | 15.8 | 64.4 |
| Fig 1 (f) | 8 | 0.14801330396 | 0.1480133036 | 0.1480133026 | 976.6 | 16.4 | 59.5 |

3-loop self-energy diagrams is given in Fig 1, with corresponding PARINT performance results in Table 1. In order to compare our integral approximations with Laporta's [1], we set all masses $m_r = 1$ and $s = 1$. Note that Laporta's method is based on the numerical solution of systems of difference equations. For our numerical integration we transform the unit simplex domain of Eq (3) to the $(N-1)$-dimensional unit cube, and apply an integration rule of polynomial degree of precision 9 (from [14]) over the individual subregions resulting in the adaptive partitioning. The approximations thus obtained are more accurate than those generated with the multivariate simplex rules in PARINT, without the transformation. We set the absolute tolerance to $t_a = 5 \times 10^{-10}$ and the maximum number of integrand evaluations to $10B = 10^{10}$ (which was reached in producing the results of Table 1).

The results in Table 1 are given for $p = 1$ and for $p = 64$ MPI processes. $T_1$ is the time with one process and $T_{64}$ is the parallel time on our cluster with $p = 64$ processes (distributed over four 16-core, 2.6 GHz compute nodes). The speedup $S_{64} = T_1/T_{64}$ is (sequential time)/(parallel time). Note that superlinear speedups ($S_{64}$) are obtained in some cases, where the speedup exceeds the number of processes. This is partially due to the fact that the timing is done within PARINT, after the processes are started. The function evaluations are distributed over all the processes. Furthermore, the load balancing option is turned on, as well as letting the controller process act as a worker and participate in the region partitioning. However the adaptive partioning reaches somewhat more accuracy sequentially. Each

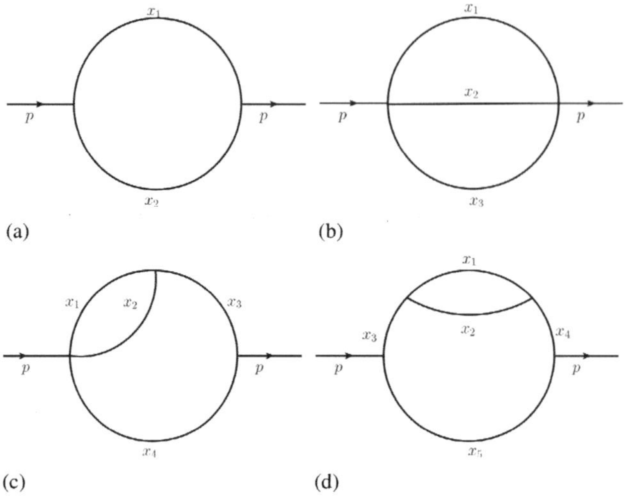


**Figure 2.** UV diagrams: (a) 1-loop self energy $N = 2$ (Laporta [1] Fig 2(a)), (b) *sunrise-sunset* $N = 3$(Laporta [1] Fig 2(b)), (c) *"lemon"* $N = 4$ (Laporta [1] Fig 2(c)), (d) *"half-boiled egg"* (Kato [16] Fig 2)

process has its own priority queue, keyed with the absolute error estimates over their region. Thus unnecessary work may be done which increases with the number of processes. The presentation of this paper [15] contains results for $t_a = 5 \times 10^{-8}$ and a maximum of 5B evaluations, which are run in about half the time.

## 3. Ultra-violet (UV) singularities and asymptotics

### *3.1. Integrals*

By replacing $C = U$ and $D = UV$ in Eq (3), the general integral is written as

$$\mathcal{I} = \frac{\Gamma\left(N - \frac{nL}{2}\right)}{(4\pi)^{nL/2}}(-1)^N \int_0^1 \prod_{r=1}^{N} dx_r \, \frac{\delta(1-\sum x_r)}{U^{n/2}(V-i\varrho)^{N-nL/2}}. \tag{4}$$

In the form of Eq (4), IR divergence occurs through a singularity arising when $V$ vanishes at the boundaries of the domain. This problem can be addressed by dimensional regularization with $n = 4 + 2\varepsilon$, which we implemented numerically in [13, 12, 7] by an extrapolation as $\varepsilon \to 0$ ($\varepsilon > 0$).

UV divergence occurs when $U$ vanishes at the boundaries. The $\Gamma$-function in (4), if it is divergent, may contribute to UV divergence. In this paper we treat UV divergence by a dimensional regularization with $n = 4 - 2\varepsilon$, implemented by a numerical extrapolation as $\varepsilon \to 0$ after an iterated integration with DQAGS from QUADPACK [3, 7] (to handle the boundary singularities).

Fig 2 depicts a 1-loop and three 2-loop self-energy diagrams with $N = 2, 3, 4$ and 5 internal lines. We refer to the 2-loop self energy diagrams (b), (c) and (d) as the *sunrise-sunset*, *lemon* and *half-boiled egg* diagram, respectively. Analytic results for the integrals have been derived by many authors. We use the formulas of Kato [16] for the functions $U$ and $V$ in (4). Let $V = \sum_r x_r m_r^2 - \frac{s}{U}W,\ \ s = q^2 = 1,$ all masses $m_r = 1$ and $n = 4 - 2\varepsilon$.

– In the case of the *1-loop self-energy diagram* (Fig 2(a)), with $L = 1, N = 2$, the integrand has $U = 1$ and $W = x_1 x_2 = x_1(1 - x_1)$, so $V = x_1^2 - x_1 + 1$; we denote

$$I_1 = \Gamma(\varepsilon) \int_0^1 \frac{1}{(x_1^2 - x_1 + 1)^{\varepsilon}} \, dx_1. \tag{5}$$

– For the *sunrise-sunset diagram* (Fig 2(b)) we have $L = 2, N = 3$ and

$$I_s = (-1)\Gamma(-1+2\varepsilon) \int_0^1 dx_1 dx_2 dx_3\, \delta(1 - \sum_r x_r)\, \frac{(V - i\rho)^{1-2\varepsilon}}{U^{2-\varepsilon}} \tag{6}$$

with $U = x_1x_2 + x_2x_3 + x_3x_1$, $W = x_1x_2x_3$.
– The integral for the *lemon diagram* (Fig 2(c)) with $L = 2, N = 4$ is

$$I_l = \Gamma(2\varepsilon) \int_0^1 dx_1 dx_2 dx_3 dx_4\, \delta(1 - \sum_r x_r)\, \frac{1}{U^{2-\varepsilon}\,(V - i\rho)^{2\varepsilon}} \tag{7}$$

where $U = x_{12}x_{34} + x_1x_2$, $W = x_4(x_1x_2 + x_2x_3 + x_3x_1)$.
– The *half-boiled egg diagram* (Fig 2(d)) with $L = 2, N = 5$ gives rise to

$$I_h = (-1)\Gamma(1+2\varepsilon) \int_0^1 dx_1 dx_2 dx_3 dx_4 dx_5\, \delta(1 - \sum_r x_r)\, \frac{1}{U^{2-\varepsilon}\,(V - i\rho)^{1+2\varepsilon}} \tag{8}$$

where $U = x_{12}x_{345} + x_1x_2$, $W = x_5(x_{12}x_{34} + x_1x_2)$.

### 3.2. Asymptotic expansions for UV integrals

The integrals in Eqs (5)-(8) are expanded with respect to the dimensional regularization parameter $\varepsilon$. The expansions are of the form of Eq (2),

$$S(\varepsilon) \sim \sum_{k \ge K} C_k \varepsilon^k \quad \text{as } \varepsilon \to 0, \tag{9}$$

and we use linear extrapolation to approximate the coefficients of the leading terms. For the *1-loop, sunrise-sunset, and lemon* integrals we expand $S(\varepsilon) = I(\varepsilon)\Gamma(1+\varepsilon)^{-L}$ where the integral $I$ is $I_1, I_s$ and $I_l$, respectively. Section 4 compares corresponding extrapolation results with those given by Laporta [1]:

$$I_1(\varepsilon)\,\Gamma(1+\varepsilon)^{-1} \sim \sum_{k\ge -1} C_k\varepsilon^k = \varepsilon^{-1} + 0.186200635766 + 0.021156303568\,\varepsilon + 0.00172674535\,\varepsilon^2 \ldots \tag{10}$$

$$I_s(\varepsilon)\,\Gamma(1+\varepsilon)^{-2} \sim \sum_{k\ge -2} C_k\varepsilon^k = -1.5\,\varepsilon^{-2} - 4.25\,\varepsilon^{-1} - 7.375 - 17.22197253479\,\varepsilon \ldots \tag{11}$$

$$I_l(\varepsilon)\,\Gamma(1+\varepsilon)^{-2} \sim \sum_{k\ge -2} C_k\varepsilon^k = 0.5\,\varepsilon^{-2} + 0.6862006357658\,\varepsilon^{-1} - 0.6868398873414 + 1.486398391913\,\varepsilon \ldots \tag{12}$$

Note that the value of $K$ in (9) corresponds with the index of the first coefficient $C_K$ in the expansion. In that case we find that, if $K$ is replaced by $K-1$ for the extrapolation, then the first coefficient converges to $C_{K-1} = 0$.

The *half-boiled egg diagram* is not covered in [1]. In order to compare with the leading order terms of [16] we expand

$$\hat{I}_h(\varepsilon) = I_h(\varepsilon)/((-1)\Gamma(1+2\varepsilon)) \sim \sum_{k\ge -1} \hat{C}_k\varepsilon^k \tag{13}$$

with $I(\varepsilon)$ as in Eq (8). The analytic results in [16] are:

$$\hat{C}_{-1} = J_1 = \int_0^1 \frac{\rho'}{M_0^2 - sG_0}\, d\rho' \tag{14}$$

$$\hat{C}_0 = -\frac{3}{2}J_1 - 2J_2 + I_B \tag{15}$$

**Table 2.** Results UV *1-loop self-energy* integral (on Mac Pro, 2.6 GHz Intel Core i7, 16GB memory, OS X), err. tol. $t_r = 10^{-14}$, $T(s)$ = Time (elapsed user time) (s); $\varepsilon = 1/b_\ell$ (starting at 1/6), $E_r$ = integration estim. rel. error

| | INTEGRAL $I_1$ | | EXTRAPOLATION | | | |
|---|---|---|---|---|---|---|
| $b_\ell$ | $E_a$ | T(S) | RESULT $C_{-1}$ | RESULT $C_0$ | RESULT $C_1$ | RESULT $C_2$ |
| 6 | 5.0e-15 | 1.2e-5 | | | | |
| 8 | 5.0e-15 | 2.0e-6 | 0.99954860270457280 | 0.192483548522 | | |
| 12 | 5.0e-15 | 2.0e-6 | 1.00000307032894753 | 0.186121001781 | 0.021814445970 | |
| 16 | 5.0e-15 | 2.0e-6 | 0.99999998779988053 | 0.186201147537 | 0.021148619692 | 0.00177553674 |
| 24 | 5.0e-15 | 2.0e-6 | 1.00000000002646638 | 0.186200634020 | 0.021156346896 | 0.00172623915 |
| 32 | 5.0e-15 | 2.0e-6 | 0.99999999999996125 | 0.186200635770 | 0.021156303426 | 0.00172674805 |
| 48 | 5.0e-15 | 2.0e-6 | 0.99999999999999878 | 0.186200635766 | 0.021156303565 | 0.00172674537 |
| 64 | 5.0e-15 | 2.0e-6 | 1.00000000000000000 | 0.186200635766 | 0.021156303578 | 0.00172674499 |
| | *Eq* (10): | | 1.0 | 0.186200635766 | 0.021156303568 | 0.00172674535 |

with

$$J_2 = \int_0^1 \frac{\rho' \log(M_0^2 - sG_0)}{M_0^2 - sG_0}\, d\rho' \tag{16}$$

$$I_B = \int_0^1 d\rho \int_0^1 d\xi \int_0^1 d\rho' (1-\rho)^2 \rho' \frac{M_0^2 - F^2 M^2 - s(G_0 - FG)}{\rho F (FM^2 - sG)(M_0^2 - sG_0)} \tag{17}$$

where $s = 1$ and

$$\begin{aligned} &F = 1 - \rho + \rho\xi(1-\xi), \;\; F_0 = F(\rho = 0) = 1 \\ &G = (1-\rho)(1-\rho')((1-\rho)\rho' + \rho\xi(1-\xi)), \;\; G_0 = G(\rho = 0) = (1-\rho')\rho' \\ &M^2 = \sum_r x_r m_r^2, \; M_0^2 = M^2(\rho = 0) = \rho' m_3^2 + (1-\rho') m_5^2 \end{aligned} \tag{18}$$

The latter holds under the assumption $m_3 = m_4$. We further have $M_0^2 = M^2 = 1$ in (18) in view of $m_r = 1,\ 1 \le r \le 5$. Note that $\hat{C}_{-1} = C_{-1}$ in the expansion of $I_h = \sum_{k \ge -1} C_k \varepsilon^k$, but $\hat{C}_0 \ne C_0$.

## 4. Numerical extrapolation results for UV singularities

Tables 2 and 3 show the convergence of the extrapolation method for the integrals $I_1$ and $I_l$ of Eqs (5) and (7), respectively. These were run on a Mac Pro, 2.6 GHz Intel Core i7, with 16 GB memory, under OS X. The elapsed user time $T(s)$ (in seconds) is listed for each integration. The time for the extrapolation is negligible compared to that of the integration. We use a standard linear system solver to solve very small systems (of sizes $2 \times 2$ up to around $15 \times 15$ for the cases in this paper). As the sequence of the extrapolation parameter $\varepsilon_\ell$ we used $\{1/b_\ell\}$ where $\{b_\ell\}$ is the Bulirsch sequence [8] started at an early index. Alternatively a slowly decreasing geometric sequence could be used - we plan on further testing with different sequences.

The convergence results in Tables 2 and 3 show excellent agreement with the expansion coefficients in [1] (see Eqs (10) and (12)). Throughout the extrapolation we keep track of the difference with the previous result as a measure of convergence. Increases of the distance between successive extrapolation results are an indicator that the convergence is no longer improving and the process can be terminated.

For the *half-boiled egg integral* we can compute $J_1$ in the analytic expressions of Eq (14) and $J_2$ of Eq (16) using DQAGE from QUADPACK [3, 7], yielding $J_1 \approx 0.6045997880780727$ with absolute error estimate 1.6e-15, and $J_2 \approx -0.11708165598778085$ with absolute error estimate 1.3e-17 on Mac Pro. In view of boundary singularities we use iterated integration by DQAGS for $I_B$ of Eq (17) and obtain $I_B \approx 0.4970393699155826$ with outer absolute error estimate 1.22e-15 (note that this error estimate does not include contributions from the inner integral error estimates). Then, using Eq (14) and Eq (15) we find $\hat{C}_{-1} = J_1 \approx 0.604599788078210687$ and $\hat{C}_0 \approx -0.175697000225964850$ for the first two coefficients of the expansion in Eq (13). As shown in Table 4, good approximations are generated by linear extrapolation.

**Table 3.** Results UV *lemon* integral (on Mac Pro), err. tol. $t_r = 10^{-10}$ (outer), $5 \times 10^{-11}$ (inner two), $T(s)$ = Time (elapsed user time) (s); $\varepsilon = 1/b_\ell$ (starting at 0.25), $E_r$ = outer integration estim. rel. error

| Integral $I_l$ | | | Extrapolation | | | |
|---|---|---|---|---|---|---|
| $b_\ell$ | $E_r$ | T(s) | Result $C_{-2}$ | Result $C_{-1}$ | Result $C_0$ | Result $C_1$ |
| 4 | 3.5e-11 | 0.36 | | | | |
| 6 | 8.8e-11 | 0.34 | 0.5130221162587 | 0.52467607220 | | |
| 8 | 2.9e-12 | 0.34 | 0.5031467341833 | 0.62342989295 | -0.237009170 | |
| 12 | 3.4e-12 | 0.40 | 0.5004379328119 | 0.67218831764 | -0.518724512 | 0.5200899 |
| 16 | 1.5e-11 | 0.41 | 0.5000485801347 | 0.68386889795 | -0.643317369 | 1.0807577 |
| 24 | 4.7e-11 | 0.39 | 0.5000037328535 | 0.68593187289 | -0.679195194 | 1.3749558 |
| 32 | 4.1e-11 | 0.38 | 0.5000002195177 | 0.68617780639 | -0.685884585 | 1.4654594 |
| 48 | 1.4e-11 | 0.43 | 0.5000000087538 | 0.68619930431 | -0.686757991 | 1.4837301 |
| 64 | 1.3e-11 | 0.44 | 0.5000000002937 | 0.68620057333 | -0.686834471 | 1.4861463 |
| 96 | 3.2e-11 | 0.31 | 0.5000000000039 | 0.68620063534 | -0.686839872 | 1.4863967 |
| | *Eq* (12): | | 0.5 | 0.68620063577 | -0.686839887 | 1.4863984 |

**Table 4.** Results UV *half-boiled egg* integral (on Mac Pro), err. tol. $t_r = 10^{-12}$ (outer), $5 \times 10^{-13}$ (inner three), $T(s)$ = Time (elapsed user time) (s); $\varepsilon = 1/b_\ell$ (starting at 1.0), $E_r$ = outer integration estim. rel. error

| Integral $\hat{I}_h$ | | | Extrapolation | | | |
|---|---|---|---|---|---|---|
| $b_\ell$ | $E_r$ | T(s) | Result $\hat{C}_{-1}$ | Result $\hat{C}_0$ | Result $\hat{C}_1$ | Result $\hat{C}_2$ |
| 1 | 4.2e-13 | 7.3 | | | | |
| 2 | 3.9e-14 | 11.4 | 0.61217953237003 | -0.26893337928 | | |
| 3 | 2.6e-13 | 10.8 | 0.62192201629541 | -0.29816083105 | -0.019484967 | |
| 4 | 5.2e-13 | 10.3 | 0.60843456496759 | -0.21723612309 | -0.128876997 | 0.0809247 |
| 6 | 1.1e-13 | 14.9 | 0.60506615959771 | -0.18355206939 | -0.246771185 | 0.2493450 |
| 8 | 7.9e-13 | 9.2 | 0.60464393463841 | -0.17679647004 | -0.28688256 | 0.3591235 |
| 12 | 9.6e-13 | 12.5 | 0.60460287243867 | -0.17581097725 | -0.296039426 | 0.4010069 |
| 16 | 2.3e-13 | 19.4 | 0.60459996124784 | -0.17570617438 | -0.297527045 | 0.4117667 |
| 24 | 8.7e-13 | 19.1 | 0.60459979484875 | -0.17569752162 | -0.297707921 | 0.4137422 |
| 32 | 7.7e-13 | 24.9 | 0.60459978828848 | -0.17569702304 | -0.297723238 | 0.4139912 |
| 32 | 4.1e-13 | 33.2 | 0.60459978807821 | -0.17569700033 | -0.297724241 | 0.4140148 |
| 32 | 6.9e-13 | 20.7 | 0.60459978807543 | -0.17569699990 | -0.297724269 | 0.4140158 |
| | Eqs (14)-(15): | | 0.60459978807807 | -0.17569700023 | | |

## 5. Conclusions

This paper presents new results for 2-loop self-energy diagrams (with 2, 3, 4 and 5 internal lines) with UV terms. These are computed with iterated integration using DQAGS from QUADPACK, and linear extrapolation, which delivers a novel numerical method for dimensional regularization of UV singularities. DQAGS is successful at treating the boundary singularities. The extrapolation yields accurate approximations for the leading term coefficients of the asymptotic expansion in the regularization parameter. The results so far have been verified with expansions by Laporta [1] and Kato [16], and we plan on testing the procedure for more complex diagrams.

New results are further obtained with the adaptive multivariate integration code from the parallel/distributed PARINT package, for 3-loop self-energy diagrams without IR or UV singularities. The adaptive partitioning strategy is capable of dealing with higher dimensions than the iterated strategies. On the other hand, while it handles irregular integrand behavior to some extent, it cannot be expected to adequately partition higher-dimensional spaces in the vicinity of severe singularities. Future work on PARINT includes testing and incorporation of special summation methods such as Kahan summation [17, 18, 19] (in view of the large numbers of function evaluations that can be performed especially on distributed processors). Other work for both the iterated and standard multivariate integration will be on further efficient parallelizations of the integration work performed throughout the extrapolation sequence.

**Acknowledgments**
We acknowledge the support from the National Science Foundation under Award Number 1126438. This work is further supported by Grant-in-Aid for Scientific Research (24540292) of JSPS, and the Large Scale Simulation Program No. 13/14-13 of KEK.

## References

[1] Laporta S 2000 *Int. J. Mod. Phys. A* **15** 5087–5159 arXiv:hep-ph/0102033v1
[2] de Boor C 1971 *Mathematical Software* ed Rice J R (Academic Press, New York) pp 417–449
[3] Piessens R, de Doncker E, Überhuber C W and Kahaner D K 1983 *QUADPACK, A Subroutine Package for Automatic Integration* (*Springer Series in Computational Mathematics* vol 1) (Springer-Verlag)
[4] de Doncker E and Yuasa F 2013 *Journal of Physics: Conf. Series* **410** 012047 doi:10.1088/1742-6596/410/1/012047
[5] de Doncker E, Yuasa F and Assaf R 2013 *Journal of Physics: Conf. Ser.* **454** doi:10.1088/1742-6596/454/1/012082
[6] de Doncker E and Yuasa F 2014 *XV Adv. Comp. and Anal. Tech. in Phys. Res., Journal of Physics: Conference Series (ACAT 2013)* **523** doi:10.1088/1742-6596/523/1/012052
[7] de Doncker E, Fujimoto J, Hamaguchi N, Ishikawa T, Kurihara Y, Shimizu Y and Yuasa F 2011 *Journal of Computational Science (JoCS)* **3** 102–112 doi:10.1016/j.jocs.2011.06.003
[8] Bulirsch R 1964 *Numerische Mathematik* **6** 6–16
[9] Wynn P 1956 *Mathematical Tables and Aids to Computing* **10** 91–96
[10] Shanks D 1955 *J. Math. and Phys.* **34** 1–42
[11] de Doncker E, Kaugars K, Cucos L and Zanny R 2001 *Proc. of Computational Particle Physics Symposium (CPP 2001)* pp 110–119
[12] de Doncker E, Yuasa F and Kurihara Y 2012 *Journal of Physics: Conf. Ser.* **368**
[13] de Doncker E, Fujimoto J, Hamaguchi N, Ishikawa T, Kurihara Y, Ljucovic M, Shimizu Y and Yuasa F 2010 Extrapolation algorithms for infrared divergent integrals arXiv:hep-ph/1110.3587; PoS (CPP2010)011
[14] Berntsen J, Espelid T O and Genz A 1991 *ACM Trans. Math. Softw.* **17** 437–451
[15] de Doncker E, Yuasa F, Kato K, Ishikawa T and Olagbemi O Automatic numerical integration methods for Feynman integrals through 3-loop XV Adv. Comp. and Anal. Tech. in Phys. Res. (ACAT 2014), https://indico.cern.ch/event/258092/timetable/#all.detailed
[16] Kato K 2010 Note on 2-loop self-energy scalar integrals Tech. rep. Department of Physics, Kogakuin University, Japan Private communication
[17] Knuth D E 1998 *The Art of Computer Programming, Volume 2, Seminumerical Algorithms* 3rd ed (Addison Wesley) ISBN 0-201-89684-2
[18] Higham N J 2002 *Accuracy and Stability of Numerical Algorithms 2nd ed.* (SIAM) iSBN 978-0-898715-21-7
[19] de Doncker E, Kapenga J and Assaf R 2014 *Numerical Computations with GPUs* ed Kindratenko V (Springer) iSBN 978-3-319-06548-9

Journal of Physics: Conference Series **608** (2015) 012072 doi:10.1088/1742-6596/608/1/012072

# Event simulation for colliders – A basic overview

**Christian Reuschle**
Institute for Theoretical Physics, Karlsruhe Institute of Technology, Karlsruhe, Germany
E-mail: christian.reuschle@kit.edu

**Abstract.** In this article we will discuss the basic calculational concepts to simulate particle physics events at high energy colliders. We will mainly focus on the physics in hadron colliders and particularly on the simulation of the perturbative parts, where we will in turn focus on the next-to-leading order QCD corrections.

## 1. Introduction

The Standard Model of Particle Physics is very successful to describe the phenomena that we observe in high energy colliders [1]. In July 2012 the ATLAS and CMS experiments at CERN finally announced to have found the last missing piece of the Standard Model – the Higgs boson [2, 3]. Measurements of the properties of the new boson, in order to identify it with the Standard Model Higgs boson, and searches for new physics beyond the Standard Model are ongoing since. In order to compare those measurements to our theoretical concepts, precise predictions for the corresponding observables are needed. However, the situation at hadron colliders is complicated through a large background of QCD radiation, and sophisticated computational techniques, theoretically as well as experimentally, are needed all the more. On the theory side Monte Carlo programs are thereby the tool of choice, in order to combine all the necessary computational techniques. In this article we will briefly describe some of the concepts that are thereby used. Due to the mixed audience, we will constrain ourselves to a rather basic description. For a more thorough description we would like to point the reader to a list of references throughout this article, which are better suited to explain certain aspects in greater detail, and we would like to apologize in advance for any relevant reference that we may miss to mention.

## 2. Hadronic collisions

In this section we will sketch the physics behind general purpose event generators. Excellent general reviews on the following can be found in [4–6], to which we would like to point the reader for a more thorough description and further references if required.

The current theoretical picture of a hadronic collision is motivated by the properties of asymptotic freedom of QCD at high energy scales and confinement at low scales. Due to the property of asymptotic freedom we may take on the parton picture, i.e. we assume that we can resolve gluons and quarks inside the hadrons, with a resolution scale $\mu_F$, as depicted in Fig.1. Correspondingly, the total cross section for the scattering of two hadrons at a center-of-mass energy (CME) $\sqrt{s}$ can be factorized according to the following formula:

$$\sigma(s) = \sum_{X_\mathrm{h}} \sum_{a,b} \int dx_1 dx_2 \; f_a(x_1; \mu_F) f_b(x_2; \mu_F) \; \hat{\sigma}_{ab \to X_\mathrm{p}}(\hat{s}; \{a, b, X_\mathrm{p}\}; \mu_F) \otimes PS_{X_\mathrm{p}} \otimes H_{X_\mathrm{p} \to X_\mathrm{h}} \quad (1)$$

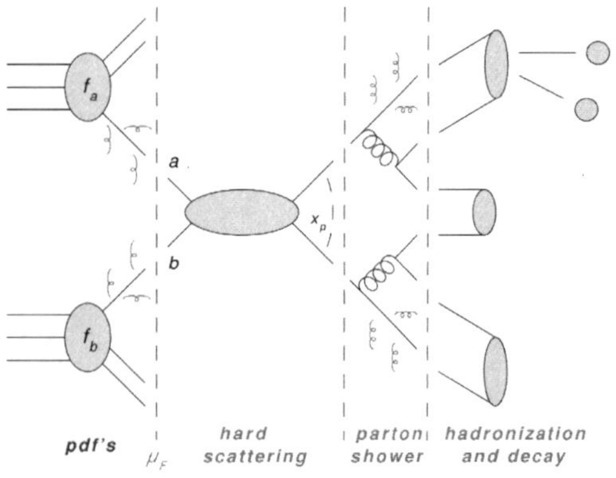


**Figure 1.** *The scattering of two hadrons with PDF $f_a(x_1, \mu_F)$ and $f_b(x_2, \mu_F)$. The picture portrays a "snapshot" of the evolution of the hadrons during the scattering process, where we pick an arbitrary scale $\mu_F$ to factorize off the actual hard interaction, which is described by the scattering between the partonic constituents a and b. The consecutive radiation of soft and collinear partons off the partons involved in the hard scattering is governed by a parton shower. The partons in the parton shower hadronize at a scale $\Lambda^2_{QCD}$, and may subsequently decay. Not depicted is the underlying event.*

Eq.1 consists of a non-perturbative and a perturbative part. In the non-perturbative part the parton density functions (PDF) $f_a(x_1; \mu_F)/f_b(x_2; \mu_F)$ describe the probabilities at a scale $\mu_F$ to "find" partons of type $a/b$ with longitudinal momentum fractions $x_1/x_2$ inside the hadrons 1/2. These probabilities cannot be calculated from first principles but can be determined from experiment. In the perturbative part, which can be completely calculated from first principles in perturbation theory, the hard (or partonic) scattering cross section $\hat{\sigma}_{ab \to X_p}(\hat{s}; \{a, b, X_p\}; \mu_F)$ describes the probability of two incoming partons of type $a$ and $b$ with a partonic CME (or hard scale) $\sqrt{\hat{s}} = \sqrt{x_1 x_2 s}$ to yield a partonic final state configuration $X_p$.

Due to the property of confinement, quarks and gluons cannot be observed separately in nature and exclusive hadronic final states $X_h$ have to be simulated: A parton shower evolves an inclusive partonic cross section into an exclusive partonic final state by simulating a cascade of soft/collinear QCD radiation, through consecutive soft and collinear parton splittings, in which each splitting product receives consecutively less energy. The parton shower evolution is determined perturbatively. At a certain scale $\Lambda_{QCD} \ll \sqrt{\hat{s}}$, approximately at the order of $\Lambda_{QCD} \sim \mathcal{O}(1\text{GeV})$, the partons in the parton shower hadronize, and may subsequently decay. The hadronization process cannot be determined from first principles, but needs to be modeled. In addition to the primary interaction, the hadron remnants, after extraction of the partons $a$ and $b$, can undergo further soft scatterings, so called multiple parton interactions (MPI), which form part of the accompanying underlying event and need to be modeled as well.

In order not to be overly sensitive to the details of such models, which usually dominate the theory uncertainties, we depend on observables that are insensitive to physics in the soft/collinear limit. Such infrared-safe observables do not change in the limit in which additional soft or collinear particles are imposed. An important class of infrared-safe observables is thereby the class of jet observables, or simply jets. Jets can be imagined as collimated bunches of soft and collinear partons inside a "cone" of a certain "radius" around a central parton, determined by suitably chosen cone or cluster algorithms to specify the possible partonic configurations inside the jets. Several formulations to define such jet algorithms are available, most of which are infrared safe [7].

In the remainder of the article we will discuss the calculational concepts needed for the computation of the perturbative contributions to multi-jet production: Fixed order calculations, needed to describe the hard scattering, will be discussed in section 3. The physics behind parton showers and how to interface them to fixed order calculations will be sketched in sections 4 and 5. The non-perturbative contributions will not be discussed any further.

Journal of Physics: Conference Series **608** (2015) 012072 doi:10.1088/1742-6596/608/1/012072

### 3. Fixed order calculations

The hard scattering for a certain partonic (sub-)process is determined predominantly by the corresponding hard matrix element, which describes the probability for the transition between the corresponding partonic initial and final state and can be calculated from first principles in perturbation theory in the strong (running) coupling constant $\alpha_s(Q)$, which exhibits small values at large scales $Q$. The corresponding fixed order calculation is evaluated at a certain (fixed) hard scale, and in the following, of which excellent reviews can be found in [8, 9], we will thus suppress the scale argument in the strong coupling constant. An (infrared-safe) observable $O$ for $n$ partons in the final state, for example, is determined as follows:

$$\langle O\rangle \propto \sum_{a,b} \int dx_1 dx_2\, f_a(x_1) f_b(x_2) \sum_n \int d\phi_n\, O(p_1, ..., p_n)\, |\mathcal{A}_{n+2}|^2 \ , \tag{2}$$

where $|\mathcal{A}_{n+2}|^2$ denotes the hard matrix element for the scattering of $n+2$ partons ($2 \to n$), which is computed in turn from the complex valued hard scattering amplitude $\mathcal{A}_{n+2}$. Summation and phase-space integration over all possible final states with $n$ partons is implied. For an observable whose leading order (LO) prediction is given by an $n$-parton tree-level amplitude $\mathcal{A}_n^{(0)}$, the following expansions, up to relative order $\mathcal{O}(\alpha_s)$, are relevant for the calculations of the next-to-leading order (NLO) prediction:

$$|\mathcal{A}_n|^2 = |\mathcal{A}_n^{(0)}|^2 + \alpha_s 2Re(\mathcal{A}_n^{(0)*}\mathcal{A}_n^{(1)}) \quad \text{and} \quad |\mathcal{A}_{n+1}|^2 = \alpha_s |\mathcal{A}_{n+1}^{(0)}|^2 \ , \tag{3}$$

where we have omitted the overall LO factor $\alpha_s^{n-2}$. The associated NLO contributions originate then from two different regions with respect to the phase space: One (the virtual) comes from the interference term of the virtual one-loop amplitude $\mathcal{A}_n^{(1)}$ with the corresponding tree-level matrix element with $n$ partons, the other (the real) from the squared $(n+1)$-parton tree-level amplitude, with one additional parton in the final state, compared to the LO contribution. Note that the two NLO contributions are defined on phase spaces of different dimensionality. We can thus sum up the computation of the NLO contribution to an observable $O$ in two terms as

$$\langle O\rangle^{NLO} = \int_n O_n V + \int_{n+1} O_{n+1} R \ , \tag{4}$$

where we have used a very condensed notation, in which $\int_n \equiv \int d\phi_n$ and where $V$ and $R$ denote the virtual and real contributions (encoding the corresponding observables and matrix elements) respectively. The amplitudes in $V$ and $R$ are traditionally computed from all possible corresponding Feynman diagrams, of which Fig.2 shows two examples. Unfortunately the two terms in Eq.4 are ill defined if taken separately, since each contains singularities for certain regions in phase space: In the virtual part the integration over the loop momentum may lead to collinear, soft or ultraviolet singularities, e.g. if $k_i||k_j$, $|k_i| \to 0$ or $|k| \to \infty$ respectively in the example in Fig.2. In the real part the integration over the additional real emission may lead to collinear or soft singularities, e.g. if $p_3||p_5$ or $|p_5| \to 0$ respectively in the example in Fig.2. In an analytic integration using e.g. $D$-dimensional regularization, with $D = 4 - 2\epsilon$ and $|\epsilon| \ll 1$, these singularities manifest themselves as poles in $1/\epsilon$. The ultraviolet singularities in the virtual part are usually taken care of by a suitably chosen ultraviolet counterterm $V_{CT}$, such that $V$ is actually ultraviolet-finite. In addition, for infrared-safe observables it is known that the soft/collinear singular terms cancel between the real and the virtual contributions after integration, rendering their sum eventually finite [10–12].

The calculation of state-of-the art processes, i.e. processes with many partons, cannot be performed analytically anymore and we need numerical methods to perform the phase-space

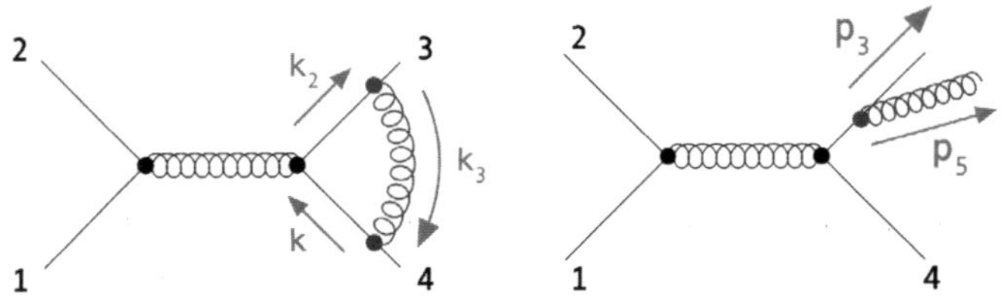


**Figure 2.** *Example diagrams for virtual (left) and real (right) corrections. The degrees of freedom of the virtual as well as of the external partons are denoted by the loop four-momenta $k_i$ and the external four-momenta $p_i$ respectively.*

integrals. The method of choice is hereby Monte Carlo integration [13, 14]. It is, however, not possible to naïvely integrate the sum of the virtual and real contributions, since the corresponding integrands live on phase spaces of different dimensionality. The cancellation of the singular terms is thus not as trivial. A standard method to deal with such a problem is the subtraction method: We can rewrite the NLO observable to read

$$\langle O\rangle^{NLO} = \int\limits_{n+1} \left(O_{n+1}R - O_n A\right) + \int\limits_{n} \left(O_n V + O_n \int\limits_{+1} A\right) \ , \tag{5}$$

where $A$ is a simple function that has the same point-wise singular behavior as $R$. Since $O_n$ is infrared safe, i.e. $O_{n+1} \to O_n$ in the soft/collinear limit, $O_n A$ acts as local counterterm to $O_{n+1}R$ in the singular regions, rendering the first bracket finite. $A$ is thereby chosen such that it can be integrated analytically over the one-parton subspace, leading to poles in $1/\epsilon$, which then cancel against the $\epsilon$-poles from $V$, rendering the second bracket finite. Both brackets can then be separately integrated over their respective phase spaces. The singular parts of the subtraction terms are universal. The finite parts, however, can be chosen freely, which results in various variants of the subtraction method, of which several automated implementations are available. Relevant references can be found in [8, 15–30].

Despite the subtraction method, the real and virtual contributions still need to be constructed and especially the results of the loop integration in the virtual contribution has to be known. To compute the virtual contribution basically three methods and variants thereof exist: Traditional tensor reduction, cut-based methods and the numerical method.

In traditional tensor reduction one uses the fact that tensor integrals of the form

$$I_{n,a}(\{p_i\},\{m_i\})^{\alpha_1\ldots\alpha_{2a}} = \int d^D k \frac{k^{\alpha_1}\ldots k^{\alpha_{2a}}}{\prod\limits_{i=1}^{n}\left(k_i^2 - m_i^2 + i\delta\right)^n} \ , \quad \text{with } D = 4 - 2\epsilon \ , \tag{6}$$

which result from all possible Feynman diagrams to a certain amplitude, can always be reduced to a linear combination of a certain set of known scalar integrals. The price to pay is the introduction of Gram determinants upon the determination of the corresponding coefficients. For example, reducing a three-point tensor integral of rank 2, leads to the introduction of an inverse Gram determinant $|G|^{-1} \propto (p_1^2 p_2^2 - (p1.p2)^2)^{-1}$, which tends to large values whenever $p_1||p_2$ and thus leads to a poor numerical behavior. However, several solutions to this type of problem exist, based on different reduction algorithms, expansion around small Gram determinants in critical regions, etc., of which several implementations are available. Relevant references can be found in [31–47].

In the cut-based methods one uses the fact that the entire amplitude can directly be decomposed into a basis of known scalar integrals. The coefficients are then obtained from tree-like objects by solving a linear system of equations numerically. These tree-like objects are universally obtained by cutting loop diagrams. For example, for a massless amplitude one needs only the coefficients in front of bubble, triangle and box integrals:

$$A_n^{(1)} = \sum_{i,j} c_{i,j} I_2^{(ij)} + \sum_{i,j,k} c_{ijk} I_3^{(ijk)} + \sum_{i,j,k,l} c_{ijkl} I_4^{(ijkl)} + R_n \ , \tag{7}$$

where the box coefficients are then for example obtained from quadruple cuts. After the box contributions have been subtracted one gets the triangle coefficients from triple cuts, and so on. Several approaches and various implementations hereof exist. Relevant references can be found in [48–66].

In the numerical method one extends the idea of the subtraction method to the virtual part by introducing local counterterms to the loop integrand in the virtual contribution as well. With $V = \int_{loop} V_{bare} + V_{CT}$ we can write

$$\langle O \rangle^{NLO} = \underbrace{\int_{n+1} \left( O_{n+1} R - O_n A \right)}_{\langle O \rangle^{NLO}_{real}} + \underbrace{\int_{n,loop} O_n \left( V_{bare} - L \right)}_{\langle O \rangle^{NLO}_{virtual}} + \underbrace{\int_n O_n \left( V_{CT} + \int_{loop} L + \int_{+1} A \right)}_{\langle O \rangle^{NLO}_{insertion}} \ , \tag{8}$$

where $L$ is a simple function that has the same point-wise singular behavior as the fully singular loop integrand $V_{bare}$ in the ultraviolet, soft and collinear regions. $L$ is thereby chosen such that it can be integrated analytically over the loop-momentum space, leading to poles in $1/\epsilon$: The explicit poles from $\int_{loop} L$ cancel against the soft/collinear $\epsilon$-poles from $\int_{+1} A$ and the ultraviolet $\epsilon$-poles from $V_{CT}$. The three contributions $\langle O \rangle^{NLO}_{real}$, $\langle O \rangle^{NLO}_{virtual}$ and $\langle O \rangle^{NLO}_{insertion}$ can then be separately integrated over their respective integration spaces, where the numerical loop integration and the numerical integration over the final state phase space in $\sigma^{NLO}_{virtual}$ are performed together in one combined Monte Carlo integration. The virtual subtraction method was originally formulated on a Feynman-diagrammatic level, but later extended to be applicable directly on the amplitude level. The actual numerical loop integration can then either be implemented for a re-parametrized form with Feynman or Schwinger parameters or in a direct approach, where the integration is over the four-dimensional loop momenta and requires an efficient method to deform the contour of the loop integration into the complex plane where necessary. Relevant references can be found in [67–82].

All methods to compute the various parts to an NLO observable depend eventually on the efficient construction and evaluation of the respective integrands, be it the tree-level constituents of the LO contribution or the NLO real contribution, or the one-loop constituents of the NLO virtual contribution. The standard for many years has been to rely on traditional Feynman diagrams, where all Feynman diagrams that contribute to a certain process are taken into account. This, however, may become quite tedious for processes with many partons, since in a naïve implementation the computational complexity grows approximately factorially with the number of partons. In addition, considering predominantly QCD processes, the algebraic manipulations due to the underlying SU($N\!=\!3$) gauge group can become quite cumbersome for processes with many partons. Many modern approaches that aim to cut down the computational complexity are based on the attempt to calculate entire amplitudes rather than single Feynman diagrams. Color decomposition offers for example a way to deal more efficiently with the underlying group theory, where one uses the fact that any QCD amplitude can in general be decomposed in a linear combination of spanning vectors in the associated group space and purely kinematical coefficients. This decomposition can be done once in the beginning of the computation, where all the group theoretical factors are computed and the linear combinations, as expansions in the number $N$ of colors, are determined. During run-time it is then only necessary to compute the kinematical coefficients for each phase-space point. This method becomes particularly efficient, if parts of the kinematical coefficients during their computation may be reused, which is possible through recursive methods. In the large-$N$ approximation one even terminates the expansions in the number $N$ of colors after the leading term, which decreases the computational effort tremendously. The kinematical coefficients can further be decomposed into more primitive objects, which exhibit a certain fixed ordering, i.e. all diagrams that are associated to them exhibit the same planar ordering with respect to their external legs. These primitive objects are particularly well suited to be constructed through recursive methods, of which various approaches are known. For certain helicity combinations even simple closed analytic expressions exist to compute the associated amplitudes, expressed in terms of particularly suitable spinor-helicity representations. Eventually one may use the benefits of the Monte Carlo approach, by additionally performing a sampling over the external quantum numbers (such as helicities or color degrees of freedom) through the Monte Carlo sum. Relevant references to the aforementioned approaches, for tree-level as well as one-loop constituents, can be found in [18, 48, 49, 65, 66, 83–119].

## 4. Parton showers

In this section we will sketch the motivation and physics behind parton showers (exemplarily for the case of time-like splittings) which are essential for the correct simulation of realistic detector events. Excellent general reviews on the following can, however, be found in [4–6, 120], to which we would like to point the reader for a more thorough description.

In order to predict detector events and observables in high energy collisions realistically, fixed order calculations are not sufficient for mainly two reasons:

1) Jet cross sections describe the inclusive production of jets: The cross section for $n$-jet production e.g. describes the production of at least $n$ jets, with a corresponding $n$-parton final state phase space. However, in order to compare to the measured detector events we need to describe the exclusive final states for exactly $n$, $n+1$, etc. jets. With fixed order calculations we will not be able to realistically describe such exclusive final states, and especially not for multi-jet configurations, due to their increasing complexity (even with modern amplitude-based methods).

2) The hard scattering is always associated with a certain hard scale, at which the corresponding fixed order calculation is evaluated. This means we need to evolve from an inclusive cross section at some hard scale $Q$ to an exclusive final state at some soft scale $\mu$, at which hadronization can take place. Due to the inherent truncation of the corresponding perturbative series and the running logarithmic behavior of the strong coupling constant, however, the evolution from one scale $Q$ to another scale $\mu$ introduces spurious logarithmic enhancements of the form $\alpha_s(\mu)^n L^m$, with $L = \log(Q^2/\mu^2)$. For a sufficiently soft scale $\mu$ the logarithms $L$ may thus easily overcome the smallness of the perturbative coupling constant $\alpha_s(\mu)$ at this scale, which leads to a divergent result for $\mu^2 \ll Q^2$. Physical cross sections, however, exhibit Sudakov dampening at soft scales. A realistic description needs thus to re-sum the large logarithms to all orders in perturbation theory, in order to properly describe the Sudakov suppression at the soft scales (cf. Fig.3).

A solution is presented by the parton shower approach, due to multiple emissions in the soft/collinear approximation: In the soft/collinear regions an $(n+1)$-parton cross section can be factorized into an $n$-parton cross section times a splitting function $P(z)$. Differentially this reads $d\sigma(\phi_{n+1}) \propto d\sigma(\phi_n)\alpha_s(t)P(z)dzdt/t$, where $t$ denotes the scale at which the splitting takes place and $z$ (respectively $1-z$) denotes the energy fraction of the splitting products, as depicted in Fig.4, and the phase-space element of the emission is given by $dzdt/t$ (for simplicity, here and in the following, we omit the azimuthal angle component of the emission). The splitting function $P(z)$ is enhanced for $z \to 0$ and/or 1 (depending on the parton types of the constituents

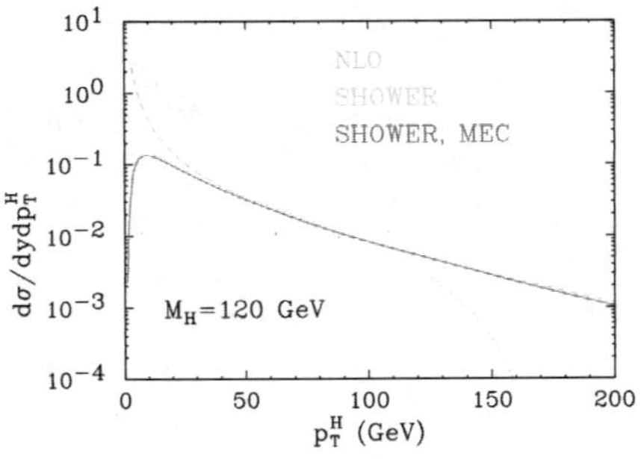


**Figure 3.** *The plot portrays the transverse momentum distribution of a Higgs boson in Higgs production at the LHC [120], calculated at fixed NLO (green), with a parton shower (cyan), and with a matrix element corrected (MEC) parton shower (red). The NLO calculation diverges in the low $p_T$-region. The parton shower exhibits the correct Sudakov suppression in the low $p_T$-region, but fails to describe the hard region correctly. The MEC parton shower describes both regions properly.*

in the splitting), and in addition we note an enhancement for small values of $t$. A parton shower algorithm implements multiple emissions in the soft/collinear approximation, and hence the production of multi-parton final states in this approximation, through the recursive application of the above factorization, which can be sketched as follows:

$$PS[d\sigma(\phi_n;t)] = \Delta(t,\mu)d\sigma(\phi_n;t) + PS[\Delta(t,t')d\sigma(\phi_n;t)\alpha_s(t')P(z')dz'dt'/t'] \,, \qquad (9)$$

where the Sudakov form factor $\Delta(t,t')$ describes the probability to evolve from a scale $t$ to a scale $t'$ without any resolvable emission, and the splitting function $P(z')$ the probability for an emission with an energy fraction $z'$. The first term in Eq.9 describes the probability to have no resolvable emission at all, between a scale $t$ and a certain cut-off scale $\mu$, whereas the second term describes the probability for at least one resolvable emission, where the integration over $z'$ and $t'$ within suitable boundaries at each recursion step is implied, which is performed through Monte Carlo integration: Choosing at each recursion step suitable values for $z'$ and $t'$ for the next emission and performing the Monte Carlo sum, where the associated weights are chosen according to the corresponding Sudakov form factors and splitting probabilities, generates eventually a corresponding exclusive final state phase space. A parton shower preserves therefore the value of the total inclusive cross section upon which it is applied, but redistributes the differential weights within the exclusive final state phase space, such that unitarity is preserved: $\mathcal{P}$(no emission)+$\mathcal{P}$(at least one emission)$=1$. Each parton has a cut-off scale $\mu \sim \Lambda_{QCD}$, below which the perturbative evolution ceases to make sense and the shower evolution is terminated. The emissions of a parton shower are naturally ordered by the values $t_n > t_{n+1} > t_{n+2}\ldots$ of the scale $t$ (also called ordering variable) at which the consecutive splittings take place, where various choices are possible (the angle between the splitting products, the transverse momentum of the splitting, etc.). Depending on those choices, a parton shower is called "angular-ordered", "$p_T$-ordered", etc. All parton showers exhibit the same behavior regarding the summation of the leading (large) logarithms. However, the choice of the ordering variable has an influence on how the next-to-leading logarithms are treated. The choice of the ordering variable will also affect how certain correlations between the splitting products are incorporated into the phase-space population. It has been shown for example, that an angular ordered shower incorporates the effects of color coherence between the splitting products correctly, but we may have small angle hard emission after large angle soft emission. A $p_T$-ordered shower on the other hand has the pleasant feature that its splitting products are ordered in "hardness", however the possibility to have large angle soft emission after small angle hard emission destroys the color coherence between the splitting products. A slightly different type of parton shower, the "dipole based parton shower", is based on the color-dipole picture of QCD and utilizes $2\rightarrow3$-splittings rather than $1\rightarrow2$-splittings. This type of parton shower has two advantages over the traditional approach: 1) In $2\rightarrow3$-splittings energy-momentum conservation and the correct mass-shell

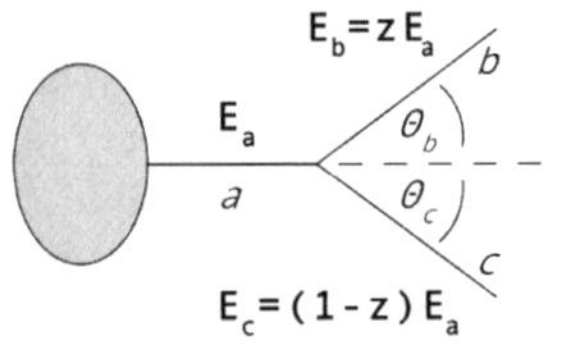


**Figure 4.** *A time-like splitting of a parton a into two partons b and c. For small angles:* $p_a^2 = t = (p_b + p_c)^2 = z(1-z)E_a^2(\Theta_b + \Theta_c)^2$*, where z parametrizes the energy fractions in the splitting. One can then show that* $d\sigma(b,c) \propto d\sigma(a)\alpha_s(t)P(z)dzdt/t$*, where we note the enhancements for small values of t but also for* $z \to 0$ *and/or* 1 *encoded in the splitting function* $P(z)$.

conditions for each participating parton can be fulfilled simultaneously. 2) It can be ordered in hardness while at the same time implementing the correct color coherence effects. One property, however, is common to all parton shower algorithms: They utilize the large-$N$ approximation in a certain sense. However, first ideas to include certain next-to-leading terms from the expansion in the number $N$ of colors have already been formulated.

If we compare fixed order calculations with the parton shower approach we note that both exhibit necessary characteristics (cf. Fig.3): A parton shower describes many-particle final states correctly to all orders in perturbation theory, but only in the soft/collinear approximation. The fixed order approach on the other hand works very well to describe hard emission, but does not work so well in the soft/collinear regions, if we want to describe a physical result, since it fails to describe the Sudakov suppression. The obvious idea is to combine the benefits of both approaches to get a more reliable description of exclusive detector events. The general strategy is thereby to correct the first few emissions from a parton shower with the help of fixed order matrix elements (in the following shortly called matrix elements). In general, the more matrix elements are involved the better: The first few emissions are then generated with the correct weights by the matrix elements (ME), whereas the remaining multi-parton soft/collinear emissions, and hence the approximate all order summation, are described by the parton shower (PS). A naïve combination of ME and PS, however, leads to overlapping contributions and double counting, and one needs to take proper care of this, for which various approaches have been engineered, known under the collective terminus of "matching" and "merging".

The simplest merging algorithm consists of correcting the first emission of a parton shower by the corresponding matrix element, for example to correct the first emission of the parton shower, applied to the leading order cross section for 2 jet production in *pp* collisions, by the LO matrix element for 3 jet production. The effect of such a matrix element correction (MEC) is portrayed in Fig.3 for the example of Higgs production at the LHC [120].

Leading order merging consists of correcting the first few emissions of a parton shower by several LO matrix elements of various jet multiplicities. However, a naïve summation of the corresponding inclusive cross sections leads to double counting. LO merging prescriptions are prescriptions on how to turn inclusive cross sections exclusive before adding them.

Next-to-leading order matching consists of correcting the first emission of a parton shower by the corresponding NLO matrix element. However, both approaches contain NLO contributions, i.e. parts of the real emission contribution: Looking at the first emission of a parton shower applied to some LO matrix element and at the real emission fixed order correction to this matrix element, these contributions will overlap in the soft/collinear regions, leading to double counting. NLO matching prescriptions are prescriptions on how to construct auxiliary NLO cross sections that return the correct results without double counting upon the application of a parton shower.

The choice of the ordering variable will also affect how we have to modify a parton shower upon the various matching and merging prescriptions. Certain matching prescriptions for example depend on an ordering in "hardness", such that if we wanted to apply them with an angular ordered shower we would have to veto any events, produced by the shower, whose first emission is harder than a certain maximum value, which is then called a "vetoed shower". However, by throwing away these events we effectively constrain the phase space that is filled by the shower.

This issue is solved by the concept of a "truncated, vetoed shower". A dipole shower on the other hand can be ordered in hardness while at the same time implementing the correct color coherence effects.

State-of-the-art nowadays are NLO multi-jet merging prescriptions, which are prescriptions on how to merge multiple NLO matrix elements and multiple LO matrix elements to a parton shower, in order to get the highest possible accuracy, of which various variants exist.

An extensive list of relevant references to the methods described in this section can be found in [30, 121–158]. References to a selection of general purpose event generators, which implement these methods can be found in [159–163].

## 5. Interfacing Monte Carlo programs

Recent developments in the automatization of fixed NLO calculations as well as NLO multi-jet merging make it of course necessary to have an interface to share the information between the two main ingredients. It is thereby assumed that the event generator (called Monte Carlo program in this context), which contains the parton shower algorithm, sets up the framework for a certain process and asks the fixed order program (called one-loop provider in this context) for the NLO matrix element during run-time. For this purpose the BLHA standard has been created, and recently extended [164, 165]. The advantage of having an automated standard is that the computation of the whole process chain is less error prone. The Monte Carlo program (MC) steers the setup of all necessary hard matrix elements, and provides for the numerical phase-space integration and the parton shower (if needed). The one-loop provider (OLP) provides the hard LO as well as NLO matrix elements on request for a certain phase-space point during the numerical integration. The initial setup and the list of required matrix elements is communicated once at the beginning of the computation, via a simple order and contract file system, the results of the individual matrix elements per phase-space point via external function calls during run-time. Besides the general purpose event generators, of which we have mentioned a selection in the previous section, there exist various Monte Carlo matrix element generators, utilizing the various fixed order approaches described in section 3, a selection of which, used for tree-level as well as one-loop calculations, can be found in [66, 109, 114, 115, 166–176]. Noteworthy hereby are those NLO programs, which may serve as independent OLP programs or come with their own parton shower capabilities [64, 66, 109, 114, 115, 166–169, 173, 174]. A selection of processes, which have been computed in a combined effort between MC and OLP programs can be found in [177–184].

## 6. Conclusion

It is probably fair to say that fixed NLO QCD calculations are now in a state in which LO calculations have been about a decade ago. Various important processes, which have been stated in an NLO "wish list" [185], have been successfully calculated during this "NLO revolution", effectively closing said wish list. New important processes have been agreed on in a new NLO "wish list" [177], with a new focus on mixed QCD and electroweak corrections but also on NNLO QCD corrections. Accordingly, methods to combine the benefits of fixed NLO QCD calculations with existing parton shower implementations have been developed and interface standards have been formulated. Several studies, using those interface standards, have been performed already. The full potential with which the new standards present us, however, has yet to be tapped into, i.e. with these new possibilities at hand we may now focus all the more on yet to be answered issues of rather physics related questions, as for example on how to assign scale uncertainties in multi-scale problems, etc., and not so much on the technical details behind the underlying calculations anymore.

## Acknowledgments
C.R. would like to thank the committees of the ACAT 2014 workshop, as well as the convenors for track 3, for the opportunity to present this overview. The work of C.R. is supported in part by the BMBF.

## References
[1] Beringer J *et al.* (Particle Data Group) 2012 *Phys.Rev.* **D86** 010001
[2] Aad G *et al.* (ATLAS Collaboration) 2012 *Phys.Lett.* **B716** 1–29 (*Preprint* 1207.7214)
[3] Chatrchyan S *et al.* (CMS Collaboration) 2012 *Phys.Lett.* **B716** 30–61 (*Preprint* 1207.7235)
[4] Buckley A, Butterworth J, Gieseke S, Grellscheid D, Hoche S *et al.* 2011 *Phys.Rept.* **504** 145–233 (*Preprint* 1101.2599)
[5] Ellis R K, Stirling W J and Webber B R 1996 *QCD and Collider Physics* (Cambridge: Cambridge University Press)
[6] Dissertori G, Knowles I and Schmelling M 2003 *Quantum Chromodynamics* (Oxford: Oxford University Press)
[7] Weinzierl S 2011 *Eur.Phys.J.* **C71** 1565 (*Preprint* 1011.6247)
[8] Weinzierl S 2006 *Eur.Phys.J.* **C45** 745–757 (*Preprint* hep-ph/0510157)
[9] Weinzierl S 2007 *PoS* **ACAT** 005 (*Preprint* 0707.3342)
[10] Bloch F and Nordsieck A 1937 *Phys.Rev.* **52** 54–59
[11] Kinoshita T 1962 *J.Math.Phys.* **3** 650–677
[12] Lee T and Nauenberg M 1964 *Phys.Rev.* **133** B1549–B1562
[13] Weinzierl S 2000 (*Preprint* hep-ph/0006269)
[14] Lepage G P 1978 *J.Comput.Phys.* **27** 192
[15] Catani S and Seymour M 1997 *Nucl.Phys.* **B485** 291–419 (*Preprint* hep-ph/9605323)
[16] Phaf L and Weinzierl S 2001 *JHEP* **0104** 006 (*Preprint* hep-ph/0102207)
[17] Catani S, Dittmaier S, Seymour M H and Trocsanyi Z 2002 *Nucl.Phys.* **B627** 189–265 (*Preprint* hep-ph/0201036)
[18] Goetz D, Schwan C and Weinzierl S 2012 *Phys.Rev.* **D85** 116011 (*Preprint* 1205.4109)
[19] Frixione S, Kunszt Z and Signer A 1996 *Nucl.Phys.* **B467** 399–442 (*Preprint* hep-ph/9512328)
[20] Somogyi G, Trocsanyi Z and Del Duca V 2005 *JHEP* **0506** 024 (*Preprint* hep-ph/0502226)
[21] Frixione S 2011 *JHEP* **1109** 091 (*Preprint* 1106.0155)
[22] Czakon M, Papadopoulos C and Worek M 2009 *JHEP* **0908** 085 (*Preprint* 0905.0883)
[23] Kosower D A 1998 *Phys.Rev.* **D57** 5410–5416 (*Preprint* hep-ph/9710213)
[24] Daleo A, Gehrmann T and Maitre D 2007 *JHEP* **0704** 016 (*Preprint* hep-ph/0612257)
[25] Gleisberg T and Krauss F 2008 *Eur.Phys.J.* **C53** 501–523 (*Preprint* 0709.2881)
[26] Frederix R, Gehrmann T and Greiner N 2008 *JHEP* **0809** 122 (*Preprint* 0808.2128)
[27] Frederix R, Frixione S, Maltoni F and Stelzer T 2009 *JHEP* **0910** 003 (*Preprint* 0908.4272)
[28] Hasegawa K, Moch S and Uwer P 2010 *Comput.Phys.Commun.* **181** 1802–1817 (*Preprint* 0911.4371)
[29] Frederix R, Gehrmann T and Greiner N 2010 *JHEP* **1006** 086 (*Preprint* 1004.2905)
[30] Platzer S and Gieseke S 2012 *Eur.Phys.J.* **C72** 2187 (*Preprint* 1109.6256)
[31] Passarino G and Veltman M 1979 *Nucl.Phys.* **B160** 151
[32] Denner A and Dittmaier S 2003 *Nucl.Phys.* **B658** 175–202 (*Preprint* hep-ph/0212259)
[33] Giele W and Glover E N 2004 *JHEP* **0404** 029 (*Preprint* hep-ph/0402152)
[34] del Aguila F and Pittau R 2004 *JHEP* **0407** 017 (*Preprint* hep-ph/0404120)
[35] Binoth T, Guillet J P, Heinrich G, Pilon E and Schubert C 2005 *JHEP* **0510** 015 (*Preprint* hep-ph/0504267)
[36] van Hameren A, Vollinga J and Weinzierl S 2005 *Eur.Phys.J.* **C41** 361–375 (*Preprint* hep-ph/0502165)
[37] Denner A and Dittmaier S 2006 *Nucl.Phys.* **B734** 62–115 (*Preprint* hep-ph/0509141)
[38] Weinzierl S 2006 345–395 (*Preprint* hep-ph/0604068)
[39] Diakonidis T, Fleischer J, Riemann T and Tausk J 2010 *Phys.Lett.* **B683** 69–74 (*Preprint* 0907.2115)
[40] Heinrich G, Ossola G, Reiter T and Tramontano F 2010 *JHEP* **1010** 105 (*Preprint* 1008.2441)
[41] Reiter T, Cullen G, Greiner N, Guffanti A, Guillet J P *et al.* 2010 *PoS* **CPP2010** 003 (*Preprint* 1011.6632)
[42] Fleischer J and Riemann T 2011 *Phys.Rev.* **D83** 073004 (*Preprint* 1009.4436)
[43] Ossola G, Papadopoulos C G and Pittau R 2008 *JHEP* **0803** 042 (*Preprint* 0711.3596)
[44] Binoth T, Guillet J P, Heinrich G, Pilon E and Reiter T 2009 *Comput.Phys.Commun.* **180** 2317–2330 (*Preprint* 0810.0992)
[45] Guillet J P, Heinrich G and von Soden-Fraunhofen J 2014 *Comput.Phys.Commun.* **185** 1828–1834 (*Preprint* 1312.3887)
[46] Denner A, Dittmaier S and Hofer L 2014 *PoS* **LL2014** 071 (*Preprint* 1407.0087)

[47] Fleischer J, Riemann T and Yundin V 2011
[48] Bern Z, Dixon L J and Kosower D A 1996 *Ann.Rev.Nucl.Part.Sci.* **46** 109–148 (*Preprint* hep-ph/9602280)
[49] Ellis R K, Kunszt Z, Melnikov K and Zanderighi G 2012 *Phys.Rept.* **518** 141–250 (*Preprint* 1105.4319)
[50] Bern Z, Dixon L J, Dunbar D C and Kosower D A 1995 *Nucl.Phys.* **B435** 59–101 (*Preprint* hep-ph/9409265)
[51] Ossola G, Papadopoulos C G and Pittau R 2007 *Nucl.Phys.* **B763** 147–169 (*Preprint* hep-ph/0609007)
[52] Anastasiou C, Britto R, Feng B, Kunszt Z and Mastrolia P 2007 *JHEP* **0703** 111 (*Preprint* hep-ph/0612277)
[53] Britto R and Feng B 2007 *Phys.Rev.* **D75** 105006 (*Preprint* hep-ph/0612089)
[54] Kilgore W B 2007 (*Preprint* 0711.5015)
[55] Ellis R K, Giele W and Kunszt Z 2008 *JHEP* **0803** 003 (*Preprint* 0708.2398)
[56] Forde D 2007 *Phys.Rev.* **D75** 125019 (*Preprint* 0704.1835)
[57] Ossola G, Papadopoulos C G and Pittau R 2008 *JHEP* **0805** 004 (*Preprint* 0802.1876)
[58] Berger C, Bern Z, Dixon L, Febres Cordero F, Forde D *et al.* 2008 *Phys.Rev.* **D78** 036003 (*Preprint* 0803.4180)
[59] Giele W T, Kunszt Z and Melnikov K 2008 *JHEP* **0804** 049 (*Preprint* 0801.2237)
[60] Ellis R K, Giele W T, Kunszt Z and Melnikov K 2009 *Nucl.Phys.* **B822** 270–282 (*Preprint* 0806.3467)
[61] van Hameren A, Papadopoulos C and Pittau R 2009 *JHEP* **0909** 106 (*Preprint* 0903.4665)
[62] Badger S, Biedermann B and Uwer P 2012 *J.Phys.Conf.Ser.* **368** 012055 (*Preprint* 1112.0412)
[63] Peraro T 2014 *Comput.Phys.Commun.* **185** 2771–2797 (*Preprint* 1403.1229)
[64] Bern Z, Dixon L, Febres Cordero F, Hoeche S, Ita H *et al.* 2014 *J.Phys.Conf.Ser.* **523** 012051 (*Preprint* 1310.2808)
[65] Badger S, Biedermann B and Uwer P 2011 *Comput.Phys.Commun.* **182** 1674–1692 (*Preprint* 1011.2900)
[66] Badger S, Biedermann B, Uwer P and Yundin V 2013 *Comput.Phys.Commun.* **184** 1981–1998 (*Preprint* 1209.0100)
[67] Nagy Z and Soper D E 2003 *JHEP* **0309** 055 (*Preprint* hep-ph/0308127)
[68] Assadsolimani M, Becker S and Weinzierl S 2010 *Phys.Rev.* **D81** 094002 (*Preprint* 0912.1680)
[69] Assadsolimani M, Becker S, Reuschle C and Weinzierl S 2010 *Nucl.Phys.Proc.Suppl.* **205-206** 224–229 (*Preprint* 1006.4609)
[70] Becker S, Reuschle C and Weinzierl S 2010 *JHEP* **1012** 013 (*Preprint* 1010.4187)
[71] Becker S, Goetz D, Reuschle C, Schwan C and Weinzierl S 2011 *PoS* **RADCOR2011** 008 (*Preprint* 1112.3521)
[72] Becker S, Reuschle C and Weinzierl S 2012 *JHEP* **1207** 090 (*Preprint* 1205.2096)
[73] Becker S, Goetz D, Reuschle C, Schwan C and Weinzierl S 2012 *PoS* **LL2012** 039 (*Preprint* 1209.2846)
[74] Nagy Z and Soper D E 2006 *Phys.Rev.* **D74** 093006 (*Preprint* hep-ph/0610028)
[75] Gong W, Nagy Z and Soper D E 2009 *Phys.Rev.* **D79** 033005 (*Preprint* 0812.3686)
[76] Becker S and Weinzierl S 2012 *Phys.Rev.* **D86** 074009 (*Preprint* 1208.4088)
[77] Becker S, Goetz D, Reuschle C, Schwan C and Weinzierl S 2012 *Phys.Rev.Lett.* **108** 032005 (*Preprint* 1111.1733)
[78] Goetz D, Reuschle C, Schwan C and Weinzierl S 2014 *PoS* **LL2014** 009 (*Preprint* 1407.0203)
[79] Becker S and Weinzierl S 2013 *Eur.Phys.J.* **C73** 2321 (*Preprint* 1211.0509)
[80] Freitas A 2012 *JHEP* **1207** 132 (*Preprint* 1205.3515)
[81] Carter J and Heinrich G 2011 *Comput.Phys.Commun.* **182** 1566–1581 (*Preprint* 1011.5493)
[82] Borowka S, Carter J and Heinrich G 2013 *Comput.Phys.Commun.* **184** 396–408 (*Preprint* 1204.4152)
[83] Dinsdale M, Ternick M and Weinzierl S 2006 *JHEP* **0603** 056 (*Preprint* hep-ph/0602204)
[84] Duhr C, Hoeche S and Maltoni F 2006 *JHEP* **0608** 062 (*Preprint* hep-ph/0607057)
[85] Schwinn C and Weinzierl S 2007 *JHEP* **0704** 072 (*Preprint* hep-ph/0703021)
[86] Giele W, Kunszt Z and Winter J 2010 *Nucl.Phys.* **B840** 214–270 (*Preprint* 0911.1962)
[87] Feng B and Luo M 2011 (*Preprint* 1111.5759)
[88] Berends F A and Giele W 1988 *Nucl.Phys.* **B306** 759
[89] Kleiss R and Kuijf H 1989 *Nucl.Phys.* **B312** 616
[90] Parke S J and Taylor T 1986 *Phys.Rev.Lett.* **56** 2459
[91] Mangano M L, Parke S J and Xu Z 1988 *Nucl.Phys.* **B298** 653
[92] Mangano M L 1988 *Nucl.Phys.* **B309** 461
[93] Parke S J and Mangano M L 1989
[94] Mangano M L and Parke S J 1988 *Nucl.Phys.* **B299** 673
[95] Kosower D A 1989 *Nucl.Phys.* **B315** 391
[96] Mangano M L and Parke S J 1991 *Phys.Rept.* **200** 301–367 (*Preprint* hep-th/0509223)
[97] Kanaki A and Papadopoulos C G 2000 (*Preprint* hep-ph/0012004)
[98] Maltoni F, Paul K, Stelzer T and Willenbrock S 2003 *Phys.Rev.* **D67** 014026 (*Preprint* hep-ph/0209271)
[99] Kilian W, Ohl T, Reuter J and Speckner C 2012 *JHEP* **1210** 022 (*Preprint* 1206.3700)

[100] Melia T 2013 *Phys.Rev.* **D88** 014020 (*Preprint* 1304.7809)
[101] Del Duca V, Dixon L J and Maltoni F 2000 *Nucl.Phys.* **B571** 51–70 (*Preprint* hep-ph/9910563)
[102] Keppeler S and Sjodahl M 2012 *JHEP* **1209** 124 (*Preprint* 1207.0609)
[103] Sjodahl M and Keppeler S 2013 *PoS* **DIS2013** 166 (*Preprint* 1307.1319)
[104] Bern Z and Kosower D A 1992 *Nucl.Phys.* **B379** 451–561
[105] Bern Z, Dixon L J and Kosower D A 1995 *Nucl.Phys.* **B437** 259–304 (*Preprint* hep-ph/9409393)
[106] Weinzierl S and Kosower D A 1999 *Phys.Rev.* **D60** 054028 (*Preprint* hep-ph/9901277)
[107] Ellis R K, Giele W, Kunszt Z, Melnikov K and Zanderighi G 2009 *JHEP* **0901** 012 (*Preprint* 0810.2762)
[108] Ita H and Ozeren K 2012 *JHEP* **1202** 118 (*Preprint* 1111.4193)
[109] Badger S, Biedermann B, Uwer P and Yundin V 2012 (*Preprint* 1209.0098)
[110] Reuschle C and Weinzierl S 2013 *Phys.Rev.* **D88** 105020 (*Preprint* 1310.0413)
[111] Reuschle C and Weinzierl S 2014 *PoS* **LL2014** 063 (*Preprint* 1406.7652)
[112] Schuster T 2014 *Phys.Rev.* **D89** 105022 (*Preprint* 1311.6296)
[113] van Hameren A 2009 *JHEP* **0907** 088 (*Preprint* 0905.1005)
[114] Cascioli F, Maierhofer P and Pozzorini S 2012 *Phys.Rev.Lett.* **108** 111601 (*Preprint* 1111.5206)
[115] Actis S, Denner A, Hofer L, Scharf A and Uccirati S 2013 *JHEP* **1304** 037 (*Preprint* 1211.6316)
[116] Kleiss R and Stirling W J 1985 *Nucl.Phys.* **B262** 235–262
[117] Dittmaier S 1998 *Phys.Rev.* **D59** 016007 (*Preprint* hep-ph/9805445)
[118] Britto R, Cachazo F and Feng B 2005 *Nucl.Phys.* **B715** 499–522 (*Preprint* hep-th/0412308)
[119] Draggiotis P, Kleiss R H and Papadopoulos C G 1998 *Phys.Lett.* **B439** 157–164 (*Preprint* hep-ph/9807207)
[120] Nason P and Webber B 2012 *Ann.Rev.Nucl.Part.Sci.* **62** 187–213 (*Preprint* 1202.1251)
[121] Gribov V and Lipatov L 1972 *Sov.J.Nucl.Phys.* **15** 438–450
[122] Altarelli G and Parisi G 1977 *Nucl.Phys.* **B126** 298
[123] Dokshitzer Y L 1977 *Sov.Phys.JETP* **46** 641–653
[124] Nagy Z and Soper D E 2006 101–123 (*Preprint* hep-ph/0601021)
[125] Giele W T, Kosower D A and Skands P Z 2008 *Phys.Rev.* **D78** 014026 (*Preprint* 0707.3652)
[126] Dinsdale M, Ternick M and Weinzierl S 2007 *Phys.Rev.* **D76** 094003 (*Preprint* 0709.1026)
[127] Schumann S and Krauss F 2008 *JHEP* **0803** 038 (*Preprint* 0709.1027)
[128] Platzer S and Gieseke S 2011 *JHEP* **1101** 024 (*Preprint* 0909.5593)
[129] Kilian W, Reuter J, Schmidt S and Wiesler D 2012 *JHEP* **1204** 013 (*Preprint* 1112.1039)
[130] Alioli S, Bauer C W, Berggren C J, Hornig A, Tackmann F J *et al.* 2013 *JHEP* **1309** 120 (*Preprint* 1211.7049)
[131] Seymour M H 1995 *Comput.Phys.Commun.* **90** 95–101 (*Preprint* hep-ph/9410414)
[132] Andre J and Sjostrand T 1998 *Phys.Rev.* **D57** 5767–5772 (*Preprint* hep-ph/9708390)
[133] Catani S, Krauss F, Kuhn R and Webber B 2001 *JHEP* **0111** 063 (*Preprint* hep-ph/0109231)
[134] Lonnblad L 2002 *JHEP* **0205** 046 (*Preprint* hep-ph/0112284)
[135] Frixione S and Webber B R 2002 *JHEP* **0206** 029 (*Preprint* hep-ph/0204244)
[136] Krauss F 2002 *JHEP* **0208** 015 (*Preprint* hep-ph/0205283)
[137] Nason P 2004 *JHEP* **0411** 040 (*Preprint* hep-ph/0409146)
[138] Nagy Z and Soper D E 2005 *JHEP* **0510** 024 (*Preprint* hep-ph/0503053)
[139] Frixione S, Nason P and Oleari C 2007 *JHEP* **0711** 070 (*Preprint* 0709.2092)
[140] Lavesson N and Lonnblad L 2008 *JHEP* **0812** 070 (*Preprint* 0811.2912)
[141] Hoche S, Krauss F, Schonherr M and Siegert F 2010 199–203 (*Preprint* 1009.1477)
[142] Hoche S, Krauss F, Schonherr M and Siegert F 2011 *JHEP* **1108** 123 (*Preprint* 1009.1127)
[143] Siegert F, Hoche S, Krauss F and Schonherr M 2010 *PoS* **ICHEP2010** 119 (*Preprint* 1011.6657)
[144] Hoche S, Krauss F, Schonherr M and Siegert F 2011 *JHEP* **1104** 024 (*Preprint* 1008.5399)
[145] Hamilton K and Nason P 2010 *JHEP* **1006** 039 (*Preprint* 1004.1764)
[146] Hoeche S, Krauss F, Schonherr M and Siegert F 2012 *JHEP* **1209** 049 (*Preprint* 1111.1220)
[147] Lonnblad L and Prestel S 2012 *JHEP* **1203** 019 (*Preprint* 1109.4829)
[148] Plaetzer S 2013 *JHEP* **1308** 114 (*Preprint* 1211.5467)
[149] Loennblad L and Prestel S 2013 *JHEP* **1303** 166 (*Preprint* 1211.7278)
[150] Lonnblad L and Prestel S 2013 *JHEP* **1302** 094 (*Preprint* 1211.4827)
[151] Frederix R and Frixione S 2012 *JHEP* **1212** 061 (*Preprint* 1209.6215)
[152] Hoeche S, Krauss F, Schonherr M and Siegert F 2013 *JHEP* **1304** 027 (*Preprint* 1207.5030)
[153] Plaetzer S 2013 (*Preprint* 1307.0774)
[154] Rubin M, Salam G P and Sapeta S 2010 *JHEP* **1009** 084 (*Preprint* 1006.2144)
[155] Nagy Z and Soper D E 2012 *JHEP* **1206** 044 (*Preprint* 1202.4496)
[156] Platzer S and Sjodahl M 2012 709–712 (*Preprint* 1206.0180)
[157] Platzer S and Sjodahl M 2012 *JHEP* **1207** 042 (*Preprint* 1201.0260)

[158] Plaetzer S 2014 *Eur.Phys.J.* **C74** 2907 (*Preprint* 1312.2448)
[159] Sjostrand T, Mrenna S and Skands P Z 2006 *JHEP* **0605** 026 (*Preprint* hep-ph/0603175)
[160] Bahr M, Gieseke S, Gigg M, Grellscheid D, Hamilton K *et al.* 2008 *Eur.Phys.J.* **C58** 639–707 (*Preprint* 0803.0883)
[161] Bellm J, Gieseke S, Grellscheid D, Papaefstathiou A, Platzer S *et al.* 2013 (*Preprint* 1310.6877)
[162] Gleisberg T, Hoeche S, Krauss F, Schonherr M, Schumann S *et al.* 2009 *JHEP* **0902** 007 (*Preprint* 0811.4622)
[163] Sjoestrand T, Ask S, Christiansen J R, Corke R, Desai N *et al.* 2014 (*Preprint* 1410.3012)
[164] Binoth T, Boudjema F, Dissertori G, Lazopoulos A, Denner A *et al.* 2010 *Comput.Phys.Commun.* **181** 1612–1622 (*Preprint* 1001.1307)
[165] Alioli S, Badger S, Bellm J, Biedermann B, Boudjema F *et al.* 2014 *Comput.Phys.Commun.* **185** 560–571 (*Preprint* 1308.3462)
[166] Arnold K, Bahr M, Bozzi G, Campanario F, Englert C *et al.* 2009 *Comput.Phys.Commun.* **180** 1661–1670 (*Preprint* 0811.4559)
[167] Baglio J, Bellm J, Campanario F, Feigl B, Frank J *et al.* 2014 (*Preprint* 1404.3940)
[168] Arnold K, Bellm J, Bozzi G, Brieg M, Campanario F *et al.* 2011 (*Preprint* 1107.4038)
[169] Kilian W, Ohl T and Reuter J 2011 *Eur.Phys.J.* **C71** 1742 (*Preprint* 0708.4233)
[170] Mangano M L, Moretti M, Piccinini F, Pittau R and Polosa A D 2003 *JHEP* **0307** 001 (*Preprint* hep-ph/0206293)
[171] Krauss F, Kuhn R and Soff G 2002 *JHEP* **0202** 044 (*Preprint* hep-ph/0109036)
[172] Gleisberg T and Hoeche S 2008 *JHEP* **0812** 039 (*Preprint* 0808.3674)
[173] Cullen G, van Deurzen H, Greiner N, Heinrich G, Luisoni G *et al.* 2014 *Eur.Phys.J.* **C74** 3001 (*Preprint* 1404.7096)
[174] Alwall J, Frederix R, Frixione S, Hirschi V, Maltoni F *et al.* 2014 *JHEP* **1407** 079 (*Preprint* 1405.0301)
[175] Campbell J M and Ellis R K 2002 *Phys.Rev.* **D65** 113007 (*Preprint* hep-ph/0202176)
[176] Alioli S, Nason P, Oleari C and Re E 2010 *JHEP* **1006** 043 (*Preprint* 1002.2581)
[177] Butterworth J, Dissertori G, Dittmaier S, de Florian D, Glover N *et al.* 2014 (*Preprint* 1405.1067)
[178] Bern Z, Dixon L, Febres Cordero F, Hche S, Ita H *et al.* 2013 *Phys.Rev.* **D88** 014025 (*Preprint* 1304.1253)
[179] Badger S, Biedermann B, Uwer P and Yundin V 2014 *Phys.Rev.* **D89** 034019 (*Preprint* 1309.6585)
[180] Bern Z, Diana G, Dixon L, Febres Cordero F, Hoeche S *et al.* 2012 *Phys.Rev.Lett.* **109** 042001 (*Preprint* 1112.3940)
[181] Hoeche S, Krauss F, Maierhoefer P, Pozzorini S, Schonherr M *et al.* 2014 (*Preprint* 1402.6293)
[182] Cullen G, van Deurzen H, Greiner N, Luisoni G, Mastrolia P *et al.* 2013 *Phys.Rev.Lett.* **111** 131801 (*Preprint* 1307.4737)
[183] Berger C, Bern Z, Dixon L J, Febres Cordero F, Forde D *et al.* 2011 *Phys.Rev.Lett.* **106** 092001 (*Preprint* 1009.2338)
[184] Campanario F, Figy T M, Pltzer S and Sjdahl M 2013 *Phys.Rev.Lett.* **111** 211802 (*Preprint* 1308.2932)
[185] Buttar C, Dittmaier S, Drollinger V, Frixione S, Nikitenko A *et al.* 2006 (*Preprint* hep-ph/0604120)

Journal of Physics: Conference Series **608** (2015) 012073 doi:10.1088/1742-6596/608/1/012073

# General formulation of the sector-improved residue subtraction

**David Heymes**

Institute for Theoretical Particle Physics and Cosmology, RWTH Aachen University, Sommerfeldstr. 16, 52074 Aachen, Germany

E-mail: dheymes@physik.rwth-aachen.de

**Abstract.** The main theoretical tool to provide precise predictions for scattering cross sections of strongly interacting particles is perturbative QCD. Starting at next-to-leading order (NLO) the calculation suffers from unphysical IR-divergences that cancel in the final result. At NLO there exist general subtraction algorithms to treat these divergences during a calculation. Since the LHC demands for more precise theoretical predictions, general subtraction methods at next-to-next-to-leading order (NNLO) are needed.

These proceedings outline the four-dimensional formulation of the sector improved residue subtraction. The subtraction scheme STRIPPER and in particular its extension to arbitrary multiplicities is explained. Therefore, it furnishes a general framework for the calculation of NNLO cross sections in perturbative QCD.

## 1. Introduction

We are interested in predicting the hadronic cross section, which is known to factorize into parton distribution functions and the partonic cross section

$$\sigma_{h_1h_2}(P_1,P_2)=\sum_{ab}\iint_0^1 \mathrm{d}x_1\mathrm{d}x_2\, f_{a/h_1}(x_1,\mu_\mathrm{F})\, f_{b/h_2}(x_2,\mu_\mathrm{F})\,\hat{\sigma}_{ab}(x_1P_1,x_2P_2;\,\alpha_s(\mu_\mathrm{R}),\,\mu_\mathrm{R},\,\mu_\mathrm{F})\,. \tag{1}$$

The summation runs over initial state partons $\{a,b\}$, i.e. massless quarks and gluons. The parton distribution function $f_{a/h_1}(x_1,\mu_\mathrm{F})$ can be understood as the probability density for finding parton $a$ inside hadron $h_1$ carrying the momentum $p_1 = x_1P_1$. Parton distribution functions are non-perturbative objects and have to be determined experimentally.

In contrast, the partonic cross section $\hat{\sigma}_{ab}$ can be calculated using perturbative QCD. Including terms up to next-to-next-to-leading order, its expansion in the strong coupling $\alpha_s$ reads

$$\hat{\sigma}_{ab}=\hat{\sigma}_{ab}^{(0)}+\hat{\sigma}_{ab}^{(1)}+\hat{\sigma}_{ab}^{(2)}\,. \tag{2}$$

The leading order contribution is known as the Born approximation and reads

$$\hat{\sigma}_{ab}^{(0)}=\hat{\sigma}_{ab}^{\mathrm{B}}=\frac{1}{2\hat{s}}\frac{1}{N_{ab}}\int \mathrm{d}\mathbf{\Phi}_n\,\langle\mathcal{M}_n^{(0)}|\mathcal{M}_n^{(0)}\rangle\,\mathrm{F}_n\,, \tag{3}$$

where $n$ is the number of final state particles and $\mathrm{d}\mathbf{\Phi}_n$ the phase space measure. The measurement functions $F_n$ defines the infrared safe observable and prevents $n$ massless partons

from becoming soft or collinear. The $i$-loop matrix element is denoted by $|\mathcal{M}_n^{(i)}\rangle$. For details of the notation we refer to [1].
Beyond leading order, we decompose the cross section according to the number of particles in the final state. At next-to-leading order (NLO) we have

$$\hat{\sigma}_{ab}^{(1)} = \hat{\sigma}_{ab}^{\mathrm{R}} + \hat{\sigma}_{ab}^{\mathrm{V}} + \hat{\sigma}_{ab}^{\mathrm{C}} \,, \tag{4}$$

with

$$\hat{\sigma}_{ab}^{\mathrm{R}} = \frac{1}{2\hat{s}}\frac{1}{N_{ab}}\int \mathrm{d}\mathbf{\Phi}_{n+1}\, \langle \mathcal{M}_{n+1}^{(0)}|\mathcal{M}_{n+1}^{(0)}\rangle\, \mathrm{F}_{n+1} \,, \quad \hat{\sigma}_{ab}^{\mathrm{V}} = \frac{1}{2\hat{s}}\frac{1}{N_{ab}}\int \mathrm{d}\mathbf{\Phi}_{n}\, 2\mathrm{Re}\,\langle \mathcal{M}_{n}^{(0)}|\mathcal{M}_{n}^{(1)}\rangle\, \mathrm{F}_{n} \,. \tag{5}$$

Starting at this order, separate parts suffer from soft and collinear (infrared) divergences [1]. They appear as poles in the regulator $\epsilon$ after setting the space-time dimension to $d = 4 - 2\epsilon$. We distinguish between explicit virtual poles, that emerge in the one-loop matrix element $|\mathcal{M}_n^{(1)}\rangle$ of the virtual contribution $\hat{\sigma}_{ab}^{\mathrm{V}}$, and real poles, that appear after integrating the phase space of the additional parton of the real contribution $\hat{\sigma}_{ab}^{\mathrm{R}}$. All poles cancel in the sum (4).
At next-to-next-to leading order (NNLO) we get

$$\hat{\sigma}_{ab}^{(2)} = \hat{\sigma}_{ab}^{\mathrm{RR}} + \hat{\sigma}_{ab}^{\mathrm{RV}} + \hat{\sigma}_{ab}^{\mathrm{VV}} + \hat{\sigma}_{ab}^{\mathrm{C1}} + \hat{\sigma}_{ab}^{\mathrm{C2}} \,, \tag{6}$$

where

$$\begin{aligned}
\hat{\sigma}_{ab}^{\mathrm{RR}} &= \frac{1}{2\hat{s}}\frac{1}{N_{ab}}\int \mathrm{d}\mathbf{\Phi}_{n+2}\, \langle \mathcal{M}_{n+2}^{(0)}|\mathcal{M}_{n+2}^{(0)}\rangle\, \mathrm{F}_{n+2} \,, \\
\hat{\sigma}_{ab}^{\mathrm{RV}} &= \frac{1}{2\hat{s}}\frac{1}{N_{ab}}\int \mathrm{d}\mathbf{\Phi}_{n+1}\, 2\mathrm{Re}\,\langle \mathcal{M}_{n+1}^{(0)}|\mathcal{M}_{n+1}^{(1)}\rangle\, \mathrm{F}_{n+1} \,, \\
\hat{\sigma}_{ab}^{\mathrm{VV}} &= \frac{1}{2\hat{s}}\frac{1}{N_{ab}}\int \mathrm{d}\mathbf{\Phi}_{n}\, \left(2\mathrm{Re}\,\langle \mathcal{M}_{n}^{(0)}|\mathcal{M}_{n}^{(2)}\rangle + \langle \mathcal{M}_{n}^{(1)}|\mathcal{M}_{n}^{(1)}\rangle\right) \mathrm{F}_{n} \,.
\end{aligned} \tag{7}$$

The double-real contribution $\hat{\sigma}_{ab}^{\mathrm{RR}}$ contains two additional massless partons in the final state that can become unresolved and lead to poles in $\epsilon$ after the phase space integration is performed. The real-virtual contribution $\hat{\sigma}_{ab}^{\mathrm{RV}}$ consists of the one-loop amplitude integrated over the $n+1$ particle phase space. In addition to virtual poles of the one-loop matrix element, it develops real poles by integrating the phase space of one unresolved particle. The double-virtual contribution $\hat{\sigma}_{ab}^{\mathrm{VV}}$ contains only explicit virtual poles in the two-loop amplitude and the squared one-loop amplitude. The sum of all contributions in (6) is finite.
In general it is not possible to perform phase space integrations analytically. Furthermore, to be able to compare a predicition with experimental data the phase space integration should be implemented in a flexible Monte-Carlo software to adapt the observable to the experimental setup easily.
Subtraction methods at NLO have been established to handle infrared singularities before numerical integrations are performed. Catani-Seymour subtraction [2] and FKS subtraction [3] are commonly used schemes.
At NNLO, subtraction schemes become more involved. Antenna subtraction [4] and $q_T$ - subtraction [5] are the most advanced proposals and have already been applied to $e^+e^- \to 3\,$jets [6], $pp \to 2\,$gluonic jets [7, 8], Higgs production [5] and vector boson pair production [9, 10] and other non-trivial examples [11, 12].
In these schemes poles cancel analytically. We illustrate this by taking the real and virtual

[1] For simplicity initial state collinear counterterms $\hat{\sigma}_{ab}^{\mathrm{C}}$ at NLO and $\hat{\sigma}_{ab}^{\mathrm{C1}}$, $\hat{\sigma}_{ab}^{\mathrm{C2}}$ at NNLO are not mentioned in the further discussion. For details see [1].

contribution of a NLO cross section in (4). The real-radiation cross section is made integrable in four dimensions by a suitable subtraction term that mimics the behaviour of the squared matrix element in the singular limits. Adding the subtracted term back and integrating it analytically over the unresolved one-particle phase space provides poles that cancel the poles of the virtual contribution. Finally, both phase space integrals are numerically integrable in four dimensions. At NNLO the procedure is similar: Subtraction terms for real unresolved particles are introduced to render the phase space integrable. Analytically integrated subtraction terms cancel the explicit poles of virtual contributions.

Here we present STRIPPER (SecToR ImProved PhasE space for Real radiation), a NNLO subtraction scheme that is completely numerical and avoids cumbersome analytic integrations. The scheme was introduced in [13] and generalized to arbitrary final states in [1]. It has been first applied to top-quark pair production [14], and subsequently to other processes of low multiplicity: Z decay [15], including some conceptual improvements, Higgs + jet [16], charmless bottom quark decay [17], top quark decay [18], single top quark production [19] and muon decay [20].

The scheme was initially formulated using conventional dimensional regularization (CDR), where momenta and spin degrees of freedom of resolved and unresolved particles are treated in $d = 4-2\epsilon$ dimensions. Unresolved particles are either virtual particles of loop contributions or real-radiated particles that can become soft or collinear. In contrast to analytic subtraction schemes, momenta of resolved particles are explicitly parameterized in $d > 4$ dimensions. The explicit dimension increases as the multiplicity of the final state rises, e.g. for top-quark pair production already five dimensions have been parameterized explicitly. It turns out that STRIPPER in CDR is not applicable for high multiplicities. In addition, tree-level Matrix elements, that appear in subtraction terms, have to be provided to several powers in $\epsilon$. Available software only provides them up to $\epsilon^0$.

Thus, it has been necessary to reformulate the scheme in 't Hooft-Veltman regularization (HV), where momenta and spin degrees of freedom of resolved particles are four-dimensional.

In these proceedings, we explain the general idea of STRIPPER in order to obtain a Laurent series in $\epsilon$ for $\hat{\sigma}_{ab}^{\mathrm{RR}}$, where each coefficient can be calculated numerically. Afterwards, we shortly point out how it can be reformulated in HV to provide a self-contained subtraction scheme for NNLO calculations. The detailed description is to be found in [1].

## 2. STRIPPER

The subtraction scheme STRIPPER is an algorithmic method to extract real singularities of different contributions in (6). Each part will be given as a Laurent series in $\epsilon$, where each coefficient has been calculated numerically. The final result, after summing the different parts, is finite.

We outline the method for $\hat{\sigma}_{ab}^{\mathrm{RR}}$ as given in (7). Since it contains two additional, potentially unresolved, partons, the phase space integral has the most complicated infrared structure.

First, we split the phase space into double-collinear and triple-collinear sectors. In a triple-collinear sector singularities are generated as three specific partons become collinear to each other and/or two of them soft. In a double-collinear sector singularities emerge as two specific pairs of partons become collinear and/or two of them soft.

In a next step, we parameterize the collinear particles in each sector separately using energies and angles. We illustrate the parametrization on the basis of a triple-collinear sector, where the three particles are in the final state: The reference momentum indicating the triple-collinear direction is denoted by $r^\mu$. The momenta of the unresolved partons are named $u_1^\mu$ and $u_2^\mu$. Each momentum is parameterized by its energy and a $(d-1)$-dimensional unit vector in spherical coordinates

$$r^\mu \equiv r^0 \begin{pmatrix} 1 \\ \hat{\boldsymbol{q}}_1 \end{pmatrix} , \quad u_1^\mu \equiv u_1^0 \begin{pmatrix} 1 \\ \hat{\boldsymbol{u}}_1 \end{pmatrix} , \quad u_2^\mu \equiv u_2^0 \begin{pmatrix} 1 \\ \hat{\boldsymbol{u}}_2 \end{pmatrix} . \tag{8}$$

The unresolved particles' energies are rescaled by their maximal value $E$, $u_i^0 = \hat{\xi}_i E$, for $i = 1, 2$. The soft limit is approached as $\hat{\xi}_1 \to 0$ and/or $\hat{\xi}_2 \to 0$. Using rotations in $(d-1)$-dimensions, the scalar products between the three given momenta take the following form

$$\hat{\boldsymbol{u}}_1 \cdot \hat{\boldsymbol{r}} = \cos\theta_1 = 1 - 2\hat{\eta}_1 \,, \quad \hat{\boldsymbol{u}}_2 \cdot \hat{\boldsymbol{r}} = \cos\theta_2 = 1 - 2\hat{\eta}_2 \,, \quad \hat{\boldsymbol{u}}_1 \cdot \hat{\boldsymbol{u}}_2 = \cos\theta_1 \cos\theta_2 + \cos\phi_2 \sin\theta_1 \sin\theta_2 \,. \tag{9}$$

This parametrization is shown in figure 1. The limit of $\hat{\eta}_i$ at zero indicates the collinear limit

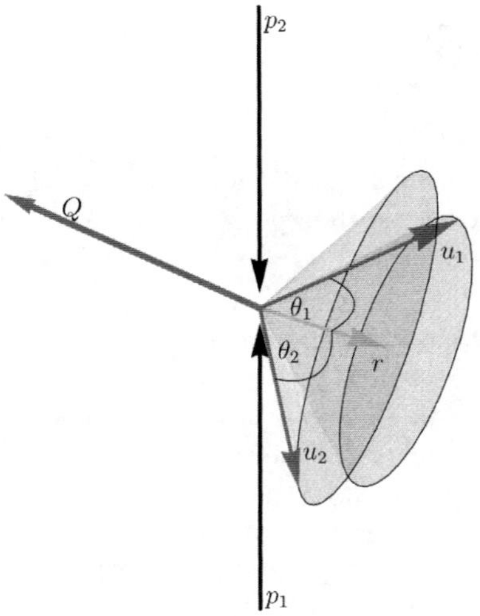


**Figure 1.** Parametrization of a triple-collinear sector, where the three partons are in the final state. The reference momentum is $r$, the unresolved momenta are denoted by $u_1$ and $u_2$. The sum of momenta of remaining resolved particles is $Q$.

of one of the unresolved partons and the reference parton. $\hat{\boldsymbol{u}}_1$ is collinear to $\hat{\boldsymbol{u}}_2$ when $\hat{\eta}_1 = \hat{\eta}_2$ and $\phi_2 = 0$. A non-linear transformation of $\phi_2$ to $\zeta$,

$$\phi_2 \to \phi_2(\hat{\eta}_1, \hat{\eta}_2, \zeta) \,, \tag{10}$$

then ensures that $\phi_2$ always vanishes as $\hat{\eta}_1 = \hat{\eta}_2$. Accordingly, all possible collinear limits are indicated only by two variables: $\hat{\eta}_1$ and $\hat{\eta}_2$. Furthermore, if we find a formulation in HV, meaning that all other momenta and the reference momentum are four-dimensional, at most six dimensions are needed to parameterize all possible scalar products consistently.

At this point, four physical variables $\{\hat{\eta}_1, \hat{\eta}_2, \hat{\xi}_1, \hat{\xi}_2\}$ parameterize all possible soft and collinear limits in a given sector. Additional sector decompositions [21] in those variables factorize all possible overlapping singularities that appear at NNLO. In practice this amounts to split each sector again. For example double-soft overlapping singularities are disentangled using

$$1 = \theta\left(\hat{\xi}_1 - \hat{\xi}_2\right) + \theta\left(\hat{\xi}_2 - \hat{\xi}_1\right) . \tag{11}$$

The decomposition of the phase space due to soft overlapping singularities is sufficient to factorize all possible limits in a double-collinear sector. In a triple-collinear sector the phase space is split into five additional sectors to factorize collinear and soft-collinear overlapping singularities. This splitting is accompanied by a transition from physical variables $\{\hat{\eta}_1, \hat{\eta}_2, \hat{\xi}_1, \hat{\xi}_2\}$ to corresponding sector variables $\{\eta_1, \eta_2, \xi_1, \xi_2\}$.

Finally, the full double-real radiation cross section can be written as a sum over different decomposed triple- and double-collinear sectors

$$\hat{\sigma}_{ab}^{\mathrm{RR}} = \sum_S \hat{\sigma}_{ab}^{\mathrm{RR},S} \,. \tag{12}$$

Each contribution has the following form

$$\hat{\sigma}_{ab}^{\mathrm{RR},S} = \int_0^1 \mathrm{d}\xi_1 \mathrm{d}\xi_2 \mathrm{d}\eta_1 \mathrm{d}\eta_2 \frac{F_S\left(\xi_1,\xi_2,\eta_1,\eta_2\right)}{\xi_1^{1-b_1\epsilon}\xi_2^{1-b_2\epsilon}\eta_1^{1-b_3\epsilon}\eta_2^{1-b_4\epsilon}}. \tag{13}$$

The function $F_S\left(\xi_1,\xi_2,\eta_1,\eta_2\right)$ is finite in the limit of vanishing arguments. All appearing singularities are factorized in the sector variables. Poles are extracted by an iterative usage of the plus distribution in each variable

$$\int_0^1 \mathrm{d}x\, \frac{f\left(x\right)}{x^{1-b\epsilon}} = \frac{f\left(0\right)}{b\epsilon} + \int_0^1 \mathrm{d}x\, \frac{f\left(x\right)-f\left(0\right)}{x^{1-b\epsilon}}\,. \tag{14}$$

We obtain a Laurent series in $\epsilon$ and all coefficients are calculated numerically. The described procedure is process independent, since it is possible to use the known universal infrared-limits of QCD amplitudes for the subtraction terms in (14).

## 3. Formulation in the 't Hooft-Veltman regularization scheme

The above procedure is carried out in CDR, i.e. resolved and unresolved momenta are $d$-dimensional. To get a transition to the 't Hooft-Veltman regularization scheme, we have to identify unresolved particles in different contributions of (6): Either one particle is unresolved, what we call the single-unresolved contribution, or two particles are unresolved, what we call the double-unresolved contribution. By adding process independent correction terms to the single-unresolved contribution and subtracting them from the double-unresolved contribution we make sure that the two contributions are finite separately. These corrections are process independent, since they depend just on the form of the splitting and soft functions and the integration region of unresolved momenta, that do not enter the matrix elements. Now we are able to identify unresolved and resolved particles unambiguously. Since each contribution is finite we set the momenta of resolved particles to $d = 4$ dimensions, the mistake is of order $\epsilon$. This makes sure that explicit parameterized dimensions are limited to six for arbitrary multiplicities. Setting also their spin degrees of freedom to the physical dimensions allows us to use matrix elements at order $\epsilon^0$.

To avoid contractions between $d$-dimensional vectors of unresolved momenta and four-dimensional matrix elements, we replace spin-correlated splitting functions in collinear subtraction terms by their azimuthal averaged counterparts.

This procedure leads to a fully general subtraction algorithm for NNLO calculations.

## 4. Summary and conclusion

In these proceedings, we presented the general formulation of STRIPPER, a subtraction scheme to calculate cross sections at NNLO in perturbative QCD for arbitrary multiplicities. The full subtraction scheme, completely described in [1], can readily be implemented in a computer code.

## Acknowledgments

This research was supported by the German Research Foundation (DFG) via the Sonderforschungsbereich/Transregio SFB/TR-9 "Computational Particle Physics". The author thanks the organizers of ACAT 2014 for remission of the conference fee.

## References

[1] M. Czakon and D. Heymes, arXiv:1408.2500 [hep-ph].
[2] S. Catani and M. H. Seymour, Nucl. Phys. B **485** (1997) 291 [Erratum-ibid. B **510** (1998) 503] [hep-ph/9605323].

[3] S. Frixione, Z. Kunszt and A. Signer, Nucl. Phys. B **467** (1996) 399 [hep-ph/9512328].
[4] A. Gehrmann-De Ridder, T. Gehrmann and E. W. N. Glover, JHEP **0509** (2005) 056 [hep-ph/0505111].
[5] S. Catani and M. Grazzini, Phys. Rev. Lett. **98**, 222002 (2007) [hep-ph/0703012].
[6] A. Gehrmann-De Ridder, T. Gehrmann, E. W. N. Glover and G. Heinrich, JHEP **0711** (2007) 058 [arXiv:0710.0346 [hep-ph]].
[7] J. Currie, A. Gehrmann-De Ridder, E. W. N. Glover and J. Pires, JHEP **1401** (2014) 110 [arXiv:1310.3993 [hep-ph]].
[8] J. Currie, A. Gehrmann-De Ridder, T. Gehrmann, N. Glover, J. Pires and S. Wells, arXiv:1407.5558 [hep-ph].
[9] F. Cascioli, T. Gehrmann, M. Grazzini, S. Kallweit, P. Maierhöfer, A. von Manteuffel, S. Pozzorini and D. Rathlev *et al.*, arXiv:1405.2219 [hep-ph].
[10] T. Gehrmann, M. Grazzini, S. Kallweit, P. Maierhfer, A. von Manteuffel, S. Pozzorini, D. Rathlev and L. Tancredi, arXiv:1408.5243 [hep-ph].
[11] G. Abelof, A. Gehrmann-De Ridder, P. Maierhofer and S. Pozzorini, JHEP **1408**, 035 (2014) [arXiv:1404.6493 [hep-ph]].
[12] X. Chen, T. Gehrmann, E. W. N. Glover and M. Jaquier, arXiv:1408.5325 [hep-ph].
[13] M. Czakon, Phys. Lett. B **693** (2010) 259 [arXiv:1005.0274 [hep-ph]].
[14] M. Czakon, Nucl. Phys. B **849** (2011) 250 [arXiv:1101.0642 [hep-ph]].
[15] R. Boughezal, K. Melnikov and F. Petriello, Phys. Rev. D **85** (2012) 034025 [arXiv:1111.7041 [hep-ph]].
[16] R. Boughezal, F. Caola, K. Melnikov, F. Petriello and M. Schulze, JHEP **1306** (2013) 072 [arXiv:1302.6216 [hep-ph]].
[17] M. Brucherseifer, F. Caola and K. Melnikov, Phys. Lett. B **721** (2013) 107 [arXiv:1302.0444 [hep-ph]].
[18] M. Brucherseifer, F. Caola and K. Melnikov, JHEP **1304** (2013) 059 [arXiv:1301.7133 [hep-ph]].
[19] M. Brucherseifer, F. Caola and K. Melnikov, arXiv:1404.7116 [hep-ph].
[20] F. Caola, A. Czarnecki, Y. Liang, K. Melnikov and R. Szafron, arXiv:1403.3386 [hep-ph].
[21] T. Binoth and G. Heinrich, Nucl. Phys. B **585** (2000) 741 [hep-ph/0004013].

Journal of Physics: Conference Series **608** (2015) 012074 doi:10.1088/1742-6596/608/1/012074

# Self-consistence of the Standard Model via the renormalization group analysis

**Fred Jegerlehner**[1,2], **Mikhail Kalmykov**[3,4] **and Bernd A. Kniehl**[3]

[1] Humboldt-Universität zu Berlin, Institut für Physik, Newtonstraße 15, 12489 Berlin, Germany

[2] Deutsches Elektronen-Synchrotron (DESY), Platanenallee 6, 15738 Zeuthen, Germany

[3] II. Institut für Theoretische Physik, Universität Hamburg, Luruper Chaussee 149, 22761 Hamburg, Germany

[4] Joint Institute for Nuclear Research, 141980 Dubna (Moscow Region), Russia

E-mail: mikhail.kalmykov@desy.de

**Abstract.** A short review of recent renormalization group analyses of the self-consistence of the Standard Model is presented.

## 1. Introduction

The recent discovery of the Higgs boson [1,2] at the LHC with mass

$$M_H = 125.03^{+0.26}_{-0.27}(\text{stat})^{+0.13}_{-0.15}(\text{syst})\ \text{GeV}\ , \quad [3]$$

$$M_H = 125.36 \pm 0.37(\text{stat}) \pm 0.18(\text{syst})\ \text{GeV}\ , \quad [4]\ ,$$

and the fact that so far no signal for new physics has been found, gives rise to the necessity to analyze in detail the self-consistency of the Standard Model. The Higgs boson is a necessary ingredient of the Standard Model (SM), required by its perturbative renormalizability [5, 6]. However, the renormalizability does not give rise to any constraints on the values of parameters of the Lagrangian, but only on the type of interactions. Some additional restrictions on the values of masses and coupling constants are coming from *unitarity*, *triviality* and *vacuum stability*. These three ideas have been used within last decades for an analysis of the self-consistency of the SM.

*Unitarity bound*: bounds on masses of fermions and Higgs bosons of any renormalizable model can be derived from considerations of the radiative corrections to decays and/or scattering processes [7, 8]. The breakdown of unitary can be avoided by adding a new particles (which implies the existence of new physics) or by the requirement that the perturbative approach must be meaningful (this implies a bound on the particles mass or its coupling constants) [9]. Unitary bounds depend on the type of process under consideration and the precise definition of the breakdown of the perturbative approach. For the Standard Model, the unitary bound for the Higgs boson mass is $M_H \lesssim 1$ TeV .

*Triviality and Landau pole*: The *triviality* constraint is related with the high energy behavior of the running couplings. Is it well known, that a running coupling $h(\Lambda)$ may suffers from a Landau pole when the corresponding $\beta$-function is positive:

$$\frac{d}{d\Lambda}h(\Lambda) = \beta_0 h^3(\Lambda) \longrightarrow h^2(\Lambda) = \frac{h_R^2}{1-\beta_0 h_R^2 \ln\frac{\Lambda}{M}}\ , \quad (1)$$

where $\Lambda$ is the ultra-violet cutoff, $M$ is a characteristics scale of the process under consideration and $h_R$ is the renormalized (running) charge at scale $M$. As follows from Eq. (1), the running coupling $h^2(\Lambda)$ diverges at scale $\Lambda_{\text{Landau}}$ defined as $\Lambda_{\text{Landau}} = M \exp\left[\frac{1}{\beta_0 h_R^2}\right]$, when $\beta_0 > 0$. In the SM, Landau poles may exists for the $U(1)$-gauge coupling, $g_1$, Yukawa couplings, $y_f$ and the Higgs self-coupling, $\lambda$. For the most important couplings the SM renormalization group (RG) equations read

$$\frac{dh}{dt} = \frac{1}{16\pi^2}\beta_h \,, \quad h \in \{g_1, g_2, g_3, y_f, \lambda\} \,,$$

where in the one-loop approximation the corresponding $\beta$-functions have the following form:

$$\beta_1 = \frac{41}{6}g_1^3 \,, \quad \beta_2 = -\frac{19}{6}g_2^3 \,, \quad \beta_3 = -7g_3^3 \,, \quad \beta_{y_t} = y_t\left[\frac{9}{2}y_t^2 - 8g_3^2 - \frac{9}{4}g_2^2 - \frac{17}{12}g_1^2\right] ,$$
$$\beta_\lambda = \lambda\left[24\lambda + 12y_t^2 - 9g_2^2 - 3g_1^2\right] - \left[6y_t^4 - \frac{9}{8}g_2^4 - \frac{3}{8}g_1^4 - \frac{3}{4}g_1^2 g_2^2\right] . \qquad (2)$$

It has been shown in [10] that the Landau pole in the Higgs self-coupling $\lambda$ would be below Planck scale if the Higgs is heavier than 180 GeV. The results of recent analyses [11,12] have confirmed that all coupling constants are free from Landau singularities and have smooth behavior in interval between $M_Z \sim 90$ GeV and $M_{Planck} = 2.435 \times 10^{18}$ GeV (see Fig. 1):

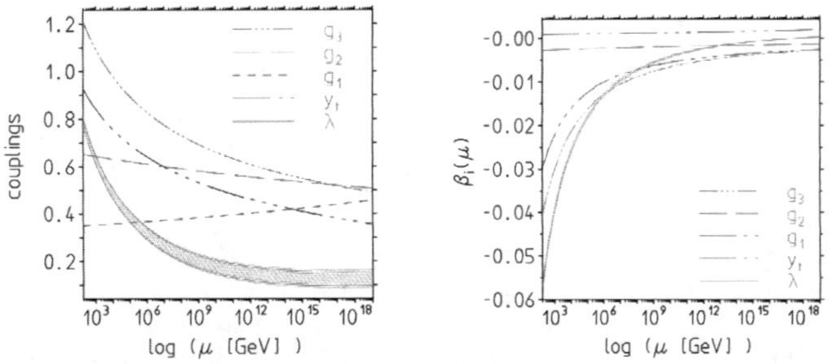


**Figure 1.** The running coupling constants(left) and corresponding $\beta$-functions (right). The plots are taken from [11].

### 1.1. *Vacuum stability and effective potential*

For the analysis of the *vacuum stability* one needs to recall some basic definitions related with the scalar potential and its generalization in Quantum Field Theory. Let us remind, that at the classical level, the Higgs potential in the SM

$$V(\phi) = \frac{1}{2}m^2\phi^2 + \frac{1}{4!}\lambda\phi^4 \,, \qquad (3)$$

is bounded from below for $\lambda > 0$ and has a trivial minimum for $m^2 > 0$ at $\phi = 0$, and non-trivial minima at $\phi^2 = \frac{-m^2}{\lambda}$ for $m^2 < 0$. At the quantum level, instead of the classical potential defined by Eq. (3), the *effective potential* should be analysed [13]. It can be written in the Landau gauge and the $\overline{\text{MS}}$ scheme as [14]

$$V(\phi) \quad \to \quad V_{eff}(\phi(t)) = -\frac{1}{2}m^2(t)\phi^2(t) + \frac{1}{4!}\lambda(t)\phi^4(t) + \frac{1}{16\pi^2}V_1 + \frac{1}{(16\pi^2)^2}V_{rest} \,, \qquad (4)$$

where (we use the notations of [15]):

$$V_1 = \sum_i \frac{k_i}{4} M_i^4(\phi(t)) \left[ \ln \frac{M_i^2(\phi(t))}{\mu^2} - C_i \right] ,$$
$$\phi(t) = \exp\{- \int_0^t \gamma(\tau) d\tau\} \phi_{clas} , \quad M_i^2(\phi(t)) = a_i h_i(t) \phi(t)^2 + b_i , \tag{5}$$

$h_i(t)$ are the running couplings, $\gamma(t)$ is the anomalous dimension of the Higgs field, and $k_i, C_i, a_i, b_i$ are numerical constants. By $V_{rest}$ we denote the higher-order contributions, which, in particular, also include the higher dimension operators (it begins at four loop) [14, 16]:

$$V_{rest} \sim \lambda \phi^4 \sum_{L>4} \left( \frac{\lambda^2 \phi^2}{m^2 + \frac{1}{2}\lambda\phi^2} \right)^{L-3} , \tag{6}$$

where $L$ is the number of loops (for more details see Section 4 of [14] or [17]).

The quantum corrections modify the shape of the effective potential such that a second minimum at large (Planck) scale may be generated. This second minimum (see Fig. (2)) is: (left plot) stable $V_{eff}(v) < V_{eff}(\phi_{\rm min})$, (middle plot) critical (two minima are degenerate in energy) $V_{eff}(v) = V_{eff}(\phi_{\rm min})$, (right plot) unstable/metastable $V_{eff}(v) > V_{eff}(\phi_{\rm min})$, electroweak vacuum and $v$ is the location of the EW minimum and $\phi_{\rm min}$ is the value of a new minimum.

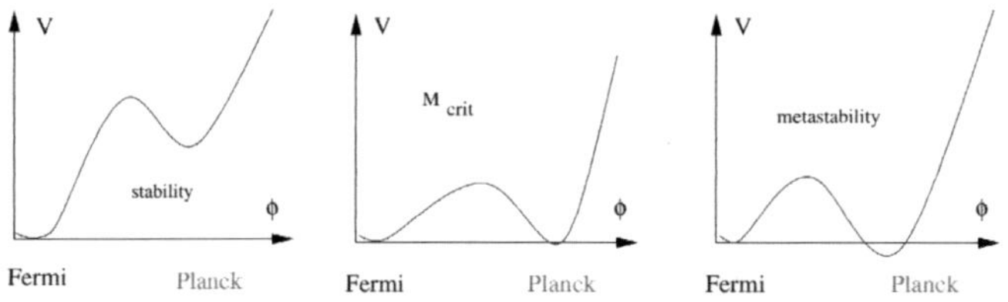


**Figure 2.** The form of the effective potential for the Higgs field $\phi$ which corresponds to the stable (left), critical (middle) and metastable (right) electroweak vacuum. The plot is taken from [18].

Depending on the values of the Higgs boson and top quark masses the lifetime of the EW vacuum can be larger or smaller then the age of the Universe [19–21]. The first case corresponds to the metastability scenario.

To get *practical* criteria of stability, the following step-by-step approximations are used [22]: (i) the potential at very high values of the field is dominated by the quartic term $V_{eff} \sim \lambda(\phi)\phi^4$ and $\lambda(\phi)$ depends on $\phi$ as the running coupling $\lambda(\mu)$ depends on the running scale $\mu$; (ii) it's looking the large value of the field $\phi = \phi_{crit}$, where $\lambda(\phi_{crit}) = 0$; (iii) combine together the previous two approximations, we get at very high values of the Higgs field $V_{eff}|_{\phi\sim\phi_{crit}} \simeq \lambda(\phi_{crit})\phi^4_{crit}$. Then, the effective potential becomes negative (unbounded from below) when $\lambda(\mu) < 0$ and the vacuum at the EW scale is not the absolute minimum. As follows from Eqs. (2), the Higgs self-coupling $\lambda$ is the only SM dimensionless coupling that can change sign with the scale variation since its beta-function, $\beta_\lambda$ contains a part which is not proportional to $\lambda$. In the considered approximation,
*the requirement that the electroweak vacuum* $<\phi>= (\sqrt{2}G_F)^{-\frac{1}{2}} = 246.22\ GeV$ *is the absolute minimum of the potential up to scale* $\Lambda$, *implies*

$$\lambda(\mu) > 0 , \quad \textit{for any} \quad \mu < \Lambda . \tag{7}$$

The role of quadratic term or higher dimensional terms (see Eq.(6)) is discussed in Section 3.

## 2. Analysis

### *2.1. Renormalization group equations and Matching Conditions*

Both the questions, the one concerning vacuum stability and the one concerning triviality, in the Standard Model are reduced to the renormalization group analysis of the Higgs self-coupling $\lambda$. The evolution of $\lambda$ includes the evolution of all coupling constants, see Eq. (2).

The starting point for solving the evolution equations is provided by the matching conditions: the relations between the running coupling constants $h_i(\mu)$ and the relevant (pseudo-) physical observables, $\sigma_i$. The simplest form for the matching conditions follows when physical masses are taken as the referring point: $\sigma_i \equiv M_i$. In this case the matching conditions have the following form [23]:

$$h_i(\mu^2) = c_i \frac{G_F}{\sqrt{2}} M_i^2 \left(1 + \delta_{\alpha_s} + \delta_{\alpha} + \delta_{\alpha\alpha_s} + \delta_{\alpha^2} + \delta_{\alpha_s^2} + \cdots \right) , \quad (8)$$

where $c_i$ are normalization constants and $\delta_{\alpha_j^k}$ are including only propagator type diagrams of order $O(\alpha_j^k)$. The evaluation of the one-loop EW matching conditions of order $O(\alpha)$ was partially done in [23–25] and has been completed in [26, 27]. The two-loop matching conditions of order $O(\alpha^2)$ for the Higgs self-coupling $\lambda$ in the limit of heavy Higgs boson have been evaluated in [28, 29], and full two-loop results were completed in [30–35]. The $O(\alpha\alpha_s)$ corrections for the top-quark Yukawa coupling within the SM have been evaluated in [36,37] and $O(\alpha^2)$ corrections are available only in the gaugeless limit [38, 39]. A state of the art evaluation attempting matching at the two-loop level has been reported in [12] (results see below).

The evaluation of the 3-loop RG equations for the SM gauge coupling was started in [40,41], and full 3-loop results for the Standard Model have been published in [42–48].

### *2.2. Results of RG analysis*

After the top-quark discovery the results of RG analyses of self-consistence of the SM are typically fitted by three parameters and can be written as follows:

$$M_{\text{min}} > \left[M_{crit} + c_1 \times (M_t - M_{\text{exp}}) + c_2 \times (\alpha_s - \alpha_{s,\text{exp}})\right] \text{ GeV} , \quad (9)$$

where $M_{crit}$ is the critical value of the Higgs boson mass, $c_1, c_2$ are some numerical coefficients and $M_{\text{exp}}$ and $\alpha_{s,\text{exp}}$ are the latest experimental values of the pole mass of the top-quark and of the strong coupling constant, respectively.

The detailed analysis of stability/metastability bounds based on the two-loop RG equations for all SM couplings and one-loop EW matching conditions (as well as two-loop QCD corrections to the top-quark Yukawa coupling $y_t$) have been presented in [49]:

$$M_{\text{min}} \geq \left[130 + 1.8 \times \left(\frac{M_t^{\text{pole}} - 173.2 \text{ GeV}}{0.9 \text{ GeV}}\right) - 0.5 \times \left(\frac{\alpha_s(M_Z) - 0.1184}{0.0007}\right) \pm 3\right] \text{ GeV} , \quad (10)$$

with $M_t = (173.2 \pm 0.9)$ GeV and $\alpha_s(M_Z) = 0.1184 \pm 0.0007$ as input parameters and where the error of 3 GeV is an estimation of unknown higher-order corrections.

The analysis of vacuum stability performed with inclusion of 3-loop RG equations and 2-loop matching conditions was presented in [50] and [51] with the following results:

$$M_{\rm min} \geq \left[128.95+2.2\times\left(\frac{M_t^{\rm pole}-172.9~{\rm GeV}}{1.1~{\rm GeV}}\right)-0.56\times\left(\frac{\alpha_s(M_Z)-0.1184}{0.0007}\right)\pm 1\right]~{\rm GeV}\,,\quad [50] \tag{11}$$

$$M_{\rm min} \geq \left[129.4+1.4\times\left(\frac{M_t^{\rm pole}-173.1~{\rm GeV}}{0.7~{\rm GeV}}\right)-0.5\times\left(\frac{\alpha_s(M_Z)-0.1184}{0.0007}\right)\pm 0.7\right]~{\rm GeV}\,,\quad [51] \tag{12}$$

*The latest published analysis* have been presented in [12]:

$$M_{\rm min} > \left[129.6+2.0\times\left(M_t^{\rm pole}-173.35~{\rm GeV}\right)-0.5\times\left(\frac{\alpha_s(M_Z)-0.1184}{0.0007}\right)\pm 0.3\right]~{\rm GeV}\,. \tag{13}$$

### *2.3. The analysis of uncertainties*

As follows from Eqs. (10)-(13) adding two-loop matching conditions and 3-loop RG equations does not changed dramatically the central value of critical mass: $130 \to\sim 129.0 \div 129.6$, but essentially reduces the theoretical uncertainties: $(3 \to 0.3)$ GeV. Adding the leading 3-loop corrections to the matching conditions for the Higgs self-coupling [52, 53] does not modified the final result, Eq. (13). The $3\sigma$ variations of $\alpha_s$ or the small variation of the value of the matching scale [54] also do not produce the significant corrections to the fitted expression.

*The main error* in Eqs. (10)-(13) is coming from the input value of top-quark mass: $\Delta M_{\rm min} \sim 2.0{\times}\Delta M_t^{\rm pole}$. Currently the most precise measurement of the top-quark mass has been reported as the world combination of the ATLAS, CDF, CMS and D0 [56] (for completeness we present also the result based only on CDF/D0 data [55]):

$$M_t = 173.18 \pm 0.94~{\rm GeV}\,,\quad [55] \tag{14}$$
$$M_t = 173.76 \pm 0.76~{\rm GeV}\,,\quad [56] \tag{15}$$

It has been pointed out in [57] that the values of the top quark mass quoted by the experimental collaborations correspond to parameters in Monte Carlo event generators in which, apart from parton showering, the partonic subprocesses are calculated at the tree level, so that a rigorous theoretical definition of the top quark mass is lacking (see more detailed discussion of this issue in [58]). In particular, the following issues in precision top mass determination at hadron colliders are relevant [59]: MC modeling, reconstruction of the top pair, unstable top and finite top width effects, bound-state effects in top pair production at hadron colliders, renormalon ambiguity in top mass definition, alternative top mass definitions, higher-order corrections, non-perturbative corrections, contributions from physics beyond the Standard Model.

To reduce the uncertainties related with undetermined differences between Monte-Carlo and pole masses (it was estimated in [57, 58] as 1 GeV) the mass of top-quark can be extracted directly from a measurement of the total top-pair production cross section $\sigma_{\rm exp}(p\bar{p} \to t\bar{t}{+}X)$. Such analysis performed in [60] with NNLO accuracy with inclusion of the full theoretical uncertainties (the scale variation as well as the (combined) PDF and $\alpha_s$ uncertainties) gives rise to the following result, $M_t^{\rm pole} = 173.3 \pm 2.8$ GeV . The central value is very close to the one in Eq. (15), but the theoretical uncertainty is much larger. Similar analyses (direct extraction of the pole mass of top-quark from measured total cross section) have been performed by a few other groups [58, 61–64] with the following results:

$$M_t^{\rm pole} = 171.4^{+5.4}_{-5.7}\ {\rm GeV} \quad [61]\,, \tag{16}$$
$$M_t^{\rm pole} = 174.3^{+4.9}_{-4.4}\ {\rm GeV} \quad [62]\,, \tag{17}$$
$$M_t^{\rm pole} = 174.2^{+3.6}_{-3.9}\ {\rm GeV} \quad [58]\,, \tag{18}$$
$$M_t^{\rm pole} = 171.2 \pm 2.4 \pm 0.7\ {\rm GeV} \quad [64]\,, \tag{19}$$

where the full NNLO QCD corrections evaluated in [65, 66] have been combined with the soft-gluon resummation at NNLL accuracy [63] and Coulomb-gluon NNLL resummation [61]. All results in Eqs. (16)-(19) have large theoretical uncertainties. To improve the current precision of the top-mass determination from the total cross section the higher order corrections as well as a reduction of PDF and $\alpha_s$ uncertainties are required [67, 68].

## 3. Some additional sources of uncertainties

### *3.1. EW contribution to the running mass of the top-quark*

In order to achieve percent level precision theoretical predictions for cross section $\sigma_{pp\to t\bar{t}}$ not only QCD NNLO radiative corrections should be applied. The EW part as well as mixed EW $\times$ QCD corrections have to be included in a systematic way. For example, the QCD interaction is not responsible for the non-zero width of the top-quark, which can be understood precisely only by inclusion of the EW interactions. In any case, the EW effect [69, 70] as well as the non-zero width should be included in addition to the QCD corrections [71].

In contrast to QCD, where the mass of a quark is the parameter of the Lagrangian, the notion of $\overline{\rm MS}$-mass in EW theory is not determined entirely by the prescriptions of minimal subtraction. It depends on the value of vacuum expectation value $v(\mu^2)$ chosen as a parameter of the calculations so that the running mass is $m_t(\mu^2) = 1/\sqrt{2}y_f(\mu^2)v(\mu^2)$. It has been shown in a series of papers [72–74] that in the scheme with explicit inclusion of tadpoles [25] the RGE for the running vacuum expectation value $v^2_{MS}(\mu^2)$ which is defined as (see Section 4 in [36])

$$v^2_{MS}(\mu^2) = \frac{1}{\sqrt{2}G_F}\frac{1}{1-\overline{\Delta}r}\left[\frac{m^2_W(\mu^2)}{M^2_W}\right]\left[\frac{\alpha(M_Z)}{\alpha_{MS}(\mu^2)}\right]\left[\frac{\sin^2\theta_W^{MS}(\mu^2)}{\sin^2\theta_W^{OS}}\right], \tag{20}$$

coincides with RGE for the classical definition of tree-level scalar potential vacuum

$$\mu^2\frac{d}{d\mu^2}v^2_{MS}(\mu^2) = \mu^2\frac{d}{d\mu^2}\left[\frac{m^2(\mu^2)}{\lambda(\mu^2)}\right], \tag{21}$$

where $m^2$ and $\lambda$ are the parameters of the scalar potential, see Eq. (3). The asymptotic behavior and properties of this vacuum at low and high energies have been analysed in [75, 76]. In particular, for a current values of Higgs and top-quark masses an IR point $\mu_{IR}$ close to the value of the Z-boson mass exist such that $v^2_{MS}(\mu^2)\big|_{\mu\sim M_Z} = (\sqrt{2}G_F)^{-1/2} = 246.22$ GeV (see details in [76]). In the framework of the effective potential approach, a similar condition for the minimum of effective potential is imposed by hand (see details in [15]).

In the SM the decoupling theorem [77] is not valid in the week sector, the "decoupling by hand" prescription does not work and we have to take full SM parameter relations as they are. In this approximation we got [75] that the EW contribution is large and has opposite sign relative to the QCD contributions, so that the total SM correction is small and approximately equal to $M_t - m_t(m_t) \sim 1\div 2$ GeV (see left plot in Fig. (3)). The complete $O(\alpha^2)$ correction to the relation between pole and $\overline{\rm MS}$ masses of the top-quark are not yet available, but our numerical estimation [75] is in agreement with result of [38, 39] and it is of the order of the $O(\alpha_s^4)$ corrections [78, 79]. For the evaluation of the NNLO corrections three different schemes have been used: with explicit inclusion of tadpole [72–74], the so called $\beta$-renormalization scheme [33–35], and the method of minimization of the effective potential [12, 51, 80]. The equivalence of these three methods of evaluation of matching conditions have never been analyzed. In particular, the structure of 2-loop UV-counterterms in [72–74] and [33,34] are different; the gauge dependence of the effective potential method versus gauge independence of the scheme including the tadpoles [72–74] (the gauge (in)dependence of prediction of the critical value of the Higgs boson mass via effective potential scheme have been analysed in [81,82]), etc. The difference between these methods is of order $O(G_F^2 M_W^4)$ and may reach the value $\sim 10^{-3} \div 10^{-4}$. Another source of numerical difference is related with the choice between complex mass [35] or pole mass [72, 74] in Eq. (8). It is equivalent to systematic additional uncertainty of order $O(\Gamma^2/(4M^2))$ (see discussion in [37,83]), where $\Gamma$ is the width and $M$ is the mass of particle. This effect is the most significant for the top-quark Yukawa-coupling: $\delta y_f \sim 1/173^2 = 3\times 10^{-5}$. The unknown QCD corrections of order $O(\alpha_s^4)$ to the top-quark Yukawa-coupling is $\delta_{\alpha_s^4} y_f \sim 1 \times 10^{-3}$ [78, 79]. The two-loop EW correction $\delta_{\alpha^2} y_t(\mu = M_t) \approx \frac{5}{(4\pi)^2} = 6 \times 10^{-4}$ (see Eq.(2.49) in [12]) and three-loop corrections can be estimated as $\delta_{\alpha_s^2 \alpha} y_t(\mu = M_t) \sim [\delta_{\alpha_s^2} y_t(\mu = M_t)] \times [\delta_\alpha y_t(\mu = M_t)] = 0.011 \times 0.00135 = 1.5 \times 10^{-5}$. All these effects do not change the central value of the critical mass of the Higgs boson, Eqs. (10)-(13), but affect the value of theoretical uncertainties.

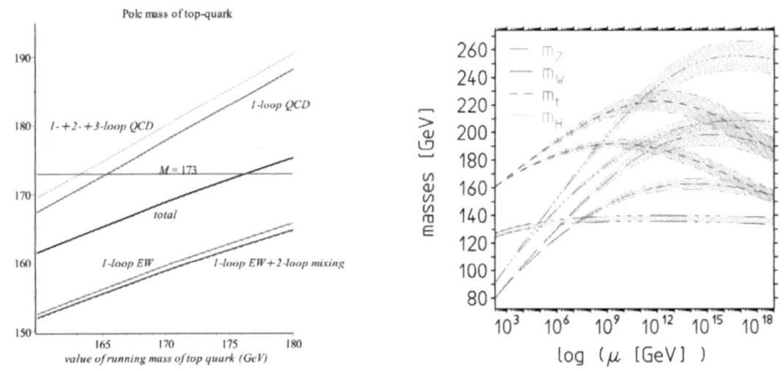


**Figure 3.** The running top-quark mass at low value of $\mu$ (left) and behavior of running masses at large values of $\mu$ (right). At the right the bands corresponds to Higgs mass in interval 124 GeV $< M_H <$ 127 GeV. The right plot is taken from [11].

### *3.2. The first order phase transition and quadratic divergences*

It was pointed out in [11] that in the region where the Higgs self-coupling is positive and close to zero, $\lambda(\Lambda) \sim 0$, the quadratic term of effective potential may start play the essential role since $\lambda(\Lambda)\phi^2(\Lambda) \ll m^2(\Lambda)$. The massive parameter $m^2$ suffers from quadratic divergences: $\Delta m^2(\Lambda) = \frac{\Lambda^2}{16\pi^2} Q_1$ , where the coefficients $Q_1$ are expressible in terms of coupling constants [84], $Q_1 = \lambda + \frac{1}{8}g_1^2 + \frac{3}{8}g_2^2 - y_t^2$ . The coefficient $Q_1$ may vanish at some high scale $\Lambda^*$ [85]. If $\Lambda \sim \Lambda^*$ then $m^2$-term may change the sign and that leads to a phase transition of the first order restoring the EW symmetry [11]. Realization of this scenario (the value of scale where $Q_1 = 0$) strongly depends on the value of top-quark mass (see discussion in [86, 87]). The Higgs inflation and hierarchy problem within this scenario have been discussed in [88, 89].

### *3.3. Inclusion of higher dimension operators*

As well known [14] the effective potential contains also the higher dimensional operators (see Eq. (6)). The impact of new physics interaction at Planck scale was analysed in [90] by adding two higher dimensional operators $\phi^6$ and $\phi^8$ suppressed by inverse powers of Planck scale $M_{Planck}$ to the Higgs potential. It has been shown that higher dimension operators may change the lifetime of the metastable vacuum, $\tau$, from $\tau = 1.49 \times 10^{714} T_U$ to $\tau = 5.45 \times 10^{-212} T_U$, where $T_U$ is the age of the Universe.

## 4. Conclusion

The main results of the LHC is the discovery of the Higgs boson. The second important result is the absence of a signal of new physics. After the Higgs boson discovery the Standard Model is completed. Since all parameters of the Standard Model are defined now experimentally, one can analyze the extrapolation of the SM up to the Planck scale. The results of the recent analysis can be summarized as follows:

- The Standard Model is a self-consistent QFT that can be extrapolated from $M_W$ to $M_{Planck}$ since all SM couplings remain perturbative (no Landau pole) in that range, see Fig. 1.
- With the current precision in $M_H(\sim 126$ GeV) , the value of top-quark mass in accordance with CDF/DO/CMS/ATLAS $M_t(\sim 173$ GeV) and 3-loop RG functions and 2-loop matching conditions, one concludes that the EW vacuum would most likely be *metastable* [12]: the *stability condition* is $M_H > (129.6 \pm 1.5)$ GeV for a given value of top-quark mass. In terms of top quark mass, the stability bound is [22] $M_t < (171.36 \pm 0.46)$ GeV.
- It was shown in [49] that the lifetime of the electroweak vacuum is longer than the age of the Universe for $M_H > 111$ GeV so that
  *Metastability of vacuum with very long lifetime cannot be used as motivation for a New Physics*
  A lot of effort has been made recently to analyzed EW vacuum during inflation [91, 92]. The condition of vacuum (meta)stability imposes the constraint on the rate of inflationary expansion that are in tension with BICEP2 result [93]. However, the recent result from the Plank collaboration [94] does not confirm the BICEP2 result [93] so that the further experimental verification is necessary.
- The result of the recent analysis [60] reveals a large theoretical uncertainties in the value of top-quark mass (see Eqs. (16)-(19)). This leads to a large uncertainty $\sim 5$ GeV in the critical value of the Higgs boson mass.
- The mass of the Higgs boson is very close to the values of "critical Higgs mass", $M_{crit}$: the "multiple point principle" [95], Higgs inflation [96–98], asymptotic safety scenario [99] The explicit realization $M_H = M_{crit}$ would be a strong indication for the absence of a new physics scale between the Fermi and Planck scales.
- To clarify the situation, more precise measurements of the coupling constants $\alpha_s, \lambda, y_t$ are needed. Unfortunately, there are no ways of directly measuring the Higgs self-coupling and top-quark Yukawa coupling with high precision in the immediate future [100–103].
- The higher order calculations are also desirable.

Therefore,
**precision determinations of parameters are more important than ever and a real challenge for experiments at the LHC and at a future ILC.**

## 5. Acknowledgments

We are thankful to S.Alekhin, A.Arbuzov, F.Bezrukov and S.Moch for the stimulating discussions. MYK would like to thank the organizers of the conference ACAT2014 and specially to Andrei Kataev, Grigory Rubtsov and Milos Lokajicek for the invitation and for creating such a stimulating atmosphere. The authors are indebted to anonymous referee for constructive and valuable comments. This work was supported in part by the German Federal Ministry for Education and Research BMBF through Grant No. 05 H12GUE, by the German Research

Foundation DFG through the Collaborative Research Center No. 676 *Particles, Strings and the Early Universe—The Structure of Matter and Space-Time.*

## References

[1] Aad G *et al.* [ATLAS Collaboration] 2012 *Phys. Lett.* B **716** 1 (Preprint 1207.7214)
[2] Chatrchyan S *et al.* [CMS Collaboration] 2012 *Phys. Lett.* B **716** 30 (Preprint 1207.7235)
[3] [CMS Collaboration] 2014 Precise determination of the mass of the Higgs boson and studies of the compatibility of its couplings with the standard model (Preprint CMS-PAS-HIG-14-009)
[4] G. Aad *et al.* [ATLAS Collaboration], *Phys. Rev.* D **90** 052004 (Preprint 1406.3827)
[5] 't Hooft G 1971 *Nucl. Phys.* B **35** 167
[6] 't Hooft G and Veltman M J G 1972 *Nucl. Phys.* B **44** 189
[7] Dicus D A and Mathur V S 1973 *Phys. Rev.* D **7** 3111
[8] Lee B W, Quigg C and Thacker H B 1977 *Phys. Rev.* D **16** 1519
[9] Vetlman M J G 1977 *Acta Phys. Polon.* B **8** 475
[10] Hambye T and Riesselmann K 1997 *Phys. Rev.* D **55** 7255 (Preprint 9610272)
[11] Jegerlehner F 2014 *Acta Phys. Polon.* B **45** 1167 (Preprint 1304.7813)
[12] Buttazzo D et al. 2013 *JHEP* **1312** 089 (Ppreprint 1307.3536)
[13] Coleman S R and Weinberg E J 1973 *Phys. Rev.* D **7** 1888
[14] Ford C et al. 1993 *Nucl. Phys.* B **395** 17 (Preprint 9210033)
[15] Casas J A et al. 1995 *Nucl. Phys.* B **436** 3 [Erratum-ibid. B **439** (1995) 466] (Preprint 9407389)
[16] Nakano H and Yoshida Y 1994 *Phys. Rev.* D **49** 5393 (Preprint 9309215)
[17] Einhorn M B and Jones D R T 2007 *JHEP* **0704** 051 (Preprint 0702295)
[18] Shaposhnikov M 2013 *PoS EPS* **-HEP2013** 155 (Preprint 1311.4979)
[19] Coleman S R 1977 *Phys. Rev.* D **15** 2929 [Erratum-ibid. D **16** (1977) 1248].
[20] Callan C G and Coleman S R 1977 *Phys. Rev.* D **16** 1762
[21] Arnold P B 1989 *Phys. Rev.* D **40** 613
[22] Espinosa J R 2013 Vacuum Stability and the Higgs Boson (Preprint 1311.1970)
[23] Sirlin A and Zucchini R 1986 *Nucl. Phys.* B **266** 389
[24] Sirlin A 1980 *Phys. Rev.* D **22** 971
[25] Fleischer J and Jegerlehner F 1981 *Phys. Rev.* D **23** 2001
[26] Böhm M, Spiesberger H and Hollik W 1986 *Fortsch. Phys.* **34** 687
[27] Hempfling R and Kniehl B A 1995 *Phys. Rev.* D **51** 1386
[28] van der Bij J and Veltman M J G 1984 *Nucl. Phys.* B **231** 205
[29] Ghinculov A and van der Bij J J 1995 *Nucl. Phys.* B **436** 30
[30] Awramik M et al. 2004 *Phys. Rev.* D **69** 053006 (Preprint 0311148)
[31] Aglietti U et al. 2004 *Phys. Lett.* B **595** 432 (Preprint 0404071)
[32] Degrassi G and Maltoni F 2004 *Phys. Lett.* B **600** 255 (Preprint 0407249)
[33] Actis S and Passarino G 2007 *Nucl. Phys.* B **777** 35 (Preprint 0612123)
[34] Actis S and Passarino G 2007 *Nucl. Phys.* B **777** 100 (Preprint 0612124)
[35] Actis S et al. 2008 *Phys. Lett.* B **670** 12 (Preprint 0809.3667)
[36] Jegerlehner F and Kalmykov M 2003 *Acta Phys. Polon.* B **34** 5335 (Preprint 0310361)
[37] Jegerlehner F and Kalmykov M 2004 *Nucl. Phys.* B **676** 365 (Preprint 0308216)
[38] Martin S P 2005 *Phys. Rev.* D **72** 096008
[39] Kniehl B A and Veretin O V 2014 *Nucl. Phys.* B **885** 459 (Preprin 1401.1844)
[40] Kazakov D I, Tarasov O V and Vladimirov A A 1979 *Sov. Phys. JETP* **50** 521
[41] Tarasov O V, Vladimirov A A and Zharkov A Y 1980 *Phys. Lett.* B **93** 429
[42] Mihaila L N, Salomon J and Steinhauser M 2012 *Phys. Rev. Lett.* **108** 151602 (Preprin 1201.5868)
[43] Mihaila L N, Salomon J and Steinhauser M 2012 *Phys. Rev.* D **86** 096008 (Preprint 1208.3357)
[44] Chetyrkin K G and Zoller M F 2012 *JHEP* **1206** 033
[45] Bednyakov A V, Pikelner A F and Velizhanin V N 2013 *JHEP* **1301** 017 (Preprint 1210.6873)
[46] Bednyakov A V, Pikelner A F and Velizhanin V N 2013 *Phys. Lett.* B **722** 336 (Preprin 1212.6829)
[47] Bednyakov A V, Pikelner A F and Velizhanin V N 2013 *Nucl. Phys.* B **875** 552 (Preprint 1303.4364)
[48] Chetyrkin K G and Zoller M F 2013 *JHEP* **1304** 091 [Erratum-ibid. **1309** 155] (Preprint 1303.2890)
[49] Elias-Miro J et al. 2012 *Phys. Lett.* B **709** 222 (Preprint1112.3022)
[50] Bezrukov F et al. 2012 *JHEP* **1210** 140 (Preprint 1205.2893)
[51] Degrassi G et al. 2012 *JHEP* **1208** 098 (Preprint 1205.6497)
[52] Martin S P 2014 *Phys. Rev.* D **89** 013003 (Preprint 1310.7553)
[53] Martin S P and Robertson D G 2014 (Preprint 1407.4336)

[54] Masina I 2012 *Phys. Rev.* D **87** 053001 (Preprint 1209.0393)
[55] T. Aaltonen *et al.* [CDF and D0 Collaborations] 2012 Phys. Rev. D **86** 092003 (Preprint 1207.1069)
[56] [ATLAS and CDF and CMS and D0 Collaborations] 2014 First combination of Tevatron and LHC measurements of the top-quark mass (Preprint 1403.4427)
[57] Hoang A H and Stewart I W 2008 *Nucl. Phys. Proc. Suppl.* **185** 220 (Preprint 0808.0222)
[58] Moch et al. 2014 High precision fundamental constants at the TeV scale (Preprint 1405.4781)
[59] Juste A et al. 2013 Determination of the top quark mass circa 2013: methods, subtleties, perspectives (Preprint 1310.0799)
[60] Alekhin S, Djouadi A and Moch S 2012 *Phys. Lett.* B **716** 214
[61] Beneke et al. 2012 *JHEP* **1207** 194 (Preprint 1206.2454)
[62] Beneke et al. 2013 *Nucl. Phys. Proc. Suppl.* **234** 101 (Preprint 1208.5578)
[63] Czakon M and Mitov M 2014 *Comput. Phys. Commun.* **185** 2930 (Preprint 1112.5675)
[64] Alekhin S, Blümlein J and Moch S 2014 *Phys. Rev.* D **89** 054028 (Preprint 1310.3059)
[65] Bärnreuther P, Czakon M and Mitov A 2012 *Phys. Rev. Lett.* **109** 132001 (Preprint 1204.5201)
[66] Czakon M, Fiedler P and Mitov A 2013 *Phys. Rev. Lett.* **110** 252004 (Preprint 1303.6254)
[67] Czakon M et al. 2013 *JHEP* **1307** 167 (Preprint 1303.7215)
[68] Ball R D *et al.* [The NNPDF Collaboration] 2014 Parton distributions for the LHC Run II (Preprint 1410.8849)
[69] Beenakker W et al. 1994 *Nucl. Phys.* **B411** 343
[70] Kühn J H, Scharf A and Uwer P 2007 *Eur. Phys. J.* **C51** 37 (Preprint 0610335)
[71] Falgari P, Papanastasiou A S and Signer A 2013 *JHEP* **1305** 156 (Preprint 1303.5299)
[72] Jegerlehner F, Kalmykov M and Veretin O 2002 *Nucl. Phys.* B **641** 285 (Preprint 0105304)
[73] Jegerlehner F, Kalmykov M and Veretin O 2003 *Nucl. Phys. Proc. Suppl.* **116** 382 (Preprint 0212003)
[74] Jegerlehner F, Kalmykov M and Veretin O 2003 *Nucl. Phys.* B **658** 49 (Preprint 0212319)
[75] Jegerlehner F, Kalmykov M and Kniehl B A 2013 *Phys. Lett.* B **722** 123 (Preprint 1212.4319)
[76] Jegerlehner F, Kalmykov M and Kniehl B A 2013 *PoS DIS* **2013** 190 (Preprint 1307.4226)
[77] Appelquist A and Carazzone J 1975 *Phys. Rev.* D **11** 2856
[78] Kataev A L and Kim V T 2010 *Phys. Part. Nucl.* **41** 946 (Preprint 1001.4207)
[79] Sumino Y 2014 *Phys. Lett.* B **728** 73 (Preprint 1309.5436)
[80] Martin S P 2004 *Phys. Rev.* D **70** 016005 (Preprint 0312092)
[81] Loinaz W and Willey R S 1997 *Phys. Rev.* D **56** 7416 (Preprint 9702321)
[82] Di Luzio L and Mihaila L 2014 *JHEP* **1406** 079 (Preprint 1404.7450)
[83] Smith M C and Willenbrock S S 1997 *Phys. Rev. Lett.* **79** 3825 (Preprint 9612329)
[84] Veltman M J G 1981 *Acta Phys. Polon.* B **12** 437
[85] Hamada Y, Kawai H and Oda K 2013 *Phys. Rev.* D **87** 053009 (Preprint 1210.2538)
[86] Masina I and Quiros M 2013 *Phys. Rev.* D **88** 093003 (Preprint 1308.1242)
[87] Jones D R T 2013 *Phys. Rev.* D **88** 098301 (Preprint 1309.7335)
[88] Jegerlehner F 2014 *Acta Phys. Polon.* B **45** 1215 (Preprint 1402.3738)
[89] Jegerlehner F 2013 The hierarchy problem of the electroweak Standard Model revisited, (Preprint 1305.6652)
[90] Branchina V and Messina E 2013 *Phys. Rev. Lett.* **111** 241801 (Preprint 1307.5193)
[91] Kobakhidze A and Spencer-Smith A 2013 *Phys. Lett.* B **722** 130 (Preprint 1301.2846)
[92] Enqvist E, Meriniemi T and Nurmi S 2014 JCAP **1407** (2014) 025 (Preprint 1404.3699)
[93] Ade P A R *et al.* [BICEP2 Collaboration] 2014 *Phys. Rev. Lett.* **112** 241101 (Preprint 1403.3985)
[94] Adam R *et al.* [Planck Collaboration] 2014 Planck intermediate results. XXX. The angular power spectrum of polarized dust emission at intermediate and high Galactic latitudes (Preprint 1409.5738)
[95] Froggatt C D and and Nielsen H B 1996 *Phys. Lett.* B **368** 96 (Preprint 9511371)
[96] Bezrukov F L and Shaposhnikov M 2008 *Phys. Lett.* B **659** 703 (Preprint 0710.3755)
[97] Hamada Y et al. 2014 *Phys. Rev. Lett.* **112** 241301 (Preprint 1403.5043)
[98] Bezrukov F and Shaposhnikov M 2014 Higgs inflation at the critical point (Preprint 1403.6078)
[99] Shaposhnikov M and Wetterich C 2010 *Phys. Lett.* B **683** 196 (Preprint 0912.0208)
[100] Dawson S et al 2013 Working Group Report: Higgs Boson (Preprint 1310.8361)
[101] Englert C et al. 2014 *J. Phys.* G **41** 113001 (Preprint 1403.7191)
[102] Bechtle P et al. 2014 Probing the Standard Model with Higgs signal rates from the Tevatron, the LHC and a future ILC (Preprint 1403.1582)
[103] Baak M *et al.* [Gfitter Group Collaboration] 2014 *Eur. Phys. J.* C **74** 3046 (Preprint 1407.3792)

Journal of Physics: Conference Series **608** (2015) 012075 doi:10.1088/1742-6596/608/1/012075

# NLO high multiplicity processes

**D. Maître**

Institute for Particle Physics Phenomenology, University of Durham, Durham DH1 3LE, UK

E-mail: daniel.maitre@durham.ac.uk

**Abstract.**
In this presentation some aspects of Next-to-Leading Order (NLO) calculations in QCD are presented. The focus is brought to aspects of such calculations for processes involving a high final-state particle multiplicity.

## 1. Introduction

Events with a large number of jets are common place at the LHC, both in Standard Model (SM) measurements and as signature or background to beyond the standard model (BSM) models. Benchmark SM processes such as vector boson and jets have been measured with up to 6 jets[1, 2, 3, 4].

Precise theoretical predictions are needed to compare with these precise measurement and also provide a good understanding of these processes when they are a background to new physics searches. A good theoretical understanding also helps designing so-called data-driven background estimation.

NLO calculations improve upon Born-level calculations by reducing the dependence of the latter on the unphysical renormalisation scale. The differential shape of distributions are described in a more precise way and the structure of jets start to be resolved at that order. The unphysical renormalisation scale dependence enters the cross section calculation through the strong coupling constant. The problem is exacerbated when a large number of jets – each yielding one power of the coupling constant – is considered. Figure 1 illustrates the increase in the renormalisation scale dependence as the power of the renormalisation constant increases. At NLO the situation is greatly improved since the virtual matrix elements they include have an additional dependence on the renormalisation scale which cancel the tree-level scale dependence to a large degree.

The uncertainty resulting from the scale dependence is usually assessed by varying the scales by factors between $1/2$ and 2 around a central scale. Figure 1 shows the relative uncertainties obtained using this procedure for the total cross section for W+jets production as a function of hte jet multiplicity from Ref. [5].

Until recently the bottleneck for NLO calculations was the calculation of the virtual matrix element. This bottleneck has been largely cleared in the last year, in what has been called the "NLO revolution". A lot of this progress has been achieved through the development of so-called unitarity techniques (for a review, see [6]). A sketch of how this method works is presented in the following section.

Journal of Physics: Conference Series **608** (2015) 012075 doi:10.1088/1742-6596/608/1/012075

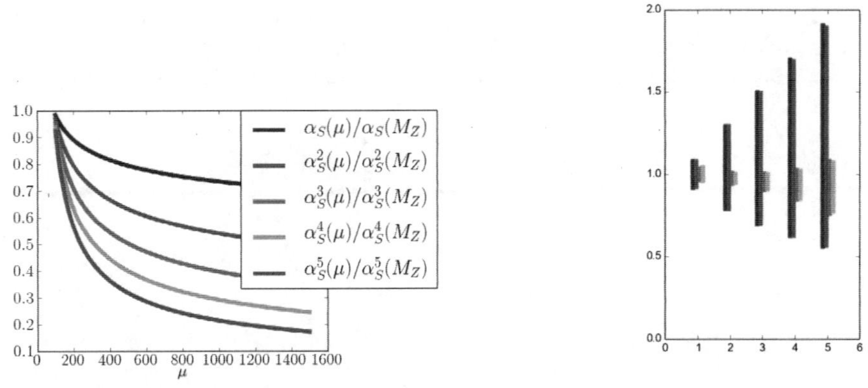


**Figure 1.** Left pane: Renormalisation scale dependence of the strong coupling constant. Each curve represents the dependence of the strong coupling constant on the renormalisation scale. The curves are normalised to their values at the Z boson mass. Right pane: Uncertainty due to renormalisation and factorisation scales for total cross section for the production of a $W$-boson as a function of the jet multiplicity, normalised by the central value. The blue and green bars represent the normalised uncertainty at LO for $W^-$ and $W^+$ respectively. The red and yellow bars show the NLO scale variation for $W^-$ and $W^+$ respectively.

## 2. Unitarity technique

To calculate NLO cross sections one has to include real emission matrix element and the virtual one-loop matrix elements for the process to account for one additional radiation and the influence of virtual particles respectively. In this section we concentrate on describing the unitarity technique approach to computing the virtual matrix element.

For this illustration we restrict ourselves to massless internal particles. In this case a general one-loop amplitude can be written in terms of known scalar two-, three- and four-point integrals.

$$\mathcal{A} = R + \sum d_i I_i^{(4)} + \sum c_i I_i^{(3)} + \sum b_i I_i^{(2)}, \tag{1}$$

where $I^{(n)}$ are the $n$-point scalar integral and the sums run over all possible propagator combinations. The scalar integrals are process independent. The first term $R$ is a rational part free of logarithm or polylogarithms. The coefficients $d_i$, $c_i$ and $b_i$ are rational functions of the spinor products. The calculation of the one-loop integral is reduced to the calculation of these coefficients.

The way unitarity techniques work is by applying so-called unitary cuts on both sides of Equation 1. A unitarity cut is the replacement under the integral:

$$\frac{1}{P^2} \to 2\pi i \delta(P^2)\,. \tag{2}$$

This replacement is done in the full loop integral on the left-hand side and in each individual scalar integral in the right-hand side. If the propagator being cut is not present in the expression the result of the cut is simply 0, unitarity cuts can thus be seen as projectors onto structures that exhibit a given set of propagators. To determine the coefficients in Equation 1 one needs to apply two, three or four cuts simultaneously. The delta-function constraints imposed by the cuts can often only be satisfied by complex momenta.

While the cuts on the right-hand side of Equation 1 will result in a set of integrals being replaced by a Jackobian factor and the other with 0, the cuts on the left-hand side will affect the propagators, transforming them into a sum of external wave-function factors. This is illustrated in Figure 2. As a result of the unitarity cut the integrand on the left-hand side of Equation 1

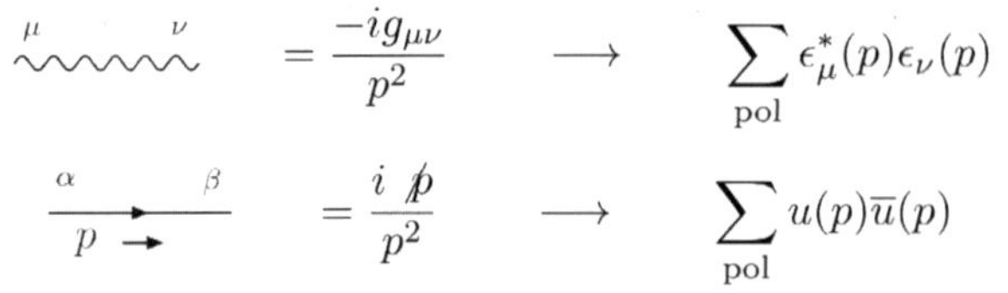


**Figure 2.** Effect on the gluon and quark propagator after application of a unitarity cut.

collapses into a product of trees. Depending on the number of cuts an integration over all momenta satisfying the cut conditions might remain. The next remaining step is to perform this integration. The nature of the one-loop integrand imposes constraints on the parametric dependence of the integrand as a function of the remaining parameter. Typically the functional form will be a rational function of the integration parameters. Through the knowledge of this form one can extract the value of the integral by evaluating the integrand at a very small number of points. To illustrate how this can be achieved we consider the example where the general form of the integrand is given by

$$i(t) = \frac{1}{t}\underbrace{\left(\frac{c_{-3}}{t^3} + \frac{c_{-2}}{t^2} + \frac{c_{-1}}{t} + c_0 + c_1 t + c_2 t^2 + c_3 t^3\right)}_{f(t)}$$

and the integration range of $t$ is a circle of unit radius in the complex plane. The result of the integral is clearly $2\pi c_0$ and the value of $c_0$ can be directly extracted from the combination of evaluations of the integrand at specific values:

$$I = \frac{2\pi i}{7}\sum_{j=0}^{6} i(e^{\frac{2\pi i j}{7}}) \; .$$

All integral coefficients can be extracted in this way starting from the box coefficients and using know coefficient to help determine lower-point coefficients. The procedure is illustrated in Figure 3. Once all coefficients of the scalar integrals are determined the last ingredient is the

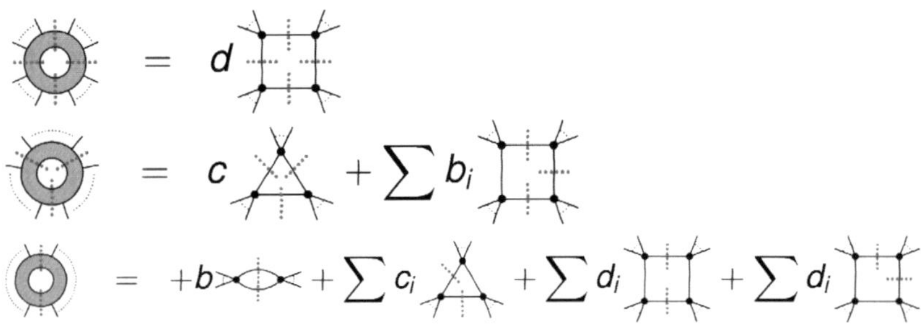


**Figure 3.** Illustration of the cut procedure to determine the scalar integral coefficients.

rational term. There are different ways of computing this term. A detailed description is beyond

Journal of Physics: Conference Series **608** (2015) 012075 doi:10.1088/1742-6596/608/1/012075

the scope of this illustration, details can be found in Ref. [6]. The unitarity method works very well at NLO and attempts are being made to apply it at NNLO (see for example ACAT 2013 proceedings [7] for a more detailed description). In this case the situation is more complicated, owing to the fact that there is no known analogous to Equation 1 at two-loop level.

## 3. NLO event files

NLO calculations for high multiplicity processes are computationally intensive and it is therefore advisable to try to reuse these calculations for as many analysis and observables as possible. The most expensive part of the calculation is the computation of the matrix element, while other steps in a typical analysis – such as the computation of the observables, parton distribution functions, jet clustering, – are relatively inexpensive. One solution to the problem is to store the computed matrix elements and the coefficient of scale dependent quantities in a file along with the phase space point. These files can then be reread for another analysis, which is less computationally expensive than the original computation. Details of a practical implementation of this strategy using ROOT files [**?**] as a storage back-end can be found in Ref. [9].

The advantages of this strategy is that analysis observables and cuts can be adapted to other analysis (provided of course that the event files have been generated for an inclusive enough phase-space for the analysis.) By storing the coefficients of factorisation and renormalisation scales logarithms allows for the inexpensive calculation of scale and pdf variation – an operation that is otherwise extremely costly. Event files may be easier to handle for experimental groups than the program that generated them. In addition, the validation of an event file is easier than the validation of an entire NLO program, which can facilitate the dissemination of NLO results. These advantages come at the price of large event files, and a balance between inclusivity and file size has to be found.

There are complications due to nature of the NLO calculation that one must be aware of when using NLO event files. The first is that NLO event files cannot be valid for arbitrary jet algoithms or arbitrary values of the parameters of a given jet algorithm such as the jet radius. The files are typically generated in such a way that they are compatible with a wide range of jet algorithms. The second issue concerns the estimation of the statistical error and has implications for the extraction of cumulative distributions and re-binning of histograms.

The issue is cause by the infrared divergences of the real matrix element. At NLO these infrared divergences are caused by the additional radiated particle going either collinear to another particle or soft. This divergence is cancelled by the infrared divergences in the virtual part, but this cancellation is happening in different phase space. To organise this cancellation across phase spaces the most fequently used strategy is to construct a subtraction term to the real matrix element that contains all the divergent behaviour of the real matrix element, while being simple enough to be integrated over the unresolved emission phase space. This integrated subtraction part is added to the virtual part to cancel its explicit infrared divergences.

The implication for the NLO event files is that the events for the real emission will contain events in the infrared regions that have associated subtraction terms. The weights of these events will be large (because of the infrared divergence) and anti-correlated (because the subtraction has been constructed for exactly that purpose.)

This artificial large correlation will lead to a severe over-estimation of the statistical integration error when using the usual formula for hte variance:

$$\begin{aligned}\sigma^2 &= \langle(\omega-\langle\omega\rangle)^2\rangle = \langle\omega^2\rangle - (\langle\omega\rangle)^2 \\ &\simeq \frac{\sum\omega_i^2}{N} - \left(\frac{\sum\omega_i}{N}\right)^2 = \frac{1}{N}\left(\sum\omega_i^2 - \frac{(\sum\omega_i)^2}{N}\right).\end{aligned}$$

To illustrate the problem, let us assume that all real weights and subtractions come in pairs, both of approximate size $\omega_+ \simeq -\omega_- \simeq L$ and that their difference (the actual physically relevant quantity) is of order $s$. In this case the error will be estimated by

$$\sigma^2 \simeq \frac{1}{N}\left(NL^2 + \frac{(Ns)^2}{N}\right) = L^2 - s^2 \simeq L^2 \tag{3}$$

Which overestimate the error. We get a better estimation by summing the contributions of the real part and the associated subtraction *before* squaring

$$\sigma^2 \simeq \frac{1}{N}\left(\sum(\omega_{i,+} + \omega_{i,-})^2 - \frac{\left(\sum \omega_i\right)^2}{N}\right). \tag{4}$$

In some cases even using NLO event files may be too inefficient. This can happen for example when one is interested in a particular distribution to perform a pdf fit or strong coupling determination. In this case one can use dedicated programs such as FastNLO [10] or ApplGrid [11] to generate an interpolation table for a specific distribution that allows to predict the distribution with a different pdf set or one that uses a different value of the strong coupling constant. FastNLO works by approximating the pdf

$$f_a(x) \simeq \sum_i f_a(x_i) E(x). \tag{5}$$

using a set of interpolating functions $E(x)$ subject to the constraint

$$\sum_i E_i(x) = 1\,, \qquad E_i(x_j) = \begin{cases} 1 & \text{if} \quad i = j \\ 0 & \text{otherwise} \end{cases} \tag{6}$$

Using this approximation one can then write any cross section or distribution as

$$\sigma = \sum_{a,b,i,j} f_a(x_i) f_b(x_j) \underbrace{\int dx_1\, dx_2\, E_i(x_1) E_j(x_2) \int d\phi\, \hat{\sigma}(x_1, x_2; \phi)}_{c_{a,b;i,j}}\,. \tag{7}$$

The coefficients $c_{a,b;i,j}$ can be computed using NLO event files by replacing the pdfs with the functions $E(x)$, which is possible since the appropriate coefficients are stored to allow for a change of pdf or factorisation scale. The use of such tables also reduces the size of the files to be handled from the hundreds of Gb to a few Mb.

We now have a situation where one can gain evaluation speed for a NLO calculation at the cost of loss of generality. First on can use NLO event files, where one is restricted to a fixed set of jet algorithms and a set of generation cut. Within these constraint, it is possible to perform any analysis and compute any observable. One can improve the speed further by choosing a fixed analysis and a fixed observable and using fastNLO.

## 4. High multiplicity

To illustrate the importance of high multiplicity processes let us consider the distribution of the total transverse energy $H_T$ in $Z/W$ production with at least one jet as measured by Atlas [12, 4] and shown in Figure 4.

We can see that the NLO description is not good for large values of $H_T$. A closer investigation shows that this is due to the fact that higher multiplicity processes are important for large values of $H_T$. This can be seen in Figure 5 from Ref. [14]. This is one example illustrating that high

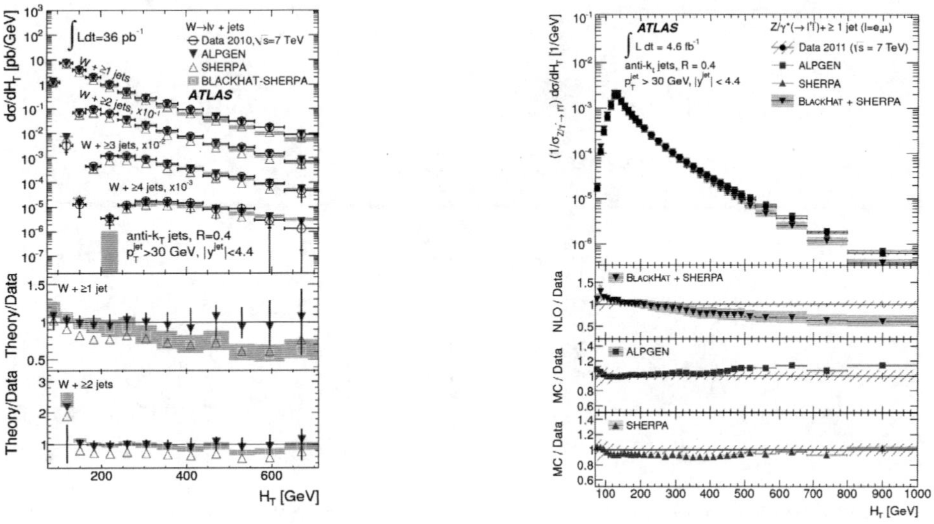


**Figure 4.** $H_T$ distribution in $W + 1$ jet (left) and $Z + 1$ jet (right) as measured by Atlas.

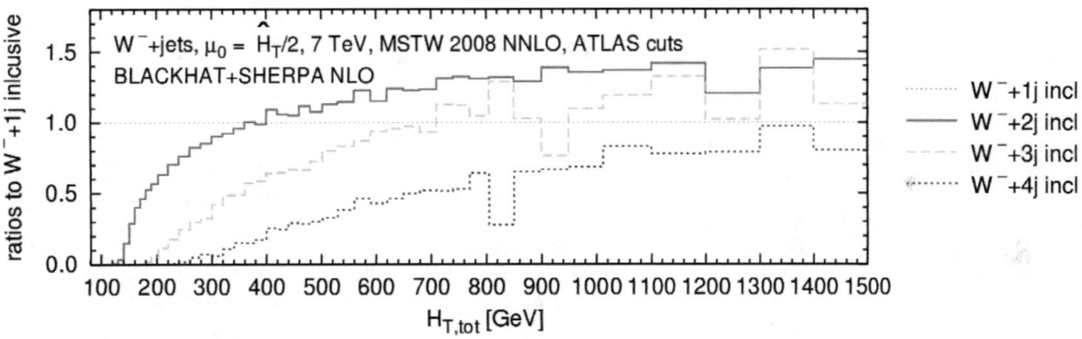


**Figure 5.**

multiplicity processes can be important in the tails of some distributions, where many searches for new physics focus their efforts.

One possible way of including higher multiplicity contributions to the distribution is add the contribution of exclusive NLO predictions for different multiplicities. By exclusive is meant here to put a cut on the transverse momentum of the additional $(n+1)$-th jet in each $V + n$ jets sample to avoid double-counting. This method is very crude and has the formal accuracy of the first sample, so from that point of view no improvement is to be expected, but for observables for which high multiplicity processes are relevant this can yield an improvement. Figure 6 taken from ref. [13] shows how higher multiplicity processes gain a larger importance as one moves towards the tail of the distribution.

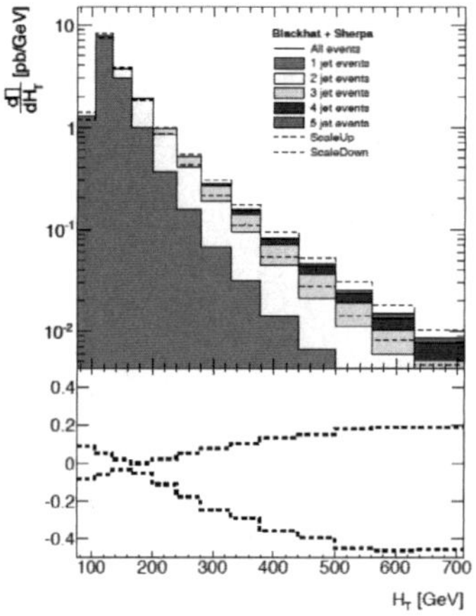


**Figure 6.** Various multiplicity contributions to the $H_T$ distribution.

## Acknowledgements
DM's work was supported by the Research Executive Agency (REA) of the European Union under the Grant Agreement number PITN–GA–2010–264564 (LHCPhenoNet).

## References
[1] V. Khachatryan *et al.* [CMS Collaboration], arXiv:1408.3104 [hep-ex].
[2] V. Khachatryan *et al.* [CMS Collaboration], arXiv:1406.7533 [hep-ex].
[3] G. Aad *et al.* [ATLAS Collaboration], JHEP **1307** (2013) 032 [arXiv:1304.7098 [hep-ex]].
[4] G. Aad *et al.* [ATLAS Collaboration], Phys. Rev. D **85** (2012) 092002 [arXiv:1201.1276 [hep-ex]].
[5] Z. Bern, L. J. Dixon, F. Febres Cordero, S. Höche, H. Ita, D. A. Kosower, D. Maître and K. J. Ozeren, Phys. Rev. D **88**, 014025 (2013) [arXiv:1304.1253 [hep-ph]].
[6] H. Ita, J. Phys. A **44** (2011) 454005 [arXiv:1109.6527 [hep-th]].
[7] S. Badger, H. Frellesvig and Y. Zhang, J. Phys. Conf. Ser. **523** (2014) 012061 [arXiv:1310.4445 [hep-ph]].
[8] R. Brun and F. Rademakers, Nucl. Instrum. Meth. A **389**, 81 (1997); I. Antcheva *et al.*, Comput. Phys. Commun. **182**, 1384 (2011).
[9] Z. Bern, L. J. Dixon, F. Febres Cordero, S. Hche, H. Ita, D. A. Kosower and D. Maitre, Comput. Phys. Commun. **185** (2014) 1443 [arXiv:1310.7439 [hep-ph]].
[10] D. Britzger *et al.* [fastNLO Collaboration], arXiv:1208.3641 [hep-ph].
[11] M. Sutton, T. Carli, D. Clements, A. Cooper-Sarkar, C. Gwenlan, G. P. Salam, F. Siegert and P. Starovoitov, PoS DIS **2010** (2010) 051.
[12] G. Aad *et al.* [ATLAS Collaboration], JHEP **1307** (2013) 032 [arXiv:1304.7098 [hep-ex]].
[13] J. Alcaraz Maestre *et al.* [SM AND NLO MULTILEG and SM MC Working Groups Collaboration], arXiv:1203.6803 [hep-ph].
[14] D. Matre and S. Sapeta, Eur. Phys. J. C **73** (2013) 2663 [arXiv:1307.2252].
[15] J. R. Andersen *et al.* [SM and NLO Multileg Working Group Collaboration], arXiv:1003.1241 [hep-ph].

# Cosmic ray propagation with CRPropa 3

**R Alves Batista[1], M Erdmann[2], C Evoli[1], K-H Kampert[3], D Kuempel[2], G Mueller[2], G Sigl[1], A Van Vliet[1], D Walz[2], T Winchen[3]**

[1] University of Hamburg, II Institut für Theoretische Physik, Hamburg, Germany
[2] RWTH Aachen University, Physikalisches Institut IIIa, Aachen, Germany
[3] University of Wuppertal, Department of Physics, Wuppertal, Germany

E-mail: walz@physik.rwth-aachen.de

**Abstract.** Solving the question of the origin of ultra-high energy cosmic rays (UHECRs) requires the development of detailed simulation tools in order to interpret the experimental data and draw conclusions on the UHECR universe. CRPropa is a public Monte Carlo code for the galactic and extragalactic propagation of cosmic ray nuclei above $\sim 10^{17}$ eV, as well as their photon and neutrino secondaries. In this contribution the new algorithms and features of CRPropa 3, the next major release, are presented. CRPropa 3 introduces time-dependent scenarios to include cosmic evolution in the presence of cosmic ray deflections in magnetic fields. The usage of high resolution magnetic fields is facilitated by shared memory parallelism, modulated fields and fields with heterogeneous resolution. Galactic propagation is enabled through the implementation of galactic magnetic field models, as well as an efficient forward propagation technique through transformation matrices. To make use of the large Python ecosystem in astrophysics CRPropa 3 can be steered and extended in Python.

## 1. Introduction

The question of the origin of ultra-high energy cosmic rays (UHECRs) continues to be of high interest. The experimental situation for cosmic rays above $10^{17}$ eV is promising, with the Telescope Array and the Pierre Auger Observatory covering the northern and southern hemisphere with large exposure. Recent measurements include the energy spectrum [1, 2], composition [3, 4, 5] and anisotropy studies [6]. Interpreting these measurements in terms of concrete astrophysical scenarios requires detailed simulations of cosmic ray propagation from the source to the observer. In these simulations the deflection of UHECRs need to be computed over several orders of magnitude in energy and length scales, ranging from hundreds of megaparsecs down to galactic kiloparsec scales. Furthermore, all relevant interactions, such as photo-disintegration, pion production and pair production, need to be included. The simulation tool should be flexible enough to cover the large parameter space of possible astrophysical scenarios, in order to constrain the origin of UHECRs in comparison with experimental data. For this purpose CRPropa 3 was developed, with the physics processes based to a large extent on the original CRPropa 2.0 [7]. In this contribution we will summarize its main new features.

## 2. Overview of CRPropa

CRPropa is a publicly available simulation software for the Monte Carlo (MC) propagation of cosmic ray nuclei, photons and neutrinos through an extragalactic and galactic environment.

CRPropa simulates interactions such as electron pair-production, pion-production and photo-disintegration of nuclei with the diffuse extragalactic background radiation, as well as nuclear decay, and it includes secondary particles created in these interactions.

The simulations can be performed either in a one-dimensional (1D) or three-dimensional (3D) mode. The 3D mode allows to define a 3D source distribution and takes into account the deflections of charged cosmic rays in extragalactic magnetic fields. In the 1D mode the cosmological evolution of the sources and background radiation as well as the adiabatic energy loss of cosmic rays can be implemented. CRPropa thereby enables the user to predict the spectra of UHECR (and of their secondaries), their composition and arrival directions. All of these features are inherited by the new version CRPropa 3 [8].

## 3. New features of CRPropa 3

CRPropa 3 features several significant advancements in the CRPropa development. The main improvement is a new modular simulation layout, that allows for multiple simulation use cases, and for easier testing, maintenance and physics extensions.

### *3.1. Code structure, steering and parallelization*

CRPropa 3 was completely rewritten with a new code structure, separating all aspects of the simulation into independent modules that correspond to individual photo-nuclear interactions, boundary conditions, observers etc. The cosmic ray particle class serves as a single interface between the simulation modules. The modules provide a method to update the cosmic ray particle according to the module's purpose. The simulation itself is a user-defined sequence of simulation modules, that are called in turn to update the cosmic ray until the propagation is either completed or aborted. Since there are no direct dependences between modules, any combination of modules can in principle be selected, allowing for multiple use cases and to study in detail individual propagation aspects.

Efficient MC propagation depends on dynamically adjusting the step size to accommodate for varying conditions, e.g. making smaller steps in regions of strong magnetic deflections. A bidding system allows all modules of a simulation to bid for the next step. The lowest bid is then selected as step size for the next iteration of the module sequence. Therefore the propagation proceeds with the largest possible step that still ensures the numerical accuracy as defined by the user.

CRPropa 3 is written in C++ and interfaced to Python using SWIG [1]. This allows to set up and steer simulations in a high level scripting language while all computations are performed with the underlying C++ code. The SWIG interface enables cross-language polymorphism, which can be used to extend a CRPropa simulation directly from the Python script that runs it. The user can for example write a custom simulation module in Python to be used in combination with the existing C++ modules. While the Python usage is the advised steering mode, backwards-compatibility to the XML steering of CRPropa 2 is provided as well.

Cosmic ray propagation is a perfectly parallel task as interactions between cosmic rays are negligible. Current multicore processors can therefore be adequately utilized by just running multiple simulation instances in parallel. However, for better memory utilization, CRPropa 3 enables shared-memory multiprocessing using OpenMP [2]. This allows to use higher resolution magnetic fields and matter distributions in the simulation. The parallelization occurs on the level of the module sequence with a dynamic distribution of cosmic rays among the available threads. As the simulation modules are stateless, only a single instance of each module is needed. The speedup is limited by the number of critical sections that are not thread safe and can only

[1] `www.swig.org`
[2] `www.openmp.org`

be executed by one thread at a time. The critical section with the largest impact is the external library SOPHIA [9], used to simulate photo-pion interactions. Thus, the speedup depends on the frequency of these interactions. The speedup for a typical extragalactic simulation is shown in fig. 1.

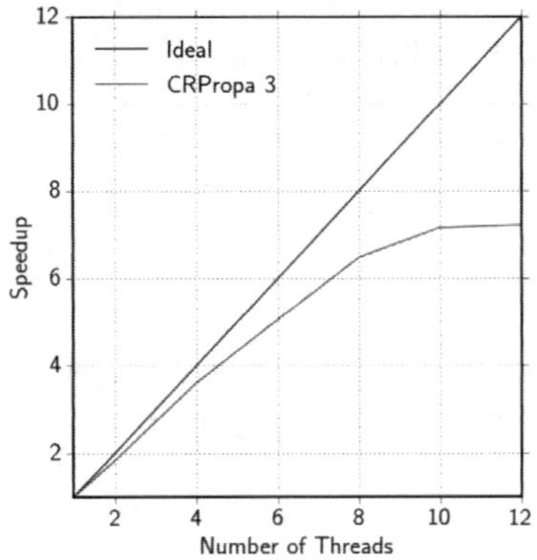


Figure 1: Speedup of CRPropa 3 in an example simulation of extragalactic propagation. The presence of non-parallelized sections limit the speedup to about 6-8 in typical simulations.

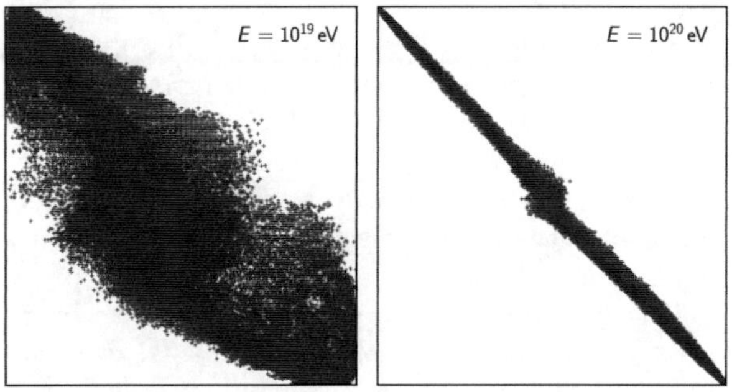


Figure 2: Representation of the non-zero matrix elements of the magnetic field lens for the JF12 model and two example energies, $10^{19}$ and $10^{20}$ eV. Non-diagonal entries correspond to cosmic ray deflections, which are more pronounced with decreasing energy. The sparsity of the matrices is used to significantly lower the memory requirement.

### *3.2. Magnetic field techniques*

Magneto-hydrodynamical simulations of structure formation provide models of the structured extragalactic magnetic field and matter distribution. The challenge of these simulations is the large difference in scale between the simulation volume of $\sim (0.1 - 1\,\mathrm{Gpc})^3$ size and the required resolution to resolve the structures down to the level of galaxies $\sim 1 - 10$kpc. Two of the techniques used in this context are smooth particles (SP) and adaptive mesh refinement (AMR). In order to directly use the resulting extragalactic magnetic field models, CRPropa provides interfaces to the SP-code Gadget [10] and the AMR-code RAMSES [11]. The trade off, however, for the small memory demand of these techniques are large lookup times, which easily become the bottleneck for tracking cosmic rays. In contrast, regular grids provide fast lookup times but easily require a terabyte memory. As a compromise CRPropa offers the use of modulated grids, which are a combination of a small high-resolution vector grid that is periodically repeated to cover a larger volume, and a large low-resolution modulation grid that carries information about the large-scale structure.

### *3.3. Galactic propagation*

Estimates for the strength of the galactic magnetic field imply that cosmic rays of rigidities $E/Z > 1\,\mathrm{EeV}$, where $E$ is the energy and $Z$ the charge number, may be considerably deflected in the Galaxy, but without propagating diffusively. Thus, due to the short galactic propagation as compared to extragalactic distances, energy loss processes are usually neglected and only magnetic deflections are considered. CRPropa 3 enables the galactic propagation by providing models of the galactic magnetic field and making use of the modular simulation layout. Both forward- and backward tracking are supported. Forward tracking is computationally expensive as the Earth is a point-like target compared to galactic distances, resulting in a very small hit probability. In the second approach, cosmic rays are propagated backwards from Earth to the galactic border. However, it is not straightforward to connect galactic backtracking with extragalactic forward tracking. As an alternative third approach, CRPropa 3 provides the lensing technique described and implemented in the PARSEC code [12]. The lensing technique uses a set of matrices for different rigidities that are constructed from backtracking simulations with e.g. CRPropa 3. The matrices transform cosmic ray directions at the galactic border into arrival directions observed at Earth. The directions are binned with a HEALpix [13] scheme into $\sim 50,000$ pixels for angular resolutions $< 1°$. The matrices thus have $\sim 50,000^2$ entries, nominally corresponding to 9 GiB in single precision. Sparse matrices are utilized to significantly lower the memory requirement to typically less than 10 MB per matrix. A representation of the non-zero elements of two example matrices is shown in fig. 2. The lenses[3] can be used to transform entire arrival distributions through matrix-vector multiplications, or, in the context of individual cosmic rays, as a lookup table of precomputed trajectories.

Generic combinations of axisymmetric or bisymmetric spiral disc and halo fields can be considered as magnetic field models. CRPropa 3 also implements the JF2012 model [14, 15] including the random large-scale and turbulent small-scale component. Additional user defined models can be easily implemented through Python extensions.

### *3.4. 4D simulations*

Cosmological effects such as the evolution of the background radiation are important when simulating the propagation of UHECRs. In fact, including these effects in 3D simulations is necessary for anisotropy studies, except for the highest cosmic ray energies, where propagation distances are sufficiently short. However, in the presence of magnetic deflections it is not possible to know *a priori* the effective propagation length and, therefore, the initial redshift of a cosmic ray that is observed at the present time. Consequently, CRPropa 3 allows for 4D simulations, in which cosmic rays propagate both spatially and in time. While computationally expensive, due to the additional loss of statistics in the time dimension, there are multiple applications for these simulations. In addition to the aforementioned anisotropy studies, effects of magnetic suppression can be directly simulated for spectrum and composition studies. 4D simulations can also be an important tool for the validation of *a posteriori* corrections, such as the magnetic suppression parametrization in [16] and [17].

## 4. Example application

In the following we present an example of a 3D simulation of cosmic ray nuclei through a structured universe and consider the galactic propagation with the lensing technique.

As a model for the structured universe we use the simulated matter distribution and magnetic fields of Dolag *et al.* [18] and Miniati *et al.* [19]. The Dolag simulation reproduces the local matter distribution by using a redshift survey as a constraint for the initial conditions. The magnetic field arises from a primordial seed field, and is compatible with rotation measures of galaxy clusters. However, the overall magnetic field strength is weaker compared to the Miniati simulation and leads to small UHECR deflections. To emphasize the cosmic ray deflections, we make use of the relation of matter density vs. magnetic field strength obtained from the Miniati simulation. With this relation we translate the Dolag matter distribution into a distribution of the magnetic field strength and use this as a modulation field for a higher resolution turbulent field, as described in 3.2. Both modulation and vector field are stored on a Cartesian grid. The modulation field covers a cubic volume of 132 Mpc edge length with a resolution of 300 kpc.

[3] Lenses for a number of galactic magnetic field models can be found on
`http://web.physik.rwth-aachen.de/Auger_MagneticFields/PARSEC/downloads.php`

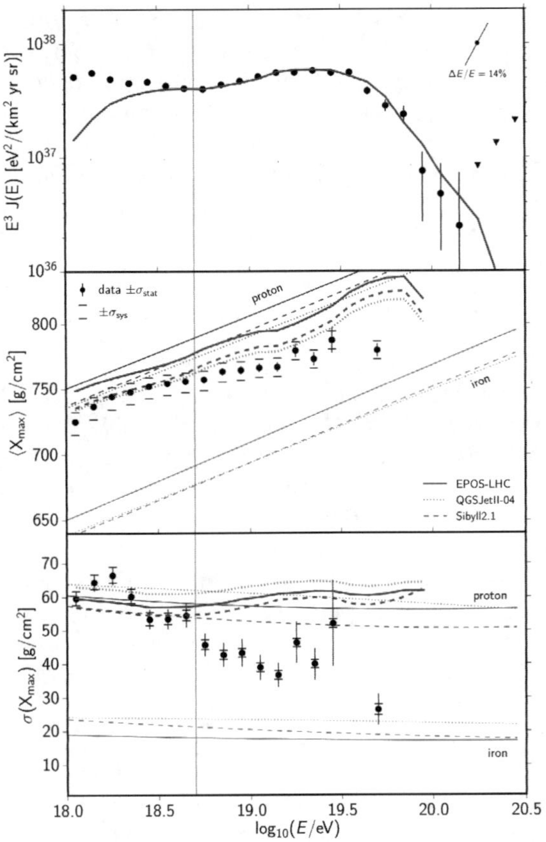


Figure 3: Energy spectrum (top) and moments of $X_{\max}$ (center and bottom) as measured by the Pierre Auger Collaboration [2, 4] and predictions from the simulated scenario (thick brown lines). The red and blue lines show the predicted $X_{\max}$ moments for a pure composition of protons and iron, respectively. Solid, dotted and dashed lines show the cosmic ray shower predictions for different hadronic interaction models.

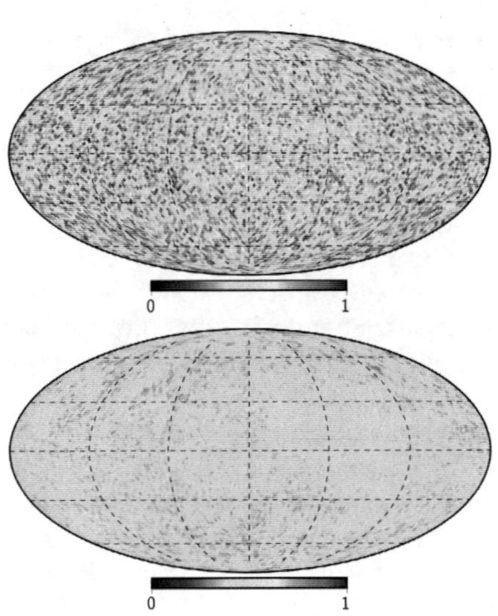


Figure 4: Distribution of arrival directions of events with energies $E > 10^{18.7}\,\mathrm{eV}$ before (top) and after galactic propagation (bottom). The events are binned in HEALpix maps and are normalized to the maximum event count of the map before galactic propagation.

The vector field has a resolution of 50 kpc and is initialized with a random turbulent realization of Kolmogorov power spectrum and a coherence length of 500 kpc.

For the cosmic ray sources, we consider a continuous distribution that follows the matter density. Using reflective boundary conditions for the cosmic rays, the contribution of sources up to 4 Gpc distance is effectively taken into account. The sources isotropically emit cosmic rays with a power-law spectrum and a rigidity dependent exponential cutoff. The differential number of cosmic rays of charge number $Z_i$ and mass number $A_i$ is given by

$$\frac{\mathrm{d}N_i}{\mathrm{d}E} \propto x_i\, A_i^{1.8-1}\, E^{-1.8}\, \exp\left(-\frac{E}{Z_i \cdot 10^{19.8}\,\mathrm{eV}}\right) \tag{1}$$

where $x_i$ is the relative abundance at equal energy per nucleon in absence of the cutoff. The spectral index and cutoff rigidity are chosen as an example to approximately reproduce the observed spectrum above $10^{18.7}$ eV. As representatives for a mixed cosmic ray composition, the four isotopes hydrogen, helium-4, nitrogen-14 and iron-56 are selected with relative abundances $x_i = 1, 0.5, 0.3$ and $0.1$, respectively. The resulting energy spectrum as well as the mean and spread of the composition-sensitive $X_{\rm max}$ observable are shown in fig. 3 in comparison with current measurements of the Pierre Auger Observatory [2, 4]. While a reasonable fit to the energy spectrum is achieved, the $X_{\rm max}$ moments are not well reproduced in this scenario. The distribution of arrival directions of events with energies $E > 10^{18.7}$ eV is shown in fig. 4 both before and after galactic propagation, for which the full JF12 field model is considered. While the small-scale anisotropy is seen to decrease through deflections in the galaxy, some anisotropic features remain.

## 5. Conclusion

In this contribution we have presented the public cosmic ray propagation code CRPropa 3. We summarized the advantages of the new code structure, the implementation of multi-processing and highlighted some of the new features, notably the galactic propagation, including the lensing technique, and the possibility to consider cosmological evolution in 4-dimensional simulations. We applied CRPropa to a scenario in which a mixed composition of hydrogen, helium, nitrogen and iron was propagated through a combination of structured universe models, and considered deflections in the galactic magnetic field. The simulation results were compared to measurements of energy spectrum and composition of the Pierre Auger Observatory. The arrival directions were shown as well, to demonstrate the possibility of simulating all observables accessible to UHECR experiments. More information on CRPropa can be found on https://crpropa.desy.de.

## Acknowledgments

This work was supported by the Deutsche Forschungsgemeinschaft (DFG) through the Collaborative Research Centre SFB 676 Particles, Strings and the Early Universe. RAB acknowledges the support from the Forschungs- und Wissenschaftsstiftung Hamburg through the program Astroparticle Physics with Multiple Messengers. GS was supported by the State of Hamburg, through the Collaborative Research program Connecting Particles with the Cosmos and by BMBF under grants 05A11GU1 and 05A14GU1.

## References

[1] Abu-Zayyad T *et al.* (Telescope Array Collaboration) 2013 (*Preprint* `arXiv:1305.7273`)
[2] Aab A *et al.* (Pierre Auger Collaboration) 2013 The Pierre Auger Observatory: Contributions to the 33rd International Cosmic Ray Conference (ICRC 2013)
[3] Abbasi R *et al.* (Telescope Array Collaboration) 2014 (*Preprint* `arXiv:1408.1726`)
[4] Aab A *et al.* (Pierre Auger Collaboration) 2014 *Phys.Rev.D* (*Preprint* `arXiv:1409.4809`)
[5] Aab A *et al.* (Pierre Auger Collaboration) 2014 *Phys.Rev.D* (*Preprint* `arXiv:1409.5083`)
[6] Aab A *et al.* (Telescope Array Collaboration and Pierre Auger Collaboration) 2014 *Astrophys.J.* (*Preprint* `arXiv:1409.3128`)
[7] Kampert K H *et al.* 2013 *Astropart.Phys.* **42** 41–51
[8] Batista R A *et al.* 2013 CRPropa 3.0 *33rd International Cosmic Ray Conference (ICRC 2013)*
[9] Mucke A, Engel R, Rachen J, Protheroe R and Stanev T 2000 *Comput.Phys.Commun.* **124** 290–314
[10] Dolag K and Stasyszyn F An MHD Gadget for cosmological simulations (*Preprint* `arXiv:0807.3553`)
[11] Teyssier R 2002 *Astron.Astrophys.* **385** 337–364
[12] Bretz H P, Erdmann M, Schiffer P, Walz D and Winchen T 2014 *Astropart.Phys.* **54** 110–117
[13] Gorski K, Hivon E, Banday A, Wandelt B, Hansen F *et al.* 2005 *Astrophys.J.* **622** 759–771
[14] Jansson R and Farrar G R 2012 *Astrophys.J.* **757** 14
[15] Jansson R and Farrar G R 2012 *Astrophys.J.* **761** L11
[16] Mollerach S and Roulet E 2013 *JCAP* **1310** 013
[17] Batista R A and Sigl G 2014 *JCAP* (*Preprint* `arXiv:1407.6150`)
[18] Dolag K, Grasso D, Springel V and Tkachev I 2005 *JCAP* **0501** 009
[19] Sigl G, Miniati F and Ensslin T A 2003 *Phys.Rev.* **D68** 043002

# Data Processing at the Pierre Auger Observatory

**Jakub Vícha[1], Jiří Chudoba[1] for the Pierre Auger Collaboration[2]**
[1]Fyzikální ústav AV ČR, Na Slovance 1999/2, Praha 8, 18221, Czech Republic
[2]Full author list: http://www.auger.org/archive/authors_2014_08.html
E-mail: vicha@fzu.cz

**Abstract.** Cosmic–ray particles with ultra–high energies (above $10^{18}$ eV) are studied through the properties of extensive air showers which they initiate in the atmosphere. The Pierre Auger Observatory detects these showers with unprecedented exposure and precision and the collected data are processed via dedicated software codes. Monte Carlo simulations of extensive air showers are very computationally expensive, especially at the highest energies and calculations are performed on the GRID for this purpose. The processing of measured and simulated data is described, together with a brief list of physics results which have been achieved.

## 1. Introduction

The Pierre Auger Observatory [1] is the largest experiment ever built to measure the extensive air showers induced by cosmic rays of energy above $10^{18}$ eV (UHECR). The Observatory (see Fig. 1) is located on the Argentinian Pampa at an altitude of about 1 400 m a.s.l., near the city of Malargüe in the province Mendoza. It predominantly observes the southern sky. On the ground, shower detection employs water–Cherenkov detectors (see left panel of Fig. 2). These are combined with the telescopic observation (see right panel of Fig. 2) of atmospheric fluorescence light on clear moonless nights. Four fluorescence detector sites surround the surface detector, which has an enclosed area of 3 000 km$^2$.

Observations began in 2004 with a partial array and the gradual deployment of the designed Observatory was completed in April 2008 with 1 600 surface detectors and 24 fluorescence telescopes. After additional enhancements to extend the energy coverage to $\sim 10^{17}$ eV, more than 1 660 water–Cherenkov stations and 27 fluorescence telescopes are installed. The Pierre Auger Collaboration is a community of more than 500 members from 19 countries.

Besides its huge aperture (volume of atmosphere and solid angle available for shower detection), the uniqueness of the Pierre Auger Observatory lies in the simultaneous observation of cosmic ray showers with two complementary techniques. Reconstructions combining surface and fluorescence data provide an opportunity to calibrate the surface detector (SD) signal using a shower energy which is measured very precisely by the fluorescence detector (FD). Moreover, in the, so–called, hybrid mode, timing information about the shower arrival on the ground from at least one SD station is used as additional information reconstructing the shower longitudinal profile with the FD. The FD provides a very precise measurement of the properties of the shower electromagnetic component, which contains ~90% of its energy.

The processes of managing and reconstructing the shower properties from the measured data are described in the following section. The third section includes specification of the Monte Carlo simulations of UHECR showers. Selected results obtained using the measured and simulated

Journal of Physics: Conference Series **608** (2015) 012077 doi:10.1088/1742-6596/608/1/012077

data of the Pierre Auger Observatory are mentioned in Section 4. The conclusions are presented in Section 5.

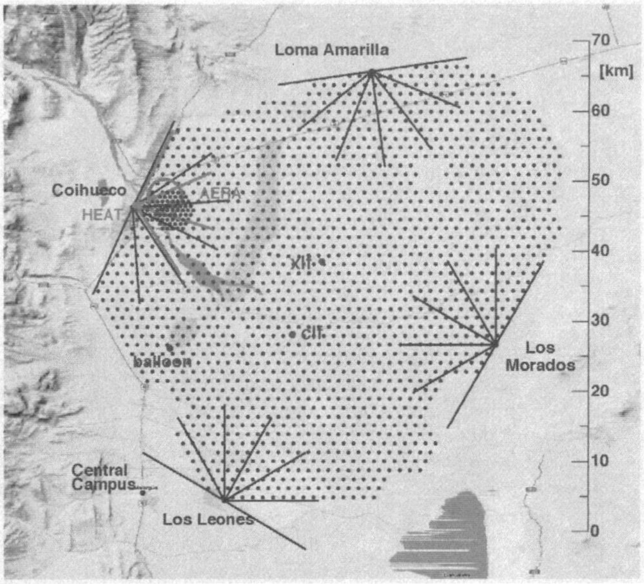


Figure 1: The view of the Pierre Auger Observatory deployment. Array of gray points represents positions of surface detector stations. The names of 4 fluorescence detector buildings are shown in blue together with illustrated azimuthal field of views of the individual fluorescence telescopes. Positions of atmospheric monitoring stations are marked in red. Locations of additional radio detection stations (AERA) and FD enhancement (HEAT) are also shown together with the position of the Central Campus.

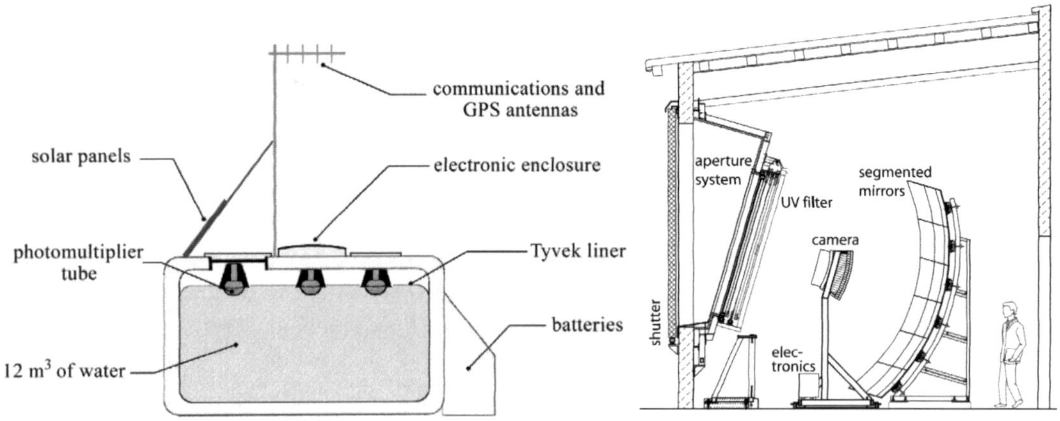


Figure 2: Schematic description of one water–Cherenkov station [1] (left panel) and one fluorescence telescope [3] (right panel).

## 2. Measured Data

Each water–Cherenkov station samples signals from three photomultipliers (PMTs) with a frequency of 40 MHz. The signal is induced by both electromagnetic particles and muons. The time stamp is obtained from GPS. When one of the local trigger conditions on the PMT signals is fulfilled, the signal traces of three PMTs are sent via the communication antenna to the communication towers placed near the fluorescence detector buildings and finally to the central data acquisition system (CDAS), when further physics triggers are tested. Altogether, the SD raw data has to pass a five–level hierarchical trigger system [2] to be stored in data centers.

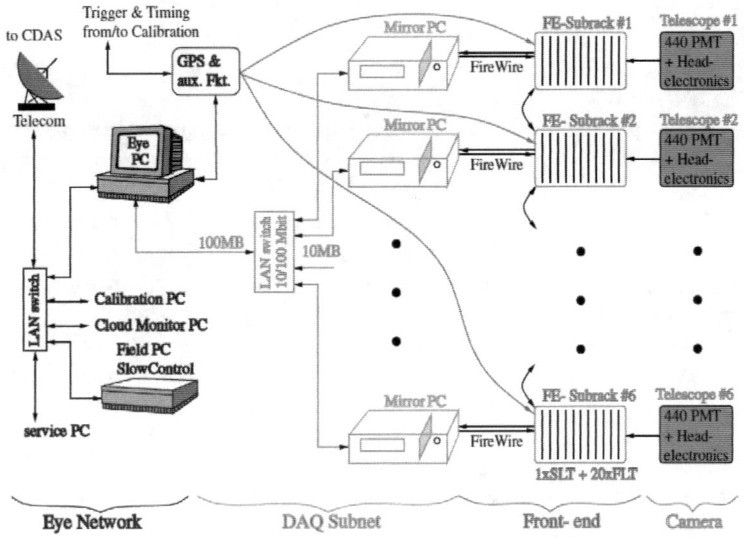


Figure 3: Schematic description of readout data flow at one fluorescence detector site. Picture comes from [3].

Each FD telescope collects fluorescence light in a camera consisting of 440 PMTs that are installed in the focal plane of the telescope. Signal traces from PMTs along a candidate shower track are also sent to the CDAS as illustrated in Fig. 3. FD raw data has to satisfy a three–level hierarchical trigger system [3] to be stored in data centers. The FD measurement is accompanied by regular monitoring of the atmosphere [4].

The SD and FD raw data from CDAS are copied to two facilities, in Lyon and Fermilab, via high speed Internet data–link. The overall size of SD and FD recorded raw data is a few TB per year. One typical shower has size ∼hundreds of kB.

The raw data in .root format are reconstructed using independent software coded in C++: Offline [5] for SD and FD and Herald [6] for SD. The Offline software is a powerful tool for the simultaneous reconstruction of SD and FD data. It has access to CDAS data and to many formats of simulated showers as well as to databases of calibration constants, atmospheric conditions etc. It has a modular structure that allows a simple implementation of additional reconstruction algorithms or detector enhancements. For the practical purpose of faster and simpler physics analyses, the package Advanced Data Summary Tree (ADST) [7] was developed to store all the relevant information including the shower properties reconstructed using the Offline. A typical ADST shower reconstructed in SD and FD has a size of ∼10 kB. The ADST data are stored in Karlsruhe on an ftp server with the overall size ∼70 GB per one production of all types of data to mid–2014.

## 3. Simulated Data

The simulation of detector signals proceeds in two steps. First, showers of typically $10^{10}$ particles for UHECR energies are simulated with the 3D simulation program CORSIKA [8]. It uses different models of hadronic interactions for high ($E_{lab}$ above 80 GeV) and low energy ($E_{lab}$ below 80 GeV) intervals. A so–called thin sampling (thinning) [9] is applied to save computational power. It assigns a higher weight to a randomly selected particle from a sample of secondary particles with total energy higher than an adjustable fraction of the primary energy, while all other secondaries are discarded. The typically used thinning level is $10^{-5}$. A typical shower generated at energy $10^{20}$ eV utilizes approximately 20 hours of CPU time of 1 core on a current processor even with the speed gained from thinning. In the second step, the simulations of detector responses of CORSIKA showers are performed using corresponding modules in the Offline. The required CPU time is smaller for lower energies but rises steeply and reaches values comparable to the CORSIKA part for energies $10^{20}$ eV.

Because there are several models to compare and many parameters to tune, we need a significant computing and storage capacity for these simulations. Local clusters available to some members of the collaboration were used. Each cluster might have different local setup and operating environment; validation of results was difficult. The unification came with the computing grid solution inspired by LHC experiments. The Prague group created a grid Virtual Organization (VO) *auger* in 2006 and started to run a common production of libraries of simulated showers using a custom production system based on shell scripts [10]. Later the group from the University of Granada took the responsibility for the official production and developed a second version of the production system with MySQL database back–end [11]. The VO *auger* became one of the biggest users of the EGI grid resources after the LHC VOs [12]. The success of grid usage was possible thanks to a similar workflow to high energy physics and thanks to resources from many centers where collaboration members were connected to. A significant fraction of CPU time was also provided by non–pledged resources. The stored output of simulated showers on the GRID is currently almost 500 TB in 36 million files.

The success in getting distributed resources brought problems with their management. It was shown that pilot jobs lead to better efficiency and higher usage. Also basic tools provided by the grid middleware for data management are not sufficient for large collections of the size of the VO *auger* data. The DIRAC interware [13] was tested first as a replacement for the file catalog and later also for the production system. We plan to do a complete migration in 2015.

## 4. Physics with Measured and Simulated Data

Following features of UHECR showers were observed with the Pierre Auger Observatory data and their comparison with MC predictions so far:

- **Spectrum:** The so called ankle at energy $10^{18.4}$ eV and steep suppression beyond $10^{19.6}$ eV are observed with high statistical significance [14].
- **Mass Composition:** A light composition occurs at energies up to the ankle. A transition towards heavier nuclei with increasing energy seems to be present at energies beyond the ankle [15].
- **p–p cross–section at 57 TeV:** The total p–p cross–section was inferred [16]. Its value lies approximately in the range that is adopted by the models of hadronic interactions.
- **Muon Excess:** The muon excess in measured data indicates a lack of understanding for all models of hadronic interactions [17, 18, 19].
- **Photons and Neutrinos Searches:** No observation of a photon or neutrino shower has been identified yet. Limits on fluxes of photons exclude Top–down models [20].
- **Anisotropy:** There is no excess of neutrons from the Galactic plane at EeV energies [21]. The only anisotropy at the highest energies (above ~55 EeV) that is stable in time is

observed in the 20° vicinity of the closest AGN: Cen A [22]. The level of correlation with known nearby AGNs from the VCV catalog decreased with time [22] and does not increase with energy [23].

An upgrade of the Pierre Auger Observatory with muon detectors is planned to be built in the future. Reliable determination of the mass composition, especially at the highest energies, with large enough statistics seems to be the crucial point in further studies of UHECR open questions (origins of the spectrum features, UHECR sources, lack of muons in MC simulations). Measurement of the muon component with a large surface array (full duty cycle) could be one way to answer these questions. However, the observed inconsistency of model predictions and measured data for the muon component is an issue that needs to be fixed simultaneously. Accompanying measurements with more types of muon detectors combined with the measurement of the electromagnetic component could be not only beneficial, but also necessary.

## 5. Conclusions

Processing of measured as well as simulated data at the Pierre Auger Observatory is well established. Important findings about cosmic rays of ultra–high energy and hadronic interactions at energies beyond accelerator capabilities came from the unprecedentedly large statistics collected by the observatory during ten years of operation. An upgrade of the observatory with muon detectors could bring more light into the presented results. The modular structure of the Offline software provides an easy implementation of such further detector enhancements. The difficulties to perform large productions of MC libraries of cosmic–ray showers for the Pierre Auger Observatory on local computer centers has led to the unprecedented usage of the GRID infrastructures in the field of astroparticle physics.

## Acknowledgment

This work is funded by the Czech Science Foundation grant 14-17501S.

## References

[1] Abraham J *et al.* 2004 *Nucl. Instrum. Meth. A* **523** pp. 50-95
[2] Abraham J *et al.* 2010 *Nucl. Instrum. Meth. A* **613** pp. 29-39
[3] Abraham J *et al.* 2010 *Nucl. Instrum. Meth. A* **620** pp. 227-251
[4] Abraham J *et al.* 2010 *Astropart. Phys.* **33** pp. 108-129 (2010)
[5] Argiro S *et al.* 2007 *Nucl. Instrum. Meth. A* **580** pp. 1485-1496
[6] CDAS Herald https://auger.colostate.edu/private/herald/
[7] The Auger Observer http://augerobserver.fzk.de
[8] Heck D *et al.* 1998 *Report FZKA* **6019**
[9] Hillas A M 1997 *Nucl. Phys. Proc. Suppl.* **52B** p. 29
[10] Chudoba J *et al.* 2010 *J. Phys. Conf. Ser.* **219** 072033
[11] Bahilo J L 2012 *J. Phys. Conf. Ser.* **368** 012015
[12] EGI accounting portal http://accounting.egi.eu
[13] DIRAC interware http://diracgrid.org
[14] Schulz A *for the Pierre Auger Collab.* 2013 *Proc. of the 33rd ICRC* arXiv:1307.5059 [astro-ph.HE]
[15] Aab A *et al.* 2014 submitted to *Phys. Rev. D* arXiv:1409.4809 [astro-ph.HE]
[16] Abreu P *et al.* 2012 *Phys. Rev. Lett.* **109** 062002
[17] Farrar G R *for the Pierre Auger Collab.* 2013 *Proc. of the 33rd ICRC* arXiv:1307.5059 [astro-ph.HE]
[18] Kegl B *for the Pierre Auger Collab.* 2013 *Proc. of the 33rd ICRC* arXiv:1307.5059 [astro-ph.HE]
[19] Valino I *for the Pierre Auger Collab.* 2013 *Proc. of the 33rd ICRC* arXiv:1307.5059 [astro-ph.HE]
[20] Aab A *et al.* 2014 *ApJ* **789** p. 160
[21] Aab A *et al.* 2014 *ApJ* **789** L34
[22] Abreu P *et al.* 2010 *Astropart. Phys.* **34** pp. 314-326
[23] Letessier-Selvon A *for the Pierre Auger Collab.* 2013 *Proc. of the 33rd ICRC* arXiv:1310.4620 [astro-ph.HE]

# The generalized BLM approach to fix scale-dependence in QCD: the current status of investigations

**A L Kataev** 1

1 Institute for Nuclear Research of the Academy of Sciences of Russia, Moscow 117312, Russia

E-mail: kataev@ms2.inr.ac.ru

**Abstract.** I present a brief review of the generalized Brodsky-Lepage-McKenzie (BLM) approaches to fix the scale-dependence of the renormalization group (RG) invariant quantities in QCD. At first, these approaches are based on the expansions of the coefficients of the perturbative series for the RG-invariant quantities in the products of the coefficients $\beta_i$ of the QCD $\beta$-function, which are evaluated in the MS-like schemes. As a next step all $\beta_i$-dependent terms are absorbed into the BLM-type scale(s) of the powers of the QCD couplings. The difference between two existing formulations of the above mentioned generalizations based on the seBLM approach and the Principle of Maximal Conformality (PMC) are clarified in the case of the Bjorken polarized deep-inelastic scattering sum rule. Using the conformal symmetry-based relations for the non-singlet coefficient functions of the Adler D-function and of Bjorken polarized deep-inelastic scattering sum rules $C^{\rm Bjp}_{\rm NS}(a_s)$ the $\beta_i$-dependent structure of the NNLO approximation for $C^{\rm Bjp}_{\rm NS}(a_s)$ is predicted in QCD with $n_{gl}$-multiplet of gluino degrees of freedom, which appear in SUSY extension of QCD. The importance of performing the analytical calculation of the $\rm N^3LO$ additional contributions of $n_{gl}$ gluino multiplet to $C^{\rm Bjp}_{\rm NS}(a_s)$ for checking the presented in the report NNLO prediction and for the studies of the possibility to determine the discussed $\{\beta\}$-expansion pattern of this sum rule at the $O(a_s^4)$-level is emphasised.

## 1. Introduction

It is known that the results of perturbative calculations of the physical quantities, which obey the RG equations (for the development of the RG method see e.g. [1, 2, 3]), depend on the choice of the scale and scheme of the renormalization procedure. In the case of QCD this problem is of particular importance. Indeed, calculations of the multiloop corrections to the observable physical quantities and to the related RG- functions (namely, $\beta$-function and various anomalous dimensions) are usually performed in the class of minimal subtractions (MS) schemes, and in the $\overline{\rm MS}$ -scheme [4], in particular. In this case, an error of the comparison of theoretical results with experimental data is usually determined by varying the corresponding renormalization scale $\mu^2$ within the concrete interval, say $\mu^2/k \leq \mu^2 \leq k\mu^2$, where $k$ is the conventionally chosen number, i.e. $k = 2 \div 4$ (this convention was recently used recently in [5] ). As can be seen from this work and from the studies of heavy flavour contributions to DIS sum rules [6], this interval for $k$ is indeed conventional. Say, the analysis of [6] motivates the choice of the following interval for $\mu^2$: $m_q^2 \leq \mu^2 \leq (6.5m_q)^2$, where $m_q$ are the $c$ and $b$-quark pole masses. Note, that in the process fitting the CCFR collaboration $xF_3$ structure functions data for $\nu N$ DIS [7] both ways of fixation of scale variations were considered for estimating a theoretical error-bar of the extracted expression for $\alpha_s^{\overline{\rm MS}}(M_Z)$. The first one was used in the case when the number of flavours was fixed as $n_f = 4$, while the second one was used to estimate the sensitivity of the fitted results to their transformation from $n_f = 4$ numbers of flavours to the energy region with $n_f = 5$ numbers of flavours.

Taking into account higher order perturbative corrections in the non-asymptotic QCD regime decreases, as a rule, theoretical errors of the analysed quantities and the extracted QCD parameters, which arise from the variations of scales within the conventionally chosen interval of values. In spite of this, there is quite understandable desire to formulate more concrete theoretical prescriptions for analysing scale-scheme dependence uncertainties using the RG-based language. Among the most applicable at present methods are the Principle of Minimal Sensitivity (PMS) [8], the Effective Charges (ECH) [9] and the Brodsky-Lepage-McKenzie (BLM) approaches [10]. The first two of them are based on the concepts of scheme-invariant quantities. Both ECH and PMS approaches are widely used in the concrete phenomenological studies (see e.g. [11, 12, 13, 14, 15]). In the process of these studies the gauge-invariant and vertex dependent schemes of defining the QCD coupling constant are usually used. These schemes include the original MS-scheme [16], its G-scheme [17] and the $\overline{\rm MS}$ -scheme [4] variants, and the number of other similar MS-like schemes, which are unified within the class of $R_\delta$-schemes [18]. Since all of them are related to the dimensional regularization [19] and are gauge-invariant, there are no problems with scale-scheme ambiguities of the BLM-approach, discussed [20, 21] in the case of various momentum subtraction (MOM) schemes, which in QCD depend on the gauge choice.

## 2. The generalizations of the BLM approach beyond the NLO: polarized Bjorken sum rule as a typical example

The first NNLO generalization of the BLM approach was formulated in [22]. It was shown that it is possible to absorb unambiguously all $n_f$-dependent terms from the NNLO corrections to the RG-invariant measurable quantities, evaluated in the $\overline{\rm MS}$ -scheme, by introducing the coupling dependence correction $O(\alpha_s)$ to the BLM scale $\mu^2_{BLM}$ fixed from the NLO approximations of the considered physical quantities. Note that this feature of the generalized BLM prescriptions was confirmed in the process of incomplete all-order extension of the BLM approach [23] aimed at resummation of the renormalon-type terms $(\beta_0\alpha_s)^n$ to the BLM scales of the perturbative series for the $\tau$- hadronic width and of the relations between the pole and running heavy quark masses.

Now let us now consider modern approaches of formulating all-order generalizations of the BLM method using the N$^3$LO approximation for the Bjorken sum rule of the polarized lepton-nucleon scattering as an example. This sum rule is defined as

$$S_{Bjp} = \int_0^1 g_1^{lp-ln}(x, Q^2)dx = \frac{g_A}{6}C^{Bjp}(a_s). \tag{1}$$

The function $C^{Bjp}$ contains the non-singlet (NS) and singlet (SI) contributions $C^{Bjp}(a_s) = C^{\rm Bjp}_{\rm NS}(a_s) + C^{\rm Bjp}_{\rm SI}(a_s)$ . The existence of the SI term at the $O(\alpha_s^4)$ level was demonstrated in [24]. Its concrete analytical expression is not yet fixed by direct diagrammatic calculations. Since we are interested in applications of the generalized BLM approaches of [25], which are similar to the seBLM method [26], and of the PMC approach (see recent reviews [27, 28]), we will neglect this SI-type $a_s^4$- term and consider the expression for $C^{\rm Bjp}_{\rm NS}(a_s)$.

Both methods start with the application of the the $\{\beta\}$-expansion of the perturbative $\overline{\rm MS}$ -scheme coefficients for $C^{\rm Bjp}_{\rm NS}(a_s)$. Within the seBLM-motivated approach of [25] this concrete structure of the $\{\beta\}$-expansion was obtained in [29] from the $\overline{\rm MS}$ -scheme generalizations of the original Crewther relation [30] fixed at the NNLO in [31] and at the $N^3$LO in [32] by using the $\{\beta\}$-expanded $O(a_s^4)$ representation for the $e^+e^-$ characteristic, namely, the $C_D^{NS}(Q^2)$ function [26].

The expression for $C_{\rm NS}^{\rm Bjp}(a_s)$ at the $N^3$LO level

$$C_{\rm NS}^{\rm Bjp}(a_s) = 1 + \sum_i^4 c_i a_s^i \tag{2}$$

was calculated in [32] in the $\overline{\rm MS}$ scheme, $a_s = \alpha_s/\pi$. Using the $\{\beta\}$-expansion formalism we express the coefficients $c_i$ as [29]

$$\begin{aligned}
c_1 &= c_1[0], && (3)\\
c_2 &= \beta_0(n_f)c_2[1] + c_2[0], && (4)\\
c_3 &= \beta_0^2(n_f)c_3[2] + \beta_1(n_f)c_3[0,1] \\
&\quad +\underline{\beta_0(n_f)c_3[1]} + c_3[0], && (5)\\
c_4 &= \beta_0^3(n_f)c_4[3] + \beta_1(n_f)\beta_0(n_f)c_4[1,1] \\
&\quad +\beta_2(n_f)c_4[0,0,1] + \beta_0^2 c_4[2] \\
&\quad +\underline{\beta_1(n_f)c_4[0,1]} + \underline{\beta_0(n_f)c_4[1]} + c_4[0], && (6)
\end{aligned}$$

while certain values of the elements in the RHS are fixed [29, 25] following to generalized Crewther relation. Here $n_f$ is the number of fermion flavours and $\beta_i(n_f)$ are the coefficients of the QCD $\beta$-function in the MS-like schemes, which are defined as

$$\mu^2 \frac{da_s}{d\mu^2} = \beta(a_s) = -\sum_{i \geq 0} \beta_i(n_f) a_s^{i+2} \tag{7}$$

In the $\overline{\rm MS}$ -scheme with one scale $\mu^2 = Q^2$ the analytical expressions for the NNLO $\{\beta\}$-approximation for $C_{\rm NS}^{\rm Bjp}(a_s)$ contain the *underlined* terms in Eq.(5),(6). They are **absent** in a similar $\{\beta\}$-representation of the $\overline{\rm MS}$ perturbative expression for $C_{\rm NS}^{\rm Bjp}(a_s)$ (see e.g. Eq.(160) review [27]). The reason for this is more technical than theoretical. As was shown in [26], at the NNLO it is possible to get the $\{\beta\}$-expansion for the $e^+e^-$ $C_D^{NS}(a_s)$-function with the concrete coefficients of the $\{\beta\}$-expanded pattern. It was possible to fix these terms only using an additional to $n_f$ degree of freedom, namely, the number $n_g$ of multiplets of massless gluino. Its analytical contribution to the NNLO correction of the $C_D^{NS}(a_s)$-function is known from the results of [32], while the additional contributions of $n_{gl}$ to the $\beta_0(n_f)$ and $\beta_1(n_f)$-functions of QCD with gluino are known from [34]. Since $\beta_0(n_f, n_{gl})$, $\beta_1(n_f, n_{gl})$ are liner in both $n_f$ and $n_{gl}$, this allows us to separate the contributions of $\beta_1$ and $\beta_0$ to the NNLO correction of the $D^{NS}(a_s)$-function and obtain the $\{\beta\}$-expansion pattern of the $O(a_s^3)$ coefficient with 4 terms similar to the ones entering in Eq.(5).

Note that without this additional information it is impossible to extract the contributions of $\beta_1(n_f)$ and $\beta_0(n_f)$ to the NNLO corrections of physical quantities from the ordinary $n_f$-expansion of these terms. Indeed, the NNLO correction contains three terms. In the case of polarized $C_{\rm NS}^{\rm Bjp}(a_s)$ the $O(a_s^3)$ coefficient has the following form:

$$c_3(n_f) = c_{3,0} + c_{3,1}n_f + c_{3,2}n_f^2 \; . \tag{8}$$

To avoid rather delicate and complicated studies, the authors of [27, 28] prefer to neglect in the NNLO corrections of their initial $\overline{\rm MS}$ -expressions the term, proportional to the single power of

$\beta_0$. The price for that is the corruption of the structure of the generalized Crewther relations in the $\overline{\text{MS}}$ -scheme (see [25] and [35] for a more detailed discussion of this subject). Note that the generalized Crewther relations result from the fundamental properties of the conformal symmetry and its violation by the conformal anomaly in QCD (for the discussions see [36]).

Unfortunately, the absence of any information about the calculated effects of the manifestation of additional degrees of freedom (like the contributions of $n_{gl}$-multiplet of gluinos) in the analytical expression of the $O(a_s^4)$ correction to $C_D^{NS}$, evaluated in [32] in the case of $SU(N_c)$ colour gauge group, does not allow one to extract analytical expressions for the coefficients $c_4[1,1]$, $c_4[0,0,1]$, $c_4[2]$, $c_4[0,1]$, $c_4[1]$ and $c_4[0]$ using the ideas proposed in [26] (apart from its $C_F^4$ contribution to $c_4[0]$, first determined in [37] from the application of the original Crewther relation [30], and the leading term of the $\beta_0$-expansion with the coefficient $c_4[3]$, which is known from the calculations of [31]). Like the coefficients $c_3[2]$, $c_3[0,1]$, $c_3[1]$ and $c_3[0]$ analytically defined in [29], these still unknown terms are crucial for demonstrating the numerical difference between different generalizations of the BLM approach, namely between the PMC approach, discussed e.g. in [27],[28], and single-scale $\{\beta\}$-expanded generalization of the BLM approach, studied in [25]. In view of the lack of knowledge of the $\{\beta\}$-expanded expression of the coefficient $c_4$ in Eq.(6), the single-scale analysis of the $O(a_s^3)$-approximation of Eq.(2) was considered only. Let us have a look at it from another point of view, namely, transforming it to the case of multiple scales, as was proposed in [18].

## 3. The scale-dependence of the BLM generalizations at the NNLO and beyond

Consider first the $\{\beta\}$-expanded form of the $O(a_s^4)$ approximation for the Bjorken polarized sum rule defined in the $\overline{\text{MS}}$ -scheme by Eqs.(2)-(5) and transform it to the multiple-scale case using the solution of the RG-equation presented in [18] and written down in the form $R_\delta$ relation, defined in [18] as

$$a_s(Q^2) = a_s(Q_\delta^2) + \sum_{n=1}^{\infty} \frac{1}{n!}\frac{d^n a_s(Q^2)}{d\ln Q^2}^n \Big|_{Q^2=Q_\delta^2} (-\delta)^n \tag{9}$$

where $\ln Q^2/Q_\delta^2 = -\delta$. This transformation relation leads to the $O(a_s^4)$ multiscale approximation of $C_{\rm NS}^{\rm Bjp}(a_s)$, which can be obtained from the straightforward RG-transformations of the power-series for $C_{\rm NS}^{\rm Bjp}(a_s)$ ( see e.g. [18]), namely,

$$\begin{aligned}
C_{\rm NS}^{\rm Bjp}(a_s) &= 1 + c_1[0]a_s(Q_1^2) + \Big[\beta_0 c_2[1] + c_2[0] + \beta_0 c_1[0]\delta_1\Big] a_s^2(Q_2^2) & (10)\\
&+ \Big[\Big(\beta_0^2 c_3[2] + \beta_1 c_3[0,1] + \underline{\beta_0 c_3[1]} + c_3[0]\Big) + \Big(\beta_0^2\delta_1^2 + \beta_1\delta_1\Big)c_1[0] \\
&+ 2\beta_0\Big(\beta_0 c_2[1] + c_2[0]\Big)\delta_2\Big] a_s^3(Q_3^2) & (11)\\
&+ \Big[\Big(\beta_0^3 c_4[3] + \beta_1\beta_0 c_4[1,1] + \beta_2 c_4[0,0,1] + \beta_0^2 c_4[2] + \underline{\beta_1 c_4[0,1]} + \underline{\beta_0 c_4[1]} + c_4[0]\Big) \\
&+ \Big(\beta_0^3\delta_1^3 + \frac{5}{2}\beta_1\beta_0\delta_1^2 + \beta_2\delta_1\Big)c_1[0] + \Big(3\beta_0^2\delta_2^2 + 2\beta_1\delta_2\Big)\Big(\beta_0 c_2[1] + c_2[0]\Big) \\
&+ 3\beta_0\Big(\beta_0^2 c_3[2] + \beta_1 c_3[0,1] + \underline{\beta_0 c_3[1]} + c_3[0]\Big)\delta_3\Big] a_s^4(Q_4^2) + O(a_s^5) & (12)
\end{aligned}$$

Here the scales are defined as $Q_K^2 = Q^2 exp(\delta_k)$. Note that the $\{\beta\}$-dependent structures of the $\delta_k$ expressions of the terms in Eqs.(10), (11) and Eq.(12) **coincide** with the introduced in [26] $\{\beta\}$-expanded structure of Eqs.(4)-(6) introduced in [26]. This feature was already observed in [18] and [27]. This fact is not **accidental** at all, but follows from the general principles of the RG method. Moreover, contrary to the claims of refs.[18], it is impossible to neglect in this expansion the terms proportional to $\beta_0 a_s^3(Q^2)$ in Eq.(5) (namely, to put **to zero** the $c_3[1]$ coefficient in Eq.(5) of the $\{\beta\}$-expanded expression for $C_{\rm NS}^{\rm Bjp}(a_s)$. Indeed, this will automatically lead to disappearance of the $3\beta_0^2 r_{4,2}$ in Eq.(6) of [18], which corresponds to the $3\beta_0^2 c_3[1]\delta_3 a_s^4(Q_4^2)$-term, underlined in part in Eq.(12).

To conclude, within the multiple-scale considerations of [18],[27], [28] the neglected and underlined in Eqs.(11),(12) terms will affect the results for the scales $Q_3^2$ and $Q_4^2$, obtained and discussed in [18],[27], [28]. They should be corrected by taking into account the $\{\beta_i\}$-dependent terms, underlined in Eq.(11),(12).

## 4. The prediction of the $\{\beta\}$-dependent NNLO expression for the Bjorken polarized sum rule with $n_{gl}$ multiplet of gluinos

One of the ways to confirm the $\{\beta\}$-dependent structure of the NNLO expression for $C_{\rm NS}^{\rm Bjp}(a_s)$ from Eqs.(3)-(5), which is in agreement with the analytical result of the direct NNLO QCD calculation of [38], is to evaluate analytically at the $a_s^3$-level additional contributions from the $n_{gl}$ multiplets of SUSY QCD gluinos, as was done [33] in the case of the $e^+e^-$ characteristic $C_D^{NS}(Q^2)$. After this, it will be extremely interesting to use the ideas of [26] and combine this possible new result with the analytical expressions for first two coefficients of the RG $\beta$-function $\beta_0(n_f, n_{gl})$ and $\beta_1(n_f, n_{gl})$. We believe, this possible study will coincide with the prediction made in [25] namely, with the following $O(a_s^3)$ approximation for $C_{\rm NS}^{\rm Bjp}(a_s)$:

$$\begin{aligned} C_{\rm NS}^{\rm Bjp}(a_s) &= 1 - \frac{3}{4}C_F a_s + \Big( - \frac{3}{2}C_F\beta_0(n_f,n_{gl}) + \frac{21}{32}C_F^2 - \frac{1}{16}C_F C_A\Big)a_s^2 \\ &+ \Big[ - \beta_0^2(n_f,n_{gl})\frac{115}{24}C_F - \beta_1(n_f,n_{gl})\Big(\frac{59}{16} + 3\zeta_3\Big)C_F \\ &+ \beta_0(n_f,n_{gl})\Big(\Big(\frac{83}{24} - \zeta_3\Big)C_F^2 + \Big(\frac{215}{192} - 6\zeta_3 + \frac{5}{2}\zeta_5\Big)C_F C_A\Big) \\ &+ -\frac{3}{128}C_F^3 - \frac{65}{64}C_F^2 C_A - \Big(\frac{523}{768} - \frac{27}{8}\zeta(3)\Big)C_F C_A^2\Big]a_s^3 + O(a_s^4) \end{aligned} \quad (13)$$

where

$$\beta_0(n_f,n_{gl}) = \frac{11}{12}C_A - \frac{1}{3}\Big(T_F N_F + \frac{1}{2}n_{gl}C_A\Big) \quad (14)$$

$$\beta_1(n_f,n_{gl}) = \frac{17}{24}C_A^2 - \frac{5}{12}C_A\Big(T_F N_F + \frac{1}{2}n_{gl}C_A\Big) - \frac{1}{4}\Big(T_F N_F C_F + \frac{1}{2}n_{gl}C_A^2\Big) \quad (15)$$

are the coefficients of the corresponding $\beta$-function, defined by Eq.(7). The prediction of Eq.(7) was obtained in [25] using a similar expression for $C_D^{NS}(Q^2)$, which result from the calculations of [33], detailed considerations of [26] and following from the conformal symmetry original Crewther relation [30] between the $C_D^{NS}(Q^2)$ and $C_{\rm NS}^{\rm Bjp}(Q^2)$ coefficient functions. It was also checked in [25] that the same expression can be obtained from the $O(a_s^3)$- generalization of the Crewther relation, discovered in [31]. In this relation the defined by the conformal symmetry term is modified by the to the conformal symmetry breaking term , namely by the factorized $\beta$-function of QCD with $n_{gl}$ multiplets of massless gluino. We are sure that a direct check of the prediction of Eq.(13) may give an additional argument in favour of the used in our works form of $\{\beta\}$-expansion at the level of NNLO corrections, considered in this report. Moreover, possible N$^3$LO evaluation of $C^{NS}(a_s)$ in QCD with $n_{gl}$ gluino multiplet should clarify how to extract still unknown analytical coefficients of the $\{\beta\}$-dependence pattern of the $O(a_s^4)$ -correction to $C^{NS}(a_s)$.

Journal of Physics: Conference Series **608** (2015) 012078 doi:10.1088/1742-6596/608/1/012078

## 5. Theoretical advantages of one-scale seBLM/PMC and its phenomenological troubles

Let us transform the general multiple-scale NNLO approximation for $C_{\rm NS}^{\rm Bjp}(a_s)$, defined in Eq.(10),(11), to the single-scale approximation, studied in [25]. This can be simply done by fixing $\delta_1$=$\delta_2$=$\delta$. Absorbing now the $\beta_0$- dependent term into the scale $\delta_1$ at the $O(a_s^2)$ level we obtain the standard BLM expression for $C_{\rm NS}^{\rm Bjp}$, namely

$$C_{\rm NS}^{\rm Bjp}(a_s) = 1 + c_1[0]a_s(Q^2_{BLM}) + c_2[0]a_s^2(Q^2_{BLM}) + O(a_s^3) \quad . \tag{16}$$

Taking into account that $c_1[0] = -\frac{3}{4}C_F = -1$ and $c_2[1] = -\frac{3}{2}C_F = -2$ we get the value of the standard BLM scale $Q^2_{BLM} = Q^2\exp(-c_2[1]/c_1[0]) = Q^2\exp[-2] = Q^2 0.135$. The $c_2[0]$ term was obtained in [25]. Its expression is $c_2[0] = -\frac{21}{32}C_F^2 + \frac{1}{16}C_F C_A = -11/12 = -0.91(6)$.

In the work of Ref.[25] we absorb into the BLM scale of the NNLO $O(a_s^3(Q^2_{BLM})$ single-scale coefficient from Eq.(11) the terms proportional to $\beta_0^2 c_3[2]$, $\beta_0 c_3[0,1]$, $\beta_0 c_3[1]$ as well. To avoid lengthy discussions, given in the original work [25] we will consider the case of $N_F = 3$ number of massless flavours and present only part of analytical and numerical expressions of the corresponding coefficients of the $\{\beta\}$-expansion procedure, which define the BLM scale of the NNLO approximation for $C_{\rm NS}^{\rm Bjp}(a_s)$. They are taken from the results of [25] and read $c_3[2] = -\frac{151}{24}C_F = -115/18 = -6.38(8)$, $c_3[0,1] = (-\frac{59}{16} + 3\zeta_3)C_F = (-\frac{59}{12} + 4\zeta_3) = -0.108$ and $c_3[1] = (\frac{83}{24} - \zeta_3)C_F^2 + (\frac{215}{192} - 6\zeta_3 + \frac{5}{2}\zeta_5)C_F C_A = \frac{4591}{432} - \frac{232}{9}\zeta_3 + 10\zeta_5 = -9.989$. Using these numbers and the expressions for $c_1[0]$, $c_2[0]$ and $c_2[1]$, as given above, we obtain the expression for the NNLO BLM scale for $C_{\rm NS}^{\rm Bjp}(a_s)$, given in [25] in a bit different normalization

$$Q^2_{NNLO} = Q^2_{BLM} exp[-7.32\beta_0 a_s(Q^2_{BLM})] = Q^2 exp[-2 - 7.32\beta_0 a_s(Q^2_{BLM})] \quad . \tag{17}$$

. The expressions for the $O(a_s^3(Q^2_{NNLO}))$ coefficient to $C_{\rm NS}^{\rm Bjp}(a_s)$ read [29]:

$$c_3[0] = -\frac{3}{128}C_F^3 - \frac{65}{64}C_F^2 C_A - \left(\frac{523}{768} - \frac{27}{8}\zeta_3\right)C_F C_A^2 \approx 35.034 \tag{18}$$

where the numerical value is given in the case of $SU(N_c = 3)$ colour gauge group. Thus, the final result we are interested in reads

$$C_{\rm NS}^{\rm Bjp}(a_s) = 1 - a_s(Q^2_{NNLO}) - 0.91(6)a_s^2(Q^2_{NNLO}) + 35.034a_s^3(Q^2_{NNLO}) + O(a_s^4) \tag{19}$$

Considering the one-scale PMC-type expression we conclude that

- its analytical coefficients extracted from the $\{\beta\}$-expansion procedure of [26] and the $\overline{\rm MS}$ -scheme generalizations of the Crewther relation of [31], [29] do not depend on the number of flavours and on the terms proportional to the QCD $\{\beta\}$-function. Combined with similar expressions for the coefficients $d_i[0]$ of the $\{\beta\}$- expansion representation of the $\overline{\rm MS}$ -scheme expression for the $e^+e^-$ characteristic $C_D^{NS}(a_s)$ these coefficients satisfy the scheme-independent relations, which follow from the original Crewther relation of [30], which is based on the conformal symmetry.
- Unfortunately, in view of not small $O(a_s^3)$-term application of the single-scale realisation of the PMC approach of Eq.(19), which is very similar to the seBLM method, spoils the satisfactory convergence of the $O(a_s^3)$ $\overline{\rm MS}$ -scheme approximation of the polarized Bjorken sum rule perturbative QCD expression, while the NNLO generalization of the BLM-scale of Eq.(17) moves the applicability of the resulting perturbative expressions to the region of higher energies.

## 6. Conclusions

At the current stage of our studies, presented in detail in [25], we discover several phenomenological disadvantages of the applicability of the BLM approach, generalized to $O(a_s^3)$-level even following the application of the $\{\beta\}$-expansion formalism, proposed in [26] and studied in [29]. We raise the question of the inappropriate use of this formalism within the Principle of Maximal Conformality considered in [18, 27, 28] and propose the calculating test, which may give extra argument in favour of the self-consistency of the structure of the $\{\beta\}$-expansion approach, as used by us at the $O(a_s^3)$-level. We would like also to mention that it is possible to invent the way how to make the generalized BLM approach more suitable for high-energy studies. However, we are still unable to give any straightforward theoretically motivated prescription for its formulation. At present, this work, which was started in [25], is based on some empirical observations and necessitates more careful analysis of $O(a_s^4)$-contributions to the RG-invariant quantities, still unknown within the used version of the theoretically consistent $\{\beta\}$-expansion approach. Note, that our studies are also of importance in view of the necessity of better understanding of the applications of the NNLO generalizations of the BLM approach to other imprtant observables, like the event-shape distributions [39], which are measured precisely at $e^+e^-$-colliders. .

## 7. Acknowledgements

I would like to thank Sergey Mikhailov for our productive collaboration and for useful discussions of the material of this talk. This work, as well as the participation at the ACAT-20014 Workshop was supported in part by RSCF, Grant N 14-22-00161. I am also grateful to the organizers of this Workshop and in particular to Milos Lokajicek for the warm hospitality in Prague and partial financial support.

## References

[1] Stueckelberg E C G and Petermann A 1951 *Helv. Phys. Acta* **24** 317
[2] Bogolyubov N N and and Shirkov D V 1956 *Nuov. Cim.* **3** 845
[3] Itzykson C and J. B. Zuber J B *Quantum Field Theory* 1980 (New York, McGRAW-HILL)
[4] Bardeen W A, Buras A J, Duke W and Muta T 1978 *Phys. Rev.* D **18** 3998
[5] Alekhin S, Blumlein J, Daum K, Lipka K and Moch S 2013 *Phys. Lett.* B **720** 172 (Preprint hep-ph/1212.2355)
[6] Blumlein J and van Neerven W L 1999 *Phys. Lett.* B **450** 417 (Preprint hep-ph/9811351)
[7] Kataev A L, Parente G and Sidorov A V 2003 *Phys. Part.Nucl.* **34** 20 (Preprint hep-ph/0106221)
[8] Stevenson P M 1981 *Phys. Rev.* D **23** 2916
[9] Grunberg G 1984 *Phys. Rev.* D **29** 2315 (Preprint Ecole Polytechnique 82-0721)
[10] Brodsky S J , Lepage G P and Mackenzie B P 1983 *Phys.Rev.* D **28** 228.
[11] Chyla J, Kataev A and Larin S A 1991 *Phys.Lett.* B **267** 269 (Preprint UM-TH-91-06, DESY 91-059)
[12] Chyla J and Kataev A L 1992 *Phys. Lett.* B **297** 385 (Preprint CERN-TH-6604-92, hep-ph/9209213)
[13] Maxwell C J and Morgan K E 2012 *Nucl. Phys.* B **858** 405 (Preprint hep-ph/1108.6204)
[14] Stevenson P M 2013 *Nucl.Phys.* B **868** 38 (Preprint hep-ph/1210.7001)
[15] P. Posolda, *J. Phys.* G **41** (2014) 095007.
[16] 't Hooft G 1973 *Nucl. Phys.* B **61** 455
[17] Chetyrkin K G, Kataev A L and Tkachov F V 1980 *Nucl. Phys.* B **174** 345
[18] Mojaza M, Brodsky S J and Wu X G 2013 *Phys. Rev. Lett.* **110** 192001 (Preprint hep-ph/1212.0049)
[19] 't Hooft G and Veltman M J G 1972 *Nucl. Phys.* B **44** 189.
[20] Celmaster W and Stevenson P M 1983 *Phys. Lett.* B **125** 493.
[21] Chyla J 1995 *Phys. Lett.* B **356** 341 (Preprint hep-ph/9505408).
[22] Grunberg G and Kataev A L 1992 *Phys. Lett.* **279** 352.
[23] Ball P, Beneke M and Braun V M 1995 *Phys. Rev.* D **52** 3929 (Preprint hep-ph/9503492)
[24] Larin S A *Phys. Lett.* 2013 B **723** 348
[25] Kataev A L and Mikhailov S V 2015 Generalization o0f the Brodsky-Lepage-Mackenzie optimization within the $\{\beta\}$-expansion and the Principle of Maximal Conformality, *Phys. Rev.* D **91** 014007; (Preprint INR-TH-9-2014; hep-ph/1408.0122)

[26] Mikhailov S V 2007 Generalization of BLM procedure and its scales in any order of pQCD: a practical approach J High Energy Physics *JHEP* **0706** 009 (Preprint JINR-E2-2005-25, hep-ph/0411397)
[27] Wu X G, Brodsky S J and Mojaza M 2013 *Prog. Part. Nucl. Phys.* **72** 44 (Preprint hep-ph/1302.0599)
[28] Wu X G, Ma Y, Wang S Q, Fu H B, Ma Y Y , Brodsky S J and Mojaza M 2014 Renormalization group invariance and optimal renormalization scale-setting Preprint SCAL-PUB-1593, hep-ph/1405.3196
[29] Kataev A L and Mikhailov S V 2012 *Theor. Math. Phys.* **170** 139 (Preprint hep-ph/1011.5248)
[30] Crewther R J 1972 *Phys. Rev. Lett.* **28** 1421.
[31] Broadhurst D J and Kataev A L 1993 *Phys. Lett.* B **315** 179 (Preprint hep-ph/9308274)
[32] Baikov P A, Chetyrkin K G and Kuhn J H 2010 *Phys. Rev. Lett.* **104** 132004 (Preprint hep-ph/1001.3606)
[33] Chetyrkin K G 1997 *Phys.Lett.* B **391** 402 (Preprint hep-ph/9608480)
[34] Clavelli L, Coulter P W and Surguladze L R 1997 *Phys. Rev.Lett.* D **55** 4268 (Preprint hep-ph/9611355)
[35] Kataev A L and Mikhailov S V 2014 $\{\beta\}$-expansion in QCD, its conformal symmetry limit: theory+applications, Preprint hep-ph/1410.0554
[36] Braun V M, Korchemsky G P and Mueller D 2003 *Prog. Part. Nucl. Phys.* **51** 311 (Preprint hep-ph/0306057)
[37] Kataev A L 2008 *Phys. Lett.* B **668** 350 (Preprint hep-ph/0808.3121)
[38] Larin S A and Vermaseren J A M 1991 *Phys. Lett.* B **259** 345.
[39] Gehrmann T , Hfliger N and Monni P F 2014 *Eur. Phys.J.* C **74** 2896 (Preprint hep/ph:1401.6809)

Journal of Physics: Conference Series **608** (2015) 012079 doi:10.1088/1742-6596/608/1/012079

# On calculations within the nonlinear sigma model

**Karol Kampf**

Institute of Particle and Nuclear Physics, Charles University,
V Holešovičkách 2, Prague, Czech Republic.

E-mail: `karol.kampf@mff.cuni.cz`

**Abstract.**
The chiral $SU(N)$ nonlinear sigma model represents one of the simplest case of effective field theory. For the last several decades it has played an extremely important role not only for the low energy phenomenology but also in many other areas of theoretical physics. In this talk we will focus on the tree-level scattering amplitudes of the $n$-Goldstone bosons. It will be shown that it can be reconstructed using the BCFW-like recursion relations. This method, which does not rely on the Lagrangian description, is much more efficient than the standard Feynman diagram techniques.

## 1. Introduction

It is well known that if we want to study the properties of quantum chromodynamics (QCD) at low energies we cannot directly use this theory as for example in the case of quantum electrodynamics. Main reason is that the degrees of freedom of QCD, quarks and leptons are not the ones we measure in experiments. One of the possible way how to tackle this problem is an effective field theory approach. This is historically connected with the name current algebra and non-linear sigma model developed later into the systematic theory called nowadays Chiral perturbation theory (ChPT) [1]. This theory describes the behaviour of the lightest pseudoscalar mesons (i.e. pions, kaons and eta) at low energies. The natural expansion parameters are thus momenta and masses of these particles and the standard choice is the same importance of them. A rough estimate of the theoretical error leads to the value of 30% for each step in the chiral series. In order to be competitive with an experimental error of the order of ten percent we are forced to provide the given calculation up to next-to-next-to-leading order (NNLO). This is true not only for the even-intrinsic-parity sector but in some cases also for the odd anomalous sector which arises due to the chiral anomaly (see e.g.[2]). In both sectors we are thus facing not only the complicated inevitable two-loop calculations but also an increase of the number of unknown low-energy constants (LECs). Their values can be estimated using the influence of resonances from the so-called intermediate energy region (for the systematic studies at NLO for both sectors see [3]). The typical calculation within ChPT thus includes complicated (usually two loop order) calculations and the part devoted to the estimation of used parameters. In this work we will further discuss two different strategies for study of the non-linear sigma model properties. Both of them will rely only on the leading order Lagrangian of ChPT and we will thus avoid the phenomenological problem connected with the estimation of LECs.

## 2. Leading logarithms

The first way is connected with the concept of (non)renormalizability in the effective field theories. Details can be found in the [4], here we will merely summarized main points.

The leading logs (LL) are the logarithms with highest power at the given order, i.e. $\mathrm{LL}^1$ at the one-loop order, $\mathrm{LL}^2$ at the two-loop order and so on. As already mentioned they are parameter-free and can be calculated, in principle, to all orders from the one-loop diagrams only. In [4] the procedure of the one-loop calculations was automatized and high orders of LL were presented. The actual automatization of the calculation was performed within cshell using C++ and the most important part, the algebraic treatment was done in FORM [5].

It is clear that main motivation for LLs is theoretical as it can hint to some deeper understanding of calculation within effective field theories. The eventual resummation of LL then leads to the same effect as is the concept of running coupling constant for the renormalizable theories. However, their application can be also phenomenological. As already stated, they are not proportional to some unknown parameters. We can define LL using the lowest-order parameters of the Lagrangian as

$$L = \frac{M^2}{16\pi^2 F^2} \log \frac{\mu^2}{M^2}\,, \tag{1}$$

or using the physical mass and decay constant

$$L_\pi = \frac{M_\pi^2}{16\pi^2 F_\pi^2} \log \frac{\mu^2}{M_\pi^2}\,. \tag{2}$$

The general notation of the expansion is as follows

$$O_{\text{phys}} = O_0(1 + \sum_i a_i L^i), \qquad O_{\text{phys}} = O_0(1 + \sum_i c_i L_\pi^i)\,. \tag{3}$$

The high orders of LL of several quantities relevant for ChPT were already calculated: expansion of mass and decay constant, vacuum expectation value, vector and scalar formfactor, $\pi\pi$ scattering and also $\gamma\gamma \to \pi\pi$ process. Concerning the anomalous processes, LL for the quantities connected with $\pi\gamma \to \pi\pi$ and $\pi^0 \to \gamma\gamma$ were calculated up to the 6th and 7th loop respectively. The similar program can be extended in different areas of particle physics, as done recently for the nucleon mass [6]. Let us also stress that though the calculation is automatic now, due to the time and computer limitations, it is still impossible to extend it to very high orders (see discussion in [7]).

## 3. Tree-level amplitudes

We have started the introduction with the phenomenologically relevant calculations within the non-linear sigma model. This includes one and two-loop diagram calculations. Further extension, in principle up to all orders, was performed using the renormalization group properties. Our next object of interest will go in the other direction. Instead of increasing number of loops we will increase number of legs. To simplify the problem we will study only tree-level diagrams and we stay in the massless limit. Let us first briefly summarize the method used in the case of the renomalizable theory which we will generalize to our model. The toy model will be the pure gluonic sector of QCD.

### *3.1. Motivation: gluon amplitudes and BCFW reconstruction*

The standard method to calculate the $n$-gluon amplitudes relies on the Feynman diagrams. We have to combine two possible elementary vertices (in the case of the studied model) and draw all possible diagrams. The number of diagrams grows exponentially with the number of external legs. What is interesting, however, after long and tedious calculation some results known to be extremely simple. In the so-called helicity formalism we have for all tree-level maximally helicity violating gluonic amplitudes close formula [8]

$$A_n(--+\ldots+) = \frac{\langle 12\rangle^4}{\langle 12\rangle\langle 23\rangle\ldots\langle n1\rangle}, \qquad \text{with} \qquad \langle ij\rangle \sim \sqrt{|2p_i\cdot p_j|}\,. \tag{4}$$

The standard method via Feynman diagrams thus leads to the big cancellations among different diagrams and is thus not very well suited for this kind of calculations. We will briefly recapitulate a new alternative method which relies on the BCFW (Britto, Cachazo, Feng and Witten) recursion relations. It reconstructs the on-shell tree-level amplitudes from its poles using the simple analytic properties [9]. First step is the possibility to define a colour-ordered stripped amplitude

$$M^{a_1\ldots a_n}(p_1,\ldots p_n) = \sum_{\sigma/Z_n} \mathrm{Tr}(t^{a_{\sigma(1)}}\ldots t^{a_{\sigma(n)}})M_\sigma(p_1,\ldots,p_n)\,. \tag{5}$$

The existence of ordering is important for the following discussion of the pole structure. The only possible poles are:

$$P_{ij}^2 = (p_i + p_{i+1} + \ldots + p_{j-1} + p_j)^2\,. \tag{6}$$

The one particle unitarity is telling us that if we go with one of the above poles to zero the amplitude will factorize into two parts (left/right), schematically:

$$\lim_{P_{1j}^2\to 0} M(1,2,\ldots n) = \sum_{h_l} M_L(1,2\ldots j,l)\cdot\frac{i}{P_{1j}^2}\cdot M_R(l,j+1,\ldots n)\,. \tag{7}$$

This can be trivially verified using e.g. the Feynman diagram representation.

In order to reconstruct the amplitude from its poles in complex plane we will make use of the following continuation. We will shift *two* (usually adjacent) external momenta

$$p_i \to p_i + zq, \qquad p_j \to p_j - zq\,, \tag{8}$$

so that the new $p_i$ and $p_j$ remain on-shell. The amplitude thus becomes a meromorphic function $A(z)$ of the complex parameter $z$. Taking $z = 0$ we should get the original function. It is important to notice that we will have only simple poles coming from propagators $P_{ab}(z)$. Employing the Cauchy's theorem one gets

$$\frac{1}{2\pi\mathrm{i}}\int\frac{dz}{z}A(z) = A(0) + \sum_k \frac{\mathrm{Res}\,(A,z_k)}{z_k}\,. \tag{9}$$

Assuming that $A(z)$ vanishes for $z\to\infty$ we get for the original amplitude

$$A = A(0) = -\sum_k \frac{\mathrm{Res}\,(A,z_k)}{z_k}\,. \tag{10}$$

The propagator $P_{ab}^2(z) = 0$ if one and only one of the two shifted momenta is in $(a, a+1,\ldots,b)$. Assuming the solution is $z_{ab}$ we get the factorization into two subamplitudes

$$\mathrm{Res}(A,z_{ab}) = \sum_s A_L^{-s}(z_{ab})\frac{i}{2(q\cdot P_{ab})}A_R^s(z_{ab})\,, \tag{11}$$

(for the allowed helicities). Using Cauchy's formula, we have finally

$$A = \sum_{k,s} A_L^{-s_k}(z_k)\frac{i}{P_k^2}A_R^{s_k}(z_k)\,. \tag{12}$$

*3.2. BCFW-like reconstruction of non-linear sigma model*

Let us turn back to the non-linear sigma model. Such a model corresponds in general to the spontaneous symmetry breaking of the chiral group $G_L \times G_R$ where $G_{L,R} = G$ to its diagonal subgroup $G_V = G$. In the following we will focus only on the group $G = SU(N)$ (which is the case of ChPT). The dynamics of the Goldstone bosons is governed by the Lagrangian (note we are in the massless limit)

$$\mathcal{L} = \frac{F^2}{4}\langle \partial_\mu U \partial^\mu U^{-1} \rangle \,, \tag{13}$$

where $\langle \ldots \rangle$ stands for the trace in the flavour space. In order we can use the analyticity properties for the amplitude reconstruction also in the case of the non-linear sigma model one important generalization is needed. The Cauchy theorem expressed by (10) was based on an assumption that the amplitude vanishes for $z \to \infty$. However, in a general case $A(z) \approx z^k$. Then $k+1$ subtractions in $a_i$ points are needed:

$$A(0) = -\sum_i \frac{\text{Res}(A, z_i)}{z_i} \prod_{j=1}^{k+1} \frac{a_j}{a_j - z_i} + \sum_{j=1}^{k+1} A(a_j) \prod_{l=1, l\neq j}^{k+1} \frac{a_l}{a_l - a_j} \,. \tag{14}$$

We will present here a reconstruction of the semi-on-shell amplitude $M_n$,

$$M_n(p_1, p_2, \ldots, p_{n+1}) = p_{n+1}^2 J_n(1, \ldots, n) \,, \tag{15}$$

where the semi-off-shell $J_n$ is the amplitude with the $(n+1)$th particle off-shell. The main reason why we have focus on the off-shell amplitude lays in the fact that the unphysical objects have different behaviour in different parametrization. For different way how to proceed see also [10]. The complex deformation which we propose is the all even-line shift, i.e.

$$p_i \to p_i(z): \qquad p_{2i+1}(z) = p_{2i+1} \quad \text{and} \quad p_{2i}(z) = z p_{2i} \,. \tag{16}$$

The shifted amplitude (15) will be denoted as

$$M_n(z) \equiv M_{2n+1}(p_1, zp_2, p_3, \ldots zp_{2n}, p_{2n+1}) \,. \tag{17}$$

The advantage of working with the off-shell amplitude is that in a specific parametrization we can show that $M_n(z) = O(z^0)$ (as $z \to \infty$). Once subtracted reconstruction formula (14) must be thus employed

$$M_n(z) = M_n(0) + \sum_P \frac{\text{Res}(M_n, z_P)}{z - z_P} \frac{z}{z_P} \,. \tag{18}$$

The amplitude for the non-linear sigma model can be reconstructed similarly to the BCFW reconstruction of gluon amplitudes. The most important practical advantage consists in the reduction of the number of diagrams. It can be shown that the number of poles, i.e. terms in (18) is at most $\sim n^2$

The recursive relations represent an equivalent description of the nonlinear $SU(N)$ sigma model to the standard Lagrangian formulation. Apart from its computational efficiency we can use it also in the investigation of the further properties. One such an application is the proof of the double Adler zero of the soft limit as conjectured in [11]. For details and formal proof see [12].

## 4. Summary and outlook

We have presented three different directions used within the non-linear sigma model. First, relevant for the phenomenology relies on a systematic calculation of the studied quantity perturbatively up to a given order. This order is dictated by the theoretical and experimental details of the problem. In practice it usually includes two-loop diagrams and they represent the most difficult part of the calculation. The question if even the NNLO order is enough is partly studied in the second direction: calculation of the leading logarithms. We were capable to automatized their calculations for different objects and showed some results up the seven order. This might be useful in resummation of the series and can thus lead to the similar concept as is the running coupling constant known for the renormalizable theories. In the third direction we stress the importance of the tree-level diagrams. We have introduced the BCFW-like reconstruction for the $SU(N)$ non-linear sigma model i.e. for model with infinite number of interaction vertices. New techniques developed primary for renormalizable models can be thus used for a broader class of theories than was expected before. One gains much faster way how to calculate amplitudes and new view on these quantities. The latter is useful in obtaining better understanding of new properties and was demonstrated on the proofs of the so-called Adler zeros. The BCFW-like reconstruction was limited so far to the tree-level. It should be possible to extend it up to one-loop level [13] or used in completely different theories (as are for example galileons, cf. [14]).

## Acknowledgements

I would like to thank my collaborators involved in parts of this presentation: Johan Bijnens, Stefan Lanz, Bachir Moussallam, Jiří Novotný and Jaroslav Trnka. This work is supported by PRVOUK P45 and grant LH 14035.

## References

[1] S. Weinberg, Physica A **96** (1979) 327; J. Gasser and H. Leutwyler, Annals Phys. **158** (1984) 142; J. Gasser and H. Leutwyler, Nucl. Phys. B **250** (1985) 465.
[2] K. Kampf and B. Moussallam, Phys. Rev. D **79** (2009) 076005 [arXiv:0901.4688 [hep-ph]]; J. Bijnens and K. Kampf, Nucl. Phys. Proc. Suppl. **207-208** (2010) 220 [arXiv:1009.5493 [hep-ph]].
[3] G. Ecker, J. Gasser, A. Pich and E. de Rafael, Nucl. Phys. B **321** (1989) 311; K. Kampf and J. Novotny, Phys. Rev. D **84** (2011) 014036 [arXiv:1104.3137 [hep-ph]].
[4] J. Bijnens and L. Carloni, Nucl. Phys. B **843** (2011) 55 [arXiv:1008.3499 [hep-ph]]. J. Bijnens, K. Kampf and S. Lanz, Nucl. Phys. B **860** (2012) 245 [arXiv:1201.2608 [hep-ph]]. J. Bijnens, K. Kampf and S. Lanz, Nucl. Phys. B **873** (2013) 137 [arXiv:1303.3125 [hep-ph]].
[5] J. A. M. Vermaseren, math-ph/0010025.
[6] J. Bijnens and A. A. Vladimirov, arXiv:1409.6127 [hep-ph].
[7] K. Kampf, PoS CD **12** (2013) 039 [arXiv:1302.7311 [hep-ph]].
[8] S. J. Parke and T. R. Taylor, Phys. Rev. Lett. **56** (1986) 2459.
[9] R. Britto, F. Cachazo and B. Feng, Nucl. Phys. B **715** (2005) 499 [hep-th/0412308]; R. Britto, F. Cachazo, B. Feng and E. Witten, Phys. Rev. Lett. **94** (2005) 181602 [hep-th/0501052].
[10] K. Kampf, J. Novotny and J. Trnka, Phys. Rev. D **87** (2013) 081701 [arXiv:1212.5224 [hep-th]].
[11] N. Arkani-Hamed, F. Cachazo and J. Kaplan, JHEP **1009** (2010) 016 [arXiv:0808.1446].
[12] K. Kampf, J. Novotny and J. Trnka, JHEP **1305** (2013) 032 [arXiv:1304.3048 [hep-th]].
[13] K. Kampf, J. Novotny and J. Trnka, *in preparation*
[14] K. Kampf and J. Novotny, JHEP **1410** (2014) 006 [arXiv:1403.6813 [hep-th]].